Addition Reactions with Unsaturated Hydrocarbons

Addition Reactions with Unsaturated Hydrocarbons

Ruimao Hua

WILEY-VCH

Author

Prof. Ruimao Hua
Tsinghua University
Department of Chemistry
30 Shuang Qing Lu
Haidian District
100084 Beijing
China

Cover Image: © Sebastian Duda/Shutterstock, Reaction equation recreated by Wiley

Library of Congress Card No.: applied for

British Library Cataloguing-in-Publication Data
A catalogue record for this book is available from the British Library.

Bibliographic information published by the Deutsche Nationalbibliothek
The Deutsche Nationalbibliothek lists this publication in the Deutsche Nationalbibliografie; detailed bibliographic data are available on the Internet at <http://dnb.d-nb.de>.

Print ISBN: 978-3-527-34189-4
ePDF ISBN: 978-3-527-80483-2
ePub ISBN: 978-3-527-80482-5
oBook ISBN: 978-3-527-80480-1

Typesetting Straive, Chennai, India
Printing and binding CPI Group (UK) Ltd, Croydon, CR0 4YY

Printed on acid-free paper

C066280_210222

Contents

Preface

The proposal of this book entitled *Addition Reactions with Unsaturated Hydrocarbons* was approved by Wiley-VCH GmbH in November 2015. It was expected to be published after two years; however, it took me much more time to finish the writing work because of the recent great development of chemistry in fundamental studies and applications of unsaturated hydrocarbons, and an enormous amount of literature must be read.

Modern synthetic methods require to be green procedures with atom/step economy and high chemo-/regio-/stereoselectivity, and the addition and cycloaddition reactions of unsaturated hydrocarbons such as alkynes [1], alkenes [2], allenes [3], enynes, and diynes [4] are showing great advances in these features, which have been well developed and applied in the synthesis of carbo- and heterocycles [5], in the construction of challenging skeletons of complex molecules and polymers [6], and other transformations [7].

This book contains 10 chapters and is devoted mainly to detailed introduction and discussions of the reactions of unsaturated hydrocarbons as the versatile building blocks in a wide variety of important organic synthesis. Chapters 1 and 2 describe the homo- and cross-dimerization of alkynes and alkenes, the addition reactions of C(sp)—H bond to unsaturated compounds to give enynes, and alkynylated compounds, which are also important unsaturated hydrocarbons for further transformations. Chapters 3 and 4 summarize the recent development of hydrofunctionalization of alkynes and alkenes to provide the atom economic approach to functionalized alkenes and alkanes. Chapters 5 and 6 describe the double functionalization of alkynes and alkenes by addition reactions of element–element bonds or carbon–element bonds. Chapters 7 and 8 include a number of annulations or cycloadditions of alkynes and alkenes for the synthesis of carbocycles and heterocycles in one-pot reactions. In addition, Chapter 9 summarizes the synthesis of carbonyl compounds from the hydration of alkynes, the addition reactions of alkynes and alkenes with carbonyl compounds, and the carbonylation of alkynes and alkenes with carbon monoxide. Finally, in Chapter 10, it introduces a few examples of the reported procedures on the synthesis of natural products involving alkyne transformation as the key step. More than 600 schemes are shown to describe the typical reactions starting from unsaturated hydrocarbons, and over

3600 literatures are cited, including about 500 review papers to allow greater access to more original papers for readers.

The new chemistry of unsaturated hydrocarbons will continue to grow and expand because of their numerous and apparently endless applications to develop the efficient synthetic methodologies for the synthesis of a vast array of chemicals. I believe this book has a comprehensive coverage of important and up-to-date reactions of unsaturated hydrocarbons, and it is hoped that the book will play multiple-purpose usage for students and chemists in both the universities and industrial laboratories.

Tsinghua University
Beijing, China

Dr. Ruimao Hua

References

Selected review papers on the applications of the addition and cycloaddition reactions of unsaturated hydrocarbons.

1 (a) Pillai, S.M., Ravindranathan, M., and Sivaram, S. (1986). Dimerization of ethylene and propylene catalyzed by transition-metal complexes. *Chem. Rev.* 86: 353–399. (b) Asao, N. and Yamamoto, Y. (2000). Lewis acid-catalyzed hydrometalation and carbometalation of unactivated alkynes. *Bull. Chem. Soc. Jpn.* 73: 1071–1087. (c) Trost, B.M. and Ball, Z.T. (2005). Addition of metalloid hydrides to alkynes: hydrometallation with boron, silicon, and tin. *Synthesis* 853–887. (d) Negishi, E.-i., Wang, G., Rao, H., and Xu, Z. (2010). Alkyne elementometalation-Pd-catalyzed cross-coupling. Toward synthesis of all conceivable types of acyclic alkenes in high yields, efficiently, selectively, economically, and safely: "green" way. *J. Org. Chem.* 75: 3151–3182. (e) Wille, U. (2013). Radical cascades initiated by intermolecular radical addition to alkynes and related triple bond systems. *Chem. Rev.* 113: 813–853. (f) Salvio, R., Moliterno, M., and Bella, M. (2014). Alkynes in organocatalysis. *Asian J. Org. Chem.* 3: 340–351. (g) Dorel, R. and Echavarren, A.M. (2015). Gold(I)-catalyzed activation of alkynes for the construction of molecular complexity. *Chem. Rev.* 115: 9028–9072. (h) Fang, G. and Bi, X. (2015). Silver-catalyzed reactions of alkynes: recent advances. *Chem. Soc. Rev.* 44: 8124–8173. (i) Boyarskiy, V.P., Ryabukhin, D.S., Bokach, N.A., and Vasilyev, A.V. (2016). Alkenylation of arenes and heteroarenes with alkynes. *Chem. Rev.* 116: 5894–5986. (j) Zheng, Y. and Zi, W. (2018). Transition-metal catalyzed enantioselective hydrofunctionalization of alkynes. *Tetrahedron Lett.* 59: 2205–2213. (k) Liu, W. and Kong, W. (2020). Ni-Catalyzed stereoselective difunctionalization of alkynes. *Org. Chem. Front.* 7: 3941–3955. (l) Leung, C.H., Baron, M., and Biffis, A. (2020). Gold-catalyzed intermolecular alkyne hydrofunctionalizations-mechanistic insights. *Catalysts* 10: 1210. (m) Saptal, V.B., Wang, R., and Park, S. (2020). Recent advances in transition metal-free catalytic hydroelementation (E = B, Si, Ge, and Sn) of alkynes. *RSC Adv.* 10: 43539–43565.

2 Costa, D.C.S. (2020). Additions to non-activated alkenes: recent advances. *Arab. J. Chem.* 13: 799–834.

3 (a) Yu, S. and Ma, S. (2012). Allenes in catalytic asymmetric synthesis and natural product syntheses. *Angew. Chem. Int. Ed.* 51: 3074–3112. (b) Muñoz, M.P. (2012). Transition metal-catalysed intermolecular reaction of allenes with oxygen nucleophiles: a perspective. *Org. Biomol. Chem.* 10: 3584–3594. (c) Yang, W. and Hashmi, A.S.K. (2014). Mechanistic insights into the gold chemistry of allenes. *Chem. Soc. Rev.* 43: 2941–2955. (d) Alonso, J.M., Quirós, M.T., and Muñoz, M.P. (2016). Chirality transfer in metal-catalysed intermolecular addition reactions involving allenes. *Org. Chem. Front.* 3: 1186–1204. (e) Santhoshkumar, R. and Cheng, C.-H. (2018). Fickle reactivity of allenes in transition-metal-catalyzed C—H functionalizations. *Asian J. Org. Chem.* 7: 1151–1163. (f) Liu, L., Ward, R.M., and Schomaker, J.M. (2019). Mechanistic aspects and synthetic applications of radical additions to allenes. *Chem. Rev.* 119: 12422–12490. (g) Xiong, Y., Sun, Y., and Zhang, G. (2018). Recent advances on catalytic asymmetric difunctionalization of 1,3-dienes. *Tetrahedron Lett.* 59: 347–355.

4 (a) García-Rubín, S., Varela, J.A., Castedo, L., and Saá, C. (2008). Recent advances in "formal" ruthenium-catalyzed [2+2+2] cycloaddition reactions of diynes to alkenes. *Chem. Eur. J.* 14: 9772–9778. (b) Aubert, C., Fensterbank, L., Garcia, P. et al. (2011). Transition-metal-catalyzed cycloisomerizations of 1,n-allenynes and -allenenes. *Chem. Rev.* 111: 1954–1993. (c) Raviola, C., Protti, S., Ravelli, D., and Fagnoni, M. (2016). (Hetero) aromatics from dienynes, enediynes and enyne–allenes. *Chem. Soc. Rev.* 45: 4364–4390. (d) Xuan, J. and Studer, A. (2017). Radical cascade cyclization of 1,*n*-enynes and diynes for the synthesis of carbocycles and heterocycles. *Chem. Soc. Rev.* 46: 4329–4346. (e) Babazadeh, M., Soleimani-Amiri, S., Vessally, E. et al. (2017). Transition metal-catalyzed [2+2+2] cycloaddition of nitrogen-linked 1,6-diynes: a straightforward route to fused pyrrolidine systems. *RSC Adv.* 7: 43716–43736. (f) Xu, C.-H., Li, Y., Li, J.-H. et al. (2019). Recent advances in radical-mediated [2+2+*m*] annulation of 1,*n*-enynes. *Sci. Chin. Chem.* 62: 1463–1475.

5 (a) Schore, N.E. (1988). Transition metal-mediated cycloaddition reactions of alkynes in organic synthesis. *Chem. Rev.* 88: 1081–1119. (b) Welker, M. (1992). [3+2] Cycloaddition reactions of transition-metal 2-alkynyl and η^1-allyl complexes and their utilization in five-membered-ring compound syntheses. *Chem. Rev.* 92: 97–112. (c) Cacchi, S. (1999). Heterocycles via cyclization of alkynes promoted by organopalladium complexes. *J. Organomet. Chem.* 576: 42–64. (d) Yet, L. (2000). Metal-mediated synthesis of medium-sized rings. *Chem. Rev.* 100: 2963–3008. (e) Belmont, P. and Parker, E. (2009). Silver and gold catalysis for cycloisomerization reactions. *Eur. J. Org. Chem.* 2009: 6075–6089. (f) Hua, R., Abrenica, M.V.A., and Wang, P. (2011). Cycloaddition of alkynes: atom-economic protocols for constructing six-membered cycles. *Curr. Org. Chem.* 15: 712–729. (g) Garayalde, D. and Nevado, C. (2012). Gold-containing and gold-generated 1,*n*-dipoles as useful platforms toward cycloadditions and cyclizations. *ACS Catal.* 2: 1462–1479. (h) Dateer, R.B., Shaibu, B.S., and Liu, R.-S. (2012). Gold-catalyzed intermolecular [4+2] and [2+2+2] cycloadditions of ynamides with alkenes. *Angew. Chem. Int. Ed.* 51: 113–117. (i) Wang, C. and Wan, B. (2012). Recent advances in the iron-catalyzed cycloaddition reactions. *Chin.*

Sci. Bull. 57: 2338–2351. (j) Palisse, A. and Kirsch, S.F. (2012). Metal-free reactions of alkynes via electrophilic iodocarbo-cyclizations. *Org. Biomol. Chem.* 10: 8041–8047. (k) Ackermann, L. (2014). Carboxylate-assisted ruthenium-catalyzed alkyne annulations by C—H/Het—H bond functionalizations. *Acc. Chem. Res.* 47: 281–295. (l) Luo, Y., Pan, X., Yu, X., and Wu, J. (2014). Double carbometallation of alkynes: an efficient strategy for the construction of polycycles. *Chem. Soc. Rev.* 43: 834–846. (m) Ye, J. and Ma, S. (2014). Palladium-catalyzed cyclization reactions of allenes in the presence of unsaturated carbon–carbon bonds. *Acc. Chem. Res.* 47: 989–1000. (n) Adams, C.S., Weatherly, C.D., Burke, E.G., and Schomaker, J.M. (2014). The conversion of allenes to strained three-membered heterocycles. *Chem. Soc. Rev.* 43: 3136–3163. (o) Thakur, A. and Louie, J. (2015). Advances in nickel-catalyzed cycloaddition reactions to construct carbocycles and heterocycles. *Acc. Chem. Res.* 48: 2354–2365. (p) Lledó, A., Pla-Quintana, A., and Roglans, A. (2016). Allenes, versatile unsaturated motifs in transition-metal-catalysed [2 + 2 + 2] cycloaddition reactions. *Chem. Soc. Rev.* 45: 2010–2023. (q) Yang, Y., Li, K., Cheng, Y. et al. (2016). Rhodium-catalyzed annulation of arenes with alkynes through weak chelation-assisted C—H activation. *Chem. Commun.* 52: 2872–2884. (r) Domínguez, G. and Pérez-Castells, J. (2016). Alkenes in [2 + 2 + 2] cycloadditions. *Chem. Eur. J.* 22: 6720–6739. (s) Aggarwal, T., Kumar, S., and Verma, A.K. (2016). Iodine-mediated synthesis of heterocycles via electrophilic cyclization of alkynes. *Org. Biomol. Chem.* 14: 7639–7653. (t) Pathipati, S.R., van der Werf, A., and Selander, N. (2017). Indium (III)-catalyzed transformations of alkynes: recent advances in carbo-and heterocyclization reactions. *Synthesis* 49: 4931–4941. (u) Zheng, L. and Hua, R. (2018). C—H activation and alkyne annulation via automatic or intrinsic directing groups: towards high step economy. *Chem. Rec.* 18: 556–569. (v) Kuila, B.; Kaur, M.; Singh, P.; Bhargava, G. Transition-metal-catalyzed [3 + 2 + 2] cycloaddition reactions. *Eur. J. Org. Chem.* 2018, 853–868. (w) Huang, M.-H., Hao, W.-J., Li, G. et al. (2018). Recent advances in radical transformations of internal alkynes. *Chem. Commun.* 54: 10791–10811. (x) Duarah, G., Kaishap, P.P., Begum, T., and Gogoi, S. (2019). Recent advances in ruthenium (II)-catalyzed C—H bond activation and alkyne annulation reactions. *Adv. Synth. Catal.* 361: 654–672. (y) Pawlowski, R., Stanek, F., and Stodulski, M. (2019). Recent advances on metal-free, visible-light-induced catalysis for assembling nitrogen-and oxygen-based heterocyclic scaffolds. *Molecules* 24: 1533. (z) Doerksen, R.S., Hodík, T., Hu, G. et al. (2021). Ruthenium-catalyzed cycloadditions to form five-, six-, and seven-membered rings. *Chem. Rev.* 121: 4045–4083.

6 (a) Tour, J.M. (1996). Conjugated macromolecules of precise length and constitution: organic synthesis for the construction of nanoarchitectures. *Chem. Rev.* 96: 537–554. (b) Martin, R.E. and Diederich, F. (1999). Linear monodisperse π-conjugated oligomers: model compounds for polymers and more. *Angew. Chem. Int. Ed.* 38: 1350–1377. (c) Nielsen, M.B. and Diederich, F. (2005). Conjugated oligoenynes based on the diethynylethene unit. *Chem. Rev.* 105: 1837–1868. (d) Shi Shun, A.L.K. and Tykwinski, R.R. (2006). Synthesis of naturally occurring polyynes. *Angew. Chem. Int. Ed.* 45: 1034–1057. (e) Rivera-Fuentes, P. and Diederich, F. (2012). Allenes in molecular materials. *Angew. Chem. Int. Ed.* 51:

2818–2828. (f) Galkin, K.I. and Ananikov, V.P. (2016). Alkynes as a versatile platform for construction of chemical molecular complexity and realization of molecular 3D printing. *Russ. Chem. Rev.* 85: 226–247.

7 (a) Rossi, R.A. and Martín, S.E. (2006). Syntheses and applications of organostannanes bonded to elements of groups XIV, XV, and XVI. *Coord. Chem. Rev.* 250: 575–601. (b) Patil, N.T., Kavthe, R.D., and Shinde, V.S. (2012). Transition metal-catalyzed addition of *C*-, *N*- and *O*-nucleophiles to unactivated C—C multiple bonds. *Tetrahedron* 68: 8079–8146. (c) Koschker, P. and Breit, B. (2016). Branching out: rhodium-catalyzed allylation with alkynes and allenes. *Acc. Chem. Res.* 49: 1524–1536. (d) Holmes, M., Schwartz, L.A., and Krische, M.J. (2018). Intermolecular metal-catalyzed reductive coupling of dienes, allenes, and enynes with carbonyl compounds and imines. *Chem. Rev.* 118: 6026–6052. (e) Beletskaya, I.P., Nájera, C., and Yus, M. (2020). Catalysis and regioselectivity in hydrofunctionalization reactions of unsaturated carbon bonds. Part I. *Russ. Chem. Rev.* 89: 250–274. (f) Li, G., Huo, X., Jiang, X., and Zhang, W. (2020). Asymmetric synthesis of allylic compounds via hydrofunctionalisation and difunctionalisation of dienes, allenes, and alkynes. *Chem. Soc. Rev.* 49: 2060–2118.

Acknowledgments

My research interests are on the development of organic synthetic methodologies starting from unsaturated hydrocarbons, which have been continued for over 20 years. It is one of my desires to publish a comprehensive book devoted to the chemistry of unsaturated hydrocarbons, thus it is very pleasing that the book of *Addition Reactions with Unsaturated Hydrocarbons* is being successfully published with the assistance and cooperation of Wiley-VCH GmbH, and the contributions of a number of literatures' authors are greatly appreciated.

I would first like to acknowledge four reviewers and two consultants for their exceedingly valuable comments and suggestions on improving the proposal of the book at the first stage.

For writing work, I downloaded more than 6000 literatures, among them, 3600 were cited in this book, and I greatly thank these authors for their excellent work published in academic journals.

My special thanks goes to the contribution of my group students, who graduated or are now working in my group; their achievements have also been cited in this book. I must also thank the National Natural Science Foundation of China for supporting my research projects on the alkyne chemistry over years (Grant Nos. 21972072, 21673124, 21473097, 21032004, 20972084, 20873073, and 20473043).

I am deeply indebted to the staff of Wiley-VCH: Dr. Lifen Yang, program manager; Ms. Shirly Samuel, project editor; and Ms. Katherine Wong, senior managing editor for their indispensable guidance and patience during my writing work.

Lastly, I wish to express my special thanks to my wife (senior research engineer in analytical chemistry) and my elder son (PhD candidate in chemistry) for their help in proofreading the manuscripts.

Ruimao Hua

Abbreviations

acac	acetylacetonate
AIBN	azobis(isobutyronitrile)
9-BBN	9-borabicyclo[3.3.1]nonane
BEMP	2-*t*-butylimino-2-diethylamino-1,3-dimethylperhydro-1,3,2-diazaphosphorine
BINAP	bis(diphenylphosphino)-1,1′-binaphthalene
BIPHEPOS	6,6′-[(3,3′-di-*t*-butyl-5,5′-dimethoxy-1,1′-biphenyl-2,2′-diyl)bis(oxy)]bis-(dibenzo[d,f][1,3,2]dioxaphosphepin)
bmim	1-*n*-butyl-3-methylimidazolium
Boc	*t*-butoxycarbonyl
Boz	*N*-benzoyloxymethyl
BQ	benzoquinone
cat	catecholato, 1,2-$O_2C_6H_4$
cod	1,5-cyclooctadiene
cot	1,3,5-cyclooctatriene
p-cymene	*p*-isopropyltoluene
Cp*	pentamethylcyclopentadienyl
DABCO	1,4-diazabicyclo[2.2.2]octane
dba	dibenzylideneacetone
dcypb	1,4-bis(dicyclohexylphosphino)butane)
DIBAL/DIBAL-H	diisobutyl aluminum hydride
dipimp	2-(2,6-diisopropylphenyl)iminomethylpyridine
DMAD	dimethyl acetylenedicarboxylate
DME	1,2-dimethoxyethane
dppb	1,4-bis(diphenylphosphino)butane
dppe	1,2-bis(diphenylphosphino)ethane
dppf	1,1′-ferrocenediyl-bis(diphenylphosphine)
dppm	bis(diphenylphosphino)methane
dppp-C3	1,3-bis(diphenylphosphino)propane
EDA	ethylenediamine
HBpin	pinacolborane
HFIP	hexafluoroisopropanol
HOTf	trifluoromethanesulfonic acid
IDTB	1,3-bis(2,5-di-*t*-butylphenyl)-imidazol-2-ylidene

Imes	*N*,*N*′-bis(2,4,6-trimethylphenyl)imidazole-2-ylidene
IPr	*N*,*N*′-bis(2,6-diisopropylphenyl)imidazole-2-ylidene
LPO	dilauroyl peroxide
PMB	*p*-methoxybenzyl
nbd	norbornadiene
NHC	*N*-heterocyclic carbine
NIS	*N*-iodosuccinimide
NTf_2	triflimide
PCC	Pyridinium chlorochromate
Phen	1,10-phenanthroline
pin	pinacolato
PTSA	*p*-toluenesulfonic acid
TBS	*t*-butyldimethylsilyl
TC	bis[(2-diphenylphosphino)phenyl]ether
TDMPP	tris(2,6-dimethoxyphenyl)phosphine
TFA	trifluoroacetic acid, trifluoroacetate
TMU	tetramethylurea
Tp*	hydrotris(3,5-dimethylpyrazolylborate)
$Tpm^{*,Br}$	tris(3,5-dimethyl-4-bromopyrazolyl)methane
p-TS	*p*-toluenesulfonamide
Xantphos	4,5-bis-(diphenylphosphino)-9,9-dimethylxanthene

1

Dimerization of Alkynes and Alkenes

Homo- or cross-dimerization of alkynes and/or alkenes is a straightforward, atom economical method to approach the conjugated enynes, 1,3-dienes, or higher alkenes. Particularly, the homo- and cross-dimerization of terminal alkynes can afford conjugated enynes, which are the important moieties in organic materials [1], biologically important molecules [2], and the versatile synthetic intermediates [3]. The functional 1,3-dienes from the cross-dimerization of alkynes with alkenes via hydrovinylation are also important intermediates in organic syntheses [4].

The catalytic dimerization of terminal alkynes can yield theoretically four unsaturated products: 2,4-disubstituted enyne (via head to tail or Markovnikov), *E*- and *Z*-1,4-disubstituted enynes (via head to head or *anti*-Markovnikov), and butatriene (Scheme 1.1).

The transition metal-catalyzed formation of 2,4- and 1,4-disubstituted enynes via alkyne dimerization can be explained by a series of conventional reaction steps: oxidative addition of C≡H, insertion of C≡C of the second alkyne molecule, and then reductive elimination [5].

1.1 Markovnikov Dimerization of Terminal Alkynes

The regioselective Markovnikov dimerization (head to tail) of terminal alkynes will afford *gem*-enynes (2,4-disubstituted enynes). $[(Rh(PMe_3)_2Cl]_2$ has been found to be the efficient catalyst to catalyze the insertion of terminal alkynes into benzene C—H bonds under irradiation, and it also shows catalytic activity for the dimerization of terminal alkynes with the *gem*-enynes as the major products in most cases (Scheme 1.2) [6].

$[Rh(cod)Cl]_2$/dppf can efficiently catalyze the dimerization of *N*-protected propargylamines to regioselectively give *gem*-enynes, which can subsequently undergo intramolecular hydroamination reaction to afford 2-(aminomethyl)pyrrole derivatives in the presence of $AuCl_3$ (Scheme 1.3) [7]. $[Rh(cod)Cl]_2/PPh_3$ was then applied in the cross-dimerization between aromatic alkynes and propargylic alcohols, ethers, or amides with high chemo- and regioselectivity [8].

Nolan and coworker studied the catalytic activity of palladium/imidazolium system in the dimerization of aromatic and aliphatic terminal alkynes [9]. It was found

Addition Reactions with Unsaturated Hydrocarbons, First Edition. Ruimao Hua.

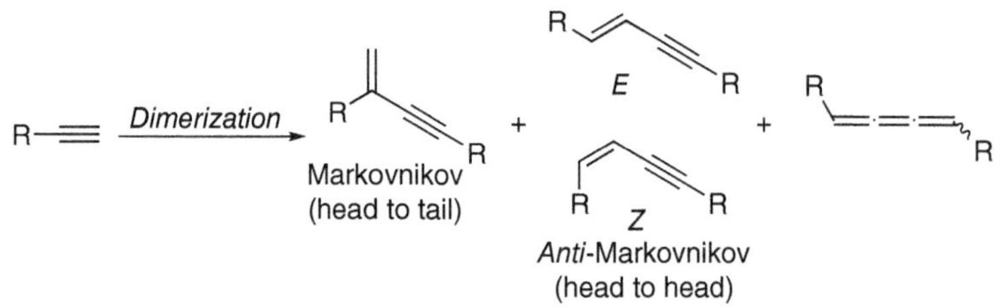

Scheme 1.1 Dimerization products of terminal alkynes.

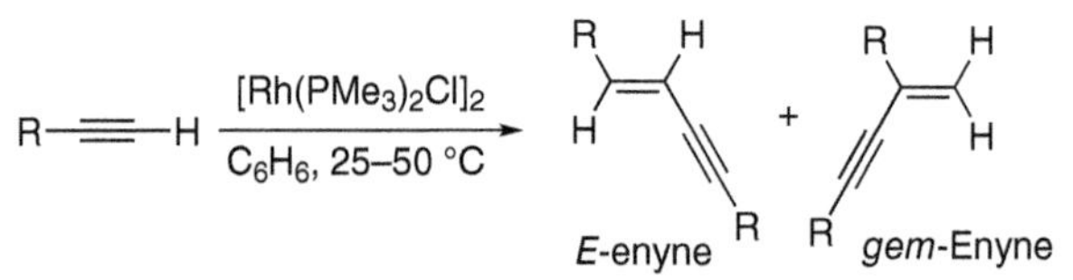

R = *n*-Pr, *E*-enyne 37%; *gem*-enyne 63%
R = *t*-Bu, *E*-enyne 100%; *gem*-enyne 0%
R = Ph, *E*-enyne 0%; *gem*-enyne 100%
R = *p*-MeOC$_6$H$_4$, *E*-enyne 0%; *gem*-enyne 100%

Scheme 1.2 [Rh(PMe$_3$)$_2$Cl]$_2$-catalyzed dimerization of terminal alkynes.

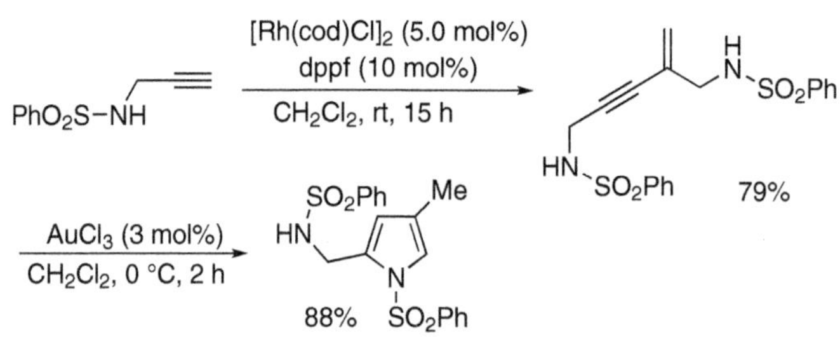

Scheme 1.3 Synthesis of trisubstituted pyrroles by cyclization of *N*-functionalized *gem*-enynes.

that Pd(OAc)$_2$/IMes·HCl is a highly efficient and regio- and stereoselective catalytic system depending on the use of the base and alkynes to control the product distribution. As shown in Scheme 1.4, in the case of K$_2$CO$_3$ used, the dimerization of 1-heptyne and phenyl acetylene affords *gem*-enyne and *E*-enyne, respectively, as the major products. However, in the former reaction, replacement of K$_2$CO$_3$ with Cs$_2$CO$_3$ results in a significant increase of *E*-enyne from the head-to-head dimerization as the predominant products.

In the presence of PdCl$_2$(PPh$_3$)$_2$/CuI/PPh$_3$, the aromatic terminal alkynes undergo the trimerization to first afford (*Z*)-1,3-diaryl-2-arylethynyl-1,3-butadienes in moderate to excellent yields with high regioselectivity, and the structures and stereochemistry of dienyne were confirmed by X-ray crystal analyses [10]. As shown in Scheme 1.5, upon heating phenyl acetylene in dimethylfomamide (DMF) or acetonitrile with a mixture of PdCl$_2$(PPh$_3$)$_2$, PPh$_3$, NEt$_3$, and CuI, the angular dienyne could be obtained with high regioselectivity in 87 and 94% yield, respectively. This

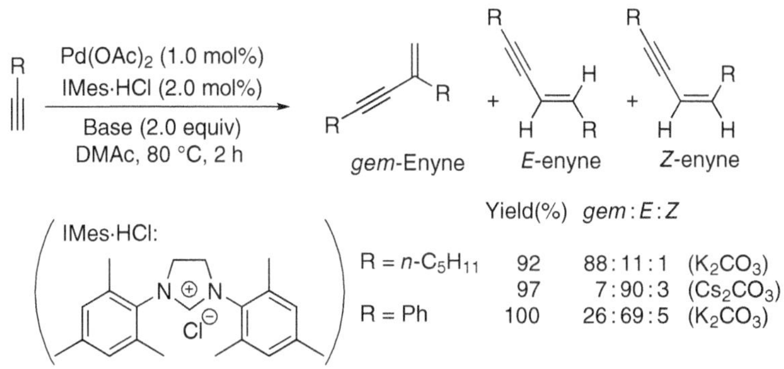

Scheme 1.4 Dimerization of terminal alkynes to enynes catalyzed by $Pd(OAc)_2$/IMes·HCl/base system.

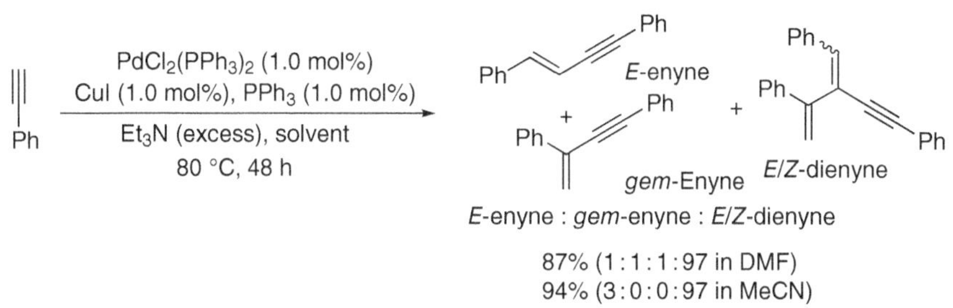

Scheme 1.5 Palladium-catalyzed synthesis of (*Z*)-dienynes via regioselective trimerization of alkynes.

protocol indicates that *gem*-enynes are much more reactive than aromatic terminal alkynes toward hydroalkynylation to yield dienynes.

Han and coworker reported a Brønsted acid $Ph_2P(O)OH$-promoted reaction of $Pd(PEt_3)_4$ with terminal alkynes to successfully isolate and characterize the complex of alkenyl(alkynyl)-palladiums from the oxidative addition of C—H bond of terminal alkyne to Pd(0) and subsequent ligand exchange reactions (Scheme 1.6). Then, a selective head-to-tail dimerization of terminal alkynes efficiently affording *gem*-enynes was developed by using Pd(0)/Brønsted acid (Scheme 1.7) [11].

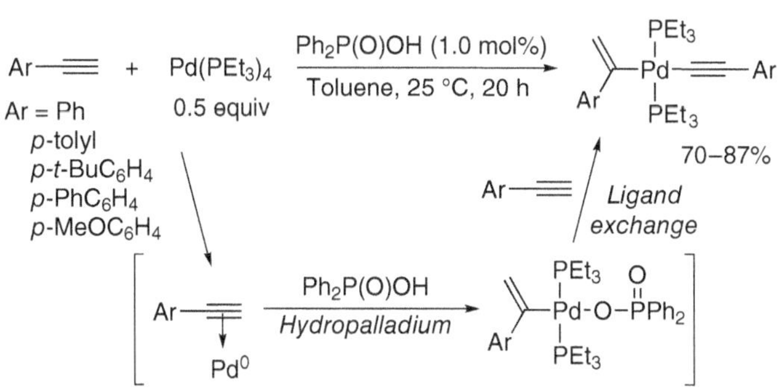

Scheme 1.6 $Ph_2P(O)OH$-catalyzed formation of alkenyl (alkynyl) palladium.

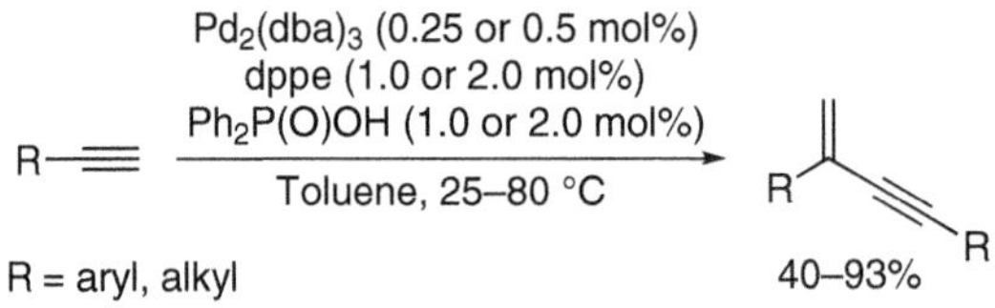

Scheme 1.7 Palladium(0)/$Ph_2P(O)OH$-catalyzed head-to-tail dimerization of terminal alkynes.

Guo and coworker reported an interesting counterion-controlled reactivity shift between dimerization and hydration of aromatic terminal alkynes catalyzed by $Pd(PPh_3)_4$ (Scheme 1.8) [12]. It is proposed that the use of acetate as counterion favors the formation of an alkenyl alkynyl palladium intermediate, resulting in the formation of 1,3-diaryl-substituted conjugated enynes via reductive elimination, while chloride is used, which is a better leaving group, leading to anion exchange on the alkenyl palladium intermediate with hydroxide to afford hydration products via reductive elimination and tautomerization (Scheme 1.9).

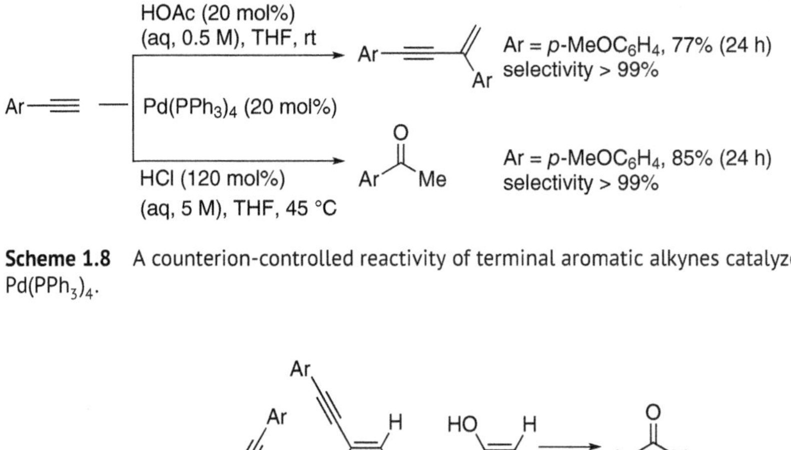

Scheme 1.8 A counterion-controlled reactivity of terminal aromatic alkynes catalyzed by $Pd(PPh_3)_4$.

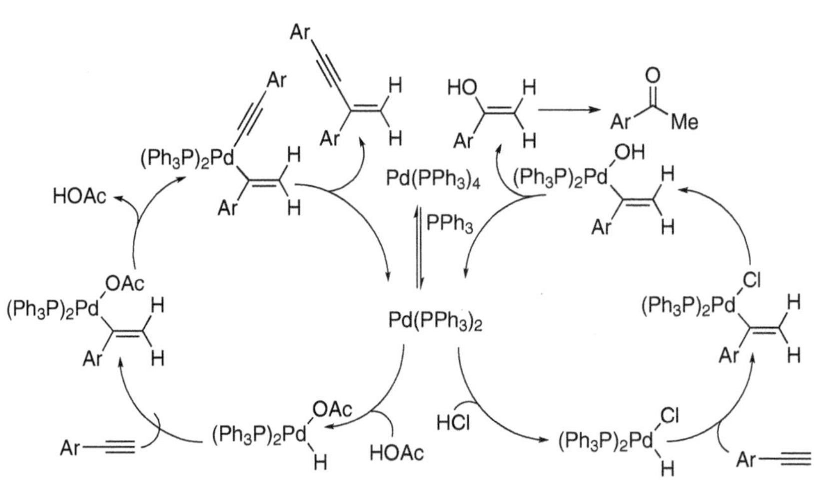

Scheme 1.9 A proposed mechanism for counterion-controlled reactivity shift of terminal aromatic alkynes.

In the presencea of t-BuXPhosAuNTf_2, the head-to-tail dimerization of aliphatic terminal alkynes occurs to give *gem*-enynes in good yields with excellent regioselectivity with the use of NaOAc as an additive [13]. Very interestingly, Hashmi and coworker have also developed a dual gold-catalyzed head-to-tail dimerization of haloalkynes (halo = Cl, Br, I) to afford *gem*-dihalogenated enynes, which are expected to be the valuable building blocks in organic synthesis. In the cases of iodoalkynes used, the dimerization occurs smoothly to give *gem*-deoxygenated enynes in good yields (Scheme 1.10) [14].

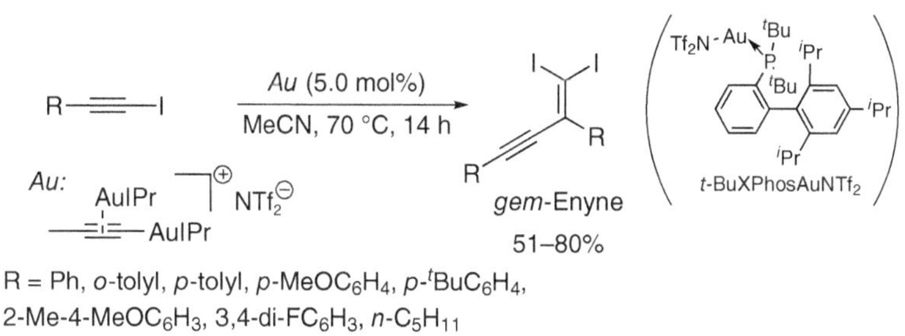

Scheme 1.10 Gold-catalyzed Markovnikov dimerization of iodoalkynes.

Methylaluminoxane (MAO) has been found to be an active catalytic precursor in benzene for the chemo- and regioselective head-to-tail dimerization of aromatic and aliphatic terminal alkynes to produce the corresponding *gem*-enynes in the excellent yields without formation of any other dimers (Scheme 1.11) [15]. Interestingly, in the case of terminal alkynes bearing an alkenyl functional group used, the *gem*-enyne can undergo an intermolecular [4+2] cycloaddition reaction to give alkynyl-substituted cyclohexene. A plausible pathway for the MAO-catalyzed dimerization of terminal alkynes is shown in Scheme 1.12; it involves a sequence of Al-alkynyl complex formation, a regioselective 1,2-head-to-tail insertion of alkyne into the Al-carbon bond, and the protonolysis of alkenyl complex.

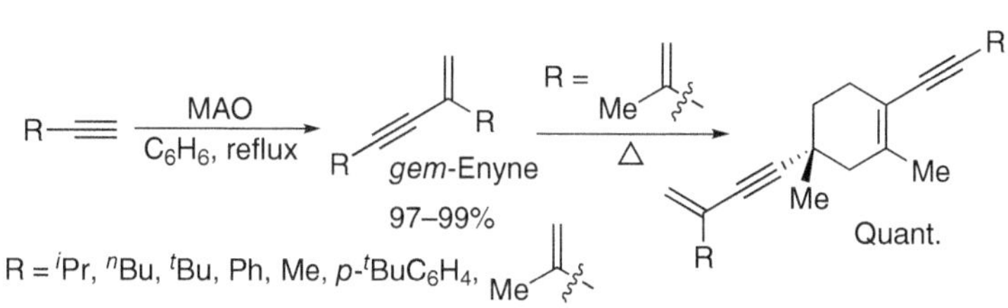

Scheme 1.11 MAO-promoted head-to-tail dimerization of terminal alkynes.

The intramolecular cyclic dimerization of α,ω-diynes is one of the atom economic synthetic ways for the construction of macrocyclic 1-en-3-ynes. Trost and coworker first reported the $Pd(OAc)_2$/tris(2,6-dimethoxyphenyl)phosphine (TDMPP)-catalyzed synthesis of the *exo*-macrocyclic 14-membered ring compound in 41% yield with high regioselectivity (Scheme 1.13) [16].

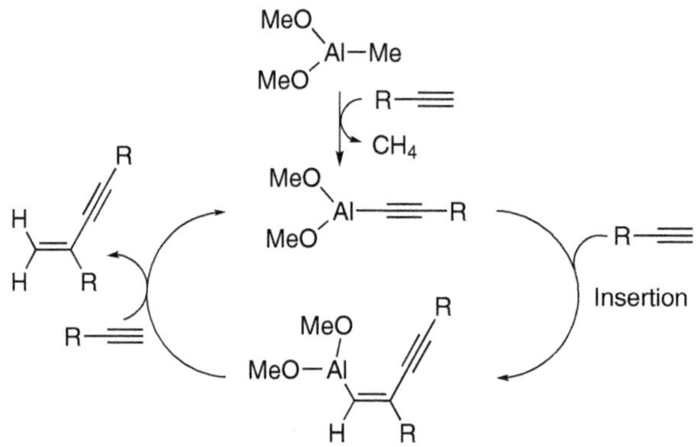

Scheme 1.12 Proposed mechanism for head-to-tail dimerization of terminal alkynes promoted by MAO.

Scheme 1.13 Palladium-catalyzed synthesis of the *exo*-macrocycles.

In the presence of rare-earth silylamides, $Ln[N(SiMe_3)_2]_3$ (Ln = Y, La, Sm), the regio- and stereoselective dimerization of terminal alkynes occurs to give enynes in high yields with the use of amines as additives [17]. It has been found that the amine additives play a crucial role to depress the oligomerization and to control the regio- and stereochemistry of the dimerization. As shown in Scheme 1.14, when primary amine of aniline was used as the additive, the dimers of (*Z*)-head-to-head enynes from aromatic terminal alkynes could be obtained as the exclusive products. In contrast, when tertiary amine of $N(SiMe_2)_3$ was employed as the additive, nearly complete formation of head-to-tail dimers from aliphatic terminal alkynes was realized.

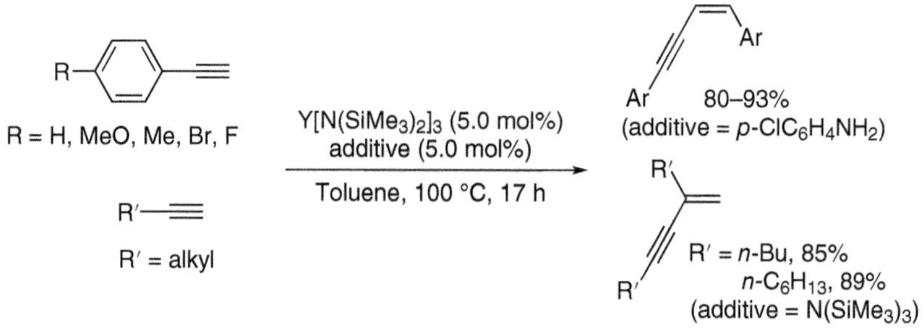

Scheme 1.14 Dimerization of terminal alkynes catalyzed by $Y[N(SiMe_3)_2]_3$/additive.

Interestingly, when α,ω-diynes were subjected to the dimerization conditions, a novel double dimerization takes place, leading to the formation of bisenynes (Scheme 1.15).

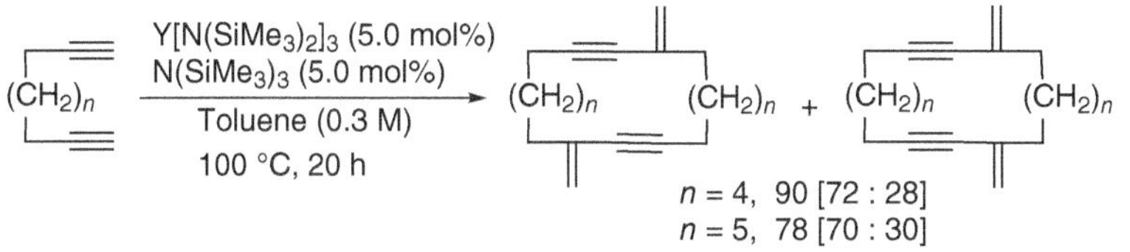

Scheme 1.15 Bisenynes formation via double dimerization of diynes.

In addition, the actinide amides $U[N(SiMe_3)_2]_3$ also show good catalytic activity for conversion of terminal alkynes into dimers, trimers, and trisubstituted benzenes; the outcome of products greatly depends on the nature of terminal alkynes and the catalyst loading [18].

In addition, other rhodium complexes [19], hafnium carboranyl complex [20], have been also confirmed to be the efficient catalysts for the dimerization of terminal alkynes to selectively give *gem*-enynes, and a series of pincer complexes of Rh(I) has been found to be the efficient catalysts for the dimerization of terminal alkynes to afford a regioisomer mixture of *E*- and *gem*-enynes [21].

Moreover, very recently, *gem*-cross-dimerization of aryl alkynes and aliphatic alkynes or gas acetylene under mild reaction conditions has been developed in the presence of $Co(OAc)_2 \cdot 4H_2O$ and phosphine ligand [22].

1.2 *Anti*-Markovnikov (Head-to-Head) Dimerization of Terminal Alkynes

Gevorgyan and coworker studied the dimerization of aromatic terminal alkynes in the presence of $[(\pi\text{-allyl})PdCl]_2$/TDMPP to give the head-to-head *E*-dimers in fair to high yields with excellent regio- and stereoselectivity, although aliphatic terminal alkynes could not produce the corresponding head-to-head dimer at all (Scheme 1.16) [23]. Further studies have revealed that both good yields and high selectivity are observed only for the dimerization of aromatic terminal

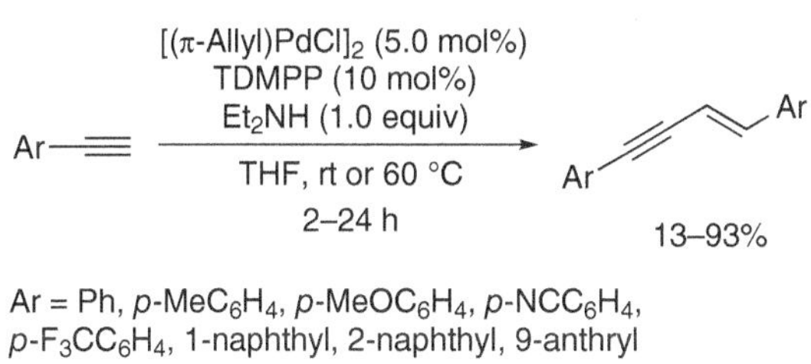

Scheme 1.16 Palladium-catalyzed head-to-head dimerization of terminal aromatic alkynes.

alkynes possessing *ortho*-hydrogen atoms. Therefore, it has been found that the introduction of one group at *ortho*-position substantially diminishes the efficiency of the head-to-head dimerization, and substitution of both *ortho*-hydrogen atoms completely inhibits the process.

However, the combination of bis-*N*-heterocyclic carbene (NHC) palladium complex (IPr—Pd—IPr) and TDMPP has been found to show highly regio- and stereoselective head-to-head dimerization not only for aliphatic terminal alkynes but also for aromatic terminal alkynes even without *ortho*-hydrogen atoms. The reaction is general for a variety of terminal alkynes possessing various functional groups such as aryl, heteroaryl, alkyl, hydroxyl, propargyl ether, and amino groups [24]. In addition, the density functional theory (DFT) calculations have revealed that the reaction proceeds via a hydropalladation pathway (Scheme 1.17). Interestingly, further studies in the same group have found that combination of several NHC-based palladium precursors with phosphine additives selectively promotes head-to-head dimerization of terminal alkynes, but the addition of carboxylate anion to the catalytic system dramatically affects the selectivity favoring the head-to-tail dimerization reaction [25]. On the basis of computational studies, it has been disclosed that the formation of anionic palladium complexes or ion pairs in the presence of carboxylate anion deactivates the hydropalladation pathway, and the head-to-tail dimerization via the carbopalladation pathway is found to be preferential for the carboxylate-assisted reaction.

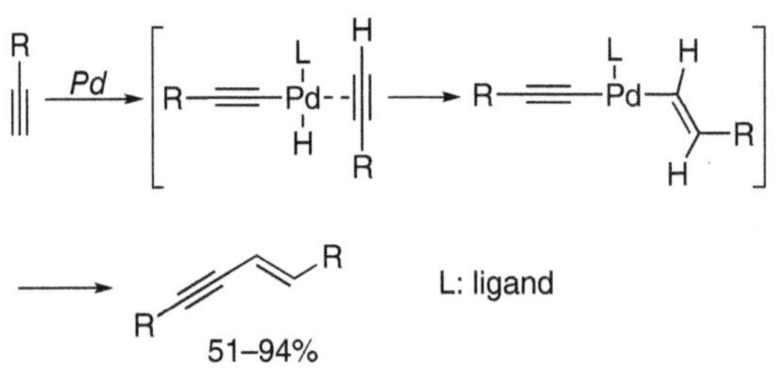

Scheme 1.17 Head-to-head dimerization of terminal acetylenes via a hydropalladation pathway.

In addition, in the presence of a catalytic amount of diethylphosphite, $Ni(cod)_2/t\text{-}Bu_3P$ [26], an iridium(III) hydride complex $IrHCl(TIMP_3)$ (Scheme 1.18) [27], copper(I) or copper(II) salts and oxides [28], and $[ReBr(CO)_3(thf)]_2$/TBAF

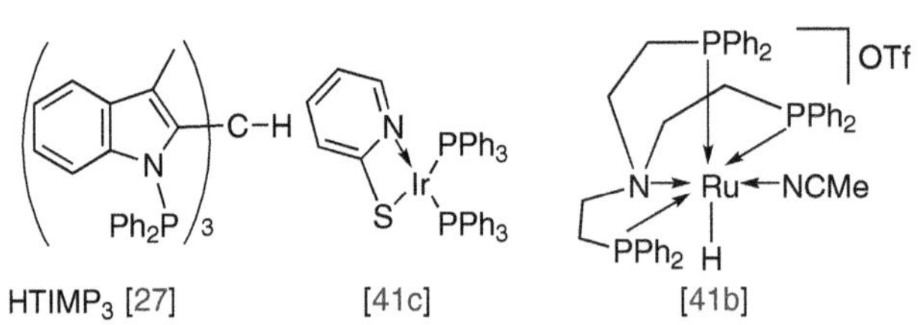

Scheme 1.18 Ligands and catalysts for dimerization of terminal alkynes.

[29] can also promote regioselective head-to-head dimerization of terminal alkynes with the *E*-enynes as major product.

A series of ruthenium complexes have been used as the efficient catalysts in the *anti*-Markovnikov dimerization of terminal alkynes. As shown in Scheme 1.19, the NHC-coordinating unsaturated 16-electron half-sandwich ruthenium complexes can be easily prepared by the reaction of $[Cp^*Ru(OMe)]_2$ with 1,3-diorganylimidazolium chloride in THF and show high catalytic activity in the dimerization of terminal alkynes without α-CH_2 moiety [30]. The regioselectivity of dimerization and the catalytic activity of catalysts greatly depend on the substituent's property of terminal alkynes. In addition, it is surprising that in the case of primary alkynes such as $PhCH_2{\equiv}CH$ used, no catalytic reaction occurs at all.

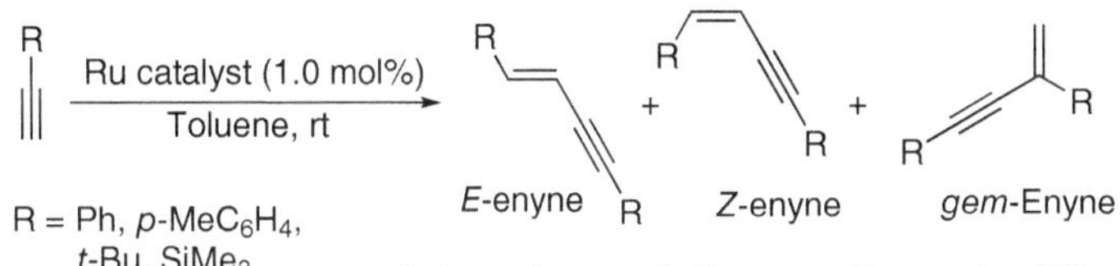

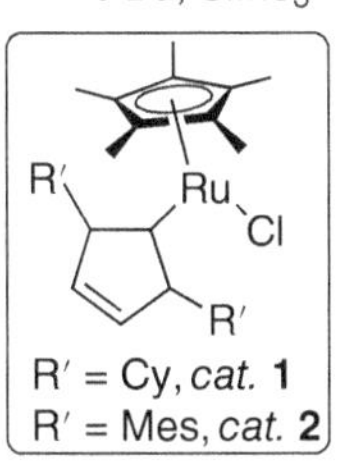

Cat.	R	*E*:*Z*:*gem*	Conversion (%)
1	Ph	76:16:8	100 (5 min)
1	*p*-MeC$_6$H$_4$	44:38:18	100 (5 min)
1	*t*-Bu	90:0:10	100 (10 min)
1	SiMe$_3$	8:0:92	100 (2 h)
2	Ph	0:10:90	95 (2 h)
2	*p*-MeC$_6$H$_4$	0:4:96	95 (2 h)
2	SiMe$_3$	0:0:100	22 (24 h)

Scheme 1.19 Ruthenium/*N*-heterocyclic carbene-catalyzed dimerization of terminal alkynes.

The thiolate-bridged diruthenium complexes $[Cp^*RuCl(\mu_2\text{-}SR)_2RuCp^*Cl]$ (R = Me, Et, *n*-Pr) have been found not only to be the effective catalysts for the head-to-head *Z*-dimerization of terminal alkynes [31] but also to show the good catalytic activity for the intramolecular cyclic dimerization of α,ω-diynes to produce the *endo*-cyclic (*Z*)-1-en-3-ynes in moderate to high yields with complete stereoselectivities (Scheme 1.20) [32].

Kirss and coworker found that ruthenium η^5-pentadienyl complex (η^5-C_5H_7)$(PPh_3)_2$RuCl shows catalytic activity in the dimerization of phenyl acetylene in benzene at 20 °C for 24 hours with nearly quantitative conversion to afford a

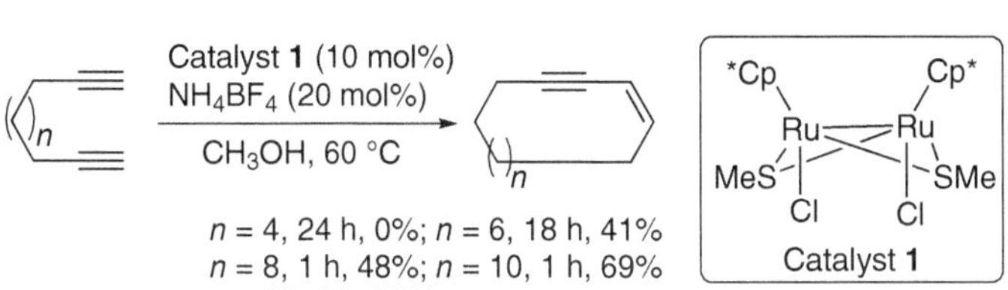

Scheme 1.20 Ruthenium-catalyzed synthesis of the *endo*-macrocycle.

mixture of enynes with an 8 : 1 : 1 ratio of *Z*/*E*/*gem*-enyne (Scheme 1.21) [33]. It was also observed that the reaction solution slowly darkened to a brown color over 30 minutes, which was assumed to be the formation of polyphenyl phosphonium salts generated in situ by cyclotrimerization of phenyl acetylene with a stoichiometric amount of (η^5-C_5H_7)(PPh_3)$_2$RuCl in the presence of KPF_6 in THF by Lin's group work (Scheme 1.22) [34].

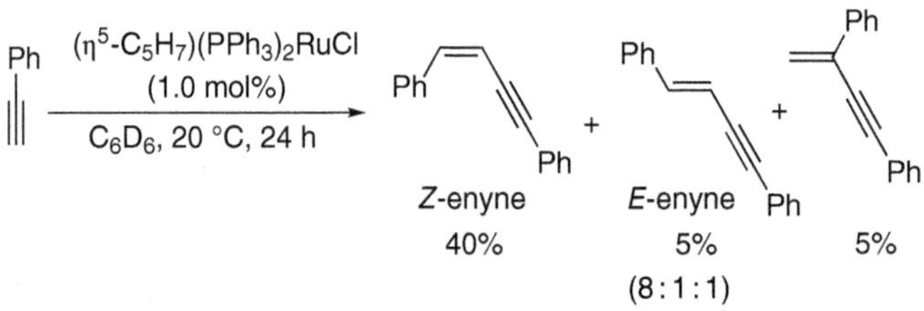

Scheme 1.21 Ruthenium-catalyzed dimerization of phenylacetylene affording *Z*-enyne as a major product.

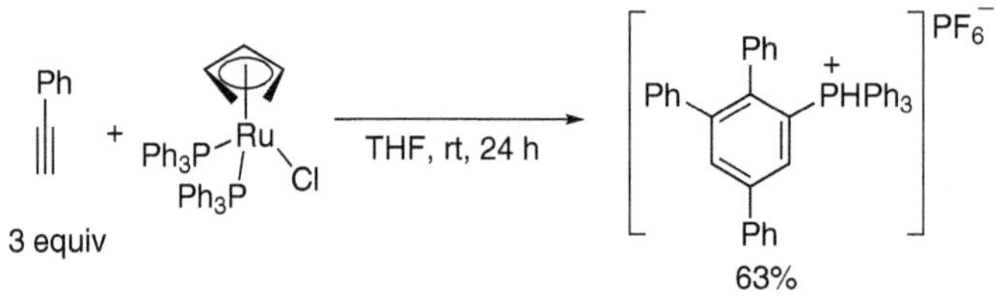

Scheme 1.22 The formation of aryl phosphonium salts.

It has been reported that in AcOH [35] or AcOH/H_2O [36] mixture (1 : 1, v/v) at room temperature, [RuCl(μ-Cl)(η^6-*p*-cymene)]$_2$ could catalyze the dimerization of aromatic alkynes to give (*E*)-1,4-diaryl-1-buten-3-ynes with excellent regio- and stereoselectivity. In addition, [RuCl(μ-Cl)(η^6-*p*-cymene)]$_2$/AcOH could promote the polyaddition of aromatic diynes affording conjugated homo- and copolymers featuring the repeat unit (—Ar—C≡C—CH=CH—) [37].

On the other hand, the (*Z*)-1,4-diphenyl-1-buten-3-yne could be obtained from the dimerization of phenyl acetylene, when a Ru(II) *cis*-dihydride [(PP_3)RuH_2] was used as catalyst [38]. Other ruthenium complexes have been also used as the efficient catalysts in the head-to-head dimerization of aromatic terminal alkynes [39].

Hou and coworker have found that the lanthanide half-metallocene complexes show high catalytic activity in C_6D_6 to catalyze the dimerization of phenylacetylene to extremely afford head-to-head (*Z*)-dimer (Scheme 1.23) [40]. The representative results of the dimerization of various aromatic terminal alkynes with the use of complex **4** as a catalyst are summarized in Scheme 1.24. However, it should be noted that a novel solvent effect on the regioselectivity was also observed. For example, the dimerization of 4-methoxyphenyl acetylene in pure toluene gave a 67 : 33 mixture of the head-to-head and head-to-tail dimers, whereas that in the presence of a small amount of THF (c. 5 equiv per 4) yielded solely the head-to-head

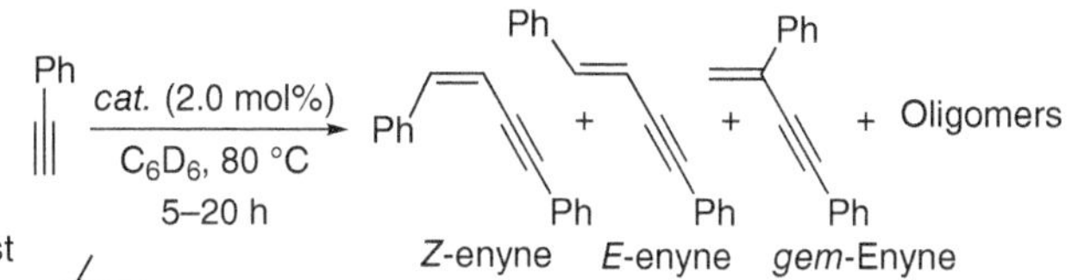

Catalyst

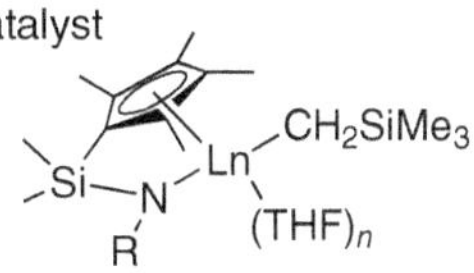

1: Ln = Y, R = Ph, $n = 2$
2: Ln = Yb, R = Ph, $n = 2$
3: Ln = Lu, R = Ph, $n = 2$
4: Ln = Lu, R = 2,4,6-$Me_3C_6H_2$, $n = 1$
5: Ln = Lu, R = 2,6-$Me_2C_6H_3$, $n = 1$
6: Ln = Lu, R = *t*-Bu, $n = 1$

Cat.	Conv.	*Z* : *E* : *gem* : *oligomers*
1	85	89 : 0 : 6 : 5
2	69	92 : 0 : 0 : 8
3	>99	100 : 0 : 0 : 0
4	>99	100 : 0 : 0 : 0
5	>99	100 : 0 : 0 : 0
6	89	72 : 0 : 13 : 15

Scheme 1.23 Dimerization of phenylacetylene catalyzed by lanthanide half-metallocene complexes.

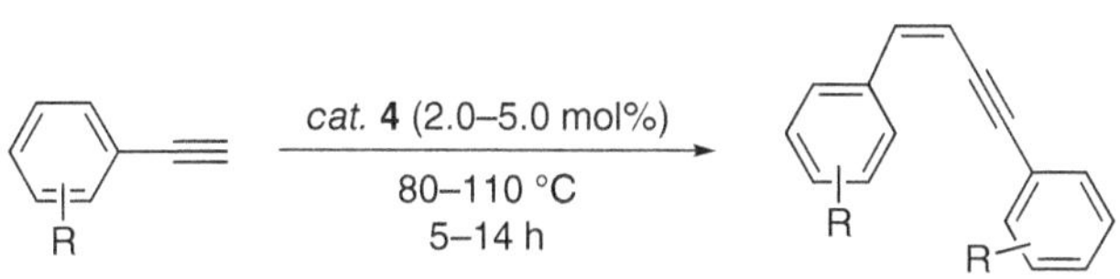

R = Me, MeO, CF_3, Cl, Br, I

Conv.: 97–99%
Select.: 95–99%

Scheme 1.24 (*Z*)-dimerization of aromatic terminal alkynes.

(*Z*)-dimer. Similarly, the dimerization of 1-octyne, an aliphatic terminal alkyne in toluene-d_8, gave the head-to-tail dimer in 76% yield, while that in THF-d_8 afforded the head-to-head *Z*-dimer in 95% yield.

In addition, the excellent regio- and stereoselective dimerization of terminal alkynes to give (*Z*)-enynes can be also achieved by using other transition metal complexes (Scheme 1.18) [41].

It is very important and interesting to develop the cheap catalyst systems for dimerization of terminal alkynes to afford enynes.

With the use of magnesium as activator, $CoBr_2$(dppe) catalyzes the dimerization of phenyl acetylene in acetonitrile, leading to linear (*E*)-enyne in 47% yield as the best result in the absence of a Lewis acid, accompanied with the formation of cyclic trimers (Scheme 1.25) [42]. In the presence of a Lewis acid, the cyclotrimerization process is favored.

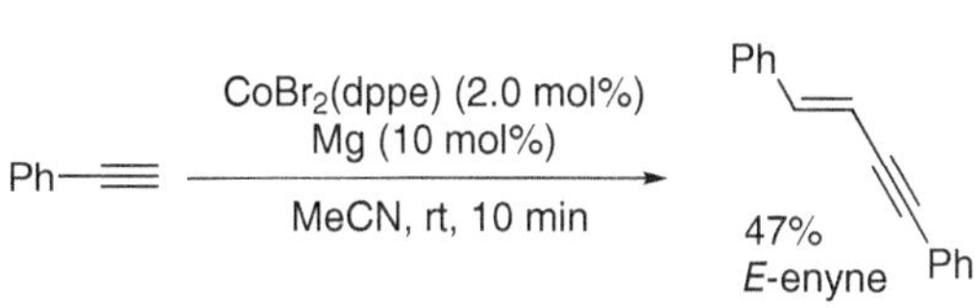

Scheme 1.25 Cobalt-catalyzed dimerization of phenylacetylene in the absence of Lewis acid.

Petit and coworker have also developed a $HCo(PMe_3)_4$-catalyzed highly regio- and stereoselective dimerization of various aromatic terminal alkynes at room temperature to afford corresponding (*E*)-1,4-enynes in good to high yields with excellent selectivity (Scheme 1.26) [43]. This method represents the first general (*E*)-selective dimerization of aromatic terminal alkynes under cobalt catalysis, and the mild catalytic conditions are tolerant of a large range of functionalized aromatic moieties. In addition, DFT calculations have revealed that the reaction proceeds via a C—H activation/hydrocobaltation pathway.

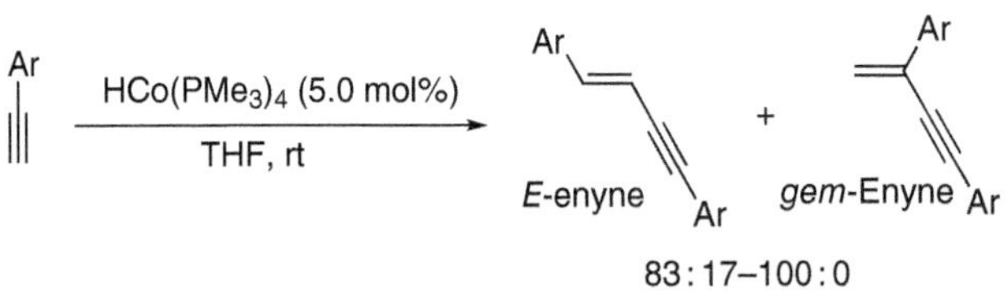

Scheme 1.26 Cobalt-catalyzed dimerization of aromatic terminal alkynes.

Iron salts are inexpensive and readily available. Dash and coworker reported the *E*-selective head-to-head dimerization of aromatic terminal alkynes catalyzed by $FeCl_3$/*N*,*N*-dimethylethylenediamine (DMEDA) in the presence of KO^tBu (Scheme 1.27) [44]. Although the regioselectivity of dimerization is not ideal, this catalytic system represents an alternative to toxic and expensive transition metals for such kind of transformation.

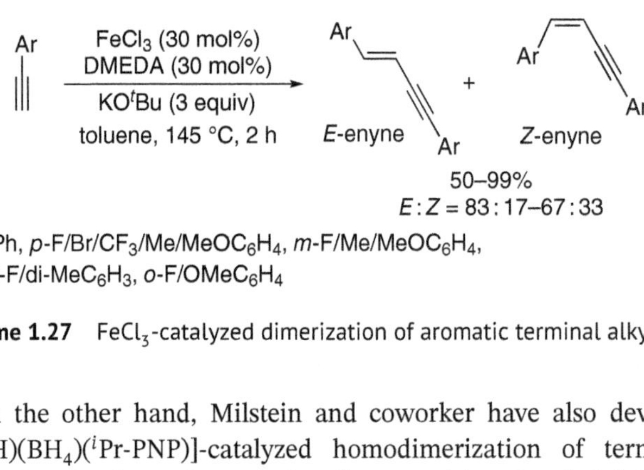

Ar = Ph, *p*-F/Br/CF_3/Me/$MeOC_6H_4$, *m*-F/Me/$MeOC_6H_4$, 3,5-di-F/di-MeC_6H_3, *o*-F/$OMeC_6H_4$

Scheme 1.27 $FeCl_3$-catalyzed dimerization of aromatic terminal alkynes.

On the other hand, Milstein and coworker have also developed an efficient $[Fe(H)(BH_4)(^iPr\text{-}PNP)]$-catalyzed homodimerization of terminal alkynes and cross-dimerization of aromatic alkynes with trimethylsilylacetylene at room temperature, without the need for a base or other additives, to give *Z*-enynes in good to high yields, and the homodimerization of trimethylsilylacetylene afforded *gem*-enyne (Scheme 1.28, **AA**) [45]. As the complementary to *E*- and *Z*-selective iron catalyst, Song and coworker reported a *gem*-specific homo- and cross-dimerization of terminal alkynes catalyzed by a well-defined iron(II) complex containing Cp* and picolyl NHC ligands (Scheme 1.28, **BB**) [46].

Zhao and coworker also reported a catalytic system of $CuSO_4{\cdot}5H_2O$ (0.5 equiv)/diethyl phosphonate (1.1 equiv)/$HNEt_2$ (0.3 equiv) could promote the

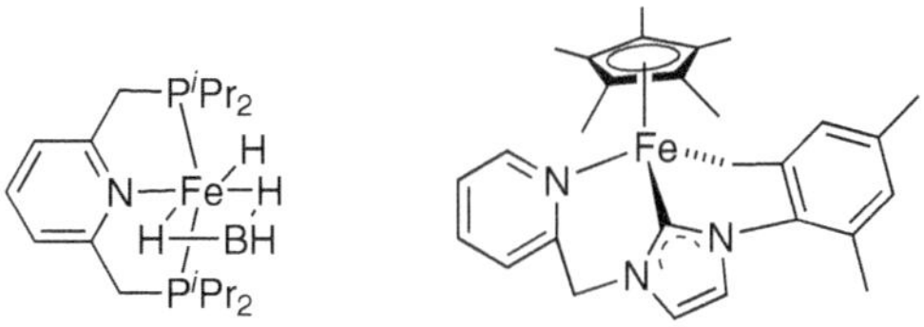

Scheme 1.28 Iron catalysts for dimerization of terminal alkynes.

regioselective head-to-head dimerization of aromatic and aliphatic terminal alkynes to give the conjugated enynes in good to high yields with (*E*)-isomer as major product [47].

In addition, NHCs as the organocatalysts have been applied in the diverse organic transformation [48], and NHCs have also shown the catalytic activity in the head-to-head dimerization of styrenes [49] and dimerization of methacrylonitrile [50].

On the other hand, the cumulated butatrienes (C=C=C=C) with two sp^2 and two sp carbon atoms are highly reactive, which are interesting structures, and their derivatives are the versatile synthons in organic transformations; thus the synthetic methodology and applications of 1,2,3-butatrienes are one of the important topics in organic chemistry [51]. The transition metal-catalyzed dimerization of terminal alkynes has been developed as one of the efficient synthetic methods for the synthesis of 1,2,3-butatrienes.

As an earlier work, Wakatsuki and coworker reported on a catalytic version of the dimerization reaction using ruthenium complexes and phosphine ligands to give 1,2,3-butatrienes [52].

In the case of $[Ir(cod)Cl]_2/PR'_3$ used as catalyst system, the outcome of dimerization of terminal alkynes greatly depends on the nature of substituent of alkynes and the use of ligands to give (*E*)-enynes, (*Z*)-enynes, or 1,2,3-butatrienes [53]. As shown in Scheme 1.29, the use of PPh_3 resulted in the selective formation of (*E*)-enynes for dimethylphenyl alkyne, while the PPr_3 complex provided linear (*Z*)-enynes for the same alkyne or 1,2,3-butatrienes for 3,3-dimethyl-1-butyne.

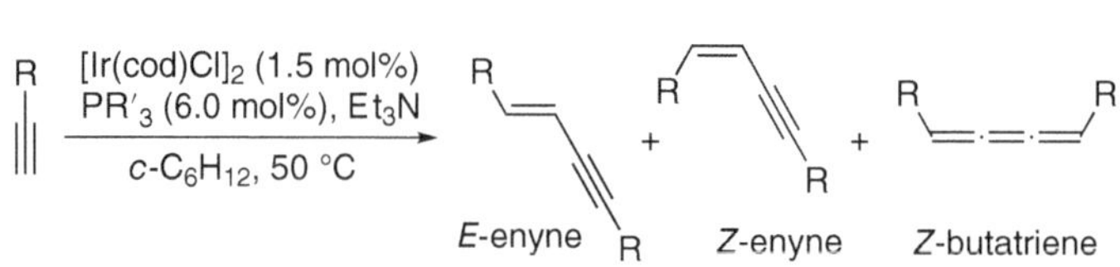

$PR'_3 = PPh_3$, R = $SiMe_2Ph$: 93% (4 h); *E*-enyne : *Z*-enyne = 96 : 4
$PR'_3 = PPr_3$, R = $SiMe_2Ph$: 75% (26 h); *E*-enyne : *Z*-enyne : *Z*-butatriene = 2 : 96 : 2
$PR'_3 = PPr_3$, R = *t*-Bu: 74% (18 h); *E*-enyne : *Z*-enyne : *Z*-butatriene = 0 : 7 : 93

Scheme 1.29 Iridium-catalyzed dimerization of terminal alkynes.

Esteruelas and coworker have reported that in the presence of Et_2NH, $OsHCl(CO)(P^iPr_3)_2$ catalyzes the dimerization of terminal alkynes (R = *t*-Bu, Cy,

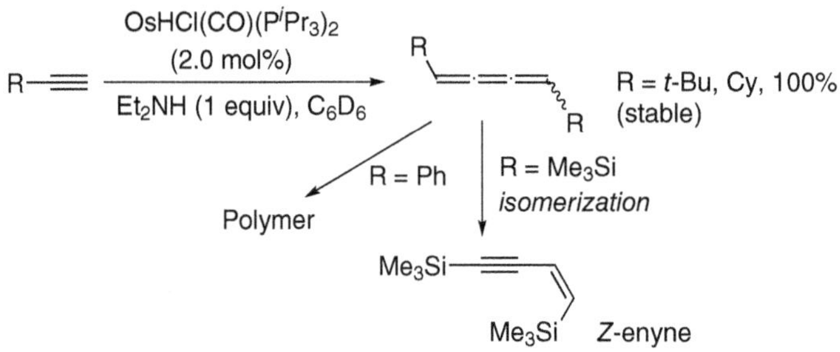

Scheme 1.30 Osmium-catalyzed synthesis of butatrienes from dimerization of terminal alkynes.

Ph, and Me_3Si) to initially give *Z*- or *E*-1,4-disubstituted butatrienes (Scheme 1.30) [54]. It has been found that when R is *t*-Bu or Cy, the corresponding butatrienes are stable under the reaction conditions, and when R is phenyl group, the butatriene undergoes polymerization. In addition, in the case of trimethylsilyl acetylene used, the butatriene isomerizes into the *Z*-enyne.

1.3 Dimerization and Cross-dimerization of Terminal Alkenes

The homo- and cross-dimerization of alkenes are important reactions for synthesis of higher alkenes, which are fundamental unsaturated hydrocarbons in the organic synthesis. The homodimerization of alkenes can afford theoretically four homodimers (Scheme 1.31).

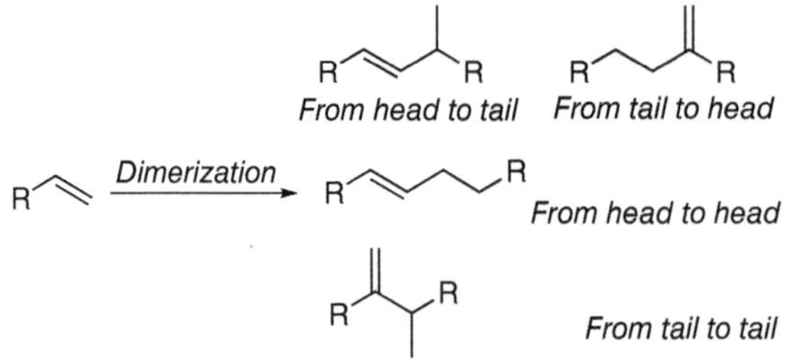

Scheme 1.31 Dimers from dimerization of terminal alkenes.

The study on the homo- and cross-dimerization of alkenes catalyzed by the transition metal complexes has been a long history, and the catalyst systems generally gave (*E*)-1,3-disubstituted-1-butenes via a head-to-tail dimerization [55].

The catalytic dimerization of vinylarenes has been well studied, and the reactions can produce three dimer's isomers depending on the reaction conditions and catalysts (Scheme 1.31) [56].

Shirakawa and coworker have reported a $Pd(OAc)_2/PPh_3/In(OTf)_3$-catalyzed head-to-tail dimerization of vinylarenes to give (*E*)-1,3-diaryl-1-butenes

(Scheme 1.32) [57]. Since lack of $In(OTf)_3$ resulted in no reaction, thus the proposed mechanism involves the activation of vinylarenes by $In(OTf)_3$ to accept nucleophilic attack of palladium(0) complexes, giving oxidative adduct equivalents, which accept insertion of another vinylarene as shown in Scheme 1.33.

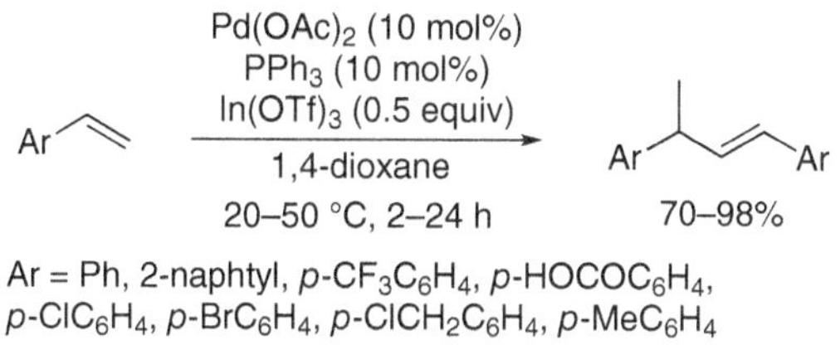

Scheme 1.32 Dimerization of vinylarenes catalyzed by $Pd(OAc)_2/In(OTf)_3$.

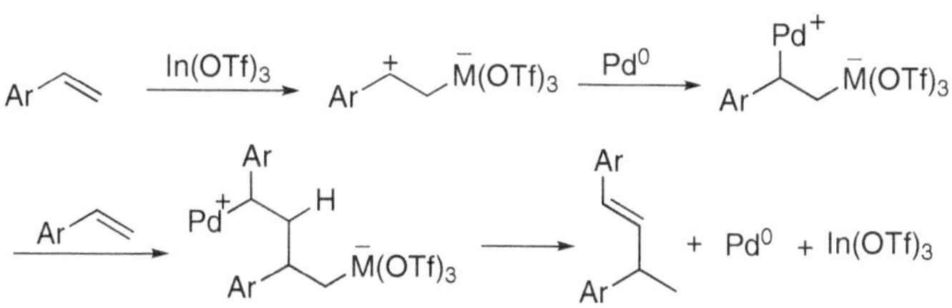

Scheme 1.33 Proposed mechanism of dimerization of vinylarenes catalyzed by $Pd(OAc)_2/In(OTf)_3$.

$NiCl_2$(dppp) also shows high catalytic activity for the head-to-tail dimerization of vinylarenes having electron-donating and electron-withdrawing groups to give (*E*)-1,3-diaryl-1-butenes in high yields [58].

The examples of tail-to-head dimerization of terminal alkenes are very few. Jun and coworker have reported the dimerization of vinylsilanes promoted by a catalyst system of $[(COE)_2IrCl]_2$ and HCl to afford exclusive tail-to-head dimers at room temperature (Scheme 1.34) [59].

The simple $RuCl_3{\cdot}nH_2O$ has been found to be the efficient catalyst in the dimerization of alkenes to afford either Markovnikov adducts or *anti*-Markovnikov adducts in a high stereospecific manner depending on the substituents of alkenes [60]. For example, the dimerization of α-methylstyrene in THF under N_2 for 24 hours produced Markovnikov dimer in 78%, and the dimerization of *N*,*N*-dimethylacrylamide with the relatively low reactivity gave *anti*-Markovnikov

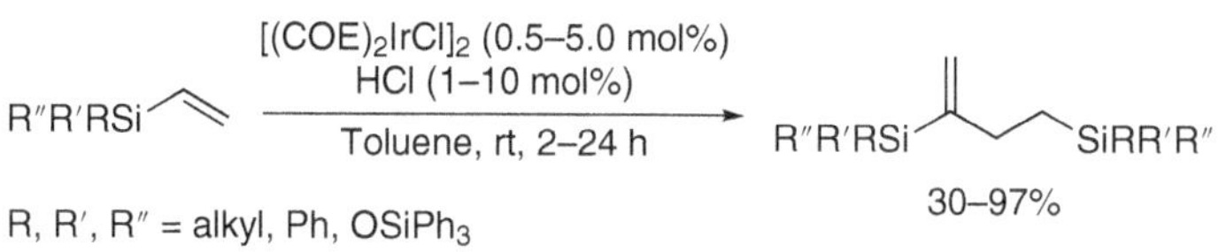

Scheme 1.34 Iridium-catalyzed dimerization of vinylsilanes.

(head-to-head) dimer in 17% yield (Scheme 1.35). However, Kondo and coworker have found that in the presence of primary alcohols, the zero-valent ruthenium complex [Ru(η^6-cot)(η^2-dmfm)$_2$] can efficiently catalyze the head-to-head dimerization of styrenes to give (*E*)-1,4-diaryl-1-butenes (Scheme 1.36) [61]. In addition, this catalyst system is also effective for the selective linear cross-dimerization of styrenes with ethylene to give (*E*)-1-aryl-1-butenes in good yields and high selectivity.

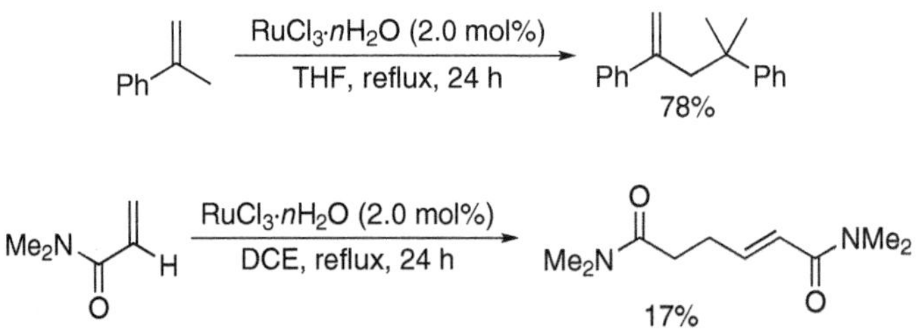

Scheme 1.35 $RuCl_3 \cdot nH_2O$-catalyzed dimerization of terminal alkenes.

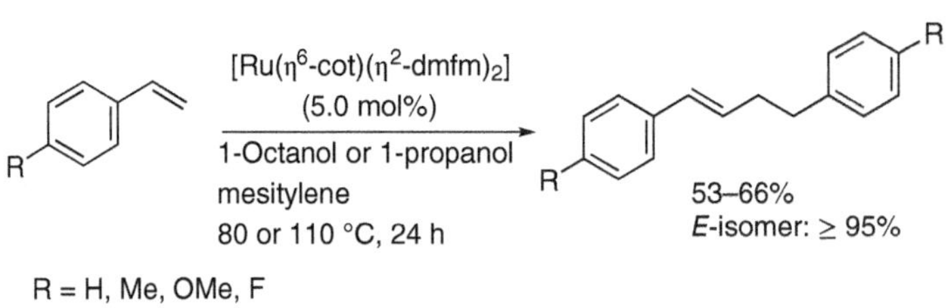

Scheme 1.36 Ruthenium-catalyzed head-to-head dimerization of styrenes.

In addition, in a primary or secondary alcoholic solvents, $RuCl_3$(tpy)/Zn or [$RuCl_2(C_6H_6)]_2$/Zn can catalyze a linear codimerization of 2-norbornenes with acrylic compounds to afford the *exo-trans*-2-norbornylacrylates as major products with regio- and stereoselectivity [62].

The cross-dimerization of terminal alkenes via the addition of C—H bond to the other alkenes is one of the efficient synthetic methods to prepare higher alkenes.

Rhodium chloride has been found to be the efficient catalyst to catalyze the addition of ethylene or propylene to dienes, such as butadiene, isoprene, and 1,3-pentadiene, to afford 1,4-dienes, and rhodium and ruthenium chlorides also catalyze the dimerization of ethylene to butenes, butadiene to 2,4,6-octatriene, and methyl acrylate to dimethyl 2-hexenedioate [63].

Yi and coworker have reported an efficient catalytic system of $RuCl(PCy_3)_2(CO)H$/$HBF_4 \cdot OEt_2$ for the hydrovinylation of alkenes [64]. For example, the reaction of styrene with excess amount of ethylene at room temperature produced the hydrovinylation product in 93% isolated yield (Scheme 1.37). Both terminal alkenes and dienes were found to give the hydrovinylation products in good to high yields.

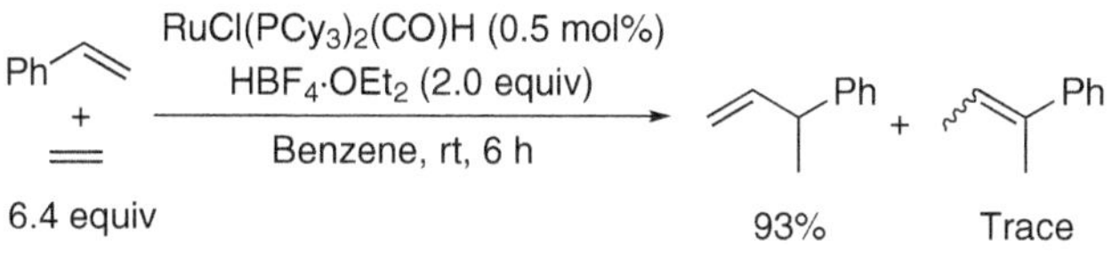

Scheme 1.37 Ruthenium-catalyzed hydrovinylation of alkenes.

The transition metal-catalyzed asymmetric hydrovinylation [65] of alkenes (cross-dimerization of alkene with ethylene) has become one of the efficient methods for synthesis of chiral alkenes.

The use of ruthenium complex with chiral (*S*,*S*)-2-methylbicyclo[3.3.1]nona-2,6-diene ligand results in the asymmetric linear cross-dimerization between methyl methacrylate and 2,5 dihydrofuran to give the cross-dimer in 74% yield in 80% ee (Scheme 1.38) [66].

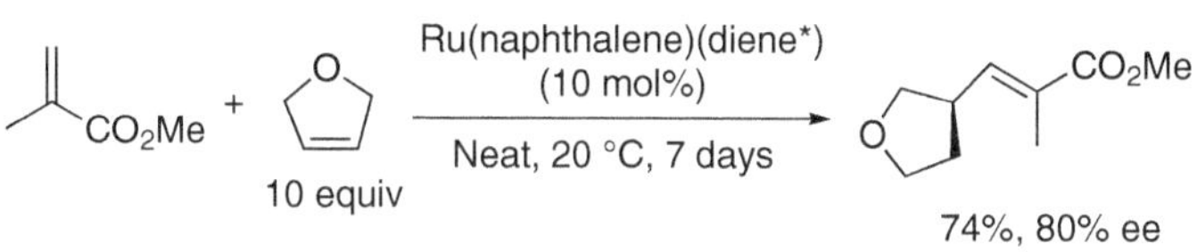

Scheme 1.38 Ruthenium-catalyzed asymmetric cross-dimerization.

The dinuclear rhodium complexes, $[(Cp^*Rh)_2(\mu\text{-}CH_2)_2(MeCN)_2](BF_4)_2$, $[(Cp^*Rh)_2(\mu\text{-}CH_2)_2(CO)_2](BF_4)_2$ [67], and Ru(η^6-naphthalene)(η^4-1,5-cod), show the catalytic activity for head-to-head dimerization of methyl methacrylate (MMA) in MeCN [68].

Ura and coworker have recently found that when $RuCl_3{\cdot}nH_2O$/Zn—Cu/EtOH is used as a catalytic system, *N*-methyl-*N*-vinylacetamide and ethyl acrylate in dimethylamine (DMA) undergo a linear cross-dimerization to give two head-to-head cross-dimers (Scheme 1.39) [69]. The ratio of two cross-dimers greatly depends on the reaction temperature; at 100 °C, the cross-dimerization affords *N*-vinyl product selectively, and at 130 °C, *N*-alkyl product is also formed. In addition, the effect of solvents is considerable to affect the outcome of two products.

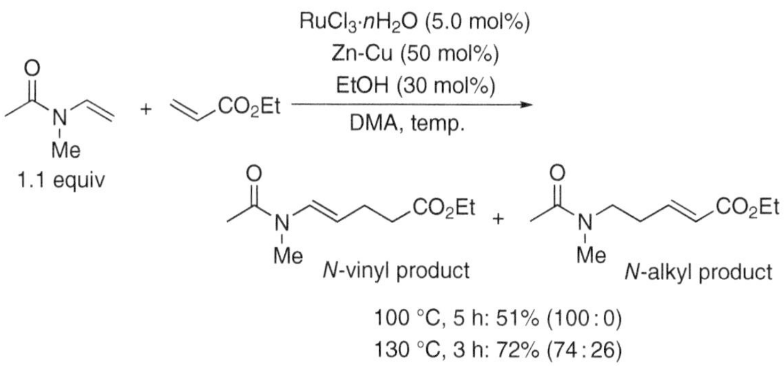

Scheme 1.39 Ruthenium-catalyzed cross-dimerization of terminal alkenes.

The results of stoichiometric and catalytic reactions have revealed that the present cross-dimerization proceeds via ruthenacyclopentane intermediates.

The first highly selective intermolecular tail-to-tail homodimerization of styrenes and cross-dimerization of styrenes with unactivated alkenes to produce branched 1,1-disubstituted alkenes were reported by Ho's group [70]. As shown in Scheme 1.40, in the presence of a catalytic amount of in situ generated [(IPr)NiH]OTf, at room temperature styrene and 1-octene (3 equiv) undergo the tail-to-tail cross-dimerization to give 2-(1-phenylethyl)-1-octene, accompanied with the formation of 2,3-diphenyl-1-butene from the tail-to-tail homodimerization of styrene in a total yield of 95%. The ratio of cross-dimer: homodimer is 90 : 10.

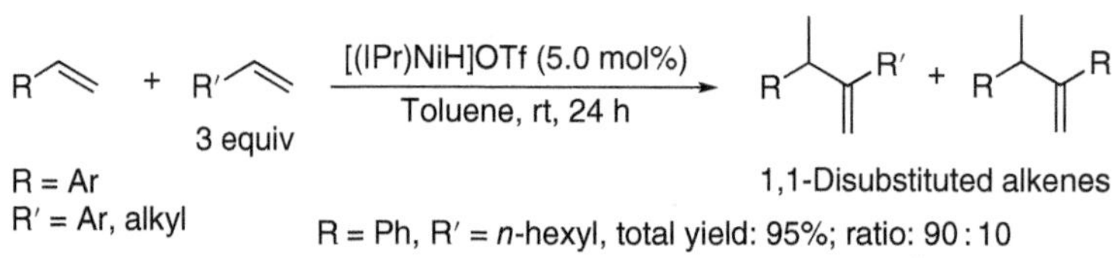

Scheme 1.40 Nickel/NHC-catalyzed intermolecular tail-to-tail hydroalkenylation of styrenes with alkenes.

Interestingly, Ritter and coworker reported an $FeCl_2$/iminopyridine-catalyzed 1,4-addition of α-alkenes to 1,3-dienes with high stereo- and regioselectivity to give linear 1,4-diene adducts [71]. For example, the addition of styrene to 2,3-dimethylbutadiene affords 1,4-diene in 94% yield, providing an efficient method for access to 1,4-dienes from 1,3-dienes (Scheme 1.41).

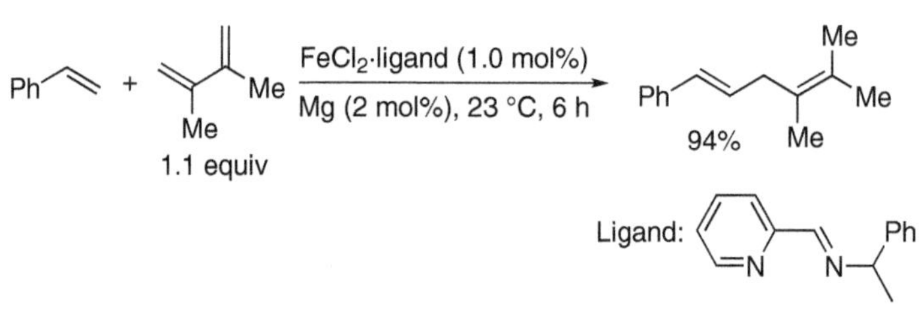

Scheme 1.41 Iron-catalyzed 1,4-addition of α-alkenes to 1,3-dienes.

Very recently, the iridium-catalyzed asymmetric hydroalkenylation of norbornenes with terminal alkenes has been reported [72].

1.4 Cross-dimerization of Different Alkynes or Alkynes with Alkenes

Development of the cross-dimerization of terminal alkynes or alkynes with alkenes is one of the challenging research topics due to the competitive homodimerization or di- and trimerization of the terminal alkynes and terminal alkenes (Scheme 1.42).

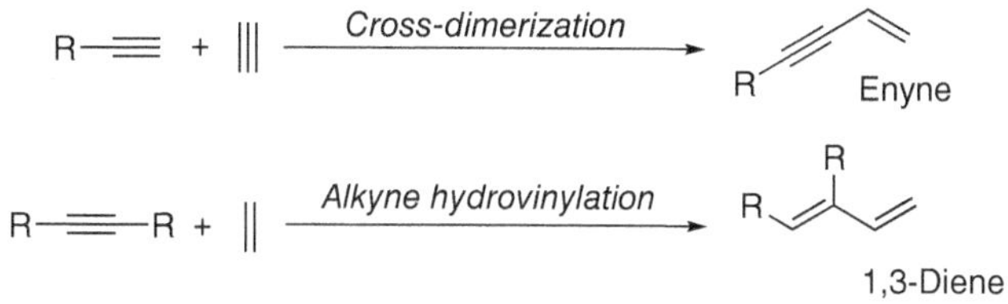

Scheme 1.42 Cross-dimerization of different alkynes or alkynes with alkenes.

In order to realize the chemoselective cross-dimerization, the catalyst systems should be able to selectively catalyze the reaction.

Trost and coworker reported the first $Pd(OAc)_2$/TDMPP-catalyzed cross-dimerization of a variety of terminal alkynes, RC≡CH (donor alkyne, R = alkyl, aryl, silyl) with electron-deficient internal alkynes and R′C≡C—EWG (acceptor alkyne or activated alkyne, R′ = alkyl, silyl; EWG = CO_2Me, COMe, SO_2Ph) affording *E*-regioisomer as single geometric isomer arising from head-to-tail cross-coupling reaction (Scheme 1.43) [73].

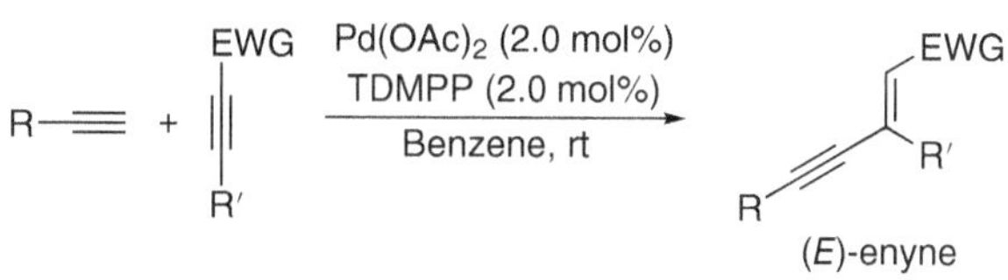

Scheme 1.43 Cross-dimerization of terminal alkynes with electron-deficient internal alkynes.

Their further studies disclosed that under the similar reaction conditions, the "unactivated" internal alkynes such as 2-butyn-1,4-diol and its diacetate derivative, 1-trimethylsilylpropargyl alcohol, could also undergo the cross-dimerization with terminal alkynes with good chemo-, regio-, and stereoselectivities [74]. Scheme 1.44 shows the cross-dimerization of trimethylsilylacetylene (5.0 equiv) to the propargyl alcohol at room temperature for 48 hours to afford the corresponding cross-dimer in 58% yield. The highly selective cross-dimerization of terminal alkynes with propargyl alcohol was applied in the synthesis of caulerpenyne analog.

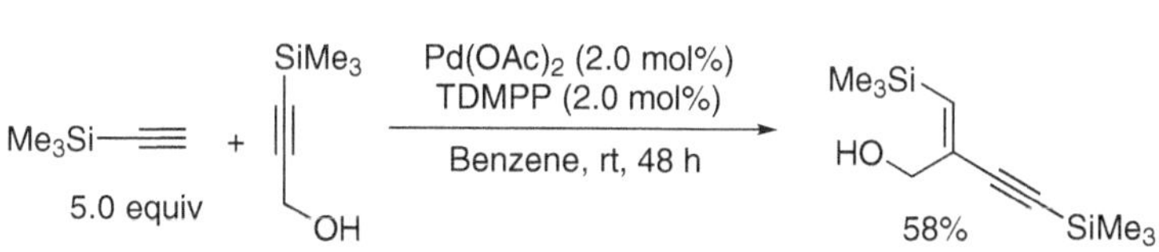

Scheme 1.44 Palladium-catalyzed cross-dimerization of terminal alkynes with unactivated internal alkynes.

The combination of CuBr (5.0 mol%) and $PdCl_2(PPh_3)_2$ (2.5 mol%) in water has been also found to be the efficient catalyst system to catalyze only the addition of terminal alkynes to electron-deficient alkynes selectively, but not the homodimerization of the terminal alkynes (Scheme 1.45) [75].

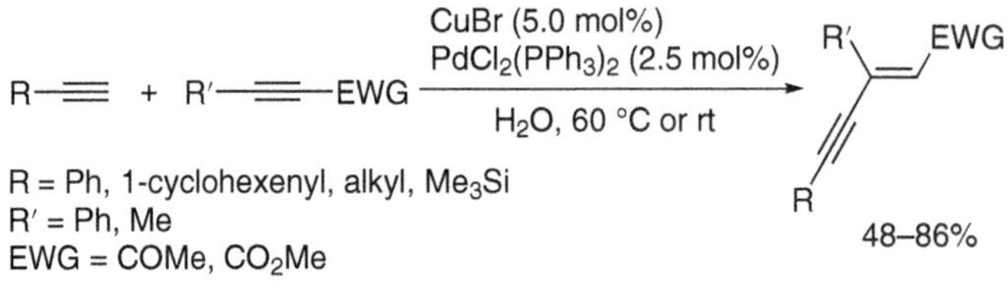

Scheme 1.45 Copper-/palladium-catalyzed cross-dimerization of terminal alkynes with activated alkynes in water.

$[IrCl(cod)]_2$/phosphines also show the catalytic activation to catalyze the cross-dimerization of electron-rich terminal alkyne with an equivalent of electron-deficient internal alkyne such as alkynyl esters and alkynyl aldehydes [76]. The regioselectivity of *cis*-adduct via the addition of C—H bond to internal alkynes markedly depends on the ligands used, and when (*rac*)-bis(diphenylphosphino)-1,1′-binaphthalene (BINAP) was employed as a ligand, the cross-dimerization occurred with excellent regioselectivity. In addition, when dppe was used as a ligand, the substituted benzene derivatives from 1 : 2 cross-cyclotrimerization of terminal alkynes with dimethyl acetylenedicarboxylate (two equivalents) could be obtained in fair yield, and the yield can be improved in the presence of a small amount of Na_2CO_3 under mild conditions (Scheme 1.46).

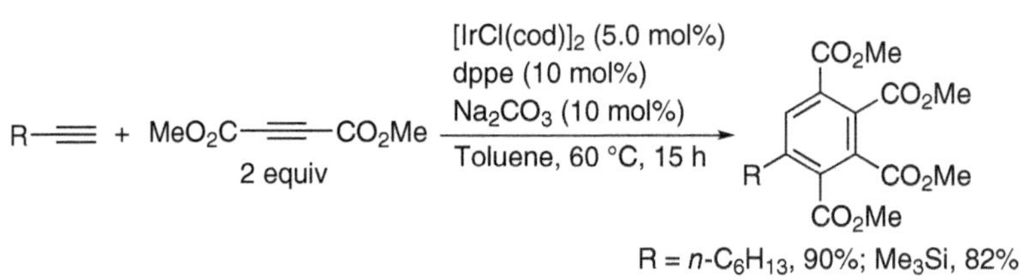

Scheme 1.46 Synthesis of substituted benzenes via 1 : 2 cross-cyclotrimerization.

Miura and coworker have described a remarkable ligand effect on the $Ni(cod)_2$-catalyzed cross-dimerization of aromatic alkynes with terminal silylacetylenes [77], and then they have also developed an interesting transition metal catalyst-controlled

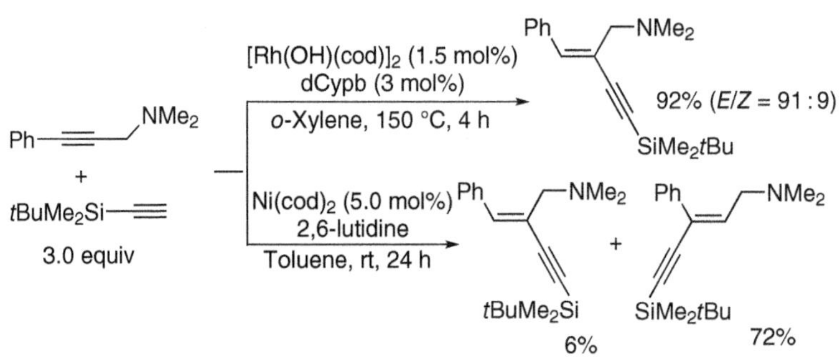

Scheme 1.47 Catalyst-dependent regioselective cross-dimerization of terminal silylacetylene to propargyl amine.

switching of regioselective cross-dimerization terminal silylacetylenes to γ-arylated propargyl amines [78]. As shown in Scheme 1.47, in the presence of $[Rh(OH)(cod)]_2$/dCypb, the reaction of *t*-butyldimethylsilyl-acetylene with dimethyl(3-phenyl-2-propynyl)amine in *o*-xylene at 150 °C (bath temperature) produced 2-alkynylallylamine as the sole regioisomer in excellent yield with good stereoselectivity. On the other hand, if the same reaction was carried out using $Ni(cod)_2$/2,6-lutidine as catalyst system in toluene at room temperature, 3-alkynylallylamine was obtained as the major adduct in good yield.

Other rhodium complexes [79] and $CoCl_2 \cdot 6H_2O$/dipime/Zn [80] have been also used in the cross-dimerization of terminal alkynes or internal alkynes with terminal alkynes.

Yi and coworker have reported a ruthenium acetylide complex $Cp^*(PPh_3)RuC{\equiv}CPh$, generated in situ from the reaction of $Cp^*(PPh_3)(Cl)Ru{=}C{=}CHPh$ with a base to be the efficient catalyst for the cross-dimerization of terminal alkynes with activated and unactivated internal alkynes to yield enynes with high regio- and stereoselectivities, when R groups of terminal alkynes are *t*-Bu, *s*-Bu, and Me_3Si (Scheme 1.48) [81]. In addition, an intermediate of β-agostic species could be isolated from the reaction of $Cp^*(PPh_3)(Cl)Ru{=}C{=}CH^tBu$ with 2-butyne in the presence of NaOMe, which showed an equally effective catalyst for the cross-dimerization (Scheme 1.49).

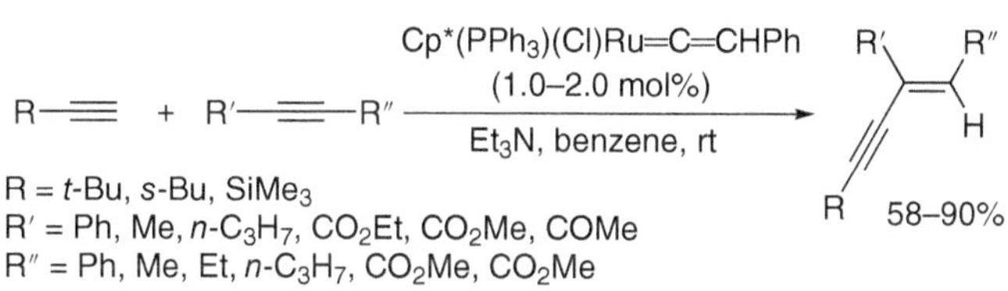

Scheme 1.48 Ruthenium-catalyzed cross-dimerization of terminal alkynes with internal alkynes.

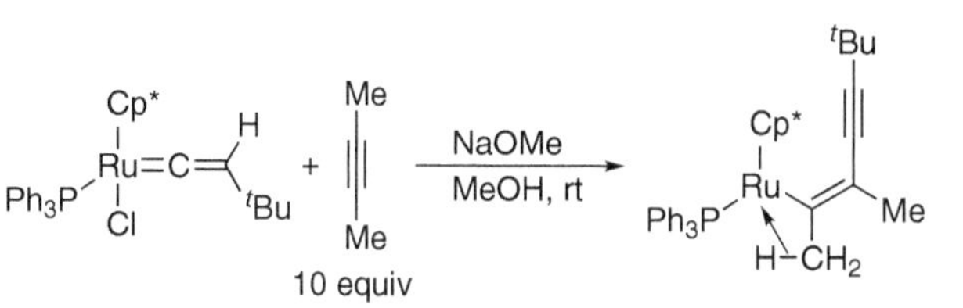

Scheme 1.49 Preparation of ruthenium β-agostic intermediate.

In addition, the use of $Pd(OAc)_2$ with some P,N-ligands shows effective activity for the homodimerization of terminal alkynes and cross-dimerization of terminal alkynes with internal alkynes with excellent yields even under solvent-free conditions. More importantly, in case of racemic propargylic alcohols employed as one of the reactants, the use of enantiomerically pure ligands resulted in the kinetic resolution of racemic propargylic alcohols [82]. For example, in the presence of

PHOX-ligand, the solvent-free reaction of 2.2 equiv of racemate 1-nonyn-3-ol with 1.0 equiv of ethyl 2-butyrate produced the cross-dimer in 43% yield with 53% ee (Scheme 1.50).

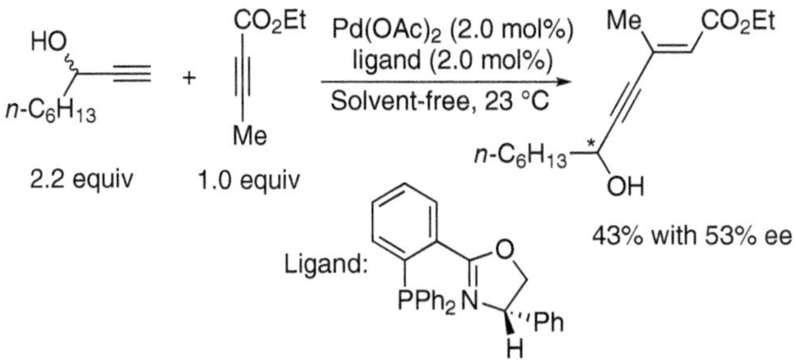

Scheme 1.50 Palladium-catalyzed cross-dimerization with kinetic resolution.

In addition, in the presence of palladium, the E-enynes formed in situ have been found to undergo a subsequent [4+2] benzannulation with 1,3-diynes to give pentasubstituted benzenes in moderate to good yields. Therefore, Yamamoto and coworker have developed a highly chemo- and regioselective formation of the benzene ring by a formal [2+2+2] sequential intermolecular trimerization of alkynes from three different acetylenic units (Scheme 1.51) [83].

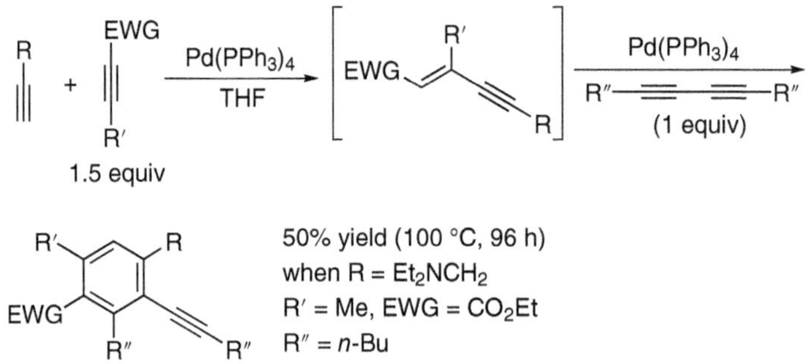

Scheme 1.51 Synthesis of pentasubstituted benzenes via formal [2 + 2 + 2] sequential cycloaddition of alkynes.

Other palladium complexes have been also found to be the efficient catalysts in the cross-dimerization of intermolecular terminal alkynes with internal alkynes [84].

It is much more interesting and challenging to develop the efficient catalyst system for the cross-dimerization between two kinds of terminal alkynes.

It has been found that the reaction of $(\eta^5\text{-}C_5Me_5)_2TiCl_2$ with two equivalents of ^{i}PrMgBr generated in situ will be the species of $(\eta^5\text{-}C_5Me_5)_2TiH$, which behaves as an efficient catalyst not only for the regioselective dimerization of terminal alkynes to 2,4-disubstituted 1-buten-3-ynes (head to tail, selectivity > 99%) in excellent yields but also for the chemoselective cross-dimerization of ethynyl-1-cyclohexene or phenylacetylene with less acidic terminal alkynes (R′C≡CH) (Scheme 1.52) [85].

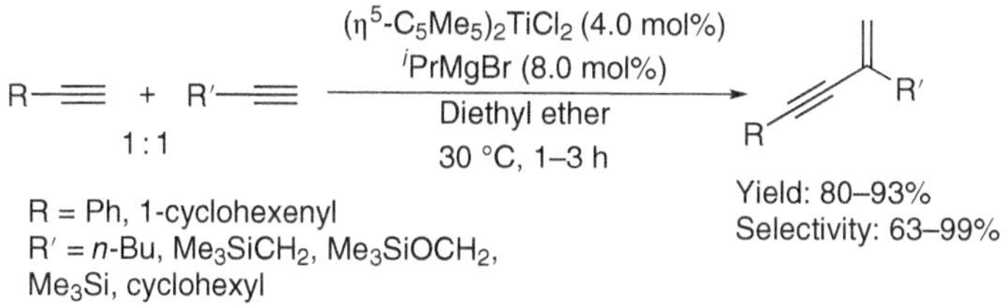

Scheme 1.52 Chemoselective head-to-tail cross-dimerization of two terminal alkynes.

However, note that when the cross-dimerization was carried out using a 1 : 1 mixture of two different normal alkyl-substituted alkynes, for example, 1-butyne and 1-hexyne, the four isomers were obtained with low chemoselectivity. The chemoselectivity can be improved by using one of the alkyne bearing secondary or tertiary alkyl substituents or Me_3Si group.

At room temperature, although $RuCl_2(=C=CHPh)(P^iPr_3)_2$/*N*-methylpyrrolidine shows high catalytic activity for the selective dimerization of electron-rich arylacetylenes to yield (*Z*)-ArCH=CH—C≡CAr in 90–97% selectivities, and the dimerization of (trimethylsilyl)acetylene affords a 15 : 85 mixture of (*E*)- and *gem*-dimerization products, but as shown in Scheme 1.53, the same catalyst system serves as good catalyst precursor for (*Z*)-selective cross-dimerization between two kinds of terminal alkynes: arylacetylenes and silylacetylenes [86]. The role of *N*-methylpyrrolidine is to abstract HCl, when a proposed catalytically active intermediate of alkynylruthenium species $[RuCl(C{\equiv}CPh)(P^iPr_3)_2]$ is formed.

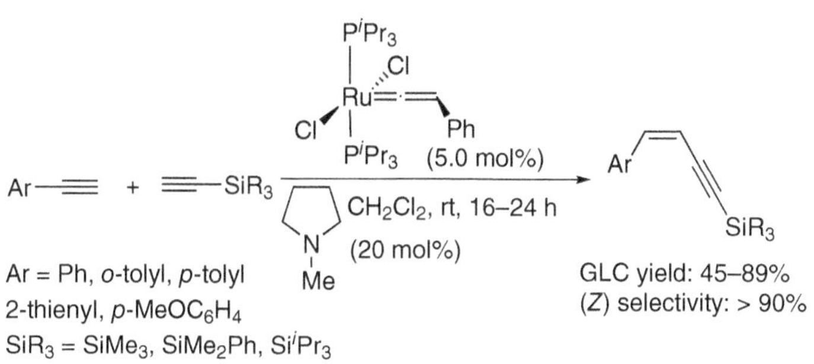

Scheme 1.53 Cross-dimerization of arylacetylenes with silylacetylenes catalyzed by vinylideneruthenium complexes.

In addition, other metal complexes have also been demonstrated to be the active precatalysts for the cross-dimerization of terminal alkynes: cationic actinide complex $[(Et_2N)_3U][BPh_4]$ with fair chemoselectivity [87], iridium(I) guanidinate complex phosphine [88], and dichlorocobalt(II) complex bearing a sterically demanding 2,9-bis(2,4,6-triisopropylphenyl)-1,10-phenanthroline [89].

On the other hand, 1,3-dienes are useful and versatile intermediates in organic synthesis via the transformation of carbon–carbon double bond and the activation of $C(sp^2)$—H bond. Therefore, it is one of the important and interesting research topics to develop the efficient synthetic methods of 1,3-dienes. The transition metal-catalyzed intermolecular cross-dimerization of alkynes with alkenes via the activation of C(sp^2)—H bond has been well studied to be a straightforward and atom-efficient approach (Scheme 1.54).

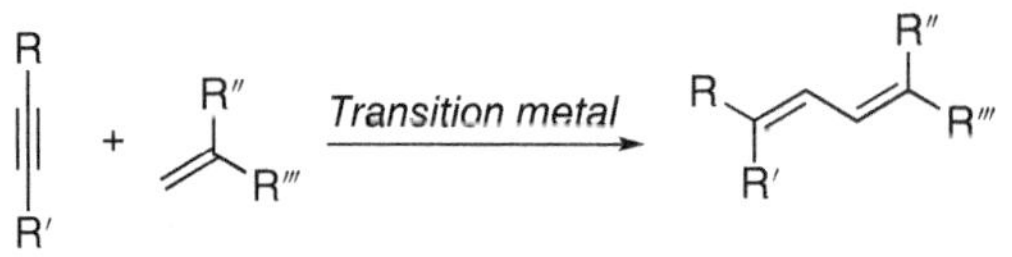

Scheme 1.54 Synthesis of 1,3-dienes via transition metal-catalyzed intermolecular enyne coupling reaction.

The first efficient catalytic linear cross-dimerization of alkynes with alkenes to give 1,3-dienes was reported in 1991 by Watanabe and coworker [90]. In the presence of Ru(cod)(cot), the intermolecular cross-dimerization of electron-rich alkynes, such as diphenylacetylene, 1-phenyl-1-propyne, 3-hexyne, and 3,3-dimethyl-1-butyne with electron-deficient α,β-unsaturated alkenes such as alkyl acrylates and *N*,*N*-dimethylacrylamide, occurs to give 1,3-dienes in fair to good yields with high regioselectivity. Scheme 1.55 shows the results from the reaction of diphenylacetylene with acrylates and acrylamide under different conditions.

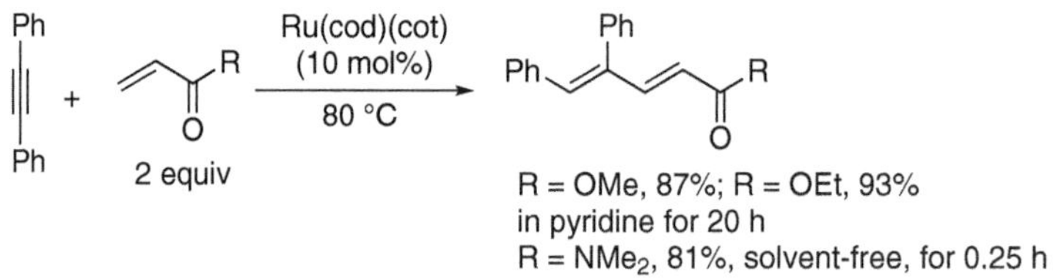

Scheme 1.55 Ruthenium-catalyzed synthesis of 1,3-dienes via cross-dimerization of alkynes with α,β-unsaturated alkenes.

With the use of bis(triphenylphosphine) iminium chloride ([PPN]Cl) or LiI as additives, $Ru_3(CO)_{12}$ catalyzes the reaction of terminal alkynes with alkyl acrylates in different chemoselective manner [91]. In the former case, γ,δ-alkynyl esters could

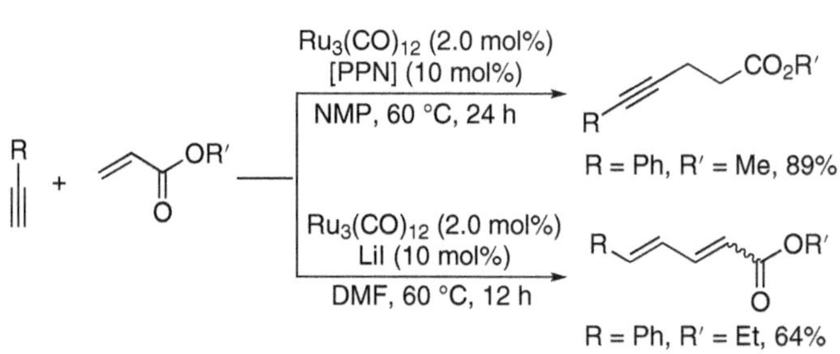

Scheme 1.56 $Ru_3(CO)_{12}$-catalyzed cross-dimerization of terminal alkynes with acrylates.

be obtained in good to high yields via a formal hydroalkynylation of alkyl acrylates; in the latter case, the linear cross-dimerization of terminal alkynes with alkyl acrylates via hydrovinylation of alkynes proceeds to give the corresponding conjugate dienes (Scheme 1.56). These two differently chemoselective carbon–carbon bond formation ways are controlled only by the nature of halide ions, either a chloride or an iodide with the use of additives.

In addition, the catalytic asymmetric hydroalkynylation of alkenes has been recently developed by Li's group [92].

$CpRu(PPh_3)_2Cl/NaPF_6$ shows the catalytic activity for the cross-dimerization of unactivated alkenes with unactivated alkynes in pyridine to provide an important method for the preparation of 1,3-dienes without electron-withdrawing group [93]. For example, the reaction of phenylacetylene with an excess amount of 1-octene yields linear and branched dienes in 65 and 12% yields, respectively (Scheme 1.57).

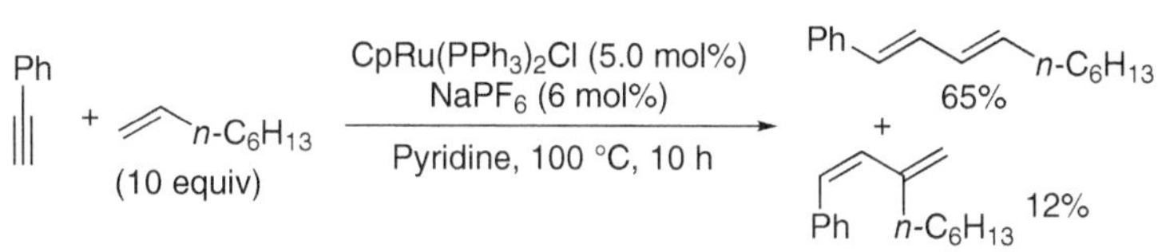

Scheme 1.57 Ruthenium-catalyzed cross-dimerization of unactivated alkene with unactivated alkyne.

Cheng and coworker have first developed a simple catalyst system for the reductive cross-dimerization of internal alkynes with electron-deficient conjugated alkenes in the presence of $CoI_2(PPh_3)_2/PPh_3/Zn$ with highly chemo-, regio-, and stereoselectivity (Scheme 1.58) [94].

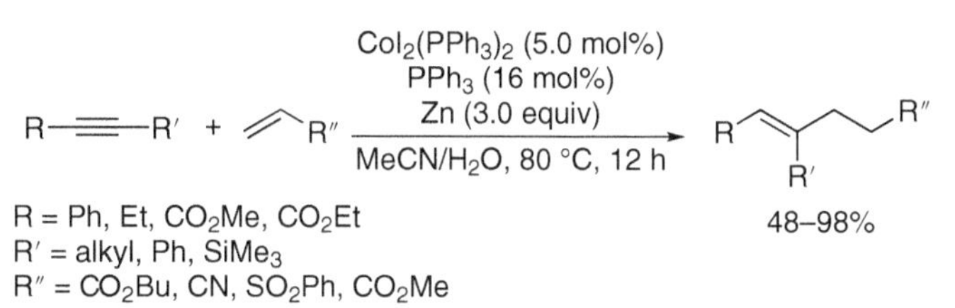

Scheme 1.58 Cobalt-catalyzed reductive cross-dimerization of internal alkynes with conjugated alkenes.

The cobalt-catalyzed Alder-ene reaction between internal alkynes and terminal alkenes to give 1,4-dienes was first reported by Hilt's group (Scheme 1.59) [95]. The reaction is proposed to involve the coordination of the two starting unsaturated compounds in the coordination sphere of the cobalt center to form a cobalt acycle intermediate and a subsequent β-hydride elimination and reductive elimination to yield 1,4-dienes.

Trost and coworker have reported a regioselective ruthenium complexes catalyzed by the cross-dimerization of silylalkynes with terminal alkenes with

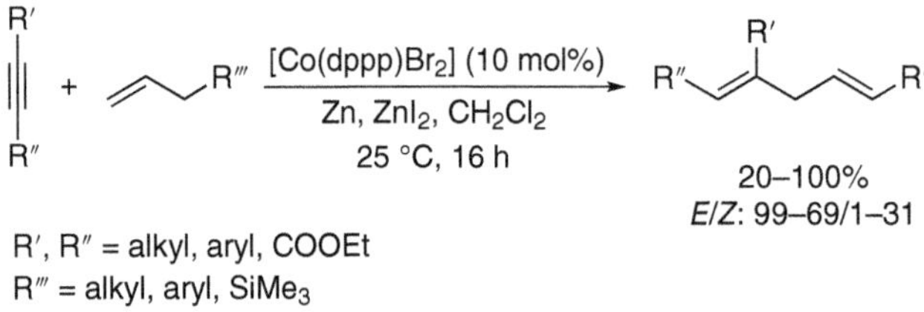

Scheme 1.59 1,4-Diene syntheses by a cobalt-catalyzed Alder-ene reaction.

complete control of regioselectivity by the silyl substituent to give geometrically defined vinylsilanes (Scheme 1.60) [96]. This protocol can be used as one of the key steps in the total synthesis of amphidinolide P [97].

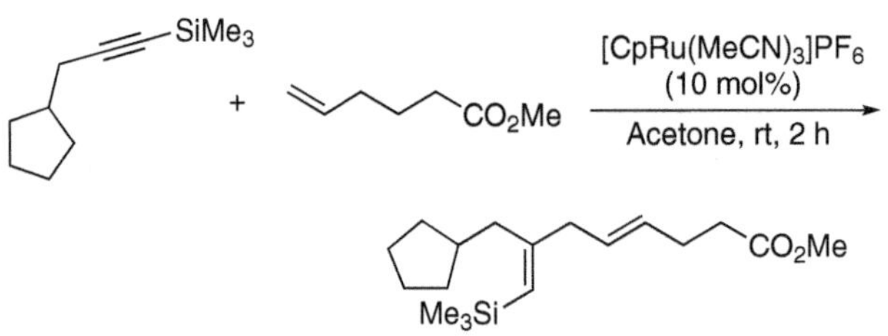

Scheme 1.60 Ruthenium-catalyzed regioselective cross-dimerization of silylalkyne with terminal alkene.

Also, $[PdCl_2(cod)]$ or $[Pd_2(dba)_3]$ [98], $[Rh(cod)_2]BF_4$ [99], $CoI_2/Zn/ZnI_2$ [100], and $[(PPh_3)_3RuH(CO)Cl]$ [101] also have been reported in the applications of cross-dimerization of alkyne with alkenes to efficiently give 1,3-dienes.

References

1 Selected reports, see: (a) Liu, Y., Nishiura, M., Wang, Y., and Hou, Z. (2006). *J. Am. Chem. Soc.* 128: 5592–5593. (b) Chen, G., Wang, L., Thompson, D.W., and Zhao, Y. (2008). *Org. Lett.* 10: 657–660.

2 Selected reports, see: (a) Sun, Z.-J., Jiang, C.-S., Chen, X.-Q. et al. (2014). *Tetrahedron* 70: 3166–3171. (b) El-Shazly, M., Barve, B.D., Korinek, M. et al. (2014). *Curr. Top. Med. Chem.* 14: 1076–1093. (c) Lan, P., White, L.E., Taher, E.S. et al. (2015). *J. Nat. Prod.* 78: 1963–1968. (d) Taher, E.S., Guest, P., Benton, A. et al. (2017). *J. Org. Chem.* 82: 211–233. (e) Wright, P.D., Veale, E.L., McCoull, D. et al. (2017). *Biochem. Biophys. Res. Commun.* 493: 444–450.

3 Selected reports, see: (a) Aubert, C., Buisine, O., and Malacria, M. (2002). *Chem. Rev.* 102: 813–834. (b) Wang, H.-Y., Zhang, W., Schienebeck, C.M. et al. (2014). *Org. Chem. Front.* 1: 386–390. (c) Ye, J. and Ma, S. (2014). *Org. Chem. Front.* 1: 1210–1224. (d) Mori, Y., Onodera, G., and Kimura, M. (2014). *Chem. Lett.* 43: 97–99. (e) Zhou, Y., Zhang, Y., and Wang, J. (2016). *Org. Biomol. Chem.* 14: 6638–6650. (f) Röse, P., Garcia, C.C.M., Pünner, F. et al. (2015). *J. Org.*

Chem. 80: 7311–7316. (g) Trost, B.M. and Masters, J.T. (2016). *Chem. Soc. Rev.* 45: 2212–2238. (h) Wei, X.-F., Xie, X.-W., Shimizu, Y., and Kanai, M. (2017). *J. Am. Chem. Soc.* 139: 4647–4650. (i) Partridge, B.M., Callingham, M., Lewis, W., and Lam, H.W. (2017). *Angew. Chem. Int. Ed.* 56: 7227–7232. (j) Yuan, S.-T., Zhou, H., Gao, L. et al. (2018). *Org. Lett.* 20: 562–565. (k) Huang, Y., del Pozo, J., Torker, S., and Hoveyda, A.H. (2018). *J. Am. Chem. Soc.* 2643–2655.

4 Selected reports, see: (a) Backvall, J.E., Chinchilla, R., Najera, C., and Yus, M. (1998). *Chem. Rev.* 98: 2291–2312. (b) Clement, N.D., Routaboul, L., Grotevendt, A. et al. (2008). *Chem. Eur. J.* 14: 7408–7420. (c) Xi, Z. (2010). *Acc. Chem. Res.* 43: 1342–1351. (d) Jadwiga Pyziak, J., Walkowiak, J., and Marciniec, B. (2017). *Chem. Eur. J.* 23: 3502–3541.

5 Kovalev, I.P., Yevdakov, K.V., Strelenko, Y.A. et al. (1990). *J. Orgonomet. Chem.* 386: 139–146.

6 Boese, W.T. and Goldman, A.S. (1991). *Organometallics* 10: 782–786.

7 Peng, H.M., Zhao, J., and Li, X. (2009). *Adv. Synth. Catal.* 351: 1371–1377.

8 Xu, H.-D., Zhang, R.W., Li, X. et al. (2013). *Org. Lett.* 15: 840–843.

9 Yang, C. and Nolan, S.P. (2002). *J. Org. Chem.* 67: 591–593.

10 Wu, Y.-T., Lin, W.-C., Liu, C.-J., and Wu, C.-Y. (2008). *Adv. Synth. Catal.* 350: 1841–1849.

11 Chen, T., Guo, C., Goto, M., and Han, L.-B. (2013). *Chem. Commun.* 49: 7498–7500.

12 Xu, C., Du, W., Zeng, Y. et al. (2014). *Org. Lett.* 16: 948–951.

13 Sun, S., Kroll, J., Luo, Y., and Zhang, L. (2012). *Synlett* 54–56.

14 Mader, S., Molinari, L., Rudolph, M. et al. (2015). *Chem. Eur. J.* 21: 3910–3913.

15 Dash, A.K. and Eisen, M. (2000). *Org. Lett.* 2: 737–740.

16 Trost, B.M., Matsubara, S., and Caringi, J.J. (1989). *J. Am. Chem. Soc.* 111: 8745–8746.

17 (a) Komeyama, K., Takehira, K., and Takaki, K. (2004). *Synthesis* 1062–1066. (b) Komeyama, K., Kawabata, T., Takehira, K., and Takaki, K. (2005). *J. Org. Chem.* 70: 7260–7266.

18 Batrice, R.J., McKinven, J., Arnold, P.L., and Eisen, M.S. (2015). *Organometallics* 34: 4039–4050.

19 Rubio-Pérez, L., Azpíroz, R., Giuseppe, A.D. et al. (2013). *Chem. Eur. J.* 19: 15304–15314.

20 Yoshida, M. and Jordan, R.F. (1997). *Organometallics* 16: 4508–4510.

21 Pell, C.J. and Ozerov, O.V. (2014). *ACS Catal.* 4: 3470–3480.

22 Chen, J.-F. and Li, C. (2020). *ACS Catal.* 10: 3881–3889.

23 Rubina, M. and Gevorgyan, V. (2001). *J. Am. Chem. Soc.* 123: 11107–11108.

24 Jahier, C., Zatolochnaya, O.V., Zvyagintsev, N.V. et al. (2012). *Org. Lett. Org.* 14: 2846–2849.

25 Zatolochnaya, O.V., Gordeev, E.G., Jahier, C. et al. (2014). *Chem. Eur. J.* 20: 9578–9588.

26 Ogoshi, S., Ueta, M., Oka, M.-a., and Kurosawa, H. (2004). *Chem. Commun.* 2732–2733.

27 Ciclosi, M., Estevan, F., Lahuerta, P. et al. (2008). *Adv. Synth. Catal.* 350: 234–236.

28 Trostyanskaya, I.G. and Beletskaya, I.P. (2017). *Tetrahedron* 73: 148–153.

29 Kawata, A., Kuninobu, Y., and Takai, K. (2009). *Chem. Lett.* 38: 836–837.

30 Baratta, W., Herrmann, W.A., Rigo, P., and Schwarz, J. (2000). *J. Organomet. Chem.* 593–594: 489–493.

31 Qü, J.-P., Masui, D., Ishii, Y., and Hidai, M. (1998). *Chem. Lett.* 1003–1004.

32 Nishibayashi, Y., Yamanashi, M., Wakiji, I., and Hidai, M. (2000). *Angew. Chem. Int. Ed.* 39: 2909–2911.

33 Daniels, M. and Kirss, R.U. (2007). *J. Organomet. Chem.* 692: 1716–1725.

34 Chen, C.-R. and Lin, Y.-C. (2014). *Organometallics* 33: 6408–6412.

35 (a) Bassetti, M., Pasquini, C., Raneri, A., and Rosato, D. (2007). *J. Org. Chem.* 72: 4558–4561. (b) Pasquini, C. and Bassetti, M. (2010). *Adv. Synth. Catal.* 352: 2405–2410.

36 Coniglio, A., Bassetti, M., García-Garrido, S.E., and Gimeno, J. (2012). *Adv. Synth. Catal.* 354: 148–158.

37 (a) Pasquini, C., Fratoddi, I., Capitani, D. et al. (2008). *J. Org. Chem.* 73: 3892–3899. (b) Pasquini, C., Fratoddi, I., and Bassetti, M. (2009). *Eur. J. Org. Chem.* 5224–5231.

38 Bianchini, C., Frediani, P., Masi, D. et al. (1994). *Organometallics* 13: 4616–4632.

39 (a) Hijazi, A., Parkhomenko, K., Djukic, J.-P. et al. (2008). *Adv. Synth. Catal.* 350: 1493–1496. (b) Pasquini, C. and Bassetti, M. (2010). *Adv. Synth. Catal.* 352: 2405–2410. (c) Kiyota, S., Soeta, H., Komine, N. et al. (2017). *J. Mol. Catal. A Chem.* 426: 419–428.

40 Nishiura, M., Hou, Z., Wakatsuki, Y. et al. (2003). *J. Am. Chem. Soc.* 125 (125): 1184–1185.

41 Selected reports, see: (a) Tazelaar, C.G.J., Bambirra, S., van Leusen, D. et al. (2004). *Organometallics* 23: 936–939. (b) Chen, X., Xue, P., Sung, H.H.Y. et al. (2005). *Organometallics* 24: 4330–4332. (c) Ogata, K. and Toyota, A. (2007). *J. Organomet. Chem.* 692: 4139–4146. (d) Platel, R.H. and Schafer, L.L. (2012). *Chem. Commun.* 48: 10609–10611. (e) Forsyth, C.D., Kerr, W.J., and Paterson, L.C. (2013). *Synlett* 587–590.

42 Hilt, G., Hess, W., Vogler, T., and Hengst, C. (2005). *J. Organomet. Chem.* 690: 5170–5181.

43 Ventre, S., Derat, E., Amatore, M. et al. (2013). *Adv. Synth. Catal.* 355: 2584–2590.

44 (a) Midya, G.C., Paladhi, S., Dhara, K., and Dash, J. (2011). *Chem. Commun.* 47: 6698–6700. (b) Midya, G.C., Parasar, B., Dhara, K., and Dash, J. (2014). *Org. Biomol. Chem.* 12: 1812–1822.

45 Rivada-Wheelaghan, O., Chakraborty, S., Shimon, L.J.W. et al. (2016). *Angew. Chem. Int. Ed.* 55: 6942–6945.

46 Liang, Q., Osten, K.M., and Song, D. (2017). *Angew. Chem. Int. Ed.* 56: 6317–6320.

47 Li, X., Chen, X.-L., Zhang, Q. et al. (2015). *RSC Adv.* 5: 5004–5009.

48 Selected reviews, see: (a) Enders, D., Niemeier, O., and Henseler, A. (2007). *Chem. Rev.* 107: 5606–5655. (b) Marion, N., Díez-González, S., and Nolan, S.P. (2007). *Angew. Chem. Int. Ed.* 46: 2988–3000. (c) Bugaut, X. and Glorius, F. (2012). *Chem. Soc. Rev.* 41: 3511–3522. (d) Flanigan, D.M., Romanov-Michailidis, F., White, N.A., and Rovis, T. (2015). *Chem. Rev.* 115: 9307–9387. (e) Menon, R.S., Biju, A.T., and Nair, V. (2015). *Chem. Soc. Rev.* 44: 5040–5052.

49 Schedler, M., Wurz, N.E., Daniliuc, C.G., and Glorius, F. (2014). *Org. Lett.* 16: 3134–3137.

50 Kato, T., Matsuoka, S.-i., and Suzuki, M. (2014). *J. Org. Chem.* 79: 4484–4491.

51 Leroyer, L., Maraval, V., and Chauvin, R. (2012). *Chem. Rev.* 112: 1310–1343.

52 (a) Wakatsuki, Y., Yamazaki, H., Kumegawa, N. et al. (1991). *J. Am. Chem. Soc.* 113: 9604–9610. (b) Wakatsuki, Y., Yamazaki, H., Kumegawa, N., and Johar, P.S. (1993). *Bull. Chem. Soc. Jpn.* 66: 987–989.

53 Ohmura, T., Yorozuya, S.-i., Yamamoto, Y., and Miyaura, N. (2000). *Organometallics* 19: 365–367.

54 Esteruelas, M.A., Herrero, J., López, A.M., and Oliván, M. (2001). *Organometallics* 20: 3202–3205.

55 Selected reports, see: (a) Barlow, M.G., Bryant, M.J., Haszeldine, R.N., and Mackie, A.G. (1970). *J. Organomet. Chem.* 21: 215–226. (b) Kaneda, K., Kiriyama, T., Hiraoka, T., and Imanaka, T. (1988). *J. Mol. Catal.* 48: 343–347. (c) Jiang, Z. and Sen, A. (1990). *J. Am. Chem. Soc.* 112: 9655–9657.

56 Selected reports, see: (a) Wu, G., Rheingold, A.L., and Heck, R.F. (1987). *Organometallics* 6: 2386–2391. (b) Wu, G., Geib, S.J., Rheingold, A.L., and Heck, R.F. (1988). *J. Org. Chem.* 53: 3238–3241. (c) Kretschmer, W.P., Troyanov, S.I., Meetsma, A. et al. (1998). *Organometallics* 17: 284–286. (d) Sui-Seng, C., Groux, L.F., and Zargarian, D. (2006). *Organometallics* 25: 571–579.

57 Tsuchimoto, T., Kamiyama, S., Negoro, R. et al. (2003). *Chem. Commun.* 852–853.

58 Yi, C. and Hua, R. (2008). *Catal. Commun.* 9: 85–88.

59 Park, J.-W., Park, S.J., and Jun, C.-H. (2012). *Org. Lett.* 14: 1468–1471.

60 Higashimura, M., Imamura, K., Yokogawa, Y., and Sakakibara, T. (2004). *Chem. Lett.* 33: 728–729.

61 Kondo, T., Takagi, D., Tsujita, H. et al. (2007). *Angew. Chem. Int. Ed.* 46: 5958–5961.

62 Ura, Y., Tsujita, H., Wada, K. et al. (2005). *J. Org. Chem.* 70: 6623–6628.

63 Alderson, T., Jenner, E.L., and Lindsey, R.V. Jr., (1965). *J. Am. Chem. Soc.* 87: 5638–5645.

64 Yi, C.S., He, Z., and Lee, D.W. (2001). *Organometallics* 20: 802–804.

65 Selected reports, see: (a) RajanBabu, T.V. (2003). *Chem. Rev.* 103: 2845–2860. (b) Rodríguez, L.-I., Rossell, O., Seco, M., and Muller, G. (2008). *Organometallics* 27: 1328–1333. (c) Ayora, I., Ceder, R.M., Espinel, M. et al. (2011). *Organometallics* 30: 115–128.

66 Hiroi, Y., Komine, N., Komiya, S., and Hirano, M. (2013). *Org. Lett.* 15: 2486–2489.

67 Kaneko, Y., Kanke, T., Kiyooka, S.-i., and Isobe, K. (1997). *Chem. Lett.* 23–24.
68 Hirano, M., Hiroi, Y., Komine, N., and Komiya, S. (2010). *Organometallics* 29: 3690–3693.
69 Fukuzawa, H., Aoyagi, N., Sato, R. et al. (2017). *Organometallics* 36: 3931–3939.
70 Ho, C.-Y. and He, L. (2010). *Angew. Chem. Int. Ed.* 49: 9182–9186.
71 Moreau, B., Wu, J.Y., and Ritter, T. (2009). *Org. Lett.* 11: 337–339.
72 Sun, X., Bai, X.-Y., Li, A.-Z., and Li, B.-J. (2021). *Organometallics* 40: 2182–2187.
73 (a) Trost, B.M., Chan, C., and Rühter, G. (1987). *J. Am. Chem. Soc.* 109: 3486–3487. (b) Trost, B.M., Sorum, M.T., Chan, C. et al. (1997). *J. Am. Chem. Soc.* 119: 698–708. (c) Trost, B.M., Hachiya, I., and McIntosh, M.C. (1998). *Tetrahedron Lett.* 39: 6445–6448.
74 Trost, B.M. and McIntosh, M.C. (1997). *Tetrahedron Lett.* 38: 3207–3210.
75 Chen, L. and Li, C.-J. (2004). *Tetrahedron Lett.* 45: 2771–2774.
76 Hirabayashi, T., Sakaguchi, S., and Ishii, Y. (2005). *Adv. Synth. Catal.* 347: 872–876.
77 Matsuyama, N., Tsurugi, H., Satoh, T., and Miura, M. (2008). *Adv. Synth. Catal.* 350: 2274–2278.
78 Matsuyama, N., Hirano, K., Satoh, T., and Miura, M. (2009). *J. Org. Chem.* 74: 3576–3578.
79 (a) Weng, W., Guo, C., Celenligil-cetin, R. et al. (2006). *Chem. Commun.* 197–199. (b) Katagiri, T., Tsurugi, H., Funayama, A. et al. (2007). *Chem. Lett.* 36: 830–831. (c) Nishimura, T., Guo, X.-X., Ohnishi, K., and Hayashi, T. (2007). *Adv. Synth. Catal.* 349: 2669–2672. (d) Ito, J.-i., Kitase, M., and Nishiyama, H. (2007). *Organometallics* 26: 6412–6417. (e) Katagiri, T., Tsurugi, H., Satoh, T., and Miura, M. (2008). *Chem. Commun.* 3405–3407.
80 Sakurada, T., Sugiyama, Y.-k., and Okamoto, S. (2013). *J. Org. Chem.* 78: 3583–3591.
81 Yi, C.S. and Liu, N. (1998). *Organometallics* 17: 3158–3160.
82 Lücking, U. and Pfaltz, A. (2000). *Synlett* 1261–1264.
83 Gevorgyan, V., Radhakrishnan, U., Takeda, A. et al. (2001). *J. Org. Chem.* 66: 2835–2841.
84 Selected reports, see: (a) Tsukada, N., Ninomiya, S., Aoyama, Y., and Inoue, Y. (2007). *Org. Lett.* 9: 2919–2921. (b) Liu, G., Kong, W., Che, J., and Zhu, G. (2014). *Adv. Synth. Catal.* 356: 3314–3318. (c) Babu, M.H., Dwivedi, V., Kant, R., and Reddy, M.S. (2015). *Angew. Chem. Int. Ed.* 54: 3783–3786. (d) Dwivedi, V., Babu, M.H., Kant, R., and Reddy, M.S. (2015). *Chem. Commun.* 51: 14996–14999.
85 Akita, M., Yasuda, H., and Nakamura, A. (1984). *Bull. Chem. Soc. Jpn.* 57: 480–487.
86 Katayama, H., Yari, H., Tanaka, M., and Ozawa, F. (2005). *Chem. Commun.* 4336–4338.
87 Wang, J., Kapon, M., Berthet, J.C. et al. (2002). *Inorg. Chim. Acta* 334: 183–192.
88 Ogata, K., Oka, O., Toyota, A. et al. (2008). *Synlett* 2663–2666.
89 Ueda, Y., Tsurugi, H., and Mashima, K. (2020). *Angew. Chem. Int. Ed.* 59: 1552–1556.

90 Mitsudo, T.-a., Zhang, S.-W., Nagao, M., and Watanabe, Y. (1991). *J. Chem. Soc. Chem. Commun.* 598–599.

91 (a) Nishimura, T., Washitake, Y., Nishiguchi, Y. et al. (2004). *Chem. Commun.* 1312–1313. (b) Nishimura, T., Washitake, Y., and Uemura, S. (2007). *Adv. Synth. Catal.* 349: 2563–2571.

92 Selected reports, see: (a) Wang, Z.-X., Bai, X.-Y., and Li, B.-J. (2017). *Synlett* 28: 509–514. (b) Wang, Z.-X. and Li, B.-J. (2019). *J. Am. Chem. Soc.* 141: 9312–9320. (c) Zhang, S.-L., Zhang, W.-W., and Li, B.-J. (2021). *J. Am. Chem. Soc.* 143: 9639–9647.

93 Murakami, M., Ubukata, M., and Ito, Y. (1998). *Tetrahedron Lett.* 39: 7361–7364.

94 (a) Wang, C.-C., Lin, P.-S., and Cheng, C.-H. (2002). *J. Am. Chem. Soc.* 124: 9696–9697. (b) Chang, H.-T., Jayanth, T.T., Wang, C.-C., and Cheng, C.-H. (2007). *J. Am. Chem. Soc.* 129: 12032–12041.

95 Hilt, G. and Treutwein, J. (2007). *Angew. Chem. Int. Ed.* 46: 8500–8502.

96 Trost, B.M., Machacek, M., and Schnaderbeck, M.J. (2000). *Org. Lett.* 2: 1761–1764.

97 Trost, B.M., Papillon, J.P.N., and Nussbaumer, T. (2005). *J. Am. Chem. Soc.* 127: 17921–17937.

98 Lindhardt, A.T., Mantel, M.L.H., and Skrydstrup, T. (2008). *Angew. Chem. Int. Ed.* 47: 2668–2672.

99 Shibata, Y., Hirano, M., and Tanaka, K. (2008). *Org. Lett.* 10: 2829–2831.

100 Mannathan, S. and Cheng, C.-H. (2010). *Chem. Commun.* 46: 1923–1925.

101 Neisius, N.M. and Plietker, B. (2009). *Angew. Chem. Int. Ed.* 48: 5752–5755.

2

Addition of C(sp)—H Bonds to Unsaturated Compounds

The addition of metal alkynylide nucleophiles [1] or terminal alkynes C(sp)—H bond into unsaturated compounds such as carbonyl compounds, imines, and α,β-unsaturated carbonyl compounds has become the important method for the preparation of propargyl alcohols/amines [2], propiolic acid derivatives, and γ,δ-alkynyl carbonyl compounds. In the case of chiral ligands used, the direct enantioselective addition reactions of prochiral unsaturated compounds can afford the corresponding chiral alkynylated products (Scheme 2.1) [3]. This chapter focuses on describing the recent research advances in the reactions of terminal alkynes as nucleophiles or the precursors of nucleophiles with carbonyl compounds, carbon dioxide, alkenes, and imines, whereas the terminal alkynes and alkynyl group as the electrophilic reagent in the addition reactions have not been included in this chapter [4].

2.1 Addition of Terminal Alkynes to Carbonyl Compounds

The addition of metal alkynylide nucleophiles or terminal alkynes C(sp)—H bond into unsaturated compounds such as carbonyl compounds is the simplest and efficient method for the preparation of chiral propargyl alcohols, and various combinations of transition metal with chiral ligands have been developed.

Scheme 2.2 shows one example of the asymmetric alkynylation of both aromatic and aliphatic aldehydes using catalytic amounts of $InBr_3$/(*R*)-1,1′-bi-2-naphthol (BINOL), and the dual activation of both substrates due to the "bifunctional character" of the In(III) catalyst enables a broad range of substrate generality with high enantioselectivity (83 to >99% ee) [5].

$ZnMe_2/Ti(O^iPr)_4$/chiral ligands was found to be the efficient catalyst system for the enantioselective alkynylation of aromatic aldehydes with phenylacetylene [6], and other similar procedures have been also reported [7]. In addition, Pu and coworker then developed a catalyst system based on the readily available Zn, ^iPrI, H_8BINOL, and $Ti(O^iPr)_4$, avoiding the use of pyrophoric $ZnEt_2$ to realize the enantioselective alkyne addition to aliphatic, aromatic, and vinyl aldehydes giving chiral propargylic alcohols at room temperature (Scheme 2.3) [8].

Addition Reactions with Unsaturated Hydrocarbons, First Edition. Ruimao Hua.

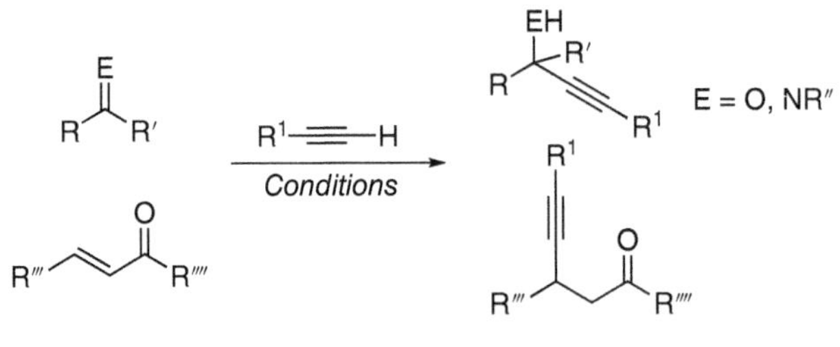

Scheme 2.1 Alkynylation of unsaturated bonds.

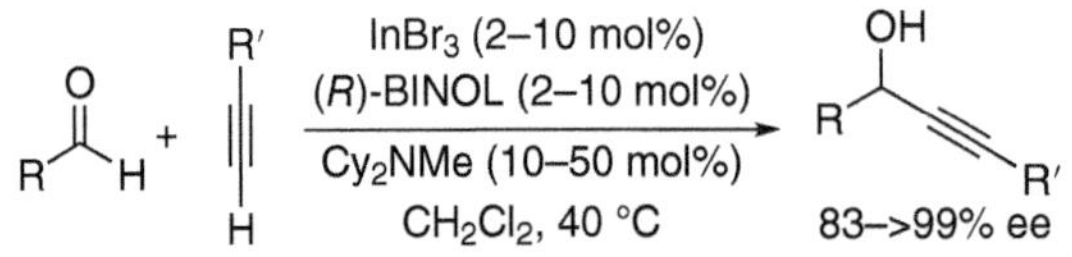

Scheme 2.2 Asymmetric alkynylation of aldehydes catalyzed by $InBr_3$/(*R*)-BINOL.

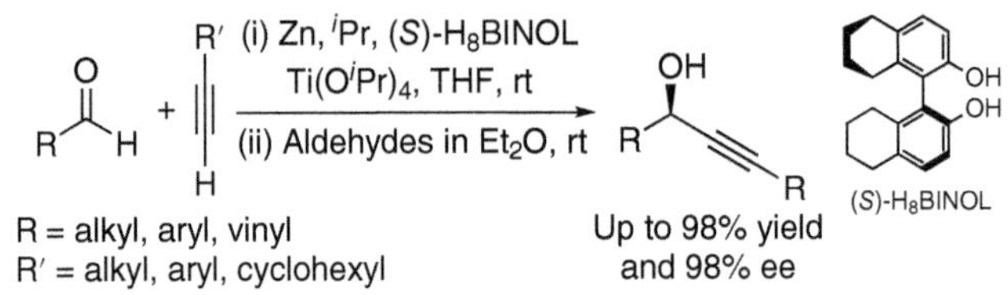

Scheme 2.3 Enantioselective alkyne addition to aliphatic, aromatic, and vinyl aldehydes.

Carreira's group has found that a mixture of $Zn(OTf)_2$/(−)-*N*-methylephedrine can promote the enantioselective addition of terminal alkynes to aldehydes [9]; the addition of α,α-dichlorinated aldehydes has the high potential applications in the synthesis of a variety of chlorinated natural products (Scheme 2.4) [10]. Also the asymmetric addition of terminal alkynes to α-ketoester can be achieved by using $Zn(OTf)_2$/chiral ligands [11].

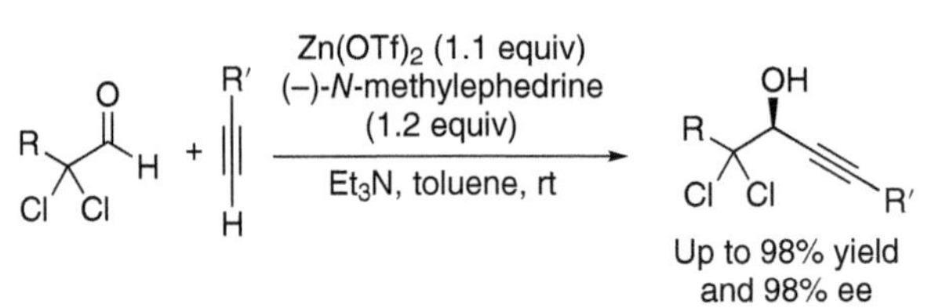

Scheme 2.4 Enantioselective addition of terminal alkynes to α,α-dichlorinated aldehydes.

With the use of diphenyl[(triisopropylsilyl)ethynyl]methanol as the alkynylating reagent, the Rh(I)/chiral-catalyzed asymmetric conjugate alkynylation of β,γ-unsaturated α-ketoesters occurs smoothly to give α-ketoesters bearing a propargylic chiral center at γ position in good yields with high enantioselectivities [12].

Meggers's group have developed the enantioselective catalytic alkynylation of various substrates such as 2-trifluoroacetyl imidazoles [13], trifluoromethyl ketones [14] and aromatic aldehydes [15] with the use of chiral cationic ruthenium/rhodium complexes.

In addition, $CuF_2 \cdot 2H_2O$/4,7-diphenyl-1,10-phenanthroline has been found to be a highly efficient catalyst for the nucleophilic alkynylation of cyclopropyl trifluoromethyl ketone affording the corresponding trifluoromethyl-substituted tertiary propargyl alcohols in moderate to excellent yields [16]. The Cu(I)-catalyzed asymmetric direct alkynylation of aromatic aldehydes [17], α-ketoesters [18] with terminal alkynes, has been also reported.

On the other hand, catalytic C(sp)—H carboxylation of terminal alkynes with carbon dioxide is one of the interesting and attractive processes for the synthesis of propiolic acids and their esters from the viewpoint of green chemistry and atom economy; since it was first reported by Inoue's group [19], the remarkable progress has been made, and these processes have been well reviewed (Scheme 2.5) [20].

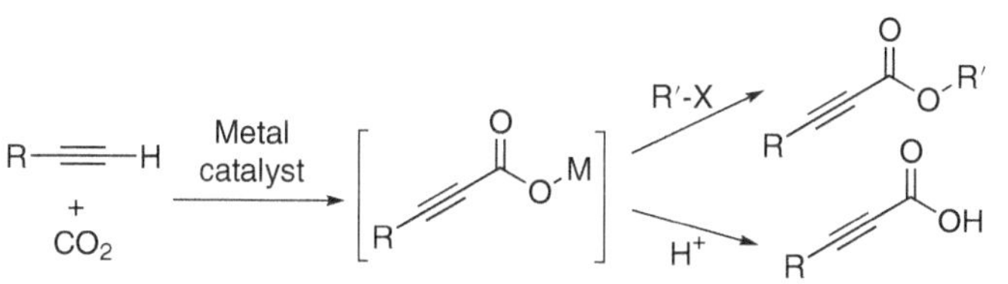

Scheme 2.5 Catalytic C−H carboxylation of terminal alkynes affording propiolic acids and esters.

As the examples described here, Suo and coworker developed an organic solvent-free, 1,8-Diazabicyclo[5.4.0]undec-7-ene (DBU)-mediated carboxylation of terminal alkynes in supercritical carbon dioxide ($ScCO_2$) to produce the corresponding propiolic acids in excellent yields (Scheme 2.6) [21]. In this process, carbon dioxide acts as both reactant and solvent for the reaction. In the presence of CuI, the carboxylation occurs smoothly to give the product with a slight increase of yields under a lower reaction temperature and pressure.

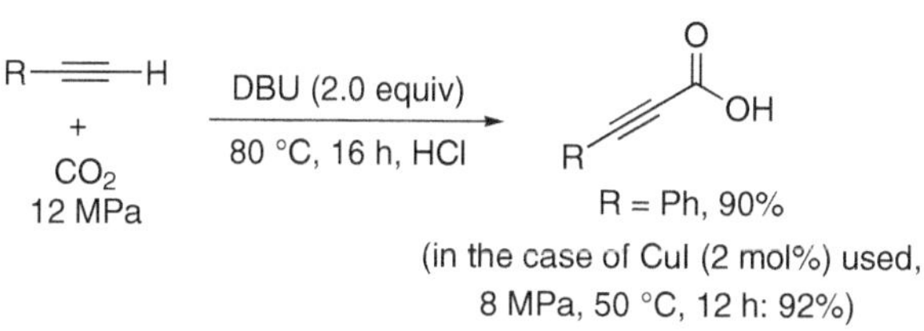

Scheme 2.6 DBU-mediated carboxylation of terminal alkynes in $ScCO_2$.

Zhang and coworker found that in the absence of transition metal catalyst, Cs_2CO_3 can efficiently promote the carboxylation of terminal alkynes with CO_2 (2.5 atm), and various propiolic acids can be prepared in good to excellent yields with a wide substrate scope and a good functional groups tolerance (Scheme 2.7) [22].

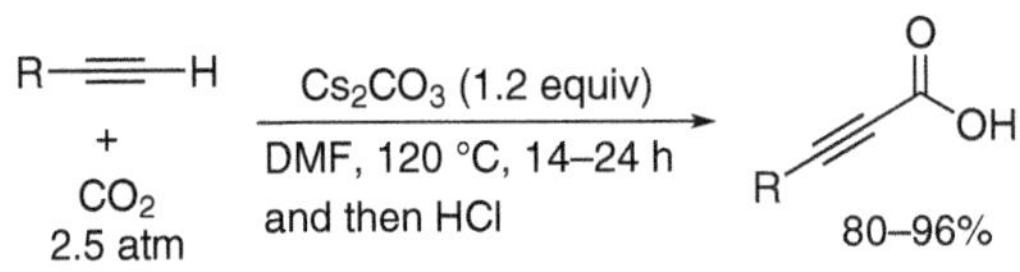

Scheme 2.7 Cs_2CO_3-promoted carboxylation of terminal alkynes with CO_2.

Ag_2WO_4 as a single-component catalyst has been found to be a bifunctional catalyst for dual activation of the terminal alkyne by silver and CO_2 by the tungstate anion to promote the carboxylation of terminal alkynes in ambient conditions in the presence of *n*-butyl iodide and cesium carbonate (Scheme 2.8) [23].

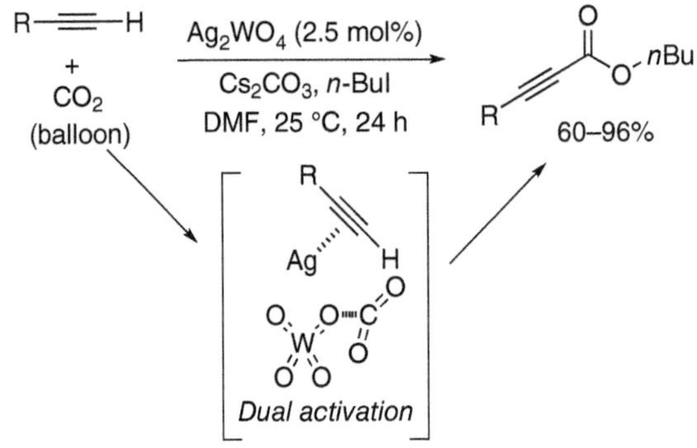

Scheme 2.8 Ag_2WO_4-catalyzed carboxylation of terminal alkynes in the presence of *n*-butyl iodide and cesium carbonate.

The carboxylation of terminal alkynes at ambient pressure can be also achieved by using AgI [24], CuI/Et_3P [25], poly(*N*-heterocyclic carbene)-supported silver nanoparticles [26], CuI with the use of ethylene carbonate as the solvent [27], bis(amidate) rare-earth metal amides [28], copper(I)-based ionic liquid [29], and activated carbon-supported CuBr [30] as catalysts or catalytic systems.

2.2 Addition of Terminal Alkynes to Alkenes

As described in Chapter 1, Section 1.4, the cross-dimerization of alkynes with alkenes via the addition reaction of C(sp^2)—H bond to alkynes can provide an efficient synthetic method for the formation of 1,3-dienes. This section will focus on the cross-dimerization of alkynes with alkenes via an alternative way by the activation of C(sp)—H bond of terminal alkynes and its hydroalkynylation of alkenes to afford internal alkynes. In these cases, terminal alkynes represent one of the useful and interesting carbon nucleophiles in Michael addition with Michael acceptors such as butenone and acrylonitrile (Scheme 2.9).

For example, Kovalev and coworker first reported that $RhCl(PMe_3)_3$ is an effective catalyst for cross-dimerization of terminal alkynes with vinyl ketones to form

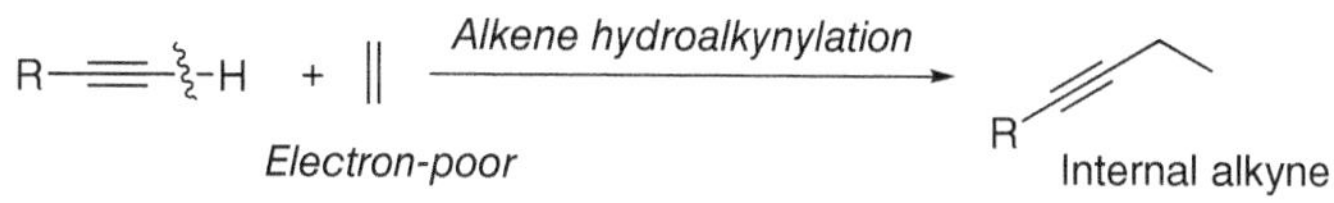

Scheme 2.9 Alkene hydroalkynylation access to internal alkynes.

γ,δ-alkynyl ketones (Scheme 2.10) [31], and Rh(acac)$(CO)_2$/phosphines [32] also show good catalytic activity for this transformation.

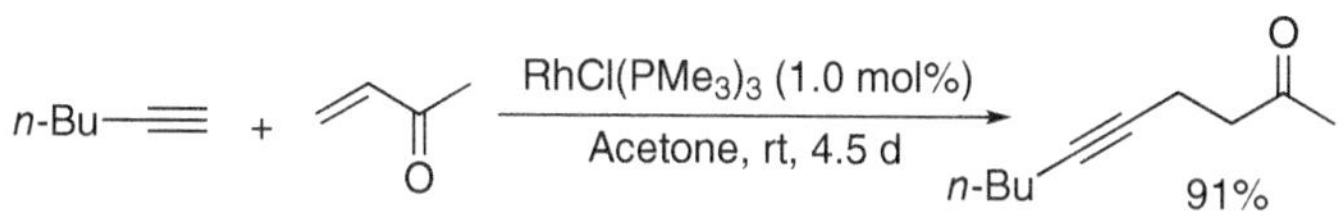

Scheme 2.10 Synthesis of γ,δ-alkynyl ketones via a rhodium-catalyzed addition of terminal alkynes to methyl vinyl ketone.

The use of the binuclear ruthenium catalyst $[Ru(O_2CH)(CO)_2(PR_3)]_2$ (R = Me, Ph) in acetonitrile at 100 °C selectively affords γ,δ-alkynyl ketones via Michael addition of terminal alkynes to butenone (Scheme 2.11) [33]. $[RuCl_2(p\text{-cymene})]_2$/pyrrolidine also shows the catalytic activity for this transformation [34].

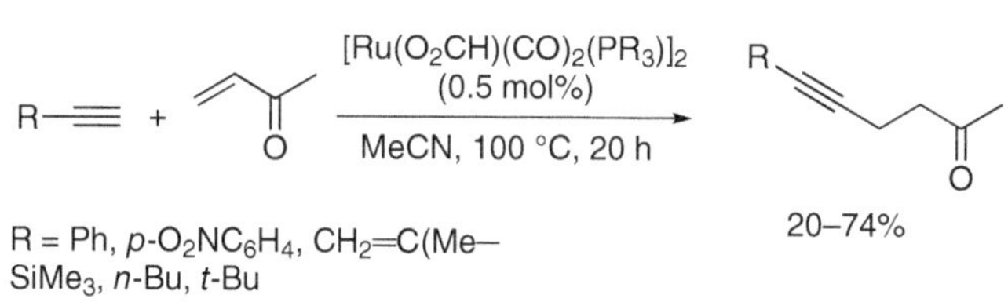

Scheme 2.11 Ruthenium-catalyzed cross-dimerization of terminal alkynes with butenone.

Cobalt [35], ruthenium [36], and palladium [37] show good catalytic activity in the addition of terminal alkynes to α,β-unsaturated ketones.

$Pd(OAc)_2/PMe_3$ has been found to be of high catalytic activity for the 1,4-addition of terminal alkynes to acrolein in water, and the addition reaction conditions are optimized to favor 1,4-addition over acrolein polymerization; a wide variety of 4-alkynals can be obtained in moderate to good yields (Scheme 2.12) [38].

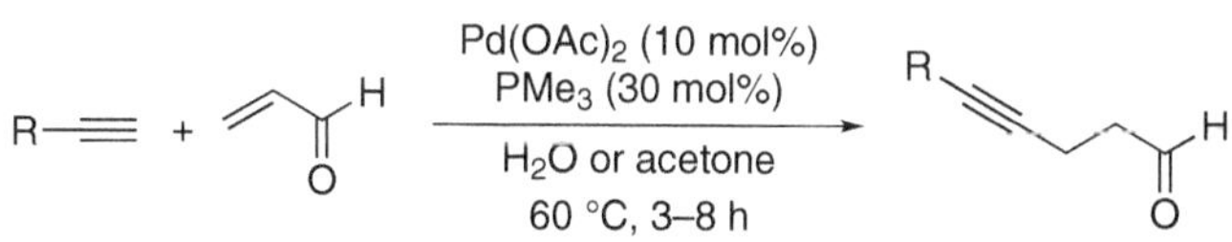

Scheme 2.12 $Pd(OAc)_2/PMe_3$-catalyzed 1,4-addition of terminal alkynes to acrolein in water.

With the use of $Ni(cod)_2$ as catalyst and PBu_3 as ligand, at room temperature, norbornene and styrenes bearing electron-withdrawing group undergo the hydroalkynylation with triisopropylsilylacetylene to produce *exo*-addition product and the internal alkynes via the addition of alkynyl group to the internal carbon of the C=C bond of styrenes (Scheme 2.13) [39]. A modified catalyst system of Ni(cod)$_2$/PMePh$_2$ shows high catalytic activity for the hydroalkynylation of styrenes, including electron-rich derivatives [40]. Note that in order to obtain the good yields of hydroalkynylation products, it is essential to use triisopropylsilylacetylene as the starting material to retard its homodimerization due to having bulky group.

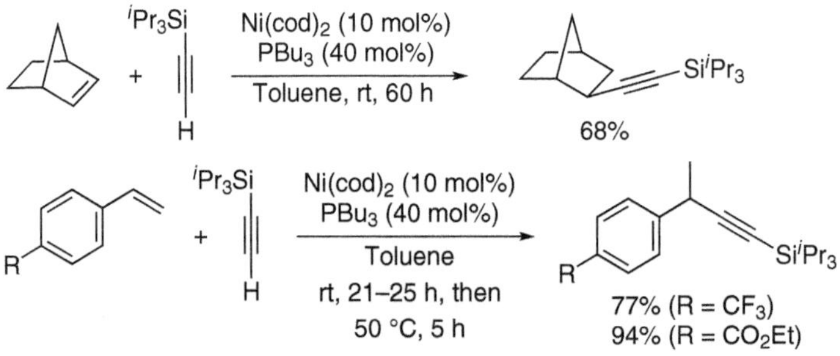

Scheme 2.13 Nickel-catalyzed hydroalkynylation of alkenes.

The same group has also developed Ni(cod)$_2$/ligand-catalyzed regio- and stereoselective hydroalkynylation of methylenecyclopropanes with retention of the cyclopropane ring for synthesis of 1-methyl-1-alkynylcyclopropanes [41], and the stereodefined alkynylcyclopropanes can be obtained by the direct addition of terminal alkynes to cyclopropenes catalyzed by the Herrmann–Beller (H–B) phosphapalladacycle catalyst [42].

Very recently, Pd(acac)$_2$/chiral ligand-catalyzed enantioselective hydroalkynylation of achiral cyclopropenes with high diastereo- and enantioselectivities has been developed by Marek group [43].

The asymmetric hydroalkynylation reactions of α,β-unsaturated compounds with terminal alkynes have been achieved by using a variety of transition metal complexes, such as rhodium [44] and copper [45], providing the efficient methods to access chiral β-alkynylated carbonyl compounds and β-alkynylated thioamides. Very interestingly, the asymmetric γ-alkynylation of α,β-unsaturated amides has been also reported in the presence of iridium complex and chiral phosphine ligand [46]. In addition, the asymmetric hydroalkynylation of norbornadiene and its derivatives with terminal alkynes has been well studied by using iridium with chiral ligand [47].

In addition, the asymmetric hydroalkynylation of diarylphosphinylallenes giving *exo*-enynes with high regio- and enantioselectivity has been also studied and developed catalyzed by [Rh(acac)(CH$_2$=CH$_2$)$_2$]/chiral/PhP(O)OH [48].

2.3 Addition of Terminal Alkynes to Imines

The direct addition of terminal alkynes to imines (aldimines and ketoimines) is one of the efficient synthetic methods for the synthesis of propargyl amines or chiral propargyl amines by asymmetric catalysis processes (Scheme 2.14) [49], and the combination of Cu(I)/chiral ligands has shown highly catalytic systems for such type of transformation [50].

Scheme 2.14 Propargyl amine from the alkyne addition to imine.

Wang and coworker have developed the enantioselective nucleophilic addition of trimethylsilylacetylene to *N*-phosphinoylimines promoted by Et_2Zn and chiral C_2-symmetric proline-derived β-amino alcohol [51], and then the direct catalytic asymmetric alkynylation of ketoimines bearing a thiophosphinoyl group by soft Lewis acid Cu(I) has been reported by Shibasaki's group [52], and a similar structure of substrates, *N*-(diphenylphosphinoyl)imines, also undergo the enantioselective addition with terminal alkynes in the presence of Et_2Zn and $Ti(O^iPr)_4$ with the use of 3,3′-di(1-diphenylmethylpiperazinyl)-methyl H_8BINOL as chiral organocatalyst to afford chiral propargylic amines via the easy removal of the *N*-(diphenylphosphinoyl) protecting groups (Scheme 2.15) [53].

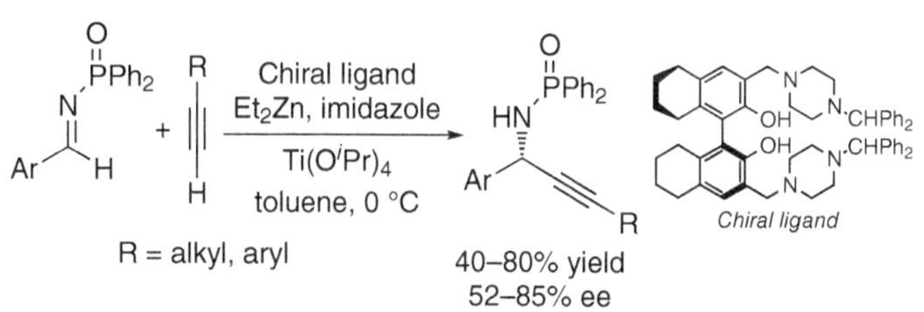

Scheme 2.15 Asymmetric addition of terminal alkynes to *N*-(diphenylphosphinoyl)imines.

Ma and coworker investigated and reported a catalytic enantioselective addition of terminal 1,3-diynes to *N*-sulfonyl aldimines providing the synthetic method for chiral diynylated carbinamines in the presence of $ZnMe_2$/chiral ligand (Scheme 2.16) [54].

On the other hand, it has been well-known that the three-component coupling reaction of aldehyde–alkyne–amine (A^3 coupling) is an alternatively efficient and convenient procedure for the synthesis of propargyl amines via the direct nucleophilic addition of the activated C—H bond of the terminal alkynes to the C=N double bond of imine or iminium ion formed in situ from aldehyde and amine (Scheme 2.17) [55].

Recently, it has been reported that $Cu(OTf)_2$ [56], dicopper complex [57], magnetic CuO nanoparticles supported on graphene oxide [58], amino acid ionic liquid bound copper Schiff base [59], CuI/PPh_3 [60], copper(I)–phosphole complex [61], Cu-MCM-41 [62], $[\{Cu(\mu-I)_2Cu\}(PPh_3)_4]$ [63], modified metal-organic framework

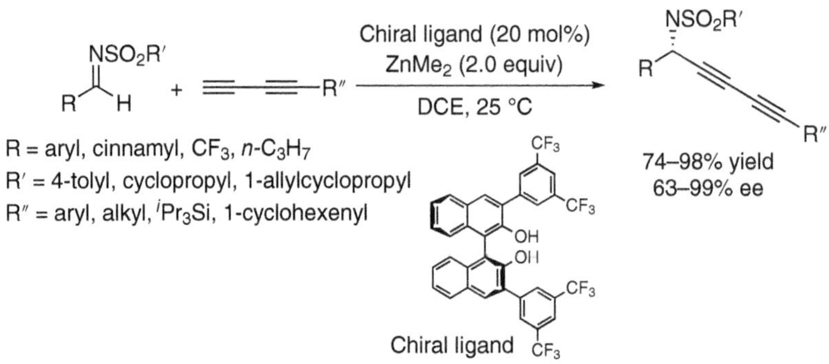

Scheme 2.16 Enantioselective addition of terminal 1,3-diynes to *N*-sulfonyl aldimines.

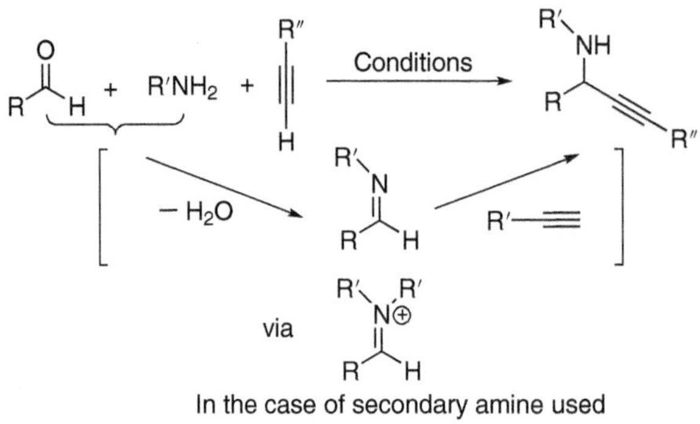

Scheme 2.17 Aldehyde–alkyne–amine (A^3) coupling and proposed mechanism approach to propargyl amines.

(MOF) [64], and cyclometalated and *N* functionalized -heterocyclic carbene (NHC) gold(I) and gold(III) complexes [65] show highly catalytic activities for A^3-coupling reactions.

In addition, copper (I) with chiral ligands has been reported to be the efficient catalyst systems for the catalytic enantioselective synthesis of propargyl amines via A^3-coupling reactions with the use of either primary amines [66] or secondary amines [67]. The A^3-coupling reactions as the versatile protocols and important steps have been also applied in various organic transformations [68].

The transition metal-catalyzed asymmetric hydroalkynylation of enamides for the synthesis of chiral propargyl amides and homopropargyl amides has been recently developed by Li's group [69].

2.4 Addition of Terminal Alkynes to Other Compounds

Transition metal-catalyzed alkynylation is the direct protocol for the synthesis of alkynylated compounds and has been well applied in organic synthesis [70]. Besides the addition of metal alkynylide nucleophiles or terminal alkynes to carbonyl compounds, imines, and α,β-unsaturated carbonyl compounds, the addition reactions to nitriles and epoxides are also very interesting and useful in organic synthesis.

The reaction of terminal alkynes with nitriles catalyzed by transition metal complexes usually undergoes the cocyclotrimerization, giving pyridine derivatives (Chapter 8) and, in some cases, the carbocyanation of alkynes via R—CN cleavage [71]. There are few examples for the hydroalkynylation of cyano group. Schemes 2.18 and 2.19 show two examples of hydroalkynylation of nitriles. One is the highly selective $Ln[N(SiMe_3)_2]_3$/n-$BuNH_2$-catalyzed addition of terminal alkynes to nitriles providing a new, economical, and mild method for the synthesis of conjugated ynones in mild to good yields via the hydrolysis of the imine adduct [72]. The other is the hydroalkynylation of 2-cyanopyridine with aromatic alkynes and a subsequent intramolecular hydroamination to afford indolizinones promoted by a basic ionic liquid of [bmim]OH under microwave irradiation and solvent-free conditions [73].

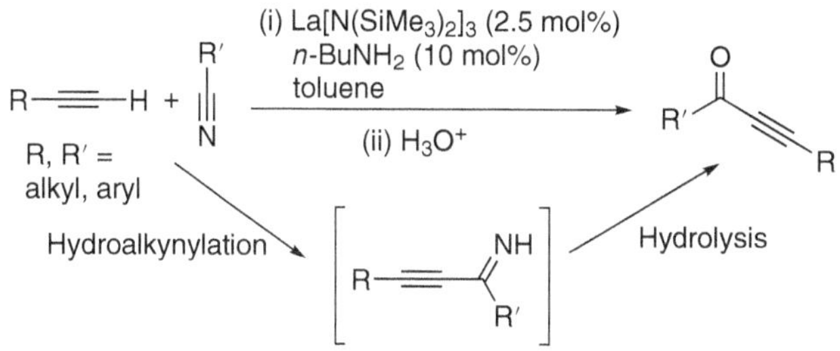

Scheme 2.18 Synthesis of conjugated ynones via hydroalkynylation of nitriles.

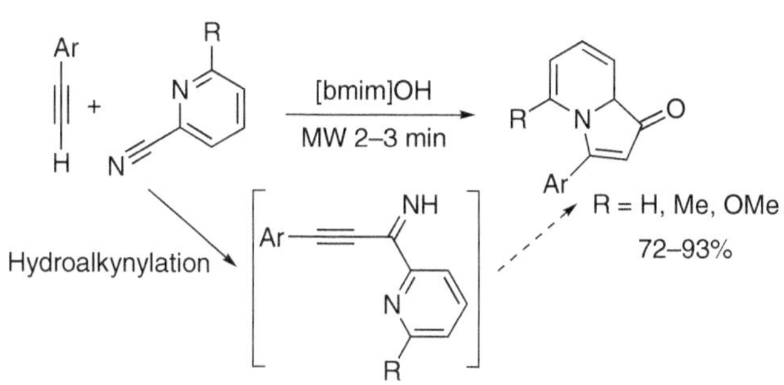

Scheme 2.19 Synthesis of indolizinones from hydroalkynylation of 2-cyanopyridine promoted by a basic ionic liquid.

The ring-opening reactions of epoxides with metal alkynylide (or generated in situ) producing homopropargyl alcohol (Scheme 2.20), as the fundamental and key step, have been well applied in the total synthesis of a variety of natural products such as volicitin [74], (+)-brefeldin A [75], soraphen A [76], (+)-aspergillide B and (+)-7-*epi*-aspergillide A [77], and (−)-aspergillide C [78]. Also, the formation of homopropargyl alcohol is one of the important transformations in the synthesis of complicated molecules [79].

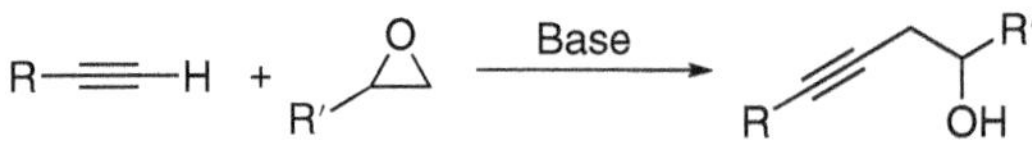

Scheme 2.20 The ring-opening reactions of epoxides with terminal alkynes affording homopropargyl alcohol.

References

1 Selected review papers, see: (a) Pu, L. and Yu, H.-B. (2001). *Chem. Rev.* 101: 757–824. (b) Pu, L. (2014). *Acc. Chem. Res.* 47: 1523–1535. (c) Kim, J.H., Ko, Y.O., Bouffard, J., and Lee, S.-j. (2015). *Chem. Soc. Rev.* 44: 2489–2507.

2 The applications of propargyl alcohols/amines in organic synthesis, review papers, see: (a) Lauder, K., Toscani, A., Scalacci, N., and Castagnolo, D. (2017). *Chem. Rev.* 117: 14091–14200. (b) Hua, R. and Nizami, T.A. (2018). *Mini-Rev. Org. Chem.* 15: 198–207.

3 Selected reviews, see: (a) Pu, L. (2003). *Tetrahedron* 59: 9873–9886. (b) Cozzi, P.J., Hilgraf, R., and Zimmermann, N. (2004). *Eur. J. Org. Chem.* 2004: 4095–4105. (c) Trost, B.M. and Weiss, A.H. (2009). *Adv. Synth. Catal.* 351: 963–983. (d) Li, C.-J. (2010). *Acc. Chem. Res.* 43: 581–590. (e) Bisai, V. and Singh, V.K. (2016). *Tetrahedron Lett.* 57: 4771–4784.

4 A comprehensive reviews for electrophilic alkynylation, see:Brand, J.P. and Waser, W. (2012). *Chem. Soc. Rev.* 41: 4165–4179.

5 Takita, R., Yakura, K., Ohshima, T., and Shibasaki, M. (2005). *J. Am. Chem. Soc.* 127: 13760–13761.

6 Li, X., Lu, G., Kwok, W.H., and Chan, A.S.C. (2002). *J. Am. Chem. Soc.* 124: 12636–12637.

7 (a) Moore, D. and Pu, L. (2002). *Org. Lett.* 4: 1855–1857. (b) Xu, M.-H. and Pu, L. (2002). *Org. Lett.* 4: 4555–4557. (c) Marshall, J.A. and Bourbeau, M.P. (2003). *Org. Lett.* 5: 3197–3199.

8 Huang, W.-C., Liu, W., Wu, X.-D. et al. (2015). *J. Org. Chem.* 80: 11480–11484.

9 (a) Frantz, D.E., Fässler, R., and Carreira, E.M. (2000). *J. Am. Chem. Soc.* 122: 1806–1807. (b) Anand, N.K. and Carreira, E.M. (2001). *J. Am. Chem. Soc.* 123: 9687–9688. (c) Boyall, D., Frantz, D.E., and Carreira, E.M. (2002). *Org. Lett.* 4: 2605–2606.

10 Molina, Y.S., Ruchti, J., and Carreira, E.M. (2017). *Org. Lett.* 19: 743–745.

11 (a) Jiang, B., Chen, Z., and Tang, X. (2002). *Org. Lett.* 4: 3451–3453. (b) Zavesky, B.P. and Johnson, J.S. (2017). *Angew. Chem. Int. Ed.* 56: 8805–8808.

12 Zhi, Y., Huang, J., Liu, N. et al. (2017). *Org. Lett.* 19: 2378–2381.

13 Zheng, Y., Harms, K., Zhang, L., and Meggers, E. (2016). *Chem. Eur. J.* 22: 11977–11981.

14 Zheng, Y., Tan, Y., Harms, K. et al. (2017). *J. Am. Chem. Soc.* 139: 4322–4325.

15 Luo, S., Zhang, X., Zheng, Y. et al. (2017). *J. Org. Chem.* 82: 8995–9005.

16 Liu, X.-D., Ma, H.-Y., Xing, C.-H., and Lu, L. (2017). *Tetrahedron Lett.* 58: 1564–1567.

17 (a) Asano, Y., Hara, K., Ito, H., and Sawamura, M. (2007). *Org. Lett.* 9: 3901–3904. (b) Asano, Y., Ito, H., Hara, K., and Sawamura, M. (2008). *Organometallics* 27: 5984–5996.

18 Schwarzer, M.C., Fujioka, A., Ishii, T. et al. (2018). *Chem. Sci.* 9: 3484–3493.

19 Fukue, Y., Oi, S., and Inoue, Y. (1994). *J. Chem. Soc. Chem. Commun.* 2091–2091.

20 (a) Huang, K., Sun, C.-L., and Shi, Z.-J. (2011). *Chem. Soc. Rev.* 40: 2435–2452. (b) Tsuji, Y. and Fujihara, T. (2012). *Chem. Commun.* 48: 9956–9964. (c) Manjolinho, F., Arndt, M., Gooßen, K., and Gooßen, L.J. (2012). *ACS Catal.* 2: 2014–2021. (d) Pinaka, A. and Vougioukalakis, G.C. (2015). *Coord. Chem. Rev.* 288: 69–97. (e) Yu, D., Teong, P.S., and Zhang, Y. (2015). *Coord. Chem. Rev.* 293 (294): 279–291. (f) Wang, S., Dua, G., and Xi, C. (2016). *Org. Biomol. Chem.* 14: 3666–3676. (g) Li, S., Sun, J., Zhang, Z. et al. (2016). *Dalton Trans.* 45: 10577–10584. (h) Yan, S.-S., Fu, Q., Liao, L.-L. et al. (2018). *Coord. Chem. Rev.* 374: 439–463.

21 Li, F.-W., Suo, Q.-L., Hong, H.-L. et al. (2014). *Tetrahedron Lett.* 55: 3878–3880.

22 Yu, D. and Zhang, Y. (2011). *Green Chem.* 13: 1275–1279.

23 Guo, C.-X., Yu, B., Xie, J.-N., and He, L.-N. (2015). *Green Chem.* 17: 474–479.

24 (a) Zhang, X., Zhang, W.-Z., Shi, L.-L. et al. (2012). *Tetrahedron* 68: 9085–9089. (b) Guo, F.-J., Zhang, Z.-Z., Wang, J.-Y. et al. (2017). *Tetrahedron* 73: 900–906.

25 Inamoto, K., Asano, N., Kobayashi, K. et al. (2012). *Org. Biomol. Chem.* 10: 1514–1516.

26 Yu, D., Tan, M.X., and Zhang, Y. (2012). *Adv. Synth. Catal.* 354: 969–974.

27 Yu, B., Diao, Z.-F., Guo, C.-X. et al. (2013). *Green Chem.* 15: 2401–2407.

28 Cheng, H., Zhao, B., Yao, Y., and Lu, C. (2015). *Green Chem.* 17: 1675–1682.

29 Xie, J.-N., Yu, B., Zhou, Z.-H. et al. (2015). *Tetrahedron Lett.* 56: 7059–7062.

30 Yu, B., Xie, J.-N., Zhong, C.-L. et al. (2015). *ACS Catal.* 5: 3940–3944.

31 Kovalev, I.P. and Nikishin, G.I. (1990). *Tetrahedron Lett.* 31: 7063–7064.

32 Lerum, R.V. and Chisholm, J.D. (2004). *Tetrahedron Lett.* 45: 6591–6594.

33 Picquet, M., Bruneau, C., and Dixneuf, P.H. (1999). *Tetrahedron* 55: 3937–3948.

34 Chang, S., Na, Y., Choi, E., and Kim, S. (2011). *Org. Lett.* 3: 2089–2091.

35 (a) Nishimura, T., Sawano, T., Ou, K., and Hayashi, T. (2011). *Chem. Commun.* 47: 10142–10144. (b) Sawano, T., Ashouri, A., Nishimura, T., and Hayashi, T. (2012). *J. Am. Chem. Soc.* 134: 18936–18939.

36 (a) Chang, S., Na, Y., Choi, E., and Kim, S. (2001). *Org. Lett.* 3: 2089–2091. (b) Nishimura, T., Washitake, Y., Nishiguchi, Y. et al. (2004). *Chem. Commun.* 1312–1313. (c) Nishimura, T., Washitake, Y., and Uemura, S. (2007). *Adv. Synth. Catal.* 349: 2563–2571. (d) Ito, J.-i., Fujii, K., and Nishiyama, H. (2013). *Chem. Eur. J.* 19: 601–605.

37 Zhou, L., Chen, L., Skouta, R. et al. (2008). *Org. Biomol. Chem.* 6: 2969–2977.

38 Wang, H., Hearne, Z., Knauber, T. et al. (2015). *Tetrahedron* 71: 5866–5870.

39 (a) Shirakura, M. and Suginome, M. (2008). *J. Am. Chem. Soc.* 130: 5410–5411. (b) Shirakura, M. and Suginome, M. (2009). *Org. Lett.* 11: 523–526.

40 Shirakura, M. and Suginome, M. (2009). *Org. Lett.* 11: 523–526.

41 Shirakura, M. and Suginome, M. (2009). *J. Am. Chem. Soc.* 131: 5060–5061.

42 Tenaglia, A., Jeune, K.L., Giordano, L., and Buono, G. (2011). *Org. Lett.* 13: 636–639.

43 Dian, L. and Marek, I. (2020). *ACS Catal.* 10: 1289–1293.

44 Selected reports, see: (a) Nishimura, T., Sawano, T., and Hayashi, T. (2009). *Angew. Chem. Int. Ed.* 48: 8057–8059. (b) Nishimura, T., Tokuji, S., Sawano, T., and Hayashi, T. (2009). *Org. Lett.* 11: 3222–3225. (c) Nishimura, T., Sawano, T., Tokuji, S., and Hayashi, T. (2010). *Chem. Commun.* 46: 6837–6839. (d) Dou, X., Huang, Y., and Hayashi, T. (2016). *Angew. Chem. Int. Ed.* 55: 1133–1137.

45 Selective reports, see: (a) Knöpfel, T.F. and Carreira, E.M. (2003). *J. Am. Chem. Soc.* 125: 6054–6055. (b) Knöpfel, T.F., Zarotti, P., Ichikawa, T., and Carreira, E.M. (2005). *J. Am. Chem. Soc.* 127: 9682–9683. (c) Yazaki, R., Kumagai, N., and Shibasaki, M. (2010). *J. Am. Chem. Soc.* 132: 10275–10277. (d) Sanz-Marco, A., García-Ortiz, A., Blay, G., and Pedro, J.R. (2014). *Chem. Commun.* 50: 2275–2278. (e) Sanz-Marco, A., Blay, G., Muñoz, M.C., and Pedro, J.R. (2016). *Chem. Eur. J.* 22: 10057–10064.

46 Wang, Z.-X., Bai, X.-Y., Yao, H.-C., and Li, B.-J. (2016). *J. Am. Chem. Soc.* 138: 14872–14875.

47 (a) Fan, B.-M., Yang, Q.-j., Hu, J. et al. (2012). *Angew. Chem. Int. Ed.* 51: 7821–7824. (b) Fan, B., Xu, J., Yang, Q. et al. (2013). *Org. Lett.* 15: 5956–5959. (c) Hu, J., Yang, Q., Xu, J. et al. (2013). *Org. Biomol. Chem.* 11: 814–820. (d) Yang, Q., Choy, P.Y., Fan, B., and Kwong, F.Y. (2015). *Adv. Synth. Catal.* 357: 2345–2350.

48 Nishimura, T., Guo, X.-X., and Hayashi, T. (2008). *Chem. Asian J.* 3: 1505–1510.

49 Selected reviews of alkynylation of imines, see: (a) Zani, L. and Bolm, C. (2006). *Chem. Commun.* 4263–4275. (b) de Armas, P., Tejedor, D., and García-Tellado, F. (2010). *Angew. Chem. Int. Ed.* 49: 1013–1016. (c) Kumagai, N. and Shibasaki, M. (2015). *Bull. Chem. Soc. Jpn.* 88: 503–517. (d) Liu, J., Dasgupta, S., and Watson, M.P. (2015). *Beilstein J. Org. Chem.* 11: 2696–2706.

50 Selected reports, see: (a) Wei, C. and Li, C.-J. (2002). *J. Am. Chem. Soc.* 124: 5638–5639. (b) Wei, C., Mague, J.T., and Li, C.-J. (2004). *PNAS* 101: 5749–5754. (c) Rueping, M., Antonchick, A.P., and Brinkmann, C. (2007). *Angew. Chem. Int. Ed.* 46: 6903–6906. (d) Lu, Y., Johnstone, T.C., and Arndtsen, B.A. (2009). *J. Am. Chem. Soc.* 131: 11284–11285. (e) de Armas, P., Tejedor, D., and García-Tellado, F. (2010). *Angew. Chem. Int. Ed.* 49: 1013–1016. (f) Yin, L., Otsuka, Y., Takada, H. et al. (2013). *Org. Lett.* 15: 698–701. (g) Ren, Y.-Y., Wang, Y.-Q., and Liu, S. (2014). *J. Org. Chem.* 79: 11759–11767. (h) Dasgupta, S., Liu, J., Shoffler, C.A. et al. (2016). *Org. Lett.* 18: 6006–6009. (i) Ling, Z., Singh, S., Xie, F. et al. (2017). *Chem. Commun.* 53: 5364–5367.

51 Zhu, S., Yan, W., Mao, B. et al. (2009). *J. Org. Chem.* 74: 6980–6985.

52 Yin, L., Otsuka, Y., Takada, H. et al. (2013). *Org. Lett.* 15: 698–701.

53 Ying, J., Wu, X.-D., Wang, D., and Pu, L. (2016). *J. Org. Chem.* 81: 8900–8905.

54 Liu, T.-L., Zhang, H.-X., Zheng, Y. et al. (2012). *Chem. Commun.* 48: 12234–12236.

55 Selected reviews, see: (a) Wei, C., Li, Z., and Li, C.-J. (2004). *Synlett* 1472–1483. (b) Yoo, W.-J., Zhao, L., and Li, C.-J. (2011). *Aldrichimica Acta* 44: 43–51. (c)

Peshkov, V.A., Pereshivko, O.P., and Van der Eycken, E.V. (2012). *Chem. Soc. Rev.* 41: 3790–3807. (d) Seidel, D. (2014). *Org. Chem. Front.* 1: 426–429. (e) Saha, T.K. and Das, R. (2018). *ChemistrySelect* 3: 147–169.

56 Meyet, C.E., Pierce, C.J., and Larsen, C.H. (2012). *Org. Lett.* 14: 964–967.

57 Chen, H.-B., Zhao, Y., and Liao, Y. (2015). *RSC Adv.* 5: 37737–37741.

58 Mirabedini, M., Motamedi, E., and Kassaee, M.Z. (2015). *Chin. Chem. Lett.* 26: 1085–1090.

59 Varyani, M., Khatri, P.K., and Jain, S.L. (2016). *Catal. Commun.* 77: 113–117.

60 Choi, Y.-J. and Jang, H.-Y. (2016). *Eur. J. Org. Chem.* 2016: 3047–3050.

61 Cammarata, J.R., Rivera, R., Fuentes, F. et al. (2017). *Tetrahedron Lett.* 58: 4078–4081.

62 Cortezano-Arellano, O., Hernández-Gasca, M.A., Ángeles-Beltrán, D. et al. (2018). *Tetrahedron Lett.* 59: 2403–2406.

63 Kashid, V.S. and Balakrishna, M.S. (2018). *Catal. Commun.* 103: 78–82.

64 Kassymova, M., de Mahieu, A., Chaemchuen, S. et al. (2018). *Catal. Sci. Technol.* 8: 4129–4140.

65 Price, G.A., Brisdon, A.K., Randall, S. et al. (2017). *J. Organomet. Chem.* 846: 251–262.

66 Selected reports, see: (a) Bisai, A. and Singh, V.K. (2006). *Org. Lett.* 8: 2405–2408. (b) Nakamura, S., Ohara, M., Nakamura, Y. et al. (2010). *Chem. Eur. J.* 16: 2360–2362. (c) Ohara, M., Hara, Y., Ohnuki, T., and Nakamura, S. (2014). *Chem. Eur. J.* 20: 8848–8851.

67 Selected reports, see: (a) Gommermann, N., Koradin, C., Polborn, K., and Knochel, P. (2003). *Angew. Chem. Int. Ed.* 42: 5763–5766. (b) Gommermann, N. and Knochel, P. (2006). *Chem. Eur. J.* 12: 4380–4392. (c) Fan, W. and Ma, S. (2013). *Chem. Commun.* 49: 10175–10177. (d) Cardoso, F.S.P., Abboud, K.A., and Aponick, A. (2013). *J. Am. Chem. Soc.* 135: 14548–14551. (e) Naeimi, H. and Moradian, M. (2014). *Tetrahedron Asymmetry* 25: 429–434. (f) Zhao, C. and Seidel, D. (2015). *J. Am. Chem. Soc.* 137: 4650–4653.

68 Review paper, see: (a) Liu, Y. (2014). *ARKIVOC* (i): 1–20. (b) Lauder, K., Toscani, A., Scalacci, N., and Castagnolo, D. (2017). *Chem. Rev.* 117: 14091–14200. (c) Visbal, R., Graus, S., Herrera, R.P., and Gimeno, M.C. (2018). *Molecules* 23: 2255.

69 (a) Bai, X.-Y., Wang, Z.-X., and Li, B.-J. (2016). *Angew. Chem. Int. Ed.* 55: 9007–9011. (b) Bai, X.-Y., Zhang, W.-W., Li, Q., and Li, B.-J. (2018). *J. Am. Chem. Soc.* 140: 506–514. (c) Zhang, W.-W., Zhang, S.-L., and Li, B.-J. (2020). *Angew. Chem. Int. Ed.* 59: 6874–6880.

70 Negishi, E.-i. and Anastasia, L. (2003). *Chem. Rev.* 103: 1979–2018.

71 Selected reports, see: (a) Nakao, Y. and Hiyama, T. (2008). *Pure Appl. Chem.* 80: 1097–1107. (b) Nakao, Y. (2012). *Bull. Chem. Soc. Jpn.* 85: 731–745.

72 Shen, Q., Huang, W., Wang, J., and Zhou, X. (2008). *Organometallics* 27: 301–303.

73 Siddiqui, I.R., Abumhdi, A.A.H., Shamim, S. et al. (2013). *RSC Adv.* 3: 10173–10176.

74 Hansen, T.V. and Stenstrom, Y. (2000). *Synth. Commun.* 30: 2549–2557.

75 (a) Suh, Y.-G., Jung, J.-K., Seo, S.-Y. et al. (2002). *J. Org. Chem.* 67: 4127–4137. (b) Wu, Y. and Gao, J. (2008). *Org. Lett.* 10: 1533–1536.

76 Trost, B.M., Sieber, J.D., Qian, W. et al. (2009). *Angew. Chem. Int. Ed.* 48: 5478–5481.

77 Sridhar, Y. and Srihari, P. (2013). *Eur. J. Org. Chem.* 2013: 578–587.

78 Srihari, P., Krishna, N.H., Sridhar, Y., and Kamal, A. (2014). *Beilstein J. Org. Chem.* 10: 3122–3126.

79 Selected reports, see: (a) Jaber, J.J., Mitsui, K., and Rychnovsky, S.D. (2001). *J. Org. Chem.* 66: 4679–4686. (b) Shin, J.-T., Hong, S.-C., Shin, S., and Cho, C.-G. (2006). *Org. Lett.* 8: 3339–3341. (c) Zhou, Y., Porco, J.A. Jr., and Snyder, J.K. (2007). *Org. Lett.* 9: 393–396.

3

Functionalized Alkenes from Hydrofunctionalization of Alkynes

The hydrofunctionalization of alkynes via the addition reactions of H–heteroatom (E) bond to carbon–carbon triple bonds is one of the efficient and atom economical processes approach to functionalized alkenes [1], which are valuable building blocks in the synthetic chemistry via the further transformation of carbon–carbon double bond and C—E bond. The addition reactions of alkynes can theoretically afford three adducts: Markovnikov-type adduct (M-adduct or α-isomer) and two *anti*-Markovnikov-type adducts (*trans*-, *cis*-adducts or (*Z*)-β-, (*E*)-β-isomers) (Scheme 3.1). Therefore, the important protocol is to develop the catalytic systems to realize the addition reactions with high regio- and stereoselectivity. This chapter focuses on the addition reactions of groups 13 (B), 14 (Si, Sn), 15 (N, P), and 16 (O, S, Se) element hydrides to alkynes.

The mechanism for transition metal-catalyzed addition reactions of E—H bond to alkyne is commonly assumed to be initiated by oxidative addition of E—H to transition metal and followed by insertion of a C—C triple bond into hydride-metal or element-metal bond and final reductive elimination of C—E or C—H bond, and the detailed information about these elementary processes have been well investigated (Scheme 3.2) [2]. The regioselectivity and stereoselectivity depend on the catalyst systems and the structures of substrates.

In addition, the asymmetric sequential double hydrofunctionalization of alkynes has recently become a powerful strategy to synthesize the chiral products, and a comprehensive review has recently reported [3].

3.1 Hydroborations of Alkynes

Alkenylboron compounds (boranes and boronate esters) are versatile synthetic intermediates in organic synthesis for the formation of C—C, C—N, and C—O bond via the activation of C—B bond and its cross-coupling reactions [4]. The hydroboration of alkynes is a straightforward and powerful method for generating alkenylboron compounds, and a few review papers have appeared [5].

The hydroboration of terminal alkynes can produce four adducts: three adducts from Markovnikov and *anti*-Markovnikov addition reactions as shown in Scheme 3.1

Addition Reactions with Unsaturated Hydrocarbons, First Edition. Ruimao Hua.

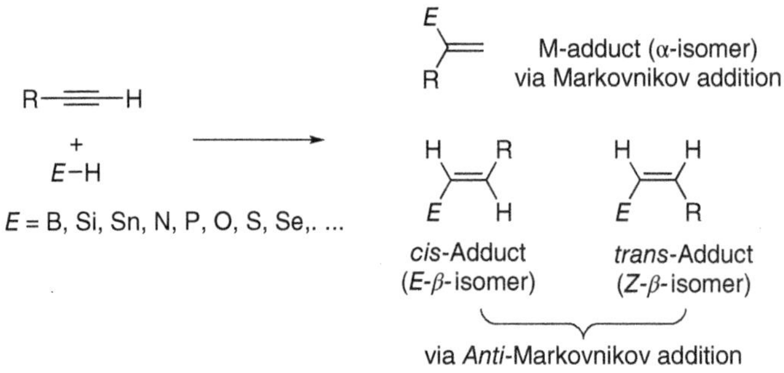

Scheme 3.1 Hydrofunctionalization of alkynes.

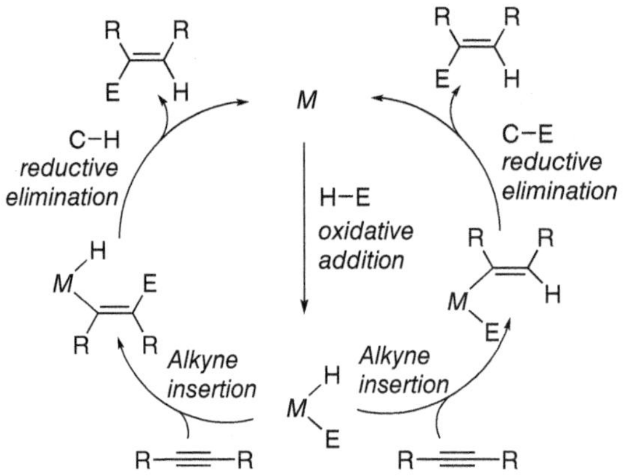

Scheme 3.2 Proposed mechanism for hydrofunctionalization of alkyne.

(E = B) and an additional adduct from *gem*-addition of B—H. The (*Z*)-isomer can be from either *trans*-addition or *gem*-addition (Scheme 3.3).

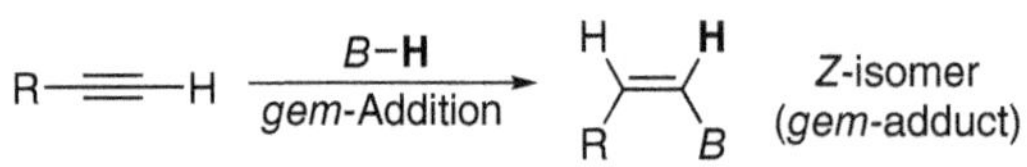

Scheme 3.3 *gem*-Adduct from the hydroboration of terminal alkyne.

For the regioselective formation of Markovnikov adduct, the recently reported procedures usually employed bis(pinacolato)diboron (B_2pin_2) as the hydroborylation reagents with the use of both base and alcohol as the required additives and found that *N*-heterocyclic carbine (NHC)–CuCl [6] and $CuBr_2$/cyclodextrin–bispyridine [7] were efficient catalyst systems.

A ligand-controlled selective hydroboration of terminal alkynes with B_2pin_2 to Markovnikov and *anti*-Markovnikov adducts of vinylboronates was developed by using $[Pd(OAc)_2]_3$ catalyst, and the high Markovnikov addition occurs with the use of trialkylphosphine as ligands, whereas the *cis*-addition of *anti*-Markovnikov reaction could be achieved by the use of *N*-heterocyclic carbine ligand [8].

On the basis of the reported procedures, the hydroboration of terminal alkynes mostly yields (*E*)-alkenylboron compounds as the major product via *cis*-addition reaction. Zirconium [9], iron [10], silver [11], gold [12], and aluminum complexes [13] have been applied in *cis*-hydroboration of terminal alkynes.

In addition, recently, a few transition metal-free catalyst systems for selective *cis*-hydroboration of terminal alkynes and internal alkynes to give (*E*)-isomers have been established, for example, *N*-heterocyclic carbenes [14], carboxylic acid [15], Lewis acid $HB(C_6F_5)_2$ [16], and base *t*-BuOLi [17].

Moreover, the regioselective *cis*-hydroboration of terminal alkynes [18] and internal alkynes [19] can be achieved by using bis(pinacolato)diboron as the starting materials catalyzed by copper salts or nanoparticles with or without the use of ligands.

For the formation of (*Z*)-alkenylboron compounds, the *gem*-addition reactions are the major ways via the metal vinylidene intermediates by migration of an acetylenic hydrogen and subsequent *gem*-addition of the borane reagent in the presence of transition metal complexes, such as rhodium [20], ruthenium [21], or via other mechanism in a Co-catalyzed hydroboration to give (*Z*)-alkenylboron compounds [22].

Transition metal complexes catalyzed by the real *trans*-hydroboration of terminal alkynes affording (*Z*)-alkenylboron compounds are rare. Recently, Lee's group has developed a copper-catalyzed stereodivergent hydroboration of aromatic terminal alkynes to afford either (*Z*)- or (*E*)-alkenylboron compounds and has first realized the highly stereoselective *trans*-hydroboration of terminal alkynes with 1,8-naphthalenediaminatoborane (HB(dan)) [23]. As shown in Scheme 3.4, CuTC catalyzes the *trans*-addition reactions to produce (*Z*)-alkenylboron compounds with excellent stereoselectivity, and with the use of SIPr—CuCl complex as the precatalyst, the *cis*-hydroboration occurs exclusively.

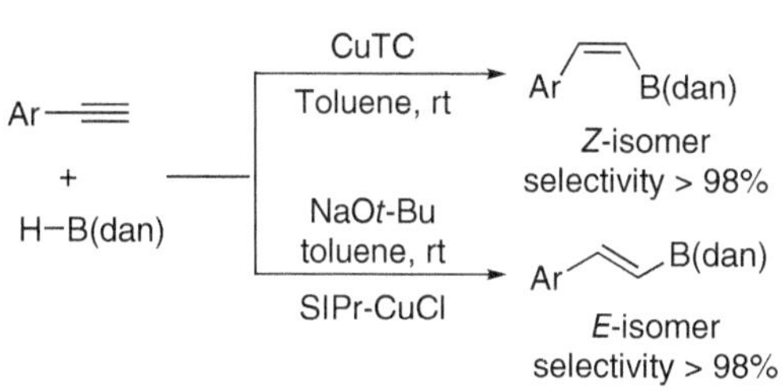

Scheme 3.4 Copper-catalyzed stereodivergent hydroboration of terminal aryl alkynes.

However, Ingleson's group has developed a *trans*-hydroboration of terminal alkynes mediated by borenium cations $[NHC(9\text{-}BBN)]^+$ to exclusively afford *Z*-alkenylboranes, and it proceeds with only catalytic amounts of $B(C_6F_5)_3$ or $[Ph_3C][B(C_6F_5)_4]$ to activate the $(NHC)_9$—BBN(H) precursor with the borenium regenerated in the hydride transfer step, and NHC can be removed from the *trans*-hydroborated products by the addition of Et_2O—BF_3 [24].

In addition, a few reaction systems have been reported for the *trans*-hydroboration of internal alkynes and in some cases with the functional group, directed to increase the regioselectivity. It includes the ruthenium-catalyzed *trans*-addition [25], radical *trans*-hydroboration with the use of *N*-heterocyclic carbene boranes [26], and trialkylphosphine-catalyzed regioselective *trans*-hydroboration of internal alkynes of alkynoic acid derivatives such as alkynoate esters and amides affording exclusive (*E*)-β-boryl acrylates and (*E*)-β-boryl acrylamides in good to excellent yields (Scheme 3.5) [27].

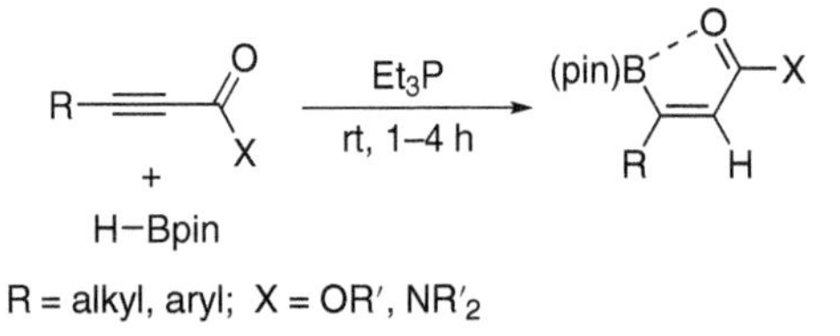

Scheme 3.5 Phosphine-catalyzed regioselective *trans*-hydroboration of alkynoates.

trans-Hydroboration of propargyl alcohol derivatives with pinacolborane has also been developed in the presence of $[Cp^*RuCl]_4$ [28].

In the cases of unsymmetrical internal alkynes used, it is a great challenge to control the highly regioselectivity of hydroboration. Recently, micro copper powder show highly regioselective catalytic activity in the hydroboration of unsymmetrical internal alkynes [29]. The activated alkenylboronates could be also prepared from the hydroboration of acetylenic esters with pinacolborane catalyzed by copper hydride complexes [30], and Tsuji and coworker have also found that with the use of different copper catalytic species, such as copper hydride and boryl copper, the regioselectivity could be controlled efficiently [31].

Very recently, $Cu(OAc)_2/P^nBu_3$-catalyzed triboration of terminal alkynes with B_2pin_2 has also been developed [32].

Boron-substituted enynes have been prepared via CuCl/P(*p*-tol)$_3$-catalyzed *cis*-hydroboration of 1,3-butadiynes with B_2pin_2 in the presence of NaO*t*-Bu and MeOH in THF with highly regio- and stereoselectivities (Scheme 3.6) [33]. With the use of NaOMe/MeOH as reaction conditions, a transition metal-free hydroboration of terminal alkynes and alkenes with B_2pin_2 has been also established [34].

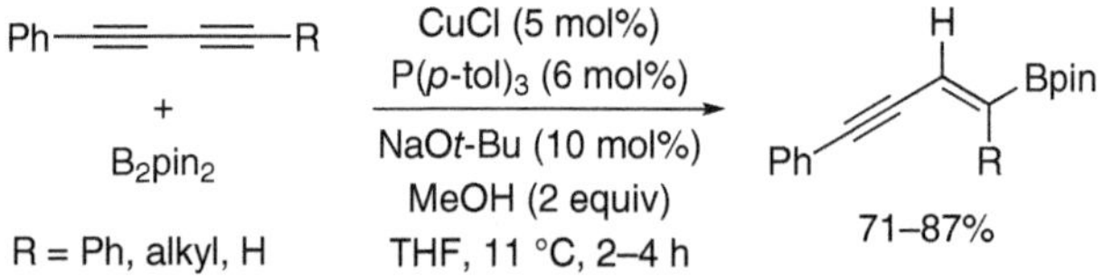

Scheme 3.6 CuCl/P(*p*-tol)$_3$-catalyzed *cis*-hydroboration of 1,3-butadiynes with B_2pin_2.

Interestingly, a cobalt-catalyzed regiodivergent stereoselective hydroboration of 1,3-diynes with HBpin to give boryl-functionalized enynes has been achieved, and the regioselectivity depends on the use of bisphosphine ligands (Scheme 3.7) [35]. With the use of $Co(acac)_2$/xantphos as the catalyst system, the hydroboration of a range of unsymmetrical and symmetrical 1,3-diynes selectively afforded enynylboronates with boron addition to the internal carbon of the 1,3-diyne unit. Whereas in the presence of $Co(acac)_2$/dppf, the addition reactions gave enynylboronats with boron addition to the external carbon of the 1,3-diyne unit.

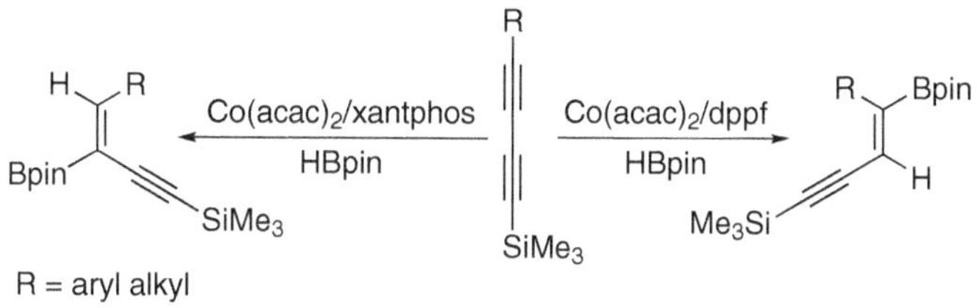

Scheme 3.7 Cobalt-catalyzed regiodivergent stereoselective hydroboration of 1,3-diynes with HBpin.

gem-Diborylalkenes and *gem*-diborylalkanes are interesting and versatile intermediates in organic synthesis; the regioselective double hydroboration of alkynes has been developed as the direct and efficient ways for these compounds (Scheme 3.8) [36].

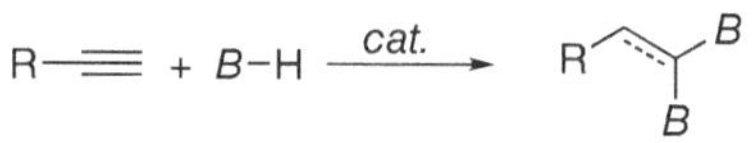

Scheme 3.8 *gem*-Diboryl alkenes/alkanes from alkynes.

The formation of *gem*-diborylalkenes usually involves C(sp)—H borylation of terminal alkyne and subsequently regioselective hydroboration of boryl-substituted internal alkyne. As the recent representative advances, NHC–zinc hydride complexes [37] and $Cu(OAc)_2$/phosphine ligand [38] were used as the catalysts in terminal alkyne C—H borylation/hydroboration affording *gem*-diborylalkenes. Zinc hydride complexes could also catalyze the formation of 1,1,1-triborylalkanes [39].

In addition, the commercially available organoborane reagent H-B-9-borabicyclo [3.3.1]nonane (BBN) was found to be an efficient catalyst for the sequential double hydroboration of terminal alkynes with HBpin to provide *gem*-diborylalkanes [40].

The metal/metal oxide nanoparticle have been extensively used in the hydroboration of alkynes [41]. The recyclable catalysts, such as copper-loaded nanocellulose sponge [42], Cu(0)-incorporated $CuFe_2O_4$ nanoparticles [43], solid polymeric carboxylic acid [44], $Ru(CO)Cl(H)(PPh_3)_3$ immobilized in ionic liquids [45], and $Ru(CO)Cl(H)(PPh_3)_3$/PEG [46], have been recently applied in the similar addition reactions.

In addition, Ge's group has developed the application of hydroboration of unsaturated hydrocarbons in the synthesis of chiral cyclic compounds. In the presence of $Co(acac)_2$/chiral bisphosphine ligands, the asymmetric hydroboration/cyclization of a variety of oxygen-, nitrogen-, carbon-, and amide-tethered 1,6-enynes with pinacolborane occurred to yield a variety of boronate esters containing chiral cyclic compounds with high to excellent enantioselectivities [47]. The same group also reported a $Cu(OAc)_2$/chiral ligand-catalyzed asymmetric hydroboration of 1,3-enynes with pinacolborane to give chiral allenylboronates [48].

As the application of hydroboration in the synthesis of heterocyclic compounds, the intramolecular *trans*-hydroboration of internal alkynes was reported for approach to cyclic aminoboranes [49] and five-membered BN-heterocycles [50].

Finally, as the same elements of 13 groups, the hydroalumination and hydrogallation of alkynes are also well-known reactions [51].

3.2 Hydrosilylation of Alkynes

Alkenyl silanes with lack of toxicity and wide functionality have been well applied as an important class of organosilicon intermediates in the synthetic chemistry for the complicated molecules, polymers, and natural products via activation of C—Si bond [52]. Therefore, the transition metal-catalyzed hydrosilylation of alkynes, which is the most straightforward and atom economical strategies for the synthesis of alkenyl silanes, has been extensively investigated in the presence of transition metal complexes, and a number of review papers have appeared [53]. The mechanism and theoretical studies have been also in detailed investigated [54]. This section focuses on summarizing the recent literatures on the development of this type of addition reactions.

The hydrosilylation of terminal alkynes can also theoretically afford three adducts via Markovnikov and *anti*-Markovnikov addition reactions as shown in Scheme 3.1 (E = Si).

For selective Markovnikov-type hydrosilylation (Scheme 3.9), the achievements have been made including the development of efficient catalyst systems by using ruthenium [55] and cobalt complexes [56].

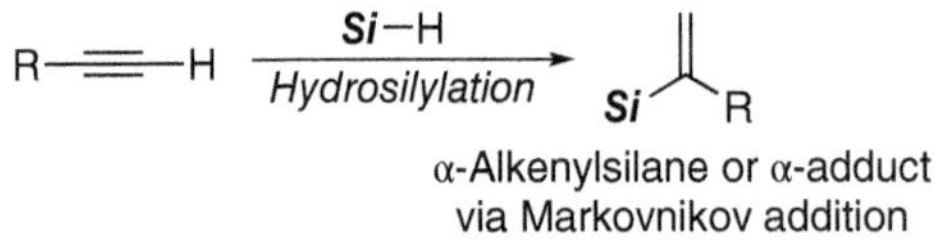

Scheme 3.9 Markovnikov hydrosilylation of terminal alkynes.

Very recently, von Wangelin and coworker have developed an interesting ligand-controlled regioselective hydrosilylation of alkynes catalyzed by $Co(OAc)_2 \cdot 4H_2O$ under mild conditions and found that the use of bipyridine ligands produces α-alkenylsilanes, whereas the use of bidentate phosphines affords (*E*)-β-alkenylsilanes [57].

The *anti*-Markovnikov *cis*-hydrosilylation of terminal alkynes can afford (*E*)-β-alkenylsilanes (Scheme 3.10). Cobalt [58], manganese [59], platinum [60], and rhodium [61] complexes have shown the high catalytic activity.

In addition, it was found that $Pd(dba)_2$/ligand-catalyzed hydrosilylation of 1,3-enynes with Me_2SiHCl proceeded to afford dienylsilanes with the silicon

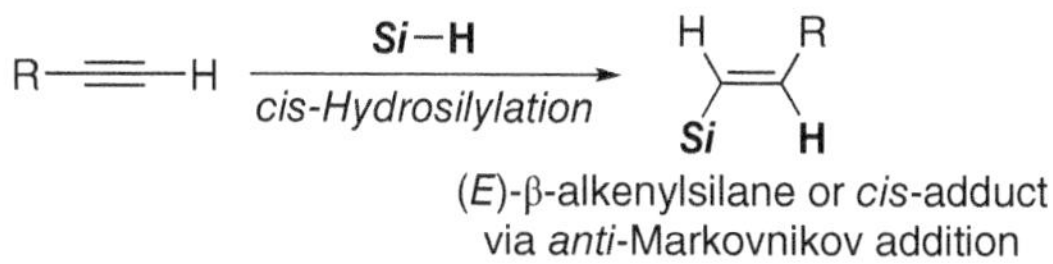

Scheme 3.10 *anti*-Markovnikov *cis*-hydrosilylation of terminal alkynes.

function added to the internal alkyne carbon atom and with (*E*)-configuration of newly formed alkene bond [62].

Catalytic enantioselective construction of silicon-stereogenic centers via desymmetrization of prochiral organosilane substrates is one of the efficient and interesting protocols [63]. Recently, the catalytic enantioselective formation of silicon-stereogenic alkenylhydrosilanes has been reported by *cis*-hydrosilylation of alkynes with dihydrosilanes in the presence of chiral platinum [64] and cobalt complexes [65] (Scheme 3.11).

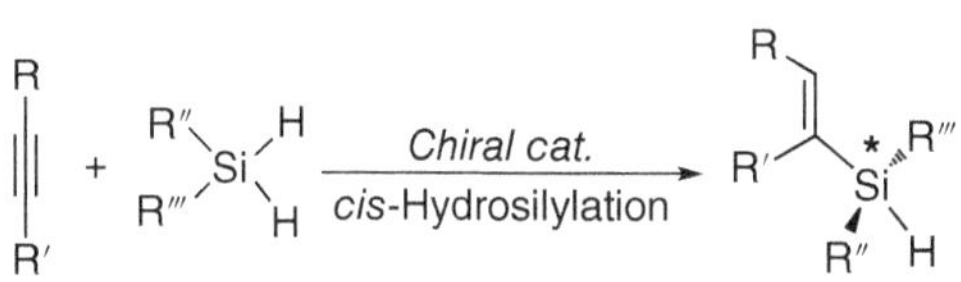

Scheme 3.11 Formation of silicon-stereogenic alkenylhydrosilanes by hydrosilylation of alkynes with dihydrosilanes.

In addition, a variety of recoverable heterogeneous catalysts show high catalytic activity for the *cis*-hydrosilylation of alkynes. For example, the impregnated platinum on magnetite have been used as an efficient catalyst for the *cis*-hydrosilylation of internal alkynes with high stereoselectivity to give €-alkenylsilanes in good yields [66]. Nanoporous gold catalyst [67], palladium [68], and platinum [69] nanoparticles stabilized with trisimidazolium tetrafluoroborates; SiO_2-supported Pd—Cu bimetallic nanoparticles also catalyze the regio- and stereoselective *cis*-hydrosilylation of both terminal alkynes and internal alkynes to give *cis*-adducts and in the case of terminal alkynes via *anti*-Markovnikov hydrosilylation under mild conditions [70]. In addition, cobalt or nickel-metalated porous organic polymers are efficient catalysts [71].

(*Z*)-β-Alkenylsilanes can be synthesized by a *trans*-hydrosilylation of terminal alkynes (Scheme 3.12) [72]. The recent representative catalyst systems including ruthenium/carbene complexes [73], ruthenium/*N* heterocyclic carbine [74], nonclassical ruthenium hydride pincer complex [75], cobalt complex [76], Lewis

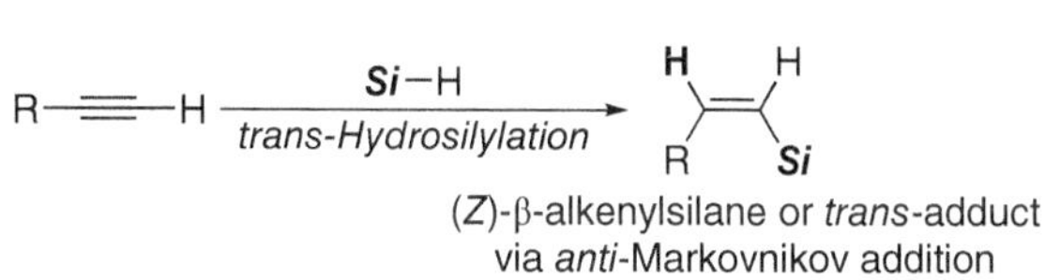

Scheme 3.12 *anti*-Markovnikov *trans*-hydrosilylation of terminal alkynes.

acids of $AlCl_3$ and $EtAlCl_2$ [77], monothiolate-bridged dirhodium complexes [78], cyclometalated rhodium(III)-triazolylidene complexes [79], zwitterionic bis-NHC rhodium(III) complex [80], and iron(II) polyhydride complex [81] have been developed. Also a visible light-initiated manganese-catalyzed *trans*-selective hydrosilylation of terminal alkynes has been reported [82].

In addition, the formation of (*E*)-β- and (*Z*)-β-alkenylsilanes can be switched via the hydrosilylation of terminal alkynes in the presence of $RhI(PPh_3)_3$ by simply changing the reaction conditions and the order of the addition of the reagents [83], by using Cp*Rh complexes of $[Cp^*Rh(BINAP)](SbF_6)_2$ and $[Cp^*RhCl_2]_2$ [84], and by using ligand- and structure-controlled ruthenium complexes [85].

In the cases of unsymmetrical internal alkynes used, one of the challenges in the *cis*-addition of a silane across the carbon–carbon triple bond is to control the regioselectivity of addition reactions (Scheme 3.13).

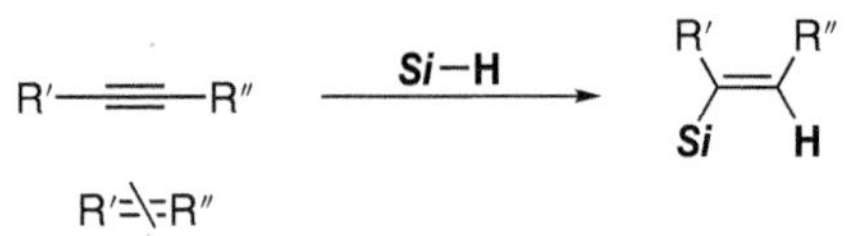

Scheme 3.13 Hydrosilylation of asymmetric internal alkynes.

An early report on the highly regioselective *cis*-hydrosilylation of unsymmetrical internal alkynes was achieved by using dimethylvinylsilyl (DMVS) group as the directing group from the reaction of propargylic and homopropargylic alcohols with dimethylvinylsilyl chloride (DMVSCl), and thus the hydrosilylation of propargylic and homopropargylic alcohol derivatives affords a single regioisomer of alkenylsilane [86].

As shown in Scheme 3.14, it has disclosed that when R′ is a carbonyl-based group such as ester, amide, and aldehyde and when R″ is an alkyl group, $PtCl_2$-catalyzed hydrosilylation of the electron-deficient alkynes occurs with excellent regioselectivity, where regioselectivity is predominantly governed by electronic effects, and the silyl group is installed to the electron-withdrawing group [87]. However, with the use of $[Cp^*Ru(MeCN)_3]PF_6$ as the catalyst, the hydrosilylation of thioalkynes at room temperature with a bulky silane of $(TMSO)_3SiH$ proceeds with excellent regioselectivity, and silyl group is attached to electron-rich carbon [88]. The similar regioselective manner can be observed with the use of various electronically and sterically distinct silanes catalyzed by $[(cod)IrCl]_2$ in DCM at room temperature [89].

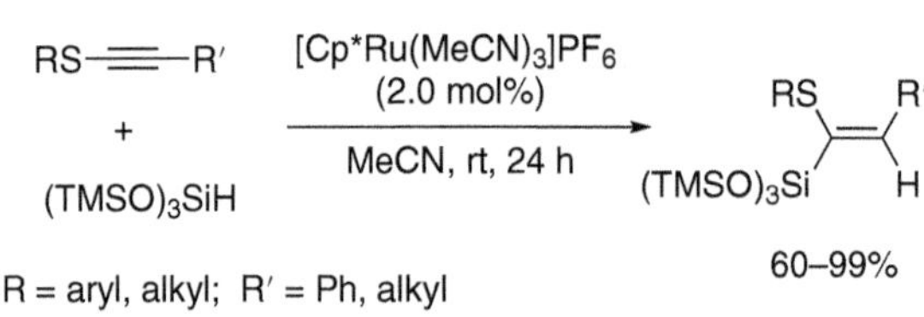

Scheme 3.14 Ruthenium-catalyzed hydrosilylation of thioalkynes with $(TMSO)_3SiH$.

The Cp*Ru complexes have been found to be efficient catalysts in the hydrosilylation of a wide variety of alkynes with good regioselectivity [90]. Wu and Sun's group has developed the first highly efficient ligand-controlled regio- and stereodivergent hydrosilylation of internal alkynes by using cationic ruthenium complexes, and the switching of the products with high regio- and stereoselectivity is resulted from the subtle variation of the catalyst ligand (Cp vs. Cp*) and the silyl group in the alkyne substrates (Scheme 3.15) [91]. The density functional theory (DFT) calculations combined with experimental evidence provide important insight into the reaction mechanism to disclose the interplay between the bulky silyl group in the alkynes and the ligand controlling the remarkable regio- and stereodivergence.

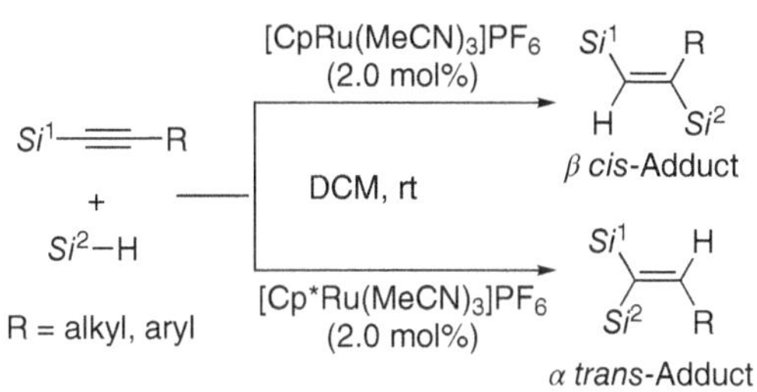

Scheme 3.15 Ligand-controlled regio- and stereodivergent hydrosilylation of internal alkynes.

Their further studies have developed a direct and efficient procedure for the synthesis of geminal disilylated terminal alkenes with excellent regioselectivity in high yields by the Markovnikov hydrosilylation of 1-silyl terminal alkynes catalyzed by $[CpRu(MeCN)_3]PF_6$ [92].

$[HCo(PMe_3)_4]$ has also been used as the catalyst in the *cis*-hydrosilylation of unsymmetrical internal alkynes with tertiary silanes and alkoxysilanes and has been found that in the cases of trimethylsilyl and aryl substituted internal alkynes employed, the *cis*-hydrosilylation occurs with highly regioselectivity, and the silyl group is installed to the aryl group [93].

With the use of $[Cp^*Ru(MeCN)_3]PF_6$ [94] and $CpRu(PPh_3)_2Cl$ [95] as catalysts, three-substituted propargyl alcohols underwent the *trans*-hydrosilylation with high regio- and stereoselectivity.

Very interestingly, Plietker and coworker have developed a stereodivergent hydrosilylation of internal alkyne catalyzed by $FeH(CO)(NO)(Ph_3P)_2$ via an aryl–aryl interactions as directing motifs [96]. Either *E*- or *Z*-selective hydrosilylation products form in excellent yields and good to excellent stereoselectivities depending on the hydrosilane employed. Very recently, iron(II)-catalyzed hydrosilylation of various alkynes by ligand-controlled divergent regioselectivity has also been developed by other group [97].

A reversed stereoselectivity has been also developed by using either $MnBr(CO)_5$/$AsPh_3$ or $Mn_2(CO)_{10}$/LPO [98].

Chang and coworker found that borane could be used as an efficient catalyst to catalyze the selective hydrosilylation of internal ynamides leading to β-silyl (*Z*)-enamides with an excellent β-regioselectivity and *anti*-stereoselectivity (*trans*-adducts) [99]. The similar procedure was then reported with the use of $[Rh(CO)_2Cl]_2$ as a catalyst and the bulky silanes as reactants; various ynamides underwent the hydrosilylation smoothly at room temperature (Scheme 3.16) [100].

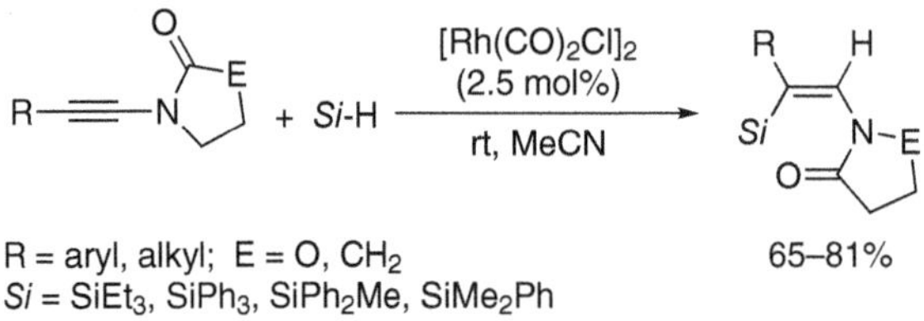

Scheme 3.16 $[Rh(CO)_2Cl]_2$-catalyzed hydrosilylation of ynamides with β-regioselectivity and *trans*-stereoselectivity.

When platinum/aminocarbene [101] and iridium/8-oxidoquinoline-2-carboxylate [102] were used as catalysts and platinum/diaminocarbene complexes were applied as photocatalysts [103] and radical reactions catalyzed by Eosin Y and thiol [104], the stereoselectivity in the hydrosilylation of terminal alkynes greatly depends on the structures and characters of substituents in alkynes and hydrosilanes.

In the CO atmosphere, alkynes could undergo the silylformylation with hydrosilanes and CO, which is a simple and efficient method for synthesis of β-silyl-α,β-unsaturated aldehyde, and a variety of rhodium complexes have been confirmed to be efficient catalysts, and the addition reactions usually occur with *cis*-addition to give (*Z*)-isomer (Scheme 3.17) [105].

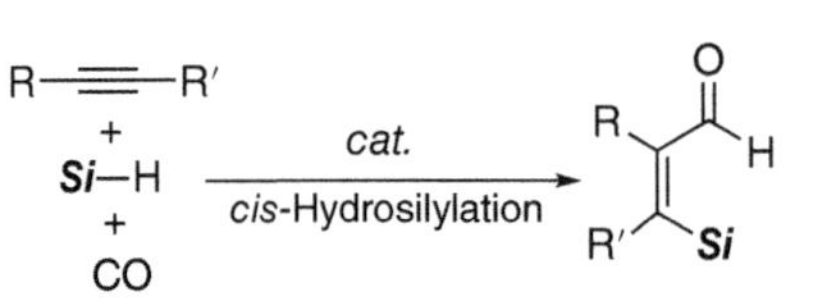

Scheme 3.17 Synthesis of β-silyl-α,β-unsaturated aldehyde by silylformylation of alkynes.

Very recently, the hydrosilylation of 1,3-diynes with highly regio- and stereoselectivity for the synthesis of mono- or bis-silylated adducts in the presence of platinum complexes has been developed (Scheme 3.18) [106]. The chemoselectivity is easily realized by using different amount of silane.

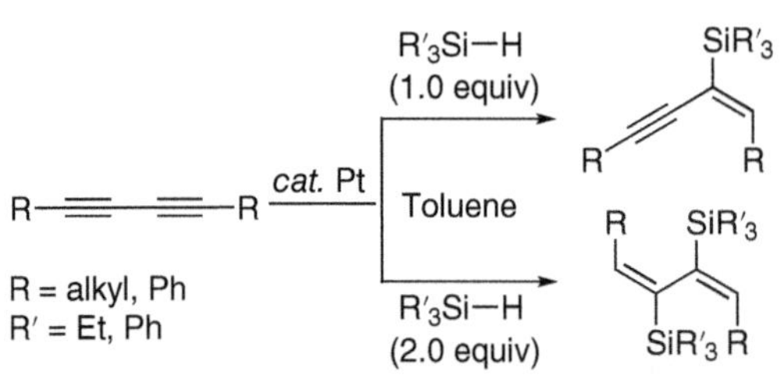

Scheme 3.18 Platinum-catalyzed regio- and stereoselective hydrosilylation of 1,3-diynes.

The catalyst system of $Co(acac)_2$/dppp has been also confirmed to be efficient for a regio- and stereoselective hydrosilylation of symmetrical and unsymmetrical 1,3-diynes, yielding the corresponding silyl-functionalized 1,3-enynes in high yields [107]. As shown in Scheme 3.19, the reaction between (buta-1,3-diyn-1-yl)silane and Ph_2SiH_2 produces the adduct from the regioselective *cis*-hydrosilylation of R-substituted triple bonds, and the high regioselectivity depends on the use of phosphine ligands.

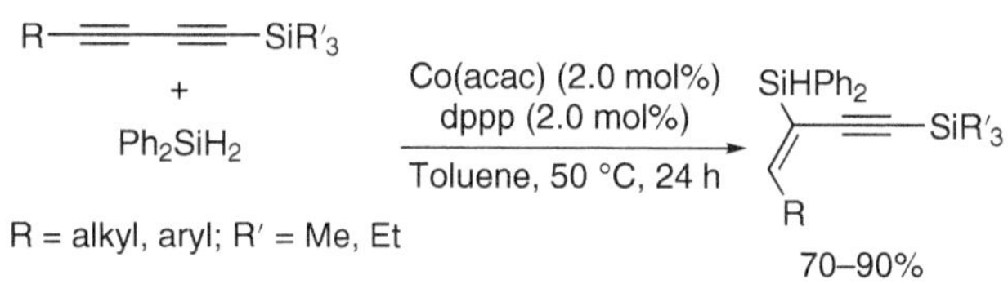

Scheme 3.19 Silyl-functionalized 1,3-enyne formation from cobalt-catalyzed regio- and stereoselective hydrosilylation of 1,3-diynes.

Very recently, iron-catalyzed dihydrosilylation of terminal alkynes access to *gem*- bis(silanes) [108] and cobalt-catalyzed asymmetric synthesis of *gem*-bis(silyl) alkanes by double hydrosilylation of aliphatic terminal alkynes have also been developed [109].

3.3 Hydrostannation of Alkynes

Transition metal-catalyzed activation of C—Sn bond of alkenylstannanes and its Stille cross-coupling reactions for C—C bond formation have been widely used in synthetic chemistry [110]. The development of the highly regio- and stereoselective reaction conditions for the hydrostannation of alkynes providing a straightforward approach to alkenylstannanes has become an important subject in modern synthetic organic chemistry, and a few comprehensive reviews have appeared [111]. As shown in Scheme 3.20, hydrostannation can afford three adducts, and the regio- and stereoselectivity depend on the use of catalysts and nature of terminal alkynes [112].

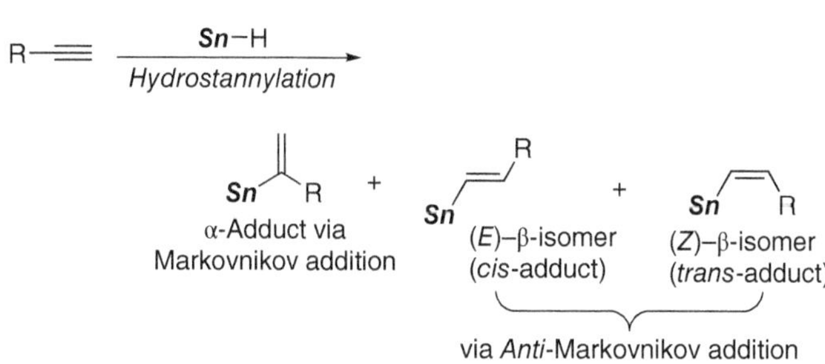

Scheme 3.20 Hydrostannation of terminal alkynes.

A pioneering work on the transition metal-catalyzed highly regioselective Markovnikov hydrostannation of terminal alkynes with Bu_3SnH was reported by Kikukawa and coworker in 1988 [113]. A few rhodium complexes catalyze hydrostannation of a variety of terminal alkynes to produce α-adduct as major products. $Pd(PPh_3)_4$ was then also reported to catalyze the Markovnikov hydrostannation of terminal alkynes efficiently [114].

One of the interesting procedures was developed by Baba's group, and they studied the reaction of $MgBr_2 \cdot OEt_2$ with n-Bu_2SnIH forming a tin hydride ate complex of $[MgBr]^+[n\text{-}Bu_2SnBrIH]^-$, which was used as the hydrostannylating reagent in the hydrostannation of aliphatic terminal alkynes in EtOAc at room temperatures without additive and catalyst to give the α-adducts [115].

Kazmaier's group has found that $Mo(CO)_3(NC\mathit{t}Bu)_3$ shows the catalytic activity for selective Markovnikov hydrostannation of various functionalized terminal alkynes [116].

Very recently, it has reported that the iron complexes of $Cp^*Fe(1,2\text{-}R_2PC_6H_4X)$ can control the Markovnikov and *anti*-Markovnikov hydrostannation of alkynes by tuning the ionic metal-heteroatom bonds (Fe—X) reactivity [117].

In addition, in the presence of H_2O, $Cu(OAc)_2$/phosphine has been used to catalyze the Markovnikov hydrostannation of terminal alkynes using a distannane or a silylstannane, and the synthetic utility of the resulting branched alkenylstannane has been demonstrated in the total synthesis of bexarotene [118].

On the other hand, the regio- and stereoselective formation of (*E*)-β-vinylstannanes via the terminal alkyne *cis*-hydrostannation has been developed with the use of molybdenum [119] and $Pd_2(dba)_3$/phosphine [120] as catalysts, under the conditions of ruthenium complex/fluorescent light [121] or under radical conditions [122].

Lewis acids, such as $ZrCI_4$ and $HfCI_4$ can catalyze the *trans*-hydrostannation of terminal and internal alkynes with n-Bu_3SnH to produce the *trans*-adducts with high regio- and stereoselectivity [123].

Oestreich's group has developed a metal-free method for selective *trans*-hydrostannation of terminal and internal alkynes with n-Bu_3SnH initiated by the trityl salt $[Ph_3C]^+[B(C_6F_5)_4]^-$ under mild conditions to give (*Z*)-isomer as major adduct, and the reaction is proposed through β-tin-stabilized vinyl cations (Scheme 3.21) [124].

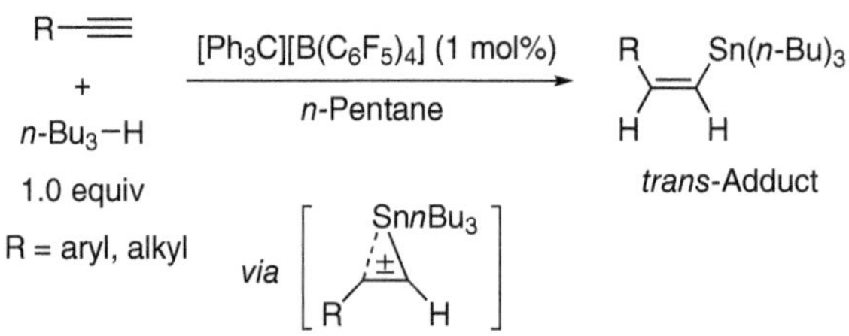

Scheme 3.21 $[Ph_3C][B(C_6F_5)_4]$-promoted *trans*-hydrostannation of alkynes.

In the cases of internal alkynes used, either *cis*-adduct or *trans*-adduct can be obtained via the high stereoselectivity hydrostannation depending on the use of catalyst systems, reaction conditions, and the substrate's structures.

Semmelhack's group has found that hexane is particularly favorable as a solvent in $Pd(OAc)_2/PCy_3$-catalyzed highly regioselective *cis*-hydrostannation of hindered internal alkynes with n-Bu_3SnH by minimizing the competing formation of hydrogen and distannane under mild conditions to give (*E*)-isomer [125].

$[(Ph_3P)CuH]_6$ effectively catalyzes the *cis*-hydrostannation of activated internal alkynes of alkynoates to produce (*E*)-α-stannylated alkenoates in good yields [126].

In the cases of α-heteroalkynes and alkynyl esters employed, the high regio- and stereoselectively addition reactions can be achieved. For example, Cai and coworker developed $Pd(PPh_3)_4$-catalyzed hydrostannation of alkynyl sulfides, alkynyl selenides, alkynyl esters, and alkynyl sulfones with n-Bu_3SnH in ionic liquid to give alkynyl sulfides, alkynyl selenides, alkynyl esters, and alkynyl sulfones via *cis*-addition and high stereoselectivity (Scheme 3.22). In addition, at room temperature, in benzene, the same catalyst could also catalyze the *cis*-hydrostannation of alkynyl esters with high regioselectivity to give (*E*)-α-stannyl-α,β-unsaturated esters, which undergo the Stille coupling reaction with alkenyl or alkynyl halides by addition of CuI and DMF to prepare 2-ethoxycarbonyl-substituted 1,3-dienes and 1,3-enynes in one-pot manner under mild conditions in good yields [127].

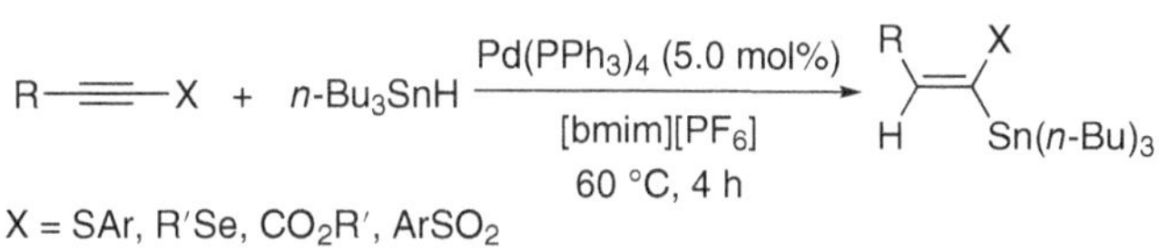

Scheme 3.22 $Pd(PPh_3)_4$-catalyzecl hydrostannation of α-heteroalkynes and alkynyl esters in ionic liquid.

The similar protocol has been used in the synthesis of alkylidenemalonates and related α,β-unsaturated esters [128].

It has also been reported that at room temperature, the glucal internal alkynes can undergo the *cis*-hydrostannation with n-Bu_3Sn—H catalyzed by $PdCl_2(PPh_3)_2$ in THF [129].

An early example of a regio- and stereocontrolled hydrostannation of internal alkynes, substituted propargyl alcohols, and derivatives was reported by Taddei's group [130]. The reaction of n-Bu_3Sn—H with different substituted propargyl alcohols by using azobis(isobutyronitrile) (AIBN) as radical initiator to give a mixture of *Z*/*E* isomers of alkenylstannane with the stannyl moiety bonded to the carbon closest to the OH or OR group, and the optimized reaction conditions could allow the isolation of pure (*Z*)-isomer.

With the use of $PdCl_2(PPh_3)_2$ as catalyst, in THF, Alami and coworker have found an *ortho*-substituent-directing regioselective addition of n-Bu_3SnH to unsymmetrical diaryl or heteroaryl alkynes providing an efficient route to stannylated stilbene derivatives (Scheme 3.23) [131]. At the same time, Marshall and coworker have found that the same catalyst can catalyze the hydrostannation of propargylic alcohols and acetates, as well as hydroxymethyl and acetoxymethyl-substituted internal alkynes with Bu_3SnH at room temperature affording the *cis*-adducts (*E*-allylic alcohols) with complete regioselectivity, and the Bu_3Sn group is affixed to the carbon proximal to the CH_2OH substituent, indicating an OH directing effect [132].

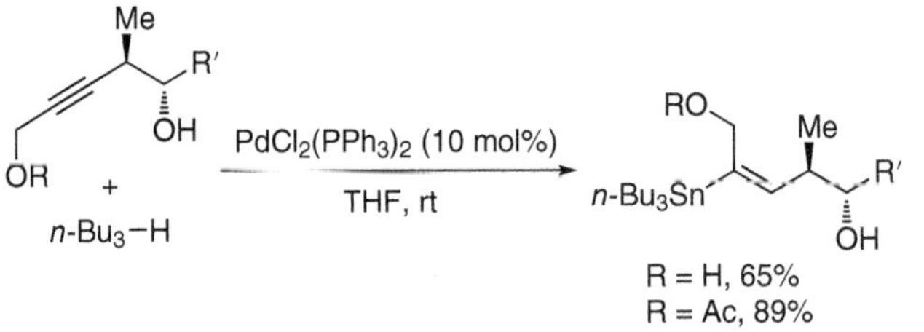

Scheme 3.23 Palladium-catalyzed *syn*-addition of hydrostannanes to alkynes.

However, an asymmetric total synthesis of (+)-inthomycin C via *O*-directed free radical propargyl alkyne hydrostannation with (*Z*)-selectivity in the presence of Et_3B has been reported [133]. In addition, when $Bu_2Sn(OTf)H$ or Bu_2SnClH is used, the Et_3B-initiated hydrostannation of propargyl alcohols (followed by butylation) occurs with high regio- and stereoselectivity, leading to (*Z*)-γ-stannylated allyl alcohols [134].

Alami's group has also then reported the $PdCl_2(PPh_3)_2$-catalyzed *ortho*-substituent-directing regioselective Markovnikov addition of *n*-Bu_3SnH to aromatic terminal alkynes [135].

In addition, remarkable differences in both regio- and stereoselectivity in radical- and non-radical-mediated hydrostannation of aryl propargylic alcohols and their derivatives under different conditions have been reported by Organ's group, and in the cases of internal propargylic alcohol derivatives used, *trans*-hydrostannation occurs with high stereoselectivity catalyzed by $B(C_6F_5)_3$ or mediated by AIBN in air [136].

Recently, [Cp*Ru]-based complexes have been found to be efficient catalysts for the selective *trans*-hydrostannation of internal alkynes to give (*Z*)-isomer [137].

Very recently, Rueping's group has reported a $MgBu_2$-catalyzed hydrostannation of internal alkynes giving *trans*-adducts, and in the cases of terminal alkynes, the addition reactions occurred with *cis*- hydrostannation to give (*E*)-β-alkenylstannane (Scheme 3.24) [138].

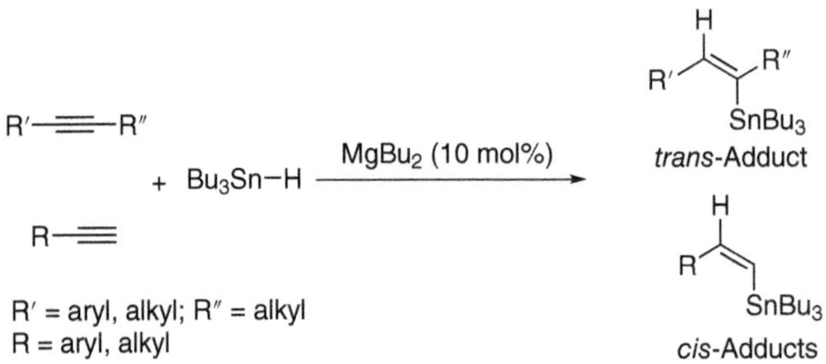

Scheme 3.24 $MgBu_2$-catalyzed hydrostannation of internal and terminal alkynes.

3.4 Hydroamination of Alkynes

Enamines [139] and imines [140] are important bulk and fine chemicals or building blocks in organic chemistry. The direct addition of amines to alkynes is an efficient and atom economical process for the preparation of these compounds, which have been summarized in detail in a few review papers [141].

The formation of enamines or imines depends on the use of types of amines, including alkylamines, anilines, and heterocyclic amines. As shown in Schemes 3.25 and 3.26, the hydroamination with ammonia (R″ = H) [142] or primary amines can give enamines, ketimines, or aldimes as products, and the hydroamination of terminal and internal alkynes with secondary amines will produce enamines.

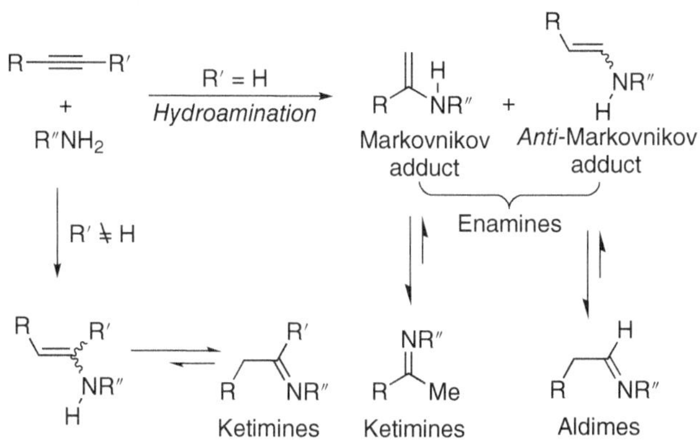

Scheme 3.25 Hydroamination of alkyne with ammonia or primary amines.

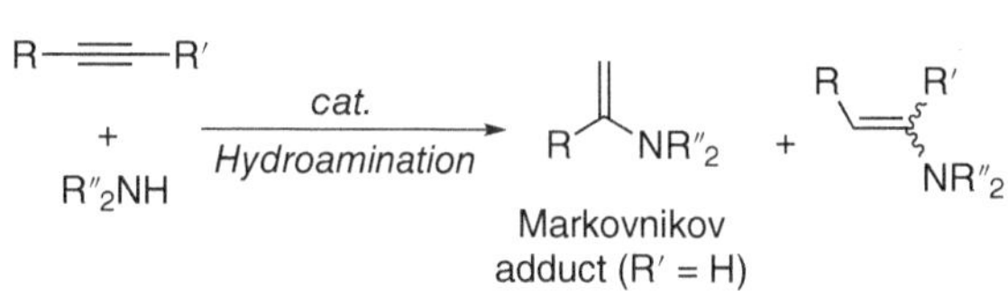

Scheme 3.26 Enamines formation from the hydroamination of alkynes with secondary amines.

3.4.1 Hydroamination of Alkynes with Primary Amines

Markovnikov hydroamination of terminal alkynes with primary amines affording enamines of ketimines can be achieved by using a variety of transition metal complexes as catalysts (Scheme 3.27).

It was reported that when titanium [143], rhodium [144], gold [145], ruthenium [146], palladium [147], zirconium [148], and platinum complexes [149] were used as the catalysts, the hydroamination with anilines, aliphatic amines, hydrazine, and hydroxylamine afforded ketimines as the major products.

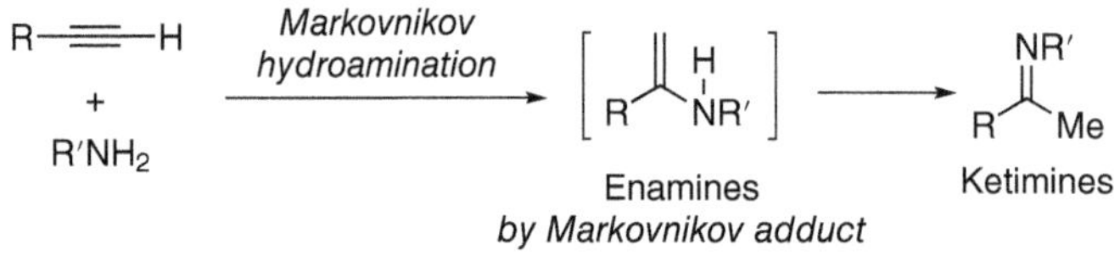

Scheme 3.27 Enamines and ketimines formation via Markovnikov hydroamination of terminal alkynes with primary amines.

In the case of $AuP^tBu_3NTf_2$ used as catalyst, the intermolecular double Markovnikov hydroamination of aromatic terminal alkynes with primary anilines occurs to give α,α′,*N*-triarylbisenamines (Scheme 3.28) [150].

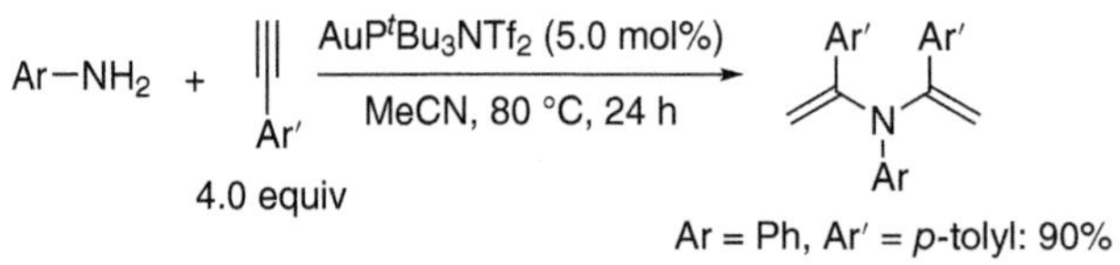

Scheme 3.28 Intermolecular double Markovnikov hydroamination of terminal alkynes.

Bis(amidate)bis(amido)Ti(IV) complexes as precatalysts show an *anti*-Markovnikov selective hydroamination of terminal alkynes with alkylamines to produce aldimine products [151] and with *N*-silylamines to afford *N*-silylenamines (Scheme 3.29) [152]. A variety of internal alkynes also underwent the addition reactions.

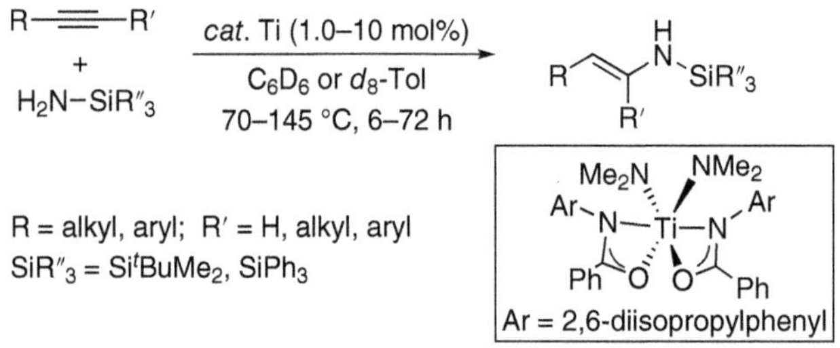

Scheme 3.29 Titanium-catalyzed *anti*-Markovnikov hydroamination of alkynes with *N*-silylamines.

The combination of *anti*-Bredt di(amino) carbene gold(I) chloride and $KBar^F$ was found to be a highly catalytic activity for the hydroamination of aryl alkynes and 1-ethynylcyclohexene with phenyl hydrazine or various anilines at room temperature to give ketimines with excellent regioselectivity of Markovnikov addition [153].

Beller's group has reported a ligand-controlled regioselectivity, titanium-catalyzed hydroamination of terminal alkynes with benzylamine, *n*-butylamine, *sec*-butylamine, and cyclooctylamine. The use of different sterically hindered phenoxy ligands results in the addition reaction switching from Markovnikov to *anti*-Markovnikov manner [154].

Under metal-free conditions, Beauchemin's group established intermolecular hydroamination procedures involving heating aqueous hydroxylamine and terminal alkynes to afford ketoximes as major products with good regioselectivity via Markovnikov addition [155], and a similar intermolecular hydroamination of a substituted hydrazine, $MeNHNH_2$, with terminal alkynes to afford hydrazones as major adducts via *anti*-Markovnikov addition (Scheme 3.30) [156].

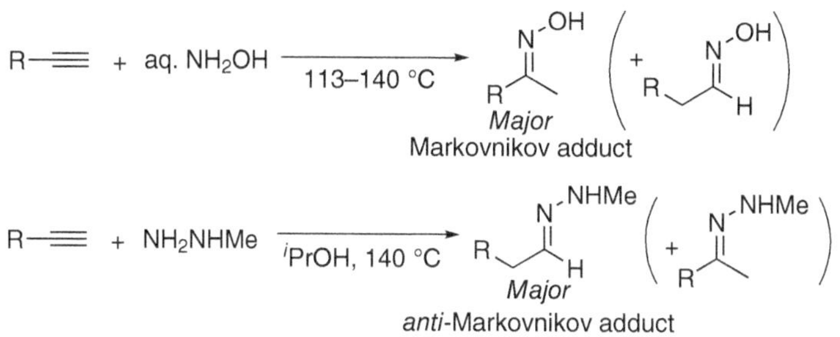

Scheme 3.30 Hydroamination of alkynes with hydroxylamine or methylated hydrazine.

In addition, Brønsted acid and HOTf could also be used as an efficient catalyst in the intermolecular hydroamination of electron-poor alkynes with primary anilines, giving enamines by *anti*-Markovnikov addition [157].

The applications of heterogeneous catalysts in hydroamination of alkynes have also been well investigated. The transition metal ion exchanged montmorillonite K-10 [158], Cu-exchanged tungstophosphoric acid [159], and Cu/Cu_2O nanoparticle [160] show high regioselectivity for the Markovnikov hydroamination of terminal alkynes with anilines giving methyl ketimines. Gold nanoparticles under visible light at ambient temperature show highly efficient activity to catalyze the hydroamination with similar selectivity [161].

3.4.2 Hydroamination of Alkynes with Secondary Amines

Hydroamination of terminal and internal alkynes with secondary amines produces enamines. The first examples of ruthenium-catalyzed Markovnikov regioselective intermolecular hydroamination of aromatic terminal alkynes with excess amount of *N*-methylanilines was reported by Uchimaru in 1999, and

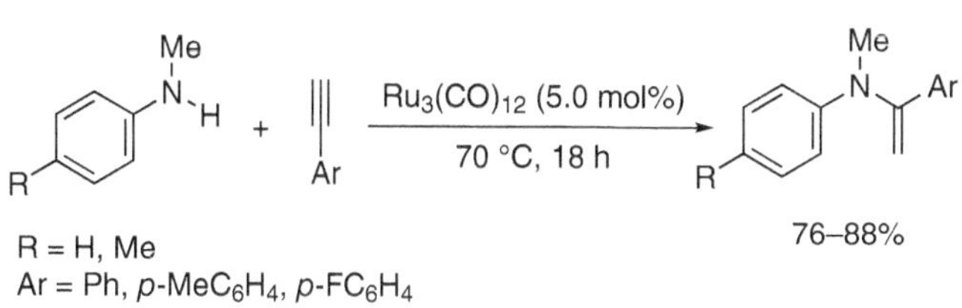

Scheme 3.31 Ruthenium-catalyzed Markovnikov intermolecular hydroamination of aromatic terminal alkynes.

N-methyl-*N*-(α-styryl)anilines were obtained in high yields under solvent-free conditions (Scheme 3.31) [162].

On the other hand, the selective synthesis of (*E*)-vinylamines could be achieved by *anti*-Markovnikov *cis*-hydroamination of terminal alkynes with secondary amines with the use of tungsten [163] and rhodium complexes [164] as catalysts.

In the presence of AuCl or cationic gold (AuCl(PPh_3/$AgSbF_6$), a *trans*-stereoselective 1,4-conjugate addition of pyrazoles to substituted propiolates occur to give (*Z*)-alkenyl acid β-pyrazolyl-(*Z*)-alkenyl acid (Scheme 3.32) [165].

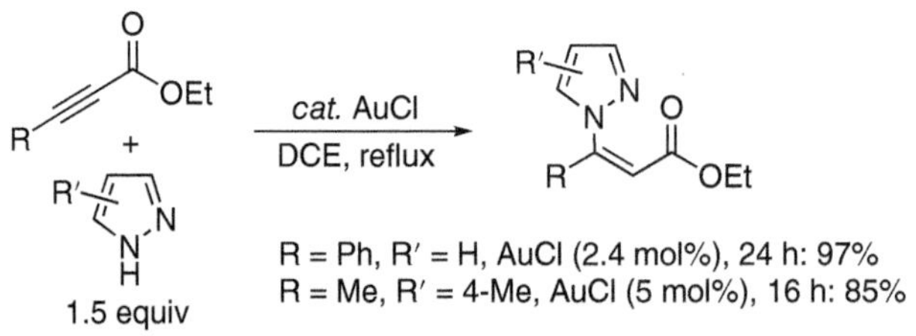

Scheme 3.32 Gold-catalyzed stereoselective 1,4-conjugate addition of pyrazoles to propiolates.

The catalyst system of Pd(0)/PhCOOH was used in the hydroamination of internal alkynes having α-CH_2 moiety with secondary amines to give the linear allylic amines via the key intermediates of allene π-allylpalladium (Scheme 3.33) [166]. When a catalyst system of $[Rh(COE)_2Cl]_2$/organic acid/chiral ligand was used, hydroamination of internal alkynes with indolines gives branched *N*-allylic indolines with high regio- and enantioselectivity [167].

Scheme 3.33 $Pd(PPh_3)_4/PhCO_2H$-catalyzecl hydroamination of internal alkynes giving allylic amines.

The *anti*-Markovnikov additions of N—H bond to terminal alkynes with various amides, lactams, anilides, ureas, bislactams, carbamates, imides, and thioamides were also reported to develop the regio- and stereoselective syntheses of enamides catalyzed by ruthenium [168] and silver complexes [169]. Gold-catalyzed *anti*-Markovnikov hydroamination of alkylidenecyclopropanes with 1-methyl-imidazolidin-2-one and other nucleophiles was also developed [170].

In addition, the regio- and stereoselective formation of (*Z*)-enamines via the *anti*-hydroamination of terminal alkynes with *N*-heterocycles could be achieved in the presence of inorganic bases, such as $CsOH{\cdot}H_2O$/NMP [171], K_3PO_4/DMSO [172], KOH/DMSO [173], and NaO*t*-Bu/DMF [174].

Vaccaro's group also developed an organic base, 2-*t*-butylimino-2-diethylamino-1,3-dimethylperhydro-1,3,2-diazaphosphorine (BEMP), catalyzed by the hydroamination of aromatic terminal alkynes with *N*-heterocycles, such as imidazole, benzimidazole, indole, 1,2,4-triazole, and pyrazole under solvent-free conditions [175]. The addition reaction occurred with high regioselectivity and stereoselectivity to give *trans*-adduct and the *Z*/*E* up to 99%. Interestingly, in the case of methyl propiolate employed, the hydroamination reaction with imidazole or 1,2,4-triazole directly affords "pincer" molecules via a regioselective double addition of imidazole or triazoles onto carbon–carbon triple bond of methyl propiolate.

Recently, under the metal-free conditions, cyclic trimeric phosphazene superbase has been used as the catalysts to catalyze the *anti*-Markovnikov stereoselective hydroamination of alkynes in good to excellent yields [176].

3.4.3 Cyclohydroamination of Alkynes

The catalytic intermolecular and intramolecular hydroamination of alkynes has become one of the efficient synthetic methods for the synthesis of nitrogen-heterocyclic compounds, which generally show important biological properties [177].

Some other examples are summarized and discussed in Chapter 8; only several representative examples are given in this section. For example, the hydroamination of alkynes with aryl hydrazines in the presence of titanium catalysts affords indole derivatives via the formation of hydrazone intermediate (Scheme 3.34) [178].

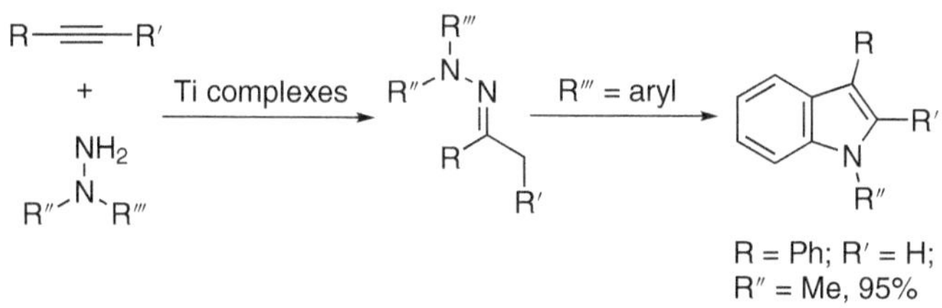

Scheme 3.34 Indoles synthesis from titanium-catalyzed hydroamination of alkynes with aryl hydrazines.

Indole ring could also be constructed by an intramolecular hydroamination of *ortho*-alkynylaniline catalyzed by iridium complexes [179] and zinc halides [180].

Yi's group studied $Ru_3(CO)_{12}/NH_4PF_6$-catalyzed intermolecular hydroamination of benzocyclic amines with an excess amount of terminal alkynes, which underwent regioselective cycloaddition reaction to give the tricyclic quinoline derivatives through the hydroamination and hydroarylation of alkynes (Scheme 3.35) [181].

In addition, 1,2-dihydroquinolines could be obtained from a $AgBF_4/HBF_4$ catalyzed by the reaction of anilines with an excess amount of terminal alkynes [182] or

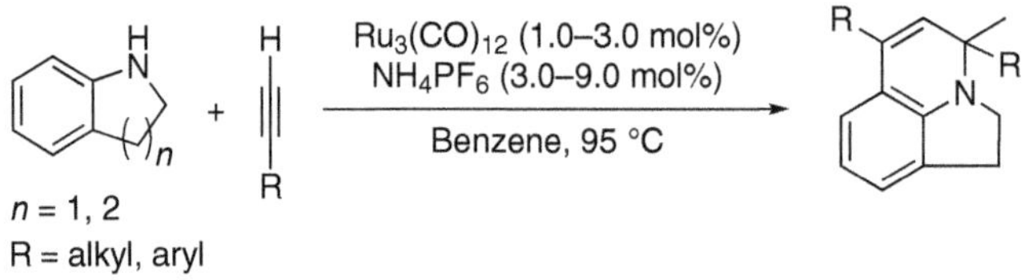

Scheme 3.35 $Ru_3(CO)_{12}$-catalyzed formation of tricyclic quinolines via regioselective hydroamination of terminal alkynes.

the reaction of two terminal alkynes with an aniline via an alkyne hydroamination followed by an alkyne insertion in the presence of $[Cp^*RhCl_2]_2$ [183].

Importantly, the intramolecular hydroamination of aminoalkynes has become one of the efficient ways to prepare the nitrogen-containing heterocycles bearing either C=N bond or C=C—N bond in the ring, depending on the reaction conditions and substrates.

Field and Messerle developed a cationic rhodium(I)-catalyzed intramolecular hydroamination of both terminal and internal alkynes affording regioselective formation of *N*-heterocycles from aliphatic aminoalkynes having C=N bond and two-substituted indoles from *o*-alkynylanilines in high yield (Scheme 3.36) [184].

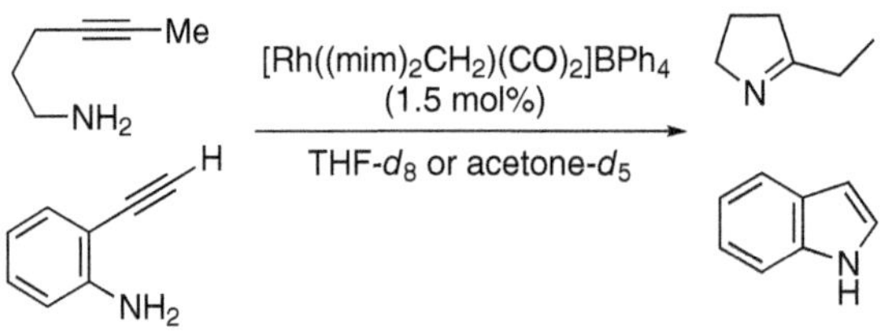

Scheme 3.36 Rhodium(I)-catalyzed intramolecular hydroaminaiton affording *N*-heterocycles.

Yamamoto's group studied the intramolecular asymmetric hydroamination of aminoalkynes to give the chiral nitrogen heterocycles. For example, in a catalyst system of $Pd_2(dba)_3·CHCl_3/PhCO_2H$/chiral ligand, the intramolecular asymmetric hydroamination of aminoalkynes has been first applied in the synthesis of five- and six-membered nitrogen heterocycles in high yields with good enantioselectivity (Scheme 3.37) [185]. The catalyst systems of Pd(0)/PhCOOH/chiral ligand have been

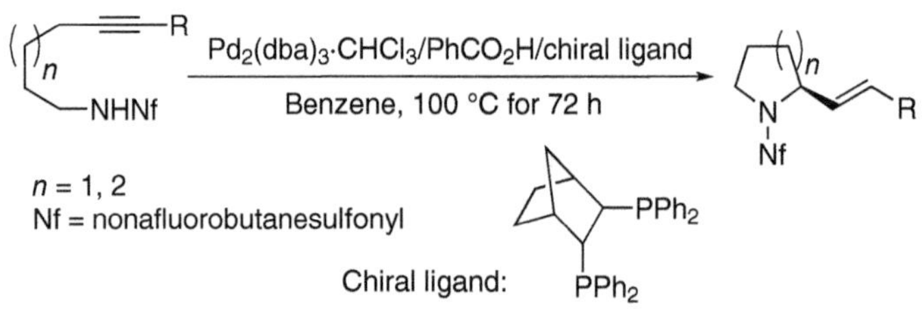

Scheme 3.37 Palladium-catalyzed intramolecular asymmetric hydroamination of aminoalkynes.

applied in the intramolecular asymmetric hydroamination of anilino-alkynes for the synthesis of two-substituted tetrahydroquinolines [186] and the intramolecular asymmetric hydroalkoxylation and hydrocarbonation of alkynes [187].

$Pd(PPh_3)_4/PPh_3$ showing the catalytic activity for the similar intramolecular hydroamination of *N*-trifluoromethanesulfonyl aminoalkynes has also been developed by the same group [188].

Other transition metal complexes, such as silver [189], copper, and zinc [190], have been also used in the intramolecular hydroamination of aminoalkynes.

Besides the previously mentioned procedures for the (cyclo)hydroamination of alkynes, organolathanide complexes have been also used as catalysts in the regioselective intermolecular hydroamination of alkynes [191].

3.4.4 Aminocarbonylation of Alkynes

The development of synthetic methods for α,β-unsaturated amides is very attractive due to their greatly important applications in academic research and industrial purposes. As shown in Scheme 3.38, an atom economic, simple, and one-pot reaction for direct synthesis of α,β-unsaturated amides is the aminocarbonylation of alkynes, and the regio- and stereoselectivity greatly depends on the use of catalysts and the reaction conditions [192].

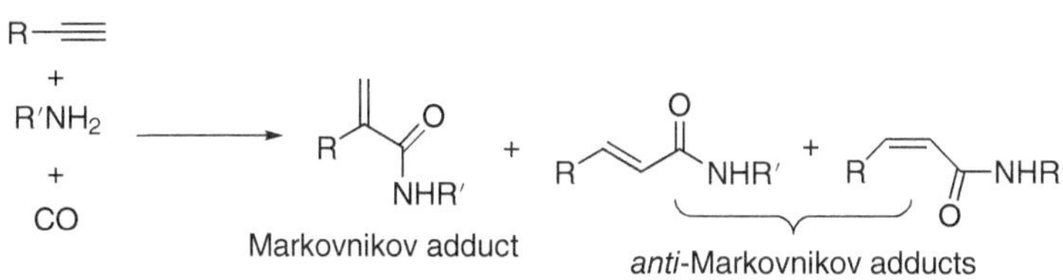

Scheme 3.38 Aminocarbonylation of terminal alkynes.

The pioneering work were reported by El Ali's and Ryu's groups. El Ali's group developed the aminocarbonylation of aromatic terminal alkynes with aniline derivatives under syngas conditions in the presence of $Pd(OAc)_2$ and dppb ligand to afford acrylamides in high yields and with good regioselectivity for the formation of Markovnikov adduct (Scheme 3.39) [193].

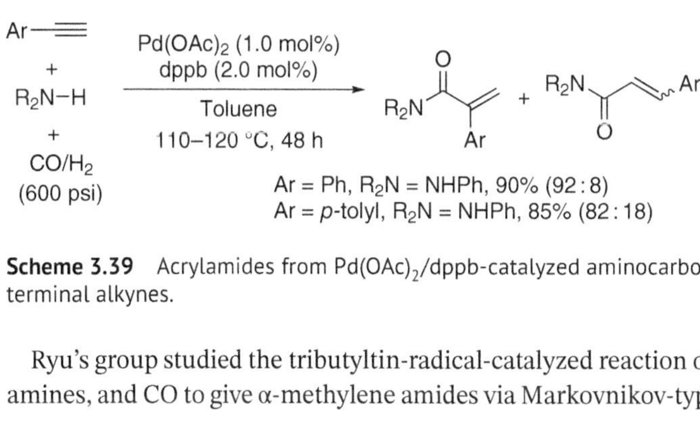

Scheme 3.39 Acrylamides from $Pd(OAc)_2$/dppb-catalyzed aminocarbonylation of aromatic terminal alkynes.

Ryu's group studied the tributyltin-radical-catalyzed reaction of terminal alkynes, amines, and CO to give α-methylene amides via Markovnikov-type addition without the use of transition metal catalysts (Scheme 3.40) [194]. The same group then developed a free-radical-mediated [2+2+1] cycloaddition of alkynes, amidines, and CO leading to five-membered α,β-unsaturated lactams [195].

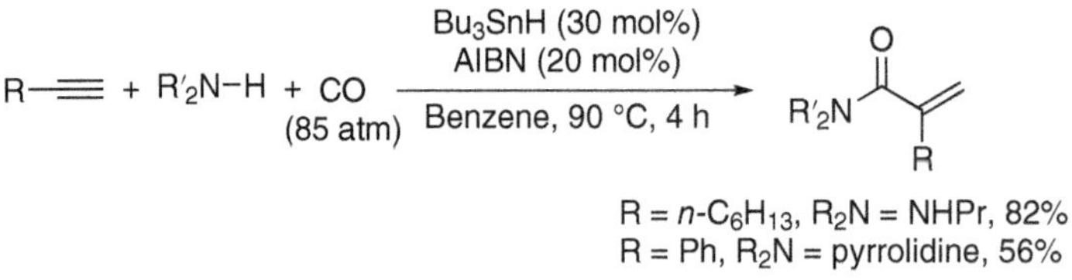

Scheme 3.40 Radical Markovnikov aminocarbonylation of terminal alkynes, amines, and CO.

In addition, ionic liquid [bmim][Tf_2N] [196], cobalt-rhodium heterobimetallic nanoparticle [197], $Fe_3(CO)_{12}$ [198], $Pd(PPh_3)_4$ [199], and nano-Pd catalyst [200] have also been used as efficient catalysts in the synthesis of α,β-unsaturated amides through the aminocarbonylation of alkynes.

An unprecedented double hydroaminocarbonylation of terminal alkynes with CO and cyclic amines catalyzed by $RhCl(PPh_3)_3$ with high chemoselectivity affording two-substituted 1,4-diamides in one-pot was reported by Hua and coworker (Scheme 3.41) [201]. The double hydroaminocarbonylation is suitable for both aliphatic and aromatic terminal alkynes to produce the corresponding 1,4-diamides in good to high isolated yields. A limitation of the procedure is that the linear secondary amines, such as Et_2NH and *n*-Bu_2NH, cannot give the similar transformation under the same reaction conditions. A proposed mechanism for the formation is shown in Scheme 3.42; it involves the hydroaminocarbonylation of alkynes affording Markovnikov and/or *anti*-Markovnikov adducts, which undergoes the second hydroaminocarbonylation to afford 1,4-diamide. Because 1,4-diamide is the exclusive product, the Markovnikov addition is considered to be the predominant reaction. As expected, the second hydroaminocarbonylation of Markovnikov adducts gives the 1,4-diamide in high yield.

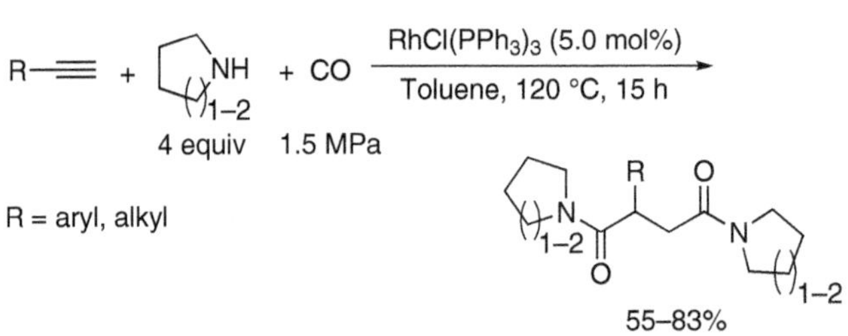

Scheme 3.41 $RhCl(PPh_3)_3$-catalyzed double hydroaminocarbonylation of terminal alkynes.

In the presence of NH_3, $Fe_3(CO)_{12}$-catalyzed double carbonylation of diaryl alkynes afforded *trans*-3,4-disubstituted succinimides, the subsequent oxidative dehydrogenation in situ yielded the corresponding 3,4-diarylmaleimides [202]. As

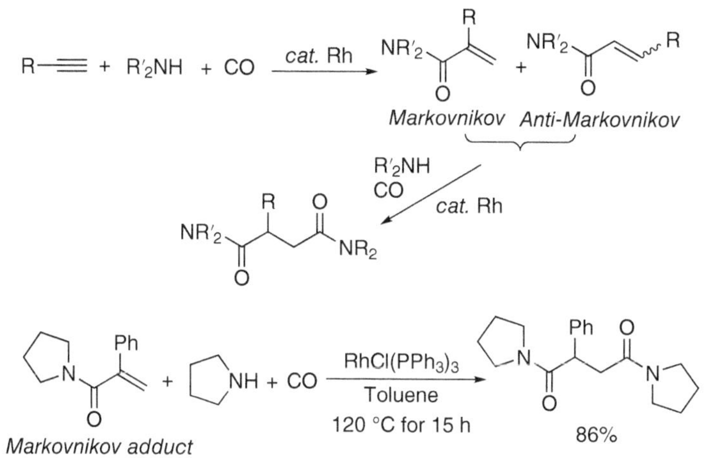

Scheme 3.42 Reaction route for the formation of 1,4-diamides.

shown in Scheme 3.43, the reaction of bisindolylacetylene gave the corresponding 3,4-bisindolylsuccinimide, and its oxidative dehydrogenation with dichloro dicyanoquinine (DDQ) produced 3,4-bisindolylmaleimide, which is an intermediate in the synthesis of the natural product arcyriarubin.

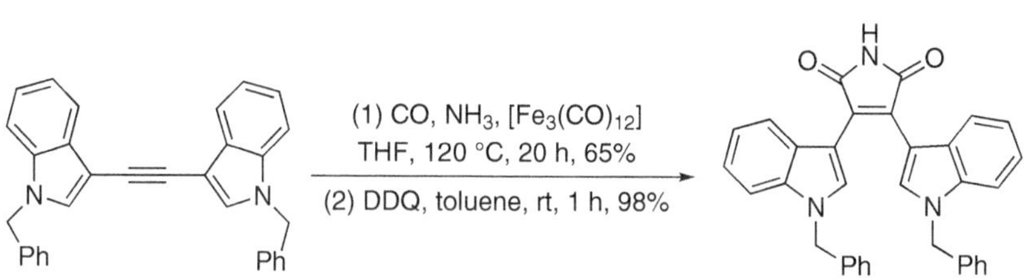

Scheme 3.43 Synthesis of arcyriarubin intermediate by aminocarbonylation.

3.5 Hydrophosphination of Alkynes and Related H–P(O) Addition

Organophosphorous compounds are of importance in organic synthesis, biochemistry, and material sciences [203]. Like other addition reactions of hydrogen-heteroatom bond to unsaturated hydrocarbons, addition reactions of P^{III}—H and P^{V}(O)—H bonds to alkynes are the direct and atom-efficient ways for the construction of P—C bonds to give organophosphorous compounds, which have been extensively studied [204].

As shown in Scheme 3.44, there are two classic additions of P^{III}—H and P^{V}(O)—H bonds to unsaturated hydrocarbons. The former is called as hydrophosphination, and the latter is called as hydrophosphinylation, hydrophosphorylation, or hydrophosphonation in the literatures, depending on the different substituent(s) of R in $P(O)R_2$—H. In this section, the latter addition reactions is simply called as $P^V(O)$ —H addition reactions.

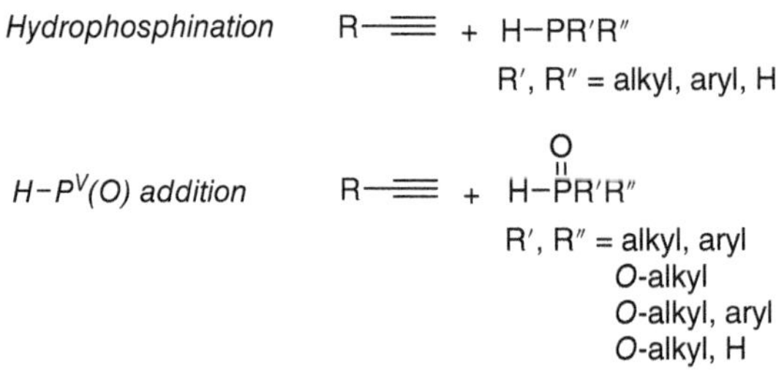

Scheme 3.44 Addition reactions of P^{III}—H and $P^V(O)$—H terminal alkynes.

3.5.1 Hydrophosphination of Alkynes

The known catalytic procedures for the hydrophosphination of terminal alkynes could produce either alkenylphosphines or in some cases afford the corresponding alkenylphosphines oxides or sulfides during workups by oxidation reactions or treatment with crystalline sulfur from the Markovnikov and *anti*-Markovnikov additions (Scheme 3.45).

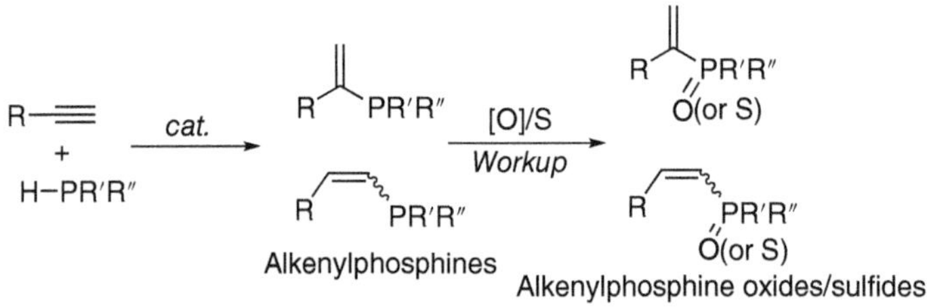

Scheme 3.45 Hydrophosphination of terminal alkynes giving alkenylphosphines or alkenylphosphine oxides/sulfides.

$Pd(PPh_3)_4$-catalyzed hydrophosphination of terminal alkynes with diphosphine-hydrosilane binary system regioselectively produced Markovnikov adducts [205]. However, with the use of diphosphine–hydrosilane, $RhCl(PPh_3)_3$-catalyzed hydrophosphination of terminal alkynes in the presence of oxygen afforded *anti*-Markovnikov-type *cis*-adducts [206].

Gaumont's group developed an interesting regio- and stereoselective hydrophosphination of terminal alkynes with the secondary phosphine-boranes ($R_2PH{\cdot}BH_3$) as hydrophosphinating under microwave irradiation affording (*Z*)-alkenylphosphine–boranes via *trans*-addition, while in the presence of palladium complex, the Markovnikov hydrophosphination occurred, and *P*-stereogenic alkenyl phosphine–boranes could be obtained with the use of chiral phosphine ligands (Scheme 3.46) [207].

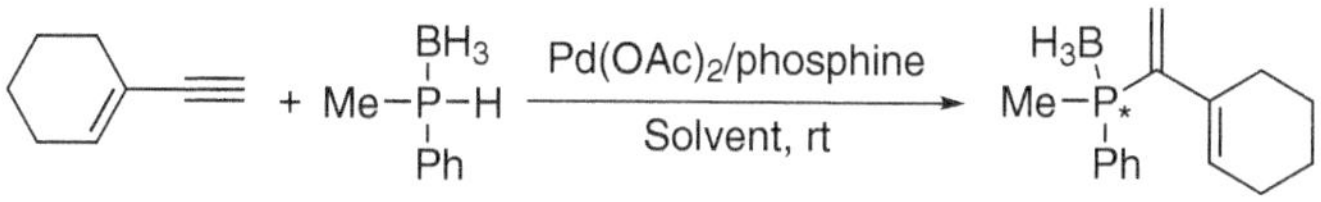

Scheme 3.46 Hydrophosphination of 1-ethynylcyclohexene with phosphine-borane.

With the use of iron(II) β-diketiminate precatalyst, the regioselectivity of hydrophosphination of terminal alkynes with H—PPh_2 are controllable from Markovnikov-selective manner in benzene to (*Z*)-selective *anti*-Markovnikov addition in dichloromethane [208].

Ytterbium [209], lanthanum [210], and cobalt complexes [211] have been used in the *cis*-hydrophosphination of terminal alkynes with H—PPh_2 to give (*E*)-alkenylphosphines.

The formation of (*Z*)-alkenylphosphines via *trans*-hydrophosphination of terminal alkynes usually occurs under transition metal-free conditions.

For example, the hydrophosphination of terminal alkynes with H—PPh_2 in the absence of a solvent and a catalyst under argon afforded (*Z*)-alkenylphosphines, or in the presence of elemental sulfur under air gave the corresponding alkenylphosphine sulfides (Scheme 3.47) [212]. Terminal alkenes could also undergo the *anti*-Markovnikov hydrophosphination under the similar conditions.

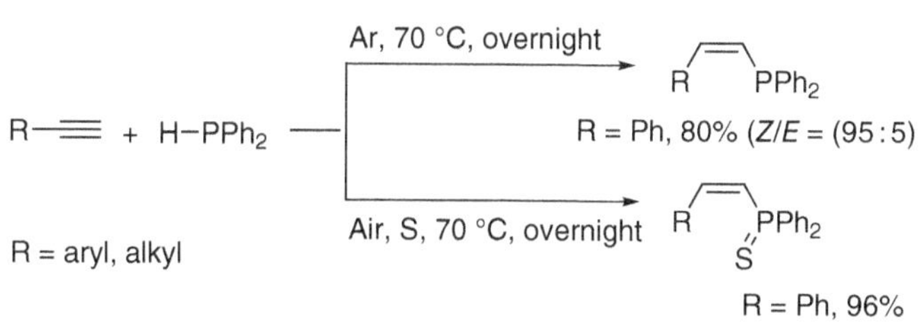

Scheme 3.47 Catalyst- and solvent-free hydrophosphination of terminal alkynes.

The first uncatalyzed hydrophosphinations of propargylic alcohols and amines with phosphine borane (H—P(*t*-Bu)$_2$BH$_3$) was also reported to give (*Z*)-isomers [213].

When propargyl alcohols were used as the substrates, RuCp*Cl(cod)- or RuCp*Cl(PPh_3)$_2$-catalyzed hydrophosphination with H-PPh_2 also led to the formation of (*Z*)-alkenylphosphines as the major adducts [214].

CuI/$CsCO_3$-catalyzed hydrophosphination of 1-alkynylphosphines with H—PPh_2 proceeded to provide an easy and efficient access to a variety of (*Z*)-1,2-diphosphino-1-alkenes or their sulfides [215].

In addition, ytterbium complexes were also found to be efficient catalysts in the hydrophosphination of terminal and internal alkynes with H—PPh_2; the regio- and stereoselectivity depends on the structures of alkynes [216]. The dual hydrophosphination of 1,3-butadiynes with H—PPh_2 to give the corresponding 1,4-bis(diphenylphosphinyl)buta-1,3-dienes in high yields after oxidative workup, and the distribution of the four possible stereoisomers sharply depended on substituents of the substrates [217].

In the presence of $Pd(OAc)_2$, the Markovnikov hydrophosphination of terminal alkynes could also take place regioselectively with tetraphenyldiphosphine to form alkenylphosphines and afforded the alkenylphosphine oxides in good yields [218].

In addition, the phosphorus heterocycles could be prepared by the catalytic intramolecular hydrophosphination/cyclization of phosphinoalkenes and phosphinoalkynes using organolanthanide precatalysts [219].

Moreover, the double hydrophosphination of alkynes with primary or secondary phosphines could provide direct access to 1,2-bis(diphosphino)ethanes (Scheme 3.48) [220]. Iron [221], zirconium [222], copper [223], and rhodium complexes [224] under different conditions have been reported to be efficient catalysts for this addition reactions.

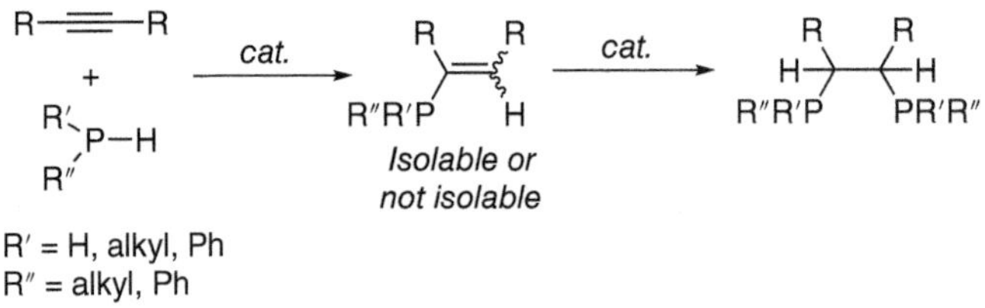

Scheme 3.48 Double hydrophosphination of alkynes with primary/secondary phosphines.

3.5.2 Addition of $P^V(O)$—H Bond to Alkynes

Although the radical or base-catalyzed additions of H—P(O)R_2 to alkynes have been well studied [225], the transition metal-catalyzed in this type of addition reactions were pioneered by Tanaka and Han's group in 1996 [226]. They first reported a palladium-catalyzed Markovnikov addition of dialkyl phosphites to terminal alkynes in THF to afford various alkenylphosphonates in excellent yields (Scheme 3.49). They also reported the oxidative addition of H—P(O)$(OR)_2$ to Pt(0) complex generating H—Pt—P(O)$(OR)_2$ species, which was proposed to be the key intermediate for the transition metal-catalyzed activation and addition of $P^V(O)$—H bond to carbon–carbon triple bond.

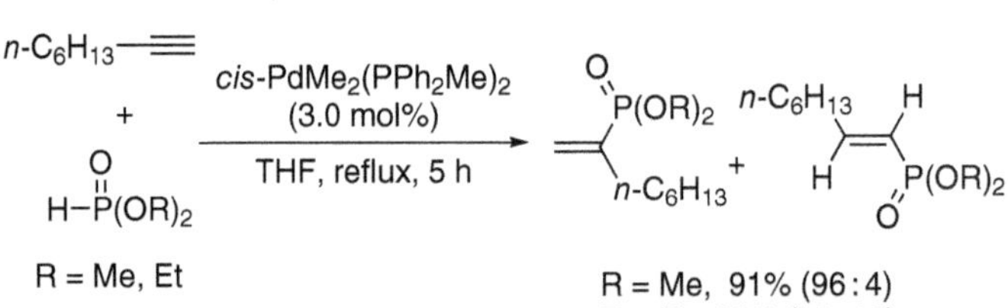

Scheme 3.49 Palladium-catalyzed Markovnikov addition of dialkyl phosphites to terminal alkynes.

The regioselectivity of addition reactions greatly depends on the use of transition metal complexes, and the structure of dialkyl phosphites. It was found that

$RhCl(PPh_3)_3$-catalyzed addition of terminal alkynes with the cyclic hydrogen phosphonate at room temperature gave the high yields of exclusively (*E*)-alkenyl phosphonates [227].

Tanaka and Han's group also developed a straightforward synthesis of enantiomerically pure *P*-chiral alkenylphosphinates via palladium-catalyzed Markovnikov addition of P^V(O)—H bond to terminal alkynes [228].

In the case of $Ni(acac)_2$/DPPE used as catalyst system, addition reaction of terminal alkynes with $(^iPrO)_2P(O)$—H proceeded with high regioselectivity to give Markovnikov adducts as the major products in the presence of either acids or bases [229].

Also the polystyrene-bound triphenylphosphine-immobilized palladium and nickel complexes were found to catalyze the Markovnikov addition efficiently to afford high yields of addition products with high regioselectivity [230].

Tanaka and Han also examined the oxidative addition of H—$P(O)Ph_2$ to $M(PEt_3)_3$ (M = Pd, Pt) at room temperature in benzene to obtain the complexes of *cis*-$MH[P(O)Ph_2][PPh_2(OH)](PEt_3)$, and the palladium complex was found to undergo an insertion reaction of alkyne to Pd–H bond [231]. Accordingly, a $Pd(PPh_3)_4$-catalyzed regio- and stereoselective *anti*-Markovnikov *cis*-addition of terminal alkynes with diphenylphosphine oxide was developed to afford (*E*)-alkenyldiphenylphosphine oxides (Scheme 3.50).

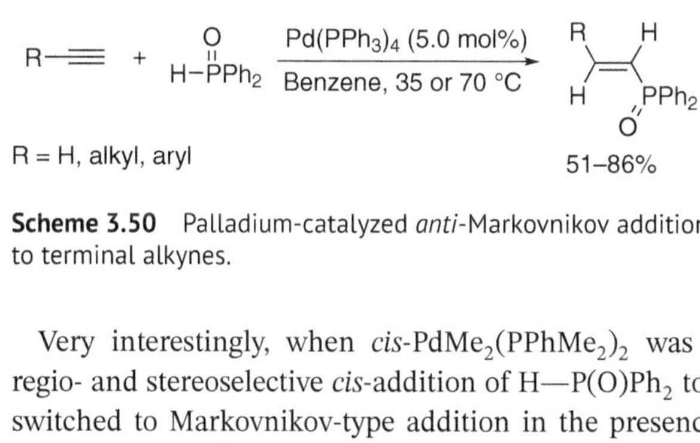

Scheme 3.50 Palladium-catalyzed *anti*-Markovnikov addition of diphenylphosphine oxide to terminal alkynes.

Very interestingly, when *cis*-$PdMe_2(PPhMe_2)_2$ was used as the catalyst, the regio- and stereoselective *cis*-addition of H—$P(O)Ph_2$ to terminal alkynes could be switched to Markovnikov-type addition in the presence of a catalytic amount of phosphinic acid [$Ph_2P(O)OH$] (Scheme 3.51) [232]. The reversal of regioselectivity was also observed in the addition of H—$P(O)(OMe)_2$ to terminal alkynes catalyzed by $Ni(cod)_2$/phosphine/$Ph_2P(O)OH$ [233].

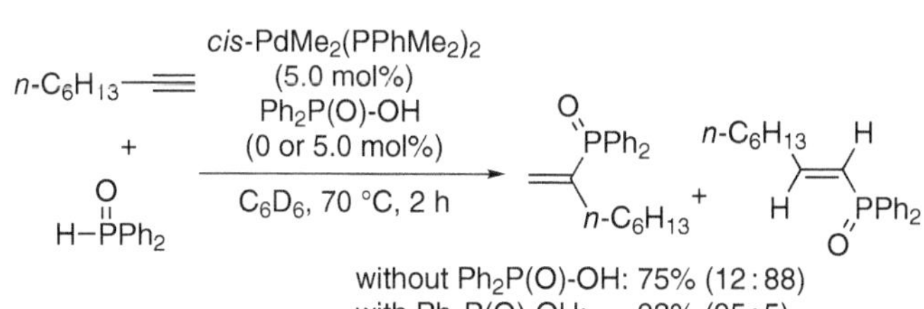

Scheme 3.51 Phosphinic acid-induced reversal of regioselectivity of hydrophosphinylation.

The same group found that, with the use of Ni(0)/$Ph_2P(O)$—OH as the catalyst system, the reactions of propargyl alcohols with $P^V(O)$—H compounds at room temperature produce high yields of phosphinoyl 1,3-dienes through an in situ dehydration process [234].

Tanaka and coworker also disclosed a ligand- and solvent-dependent regioselective $P^V(O)$—H addition to terminal alkynes with ethyl phenylphosphinate (Scheme 3.52) [235]. It has found that $Pd(OAc)_2$/dppe catalyzes regioselective Markovnikov addition of H—P(O)Ph(OEt) to terminal alkynes in toluene, while the use of t-Bu_3P as the ligand or ethanol as the solvent leads to regioselectivity reversal giving (*E*)-adduct via an *anti*-Markovnikov *cis*-hydrophosphinylation reaction. The similar regioselectivity reversal by using different ligands and solvents was also reported by other group [236].

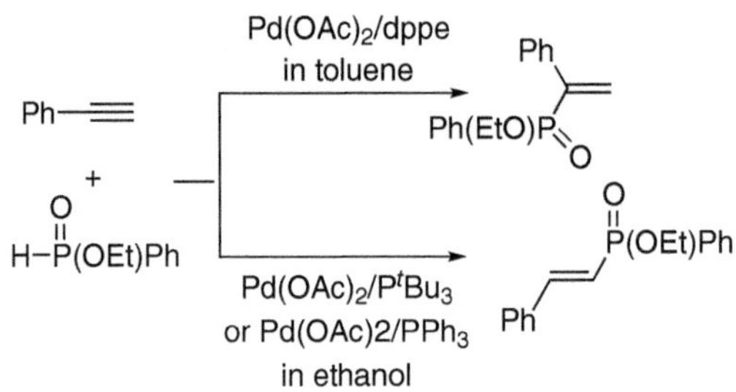

Scheme 3.52 Palladium-catalyzed regioselective addition of ethyl phenylphosphinate to terminal acetylenes.

The regioselectivity reversal could also be observed with the use of chiral enantiopure phospholane oxide; when palladium and rhodium complexes were used, Markovnikov addition and *anti*-Markovnikov addition occur, respectively [237].

The highly regioselective and stereoselective synthesis of (*E*)-alkenylphosphine oxides via *anti*-Markovnikov *cis*-addition of terminal alkynes with H-phosphine oxides was also achieved by using Rh(I) complexes [238], CuI/EDA [239] catalyst systems, $Cu(acac)_2$ in a mixture solvent of DMF/THF [240], and the recoverable catalysts of MCM-41-immobilized rhodium complex [241]. $Cu(OAc)_2$ shows activity for addition reactions of H-phosphinates to terminal ynamides giving β-aminophosphinates [242].

In the absence of ligand, $NiCl_2$ shows catalytic activity for the *cis*-addition reactions of terminal and internal alkynes with alkyl phosphinates [H-P(O)H(OR) [243]. Ananikov's group has also disclosed a unique $Ni(acac)_2$/DIBAL-catalyzed mono- and double hydrophosphorylation of terminal and internal alkynes under solvent-free and ligand-free conditions, which can be controlled by varying the catalyst loading [244]. As shown in Scheme 3.53, when equivalent of diphenylacetylene and $(^iPrO)_2P(O)H$ were used, the use of 9 mol% of $Ni(acac)_2$ and 18 mol% of DIBAL led to the complete monohydrophosphorylation to afford the *cis*-adduct of alkenylmonophosphonate in 99% yield with excellent stereoselectivity. While the application of 1 mol% of $Ni(acac)_2$, 2 mol% of DIBAL, and 2.0 equivalents of

($(^{i}PrO)_2P(O)H$ resulted in a quantitative yield of 1,2-bisphosphonate via double hydrophosphorylation. On the basis of the experiments, it is proposed that the formation of 1,2-bisphosphonate is from the transformation of monophosphonate mediated by a catalytic amount of DIBAL. In addition, in the cases of terminal alkynes used, the monohydrophosphorylation proceeds with excellent regio- and stereoselectivity affording (*E*)-alkenylphosphonate exclusively.

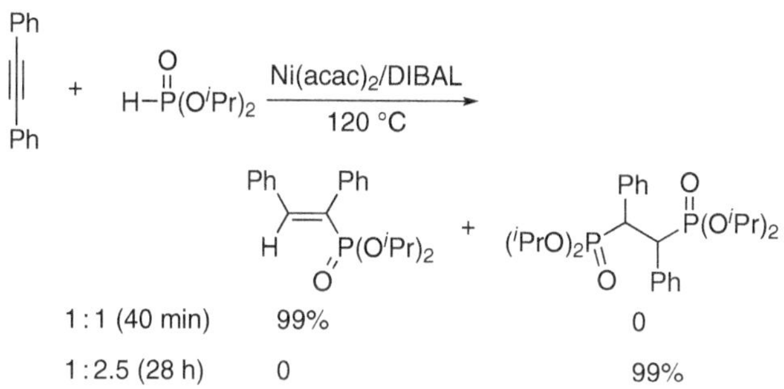

Scheme 3.53 Nickel-catalyzed mono- and bishydrophosphorylation with H-phosphonate.

Interestingly, in some cases, alkynes could undergo the double addition with $P^{V}(O)$—H bond giving 1,2-bisphosphine oxides (Scheme 3.54), which were catalyzed by $Pd(PPh_3)_4$ [245], *t*-BuOLi [246], and homogeneous/heterogeneous rhodium complexes under microwave radiation [247]. Recently, Han's group has developed an air-induced double addition of $P^{V}(O)$—H bond to terminal alkynes to provide a clean and practical method for the preparation of 1,2-bisphosphorylethanes [248]. In addition, double addition with dialkyl/aryl phosphites afforded 1,2-bisphosphonates with the use of $Pd(PPh_3)_4$ as catalyst [249].

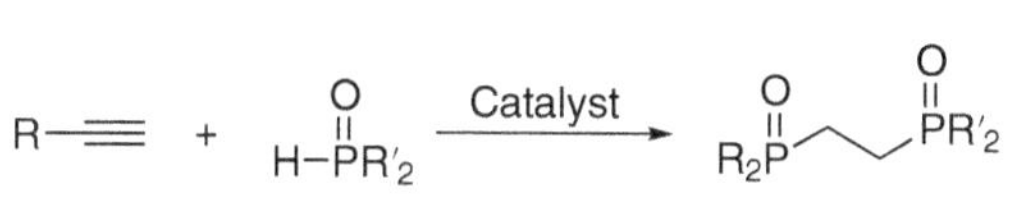

Scheme 3.54 Double addition of $P^{V}(O)$–H to terminal alkynes.

Oshima and coworker developed a highly efficient and concise diphosphanylation of terminal alkynes affording 1,2-diphosphanylethenes in good yield with high *E*-selectivity via radical addition of a tetraorganodiphosphane formed in situ by mixing a diorganophosphane and a chlorodiorganophosphane in the presence of trimethylamine [250]. As a typical reaction shown in Scheme 3.55, the reaction of a mixture of 1-dodecyne, diphenylphosphane, chlorodiphenylphosphane, triethylamine, and 1,1′-azobis(cyclohexanecarbonitrile) (V-40) in boiling benzene for 10 hours produced the adducts, isolated as phosphane sulfides in 84% yield with a 91 : 9 mixture of *E* and *Z* isomers.

The same group also reported a *cis*-hydrophosphination of terminal and internal alkynes with diphenylphosphane catalyzed by $Co(acac)_2$/BuLi, and in case of terminal alkynes used, the *E*-isomer is the major products [251].

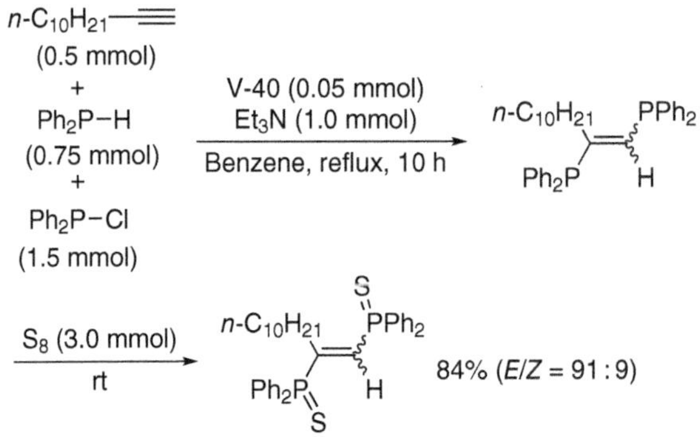

Scheme 3.55 Synthesis of (*E*)-1,2-diphosphanylethenes via radical addition of tetraorganodiphosphane to alkynes.

Han and coworker reported a palladium-catalyzed dehydrogenative *cis*-double addition of H-phosphonate to terminal alkynes leading to (*Z*)-bis(phosphinoyl) alkenes as the main products with the use of methyl acrylate as an efficient hydrogen scavenger (Scheme 3.56) [252]. They have recently studied in detailed the generality and mechanism of palladium-catalyzed hydrophosphorylation of alkynes [253].

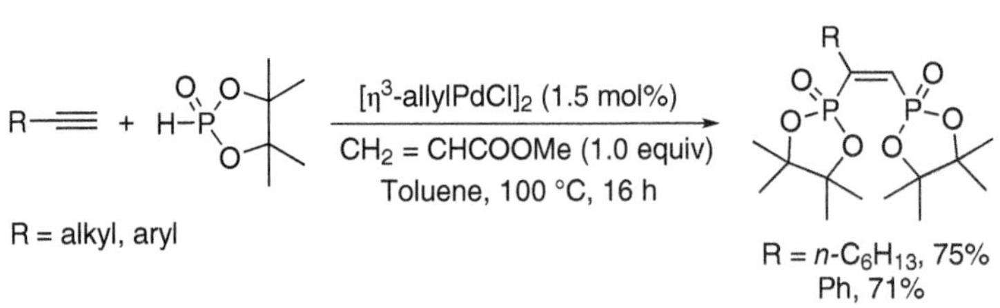

Scheme 3.56 Palladium-catalyzed dehydrogenative *cis*-double addition of H-phosphonate to terminal alkynes.

Very recently, the addition of Sb—H bond to alkynes has been reported recently as a new hydroelementation reactions [254].

3.6 Hydrothiolation of Alkynes

Transition metal-catalyzed or transition metal-free Markovnikov or *anti*-Markovnikov addition of thiols to terminal alkynes (hydrothiolation) is an important reaction for synthesizing α- and β-alkenyl sulfides [255], which have the versatile synthetic applicability via activation and transformations of C(sp^2)—S bonds [256].

The radical addition of thiols usually give *anti*-Markovnikov products as mixtures of *E* and *Z* isomers [257], and the transition metal-catalyzed hydrothiolation mostly proceed in a *cis*-addition to afford *anti*-Markovnikov (*E*)-1-alkenyl sulfides or Markovnikov adducts. In this section, only the transition metal-catalyzed hydrothiolations are reviewed.

3.6.1 Markovnikov Hydrothiolation of Alkynes

The first example of transition metal-catalyzed Markovnikov addition of aromatic thiols to alkynes gives α-alkenyl sulfides in the presence of $Pd(OAc)_2$ (Scheme 3.57) [258]. This complex also catalyzed the Markovnikov addition of thiophenol to terminal 1,3-conjugated enynes for efficient syntheses of 2-(phenylsulfinyl)-1,3-dienes [259]. Other palladium complexes were then found to be efficient catalyst systems for this transformation [260].

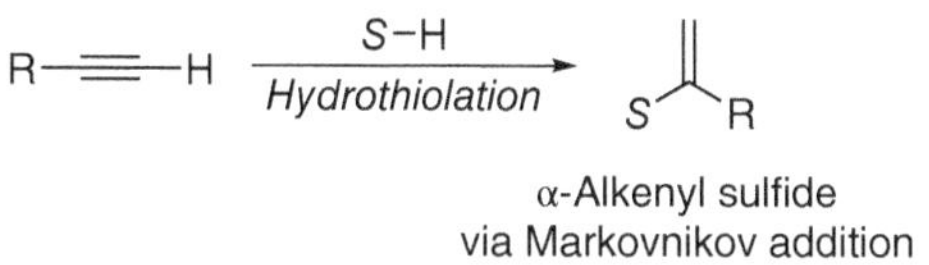

Scheme 3.57 Markovnikov hydrothiolation of terminal alkynes.

In addition, other transition metal complexes showing catalytic activity for Markovnikov intermolecular hydrothiolation of terminal alkynes include nickel [261], rhodium [262], organozirconium [263], and lanthanide- or actinide complexes [264].

The study of intermediates disclosed that in the hydrothiolation of terminal alkynes with Markovnikov regioselectivity, hydroxo–rhodium–*N*-heterocyclic carbene complexes were confirmed to be efficient catalyst precursors [265].

$Pd(OAc)_2$ and $Cu(OTf)_2$ cocatalyst system showed high catalytic activity and regioselectivity for the Markovnikov addition of heteroaromatic thiols to terminal alkynes, providing an efficient method to prepare heteroaromatic-sulfides, although the use of benzenethiol could not give the corresponding adduct [266]. A proposed mechanistic study indicates that Cu(II) plays a key role in obtaining Markovnikov adducts.

Markovnikov hydrothiolation of terminal alkynes could also be realized by using the heterogeneous catalysts [267] and the radical addition through a photoredox process [268].

Recently, nickel(II)-catalyzed Markovnikov hydrothiolation of terminal propargylic alcohols has been used in the synthesis of two-thiosubstituted-1,3-butadienes via a two-step reaction procedure including the formation of vinyl sulfides followed by their dehydration [269].

In addition, Castarlenas and coworker have developed a ligand-controlled regioselective either Markovnikov or *anti*-Markovnikov hydrothiolation of terminal alkynes catalyzed by rhodium *N*-heterocyclic carbene catalysts to give α- or β-alkenyl sulfides as majors [270].

$In(OTf)_3$ has also been reported to have the ability to selectively catalyze both Markovnikov and *anti*-Markovnikov hydrothiolation of terminal alkynes depending upon the nature of the thiol employed, and heteroaromatic thiols undergo selective Markovnikov hydrothiolation, whereas aromatic and aliphatic thiols show *anti*-Markovnikov selectivity under identical reaction conditions [271].

3.6.2 *Anti*-Markovnikov Hydrothiolation of Alkynes

Regio- and stereoselective synthesis of (*E*)-vinyl sulfides via *anti*-Markovnikov *cis*-hydrothiolation of terminal alkynes could be achieved (Scheme 3.58) by using Wilkinson's catalyst of $RhCl(PPh_3)_3$ [272], well-defined rhodium complex supported on mesoporous SBA-15 silica [273], and diphosphino-functionalized MCM-41 anchored rhodium complex [MCM-41-2P-$RhCl(PPh_3)$] [274]. Also, gold(I)/(III) complexes show good catalytic activity for the *cis*-hydrothiolation of terminal alkynes, as well as the hydrothiolation of electron-deficient alkenes to give vinyl sulfides and alkyl sulfides in excellent yields [275].

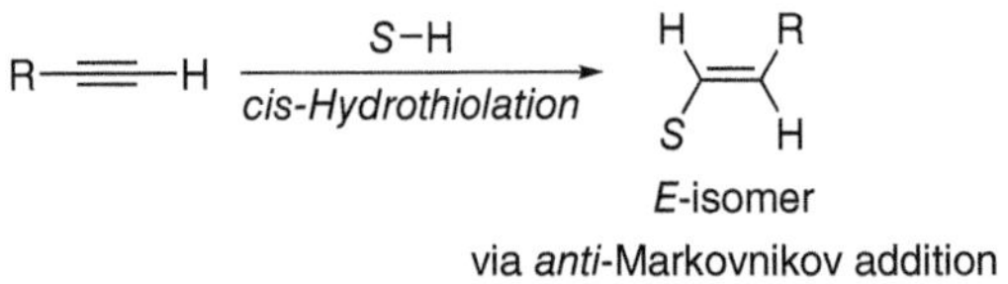

Scheme 3.58 *cis*-Hydrothiolation of terminal alkynes.

Without the use of any catalyst, the regio- and stereoselective hydrothiolation of propargyl alcohols with arylthiols could occur in water at room temperature to give (*E*)-vinyl sulfides as the major adducts [276]. The role of additives in the *anti*-Markovnikov addition of PhS-H to phenylacetylene in water at room temperature to realize the "stereoselective switch" was also studied [277].

Moreover, the use of cheap and commercially available Eosin Y as a photocatalyst could also realize the *cis*-hydrothiolation of terminal propargyl alcohols with arylthiols affording (*E*)-vinyl sulfides in good yields [278].

On the other hand, a variety of catalysis have also been developed for the regio- and stereoselective *anti*-Markovnikov *trans*-hydrothiolation of terminal alkynes affording (*Z*)-alkenyl sulfides as the major product (Scheme 3.59).

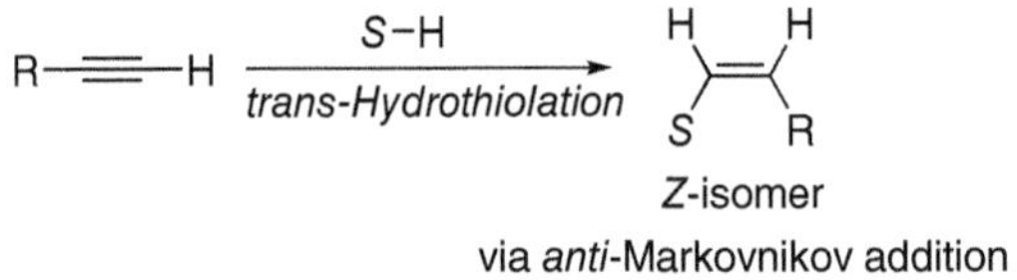

Scheme 3.59 *trans*-Hydrothiolation of terminal alkynes.

(*Z*)-alkenyl sulfides could be obtained by radical processes depending on the structures of substrates [279]. Also (*Z*)-alkenyl sulfides were generated in the presence of a catalytic amount of Cs_2CO_3 and a radical inhibitor in DMSO [280], NaOH-promoted hydrothiolation of three-aryl propargyl alcohols [281], and hydrothiolation of activated propiolate with heterocyclic thiols in MeOH under microwave irradiation conditions in absence of any catalyst or additive [282].

[Ir(PyP)$(CO)_2$]BPh_4 was reported to show the catalytic activity for *trans*-hydrothiolation of phenyacetylene to give (*Z*)-vinyl sulfides as major product

(Z/E = 90/10) [283]. $PdCl_2[P(Cy)_2(NC_5H_{10})]_2$ [284] and CuI under a CO_2 atmosphere [285] also show the selectivity.

K_3PO_4/NMP/argon has been found to be efficient base and solvent for the *anti*-Markovnikov hydrothiolation of terminal alkynes with arylthiols giving (*Z*)-vinyl sulfides as the major products [286].

Recently, under the metal-free conditions, *N*-heterocyclic carbenes (NHCs) [287] and cyclic trimeric phosphazene superbase [288] have been used as efficient catalysts to catalyze the *trans*-hydrothiolation of terminal alkynes with thiols with high stereoselectively.

The heterogeneous catalyst consisting of a Cu–NHC complex grafted to SBA-15 silica displays high activity and stereoselectivity to (*Z*)-*anti*-Markovnikov adducts in the hydrothiolation of terminal and internal alkynes [289].

In addition, in the presence of CuI and Cs_2CO_3, the *trans*-hydrothiolation of terminal alkynes could be achieved with diaryl disulfides [290]. Under the similar catalytic systems, decarboxylative thiolation of arylpropiolic acids can also selectively give (*Z*)-vinyl sulfides in high yields [291].

When 1-alkynylphosphines were used as starting materials, the $Pd(OAc)_2$-catalyzed *trans*-hydrothiolation occurred smoothly at 25 °C in EtOH to (*Z*)-1-phosphino-2-thio-1-alkenes in high yields with excellent regioselectivity and stereoselectivity (Scheme 3.60) [292]. .

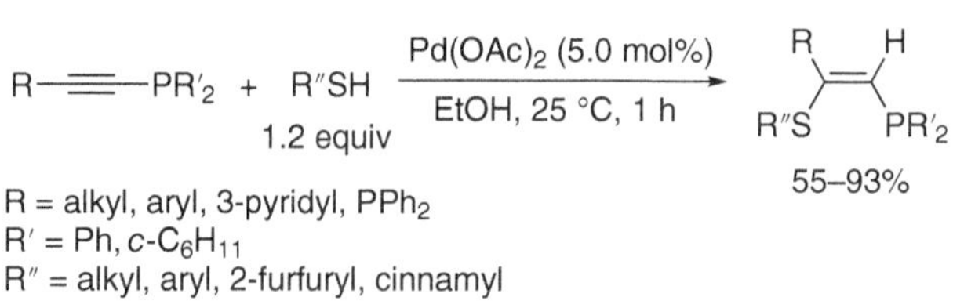

Scheme 3.60 $Pd(OAc)_2$-catalyzed hydrothiolation of 1-alkynylphosphines with thiols.

Very interestingly, Yadav's group developed an $InBr_3$-catalyzed highly regioselective Markovnikov double hydrothiolation of aliphatic terminal alkynes with various thiols under mild conditions to produce the corresponding dithioacetals in excellent yields [293]. The similar reaction could be also achieved in the presence of $Pd(OAc)_2$ and water (Scheme 3.61) [294].

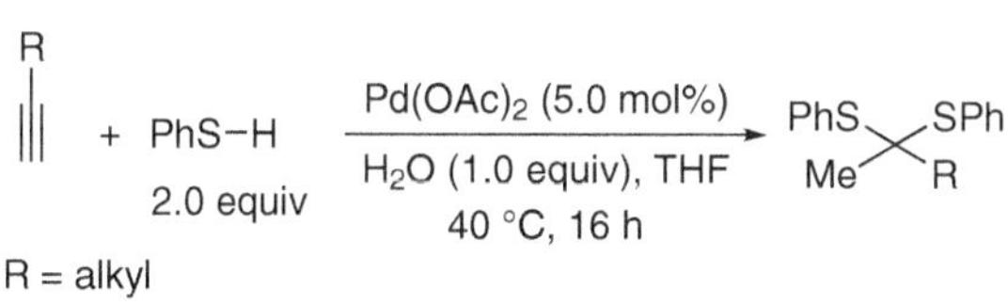

Scheme 3.61 $Pd(OAc)_2$-catalyzed dithioketal formation from regioselective double hydrothiolation of terminal alkynes.

Kobayashi and coworker then synthesized a Lewis-acidic $Ca(OSO_2C_4F_9)_2$, which shows the catalytic activity for bis-hydrothiolation of alkynes providing *anti*-Markovnikov dithioacetals in good to high yields [295].

In the presence of acetic acid, $Pd(OAc)_2$ can also catalyze the double hydroselenation of terminal alkynes with benzeneselenol with high regioselectivity to give the corresponding diselenoketals [296].

Kawatsura and coworker developed the synthesis of 1,3-oxathiine derivatives in moderate to high yields via the intermolecular hydrothiolation and sequential intramolecular cyclization of internal alkynes with thiosalicylic acid derivatives catalyzed by $Fe(acac)_2$/1,10-phenanthroline in a mixture solvent of toluene and HFIP (Scheme 3.62) [297].

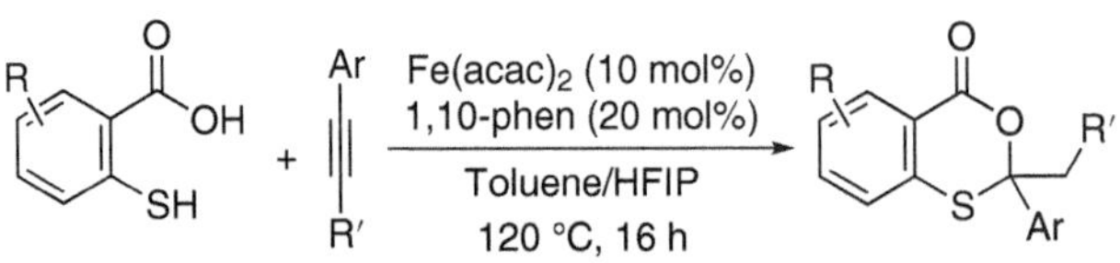

R = Me, OMe, F
R′ = CF_3, CO_2R'', COR″

Scheme 3.62 Synthesis of 1,3-oxathiines via iron-catalyzed intermolecular cyclo-coupling of internal alkynes with thiosalicylic acids.

The recoverable solid catalysts such as Al_2O_3/KF [298], native silica nanoparticles [299], elenium ionic liquid of [bmim][$SeO_2(OMe)$] [300], and $ZnIn_2S_4$ [301] have also been applied in the hydrothiolation of alkynes under different conditions, but both regio- and stereoselectivity are difficult to be expected depending on the structures of substrates.

Under CO atmosphere, terminal alkynes can undergo either thioformylation or hydrothiocarbonylation depending upon the use of transition metal catalysts (Scheme 3.63). By using AIBN as the radical initiator [302], $RhH(CO)(PPh_3)_3$ [303] could selectively catalyze the former carbonylative addition. Whereas $Pd(OAc)_2$/dppp-C_3 [304] and $Pt(PPh_3)_4$ [305] showed the catalytic activity for the Markovnikov-type hydrothiocarbonylation of terminal alkynes.

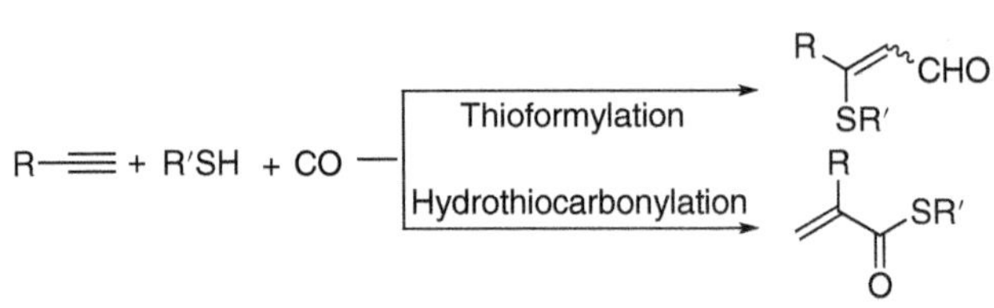

Scheme 3.63 Carbonylation-addition of alkyne with thiol and carbon monoxide.

Alper's group applied the reaction of propargylic alcohols with thiols and CO in the presence of palladium complexes to develop an efficient synthesis of thiofuranones [306].

The controlled regioselective hydrothiocarbonylation of terminal alkynes with arylthiols could occur by using the type of ligand (dppp or dppb-C_3) and the solvent (THF or CH_2Cl_2) under CO/H^+ or CO/H_2 in the presence of $Pd(OAc)_2$ [307].

In addition, $Pd(OAc)_2$-catalyzed Markovnikov thiocarbonylation of conjugated enynes with thiols and CO giving 2-(phenylthiocarbonyl)-1,3-dienes and $Pd(PPh_3)_4$-catalyzed highly stereoselective dithiocarbonylation of propargylic mesylates affording dithioesters have also been developed by Alper's group [308].

Moreover, a selective thiolative lactonization of internal alkynes bearing a hydroxyl group with CO and disulfides catalyzed by cobalt and palladium complexes, and cobalt-catalyzed carbonylative cyclization of internal alkynes with CO and thiols forming the corresponding α,β-unsaturated γ-thio-γ-lactones were then studied by Ogawa's group [309].

On the other hand, the first example of transition metal-catalyzed hydroselenation of terminal alkynes was reported by Ogawa and Sonoda's group in 1992 [310]. In the presence of a catalytic amount of $Pd(OAc)_2$, the addition reactions occur with high regioselectivity to afford the Markovnikov adducts (Scheme 3.64).

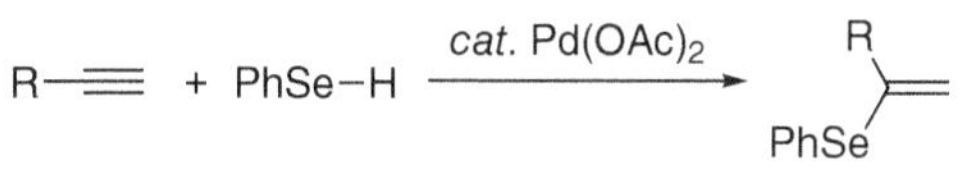

Scheme 3.64 $Pd(OAc)_2$-catalyzed Markovnikov hydroselenation of terminal alkynes.

Afterward, a variety of catalytic procedures have also been reported for the addition reactions of RSe—H to alkynes [311] and allenes [312].

3.7 Addition of *O*-nucleophiles to Alkynes

This section focuses on the intermolecular addition reactions of oxygen nucleophiles, such as alcohols, phenols, and carboxylic acids to alkynes (Scheme 3.65) [313]. The intramolecular addition of O—H bond to alkynes giving oxygen-heterocyclic compounds and the addition reactions of H_2O to alkynes (hydration) are described in Chapters 8 and 9, respectively.

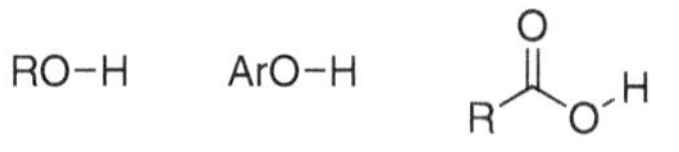

Scheme 3.65 General oxygen nucleophiles.

3.7.1 Addition of Alcohols and Phenols to Alkynes

The intermolecular hydroalkoxylation and hydrophenoxylation of alkynes with alcohol or phenols are the efficient ways for the synthesis of vinyl ethers, ketals/acetals, and oxygen-containing heterocycles (Scheme 3.66) [314].

At room temperature, $AuCl(PPh_3)/AgOTf$-catalyzed intermolecular addition of hydroxybenzotriazoles to terminal alkynes gave vinyl ethers in high yields with excellent Markovnikov regioselectivity, the adducts underwent a 3,3-sigmatropic rearrangement to afford highly functionalized benzotriazoles at 100 °C in dioxane [315].

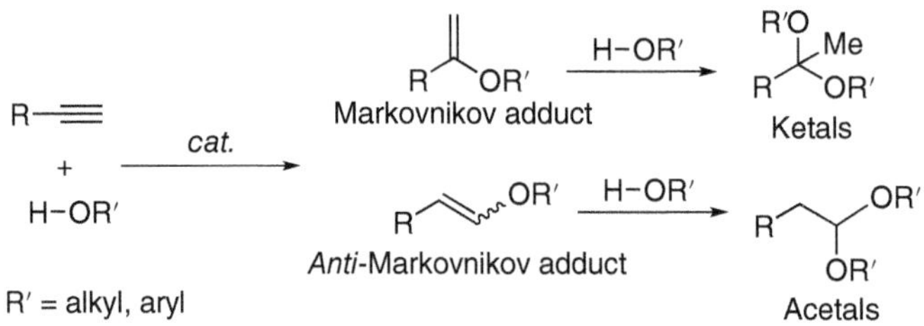

Scheme 3.66 Addition of alcohols and phenols to terminal alkynes.

Base-promoted additions of alcohols to alkynes were well reported. The addition reactions catalyzed by CsOH·H_2O in NMP [316] and by KOH in DMSO [317] gave a mixture of *trans*- and *cis*-adducts with high *anti*-Markovnikov addition. KOH/DMSO, as a superbasic system also catalyzes the addition reactions of phenols and naphthols to acetylene affording the corresponding aryl vinyl ethers [318] and *trans*-hydrophenoxylation of terminal alkynes forming (*Z*)-styrylaryl ethers [319].

The cobalt-catalyzed double hydroalkoxylation of terminal alkynes was used for the synthesis of acetals via Markovnikov regioselective addition of alcohols to terminal alkynes and the subsequent similar reaction of terminal alkenes formed in situ under slightly acidic, aerobic conditions [320]. The selective formation of ketals was also developed by using 18-crown-6 (18C6) ether adduct of sodium hexachloroiridate $[Na(18C6)]_2[IrCl_6]\cdot xH_2O(1)$ [321] and cationic gold(I) [322] complexes as catalysts.

Rhodium(I) complexes were found to be efficient catalysts for the *anti*-Markovnikov intermolecular *trans*-hydroalkoxylation of terminal alkynes to give (*Z*)-enol ethers with high regio- and stereoselectivity [323].

The addition reaction of phenols or 1,2-diphenols with activated alkynes took place smoothly in the presence of a catalytic amount of DABCO at room temperature and resulted in the formation of various alkenoic acid esters or 1,3-dioxole derivatives [324]. Hua's group then investigated the addition reaction of 1,2-diphenols with terminal and internal alkynes in the presence of $Ru_3(CO)_{12}$, establishing an efficient procedure for the synthesis of 2,2-disubstituted 1,3-benzodioxoles in good to high yields via the tandem addition of two O—H bonds of 1,2-diphenols to alkynes (Scheme 3.67) [325]. When terminal alkynes were used, the cyclic addition reactions occurred with excellent Markovnikov regioselectivity. In the cases of unsymmetrical internal alkynes used, the regioselectivity depended on the structural nature of alkynes.

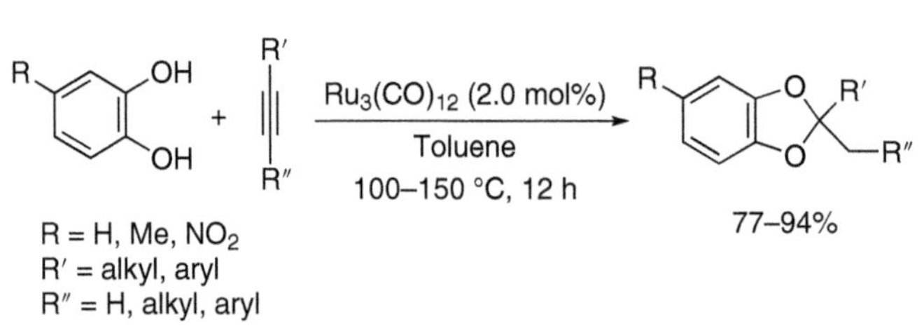

Scheme 3.67 $Ru_3(CO)_{12}$-catalyzed cycloaddition of 1,2-diphenols with alkynes.

The formation of 2,2-substituted 1,3-benzodioxoles is considered from the tandem addition of two O—H bonds of diphenols to alkyne's triple bond, and the proposed mechanism is depicted in Scheme 3.68, involving the oxidative addition of O—H bond to ruthenium complex, the insertion of alkyne, and subsequent intramolecular addition of O—H bond to C=C.

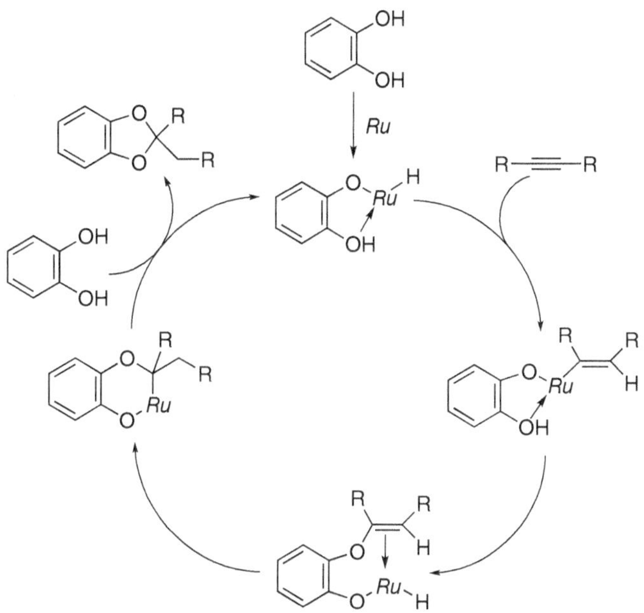

Scheme 3.68 Proposed mechanism for the cycloaddition of 1,2-diphenols to alkynes.

With the use of $TiCl_4$ as catalyst, the cyclic addition of 2,2′-dihydroxybiphenyl to terminal alkynes occurred to give dibenzo[d,f][1,3]dioxepines in good yields with excellent regioselectivity (Scheme 3.69) [326].

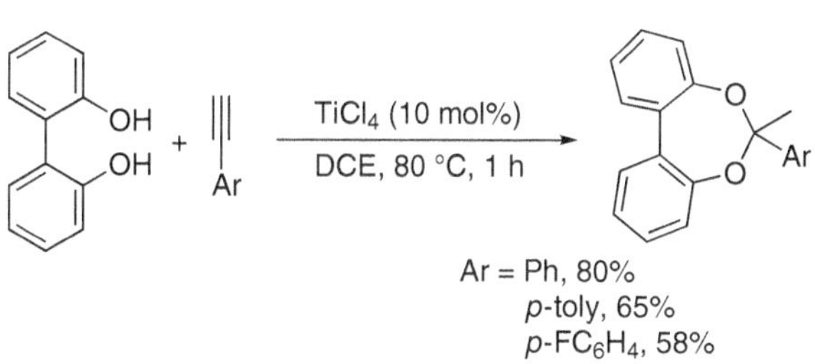

Scheme 3.69 Dibenzo[d,f][1,3]dioxepine formation by cyclic addition of 2,2″-dihydroxybiphenyl with aromatic alkynes.

Recently, 2-fluoroalkylated 1,3-benzodioxole derivatives were prepared by the addition reaction ethyl 4,4,4-trifluorobut-2-ynoate with 1,2-diphenols [327].

In the cases of internal alkynes, Au(I) complexes were found to be efficient catalysts for the intermolecular hydroalkoxlation with high stereoselectivity, and intramolecular hydroetherification of alkynyl bisphenols and dialkynyl phenols [328].

$AuCl_3$/ligand shows the catalytic activity for the intermolecular *trans*-hydrophenoxylation of internal alkynes, and the addition of phenols to unsymmetrical alkynes provides the corresponding mixture of regioisomers with appreciable selectivity [329]. Then, $[\{Au(NHC)\}_2(\mu\text{-}OH)][BF_4]$ has been found to be the catalytic activity for the same addition reactions with similar selectivity [330].

Allylic ethers could be obtained by intermolecular hydroalkoxylation of 1-phenyl-1-propyne catalyzed palladium(0) [331], and allenes catalyzed by rhodium(I) [332].

3.7.2 Addition of Acids to Alkynes

Transition metal-catalyzed addition of carboxylic acids to alkynes (hydrocarboxylation of alkynes) is a very effective tool for the synthesis of enol esters in an atom economical manner, the addition reactions can also afford three possible adducts: one Markovnikov adduct and two *anti*-Markovnikov adducts of (*E*)- and (*Z*)- isomers (Scheme 3.70) [333].

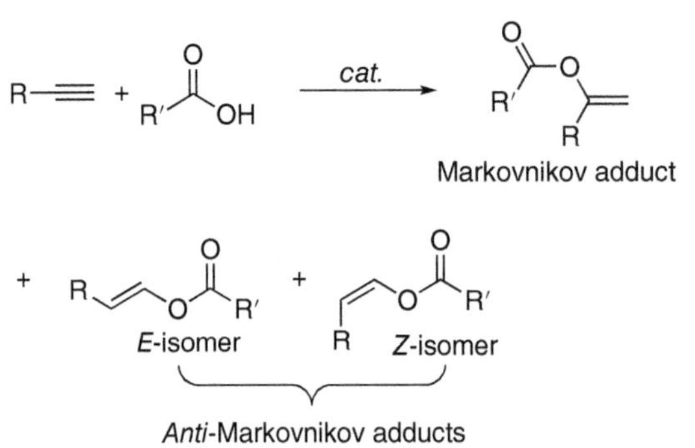

Scheme 3.70 Formation of enol esters via addition of carboxylic acids to alkynes.

Since Shvo and coworker first reported the addition of carboxylic acids to terminal and internal alkynes affording a mixture of adducts catalyzed by $Ru_3(CO)_{12}$ in 1983 [334], a variety of transition metal complexes were then employed as catalysts to optimize the regio- and stereoselectivity, including other ruthenium [335], rhodium [336], palladium [337], iridium [338], cobalt [339], and silver complexes [340].

Recently, the ruthenium complexes $[Ru(CO)_2(P(p\text{-}C_6H_4X)_3)_2(O_2CPh)_2]$ ($X = H$, Me, OMe, Cl, CF_3) have been successfully applied in the regioselective Markovnikov addition of carboxylic acids to terminal alkynes [341]. Among them, $[Ru(CO)_2(P(p\text{-}CF_3C_6H_4)_3)_2(O_2CPh)_2]$ shows stereoselective *syn*-addition of carboxylic acids to internal alkynes yielding trisubstituted enol esters with (*E*)-configuration [342].

Hua's group also developed an early transition metal complex $Re(CO)_5Br$-catalyzed addition reactions of carboxylic acids to terminal alkynes with high selectivity to afford *anti*-Markovnikov adducts under an air atmosphere; catalyst $Re(CO)_5Br$ could be partly recovered after the catalytic reactions in *n*-heptane solvent [343].

The high *anti*-Markovnikov intermolecular addition of carboxylic acids to terminal alkynes yielding *Z*-enol esters as major products was realized in the presence of $[RhCl(cod)]_2$ with the use of 2-(diphenylphosphinomethyl)pyridine as ligand, the catalyst system was applicable to a broad substrate scope and displays a wide functional group tolerance (Scheme 3.71) [344].

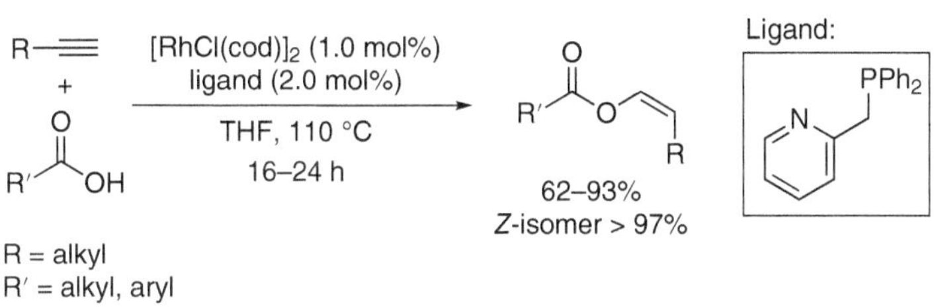

Scheme 3.71 (Z)-enol ester formation via rhodium-catalyzed *anti*-Markovnikov intermolecular addition of carboxylic acids to terminal alkynes.

A *trans*-addition of carboxylic acids to internal alkynes with a variety of aromatic and aliphatic acids catalyzed by a digold hydroxide complex has been recently developed by Nolan's group [345]. The addition reactions give various aryl- and alkyl enol esters in good to excellent yields with (*Z*)-stereospecificity and good regioselectivities under solvent-free conditions.

The *trans*-addition of carboxylic acids to internal iodoalkynes was developed for the synthesis of (*Z*)-β-iodoenol esters in the presence of $AuCl(PPh_3)/AgPF_6$ at room temperature (Scheme 3.72) [346]. The obtained (*Z*)-β-iodoenol esters can be applied in the synthesis of a broad family of 1,4-disubstituted (*Z*)-enynyl esters via Sonogashira coupling reactions with different terminal alkynes.

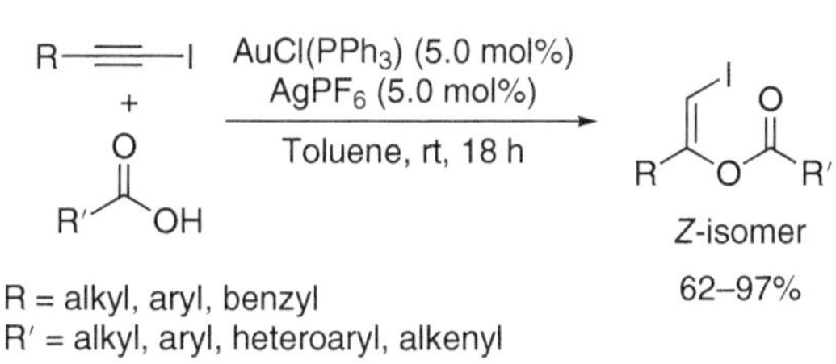

Scheme 3.72 Formation of (Z)-β-iodoenol esters via *trans*-addition of carboxylic acids to internal iodoalkynes.

In addition, by activation of P(O)O—H bond, ruthenium-catalyzed regioselective addition of diphenylphosphinic acid to terminal alkynes affording alkenyl diphenylphosphinates [347] and a similar addition reaction with the use of diphenyl phosphate catalyzed by Au(I) [348] were also reported.

References

1 Selected reviews, see: (a) Han, L.-B. and Tanaka, M. (1999). *Chem. Commun.* 395–402. (b) Yamamoto, Y. and Radhakrishnan, U. (1999). *Chem. Soc. Rev.* 28: 199–207. (c) Alonso, F., Beletskaya, I.P., and Yus, M. (2004). *Chem. Rev.* 104: 3079–3159. (d) Beller, M., Seayad, J., Tillack, A., and Jiao, H. (2004). *Angew. Chem. Int. Ed.* 43: 3368–3398. (e) Bruneau, C. and Dixneuf, P.H. (2006). *Angew. Chem. Int. Ed.* 45: 2176–2203. (f) Zeng, X. (2013). *Chem. Rev.* 113: 6864–6900. (g) Evano, G., Gaumont, A.-C., Alayrac, C. et al. (2014). *Tetrahedron* 70: 1529–1616. (h) Greenhalgh, M.D., Jones, A.S., and Thomas, S.P. (2015). *ChemCatChem* 7: 190–222. (i) Orlov, N.V. (2015). *ChemistryOpen* 4: 682–697. (j) Yoshida, H. (2016). *Chem. Rec.* 16: 419–434. (k) Wathier, M. and Love, J.A. (2016). *Eur. J. Inorg. Chem.* 2391–2402. (l) Hadlington, T.J., Driess, M., and Jones, C. (2018). *Chem. Soc. Rev.* 47: 4176–4197. (m) Chen, J.H., Guo, J., and Lu, Z. (2018). *Chin. J. Chem.* 36: 1075–1109. (n) Wen, H., Liu, G., and Huang, Z. (2019). *Coord. Chem. Rev.* 386: 138–153.

2 Chu, T. and Nikonov, G.I. (2018). *Chem. Rev.* 118: 3608–3680.

3 Cheng, Z., Guo, J., and Lu, Z. (2020). *Chem. Commun.* 56: 2229–2239.

4 Selected reviews, see: (a) Miyaura, N. and Suzuki, A. (1995). *Chem. Rev.* 95: 2457–2483. (b) Suzuki, A. (1999). *J. Organomet. Chem.* 576: 147–168. (c) Suzuki, A. (2002). *J. Organomet. Chem.* 653: 83–90. (d) Liu, C., Zhang, H., Shi, W., and Lei, A. (2011). *Chem. Rev.* 111: 1780–1824. (e) Heravi, M.M. and Hashemi, E. (2012). *Tetrahedron* 68: 9145–9178. (f) Greenhalgh, M.D., Jones, A.S., and Thomas, S.P. (2015). *ChemCatChem* 7: 190–222. (g) Neeve, E.C., Geier, S.J., Mkhalid, I.A.I. et al. (2016). *Chem. Rev.* 116: 9091–9161. (h) Royes, J., Cuenca, A.B., and Fernández, E. (2018). *Eur. J. Org. Chem.* 2728–2739. (i) Yang, X., Kalita, S.J., Maheshuni, S., and Huang, Y.-Y. (2019). *Coord. Chem. Rev.* 392: 35–48. (j) Babu, K.N., Massarwe, F., Reddy, R.R. et al. (2020). *Molecules* 25: 959.

5 Selected reviews on the hydroboration of alkynes, see: (a) Pereira, S. and Srebnik, M. (1996). *Tetrahedron Lett.* 37: 3283–3286. (b) Beletskaya, I. and Pelter, A. (1997). *Tetrahedron* 53: 4957–5026. (c) Fujihara, T., Semba, K., Terao, J., and Tsuji, Y. (2014). *Catal. Sci. Technol.* 4: 1699–1709. (d) Barbeyron, R., Benedetti, E., Cossy, J. et al. (2014). *Tetrahedron* 70: 8431–8452. (e) Yoshida, H. (2016). *ACS Catal.* 6: 1799–1811. (f) Buñuel, E. and Cárdenas, D.J. (2018). *Chem. Eur. J.* 24: 11239–11244. (g) Carreras, J., Caballero, A., and Pérez, P.J. (2019). *Chem. Asian J.* 14: 329–343. (h) Wen, Y., Deng, C., Xie, J., and Kang, X. (2019). *Molecules* 24: 101. (i) Ding, S., Xu, L., and Miao, Z. (2019). *Molecules* 24: 1325.

6 (a) Jang, H., Zhugralin, A.R., Lee, Y., and Hoveyda, A.H. (2011). *J. Am. Chem. Soc.* 133: 7859–7871. (b) Yoshida, H., Takemoto, Y., and Takaki, K. (2014). *Chem. Commun.* 8299–8302.

7 Yao, Z.-J., Hong, S., Zhang, W. et al. (2016). *Tetrahedron Lett.* 57: 910–913.

8 Ojha, D.P. and Prabhu, K.R. (2016). *Org. Lett.* 18: 432–435.

9 Wang, Y.D., Kimball, G., Prashad, A.S., and Wang, Y. (2005). *Tetrahedron Lett.* 46: 8777–8780.

10 Haberberger, M. and Enthaler, S. (2013). *Chem. Asian J.* 8: 50–54.

11 Wang, Y., Guan, R., Sivaguru, P. et al. (2019). *Org. Lett.* 21: 4035–4038.

12 Leyva, A., Zhang, X., and Corma, A. (2009). *Chem. Commun.* 4947–4949.

13 (a) Yang, Z., Zhong, M., Ma, X. et al. (2016). *J. Am. Chem. Soc.* 138: 2548–2551. (b) Bismuto, A., Thomas, S.P., and Cowley, M.J. (2016). *Angew. Chem. Int. Ed.* 55: 15356–15359.

14 Wen, K., Chen, J., Gao, F. et al. (2013). *Org. Biomol. Chem.* 11: 6350–6356.

15 Ho, H.E., Asao, N., Yamamoto, Y., and Jin, T. (2014). *Org. Lett.* 16: 4670–4673.

16 Fleige, M., Möbus, J., vom Stein, T. et al. (2016). *Chem. Commun.* 52: 10830–10833.

17 Hong, S., Zhang, W., Liu, M. et al. (2016). *Tetrahedron Lett.* 57: 1–4.

18 (a) Khan, A., Asiri, A.M., Kosa, S.A. et al. (2015). *J. Catal.* 329: 401–412. (b) Mohan, B. and Park, K.H. (2016). *Appl. Catal. A: Gen.* 519: 78–84. (c) Liu, S., Zeng, X., and Xu, B. (2016). *Tetrahedron Lett.* 57: 3706–3710.

19 Lee, J.-E., Kwon, J., and Yun, J. (2008). *Chem. Commun.* 733–734.

20 (a) Ohmura, T., Yamamoto, Y., and Miyaura, M. (2000). *J. Am. Chem. Soc.* 122: 4990–4991. (b) Cid, J., Carbó, J.J., and Fernández, E. (2012). *Chem. Eur. J.* 18: 1512–1521.

21 Gunanathan, C., Hölscher, M., Pan, F., and Leitner, W. (2012). *J. Am. Chem. Soc.* 134: 14349–14352.

22 Obligacion, J.V., Neely, J.M., Yazdani, A.N. et al. (2015). *J. Am. Chem. Soc.* 137: 5855–5858.

23 Jang, W.J., Lee, W.L., Moon, J.H. et al. (2016). *Org. Lett.* 18: 1390–1393.

24 McGough, J.S., Butler, S.M., Cade, I.A., and Ingleson, M. (2016). *J. Chem. Sci.* 7: 3384–3389.

25 Sundararaju, B. and Fürstner, A. (2013). *Angew. Chem. Int. Ed.* 52: 14050–14054.

26 Shimoi, M., Watanabe, T., Maeda, K. et al. (2018). *Angew. Chem. Int. Ed.* 57: 9485–9490.

27 (a) Nagao, K., Yamazaki, A., Ohmiya, H., and Sawamura, M. (2018). *Org. Lett.* 20: 1861–1865. (b) Fritzemeier, R., Gates, A., Guo, X. et al. (2018). *J. Org. Chem.* 83: 10436–10444.

28 Longobardi, L.E. and Fúrstner, A. (2019). *Chem. Eur. J.* 25: 10063–10068.

29 Zhao, J., Niu, Z., Fu, H., and Li, Y. (2014). *Chem. Commun.* 50: 2058–2060.

30 Lipshutz, B.H., Boškovic, Ž.V., and Aue, D.H. (2008). *Angew. Chem., Int. Ed.* 47: 10183–10186.

31 Semba, K., Fujihara, T., Terao, J., and Tsuji, Y. (2012). *Chem. Eur. J.* 18: 4179–4184.

32 Liu, X., Ming, W., Friedrich, A. et al. (2020). *Angew. Chem. Int. Ed.* 59: 304–309.

33 Li, D.X., Kim, Y.E., and Yun, J. (2015). *Org. Lett.* 17: 860–863.

34 Deng, C.M., Ma, Y.F., and Wen, Y.M. (2018). *ChemistrySelect* 3: 1202–1204.

35 Sang, H.L., Wu, C., Phua, G.G.D., and Ge, S. (2019). *ACS Catal.* 9: 10109–10114.

36 Selected reviews, see: (a) Nallagonda, R., Padala, K., and Masarwa, A. (2018). *Org. Biomol. Chem.* 16: 1050–1064. (b) Wu, C. and Wang, J. (2018). *Tetrahedron Lett.* 59: 2128–2140. (c) Royes, J., Cuenca, A.B., and Fernandez, E. (2018). *Eur. J. Org. Chem.* 2728–2739. (d) Babu, K.N., Massarwe, F., Reddy, R.R. et al. (2020). *Molecules* 25: 959.

37 Procter, R.J., Uzelac, M., Cid, J. et al. (2019). *ACS Catal.* 9: 5760–5771.

38 Gao, Y., Wu, Z.-Q., and Engle, K.M. (2020). *Org. Lett.* 22: 5235–5239.

39 Uzelac, M., Yuan, K., and Ingleson, M.J. (2020). *Organometallics* 39: 1332–1338.

40 Docherty, J.H., Nicholson, K., Dominey, A.P., and Thomas, S.P. (2020). *ACS Catal.* 10: 4686–4691.

41 A review:Verma, P.K., Shegavi, M.L., Bose, S.K., and Geetharani, K. (2018). *Org. Biomol. Chem.* 16: 857–873.

42 Zhang, C., Zhou, M., Liu, S. et al. (2018). *Carbohyd. Poly.* 191: 17–24.

43 Zeng, X., Gong, C., Guo, H. et al. (2018). *New J. Chem.* 42: 17346–17350.

44 Hoyt, C.B., Sarazen, M.L., and Jones, C.W. (2019). *J. Catal.* 369: 493–500.

45 Szyling, J., Franczyk, A., Stefanowska, K. et al. (2018). *ACS Sustain. Chem. Eng.* 6: 10980–10988.

46 Szyling, J., Franczyk, A., Stefanowska, K., and Walkowiak, J. (2018). *Adv. Synth. Catal.* 360: 2966–2974.

47 (a) Yu, S., Wu, C., and Ge, S. (2017). *J. Am. Chem. Soc.* 139: 6526–6529. (b) Wang, C. and Ge, S. (2018). *J. Am. Chem. Soc.* 140: 10687–10690.

48 Sang, H.L., Yu, S., and Ge, S. (2018). *Org. Chem. Front.* 5: 1284–1287.

49 Wang, Q., Motika, S.E., Akhmedov, N.G. et al. (2014). *Angew. Chem. Int. Ed.* 53: 5418–5422.

50 Yuan, K., Suzuki, N., Mellerup, S.K. et al. (2016). *Org. Lett.* 18: 720–723.

51 A review on the hydroalumination and hydrogallation of alkynes, see:Uhl, W. (2008). *Coord. Chem. Rev.* 252: 1540–1563.

52 Selected reports, see: (a) Blumenkopf, T.A. and Overman, L.E. (1986). *Chem. Rev.* 86: 857–873. (b) Langkopf, E. and Schinzer, D. (1995). *Chem. Rev.* 95: 1375–1408. (c) Fleming, I., Barbero, A., and Walter, D. (1997). *Chem. Rev.* 97: 2063–2192. (d) Denmark, S.E. and Ober, M.H. (2003). *Aldrichimica Acta* 36: 75–85. (e) Curtis-Long, M.J. and Aye, Y. (2009). *Chem. Eur. J.* 15: 5402–5416. (f) Denmark, S.E. and Liu, J.H.C. (2010). *Angew. Chem. Int. Ed.* 49: 2978–2986. (g) Sore, H.F., Galloway, W.R.J.D., and Spring, D.R. (2012). *Chem. Soc. Rev.* 41: 1845–1866. (h) McAdam, C.A., McLaughlin, M.G., and Cook, M.J. (2015). *Org. Chem. Front.* 2: 510–514. (i) Komiyama, T., Minami, Y., and Hiyama, T. (2017). *ACS Catal.* 7: 631–651.

53 Selected reviews, see: (a) Díez-González, S. and Nolan, S.P. (2008). *Acc. Chem. Res.* 41: 349–358. Iron complexes: (b) Zhang, M. and Zhang, A. (2010). *Appl. Organometal. Chem.* 24: 751–757. Ruthenium complexes: (c) Zaranek, M., Marciniec, B., and Pawluc, P. (2016). *Org. Chem. Front.* 3: 1337–1344. Nickel complexes: (d) Nakajima, Y., Sato, K., and Shimada, S. (2016). *Chem. Rec.* 16: 2379–2387. Cobalt complexes: (e) Sun, J. and Deng, L. (2016). *ACS Catal.* 6: 290–300. Manganese complexes: (f) Yang, X. and Wan, C. (2018). *Chem. Asian J.* 13: 2307–2315.

54 Selected reports, see: (a) Sakaki, S., Mizoe, N., Sugimoto, M., and Musashi, Y. (1999). *Coord. Chem. Rev.* 190–192: 933–960. (b) Zhao, L., Nakatani, N., Sunada, Y. et al. (2019). *J. Org. Chem.* 84: 8552–8561.

55 (a) Menozzi, C., Dalko, P.I., and Cossy, J. (2005). *J. Org. Chem.* 70: 10717–10719. (b) Kanno, K.-i., Aikawa, Y., and Kyushin, S. (2020). *Tetrahedron Lett.* 61: 152274.

56 Selected reports, see: (a) Guo, J. and Lu, Z. (2016). *Angew. Chem. Int. Ed.* 55: 10835–10838. (b) Ma, Y. and Han, Z. (2018). *J. Org. Chem.* 83: 14646–14657. (c) Zhang, S., Ibrahim, J.J., and Yang, Y. (2018). *Org. Lett.* 20: 6265–6269. (d) Kong, D., Hu, B., and Chen, D. (2019). *Chem. Asian J.* 14: 2694–2703. (e) Zong, Z., Yu, Q., Sun, N. et al. (2019). *Org. Lett.* 21: 5767–5772.

57 Wu, G., Chakraborty, U., and von Wangelin, A.J. (2018). *Chem. Commun.* 54: 12322–12325.

58 (a) Mo, Z., Xiao, J., Gao, Y., and Deng, L. (2014). *J. Am. Chem. Soc.* 136: 17414–17417. (b) Wu, C., Teo, W.J., and Ge, S. (2018). *ACS Catal.* 8: 5896–5900.

59 Carney, J.R., Dillon, B.R., Campbell, L., and Thomas, S.P. (2018). *Angew. Chem. Int. Ed.* 57: 10620–10624.

60 (a) Aneetha, H., Wu, W., and Verkade, J.G. (2005). *Organometallics* 24: 2590–2596. (b) Dong, C., Yuan, Y., Cui, Y.-M. et al. (2018). *Appl. Organomet. Chem.* 32: e4037. (c) Zak, P., Bołt, M., and Pietraszuk, C. (2018). *RSC Adv.* 8: 40016–40021.

61 (a) Morales-Cerón, J.P., Lara, P., López-Serrano, J. et al. (2017). *Organometallics* 36: 2460–2469. (b) Żak, P., Bołt, M., Kubicki, M., and Pietraszuk, C. (2018). *Dalton Trans.* 47: 1903–1910.

62 Zhou, H. and Moberg, C. (2013). *Org. Lett.* 15: 1444–1447.

63 Shintani, R.J. (2018). *Synth. Org. Chem. Jpn.* 76: 1163–1169.

64 Igawa, K., Yoshihiro, D., Ichikawa, N. et al. (2012). *Angew. Chem. Int. Ed.* 51: 12745–12748.

65 Wen, H., Wan, X., and Huang, Z. (2018). *Angew. Chem. Int. Ed.* 57: 6319–6323.

66 Cano, R., Yus, M., and Ramón, D. (2012). *ACS Catal.* 2: 1070–1078.

67 Ishikawa, Y., Yamamoto, Y., and Asao, N. (2013). *Catal. Sci. Technol.* 3: 2902–2905.

68 Planellas, M., Guo, W., Alonso, F. et al. (2014). *Adv. Synth. Catal.* 356: 179–188.

69 Fernández, G. and Pleixats, R. (2018). *ChemistrySelect* 3: 11486–11493.

70 Zhang, J.-w., Lu, G.-p., and Cai, C. (2017). *Green Chem.* 19: 2535–2540.

71 (a) Li, R.-H., An, X.-M., Yang, Y. et al. (2018). *Org. Lett.* 20: 5023–5026. (b) Zhou, Y.-B., Liu, Z.-K., Fan, X.-Y. et al. (2018). *Org. Lett.* 20: 7748–7752. (c) Li, R.-H., Zhang, G.-L., Dong, J.-X. et al. (2019). *Chem. Asian J.* 14: 149–154.

72 Selected early reports, see: (a) Tanke, R.S. and Crabtree, R.H. (1990). *J. Am. Chem. Soc.* 112: 7984–7989. (b) Ojima, I., Clos, N., Donovan, R.J., and Ingallina, P. (1990). *Organometallics* 9: 3127–3133. (c) Jun, C.-H. and Crabtree, R.H. (1993). *J. Organomet. Chem.* 447: 177–187. (d) Maruyama, Y., Yamamura, K., Nakayama, I. et al. (1998). *J. Am. Chem. Soc.* 120: 1421–1429. (e) Trost, B.M. and Ball, Z.T. (2003). *J. Am. Chem. Soc.* 125: 30–31.

73 Maifeld, S.V., Tran, M.N., and Lee, D. (2005). *Tetrahedron Lett.* 46: 105–108.

74 Mutoh, Y., Mohara, Y., and Saito, S. (2017). *Org. Lett.* 19: 5204–5207.

75 Conifer, C., Gunanathan, C., Rinesch, T. et al. (2015). *Eur. J. Inorg. Chem.* 333–339.

76 (a) Du, X., Hou, W., Zhang, Y., and Huang, Z. (2017). *Org. Chem. Front.* 4: 1517–1521. (b) Teo, W.J., Wang, C., Tan, Y.W., and Ge, S. (2017). *Angew. Chem. Int. Ed.* 56: 4328–4332.

77 (a) Asao, N., Sudo, T., and Yamamoto, Y. (1996). *J. Org. Chem.* 61: 7654–7655. (b) Sudo, T., Asao, N., Gevorgyan, V., and Yamamoto, Y. (1999). *J. Org. Chem.* 64: 2494–2499.

78 Zhao, X., Yang, D., Zhang, Y. et al. (2018). *Org. Lett.* 20: 5357–5361.

79 Sánchez-Page, B., Munarriz, J., Jiménez, M.V. et al. (2020). *ACS Catal.* 10: 13334–13351.

80 Puerta-Oteo, R., Munarriz, J., Polo, V. et al. (2020). *ACS Catal.* 10: 7367–7380.

81 Gorgas, N., Stöger, B., Veiros, L.F., and Kirchner, K. (2018). *ACS Catal.* 8: 7973–7982.

82 Liang, H., Ji, Y.-X., Wang, R.-H. et al. (2019). *Org. Lett.* 21: 2750–2754.

83 (a) Mori, A., Takahisa, E., Kajiro, H. et al. (2000). *Polyhedron* 19: 567–568. (b) Mori, A., Takahisa, E., Yamamura, Y. et al. (2004). *Organometallics* 23: 1755–1765.

84 Faller, J.W. and D'Alliessi, D.G. (2002). *Organometallics* 21: 1743–1746.

85 Katayama, H., Taniguchi, K., Kobayashi, M. et al. (2002). *J. Organomet. Chem.* 645: 192–200.

86 Kawasaki, Y., Ishikawa, Y., Igawa, K., and Tomooka, K. (2011). *J. Am. Chem. Soc.* 133: 20712–20715.

87 Rooke, D.A. and Ferreira, E.M. (2012). *Angew. Chem. Int. Ed.* 51: 3225–3230.

88 Ding, S., Song, L.-J., Wang, Y. et al. (2015). *Angew. Chem. Int. Ed.* 54: 5632–5635.

89 Song, L.-J., Ding, S., Wang, Y. et al. (2016). *J. Org. Chem.* 81: 6157–6164.

90 Trost, B.M. and Ball, Z.T. (2005). *J. Am. Chem. Soc.* 127: 17644–11765.

91 Ding, S., Song, L.-J., Chung, L.W. et al. (2013). *J. Am. Chem. Soc.* 135: 13835–13842.

92 Zhang, X., Ji, X., Xie, X., and Ding, S. (2018). *Chem. Commun.* 54: 12958–12961.

93 Rivera-Hernández, A., Fallon, B.J., Ventre, S. et al. (2016). *Org. Lett.* 18: 4242–4245.

94 (a) Rummelt, S.M., Radkowski, K., Roşca, D.-A., and Fürstner, A. (2015). *J. Am. Chem. Soc.* 137: 5506–5519. (b) Roşca, D.-A., Radkowski, K., Wolf, L.M. et al. (2017). *J. Am. Chem. Soc.* 139: 2443–2455.

95 Dai, W., Wu, X., Li, C. et al. (2018). *RSC Adv.* 8: 28261–28265.

96 Belger, C. and Plietker, B. (2012). *Chem. Commun.* 48: 5419–5421.

97 Hu, M.-Y., He, P., Qiao, T.-Z. et al. (2020). *J. Am. Chem. Soc.* 142: 16894–16902.

98 Yang, X. and Wang, C. (2018). *Angew. Chem. Int. Ed.* 57: 923–928.

99 Kim, Y., Dateer, R.B., and Chang, S. (2017). *Org. Lett.* 19: 190–193.

100 Zheng, Z., Song, W., Zhang, T. et al. (2018). *J. Org. Chem.* 83: 6210–6216.

101 Chay, R.S., Rocha, B.G.M., Pombeiro, A.J.L. et al. (2018). *ACS Omega* 3: 863–871.

102 Pérez-Torrente, J.J., Nguyen, D.H., Jiménez, M.V. et al. (2016). *Organometallics* 35: 2410–2422.

103 Gee, J.C., Fuller, B.A., Lockett, H.-M. et al. (2018). *Chem. Commun.* 54: 9450–9453.

104 Zhu, J., Cui, W.-C., Wang, S., and Yao, Z.-J. (2018). *Org. Lett.* 20: 3174–3178.

105 Selected reports, see: (a) Matsuda, I., Ogiso, A., Sato, S., and Izumi, Y. (1989). *J. Am. Chem. Soc.* 111: 2332–2333. (b) Doyle, M.P. and Shanklin, M.S. (1993). *Organometallics* 12: 11–12. (c) Doyle, M.P. and Shanklin, M.S. (1994). *Organometallics* 13: 1081–1088. (d) Monteil, F., Matsuda, I., and Alper, H. (1995). *J. Am. Chem. Soc.* 117: 4419–4420. (e) Alonso, M.A., Casares, J.A., Espinet, P. et al. (2001). *Tetrahedron Lett.* 42: 5697–5700.

106 Walkowiak, J., Salamon, K., Franczyk, A. et al. (2019). *J. Org. Chem.* 84: 2358–2365.

107 Sang, H.L., Hu, Y., and Ge, S. (2019). *Org. Lett.* 21: 5234–5237.

108 Hu, M.-Y., Lian, J., Sun, W. et al. (2019). *J. Am. Chem. Soc.* 141: 4579–4583.

109 (a) Cheng, Z., Xing, S., Guo, J. et al. (2019). *Chin. J. Chem.* 37: 457–461. (b) Guo, J., Wang, H., Xing, S. et al. (2019). *Chem* 5: 881–895. (c) Cheng, Z. (2019). *Chin. J. Chem.* 37: 632–633.

110 Selected recent reviews, see: (a) Frihed, T.G. and Fürstner, A. (2016). *Bull. Chem. Soc. Jpn.* 89: 135–160. (b) Heravi, M.M. and Mohammadkhani, L. (2018). *J. Organomet. Chem.* 869: 106–200.

111 Selected reviews, see: (a) Smith, N.D., Mancuso, J., and Lautens, M. (2000). *Chem. Rev.* 100: 3257–3282. (b) Alami, M., Hamze, A., and Provot, O. (2019). *ACS Catal.* 9: 3437–3466.

112 Selected reports, see: (a) Zhang, X.H., Guibé, F., and Balavoine, G. (1990). *J. Org. Chem.* 55: 1857–1867. (b) Ghosh, B. and Maleczka, R.E. Jr., (2011). *Tetrahedron Lett.* 52: 5285–5287. (c) Cheng, L.-J. and Mankad, N.P. (2019). *J. Am. Chem. Soc.* 141: 3710–3716. (d) Mi, J., Huo, S., Zeng, Y. et al. (2021). *Organometallics* 40: 654–662.

113 Kikukawa, K., Umekawa, H., Wada, F., and Matsuda, T. (1988). *Chem. Lett.* 881–884.

114 Cochran, J.C., Bronk, B.S., Terrence, K.M., and Phillips, H.K. (1990). *Tetrahedron Lett.* 31: 6621–6624.

115 Shibata, I., Suwa, T., Ryu, K., and Baba, A. (2001). *J. Am. Chem. Soc.* 123: 4101–4102.

116 (a) Kazmaier, U., Schauss, D., and Pohlman, M. (1999). *Org. Lett.* 1: 1017–1019. (b) Kazmaier, U., Pohlman, M., and Schauß, D. (2000). *Eur. J. Org. Chem.* 2761–2766. (c) Braune, S. and Kazmaier, U. (2002). *J. Organomet. Chem.* 641: 26–29. (d) Lin, H. and Kazmaier, U. (2007). *Eur. J. Org. Chem.* 2839–2843. (e) Jena, N. and Kazmaier, U. (2008). *Eur. J. Org. Chem.* 3852–3858. (f) Lin, H. and Kazmaier, U. (2009). *Eur. J. Org. Chem.* 12216–11227. (g) Klos, M.R. and Kazmaier, U. (2013). *Eur. J. Org. Chem.* 1726–1731.

117 Liu, J., Song, H., Wang, T. et al. (2021). *J. Am. Chem. Soc.* 143: 409–419.

118 Yoshida, H., Shinke, A., Kawanoa, Y., and Takakia, K. (2015). *Chem. Commun.* 51: 10616–10619.

119 Mandla, K.A., Moore, C.E., Rheingold, A.L., and Figueroa, J.S. (2018). *Angew. Chem. Int. Ed.* 57: 6853–6857.

120 Darwish, A., Lang, A., Kim, T., and Chong, J.M. (2008). *Org. Lett.* 10: 861–864.

121 Gupta, S., Do, Y., Lee, J.H. et al. (2014). *Chem. Eur. J.* 20: 1267–1271.

122 Dodero, V.I., Koll, L.C., Mandolesi, S.D., and Podestá, J.C. (2002). *J. Organomet. Chem.* 650: 173–180.

123 (a) Asao, N., Liu, J.-X., Sudoh, T., and Yamamoto, Y. (1995). *J. Chem. Soc., Chem. Commun.* 2405–2406. (b) Asao, N., Liu, J.-X., Sudoh, T., and Yamamoto, Y. (1996). *J. Org. Chem.* 61: 4568–4571.

124 Forster, F., López, V.M.R., and Oestreich, M. (2018). *Organometallics* 37: 2656–2659.

125 Semmelhack, M.F. and Hooley, R.J. (2003). *Tetrahedron Lett.* 44: 5737–5739.

126 (a) Leung, L.T., Leung, S.K., and Chiu, P. (2005). *Org. Lett.* 7: 5249–5252. (b) Miao, R., Li, S., and Chiu, P. (2007). *Tetrahedron* 63: 6737–6740.

127 (a) Cai, M., Wang, Y., and Hao, W. (2008). *Eur. J. Org. Chem.* 2983–2988. (b) Cai, M., Fang, X., Dai, R., and Zha, L. (2009). *Appl. Organometal. Chem.* 23: 229–234.

128 Fujiwara, S., Cadou, R., Yamaoka, Y. et al. (2015). *Eur. J. Org. Chem.* 1264–1272.

129 Shamim, A., Barbeiro, C.S., Ali, B., and Stefani, H.A. (2016). *ChemistrySelect* 1: 5653–5659.

130 Nativi, C. and Taddei, M. (1988). *J. Org. Chem.* 53: 820–826.

131 Alami, M., Liron, F., Gervais, M. et al. (2002). *Angew. Chem. Int. Ed.* 41: 1578–1580.

132 (a) Marshall, J.A. and Bourbeau, M.P. (2002). *Org. Lett.* 4: 3931–3934. (b) Marshall, J.A. and Bourbeau, M.P. (2003). *Tetrahedron Lett.* 44: 1087–1089.

133 Hale, K.J., Grabski, M., Manaviazar, S., and Maczka, M. (2014). *Org. Lett.* 16: 1164–1167.

134 (a) Miura, K., Wang, D., Matsumoto, Y. et al. (2003). *J. Org. Chem.* 68: 8730–8732. (b) Miura, K., Wang, D., Matsumoto, Y., and Hosomi, A. (2005). *Org. Lett.* 7: 503–505.

135 Hamze, A., Veau, D., Provot, O. et al. (2009). *J. Org. Chem.* 74: 1337–1340.

136 (a) Oderinde, M.S. and Organ, M.G. (2012). *Angew. Chem. Int. Ed.* 51: 9834–9837. (b) Oderinde, M.S., Hunter, H.N., Froese, R.D.J., and Organ, M.J. (2012). *Chem. Eur. J.* 18: 10821–10824. (c) Oderinde, M.S., Froese, R.D.J., and Organ, M.G. (2013). *Angew. Chem. Int. Ed.* 52: 11334–11338. (d) Oderinde, M.S., Froese, R.D.J., and Organ, M.G. (2014). *Chem. Eur. J.* 20: 8579–8583.

137 (a) Rummelt, S.M. and Fürstner, A. (2014). *Angew. Chem. Int. Ed.* 53: 3626–3630. (b) Roşca, D.-A., Radkowski, K., Wolf, L.M. et al. (2017). *J. Am. Chem. Soc.* 139: 2443–2455.

138 Magre, M., Szewczyk, M., and Rueping, M. (2020). *Org. Lett.* 22: 1594–1598.

139 Selected reviews on enamines applications, see: (a) Granik, V.G. (1984). *Russ. Chem. Rev.* 53: 383–400. (b) Erian, A.W. (1993). *Chem. Rev.* 93: 1991–2005. (c) Granik, V.G., Makarov, V.A., and Párkányi, C. (1998). *Adv. Heterocycl. Chem.* 72: 283–359. (d) Adams, J.P. (2000). *J. Chem. Soc., Perkin Trans.* 1: 125–139. (e) Cheng, Y., Huang, Z.-T., and Wang, M.-X. (2004). *Curr. Org. Chem.* 8: 325–351. (f) Xie, J.-H., Zhu, S.-F., and Zhou, Q.-L. (2011). *Chem. Rev.* 111: 1713–1760.

140 Selected reviews, see: (a) Layer, R.W. (1963). *Chem. Rev.* 63: 489–510. (b) Pratt, A.C. (1977). *Chem. Soc. Rev.* 6: 63–81. (c) Kobayashi, S. and Ishitani, H. (1999).

Chem. Rev. 99: 1069–1094. (d) Adams, J.P. (2000). *J. Chem. Soc., Perkin Trans.* 1: 125–139. (e) Cozzi, P.G., Hilgraf, R., and Zimmermann, N. (2004). *Eur. J. Org. Chem.* 4095–4105. (f) Martin, S.F. (2009). *Pure Appl. Chem.* 81: 195–204. (g) Otero, A., Chapela, M.-J., Atanassova, M. et al. (2011). *Chem. Res. Toxicol.* 24: 1817–1829. (h) Kobayashi, S., Mori, Y., Fossey, J.S., and Salter, M.M. (2011). *Chem. Rev.* 111: 2626–2704. (i) Yus, M., González-Gómez, J.C., and Francisco Foubelo, F. (2013). *Chem. Rev.* 113: 5595–5698. (j) Dong, H.-Q., Xu, M.-H., Feng, C.-G. et al. (2015). *Org. Chem. Front.* 2: 73–89. (k) Ochiai, T., Franz, D., and Inoue, S. (2016). *Chem. Soc. Rev.* 45: 6327–6344. (l) Iwanejko, J. and Wojaczyńska, E. (2018). *Org. Biomol. Chem.* 16: 7296–7314.

141 Selected reviews, see: (a) Müller, T.E. and Beller, M. (1998). *Chem. Rev.* 98: 675–703. (b) Nobis, M. and Drieûen-Hölscher, B. (2001). *Angew. Chem. Int. Ed.* 40: 3983–3985. (c) Bytschkov, I. and Doye, S. (2001). *Eur. J. Org. Chem.* 4411–4418. (d) Pohlki, F. and Doye, S. (2003). *Chem. Soc. Rev.* 32: 104–114. (e) Bytschkov, I. and Doye, S. (2003). *Eur. J. Org. Chem.* 935–946. (f) Widenhoefer, R.A. and Han, X. (2006). *Eur. J. Org. Chem.* 4555–4563. (g) Severin, R. and Doye, S. (2007). *Chem. Soc. Rev.* 36: 1407–1420. (h) Müller, T.E., Hultzsch, K.C., Yus, M. et al. (2008). *Chem. Rev.* 108: 3795–3892. (i) Chemler, S.R. (2009). *Org. Biomol. Chem.* 7: 3009–3019; (j) Patil, R.D. and Adimurthy, S. (2013). *Asian J. Org. Chem.* 2: 726–744. (k) Yim, J.C.-H. and Schafer, L.L. (2014). *Eur. J. Org. Chem.* 6825–6840. (l) Shi, S.-L. and Buchwald, S.L. (2015). *Nat. Chem.* 7: 38–44. (m) Huang, L., Arndt, M., Gooßen, K. et al. (2015). *Chem. Rev.* 115: 2596–2697. (n) Isaeva, V.I. and Kustov, L.M. (2016). *Top. Catal.* 59: 1196–1206. (o) Pirnot, M.T., Wang, Y.-M., and Buchwald, S.L. (2016). *Angew. Chem. Int. Ed.* 55: 48–57. (p) Patel, M., Saunthwal, R.K., and Verma, A.K. (2017). *Acc. Chem. Res.* 50: 240–254. (q) Tashrifi, Z., Khanaposhtani, M.M., Biglar, M. et al. (2020). *Asian J. Org. Chem.* 9: 969–991.

142 Selected reports, see: (a) Lavallo, V., Frey, G.D., Donnadieu, B. et al. (2008). *Angew. Chem. Int. Ed.* 47: 5224–5228. (b) Kovács, G., Lledós, A., and Ujaque, G. (2011). *Angew. Chem. Int. Ed.* 50: 11147–11151.

143 (a) Haak, E., Bytschkov, I., and Doye, S. (1999). *Angew. Chem. Int. Ed.* 38: 3389–3391. (b) Shi, Y., Ciszewski, J.T., and Odom, A.L. (2001). *Organometallics* 20: 3967–3969. (c) Tillack, A., Castro, I.G., Hartung, C.G., and Beller, M. (2002). *Angew. Chem. Int. Ed.* 41: 2541–2543. (d) Khedkar, V., Tillack, A., and Beller, M. (2003). *Org. Lett.* 5: 4767–4770. (e) Anderson, L.L., Arnold, J., and Bergman, R.G. (2004). *Org. Lett.* 6: 2519–2522.

144 (a) Hartung, C.G., Tillack, A., Trauthwein, H., and Beller, M. (2001). *J. Org. Chem.* 66: 6339–6343. (b) Kumaran, E. and Leong, W.K. (2012). *Organometallics* 31: 1068–1072.

145 (a) Mizushima, E., Hayashi, T., and Tanaka, M. (2003). *Org. Lett.* 5: 3349–3352. (b) Corma, A., Concepción, P., Domínguez, I. et al. (2007). *J. Catal.* 251: 39–47. (c) Kinjo, R., Donnadieu, B., and Bertrand, G. (2011). *Angew. Chem. Int. Ed.* 50: 5560–5563. (d) Gabbaï, F.P. and Yang, H. (2015). *J. Am. Chem. Soc.* 137: 13425–13432.

146 Tokunaga, M., Eckert, M., and Wakatsuki, Y. (1999). *Angew. Chem. Int. Ed.* 38: 3222–3225.

147 Virant, M., Mihelač, M., Gazvoda, M. et al. (2020). *Org. Lett.* 22: 2157–2161.
148 Born, K. and Doye, S. (2012). *Eur. J. Org. Chem.* 764–771.
149 Brunet, J.-J., Chu, N.C., Diallo, O., and Vincendeau, S. (2005). *J. Mol. Catal. A: Chem.* 240: 245–248.
150 Leyva-Pérez, A., Cabrero-Antonino, J.R., Cantín, Á., and Corma, A. (2010). *J. Org. Chem.* 75: 7769–7780.
151 Zhang, Z. and Schafer, L.L. (2003). *Org. Lett.* 5: 4733–4736.
152 Lui, E.K.J., Brandt, J.W., and Schafer, L.L. (2018). *J. Am. Chem. Soc.* 140: 4973–4976.
153 Hu, X., Martin, D., and Bertrand, G. (2016). *New J. Chem.* 40: 5993–5996.
154 Tillack, A., Khedkar, V., and Beller, M. (2004). *Tetrahedron Lett.* 45: 8875–8878.
155 Beauchemin, A.M., Moran, J., Lebrun, M.-E. et al. (2008). *Angew. Chem. Int. Ed.* 47: 1410–1413.
156 Cebrowski, P.H., Roveda, J.-G., Moran, J. et al. (2008). *Chem. Commun.* 492–493.
157 Xu, X., Zhang, X., Wang, Z., and Kong, M. (2015). *RSC Adv.* 5: 40950–40952.
158 Shanbhag, G.V., Kumbar, S.M., Joseph, T., and Halligudi, S.B. (2006). *Tetrahedron Lett.* 47: 141–143.
159 Pasha, N., Babu, N.S., Rao, K.T.V. et al. (2009). *Tetrahedron Lett.* 50: 239–242.
160 Sengupta, M., Das, S., and Bordoloi, A. (2017). *Mol. Catal.* 440: 57–65.
161 Zhao, J., Zheng, Z., Bottle, S. et al. (2013). *Chem. Commun.* 49: 2676–2678.
162 Uchimaru, Y. (1999). *Chem. Commun.* 1133–1134.
163 Kocięcka, P., Czeluśniak, I., and Szymańska-Buzar, T. (2014). *Adv. Synth. Catal.* 356: 3319–3324.
164 (a) Fukumoto, Y., Asai, H., Shimizu, M., and Chatani, N. (2007). *J. Am. Chem. Soc.* 129: 13792–13793. (b) Sakai, K., Kochi, T., and Kakiuchi, F. (2011). *Org. Lett.* 13: 3928–3931.
165 Luo, G. and Chen, L. (2015). *Tetrahedron Lett.* 56: 6276–6278.
166 Kadota, I., Shibuya, A., Lutete, L.M., and Yamamoto, Y. (1999). *J. Org. Chem.* 64: 4570–4571.
167 Chen, Q.-A., Chen, Z., and Dong, V.M. (2015). *J. Am. Chem. Soc.* 137: 8392–8395.
168 (a) Gooßen, L.J., Rauhaus, J.E., and Deng, G. (2005). *Angew. Chem. Int. Ed.* 44: 4042–4045. (b) Goossen, L.J., Blanchot, M., Brinkmann, C. et al. (2006). *J. Org. Chem.* 71: 9506–9509. (c) Goossen, L.J., Blanchot, M., Salih, K.S.M. et al. (2008). *Org. Lett.* 10: 4497–4499. (d) Gooßen, L.J., Salih, K.S.M., and Blanchot, M. (2008). *Angew. Chem. Int. Ed.* 47: 8492–8495. (e) Kuai, C., Wang, L., Cui, H. et al. (2016). *ACS Catal.* 6: 186–190.
169 Sun, J. and Kozmin, S.A. (2006). *Angew. Chem. Int. Ed.* 45: 4991–4993.
170 Timmerman, J.C., Robertson, B.D., and Widenhoefer, R.A. (2015). *Angew. Chem. Int. Ed.* 54: 2251–2254.
171 Tzalis, D., Koradin, C., and Knochel, P. (1999). *Tetrahedron Lett.* 40: 6193–6195.
172 Lu, L., Yan, H., Liu, D. et al. (2014). *Chem. Asian J.* 9: 75–78.
173 (a) Verma, A.K., Joshi, M., and Singh, V.P. (2011). *Org. Lett.* 13: 1630–1633. (b) Joshi, M., Patel, M., Tiwari, R., and Verma, A.K. (2012). *J. Org. Chem.* 77:

5633–5645. (c) Patel, M., Saunthawal, R.K., and Verma, A.K. (2014). *Tetrahedron Lett.* 55: 1310–1315. (d) Garg, V., Kumar, P., and Verma, A.K. (2017). *J. Org. Chem.* 82: 10247–10262.

174 Hentz, A., Retailleau, P., Gandon, V. et al. (2014). *Angew. Chem. Int. Ed.* 53: 8333–8337.

175 Kozell, V., Rahmani, F., Piermatti, O. et al. (2018). *Mol. Catal.* 455: 188–191.

176 Zhao, N., Lin, C., Wen, L., and Li, Z. (2019). *Tetrahedron* 75: 3432–3440.

177 Selected reports, see: (a) Müller, T.E. and Pleier, A.K. (1999). *J. Chem. Soc., Dalton Trans.* 583–587. (b) Patil, N.T. and Singh, V. (2011). *J. Organomet. Chem.* 696: 419–432. (c) Dighe, V.A. and Pujari, R.R. (2017). *Asian J. Pharmaceut. Res.* 7: 21–24. (d) Hosseinzadeh, A.R.Z. (2018). *Curr. Org. Chem.* 22: 1–24.

178 (a) Cao, C., Shi, Y., and Odom, A.L. (2002). *Org. Lett.* 4: 2853–2856. (b) Khedkar, V., Tillack, A., Michalik, M., and Beller, M. (2005). *Tetrahedron* 61: 7622–7631.

179 Lai, R.-Y., Surekha, K., Hayashi, A. et al. (2007). *Organometallics* 26: 1062–1068.

180 Okuma, K., Seto, J.-i., Sakaguchi, K.-i. et al. (2009). *Tetrahedron Lett.* 50: 2943–2945.

181 Yi, C.S., Yun, S.Y., and Guzei, I.A. (2005). *J. Am. Chem. Soc.* 127: 5782–5783.

182 Luo, Y., Li, Z., and Li, C.-J. (2005). *Org. Lett.* 7: 2675–2678.

183 Kumaran, E. and Leong, W.K. (2015). *Organometallics* 34: 1779–1782.

184 Burling, S., Leslie, D., Field, L.D., and Messerle, B.A. (2000). *Organometallics* 19: 87–90.

185 (a) Lutete, L.M., Kadota, I., and Yamamoto, Y. (2004). *J. Am. Chem. Soc.* 126: 1622–1623. (b) Narsireddy, M. and Yamamoto, Y. (2008). *J. Org. Chem.* 73: 9698–9709.

186 Patil, N.T., Wu, H., and Yamamoto, Y. (2007). *J. Org. Chem.* 72: 6577–6579.

187 Patil, N.T., Lutete, L.M., Wu, H. et al. (2006). *J. Org. Chem.* 71: 4270–4279.

188 Bajracharya, G.B., Huo, Z., and Yamamoto, Y. (2005). *J. Org. Chem.* 70: 4883–4886.

189 Beeren, S.R., Dabb, S.L., and Messerle, B.A. (2009). *J. Organomet. Chem.* 694: 309–312.

190 Burling, S., Field, L.D., Messerle, B.A., and Rumble, S.L. (2007). *Organometallics* 26: 4335–4343.

191 Ryu, J.-S., Li, G.Y., and Marks, T.J. (2003). *J. Am. Chem. Soc.* 125: 12584–12605.

192 Selected reports in palladium-catalyzed reactions, see: (a) Torii, S., Okumoto, H., Sadakane, M., and Long, H.X. (1991). *Chem. Lett.* 1673–1676. (b) El Ali, B., Tijani, J., and EI-Ghanam, A.M. (2002). *J. Mol. Catal. A: Chem.* 187: 17–33. (c) El Ali, B., Tijani, J., and EI-Ghanam, A.M. (2002). *Appl. Organometal. Chem.* 16: 369–376. (d) El Ali, B. and Tijani, J. (2003). *Appl. Organomet. Chem.* 17: 921–931. (e) Matteoli, U., Scrivanti, A., and Beghetto, V. (2004). *J. Mol. Catal. A: Chem.* 213: 183–186. (f) Sha, F. and Alper, H. (2017). *ACS Catal.* 7: 2220–2229.

193 El Ali, B., El-Ghanam, A., Fettouhi, M., and Tijani, J. (2000). *Tetrahedron Lett.* 41: 5761–5764.

194 Uenoyama, Y., Fukuyama, T., Nobuta, O. et al. (2005). *Angew. Chem. Int. Ed.* 44: 1075–1078.

195 Fukuyama, T., Nakashima, N., Okada, T., and Ryu, I. (2013). *J. Am. Chem. Soc.* 135: 1006–1008.

196 Li, Y., Alper, H., and Yu, Z. (2006). *Org. Lett.* 8: 5199–5201.

197 Park, J.H., Kim, S.Y., Kim, S.M., and Chung, Y.K. (2007). *Org. Lett.* 9: 2465–2468.

198 Driller, K.M., Prateeptongkum, S., Jackstell, R., and Beller, M. (2011). *Angew. Chem. Int. Ed.* 50: 537–541.

199 (a) Ji, X., Gao, B., Zhou, X. et al. (2018). *J. Org. Chem.* 83: 10134–10141. (b) Wang, D.-L., Guo, W.-D., Liu, L. et al. (2019). *Catal. Sci. Technol.* 9: 1334–1337.

200 Wang, H., Yuan, H., Wang, X. et al. (2020). *Adv. Synth. Catal.* 362: 2348–2353.

201 Huang, Q. and Hua, R.R. (2007). *Adv. Synth. Catal.* 349: 849–852.

202 Prateeptongkum, S., Driller, K.M., Jackstell, R. et al. (2010). *Chem. Eur. J.* 16: 9606–9615.

203 Selected reviews on the application of organophosphorous compounds, see: (a) Baumgartner, T. and Réau, R. (2006). *Chem. Rev.* 106: 4681–4727. (b) Shameem, M.A. and Orthaber, A. (2016). *Chem. Eur. J.* 22: 10718–10735.

204 Selected reviews, see: (a) Greenberg, S. and Stephan, D.W. (2008). *Chem. Soc. Rev.* 37: 1482–1489. (b) Glueck, D.S. (2010). *Top Organomet. Chem.* 31: 65–100. (c) Xu, Q. and Han, L.-B. (2011). *J. Organomet. Chem.* 696: 130–140. (d) Ananikov, V.P. and Beletskaya, I.P. (2011). *Chem. Asian J.* 6: 1423–1430. (e) Ananikov, V.P., Makarov, A.V., and Beletskaya, I.P. (2011). *Chem. Eur. J.* 17: 12623–12630. (f) Tanaka, M. (2013). *Top. Organomet. Chem.* 43: 167–202. (g) Rosenberg, L. (2013). *ACS Catal.* 3: 2845–2855. (h) Koshti, V., Gaikwad, S., and Chikkali, S.H. (2014). *Coord. Chem. Rev.* 265: 52–73. (i) Wauters, I., Debrouwer, W., and Stevens, C.V. (2014). *Beilstein J. Org. Chem.* 10: 1064–1096. (j) Bange, C.A. and Waterman, R. (2016). *Chem. Eur. J.* 22: 12598–12605.

205 Kawaguchi, S.-i., Nagata, S., Nomoto, A. et al. (2008). *J. Org. Chem.* 73: 7928–7933.

206 Kawaguchi, S.-I., Kotani, M., Ohe, T. et al. (2010). *Phosphorus, Sulfur, Silicon Relat. Elem.* 185: 1090–1097.

207 (a) Mimeau, D. and Gaumont, A.-C. (2003). *J. Org. Chem.* 68: 7016–7022. (b) Join, B., Mimeau, D., Delacroix, O., and Gaumont, A.-C. (2006). *Chem. Commun.* 3249–3251.

208 King, A.K., Gallagher, K.J., Mahon, M.F., and Webster, R.L. (2017). *Chem. Eur. J.* 23: 9039–9043.

209 (a) Takaki, K., Takeda, M., Koshoji, G. et al. (2001). *Tetrahedron Lett.* 42: 6357–6360. (b) Takaki, K., Koshoji, G., Komeyama, K. et al. (2003). *J. Org. Chem.* 68: 6554–6565.

210 Basiouny, M.M.I., Dollard, D.A., and Schmidt, J.A.R. (2019). *ACS Catal.* 9: 7143–7153.

211 Rajpurohit, J., Kumar, P., Shukla, P. et al. (2018). *Organometallics* 37: 2297–2304.

212 Moglie, Y., González-Soria, M.J., Martín-García, I. et al. (2016). *Green Chem.* 18: 4896–4907.

213 Busacca, C.A., Qu, B., Farber, E. et al. (2013). *Org. Lett.* 15: 1132–1135.

214 Jérôme, F., Monnier, F., Lawicka, H. et al. (2003). *Chem. Commun.* 696–697.

215 Kondoh, A., Yorimitsu, H., and Osh, K. (2007). *J. Am. Chem. Soc.* 129: 4099–4104.

216 Takaki, K., Komeyama, K., Kobayashi, D. et al. (2006). *J. Alloys Compd.* 408–412: 432–436.

217 Komeyama, K., Kobayashi, D., Yamamoto, Y. et al. (2006). *Tetrahedron* 62: 2511–2519.

218 Nagata, S., Kawaguchi, S.-i., Matsumoto, M. et al. (2007). *Tetrahedron Lett.* 48: 6637–6640.

219 Douglass, M.R. and Marks, T.J. (2000). *J. Am. Chem. Soc.* 122: 1824–1825.

220 Hirano, K. and Miura, M. (2017). *Tetrahedron Lett.* 58: 4317–4322.

221 (a) Kamitani, M., Itazaki, M., Tamiya, C., and Nakazawa, H. (2012). *J. Am. Chem. Soc.* 134: 11932–11935. (b) Ackley, B.J., Pagano, J.K., and Waterman, R. (2018). *Chem. Commun.* 54: 2774–2776.

222 (a) Bange, C.A. and Waterman, R. (2016). *ACS Catal.* 6: 6413–6416. (b) Bange, C.A., Conger, M.A., Novas, B.T. et al. (2018). *ACS Catal.* 8: 6230–6238.

223 Yuan, J., Zhu, L., Zhang, J. et al. (2017). *Organometallics* 36: 455–459.

224 Di Giuseppe, A., De Luca, R., Castarlenas, R. et al. (2016). *Chem. Commun.* 52: 5554–5557.

225 Recent report on the radical hydrophosphorylation of alkynes, see:Huang, T., Saga, Y., Guo, H. et al. (2018). *J. Org. Chem.* 83 (15): 8743–8749. and references cited therein.

226 Han, L.-B. and Tanaka, M. (1996). *J. Am. Chem. Soc.* 118: 1571–1572.

227 Zhao, C.-Q., Han, L.-B., Goto, M., and Tanaka, M. (2001). *Angew. Chem. Int. Ed.* 40: 1929–1932.

228 Han, L.-B., Zhao, C.-Q., Onozawa, S.-Y. et al. (2002). *J. Am. Chem. Soc.* 124: 3842–3843.

229 Ananikov, V.P., Khemchyan, L.L., Beletskaya, I.P., and Starikova, Z.A. (2010). *Adv. Synth. Catal.* 352: 2979–2992.

230 Chen, T., Zhou, Y., Guo, C., and Han, L.-B. (2013). *Chem. Lett.* 42: 1065–1067.

231 Han, L.-B., Choi, N., and Tanaka, M. (1996). *Organometallics* 15: 3259–3261.

232 Han, L.-B., Hua, R., and Tanaka, M. (1998). *Angew. Chem. Int. Ed.* 37: 94–96.

233 Han, L.-B., Zhang, C., Yazawa, H., and Shimada, S. (2004). *J. Am. Chem. Soc.* 126: 5080–5081.

234 Han, L.-B., Ono, Y., and Yazawa, H. (2005). *Org. Lett.* 7: 2909–2911.

235 Nune, S.K. and Tanaka, M. (2007). *Chem. Commun.* 2858–2860.

236 Belabassi, Y., Bravo-Altamirano, K., and Montchamp, J.-L. (2011). *J. Organomet. Chem.* 696: 106–111.

237 Duraud, A., Toffano, M., and Fiaud, J.-C. (2009). *Eur. J. Org. Chem.* 4400–4403.

238 (a) Han, L.-B., Zhao, C.-Q., and Tanaka, M. (2001). *J. Org. Chem.* 66: 5929–5932. (b) Rooy, S.V., Cao, C., Patrick, B.O. et al. (2006). *Inorg. Chim. Acta* 359: 2918–2923.

239 Niu, M., Fu, H., Jiang, Y., and Zhao, Y. (2007). *Chem. Commun.* 272–274.

240 Trostyanskaya, I.G. and Beletskay, I.P. (2014). *Tetrahedron* 70: 2556–2562.

241 Huang, Y., Hao, W., Ding, G., and Cai, M.-Z. (2012). *J. Organomet. Chem.* 715: 141–146.

242 Kinbara, A., Sato, M., Yumita, K., and Yamagishi, T. (2017). *Tetrahedron* 73: 1705–1710.

243 Ribière, P., Bravo-Altamirano, K., Antczak, M.I. et al. (2005). *J. Org. Chem.* 70: 4064–4072.

244 Khemchyan, L.L., Ivanova, J.V., Zalesskiy, S.S. et al. (2014). *Adv. Synth. Catal.* 356: 771–780.

245 Allen, A. Jr., Ma, L., and Lin, W. (2002). *Tetrahedron Lett.* 43: 3707–3710.

246 Yoshimura, A., Saga, Y., Sato, Y. et al. (2016). *Tetrahedron Lett.* 57: 3382–3384.

247 Stone, J.J., Stockland, R.A. Jr., Reyes, J.M. Jr. et al. (2005). *J. Mol. Catal. A: Chem.* 226: 11–21.

248 Guo, H., Yoshimura, A., Chen, T. et al. (2017). *Green Chem.* 19: 1502–1506.

249 Allen, A. Jr., Manke, D.R., and Lin, W. (2000). *Tetrahedron Lett.* 41: 151–154.

250 Sato, A., Yorimitsu, H., and Oshima, K. (2005). *Angew. Chem. Int. Ed.* 44: 1694–1696.

251 Ohmiya, H., Yorimitsu, H., and Oshima, K. (2005). *Angew. Chem. Int. Ed.* 44: 2368–2370.

252 Han, L.-B., Ono, Y., and Shimada, S. (2008). *J. Am. Chem. Soc.* 130: 2752–2753.

253 Chen, T., Chang-Qiu Zhao, C.-Q., and Han, L.-B. (2018). *J. Am. Chem. Soc.* 140: 3139–3155.

254 MacMillan, J.W.M., Marczenko, K.M., Johnson, E.R., and Chitnis, S.S. (2020). *Chem. Eur. J.* 26: 17134–17142.

255 Selected reviews, see: (a) Kondo, T. and Mitsudo, T.-a. (2000). *Chem. Rev.* 100: 3205–3220. (b) Ogawa, A. (2000). *J. Organomet. Chem.* 611: 463–474. (c) Weiss, C.J. and Marks, T.J. (2010). *Dalton Trans.* 39: 6576–6588. (d) Beletskaya, I.P. and Ananikov, V.P. (2011). *Chem. Rev.* 111: 1596–1636. (e) Dondoni, A. and Marra, A. (2014). *Eur. J. Org. Chem.* 3955–3969. (f) Orlov, N.V. (2015). *ChemistryOpen* 4: 682–697. (g) Giuseppe, A.D., Castarlenas, R., and Oro, L.A. (2018). *Top. Organomet. Chem.* 61: 31–68.

256 Selected reviews, see: (a) Luh, T.-Y. (1991). *Acc. Chem. Res.* 24: 257–263. (b) Dubbaka, S.R. and Vogel, P. (2005). *Angew. Chem. Int. Ed.* 44: 7674–7684. (c) Wang, L., He, W., and Yu, Z. (2013). *Chem. Soc. Rev.* 42: 599–621. (d) Modha, S.G., Mehta, V.P., and van der Eycken, E.V. (2013). *Chem. Soc. Rev.* 42: 5042–5055. (e) Munjanja, L., Brennessel, W.W., and Jones, W.D. (2015). *Organometallics* 34: 1716–1724. (f) Otsuka, S., Nogi, K., and Yorimitsu, H. (2018). *Top. Curr. Chem. (Z)* 376: 199–237. (g) Hirschbeck, V., Gehrtz, P.H., and Fleischer, I. (2018). *Chem. Eur. J.* 24: 7092–7107. (h) Lou, J., Wang, Q., Wu, P. et al. (2020). *Chem. Soc. Rev.* 49: 4307–4359.

257 Chatterjee, A., König, B., and Natarajan, P. (2020). *ChemPhotoChem* 4: 1–4. and references cited therein.

258 (a) Kuniyasu, H., Ogawa, A., Sato, K.-I. et al. (1992). *J. Am. Chem. Soc.* 114: 5902–5903. (b) Ogawa, A., Ikeda, T., Kimura, K., and Hirao, T. (1999). *J. Am. Chem. Soc.* 121: 5108–5114.

259 Bäckvall, J.-E. and Ericsson, A. (1994). *J. Org. Chem.* 59: 5850–5851.

260 (a) Degtyareva, E.S., Burykina, J.V., Fakhrutdinov, A.N. et al. (2015). *ACS Catal.* 5: 7208–7213. (b) Eremin, D.B., Boiko, D.A., Borkovskaya, E.V. et al. (2018). *Catal. Sci. Technol.* 8: 3073–3080.
261 Selected reports, see: (a) Ananikov, V.P., Orlov, N.V., and Beletskaya, I.P. (2006). *Organometallics* 25: 1970–1977. (b) Malyshev, D.A., Scott, N.M., Marion, N. et al. (2006). *Organometallics* 25: 4462–4470.
262 Selected reports, see: (a) Cao, C., Fraser, L.R., and Love, J.A. (2005). *J. Am. Chem. Soc.* 127: 17614–17615. (b) Misumi, Y., Seino, H., and Mizobe, Y. (2006). *J. Organomet. Chem.* 691: 3157–3164. (c) Fraser, L.R., Bird, J., Wu, Q. et al. (2007). *Organometallics* 26: 5602–5611. (d) Sabarre, A. and Love, J.A. (2008). *Org. Lett.* 10: 3941–3944. (e) Yang, J., Sabarre, A., Fraser, L.R. et al. (2009). *J. Org. Chem.* 74: 182–187. (f) Kankala, S., Nerella, S., Vadde, R., and Vasam, C.S. (2013). *RSC Adv.* 3: 23582–23588. (g) Palacios, L., Giuseppe, A.D., Artigas, M.J. et al. (2016). *Catal. Sci. Technol.* 6: 8548–8561.
263 Weiss, C.J. and Marks, T.J. (2010). *J. Am. Chem. Soc.* 132: 10533–10546.
264 (a) Weiss, C.J., Wobser, S.D., and Marks, T.J. (2009). *J. Am. Chem. Soc.* 131: 2062–2063. (b) Weiss, C.J., Wobser, S.D., and Marks, T.J. (2010). *Organometallics* 29: 6308–6320.
265 Palacios, L., Artigas, M.J., Polo, V. et al. (2013). *ACS Catal.* 3: 2910–2919.
266 Ma, H., Ren, X., Zhou, X. et al. (2015). *Tetrahedron Lett.* 56: 6022–6029.
267 Ananikov, V.P., Orlov, N.V., Beletskaya, I.P. et al. (2007). *J. Am. Chem. Soc.* 129: 7252–7253.
268 Wang, H., Lu, Q., Chiang, C.-W. et al. (2017). *Angew. Chem. Int. Ed.* 56: 595–599.
269 Degtyareva, E.S., Borkovskaya, E.V., and Ananikov, V.P. (2019). *ACS Sustain. Chem. Eng.* 7: 9680–9689.
270 Giuseppe, A.D., Castarlenas, R., Pérez-Torrente, J.J. et al. (2012). *J. Am. Chem. Soc.* 134: 8171–8183.
271 (a) Sarma, R., Rajesh, N., and Prajapati, D. (2012). *Chem. Commun.* 48: 4014–4016. (b) Rajesh, N. and Prajapati, D. (2014). *RSC Adv.* 4: 32108–32112.
272 (a) Shoai, S., Bichler, P., Kang, B. et al. (2007). *Organometallics* 26: 5778–5781. (b) Yoshimura, A., Nomoto, A., and Ogawa, A. (2014). *Res. Chem. Intermed.* 40: 2381–2389.
273 Yang, Y. and Rioux, R.M. (2011). *Chem. Commun.* 47: 6557–6559.
274 Zhao, H., Peng, J., and Cai, M. (2012). *Catal. Lett.* 142: 138–142.
275 Corma, A., González-Arellano, C., Iglesias, M., and Sánchez, F. (2010). *Appl. Catal. A: Gen.* 375: 49–54.
276 Jin, Z., Xu, B., and Hammond, G.B. (2010). *Eur. J. Org. Chem.* 168–173.
277 Basu, B., Biswas, K., Kundu, S., and Sengupta, D. (2014). *Org. Chem. Int.* 358932.
278 Zalesskiy, S.S., Shlapakov, N.S., and Ananikov, V.P. (2016). *Chem. Sci.* 7: 6740–6745.
279 (a) Ichinose, Y., Wakamatsu, K., Nozaki, K. et al. (1987). *Chem. Lett.* 1647–1650. (b) Benati, L., Capella, L., Montevecchi, P.C., and Spagnolo, P. (1995). *J. Chem. Soc., Perkin. Trans. 1* 1035–1038.

280 Kondoh, A., Takami, K., Yorimitsu, H., and Oshima, K. (2005). *J. Org. Chem.* 70: 6468–6473.

281 Waters, M. S.; Cowen, J. A.; McWilliams, J. C.; Maligres, P. E.; Askin, D. *Tetrahedron Lett.* 2000, 41, 141–144.

282 Rajesh, N., Sarma, R., and Prajapati, D. (2014). *RSC Adv.* 4: 7834–7837.

283 Burling, S., Field, L.D., Messerle, B.A. et al. (2003). *Dalton Trans.* 4181–4191.

284 Gerber, R. and Frech, C.M. (2012). *Chem. Eur. J.* 18: 8901–8905.

285 Riduan, S.N., Ying, J.Y., and Zhang, Y. (2012). *Org. Lett.* 14: 1780–1783.

286 Liao, Y., Chen, S., Jiang, P. et al. (2013). *Eur. J. Org. Chem.* 6878–6885.

287 Cong, Z.-S., Zhang, Y., Du, G.-F. et al. (2018). *Synth. Commun.* 48: 1838–1846.

288 Zhao, N., Lin, C., Wen, L., and Li, Z. (2019). *Tetrahedron* 75: 3432–3440.

289 Yang, Y. and Rioux, R.M. (2014). *Green Chem.* 16: 3916–3925.

290 Wang, Z.-L., Tang, R.-Y., Luo, P.-S. et al. (2008). *Tetrahedron* 64: 10670–10675.

291 Ranjit, S., Duan, Z., Zhang, P., and Liu, X. (2010). *Org. Lett.* 12: 4134–4136.

292 Kondoh, A., Yorimitsu, H., and Oshima, K. (2007). *Org. Lett.* 9: 1383–1385.

293 Yadav, J.S., Reddy, B.V.S., Raju, A. et al. (2007). *Chem. Lett.* 36: 1474–1475.

294 Mitamura, T., Daitou, M., Nomoto, A., and Ogawa, A. (2011). *Bull. Chem. Soc. Jpn.* 84: 413–415.

295 Hut'ka, M., Tsubogo, T., and Kobayashi, S. (2014). *Organometallics* 33: 5626–5629.

296 Ikeda, T., Tamai, T., Daitou, M. et al. (2013). *Chem. Lett.* 42: 1383–1385.

297 Sonehara, T., Murakami, S., Yamazaki, S., and Kawatsura, M. (2017). *Org. Lett.* 19: 4299–4302.

298 Silva, M.S., Lara, R.G., Marczewski, J.M. et al. (2008). *Tetrahedron Lett.* 49: 1927–1930.

299 Banerjee, S., Das, J., and Santra, S. (2009). *Tetrahedron Lett.* 50: 124–127.

300 Thurow, S., Ostosi, N.T., Mendes, S.R. et al. (2012). *Tetrahedron Lett.* 53: 2651–2653.

301 Li, Y., Cai, J., Hao, M., and Li, Z. (2019). *Green Chem.* 21: 2345–2351.

302 Nakatani, S., Yoshida, J.-i., and Isoe, S. (1992). *J. Chem. Soc., Chem. Commun.* 880–881.

303 (a) Ogawa, A., Takeba, M., Kawakami, J.-i. et al. (1995). *J. Am. Chem. Soc.* 117: 7564–7565. (b) Kawakami, J.-i., Takeba, M., Kamiya, I. et al. (2003). *Tetrahedron* 59: 6559–6567.

304 Xiao, W.J., Vasapollo, G., and Alper, H. (1999). *J. Org. Chem.* 64: 2084–2084.

305 (a) Ogawa, A., Kawakami, J.-i., Mihara, M. et al. (1997). *J. Am. Chem. Soc.* 119: 12380–123810. (b) Kawakami, J.-i., Mihara, M., Kamiya, I. et al. (2003). *Tetrahedron* 59: 3521–3526.

306 Xiao, W.-J. and Alper, H. (1997). *J. Org. Chem.* 62: 3422–3423.

307 El Ali, B., Tijani, J., El-Ghanam, A., and Fettouhi, M. (2001). *Tetrahedron Lett.* 42: 1567–1570.

308 (a) Xiao, W.-J., Vasapollo, G., and Alper, H. (1999). *J. Org. Chem.* 64: 2080–2084. (b) Xiao, W.-J. and Alper, H. (2005). *J. Org. Chem.* 70: 1802–1807.

309 (a) Higashimae, S., Tamai, T., Nomoto, A., and Ogawa, A. (2015). *J. Org. Chem.* 80: 7126–7133. (b) Higuchi, Y., Higashimae, S., Tamai, T., and Ogawa, A. (2013). *Tetrahedron* 69: 11197–11202.

310 Kuniyasu, H., Ogawa, A., Sato, K.-I. et al. (1992). *Tetrahedron Lett.* 33: 5525–5528.
311 Selected reports, see: (a) Comasseto, J.V. and Brandt, C.A. (1987). *Synthesis* 146–149. (b) Ananikov, V.P., Malyshev, D.A., Beletskaya, I.P. et al. (2003). *J. Organomet. Chem.* 679: 162–172. (c) Kamiya, I., Nishinaka, E., and Ogawa, A. (2005). *J. Org. Chem.* 70: 696–698. (d) Ananikov, V.P., Orlov, N.V., and Beletskaya, I.P. (2007). *Organometallics* 26: 740–750. (e) Ishii, A., Kamon, H., Murakami, K., and Nakata, N. (2010). *Eur. J. Org. Chem.* 1653–1659. (f) Ozaki, T., Kotani, M., Kusano, H. et al. (2011). *J. Organomet. Chem.* 696: 450–455. and references therein. (g) Kobiki, Y., Kawaguchi, S.-i., and Ogawa, A. (2013). *Tetrahedron Lett.* 54: 5453–5456. (h) Orlov, N.V., Chistyakov, I.V., Khemchyan, L.L. et al. (2014). *J. Org. Chem.* 79: 12111–12121.
312 Ogawa, A., Kudo, A., and Hirao, T. (1998). *Tetrahedron Lett.* 39: 5213–5216.
313 Selected reports, see: (a) Patil, N.T., Pahadi, N.K., and Yamamoto, Y. (2005). *Can. J. Chem.* 83: 569–573. (b) Weiss, C.J. and Marks, T.J. (2010). *Dalton Trans.* 39: 6576–6588. (c) Bruneau, C. (2013). *Top Organomet. Chem.* 43: 203–230.
314 Selected reports, see: (a) Wobser, S.D. and Marks, T.J. (2013). *Organometallics* 32: 2517–2528. (b) Goodwin, J.A. and Aponick, A. (2015). *Chem. Commun.* 51: 8730–8741. (c) Iio, K., Sachimori, S., Watanabe, T., and Fuwa, H. (2018). *Org. Lett.* 20: 7851–7855.
315 Kumar, M., Scobie, M., Mashuta, M.S. et al. (2013). *Org. Lett.* 15: 724–727.
316 Tzalis, D., Koradin, C., and Knochel, P. (1999). *Tetrahedron Lett.* 40: 6193–6195.
317 Trofimov, B.A., Oparina, L.A., Tarasova, O.A. et al. (2014). *Tetrahedron* 70: 5954–5960.
318 Trofimov, B.A., Oparina, L.A., Kolyvanov, N.A. et al. (2015). *Russ. J. Org. Chem.* 51: 188–194.
319 Patel, M., Saunthwal, R.K., Dhaked, D.K. et al. (2016). *Asian J. Org. Chem.* 5: 213–221.
320 Ushimaru, R., Nishimura, T., Iwatsuki, T., and Naka, H. (2017). *Chem. Pharm. Bull.* 65: 1000–1003.
321 Konkol, M., Schmidt, H., and Steinborn, D. (2007). *J. Mol. Catal. A: Chem.* 269: 119–124.
322 Teles, J.H., Brode, S., and Chabanas, M. (1998). *Angew. Chem. Int. Ed.* 37: 1415–1418.
323 Kondo, M., Kochi, T., and Kakiuchi, F. (2011). *J. Am. Chem. Soc.* 133: 32–34.
324 Fan, M.-J., Li, G.-Q., Li, L.-H. et al. (2006). *Synthesis* 2286–2292.
325 Li, M. and Hua, R. (2008). *J. Org. Chem.* 73: 8658–8660.
326 Wu, H., Yang, J., and Wang, L. (2011). *Chin. J. Chem.* 29: 1211–1215.
327 Ndong, G.J.E., Jismy, B., Ngi, S.I., and Mohamed Abarbri, M. (2018). *J. Fluor. Chem.* 212: 45–50.
328 (a) Veenboer, R.M.P., Dupuy, S., and Nolan, S.P. (2015). *ACS Catal.* 5: 1330–1334. (b) Zheng, Y., Guo, L., and Zi, W. (2018). *Org. Lett.* 20: 7039–7043.
329 Kuram, M.R., Bhanuchandra, M., and Sahoo, A.K. (2010). *J. Org. Chem.* 75: 2247–2258.
330 Oonishi, Y., Gómez-Suórez, A., Martin, A.R., and Nolan, S.P. (2013). *Angew. Chem. Int. Ed.* 52: 9767–9771.

331 (a) Kadota, I., Lutete, L.M., Shibuya, A., and Yamamoto, Y. (2001). *Tetrahedron Lett.* 42: 6207–6210. (b) Zhang, W., Haight, A.R., and Hsu, M.C. (2002). *Tetrahedron Lett.* 43: 6575–6578.

332 Liu, Z. and Breit, B. (2016). *Angew. Chem. Int. Ed.* 55: 8440–8443.

333 Francos, J. and Cadierno, V. (2017). *Catalysts* 7: 328.

334 (a) Rotem, M. and Shvo, Y. (1983). *Organometallics* 2: 1689–1691. (b) Rotem, M. and Shvo, Y. (1993). *J. Organomet. Chem.* 448: 189–204.

335 Selected reports, see: (a) Mitsudo, T., Hori, Y., and Watanabe, Y. (1985). *J. Org. Chem.* 50: 1566–1568. (b) Ruppin, C. and Dixneuf, P.H. (1986). *Tetrahedron Lett.* 27: 6323–6324. (c) Mitsudo, T.-a., Hori, Y., Yamakawa, Y., and Watanabe, Y. (1987). *J. Org. Chem.* 52: 2230–2239. (d) Doucet, H., Martin-Vaca, B., Bruneau, C., and Dixneuf, P.H. (1995). *J. Org. Chem.* 60: 7247–7255. (e) Bruneau, C. and Dixneuf, P.H. (1997). *Chem. Commun.* 507–512. (f) Melis, K., Opstal, T., and Verpoort, F. (2002). *Eur. J. Org. Chem.* 3779–3784. (g) Goossen, L.J., Paetzold, J., and Koley, D. (2003). *Chem. Commun.* 706–707. (h) Melis, K., Vos, D., Jacobs, P., and Verpoort, F. (2003). *J. Organomet. Chem.* 671: 131–136. (i) Ye, S. and Leong, W.K. (2006). *J. Organomet. Chem.* 691: 1117–1120. (j) Tripathy, J. and Bhattacharjee, M. (2009). *Tetrahedron Lett.* 50: 4863–4865. (k) Yi, C.S. and Gao, R. (2009). *Organometallics* 28: 6585–6592. (l) Tan, S.T. and Fan, W.Y. (2010). *Eur. J. Inorg. Chem.* 4631–4635. (m) Maity, B. and Koley, D. (2018). *ChemCatChem* 10: 566–580. (n) Liu, G., Zhang, X., Kuang, G. et al. (2020). *ACS Omega* 5: 4158–4166.

336 Bianchini, C., Meli, A., Peruzzini, M., and Zanobini, F. (1990). *Organometallics* 9: 1155–1160.

337 Lu, X., Zhu, G., and Ma, S. (1992). *Tetrahedron Lett.* 33: 7205–7209.

338 Nakagawa, H., Okimoto, S., Sakaguchi, S., and Ishii, Y. (2003). *Tetrahedron Lett.* 44: 103–106.

339 Chen, J.-F. and Li, C. (2018). *Org. Lett.* 20: 6719–6724.

340 Yin, J., Bai, Y., Mao, M., and Zhu, G. (2014). *J. Org. Chem.* 79: 9179–9185.

341 Jeschke, J., Gäbler, G., and Lang, H. (2016). *J. Org. Chem.* 81: 476–484.

342 Jeschke, J., Engelhardt, T.B., and Lang, H. (2016). *Eur. J. Org. Chem.* 1548–1554.

343 Hua, R. and Tian, X. (2004). *J. Org. Chem.* 69: 5782–5784.

344 (a) Lumbroso, A., Vautravers, N.R., and Breit, B. (2010). *Org. Lett.* 12: 5498–5501. (b) Wei, S., Pedroni, J., Meißner, A. et al. (2013). *Chem. Eur. J.* 19: 12067–12076.

345 Dupuy, S., Gasperini, D., and Nolan, S.P. (2015). *ACS Catal.* 5: 6918–6921.

346 González-Liste, P.J., Francos, J., García-Garrido, S.E., and Cadierno, V. (2017). *J. Org. Chem.* 82: 1507–1516.

347 Hua, R. and Tananka, M. (1998). *Chem. Lett.* 431–432.

348 Lee, P.H., Kim, S., Park, A. et al. (2010). *Angew. Chem. Int. Ed.* 49: 6806–6809.

4

Hydrofunctionalization of Carbon–Carbon Double Bonds

Similar to the addition reactions of H—heteroatom (E) bond to alkynes, the addition of alkenes with H—E bond has also been developed as one of the efficient and atom economical processes approach to functionalized alkanes [1]. In this chapter, the recent hydrofunctionalization of allenes [2] and 1,3-dienes [3] will also be introduced in each section.

The mechanism for the hydrofunctionalization of alkenes is essentially similar to that of alkynes as shown in Scheme 3.2 (Chapter 3), by replacement of alkyne with alkene.

Some catalyst systems were found to show versatile catalytic activity to promote the addition reactions of different H—E bond to alkenes, such as $FeCl_3$/$AgSbF_6$ [4], $FeBr_2$/phosphine/phosphinite-diimine ligand [5], and manganese(II) bis(imino)-pyridine compounds [6]. Organolanthanide complexes show catalytic activity in the stereoselective intramolecular cyclo-hydroamination of aminoalkenes and cyclo-hydrophosphination of phosphinoalkenes [7].

Intermolecular hydrofunctionalization of 1,3-dienes usually gives a mixture of adducts; the high regioselectivity depends on not only the use of catalyst systems and reaction conditions but also the structures and natures of 1,3-dienes [8].

4.1 Hydroboration of Alkenes

The cross-coupling reaction of alkylboron compounds (boranes and boronate esters) is important and useful reaction via activation of C(sp^3)—B bond in the organic synthesis [9]. The *B*-alkyl Suzuki–Miyaura reaction involving an sp^3 carbon in the coupling event is important in the formation of C(sp^3)—C(sp^2) bond, which has been well applied in the synthesis of natural products [10].

Alkylboron compounds are readily prepared by hydroboration of alkenes [11]; the hydroboration of allenes can also produce either alkylborons or alkenylborons under different conditions [12].

Addition Reactions with Unsaturated Hydrocarbons, First Edition. Ruimao Hua.

4.1.1 Markovnikov Hydroboration of Alkenes

For highly selective Markovnikov hydroboration of terminal alkenes (Scheme 4.1), recently a variety of transition metal complexes have been found to be efficient catalysts, such as copper [13], cobalt [14], manganese [15], nickel [16], and iron complexes [17]. Very recently, a review paper has summarized the Markovnikov hydroboration of terminal alkenes catalyzed by Earth-abundant metal complexes [18].

Scheme 4.1 Markovnikov hydroboration of terminal alkenes.

4.1.2 *Anti*-Markovnikov Hydroboration of Alkenes

The regioselective *anti*-Markovnikov hydroboration of alkenes with pinacolborane (HBpin) (Scheme 4.2) can be achieved by using iron [19], silver [20], ruthenium [21], and cobalt complexes [22] as catalysts.

Scheme 4.2 *Anti*-Markovnikov hydroboration of terminal alkenes.

Copper [23], iron [24], and nickel [25] show high catalytic activity for the *anti*-Markovnikov hydroboration of arylalkenes with $B_2(pin)_2$ under mild conditions. Note that the hydroboration of aliphatic terminal alkenes with $B_2(pin)_2$ usually gives the adducts in unsatisfactory regioselectivity.

Recently, iron- and cobalt-catalyzed hydroboration of alkenes using *t*-BuONa as a precatalyst activator has been reported [26], and without using any activator, the commercially available $M(BF_4)_2{\cdot}6H_2O$ (M = Fe, Co) has also been found to efficiently catalyze the hydroboration of aryl and alkyl alkenes with HBpin enabled by counterion dissociation [27].

In addition, bis(imino)pyridine iron dinitrogen complexes can promote the *anti*-Markovnikov hydroboration of terminal, internal, and geminal alkenes with high activity and selectivity [28].

Recently, aluminum-catalyzed hydroboration of alkynes has been reported (see Chapter 3, Section 3.2), and a $LiAlH_4$-catalyzed hydroboration of alkenes with HBpin has also been developed by Thomas's group [29]. The present hydroboration is proposed to proceed by alkene hydroalumination giving an alkyl aluminum species, which undergoes σ-bond metathesis with HBpin to drive turnover of the catalytic cycle (Scheme 4.3).

Very recently, a phenalenyl-based nickel catalyst for the hydroboration of terminal alkenes under ambient conditions has been developed, and both Markovnikov and *anti*-Markovnikov adducts could be obtained with high regioselectivity

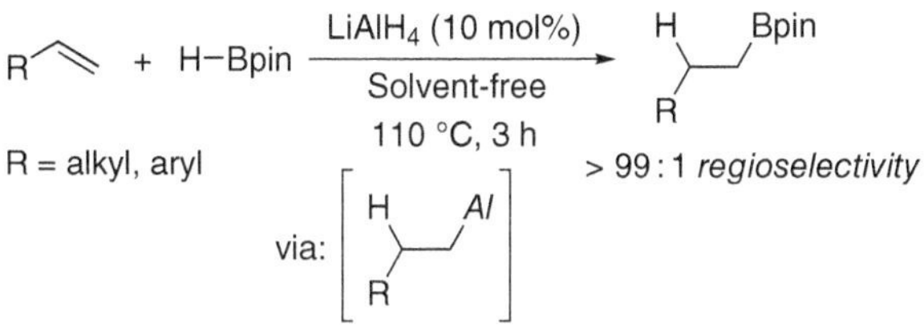

Scheme 4.3 $LiAlH_4$-catalyzed hydroboration of alkenes with HBpin.

depending on the substrates [30]. Hydroboration of arylalkenes with HBpin afforded Markovnikov adduct, whereas the hydroboration of aliphatic alkenes produced the *anti*-Markovnikov adducts.

It has been found that the use of Lewis acids, such as $B(C_6F_5)_3$ as cocatalysts, has a dramatic effect on both reaction rates and regioselectivity in a Rh(cod)(dppb)BF_4-catalyzed hydroboration of terminal and internal alkenes with pinacolborane, indicating the importance of $B(C_6F_5)_3$ in the heterolytic cleavage of the B—H bond of HBpin [31].

Although $B(C_6F_5)_3$ cannot catalyze the hydroboration of terminal alkenes, tris[3,5-bis(trifluoromethyl)phenyl]borane (BAr^F3) enables the catalytic hydroboration of various alkenes with HBpin without the assistance of an external additive [32].

In addition, Chirik and coworker have designed and synthesized cobalt alkyl complexes bearing readily available and redox-active 2,2′ : 6′,2″-terpyridine and α-diimine ligands, which show effective catalytic activity for the isomerization–hydroboration of sterically hindered alkenes of tri-, tetra-, and geminally substituted alkenes with HBpin, giving formal *anti*-Markovnikov adducts [33].

Interestingly, a zirconium-catalyzed formation of 1,1-diborylalkanes from the reaction of terminal alkenes with HBpin has been reported recently, and the formal *anti*-Markovnikov reaction with the formation of C—B bond of alkene terminal carbon is proposed in the plausible mechanism [34].

4.1.3 Hydroboration of Allenes and 1,3-dienes

The reported procedures for intermolecular hydroboration of allenes with HBpin can afford *E*/*Z*-allylboronates (Scheme 4.4).

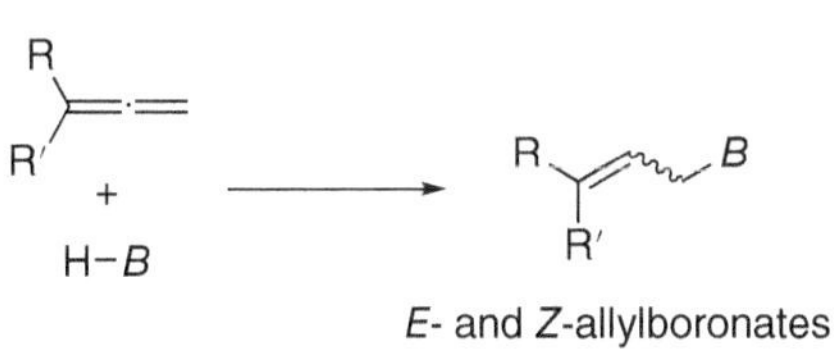

Scheme 4.4 Intermolecular hydroboration of allenes.

Transition metal-catalyzed (platinum [35] and cobalt complexes [36]) hydroboration of allenes with HBpin selectively gives *Z*-allylboronates, copper complexes [37] afforded *E*-allylboronates as major adducts.

On the other hand, the hydroboration of substituted 1,3-dienes can produce several possible adducts resulting from the chemoselectivity under different conditions with the use of different substrates. A recent report on the selective hydroboration of 2-substituted 1,3-dienes to access homoallylic boronates is cited, which gives an overview for this type of addition reactions [38].

4.1.4 Asymmetric Hydroboration of Alkenes

Transition metal-catalyzed asymmetric hydroboration of alkenes has been shown to be one of the most efficient methods for the synthesis of chiral boronic esters [39], which are important and useful building blocks in asymmetric synthesis [40]. In the case of terminal alkene used, the chiral organoboronates can be obtained by both asymmetric Markovnikov and asymmetric *anti*-Markovnikov hydroboration reactions catalyzed by transition metal complexes with the use of chiral ligands (Scheme 4.5). Recently, copper [41], iridium [42], cobalt [43], rhodium [44], and iron complexes [45] have been applied as the efficient catalysts in the enantioselective hydroboration with HBpin and enantioselective hydroboration by copper [46] and rhodium complexes [47] catalyzed with (pin)B–B(pin). The internal alkenes could also undergo the enantioselective hydroboration in the presence of rhodium [48] and copper complexes [49].

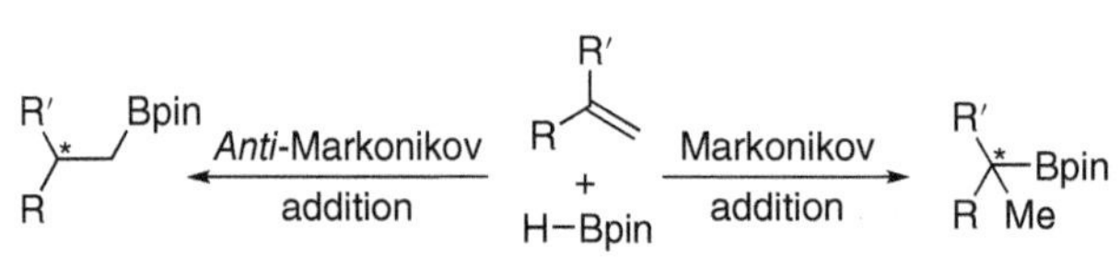

Scheme 4.5 Enantioselective hydroboration of 1,1-disubstituted alkenes.

The enantioselective hydroboration of internal alkenes of cyclopropenes with catecholborane in the presence of rhodium complexes has been also developed [50].

In addition, the asymmetric hydroboration of substituted alkenes has become the convenient routes for synthesis of chiral alcohols in situ [51].

Very recently, Li's group has developed a rhodium(I)-catalyzed regiodivergent and enantioselective hydroboration of enamides affording α- and β-aminoboronic esters with high regio-, diastereo-, and enantioselectivity [52].

For hydroboration of substituted 1,3-dienes, with the use of different transition metal complexes such as palladium [53], iron [54], copper [55], cobalt [56], and iridium [38] as catalysts, the addition reactions occurred with different regioselectivity to give either allylic boronates or homoallylic boronates.

The metal/metal oxide nanoparticles have been extensively used in the hydroboration of alkenes [57].

gem-Bis(boryl)alkanes can undergo various transformations via activation of $C(sp^3)$—B bond under different conditions (Scheme 4.6) [58], which could be prepared from the catalytic dihydroboration of alkynes [59], hydroboration of alkenylboronates [60], and diborylation of alkenes with B_2pin_2 [61]. Recently, a cobalt-catalyzed diborylation of 1,1-disubstituted vinylarenes with HBpin affording branched *gem*-bis(boryl)alkanes has also been reported [62].

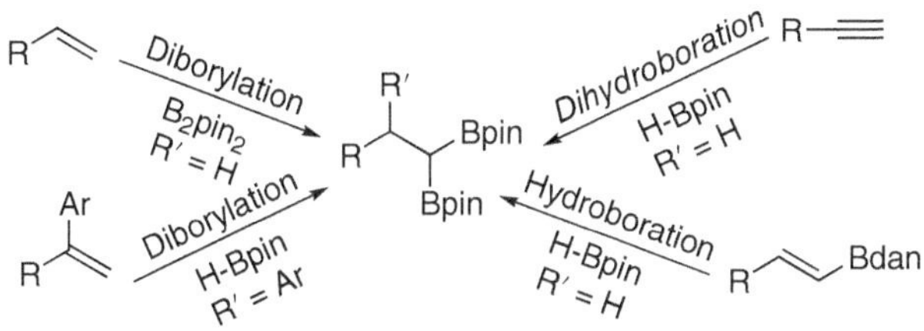

Scheme 4.6 Synthesis of *gem*-bis(boryl)alkanes.

4.2 Hydrosilylation of Carbon–Carbon Double Bonds

This section covers the advances in the development of hydrosilylation of alkenes, allenes, and 1,3-dienes, which are also important reactions for the synthesis of alkylsilanes, vinylsilanes, allylsilanes, and homoallylsilanes [63].

4.2.1 Markovnikov and *Anti*-Markovnikov Hydrosilylation of Alkenes

When terminal alkenes are used, the hydrosilylation reactions can give α-adduct by Markovnikov-type addition and/or β-adduct by *anti*-Markovnikov-type addition (Scheme 4.7). Similar to the alkyne hydrosilylation, a variety of transition metal complexes show not only the efficient catalytic activity but also high regioselectivity.

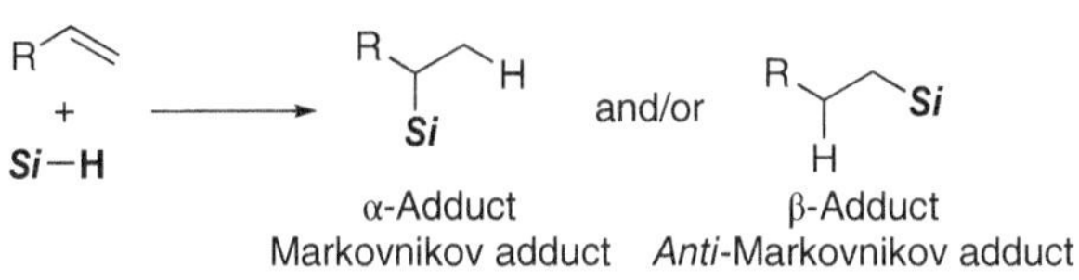

Scheme 4.7 Hydrosilylation of terminal alkenes.

Very recently, a review has summarized the progress of Markovnikov hydrosilylation of alkenes under different conditions [64]. A few new catalyst systems have been recently developed for the Markovnikov hydrosilylation [65].

For *anti*-Markovnikov hydrosilylation of alkenes, various transition metal complexes have been confirmed to be the efficient catalysts or precatalysts, including platinum [66], cobalt [67], iridium [68], rhodium [69], iron [70], nickel [71], ruthenium [72], and lanthanide–imine complexes [74] as well as other metal complexes of dimeric β-diketiminato magnesium hydride [74], cationic aluminum [75], and dicationic bismuth(III) Lewis acid [76].

In addition, the simple inorganic compounds of cobalt(0) and iron(0) isocyanides have also been used as catalysts for alkene hydrosilylation with hydrosiloxanes affording β-adducts [77].

Recently, nickel complexes have been applied for the hydrosilylation of alkenes showing excellent regioselectivity. For example, the reaction of air-stable nickel(II) bis(carboxylates) with α-diimine ligand generates in situ a highly active catalyst showing *anti*-Markovnikov selectivity for the hydrosilylation of alkenes with a variety of industrially relevant tertiary alkoxy- and siloxy-substituted silanes [78].

$[Ir(cod)Cl]_2$ was used in the *anti*-Markovnikov hydrosilylation of terminal alkenes with ethynylsilanes, which are substrates bearing more reactive alkynyl group (Scheme 4.8) [79]. It also catalyzed the hydrosilylation of unactivated terminal alkenes with excellent *anti*-Markovnikov regioselectivity [80]. In addition, iridium complexes of $[Ir(cod)X]_2$ (X = Cl, SPh) could catalyze the *anti*-Markovnikov hydrosilylation of various sulfur-containing alkenes with $(RO)_3SiH$; the present catalysis was applicable to the chemoselective hydrosilylation of thioacetate, which enabled the preparation of an industrially important silane coupling agent (Scheme 4.9) [81].

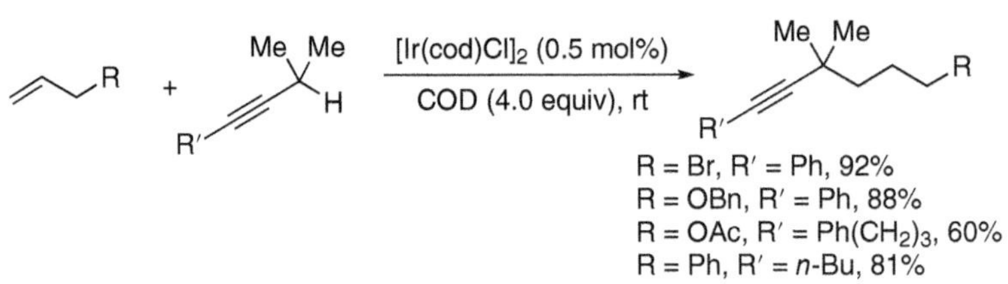

Scheme 4.8 Iridium-catalyzed *anti*-Markovnikov hydrosilylation of alkenes with ethynylsilanes.

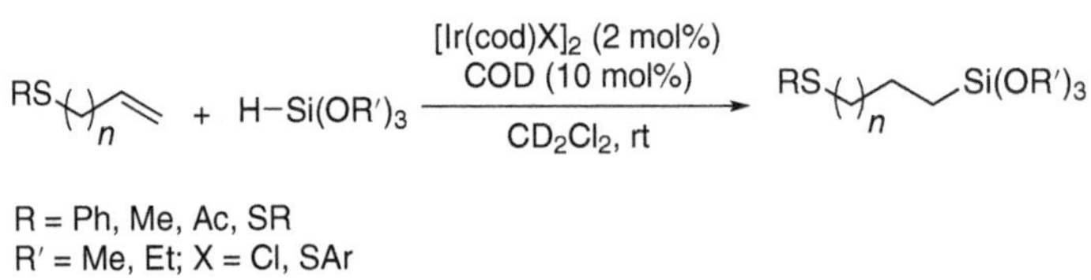

Scheme 4.9 Iridium-catalyzed *anti*-Markovnikov hydrosilylation of sulfur-containing alkenes with $(RO)_3SiH$.

The recoverable heterogeneous catalysts have also been well studied in the hydrosilylation of alkenes [82]. For example, MCM-41-supported mercapto platinum complex [83], silica-supported nitrogenous platinum complex [84], graphite oxide-supported Karstedt heterogeneous catalyst [85], Pt-Ni alloy [86], graphene-supported platinum [87], assembled Ni(II)-carboxylate metal-organic frameworks (MOFs) [88], mesoporous organosilica microspheres doped with Pt nanoparticles [89], cobalt ion-doped titanium oxide catalyst [90], and oxide-supported Pt-ligand single-site catalysts [91] were shown to be highly active for *anti*-Markovnikov hydrosilylation of terminal alkene to give β-adducts as the major products.

Cobalt complexes are efficient catalysts for the activation of Si—H bond and hydrosilylation of alkenes [92]. Recently, the regioselectivity of terminal alkene hydrosilylation can be switched from Markovnikov to *anti*-Markovnikov by using different ligands in the presence of cobalt complexes [93] or using well-defined low-valent cobalt catalysts $Co(PMe_3)_4$ and $Co(PMe_3)_3Cl$ [94].

The hydrosilylation of alkenes can also be achieved by using alkali metal and alkaline earth metal complexes as catalysts [95].

Under metal-free conditions, several procedures have been developed for hydrosilylation of alkenes. For example, Lewis acids were used as the catalysts in the hydrosilylation of cyclic alkenes and linear alkenes with trialkylsilanes giving the corresponding (trialkylsilyl)alkanes in fair to good yields [96].

Roberts and coworker developed the radical *anti*-Markovnikov hydrosilylation of terminal alkenes catalyzed by thiols [97], and then they reported the enantioselective radical hydrosilylation of prochiral alkenes using optically active thiols catalysts [98].

In addition, $B(C_6F_5)_3$ can be used as the catalyst for hydrosilylation [99] and hydrogermylation of alkenes[100]. Et_3B-induced radical hydrogermylation of terminal and internal alkenes has been achieved by using tri-2-furanylgermane to give the corresponding adducts in good to excellent yields [101].

In the cases of internal alkene used, the hydrosilylation can also efficiently take place with the use of the rare-earth complexes of Sm [102] and $Cu(OAc)_2$/ligand [103].

Similar to alkynes, in the CO atmosphere, the alkenes could also undergo the silylformylation reaction, and a tandem intramolecular silylformylation-allyl(crotyl) silylation was developed to prepare polyketide fragments in the presence of $Rh(acac)(CO)_2$ [104].

4.2.2 Hydrosilylation of Allenes

Hydrosilylation of allenes can afford either vinylsilanes or allylsilanes depending on the regioselectivity of addition reactions (Scheme 4.10).

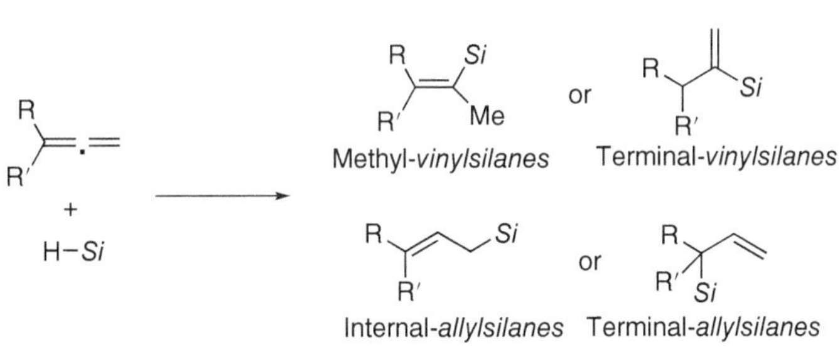

Scheme 4.10 Hydrosilylation of terminal allenes.

It was reported that with the use of stoichiometric amounts of $Co_2(CO)_8$, the hydrosilylation of sugar-substituted terminal allenes produced a mixture of two regio-isomeric vinylsilanes [105]. Without any ligand, Au/TiO_2 nanoparticles also showed the catalytic activity for regioselective hydrosilylation of terminal allenes to give terminal vinylsilanes as major products [106].

The hydrosilylation of allenes, in the presence of nickel(0) or palladium(0) complexes bearing *N*-heterocyclic carbine ligands generated in situ, occurs to give either terminal vinylsilanes or terminal allylsilanes, and the switchable regioselectivity is resulted from the use of the bulkiness of ligand [107]. The formation of terminal allylsilanes could be achieved by the hydrosilylation of 1,1-disubstituted allenes with a palladium catalyst in high yields and exceptional regioselectivity, and the use of chiral ligand afforded a chiral allylsilane in excellent enantioselectivity [108].

With the use of $AlCl_3$, a Lewis acid-catalyzed hydrosilylation of 1,3-disubstituted allenes afforded (*E*)-alkenylsilanes by silyl group added to sp C of allenes [109]. In addition, recently a regiodivergent and stereoselective hydrosilylation of 1,3-disubstituted allenes has been developed by using palladium and nickel complexes, respectively [110].

Interestingly, the regioselectivity can also be controlled by using different silanes with different numbers of substituted groups catalyzed by palladium complex [111].

4.2.3 Hydrosilylation of 1,3-dienes

1,3-Dienes undergo 1,4- or 1,2-hydrosilylation reactions affording allylsilanes and homoallylsilanes, and the regioselectivity depends on the catalyst and reaction conditions (Scheme 4.11).

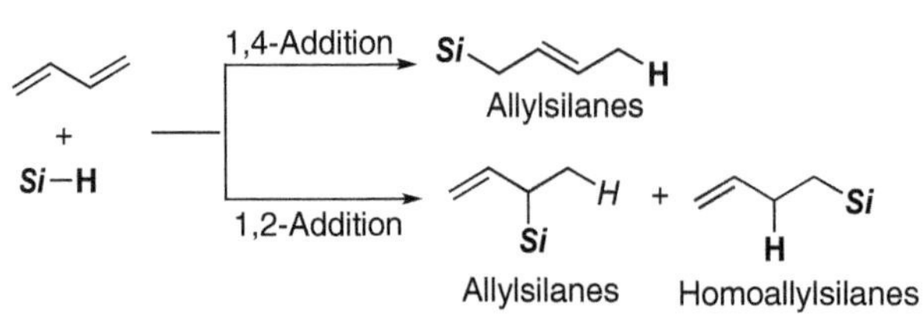

Scheme 4.11 Hydrosilylation of 1,3-dienes.

Recently, the regioselective 1,4-hydrosilylation has been achieved by using cobalt [112] and iron complexes [113], while 1,2-hydrosilylation occurs with the use of platinum [114] and cobalt complexes [115].

Recently, the regio- and enantioselective Markovnikov 1,2-hydrosilylation of 1,3-dienes catalyzed by chiral cobalt complexes has also been developed [116].

4.2.4 Asymmetric Hydrosilylation of Alkenes

Chiral silanes can be prepared by the asymmetric hydrosilylation of 1,1-/1,2-disubstituted or trisubstituted alkenes. In the case of terminal alkenes, as shown in Scheme 4.12, chiral silanes can be obtained from either Markovnikov addition or *anti*-Markovnikov addition.

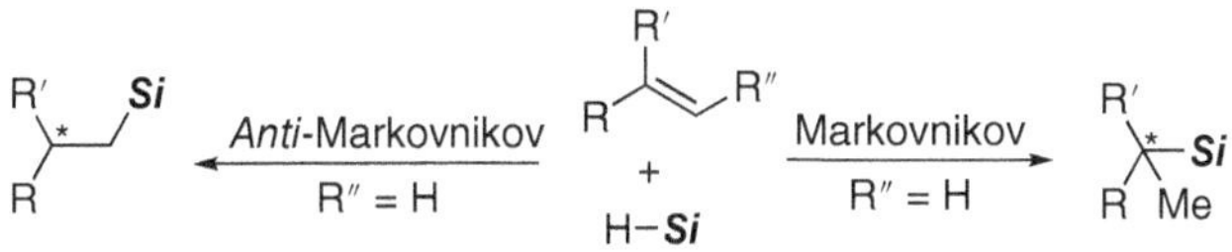

Scheme 4.12 Enantioselective hydrosilylation of alkenes.

With the use of chiral organolanthanides [117], palladium [118] and copper [119], the enantioselective Markovnikov hydrosilylation of terminal alkenes has been reported.

On the other hand, iron [120], platinum [121], rhodium complexes [122], and single-atom platinum [123] have also been used as the efficient catalysts with highly enantioselective *anti*-Markovnikov hydrosilylation of terminal alkenes.

With the use of $Pt(dba)_3$/chiral ligand, the enantioselective hydrosilylation of alkenylboronates occurs with high regio- and enantioselectivity to provide a convenient route to chiral geminal silylboronates, which are useful reagents in stereoselective synthesis (Scheme 4.13) [124].

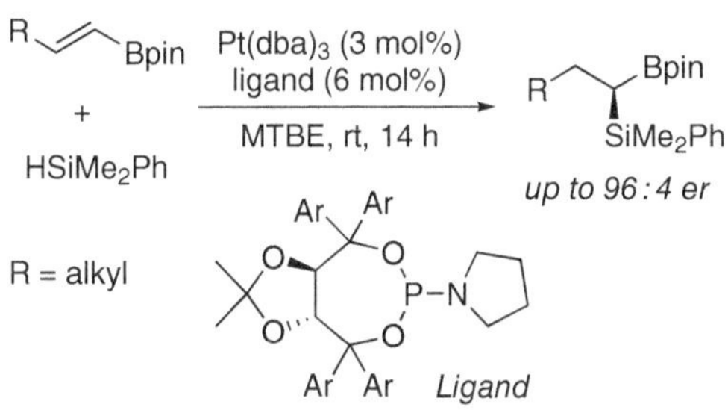

Scheme 4.13 $Pt(dba)_3$-catalyzed enantioselective hydrosilylation of alkenylboronates affording geminal silylboronates.

The enantioselective construction of silicon-stereogenic silanes could be also achieved by regio- and enantioselective *anti*-Markovnikov hydrosilylation of terminal alkenes with dihydrosilanes catalyzed by scandium complexes [125].

Very recently, the diastereoselective and enantioselective hydrosilylation of achiral cyclopropenes has been developed in the presence of chiral cobalt complex [126].

In addition, enantioselective hydrosilylation of alkenes also occurs under transition metal-free conditions [127].

4.3 Hydrostannation of Carbon–Carbon Double Bonds

The hydrostannation of alkenes and allenes can afford organostannanes, although relative few examples have been reported, which are useful intermediates in organic synthesis via activation of $C(sp^3)$—Sn and $C(sp^2)$—Sn bonds [128].

The hydrostannation of alkenes can give α-adduct by Markovnikov-type addition and/or β-adduct by *anti*-Markovnikov-type addition (Scheme 4.14). Similar to the alkyne hydrostannation, a variety of transition metal complexes show not only the efficient catalytic activity but also high regioselectivity.

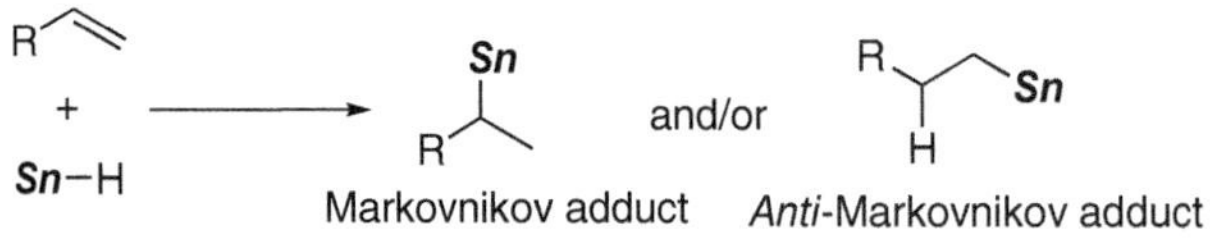

Scheme 4.14 Hydrostannation of alkenes.

For *anti*-Markovnikov hydrostannation of alkenes, a pioneering work was reported by Lautens and coworker with the use of heterogeneous palladium catalyst to catalyze a regioselective hydrostannation of a variety of alkenes, such as allylic alcohols/amines, acrolein diethyl acetal, vinylacetic acid, acrylonitrile, *N*-vinylphthalimide, and 1,2-dihydronaphthalene [129]. Scheme 4.15 shows several examples of the addition reactions of terminal alkenes affording *anti*-Markovnikov adducts in excellent regioselectivity.

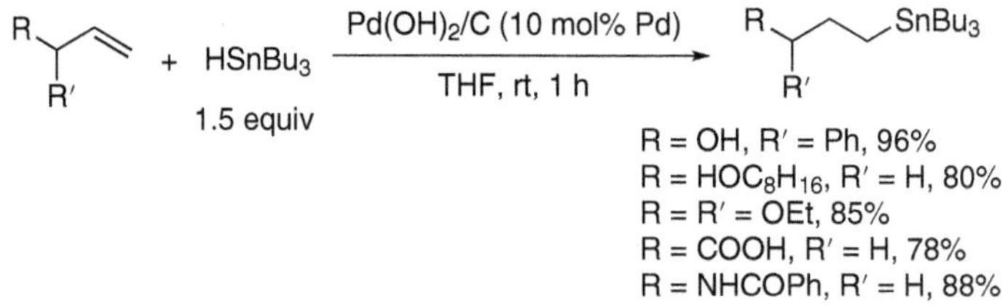

Scheme 4.15 Heterogeneous palladium-catalyzed regioselective hydrostannation of alkenes.

The photochemically activated [$Mo(CO)_6$] and [$Mo(CO)_4(\eta^4$-nbd)] are also shown to be good catalysts for the hydrostannation of norbornadiene (nbd) with Bu_3SnH and Ph_3SnH to give stannylnorbornene in excellent yields [130].

Hydrostannation of allenes can give either allylstannanes or vinyl stannanes depending on the regioselective addition of Sn—H bond across allene, and the outcome strongly depends on the conditions employed and the substrates (Scheme 4.16) [131].

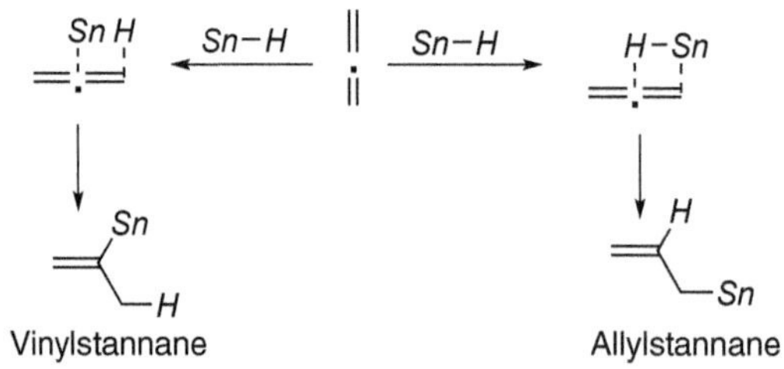

Scheme 4.16 Hydrostannation of allenes giving allylstannanes and vinylstannanes.

The palladium-catalyzed hydrostannation of a number of monosubstituted allenes, in most cases, afforded allylstannanes [132].

When Lewis acid, $B(C_6F_5)_3$ [132], and a heterogeneous $Pd(OH)_2/C$ [133] were used as the catalysts, the hydrostannation with Bu_3SnH selectively produces vinylstannane.

In addition, without any promoters and catalysts, in THF at room temperature, the hydrostannation of monosubstituted allenes with Bu_2SnIH (in situ generated by the redistribution between Bu_2SnI_2 and Bu_2SnH_2) took place to give vinylstannane as major products [134].

Moreover, allylstannanes could also be obtained by the highly regio- and stereoselective hydrostannation of 1,3-dienes in the presence of $Pd(PPh_3)_4$ [135].

The first $[Rh(cod)Cl]_2$-catalyzed enantioselective hydrostannation of cyclopropenes was reported by Gevorgyan's group in 2004; the addition reactions provide a straightforward approach to optically active cyclopropylstannanes [136].

4.4 Hydroamination of Carbon–Carbon Double Bonds

Direct N—H addition of amines and amides to alkenes is the most efficient approach toward the synthesis of higher substituted amines and amides. A few reviews have summarized the development of these types of addition reactions and their applications in the synthesis of amines, amides, and nitrogen-heterocyclic compounds [137]. A review on the hydroamination of nonactivated alkenes with ammonia has appeared very recently [138]. This section will mostly describe the typical and recently reported procedures.

4.4.1 Markovnikov Hydroamination of Alkenes

The regioselective Markovnikov intermolecular hydroamination (Scheme 4.17) can be achieved in the presence of a variety of transition metal complexes, such as palladium [139], platinum [140], copper [141], gold [142], rhodium complexes [143], and lanthanide salts [144].

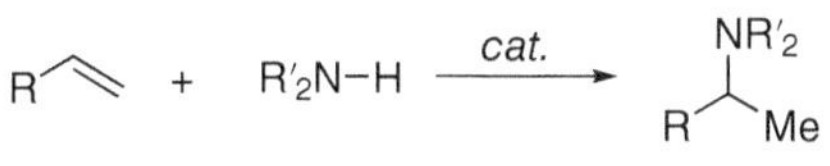

Scheme 4.17 Markovnikov hydroamination of alkenes.

The visible light-induced copper-catalyzed intermolecular Markovnikov hydroamination of alkenes has also been developed recently [145].

In the presence of CO, the hydroaminocarbonylation of alkenes can produce branched and linear amides via Markovnikov and *anti*-Markovnikov hydroaminocarbonylation of terminal alkenes (Scheme 4.18).

Beller's group first developed the $PdCl_2$-catalyzed formation of branched amides as the major adducts via Markovnikov-type hydroaminocarbonylation

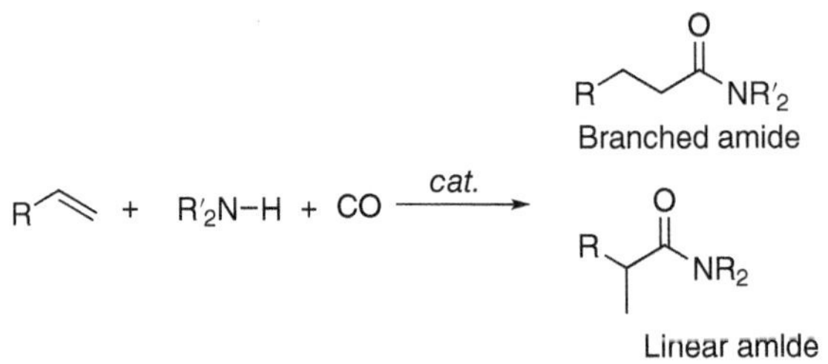

Scheme 4.18 Hydroaminocarbonylation of terminal alkenes.

with aliphatic amines by using a specific 2-phosphino-substituted pyrrole ligand [146]. The early work from the same group, however, developed a rhodium(I)-catalyzed hydroaminocarbonylation of alkenes with aliphatic amines [147], palladium(II)-catalyzed hydroamidocarbonylation of alkenes [148], and 1,3-dienes [149] to imides with good regioselectivity for linear products.

Palladium(0) complexes, like $Pd(t\text{-}Bu_3P)_2$ and $Pd(PPh_3)_4$, could also catalyze the hydroaminocarbonylation of terminal alkenes with ammonium chloride to give branched primary amides, and the use of PdI_2/phosphine as catalyst system resulted in the formation of linear amide as the major products [150].

In addition, a highly ligand-controlled regioselective palladium-catalyzed hydroaminocarbonylation of aromatic alkenes with aminophenols [151] and palladium-catalyzed highly regioselective hydroaminocarbonylation of aromatic alkenes to branched amides [152] were then developed.

Besides the transition metal complexes, calcium(II) triflimide, a main group salt, has been found to show the high catalytic activity for the hydroamination of alkenes. Leboeuf and coworker have recently disclosed that a combination of a $Ca(NTf_2)_2$ and $n\text{-}Bu_4NPF_6$ in hexafluoroisopropanol (HFIP) is a highly efficient promoter system for the intra- and intermolecular hydroamination of unactivated alkenes, and a vast array of functional groups is tolerated at the nitrogen and alkene moieties [153]. As shown in Scheme 4.19, the intermolecular hydroamination of styrene derivatives with *N*-tosylamine derivatives produces the Markovnikov adducts, and the intramolecular hydroamination of amidoalkenes affords the corresponding nitrogen-containing cyclic compounds in good to high yields. The use of HFIP as solvent and NTf_2 as ligand has been proved to be crucial for this transformation.

The hydroamination of unactivated alkenes with sulfonamides has also been achieved with the use of $FeCl_3/AgSbF_6$ as the catalyst system [4].

An iridium(III)-catalyzed hydroamination of alkenes with *N*-alkyl amides via transient amidyl radical intermediates formed by proton-coupled electron transfer activation of the strong N—H bonds in *N*-alkyl amides has been developed recently [154].

In addition, the catalyst system of $[(PhO)_3P]AuCl/AgOTf$ shows the high catalytic activity not only for the Markovnikov hydroamination of vinylarenes but also for the hydroamination of internal alkenes and 1,3-dienes [155].

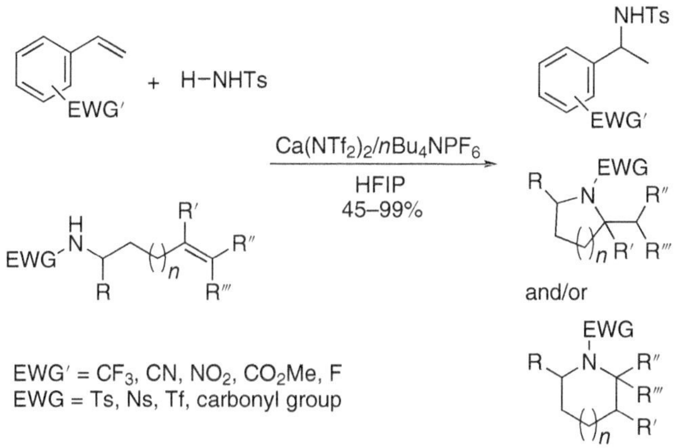

Scheme 4.19 Calcium(II)-catalyzed inter- and intramolecular hydroamination of alkenes.

4.4.2 *Anti*-Markovnikov Hydroamination of Alkenes

Transition metal complexes of ruthenium [156], rhodium [157], iridium [158], and palladium [159] have been reported to be the efficient catalysts to catalyze the *anti*-Markovnikov hydroamination of terminal alkenes with high regioselectivity (Scheme 4.20). Organolanthanide-catalyzed regioselective *anti*-Markovnikov hydroamination of alkenes and the mechanism have also been reported [160].

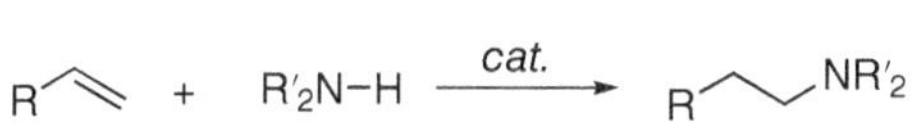

Scheme 4.20 *Anti*-Markovnikov hydroamination of alkenes.

As the representative progress on the development of *anti*-Markovnikov hydroamination of alkenes, recently, Wang and Xiao have reported an iron-catalyzed *anti*-Markovnikov hydroamination and hydroamination of allylic alcohols affording γ-amino and γ-amido alcohols with high regioselectivity (Scheme 4.21) [161]. This catalytic system with the use of the pincer Fe—PNP complex and a weak base in nonpolar solvent produces exclusive *anti*-Markovnikov adducts with a broad substrate scope and good functional group tolerance. When chiral substrates are used, the stereochemistry and enantiomeric excess are retained.

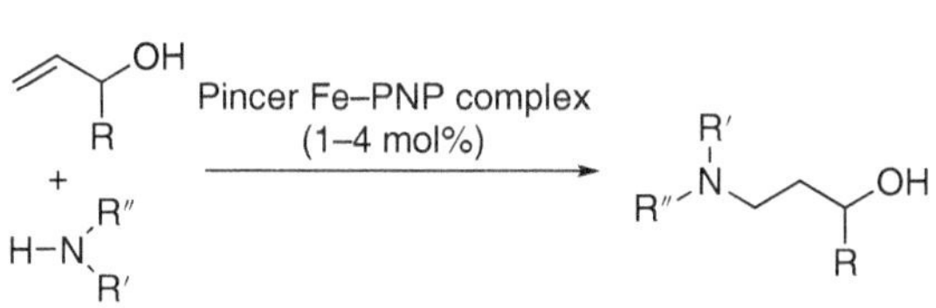

Scheme 4.21 Iron-catalyzed *anti*-Markovnikov hydroamination of allylic alcohols.

In addition, the heavier group two complexes $[M\{N(SiMe_3)_2\}_2]_2$ and $[M\{CH(SiMe_3)_2\}_2(THF)_2]$ (M = Ca, Sr) are shown to be effective precatalysts for the intermolecular *anti*-Markovnikov hydroamination of vinylarenes under mild conditions [162].

Recently, HOTf, a Brønsted acid, has been found to be the efficient catalyst in the intermolecular hydroamination of electron-poor alkenes with anilines under mild conditions giving *anti*-Markovnikov adducts [163].

Moreover, without catalyst, under solvent-free conditions, the addition reactions of acrylamides with secondary amines at 120 °C occur to give 3-aminopropionamides in good to excellent yields [164]. A computational study on the non-catalytic hydroamination of alkenes has appeared [165].

Chung and coworker first reported the heterogeneous catalytic formation of *N*-phenyl alkyl amides in reasonable to high yields from the *anti*-Markovnikov hydroaminocarbonylation of aliphatic alkenes, anilines, and carbon monoxide with the use of cobalt on charcoal as catalyst; the reaction occurred with high chemoselectivity (Scheme 4.22) [166]. It has been found that under the optimized condition, other cobalt compounds such as $Co_2(CO)_8$, $Co_4(CO)_{12}$, $CoCl(PPh_3)_3$, and $CoBr_2$ showed low catalytic activity, and internal alkenes and cycloalkenes also underwent the reaction albeit in relatively low yields.

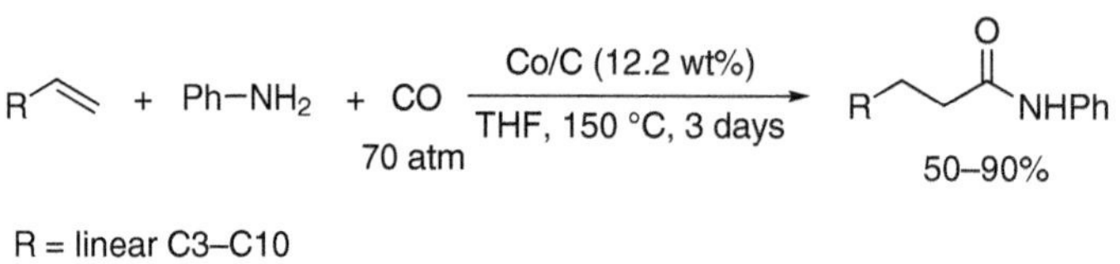

Scheme 4.22 Synthesis of *N*-phenyl alkyl amides from alkene and aniline.

The recoverable heterogeneous catalysts of zeolites was then also used for the hydroamination of alkenes [167].

4.4.3 Hydroamination of Allenes and 1,3-dienes

The intermolecular hydroamination of allenes can give enamines, imines, or allylamines depending on the use of catalysts and the structures of substrates (Scheme 4.23). The use of early transition metal catalysts such as zirconium and titanium complexes usually leads to the formation of imines [168]. Palladium [169], gold [170], and copper complexes [171]-catalyzed intermolecular hydroamination of allenes mostly afford allylic amines.

Intramolecular hydroamination of aminoallenes can also provide a conveniently available entry to nitrogen-containing heterocycles, mostly bearing a vinyl substituent [172]. Recently, $[Cu(NCMe)_4]PF_6$-catalyzed hydroamination of *N*-allenylazoles at room temperature has been developed, providing the simple synthetic method of *N*-vinylazoles bearing an amino substituent at the allylic position with excellent regio- and *E*-stereoselectivity [173].

In addition, in the presence of $Pd(PPh_3)_4$, 1,2-dienes can undergo the amidation with amines and CO under mild conditions to afford enamide derivatives in good

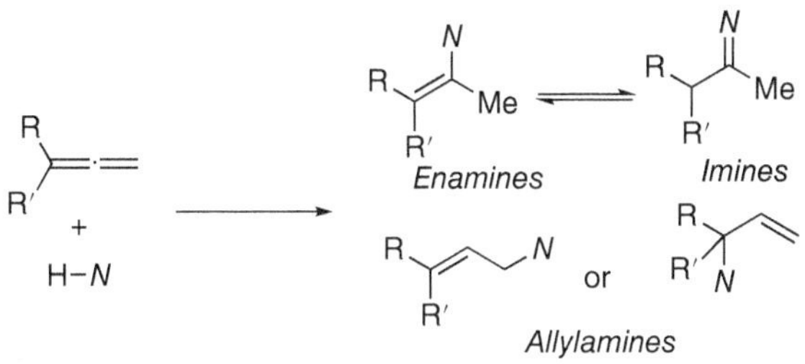

Scheme 4.23 Intermolecular hydroamination of allenes.

to high yields with excellent regioselectivity (Scheme 4.24) [174]. The oxygen nucleophiles such as phenols and oximes also undergo the similar transformation to give methacrylamides or methacrylate esters and their derivatives in good to excellent yield. The formation of hydridopalladium(II) species in situ is proposed as the key intermediate by oxidative addition of palladium(0) to acetic acid or acidic hydroxyl substrates.

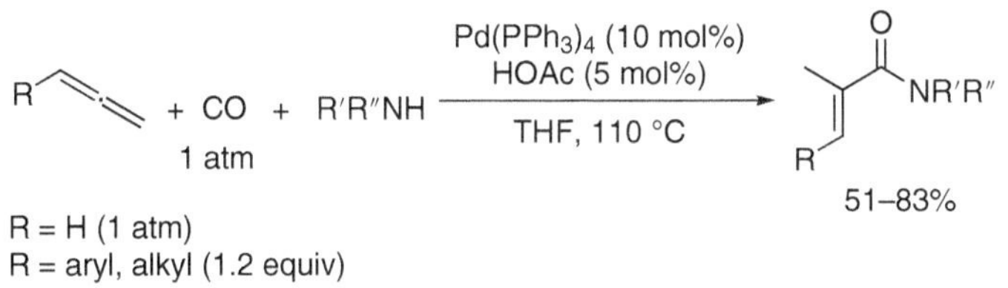

Scheme 4.24 Palladium-catalyzed amidation of allenes with amines and CO.

On the other hand, intermolecular hydroamination of 1,3-dienes with 1,2-Markovnikov [175] and 1,4-addition regioselectivity [176] could afford allylic amines, which have been greatly achieved in the presence of a variety of transition metal complexes (Scheme 4.25a) [177].

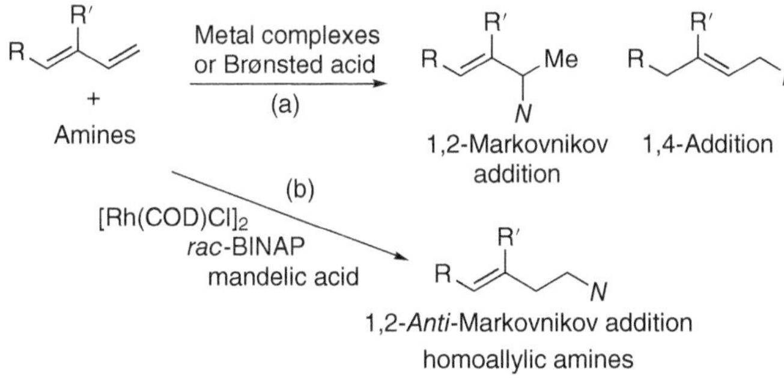

Scheme 4.25 Intermolecular hydroamination of 1,3-dienes.

The rhodium-catalyzed *anti*-Markovnikov hydroamination of 1,3-dienes generating homoallylic amines has also been reported (Scheme 4.25b) [178].

$Ni(cod)_2$/dppf/acid showed highly catalytic activity for the hydroamination of 1,3-dienes, focusing on cyclic 1,3-dienes with alkylamines to form allylic amines [179].

The enantioselective addition of 2-pyridones to terminal allenes has been reported with the use of $[Rh(cod)Cl]_2$/chiral phosphine as catalyst system, and the N—H addition occurs with high chemo- and regioselectivity to give branched *N*-allyl 2-pyridones [180]. In addition, the highly enantioselective formation of chiral allylamines has been developed through the intermolecular hydroamination of 1,3-disubstituted allenes catalyzed by chiral phosphinothioureas [181] and by the chirality transfer in the presence of $AuBr_3$ [182].

The regioselective 1,4-hydroamination [183] and 1,4-hydroaminocarbonylation [184] of 1,3-dienes represent the efficient and atom economic procedures for the formation of allylic amines and β,γ-unsaturated amides, respectively, which have been developed in the presence of palladium complexes.

$Bi(OTf)_3/Cu(MeCN)_4PF_6$ could efficiently promote the intermolecular hydroamination of 1,3-dienes with various carbamates, sulfonamides, and carboxamides to afford allylic amines in good yield in 1,4-dioxane [185].

The lanthanoid-catalyzed intramolecular hydroamination of 1,3-dienes could afford alkenylpyrrolidine and -piperidine derivatives in varying ratios of regio- and *E*/*Z*-isomers, and, in most cases, the ratio depends on the organolanthanoid used [186].

The asymmetric intramolecular hydroaminations of 1,3-dienes catalyzed by a dithiophosphoric acid [187] and other Brønsted acids [188] have also been investigated.

4.4.4 Asymmetric Hydroamination of Alkenes

Enantioselective intermolecular hydroamination of alkenes and intramolecular hydroamination of aminoalkenes have been well applied in the synthesis of chiral amines [189].

As a pioneering work, Hartwig's group reported an efficient, palladium-catalyzed intermolecular Markovnikov hydroamination of vinylarenes with aromatic amines to give sec-phenethylamines in the presence of acid cocatalyst; the use of chiral palladium catalyst resulted in enantioselective hydroamination [190]. For example, the reaction of trifluoromethylstyrene with aniline catalyzed by $[((R)\text{-BINAP})Pd(OSO_2CF_3)_2]$ at 25 °C gave the adduct in 81% yield and 81% enantioselectivity (Scheme 4.26). They have also developed iridium/chiral ligand-catalyzed addition of the N—H bonds of indoles to simple terminal alkenes, and this strategy for indole alkylation occurs exclusively at nitrogen with Markovnikov selectivity [191].

Copper-catalyzed intermolecular asymmetric hydroamination of unactivated internal alkenes aliphatic amines was also developed by Buchwald's group in 2015 [192].

The first rhodium-catalyzed asymmetric intramolecular hydroamination of aminoalkenes affording chiral 2-methylpyrrolidines with high enantioselectivity

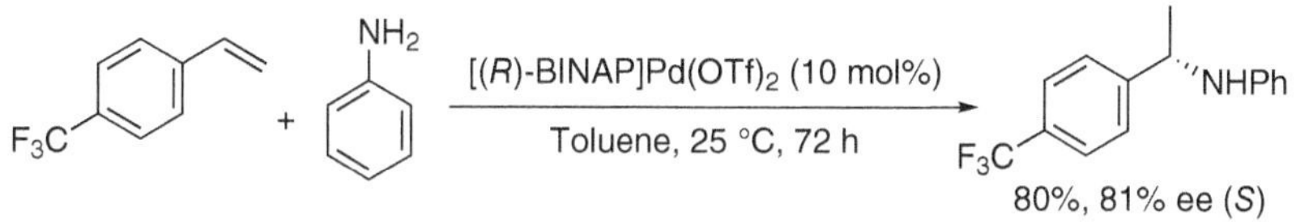

Scheme 4.26 Palladium-catalyzed enantioselective intermolecular hydroamination of vinylarenes with aromatic amines.

was developed by using chiral dialkylbiaryl phosphine ligands (Scheme 4.27) [193]. The rhodium-catalyzed intermolecular asymmetric hydroamination of allyl amines affords 1,2-diamines in good yields with excellent enantioselectivity by Markovnikov regioselectivity [194].

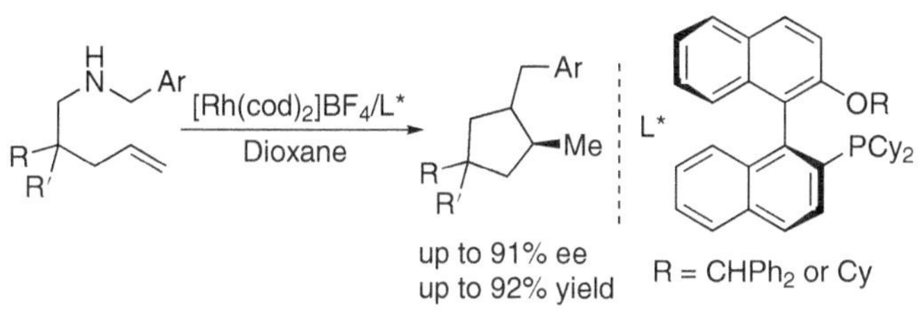

Scheme 4.27 Rhodium-catalyzed asymmetric intramolecular hydroamination of aminoalkenes.

The highly enantioselective chiral Brønsted acids have also been used as efficient catalysts for the intramolecular hydroamination of aminoalkenes to construct a series of chiral (spirocyclic)pyrrolidines with an α-tetrasubstituted carbon stereocenter [195].

In addition, the chiral rare-earth metal complexes have been well applied in the asymmetric hydroamination [196].

Very recently, Malcolmson and coworker have developed an interesting palladium-catalyzed selective 1,4-hydroamination of a wide range of mono- and disubstituted enynes affording chiral allenes with pendant allylic amines [197].

The enantioselective intermolecular hydroamination of 1,3-dienes catalyzed by transition metal complexes and organocompounds has also been developed [198].

4.4.5 Nitrogen Heterocycles from Intramolecular Hydroamination of Alkenes

The intramolecular hydroamination of alkenes has been developed as one of the efficient synthetic routes for five- and six-membered *N*-heterocyclic compounds in the presence of transition metal complexes, such as palladium [199], copper [200], titanium [201], cobalt [202], and zinc [203]. Note that most of the reported procedures give the Markovnikov adducts, namely, α-methyl-substituted *N*-heterocyclic compounds.

For example, Che and coworker studied the $(PPh_3)AuCl/AgOTf$-catalyzed intramolecular hydroamination of terminal alkenes with Markovnikov selectivity affording five- and six-membered *N*-heterocyclic compounds, and the use of microwave radiation as a heat source could shorten the reaction time (Scheme 4.28) [204]. A mixture of (phosphine)AuCl and AgOTf could also catalyze the Markovnikov intramolecular hydroamination of unactivated alkenes with primary and secondary ammonium salts [205].

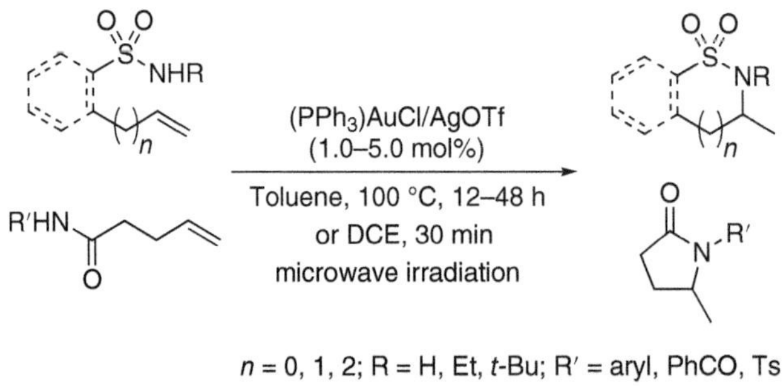

Scheme 4.28 Gold(I)-catalyzed intramolecular hydroamination of terminal alkenes.

The intramolecular hydroaminations of 1-amino-2,2-di-substituted-4-pentenes and 1-amino-2,2-di-substituted-5-hexenes access to 2-methyl-4,4-di-substituted pyrrolidines and 2-methyl-5,5-di-substituted piperidines (Scheme 4.29) have been well investigated with the use of ZnI_2/8-hydroxyquinoline [206], rare-earth metal diisopropylamides [207], calcium [208], gold [209], iridium [210], and group IV metal complexes [211] as the efficient catalysts. The $Bi(OTf)_3$-mediated intramolecular hydroamination of 2-aminostilbenes afforded 2-arylindolines (Scheme 4.29) [212].

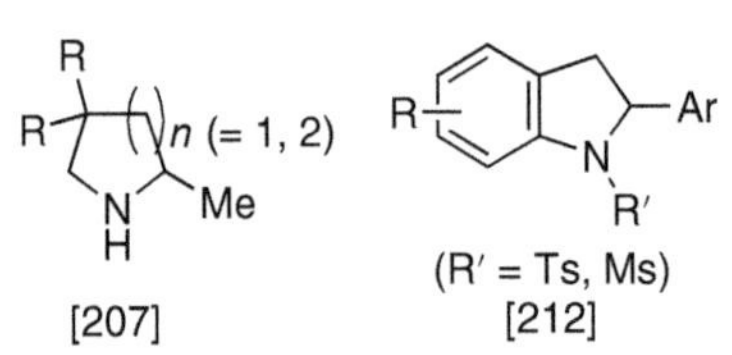

Scheme 4.29 Nitrogen-heterocyclic compounds from intramolecular hydroamination of aminoalkenes.

On the other hand, the catalytic systems for intramolecular *anti*-Markovnikov hydroamination of alkenes are few (Scheme 4.30). The first example of intramolecular *anti*-Markovnikov hydroamination of styrenes was reported by Hartwig's group in 2006 with the use of rhodium as catalysts [213], which was then developed by using catalytic amounts of thiophenol and an organic photocatalyst promoted by visible light [214].

The $PdCl_2(PPh_3)_2$-catalyzed intramolecular hydroaminocarbonylation of 2-vinylbenzylamines in the absence of acidic or any other additives could afford a variety of six-membered lactams in good to excellent yields with high regioselectivity (Scheme 4.31) [215].

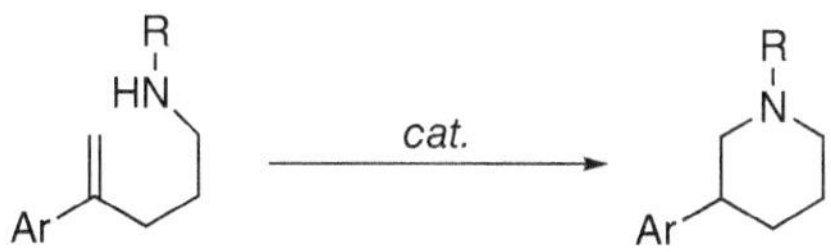

Scheme 4.30 Intramolecular *anti*-Markovnikov hydroamination of alkenes.

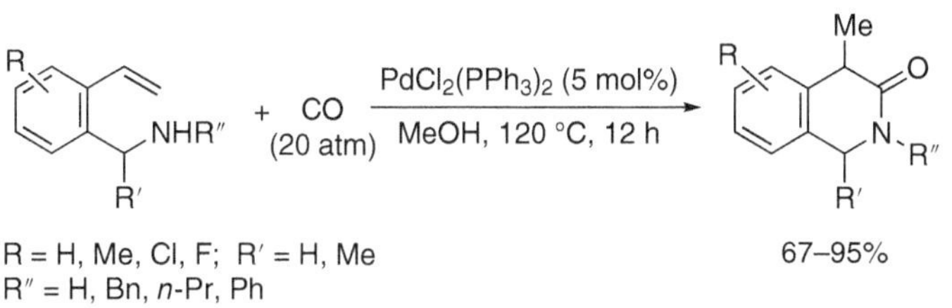

Scheme 4.31 Palladium-catalyzed intramolecular hydroaminocarbonylation of 2-vinylbenzylamines.

4.5 Hydrophosphination of Alkenes and Related P^V(O)–H Addition

Hydrophosphination and asymmetric hydrophosphination of alkenes can synthesize tertiary phosphines, which are an important class of phosphorus compounds widely applied as ligands for transition metal complexes and reagents in organic synthesis (Scheme 4.32). Therefore, hydrophosphination and asymmetric hydrophosphination of alkenes have been well developed, which are easily achieved under mild thermal activation, in the presence of radical initiators, in the acid- or base-catalyzed process, or in the presence of transition metal catalysts, and several reviews have appeared [216].

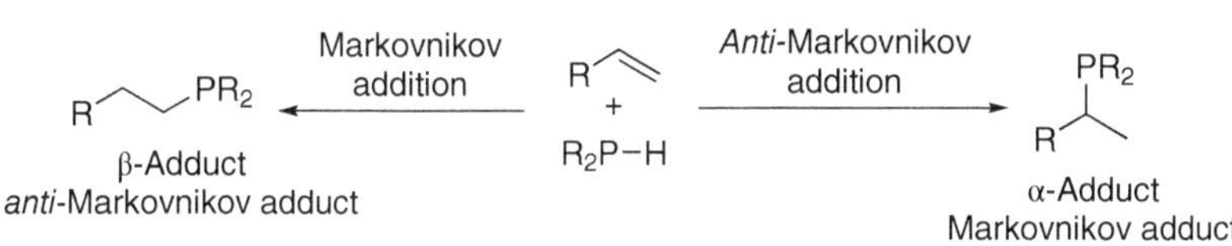

Scheme 4.32 Alkene hydrophosphination giving Markovnikov and *anti*-Markovnikov adducts.

The first examples of intermolecular hydrophosphination of arylalkenes catalyzed by Ni[P(OEt)$_3$]$_4$ and Pd(MeCN)$_2$Cl$_2$ complexes were reported by Beletskaya' group [217]. The addition reactions of Ph$_2$PH with electron-rich and -poor arylalkenes, such as styrene, 4-vinylpyridine, 2-vinylpyridine, 4-methoxystyrene, 2-methoxystyrene, and 5-vinyl-2-methylpyridine, proceed with good to high yields and excellent regioselectivity to give only *anti*-Markovnikov adducts (Scheme 4.33). Other nickel complexes have also shown catalytic activity for these addition reactions [218].

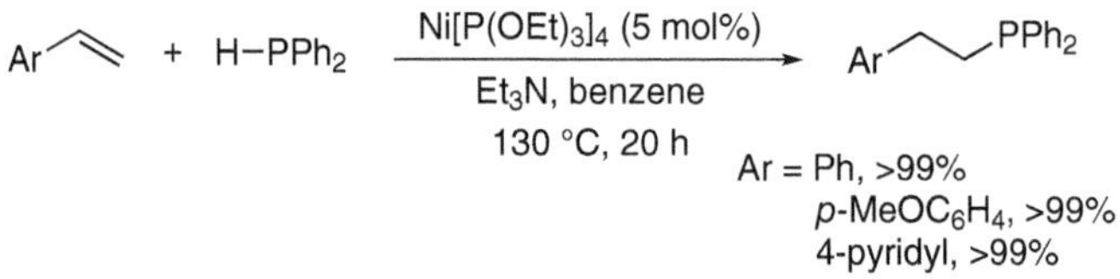

Scheme 4.33 Nickel-catalyzed hydrophosphination of aryl alkenes with Ph_2PH affording *anti*-Markovnikov adducts.

Copper [219], europium [220], platinum [221], iron [222], nickel [223], ytterbium [224], and ruthenium complexes [225] also show catalytic activity for *anti*-Markovnikov hydrophosphination of alkenes with high regioselectivity.

It was found that the simple tin derivatives, $Cp^*_2SnCl_2$ and Ph_2SnCl_2, could also catalyze the hydrophosphination of alkenes with diphenylphosphine to give adducts under an H_2 atmosphere [226]. In addition, β-diketiminato calcium complexes [227] were also used as catalysts to catalyze intermolecular hydrophosphination of alkenes.

In the presence of zirconium complex, the highly selective *anti*-Markovnikov hydrophosphination of terminal alkenes and dienes with primary phosphines afforded either the secondary or the tertiary phosphine (via double hydrophosphination) products, depending on reaction conditions (Scheme 4.34) [228]. A chiral, air-stable primary phosphine, (*R*)-[2′-methoxy(1,1′-binapthalen)-2-yl]phosphine could also proceed the addition reactions catalyzed by the same catalyst [229]. Other zirconium complexes were then prepared and used as catalysts in the same transformation [230].

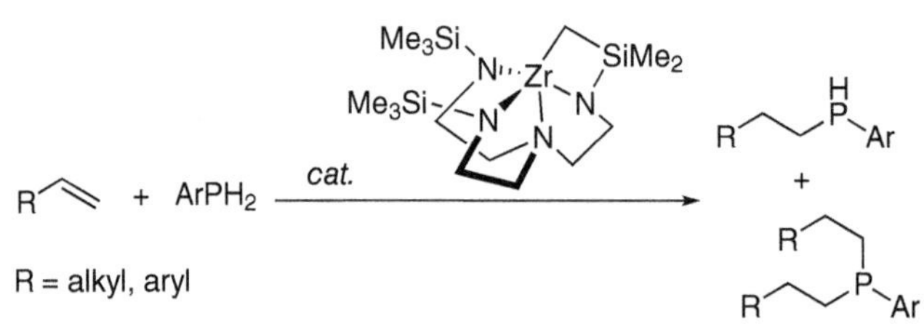

Scheme 4.34 Zirconium-catalyzed hydrophosphination of alkenes with primary phosphines.

Very interesting, a few examples on the solvent- and catalyst-free hydrophosphination of alkenes have also been reported [231].

Very interestingly, the regioselectivity of hydrophosphination of styrenes can be switched to give α- or β-adducts by using simple $FeCl_3$ or $FeCl_2$ salts as promoters in MeCN [232].

The alkyldiarylphosphines and aryldialkylphosphines could also be prepared by a regioselective hydrophosphination of alkenes with secondary phosphine boranes at room temperature or 60 °C or under microwave conditions, without using any catalyst due to the strong P—H activation induced by the borane [233].

Recently, photocatalyses have also been developed for the hydrophosphination of activated and unactivated alkenes [234].

In addition, Ti [235], complexes-catalyzed hydrophosphination of 1,3-dienes, cyclic dienes, and trienes have also been reported.

In the case of $P^V(O)$—H, addition reactions to alkenes, palladium [236], manganese [237], and nickel complexes [238] show efficient catalytic activity with dialkyl *H*-phosphonates to give *anti*-Markovnikov adducts as the major products. Visible light photocatalysis has also become the efficient procedure for the addition reactions [239].

Recently, the direct radical oxyphosphorylation of styrenes with *H*-diarylphosphine oxides under air atmosphere mediated by $Mn(OAc)_3$ in MeOH provides an efficient route to β-ketophosphine oxides [240], and the similar reaction with the use of alkenes and dioxygen can be achieved in the presence of CuCN [241] or catalyzed by trialkyl phosphines in the case of electron-deficient alkenes used [242].

Zhu and coworker have also developed a metal-free and visible light-promoted aerobic oxyphosphorylation of alkenes with high regioselectivity approach to β-ketophosphine oxides (Scheme 4.35) [243].

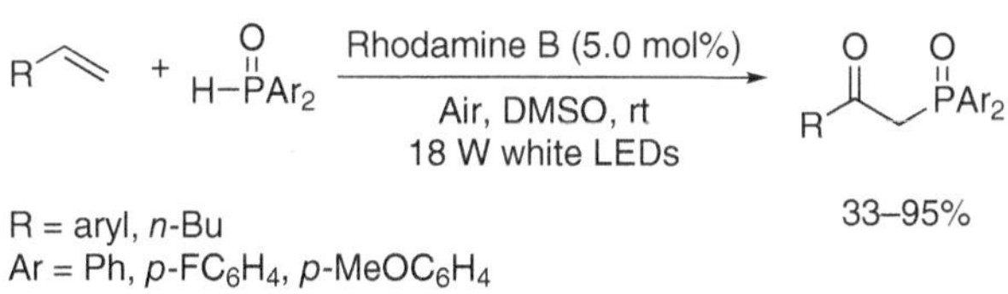

Scheme 4.35 Synthesis of β-ketophosphine oxides by visible light-promoted aerobic oxyphosphorylation of alkenes.

1,2-Bisphosphorylethanes are very useful and interesting compounds, because they can be easily reduced to the corresponding trivalent 1,2-bisphosphinoethanes, which are important ligands for transition metal complexes. One of the efficient synthetic methods is the regioselective double addition of H—P(O)R_2 to alkynes as described in Chapter 3, Section 3.5.2. The other way for the synthesis is the *anti*-Markovnikov addition of H—P(O)R_2 to vinylphosphoryl compounds (Scheme 4.36).

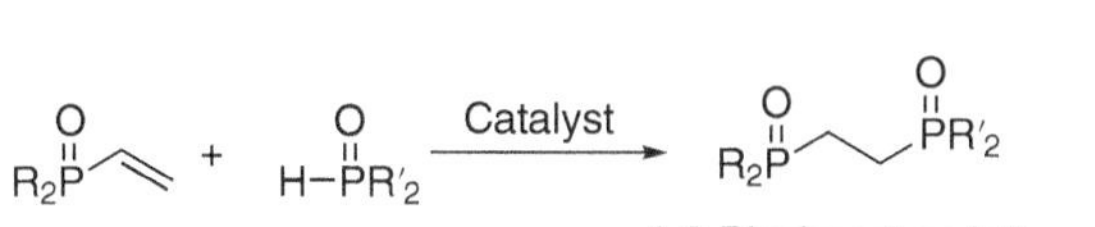

Scheme 4.36 1,2-Bisphosphoryl compounds from addition of *H*-phosphine oxide to vinylphosphoryl compounds.

Han's group first developed the stereospecific radical or base-catalyzed addition of menthyl phenylphosphinate to vinylphosphonate affording optical 1,2-bisphosphorylethanes [244] and then developed a PMe_3-catalyzed addition of dimethyl phosphite (R′ = OMe) to dimethyl vinylphosphonate (R = OMe) [245] and optically active menthylO(Ph)P(O)H to electron-deficient alkenes [246] in THF at room temperature to afford the corresponding *anti*-Markovnikov adducts in high yields. The same group also reported an air-induced *anti*-Markovnikov addition of secondary phosphine oxides and H-phosphinates to alkenes [247].

The radical [248] or base-mediated [249] addition of $P^V(O)$—H to terminal and internal alkenes has also been studied by several other groups.

When activated internal alkenes were used, the addition of $P^V(O)$—H could occur without catalyst and additive under microwave irradiation [250] or at room temperature [251].

In addition, allylic *H*-phosphinates could be synthesized by the reactions of hypophosphorous compounds with allenes, dienes, and allylic electrophiles [252].

On the other hand, asymmetric hydrophosphination of alkenes (addition of P^{III}—H to alkenes) is the most atomically economical and straightforward approach to the construction of chiral phosphine ligands.

A pioneering work for the enantioselective synthesis of chiral phosphines by transition metal-catalyzed asymmetric hydrophosphination of activated alkenes such as acrylonitrile and acrylate esters in the presence of platinum complex was reported by Glueck and coworker in 2000 (Scheme 4.37) [253].

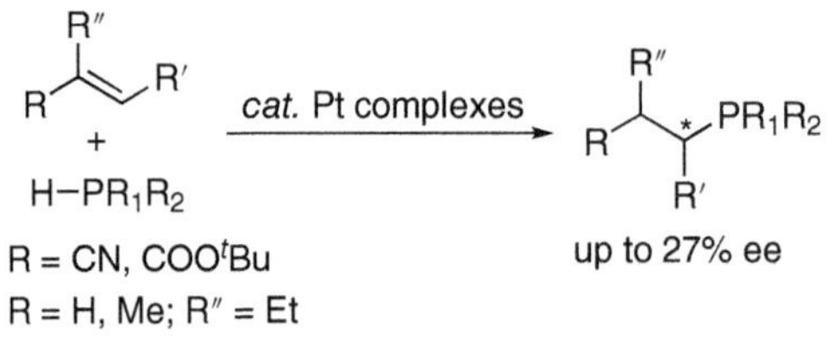

Scheme 4.37 Asymmetric hydrophosphination of electron-deficient alkenes.

A highly enantioselective asymmetric hydrophosphination of methacrylonitrile was then developed by Togni's group with the use of nickel complexes as catalysts in 2004, and the ee value was up to 94% [254].

In the cases of α,β-unsaturated carbonyl compounds used, the regioselective 1,4-hydrophosphination affording β-chiral phosphines has been extensively investigated in the presence of palladium complexes [255] and organocatalysts [256] (Scheme 4.38).

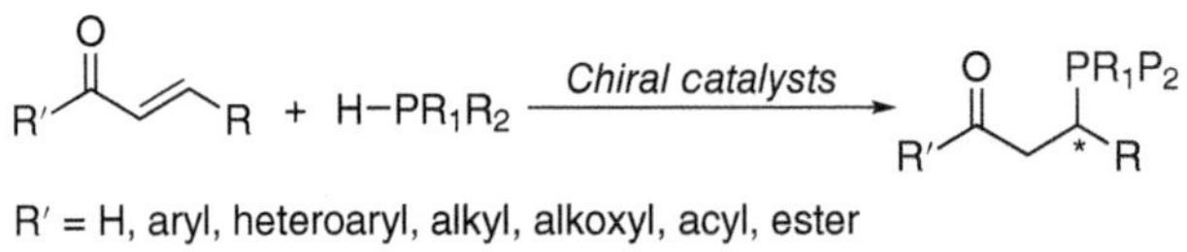

Scheme 4.38 Regioselective 1,4-hydrophosphination of α,β-unsaturated carbonyl compounds.

A palladacycle-catalyzed diastereo- and enantioselective stepwise double hydrophosphination of bis(enones) with $PhPH_2$ has been developed to construct chiral tertiary bulky *P*-heterocycles in one-pot manner with high yields (Scheme 4.39) [257].

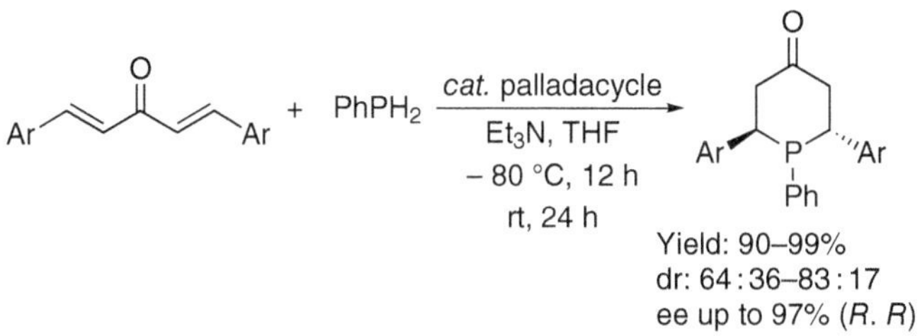

Scheme 4.39 Palladacycle-catalyzed double hydrophosphination of bis(enones) forming *P*-heterocycles.

Palladium-catalyzed asymmetric addition of diarylphosphines to nitroalkenes [258], to α,β-unsaturated sulfonic esters [259], to 3-methyl-4-nitro-5-alkenylisoxazoles [260], to electron-deficient dienes [261], and to (4-aryl-1,3-butadienylidene) bis(phosphonates) via 1,6-addition [262] was also reported by Duan and other groups.

Recently, palladium(II)-catalyzed asymmetric hydrophosphination of oxa- and azabicyclic alkenes with diarylphosphines offering a direct and atom efficient access to chiral tertiary phosphines has also been developed [263].

In addition, the asymmetric synthesis of 1,2-diphosphines by hydrophosphination of functionalized allenes has also been well developed [264].

Han and coworker also developed the asymmetric addition of cyclic hydrogen phosphonates to norbornenes catalyzed by palladium complexes [265].

4.6 Hydrothiolation of Carbon–Carbon Double Bonds

Hydrothiolation of alkenes is the most important reaction for the synthesis of thioethers with the formation of C—S bonds, which are versatile applications in organic synthesis through further transformation of C—S bonds [266]

Compared with the hydroboration and hydrosilylation of alkenes, the procedures for hydrothiolation are relatively few. The addition can also give α- or β-adducts by either Markovnikov or *anti*-Markovnikov-type addition (Scheme 4.40). In this section, only a few representative procedures are introduced.

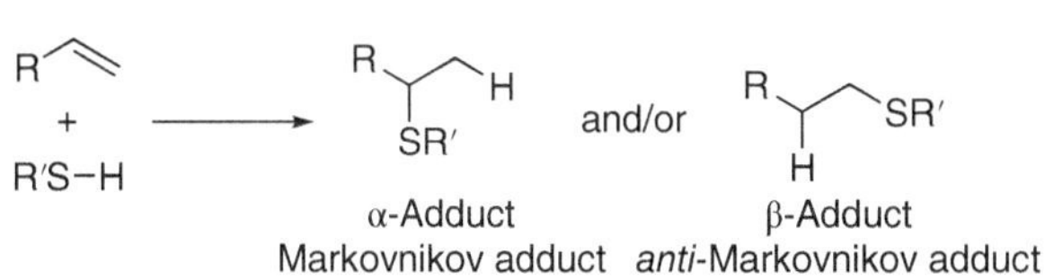

Scheme 4.40 Hydrothiolation of terminal alkenes.

Markovnikov hydrothiolation of alkenes has been well established, and various thioethers are conventionally available by this transformation, which are often promoted by transition metal catalysts, such as palladium [267], copper [268], palladium/copper [269], ZnI_2/Brønsted acid [270], Mont K 10 clay [271], and organic compounds [272].

Recently, a review on the *anti*-Markovnikov-selective hydrothiolation of alkenes has appeared, and all the pioneering work and recent procedures have been summarized [273].

Very recently, an interesting and novel formic acid-assisted rapid and efficient *anti*-Markovnikov hydrothiolation of styrenes has been reported to give adducts in good to excellent yield [274]. Other interesting metal-free, organic compounds-promoted hydrothiolation of alkenes [275] or without catalysts and additives [276] has been reported.

PPh_3AuNTf_2 can catalyze *anti*-Markovnikov hydrothiolation of unactivated alkenes with benzenethiols and aliphatic thiols to give adducts in good yields with high regioselective manner [277]. $[Au_2(NTf_2)_2(\mu\text{-dppf})]$-promoted *anti*-Markovnikov hydrothiolation of alkenes with methanethiol generated ex situ was suitable for a broad range of alkenes affording the corresponding hydrothiolated adduct in good to excellent yields [278].

In addition, the recoverable solid catalyst of $ZnIn_2S_4$ under visible light irradiation has also been applied in the hydrothiolation of alkenes [279].

On the other hand, hydrothiolation of terminal allenes can be achieved in the presence of transition metal complexes or radical initiators affording four types of regioisomers, vinylic, and/or allylic sulfides resulting from the different regioselectivity, depending on the substrates and reaction conditions (Scheme 4.41).

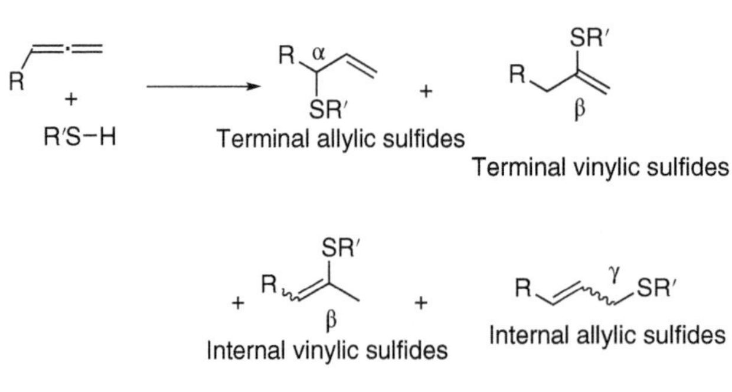

Scheme 4.41 Hydrothiolation of terminal allenes.

On the basis of the reported procedures, the radical hydrothiolation of monosubstituted allenes usually produces a regioisomeric mixture of adducts, resulting in the radical reaction being synthetically less useful.

However, transition metal-catalyzed addition reactions feature high regioselectivity. For example, $Pd(OAc)_2$-catalyzed addition of benzenethiol to monalkyl-substituted allenes gave terminal vinylic sulfides in good yields with high regioselectivity [280]; $[Rh(cod)Cl]_2$/chiral ligand-catalyzed addition reaction afforded optically active terminal vinylic sulfides [281].

Very recently, boron Lewis acid-catalyzed regioselective hydrothiolation of 1-aryl-1,3-dienes with thiols has also been developed, and either 1,4- or 3,4-regioselective addition reactions could be achieved in the presence of either $B(C_6F_5)_3$ or $BF_3{\cdot}Et_2O$ to give secondary and tertiary allylic sulfides [282]. The synthesis of allylic sulfides with high chemo-, regio-, and enantiocontrol has been achieved in the presence of $Rh(cod)_2SbF_6$/chiral ligand [283], and the counterion-controlled regioselectivity has been disclosed [284].

Alper's group [285] and others [286] have developed the regioselective thiocarbonylation of alkenes, allenes, and 1,3-dienes with thiols and CO in the presence of transition metal complexes to synthesize β,γ-unsaturated thiol esters via 1,4-addition (Scheme 4.42).

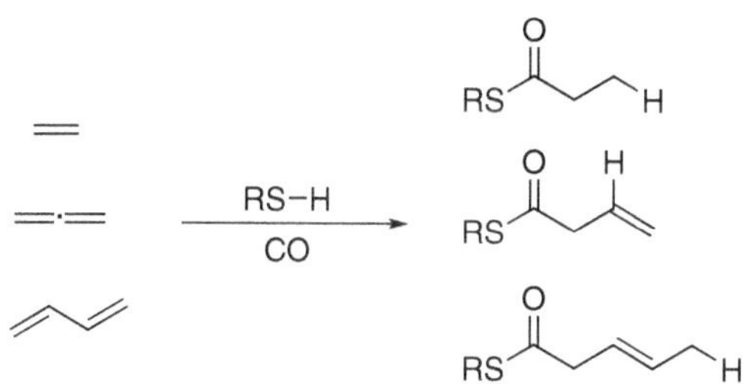

Scheme 4.42 Palladium-catalyzed hydrothiocarbonylation of alkenes, allenes, and 1,3-dienes.

The optically active β,γ-unsaturated thiol esters was also first reported by Alper's group via an enantioselective $Pd(OAc)_2$/chiral ligand-catalyzed thiocarbonylation of prochiral 1,3-conjugated dienes with thiols and CO [287].

Very recently, Dong and coworker have developed a rhodium(I)-catalyzed enantioselective hydrothiolation of cyclopropenes to give either cyclopropyl sulfides or allylic sulfides, and the chemoselectivity was controlled by the simple choice of bisphosphine ligands (Scheme 4.43) [288].

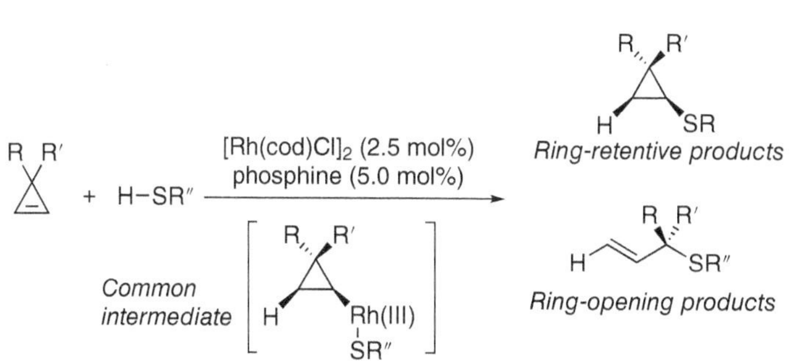

Scheme 4.43 Rhodium(I)-catalyzed enantioselective hydrothiolation of cyclopropenes.

In addition, $Pd(OAc)_2$-catalyzed hydroselenation of allenes affording vinylic selenides [289] and Markovnikov hydroselenation of *N*-vinyl lactams and internal *N*-vinylpyrrolidones [290] were also developed.

4.7 Addition of *O*-nucleophiles to Alkenes

This section focuses on the addition of *O*-nucleophiles, such as H_2O, alcohols, phenols, and acids, to alkenes affording alcohols and ethers [291].

The addition of water to terminal alkene can afford either primary alcohols or secondary/tertiary alcohols, depending on the regioselectivity (Scheme 4.44).

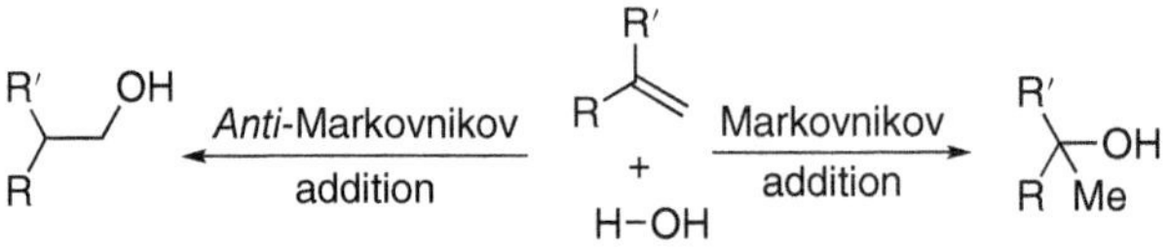

Scheme 4.44 Alcohol from addition of water to terminal alkenes.

Because most addition reactions were achieved under acidic conditions, the Markovnikov additions forming secondary/tertiary alcohols are the main manners. A wool–palladium complex was also reported to efficiently catalyze the hydration of styrene giving α-methylbenzyl alcohol in high yield [292].

A few catalytic systems have been developed recently for the *anti*-Markovnikov addition affording primary alcohols, including Shvo's catalyst/$PdCl_2(MeCN)_2/CuCl_2$ [293], visible light-mediated with an organic photoredox catalyst [294].

Very recently, a theoretical study on the direct *anti*-Markovnikov addition of water to alkenes to form primary alcohols has been reported [295].

In addition, Au(I)/NHC complexes show catalytic activity to catalyze the hydration of terminal allenes to form allylic alcohols in modest yields with high regioselectivity installing OH group to the terminal carbon of the allenyl moiety [296].

$Ru_3(CO)_{12}/CF_3COOH$ can catalyze the hydrative dimerization of allenes affording a mixture of γ,δ-unsaturated ketones and methyl ketones in moderate combined yields [297].

The addition reactions of alcohols and phenols to alkenes are the green synthetic routes of ethers and aryl ethers (Scheme 4.45).

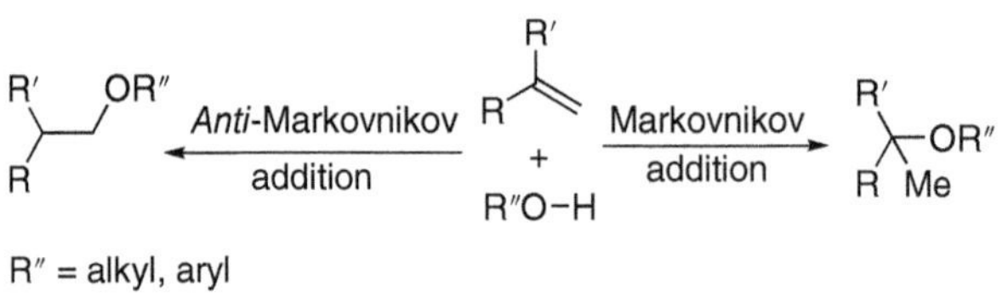

Scheme 4.45 Addition of alcohol and phenol to terminal alkene.

Intermolecular Markovnikov hydroalkoxylation of unactivated alkenes was catalyzed by Au(I)/Ag(I) in the presence of electron-deficient phosphine ligands [298] and $FeCl_3$ iron(III) in the presence of TsOH [299].

Recently, $Ph_3PAuOTf$ shows high catalytic activity for the addition of both phenols and carboxylic acids to alkenes [300], and a theoretical investigation on the overall reaction mechanism for the Au(I)-catalyzed nucleophilic addition to alkenes, including phenol to alkenes, was also reported [301].

Recently, $[Ir(cod)Cl]_2$/chiral ligand-catalyzed intermolecular additions of phenols to unactivated terminal alkenes occurred with Markovnikov selectivity to give aryl ethers in good yields [302]. The adducts having the phenoxy group at the second-position of the alkyl chain could be obtained with ee of 8–60%.

Marks and coworker reported an efficient intramolecular hydroalkoxylation of unactivated alkenols catalyzed by recyclable lanthanide triflate ionic liquids. The intramolecular hydroalkoxylation of primary/secondary and aliphatic/aromatic hydroxyalkenes provided an efficient access to furans, pyrans, spirobicyclic furans, spirobicyclic furans/pyrans, benzofurans, and isochromans [303].

The addition of carboxylic acids to alkenes is a green synthesis of carboxylic esters (Scheme 4.46).

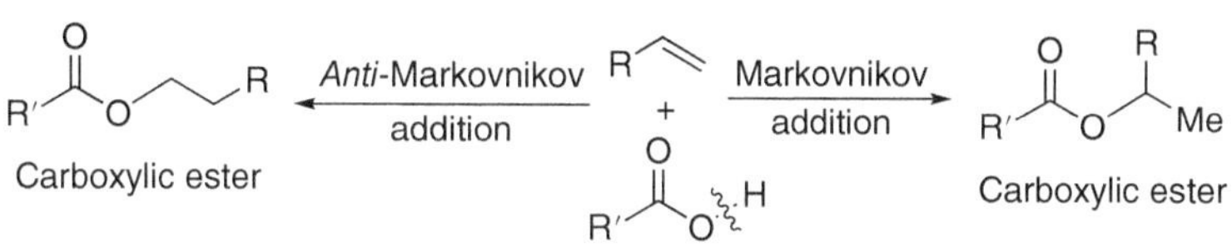

Scheme 4.46 Addition of carboxylic acid to alkenes.

$Fe(OTf)_3$, in situ formed from $FeCl_3$ and TfOH, or $FeCl_3$/AgOTf showed high catalytic activity for the intermolecular addition of carboxylic acids to various alkenes to yield carboxylic esters, including unstable esters of acrylates [304]. In the case of linear terminal alkene used, the addition reaction produced an isomeric mixture of esters.

$[(p\text{-Cymene})RuCl_2]_2$/phosphine/AgOTf could catalyze the addition reactions of 2-phenylbenzoic acid to unactivated alkenes giving Markovnikov adducts with excellent regioselectivity [305].

Note that $Cu(OTf)_2$, an inexpensive, air- and moisture-stable inorganic salt, shows catalytic activity for the O—H addition of aliphatic and aromatic acids and alcohols to norbornene [306].

In the presence of $(PPh_3)AuCl$/AgOTf as a catalyst, the intramolecular additions of hydroxyl groups to tethered conjugated dienes proceeded smoothly at 60 °C to produce substituted tetrahydrofurans and tetrahydropyrans in good yields. Scheme 4.47 shows two examples for this intramolecular additions for the formation of 5-*exo*/6-*exo* 1,4-addition cyclic ether products [307].

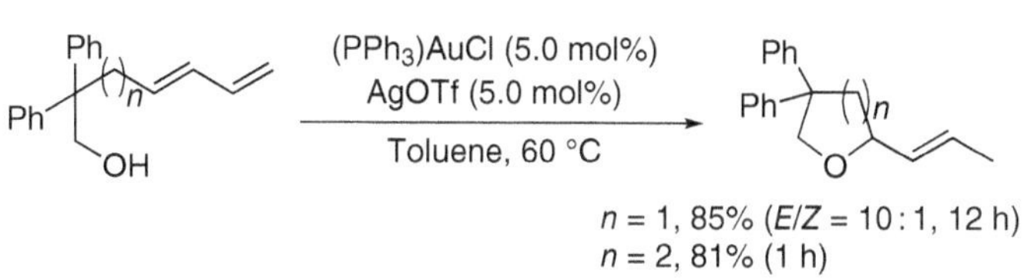

Scheme 4.47 Intramolecular hydroalkoxylation of conjugated dienyl alcohols.

References

1 Selected reviews, see: (a) Yamamoto, Y. and Radhakrishnan, U. (1999). *Chem. Soc. Rev.* 28: 199–207. (b) Bytschkov, I. and Doye, S. (2003). *Eur. J. Org. Chem.* 935–946. (c) Beller, M., Seayad, J., Tillack, A., and Jiao, H. (2004). *Angew. Chem. Int. Ed.* 43: 3368–3398. (d) Bruneau, C. and Dixneuf, P.H. (2006). *Angew. Chem. Int. Ed.* 45: 2176–2203. (e) Yadav, J.S., Antony, A., Rao, T.S., and Reddy, B.V.S. (2011). *J. Organomet. Chem.* 696: 16–36. (f) Zeng, X. (2013). *Chem. Rev.* 113: 6864–6900. (g) Greenhalgh, M.D., Jones, A.S., and Thomas, S.P. (2015). *ChemCatChem* 7: 190–222. (h) Yoshida, H. (2016). *Chem. Rec.* 16: 419–434. (i) Wathier, M. and Love, J.A. (2016). *Eur. J. Inorg. Chem.* 2391–2402. (j) Coombs, J.R. and Morken, J.P. (2016). *Angew. Chem. Int. Ed.* 55: 2636–2649. (k) Crossley, S.W.M., Obradors, C., Martinez, R.M., and Shenvi, R.A. (2016). *Chem. Rev.* 116: 8912–9000. (l) Obligacion, J.V. and Chirik, P.J. (2018). *Nat. Rev. Chem.* 2: 15–34. (m) Chen, J.H. and Lu, Z. (2018). *Org. Chem. Front.* 5: 260–272. (n) Chen, J.H., Guo, J., and Lu, Z. (2018). *Chin. J. Chem.* 36: 1075–1109. (o) Wen, Y., Deng, C., Xie, J., and Kang, X. (2019). *Molecules* 24: 101. (p) Ding, S., Xu, L., and Miao, Z. (2019). *Molecules* 24: 1325. (q) Alyabyev, S.B. and Beletskaya, I.P. (2020). *Russ. Chem. Rev.* 89: 491–536.

2 Muñoz, M.P. (2014). *Chem. Soc. Rev.* 43: 3164–3183.

3 McNeill, E. and Ritter, T. (2015). *Acc. Chem. Res.* 48: 2330–2343.

4 Marcyk, P.T. and Cook, S.P. (2019). *Org. Lett.* 21: 1547–1550.

5 Gilbert-Wilson, R., Chu, W.-Y., and Rauchfuss, T.B. (2015). *Inorg. Chem.* 54: 5596–5603.

6 Carney, J.R., Dillon, B.R., Campbell, L., and Thomas, S.P. (2018). *Angew. Chem. Int. Ed.* 57: 10620–10624.

7 Douglass, M.R., Ogasawara, M., Hong, S. et al. (2002). *Organometallics* 21: 283–292.

8 Adamson, N.J. and Malcolmson, S.J. (2020). *ACS Catal.* 10: 1060–1076.

9 Selected reviews, see: (a) Miyaura, N. and Suzuki, A. (1995). *Chem. Rev.* 95: 2457–2483. (b) Suzuki, A. (1999). *J. Organomet. Chem.* 576: 147–168. (c) Suzuki, A. (2002). *J. Organomet. Chem.* 653: 83–90. (d) Liu, C., Zhang, H., Shi, W., and Lei, A. (2011). *Chem. Rev.* 111: 1780–1824. (e) Heravi, M.M. and Hashemi, E. (2012). *Tetrahedron* 68: 9145–9178.

10 (a) Chemler, S.R., Trauner, D., and Danishefsky, S.J. (2001). *Angew. Chem. Int. Ed.* 40: 4544–4568. (b) Yang, X., Kalita, S.J., Maheshuni, S., and Huang, Y.-Y. (2019). *Coord. Chem. Rev.* 392: 35–48.

11 Selected reviews, see: (a) Burgess, K. and Ohlmeyer, M.J. (1991). *Chem. Rev.* 91: 1179–1191. (b) Pereira, S. and Srebnik, M. (1996). *Tetrahedron Lett.* 37: 3283–3286. (c) Beletskaya, I. and Pelter, A. (1997). *Tetrahedron* 53: 4957–5026. (d) Zeng, X. (2013). *Chem. Rev.* 113: 6864–6900. (e) Neeve, E.C., Geier, S.J., Mkhalid, I.A.I. et al. (2016). *Chem. Rev.* 116: 9091–9161. (f) Buñuel, E. and Cárdenas, D.J. (2018). *Chem. Eur. J.* 24: 11239–11244. (g) Chen, J.-B. and Whiting, A. (2018). *Synthesis* 50: 3843–3861. (h) Hemming, D., Fritzemeier, R., Westcott, S.A. et al. (2018). *Chem. Soc. Rev.* 47: 7477–7494. (i) Carreras, J.,

Caballero, A., and Pérez, P.J. (2019). *Chem. Asian J.* 14: 329–343. (j) Wen, H., Liu, G., and Huang, Z. (2019). *Coord. Chem. Rev.* 386: 138–153.

12 Selected recent reports, see: (a) Nagashima, Y., Sasaki, K., Suto, T. et al. (2018). *Chem. Asian J.* 13: 1024–1028. (b) García, L., Sendra, J., Miralles, N. et al. (2018). *Chem. Eur. J.* 24: 14059–14063.

13 Iwamoto, H., Kubota, K., and Ito, H. (2016). *Chem. Commun.* 52: 5916–5919.

14 (a) Reilly, S.W., Webster, C.E., Hollis, T.K., and Valle, H.U. (2016). *Dalton Trans.* 45: 2823–2828. (b) Zhang, G., Wu, J., Wang, M. et al. (2017). *Eur. J. Org. Chem.* 5814–5818. (c) Peng, J., Docherty, J.H., Dominey, A.P., and Thomas, S.P. (2017). *Chem. Commun.* 53: 4726–4729. (d) Tamang, S.R., Bedi, D., Shafiei-Haghighi, S. et al. (2018). *Org. Lett.* 20: 6695–6700. (e) Verma, P.K., Sethulekshmi, A.S., and Geetharani, K. (2018). *Org. Lett.* 20: 7840–7845. (f) Zhang, G., Wu, J., Li, S. et al. (2018). *Org. Lett.* 20: 7893–7897.

15 Zhang, G., Zeng, H., Wu, J. et al. (2016). *Angew. Chem. Int. Ed.* 55: 14369–14372.

16 Touney, E.E., Hoveln, R.V., Buttke, C.T. et al. (2016). *Organometallics* 35: 3436–3439.

17 (a) MacNair, A.J., Millet, C.R.P., Nichol, G.S. et al. (2016). *ACS Catal.* 6: 7217–7221. (b) Chen, X., Cheng, Z., and Lu, Z. (2017). *Org. Lett.* 19: 969–971.

18 Fan, W., Li, L., and Zhang, G. (2019). *J. Org. Chem.* 84: 5987–5996.

19 (a) Zhang, L., Peng, D., Leng, X., and Huang, Z. (2013). *Angew. Chem. Int. Ed.* 52: 3676–3680. (b) Greenhalgh, M.D. and Thomas, S.P. (2013). *Chem. Commun.* 49: 11230–11232. (c) Zheng, J., Sortais, J.-B., and Darcel, C. (2014). *ChemCatChem* 6: 763–766. (d) Tseng, K.-N.T., Kampf, J.W., and Szymczak, N.K. (2015). *ACS Catal.* 5: 411–415.

20 Wang, Y., Guan, R., Sivaguru, P. et al. (2019). *Org. Lett.* 21: 4035–4038.

21 Kisan, S., Krishnakumar, V., and Gunanathan, C. (2017). *ACS Catal.* 7: 5950–5954.

22 (a) Zhang, L., Zuo, Z., Leng, X., and Huang, Z. (2014). *Angew. Chem. Int. Ed.* 53: 2696–2700. (b) Pang, M., Wu, C., Zhuang, X. et al. (2018). *Organometallics* 37: 1462–1467.

23 (a) Mun, S., Lee, J.-E., and Yun, J. (2006). *Org. Lett.* 8: 4887–4889. (b) Hong, S., Liu, M., Zhang, W. et al. (2015). *Tetrahedron Lett.* 56: 2297–2302. (c) Liu, S., Zeng, X., and Xu, B. (2016). *Tetrahedron Lett.* 57: 3706–3710.

24 Liu, Y., Zhou, Y., Wang, H., and Qu, J. (2015). *RSC Adv.* 5: 73705–73713.

25 Li, J.-F., Wei, Z.-Z., Wang, Y.-Q., and Ye, M. (2017). *Green Chem.* 19: 4498–4502.

26 Docherty, J.H., Peng, J., Dominey, A.P., and Thomas, S.P. (2017). *Nat. Chem.* 9: 595–600.

27 Agahi, R., Challinor, A.J., Carter, N.B., and Thomas, S.P. (2019). *Org. Lett.* 21: 993–997.

28 Obligacion, J.V. and Chirik, P.J. (2013). *Org. Lett.* 15: 2680–2683.

29 Bismuto, A., Cowley, M.J., and Thomas, S.P. (2018). *ACS Catal.* 8: 2001–2005.

30 Vijaykumar, G., Bhunia, M., and Mandal, S.K. (2019). *Dalton Trans.* 48: 5779–5784.

31 Lata, C.J. and Crudden, C.M. (2010). *J. Am. Chem. Soc.* 132: 131–137.

32 Yin, Q., Kemper, S., Klare, H.F.T., and Oestreich, M. (2016). *Chem. Eur. J.* 22: 13840–13844.

33 Palmer, W.N., Diao, T., Pappas, I., and Chirik, P.J. (2015). *ACS Catal.* 5: 622–626.

34 Wang, X., Cui, X., Li, S. et al. (2020). *Angew. Chem. Int. Ed.* 59: 13608–13612.

35 Yamamoto, Y., Fujikawa, R., Yamada, A., and Miyaura, N. (1999). *Chem. Lett.* 28: 1069–1070.

36 Li, C., Yang, Z., Wang, L. et al. (2020). *Angew. Chem. Int. Ed.* 59: 6278–6283.

37 Semba, K., Shinomiya, M., Fujihara, T. et al. (2013). *Chem. Eur. J.* 19: 7125–7132.

38 Fiorito, D. and Mazet, C. (2018). *ACS Catal.* 8: 9382–9387.

39 Selected reviews, see: (a) Crudden, C.M. and Edwards, D. (2003). *Eur. J. Org. Chem.* 4695–4712. (b) Carroll, A.M., O'Sullivan, T.P., and Guiry, P.J. (2005). *Adv. Synth. Catal.* 347: 609–631. (c) Collins, B.S.L., Wilson, C.M., Myers, E.L., and Aggarwal, V.K. (2017). *Angew. Chem. Int. Ed.* 56: 11700–11733.

40 Selected reviews, see: (a) Brown, H.C. and Singaram, B. (1988). *Acc. Chem. Res.* 21: 287–293. (b) Sandford, C. and Aggarwal, V.K. (2017). *Chem. Commun.* 53: 5481–5494.

41 (a) Noh, D., Chea, H., Ju, J., and Yun, J. (2009). *Angew. Chem. Int. Ed.* 48: 6062–6064. (b) Xi, Y. and Hartwig, J.F. (2016). *J. Am. Chem. Soc.* 138: 6703–6706. (c) Xi, Y. and Hartwig, J.F. (2017). *J. Am. Chem. Soc.* 139: 12758–12772. (d) Jang, W.J., Song, S.M., Moon, J.H. et al. (2017). *J. Am. Chem. Soc.* 139: 13660–13663.

42 (a) Mazet, C. and Gérard, D. (2011). *Chem. Commun.* 47: 298–300. (b) Magre, M., Biosca, M., Pàmies, O., and Diéguez, M. (2015). *ChemCatChem* 7: 114–120.

43 (a) Zhang, L., Zuo, Z., Wan, X., and Huang, Z. (2014). *J. Am. Chem. Soc.* 136: 15501–15504. (b) Chen, J., Xi, T., Ren, X. et al. (2014). *Org. Chem. Front.* 1: 1306–1309. (c) Chen, X., Cheng, Z., and Lu, Z. (2019). *ACS Catal.* 9: 4025–4029.

44 (a) Shoba, V.M., Thacker, N.C., Bochat, A.J., and Takacs, J.M. (2016). *Angew. Chem. Int. Ed.* 55: 1465–1469. (b) Hoang, G.L., Zhang, S., and Takacs, J.M. (2018). *Chem. Commun.* 54: 4838–4841.

45 Chen, J., Xi, T., and Lu, Z. (2014). *Org. Lett.* 16: 6452–6455.

46 (a) Wang, Z., He, X., Zhang, R. et al. (2017). *Org. Lett.* 19: 3067–3070. (b) Iwamoto, H., Imamoto, T., and Ito, H. (2018). *Nat. Commun.* 9: 2290. (c) Han, S., Shen, X., Kong, D. et al. (2019). *J. Org. Chem.* 84: 4318–4329.

47 Smith, J.R., Collins, B.S.L., Hesse, M.J. et al. (2017). *J. Am. Chem. Soc.* 139: 9148–9151.

48 (a) Smith, S.M., Thacker, N.C., and Takacs, J.M. (2008). *J. Am. Chem. Soc.* 130: 3734–3735. (b) Smith, S.M. and Takacs, J.M. (2010). *J. Am. Chem. Soc.* 132: 1740–1741. (c) Smith, S.M. and Takacs, J.M. (2010). *Org. Lett.* 12: 4612–4615. (d) Smith, S.M., Hoang, G.L., Pal, R. et al. (2012). *Chem. Commun.* 48: 12180–12182. (e) Chakrabarty, S. and Takacs, J.M. (2017). *J. Am. Chem. Soc.* 139: 6066–6069.

49 Lee, Y. and Hoveyda, A.H. (2009). *J. Am. Chem. Soc.* 131: 3160–3161.

50 Rubina, M., Rubin, M., and Gevorgyan, V. (2003). *J. Am. Chem. Soc.* 125: 7198–7199.

51 A highlight paper, see:Thomas, S.P. and Aggarwal, V.K. (2009). *Angew. Chem. Int. Ed.* 48: 1896–1898.

52 Bai, X.-Y., Zhao, W., Sun, X., and Li, B.-J. (2019). *J. Am. Chem. Soc.* 141: 19870–19878.

53 (a) Satoh, M., Nomoto, Y., Miyaura, N., and Suzuki, A. (1989). *Tetrahedron Lett.* 30: 3789–3792. (b) Zhu, C., Yang, B., Qiu, Y., and Bäckvall, J.-E. (2016). *Chem. Eur. J.* 22: 2939–2943.

54 Wu, J.Y., Moreau, B., and Ritter, T. (2009). *J. Am. Chem. Soc.* 131: 12915–12917.

55 (a) Sasaki, Y., Zhong, C., Sawamura, M., and Ito, H. (2010). *J. Am. Chem. Soc.* 132: 1226–1227. (b) Liu, Y., Fiorito, D., and Mazet, C. (2018). *Chem. Sci.* 9: 5284–5288.

56 (a) Obligacion, J.V. and Chirik, P.J. (2013). *J. Am. Chem. Soc.* 135: 19107–19110. (b) Ibrahim, A.D., Entsminger, S.W., and Fout, A.R. (2017). *ACS Catal.* 7: 3730–3734.

57 A review: Verma, P.K., Shegavi, M.L., Bose, S.K., and Geetharani, K. (2018). *Org. Biomol. Chem.* 16: 857–873.

58 Selected reports, see: (a) Sun, C., Potter, B., and Morken, J.P. (2014). *J. Am. Chem. Soc.* 136: 6534–6537. (b) Xu, S., Shangguan, X., Li, H. et al. (2015). *J. Org. Chem.* 80: 7779–7784. (c) Kim, J., Park, S., Park, J., and Cho, S.H. (2016). *Angew. Chem. Int. Ed.* 55: 1498–1501. (d) Potter, B., Edelstein, E.K., and Morken, J.P. (2016). *Org. Lett.* 18: 3286–3289. (e) Huang, Q. and Zard, S.Z. (2018). *Org. Lett.* 20: 5304–5308. (f) Babu, K.N., Massarwe, F., Reddy, R.R. et al. (2020). *Molecules* 25: 959.

59 (a) Lee, S., Li, D., and Yun, J. (2014). *Chem. Asian J.* 9: 2440–2443. (b) Zuo, Z. and Huang, Z. (2016). *Org. Chem. Front.* 3: 434–438.

60 (a) Lee, J.C.H., McDonald, R., and Hall, D.G. (2011). *Nat. Chem.* 3: 894–899. (b) Feng, X., Jeon, H., and Yun, J. (2013). *Angew. Chem. Int. Ed.* 52: 3989–3992.

61 Li, L., Gong, T., Lu, X. et al. (2017). *Nat. Commun.* 8: 345.

62 Teo, W.J. and Ge, S. (2018). *Angew. Chem. Int. Ed.* 57: 1654–1658.

63 Selected reviews, see: (a) Marciniec, B. (2005). *Coord. Chem. Rev.* 249: 2374–2390. (b) Díez-González, S. and Nolan, S.P. (2008). *Acc. Chem. Res.* 41: 349–358. (c) Zhang, M. and Zhang, A. (2010). *Appl. Organometal. Chem.* 24: 751–757. (d) Troegel, D. and Stohrer, J. (2011). *Coord. Chem. Rev.* 255: 1440–1459. (e) Nakajima, Y. and Shimada, S. (2015). *RSC Adv.* 5: 20603–20616. (f) Sun, J. and Deng, L. (2016). *ACS Catal.* 6: 290–300. (g) Zaranek, M., Marciniec, B., and Pawluc, P. (2016). *Org. Chem. Front.* 3: 1337–1344. (h) Nakajima, Y., Sato, K., and Shimada, S. (2016). *Chem. Rec.* 16: 2379–2387. (i) Du, X. and Huang, Z. (2017). *ACS Catal.* 7: 1227–1243. (j) Obligacion, J.V. and Chirik, P.J. (2018). *Nat. Rev. Chem.* 2: 15–34. (k) Yang, X. and Wan, C. (2018). *Chem. Asian J.* 13: 2307–2315. (l) Carney, J.R., Dillon, B.R., Campbell, L., and Thomas, P.S. (2018). *Angew. Chem. Int. Ed.* 57: 10620–10624. (m) Liu, D., Liu, B., Pan, Z. et al. (2019). *Sci. China Chem.* 62: 571–582.

64 Zaranek, M. and Pawluc, P. (2018). *ACS Catal.* 8: 9865–9876.

65 (a) Pindwal, A., Patnaik, S., Everett, W.C. et al. (2017). *Angew. Chem. Int. Ed.* 56: 628–631. (b) Xie, S., Li, X., Sun, H. et al. (2020). *Organometallics* 39: 2455–2463.
66 Selected reports, see: (a) Markó, I.E., Stérin, S., Buisine, O. et al. (2002). *Sciene* 298: 204–206. (b) Buisine, O., Berthon-Gelloz, G., Brière, J.-F. et al. (2005). *Chem. Commun.* 3856–3858. (c) Meister, T.K., Riener, K., Gigler, P. et al. (2016). *ACS Catal.* 6: 1274–1284. (d) Wu, H., Zheng, C., Chen, N. et al. (2017). *Tetrahedron Lett.* 58: 1576–1578.
67 Selected reports, see: (a) Schuster, C.H., Diao, T., Pappas, I., and Chirik, P.J. (2016). *ACS Catal.* 6: 2632–2636. (b) Ibrahim, A.D., Entsminger, S.W., Zhu, L., and Fout, A.R. (2016). *ACS Catal.* 6: 3589–3593. (c) Verhoeven, D.G.A., Kwakernaak, J., van Wiggen, M.A.C. et al. (2019). *Eur. J. Inorg. Chem.* 660–667.
68 (a) Xie, X., Zhang, X., Yang, H. et al. (2019). *J. Org. Chem.* 84: 1085–1093. (b) Gao, W. and Ding, S. (2020). *Synthesis* 52: 3549–3563.
69 Selected reports, see: (a) Marciniec, B., Krzyzanowski, P., Walczuk-Guściora, E., and Duczmal, W. (1999). *J. Mol. Catal. A: Chem.* 144: 263–271. (b) Itami, K., Mitsudo, K., Nishino, A., and Yoshida, J.-i. (2002). *J. Org. Chem.* 67: 26455–22652. (c) Żak, P., Bołt, M., Kubicki, M., and Pietraszuk, C. (2018). *Dalton Trans.* 47: 1903–1910.
70 (a) Tondreau, A.M., Atienza, C.C.H., Weller, K.J. et al. (2012). *Science* 335: 567–570. (b) Kamata, K., Suzuki, A., Nakai, Y., and Nakazawa, H. (2012). *Organometallics* 31: 3825–3828. (c) Greenhalgh, M.D., Frank, D.J., and Thomas, S.P. (2014). *Adv. Synth. Catal.* 356: 584–590.
71 (a) Buslov, I., Becouse, J., Mazza, S. et al. (2015). *Angew. Chem. Int. Ed.* 54: 14523–14526. (b) Rock, C.L. and Trovitch, R.J. (2019). *Dalton Trans.* 48: 461–467.
72 (a) Glaser, P.B. and Tilley, T.D. (2003). *J. Am. Chem. Soc.* 125: 13640–13641. (b) Lipke, M.C., Poradowski, M.-N., Raynaud, C. et al. (2018). *ACS Catal.* 8: 11513–11523.
73 Takaki, K., Sonoda, K., Kousaka, T. et al. (2001). *Tetrahedron Lett.* 42: 9211–9214.
74 Garcia, L., Dinoi, C., Mahon, M.F. et al. (2019). *Chem. Sci.* 10: 8108–8118.
75 Jakobsson, K., Chu, T., and Nikonov, G.I. (2016). *ACS Catal.* 6: 7350–7356.
76 Balasubramaniam, S., Kumar, S., Andrews, A.P. et al. (2019). *Eur. J. Inorg. Chem.* 3265–3269.
77 Sanagawa, A. and Nagashima, H. (2018). *Organometallics* 37: 2859–2871.
78 Pappas, I., Treacy, S., and Chirik, P.J. (2016). *ACS Catal.* 6: 4105–4109.
79 Muchnij, J.A., Kwaramba, F.B., and Rahaim, R.J. (2014). *Org. Lett.* 16: 1330–1333.
80 Xie, X., Zhang, X., Yang, H. et al. (2019). *J. Org. Chem.* 84: 1085–1093.
81 Srinivas, V., Nakajima, Y., Sato, K., and Shimada, S. (2018). *Org. Lett.* 20: 12–15.
82 Schmid, G., West, H., Mehles, H., and Lehnert, A. (1997). *Inorg. Chem.* 36: 891–895.
83 Hu, R., Zha, L., and Cai, M. (2010). *Catal. Commun.* 11: 563–566.

84 Li, J., Yang, C., Zhang, L., and Ma, T. (2011). *J. Organomet. Chem.* 696: 1845–1849.

85 Rao, F., Deng, S., Chen, C., and Zhang, N. (2014). *Catal. Commun.* 46: 1–5.

86 Li, Q., Ji, S., Li, M., and Duan, X. (2018). *Sci. China Mater.* 61: 1339–1344.

87 Kong, C.J., Gilliland, S.E. III, Clark, B.R., and Gupton, B.F. (2018). *Chem. Commun.* 54: 13343–13346.

88 Zhang, Z., Bai, L., and Hu, X. (2019). *Chem. Sci.* 10: 3791–3795.

89 (a) Zai, H., Zhao, Y., Chen, S. et al. (2018). *Nano Res.* 11: 2544–2552. (b) Pandarus, V., Ciriminna, R., Gingras, G. et al. (2019). *Green Chem.* 21: 129–140.

90 Mitsudome, T., Fujita, S., Sheng, M. et al. (2019). *Green Chem.* 21: 4566–4570.

91 Chen, L., Ali, I.S., Sterbinsky, G.E. et al. (2019). *ChemCatChem* 11: 2843–2854.

92 (a) Gandon, V., Agenet, N., Vollhardt, K.P.C. et al. (2009). *J. Am. Chem. Soc.* 131: 3007–3015. (b) Chen, C., Hecht, M.B., Kavara, A. et al. (2015). *J. Am. Chem. Soc.* 137: 13244–13247. (c) Kim, D., Chen, C., Mercado, B.Q. et al. (2020). *Organometallics* 39: 2415–2424.

93 (a) Wang, C., Teo, W.J., and Ge, S. (2017). *ACS Catal.* 7: 855–863. (b) Gao, Y., Wang, L., and Deng, L. (2018). *ACS Catal.* 8: 9637–9646.

94 Xie, S., Dong, Y., Du, X. et al. (2021). *Organometallics* 40: 286–293.

95 Selected reports, see: (a) Buch, F., Brettar, J., and Harder, S. (2006). *Angew. Chem. Int. Ed.* 45: 2741–2745. (b) Ruspic, C., Spielmann, J., and Harder, S. (2007). *Inorg. Chem.* 46: 5320–5326. (c) Leich, V., Spaniol, T.P., Maron, L., and Okuda, J. (2014). *Chem. Commun.* 50: 2311–2314. (d) Leich, V., Lamberts, K., Spaniol, T.P., and Okuda, J. (2014). *Dalton Trans.* 43: 14315–14321. (e) Leich, V., Spaniol, T.P., and Okuda, J. (2016). *Organometallics* 35: 1179–1182. (f) Zaranek, M., Witomska, S., Patroniak, V., and Pawluc, P. (2017). *Chem. Commun.* 53: 5404–5407. (g) Schuhknecht, D., Leich, V., Spaniol, T.P., and Okuda, J. (2018). *Chem. Eur. J.* 24: 13424–13427. (h) Garcia, L., Dinoi, C., Mahon, M.F. et al. (2019). *Chem. Sci.* 10: 8108–8118. (i) Garcia, L., Mahon, M.F., and Hill, M.S. (2019). *Organometallics* 38: 3778–3785. (j) Schuhknecht, D., Spaniol, T.P., Maron, L., and Okuda, J. (2020). *Angew. Chem. Int. Ed.* 59: 310–314.

96 Song, Y.-S., Yoo, B.R., Lee, G.-H., and Jung, I.N. (1999). *Organometallics* 18: 3109–3115.

97 Dang, H.-S. and Roberts, B.P. (1995). *Tetrahedron Lett.* 36: 2875–2878.

98 Haque, M.B. and Roberts, B.P. (1996). *Tetrahedron Lett.* 37: 9123–9126.

99 Rubin, M., Schwier, T., and Gevorgyan, V. (2002). *J. Org. Chem.* 67: 1936–1940.

100 Keess, S. and Oestreich, O.M. (2017). *Org. Lett.* 19: 1898–1901.

101 Tanaka, S., Nakamura, T., Yorimitsu, H. et al. (2000). *Org. Lett.* 2: 1911–1914.

102 Liu, J., Chen, W., Li, J., and Cui, C. (2018). *ACS Catal.* 8: 2230–2235.

103 Wang, H., Zhang, G., Zhang, Q. et al. (2020). *Chem. Commun.* 56: 1819–1822.

104 Zacuto, M.J., O'Malley, S.J., and Leighton, J.L. (2003). *Tetrahedron* 59: 8889–8900.

105 Huang, G. and Isobe, M. (2001). *Tetrahedron* 57: 10241–10246.

106 Kidonakis, M. and Stratakis, M. (2015). *Org. Lett.* 17: 4538–4541.

107 (a) Miller, Z.D., Li, W., Belderrain, T.R., and Montgomery, J. (2013). *J. Am. Chem. Soc.* 135: 15282–15285. (b) Xie, H., Zhao, L., Yang, L. et al. (2014). *J. Org. Chem.* 79: 4517–4527.

108 Li, K., Nie, M., and Tang, W. (2020). *Green Synth. Catal.* 1: 171–174.

109 Sudo, T., Asao, N., Gevorgyan, V., and Yamamoto, T. (1999). *J. Org. Chem.* 64: 2494–2499.

110 Miller, Z.D., Dorel, R., and Montgomery, J. (2015). *Angew. Chem. Int. Ed.* 54: 9088–9091.

111 Tafazolian, H. and Schmidt, J.A.R. (2015). *Chem. Commun.* 51: 5943–5946.

112 Hilt, G., Lüers, S., and Schmidt, F. (2004). *Synthesis* 634–638.

113 Wu, J.Y., Stanzl, B.N., and Ritter, T. (2010). *J. Am. Chem. Soc.* 132: 13214–13216.

114 Parker, S.E., Börgel, J., and Ritter, T. (2014). *J. Am. Chem. Soc.* 136: 4857–4860.

115 (a) Raya, B., Jing, S., Balasanthiran, V., and RajanBabu, T.V. (2017). *ACS Catal.* 7: 2275–2283. (b) Sang, H.L., Yu, S., and Ge, S. (2018). *Chem. Sci.* 9: 973–978.

116 Wen, H., Wang, K., Zhang, Y. et al. (2019). *ACS Catal.* 9: 1612–1618.

117 Fu, P.-F., Brard, L., Li, Y., and Marks, T.J. (1995). *J. Am. Chem. Soc.* 117: 7157–7168.

118 Selected reports, see: (a) Uozumi, Y. and Hayashi, T. (1991). *J. Am. Chem. Soc.* 113: 9887–9888. (b) Hayashi, T., Hirate, S., Kitayama, K. et al. (2001). *J. Org. Chem.* 66: 1441–1449. (c) Jensen, J.F., Svendsen, B.Y., la Cour, T.V. et al. (2002). *J. Am. Chem. Soc.* 124: 4558–4559. (d) Magistrato, A., Woo, T.K., Togni, A., and Rothlisberger, U. (2004). *Organometallics* 23: 3218–3227.

119 Gribble, M.W. Jr., Pirnot, M.T., Bandar, J.S. et al. (2017). *J. Am. Chem. Soc.* 139: 2192–2195.

120 (a) Chen, J., Cheng, B., Cao, M., and Lu, Z. (2015). *Angew. Chem. Int. Ed.* 54: 4661–4664. (b) Nikonov, G.I. (2015). *ChemCatChem* 7: 1918–1919. (c) Challinor, A.J., Calin, M., Nichol, G.S. et al. (2016). *Adv. Synth. Catal.* 358: 2404–2409. (d) Cheng, B., Lu, P., Zhang, H. et al. (2017). *J. Am. Chem. Soc.* 139: 9439–9442. (e) Cheng, B., Liu, W., and Lu, Z. (2018). *J. Am. Chem. Soc.* 140: 5014–5017.

121 Yamamoto, K., Hayashi, T., Zembayashi, M., and Kumada, M. (1976). *J. Organomet. Chem.* 118: 161–181.

122 Yamamoto, K., Hayashi, T., Uramoto, Y. et al. (1976). *J. Organomet. Chem.* 118: 331–348.

123 Chen, Y., Ji, S., Sun, W. et al. (2018). *J. Am. Chem. Soc.* 140: 7407–7410.

124 Szymaniak, A.A., Zhang, C., Coombs, J.R., and Morken, J.P. (2018). *ACS Catal.* 8: 2897–2901.

125 Zhan, G., Teng, H.-L., Luo, Y. et al. (2018). *Angew. Chem. Int. Ed.* 57: 12342–12346.

126 Dian, L. and Marek, I. (2020). *Org. Lett.* 22: 4914–4918.

127 Selected reports, see: (a) Haque, M.B., Roberts, B.P., and Tocher, D.A. (1998). *J. Chem. Soc., Perkin Trans. 1* 2881–2889.

128 Selected reports, see: (a) Allred, G.D. and Liebeskind, L.S. (1996). *J. Am. Chem. Soc.* 118: 2748–2749. (b) Shirakawa, E. and Hiyama, T. (1999). *J. Organomet. Chem.* 576: 169–178. (c) Roy, U.K. and Roy, S. (2010). *Chem.*

Rev. 110: 2472–2535. (d) Li, H., Jin, R., Li, Y. et al. (2017). *Sci. Rep.* 7: 16559. (e) Ghazi, D., Zahraa Rasheed, Z., and Yousif, E. (2018). *Arc. Org. Inorg. Chem. Sci.* 3: 344–352.

129 Lautens, M., Kumanovic, S., and Meyer, C. (1996). *Angew. Chem. Int. Ed. Engl.* 35: 1329–1330.

130 Zyder, M. and Szymańska-Buzar, T. (2010). *J. Organomet. Chem.* 695: 1734–1737.

131 Kazmaier, U. and Klein, M. (2005). *Chem. Commun.* 501–503.

132 Selected reports, see: (a) Ichinose, Y., Oshima, K., and Utimoto, K. (1988). *Bull. Chem. Soc. Jpn.* 61: 2693–2695. (b) Mitchell, T.N. and Schneider, U. (1991). *J. Organomet. Chem.* 405: 195–199. (c) Koerber, K., Gore, J., and Vatele, J.-M. (1991). *Tetrahedron Lett.* 32: 1187–1190. (d) Gevorgyan, V., Liu, J.-X., and Yamamoto, Y. (1997). *J. Org. Chem.* 62: 2963–2967.

133 Lautens, M., Ostrovsky, D., and Tao, B. (1997). *Tetrahedron Lett.* 38: 6343–6346.

134 Hayashi, N., Kusano, K., Sekizawa, S. et al. (2007). *Chem. Commun.* 4913–4915.

135 Miyake, H. and Yamamura, K. (1992). *Chem. Lett.* 507–508.

136 Rubina, M., Rubin, M., and Gevorgyan, V. (2004). *J. Am. Chem. Soc.* 126: 3688–3689.

137 Selected reviews, see: (a) Müller, T.E. and Beller, M. (1998). *Chem. Rev.* 98: 675–703. (b) Nobis, M. and Drieûen-Hölscher, B. (2001). *Angew. Chem. Int. Ed.* 40: 3983–3985. (c) Hong, S. and Marks, T.J. (2004). *Acc. Chem. Res.* 37: 673–686. (d) Widenhoefer, R.A. and Han, X. (2006). *Eur. J. Org. Chem.* 4555–4563. (e) Müller, T.E., Hultzsch, K.C., Yus, M. et al. (2008). *Chem. Rev.* 108: 3795–3892. (f) Chemler, S.R. (2009). *Org. Biomol. Chem.* 7: 3009–3019. (g) Hesp, K.D. and Stradiotto, M. (2010). *ChemCatChem* 2: 1192–1207. (h) Huang, L., Arndt, M., Gooßen, K. et al. (2015). *Chem. Rev.* 115: 2596–2697. (i) Bernoud, E., Lepori, C., Mellah, M. et al. (2015). *Catal. Sci. Technol.* 5: 2017–2037. (j) Isaeva, V.I. and Kustov, L.M. (2016). *Top. Catal.* 59: 1196–1206. (k) Pirnot, M.T., Wang, Y.-M., and Buchwald, S.L. (2016). *Angew. Chem. Int. Ed.* 55: 48–57. (l) Patel, M., Saunthwal, R.K., and Verma, A.K. (2017). *Acc. Chem. Res.* 50: 240–254. (m) Hannedouche, J. and Schulz, E. (2018). *Organometallics* 37: 4313–4326.

138 Streiff, S. and Jérôme, F. (2021). *Chem. Soc. Rev.* 50: 1512–1521.

139 Kang, O.-Y., Kim, B.E., Park, S.J. et al. (2018). *Asian J. Org. Chem.* 7: 451–457.

140 Brunet, J.-J., Chu, N.C., and Diallo, O. (2005). *Organometallics* 24: 3104–3110.

141 Taylor, J.G., Whittall, N., and Hii, K.K. (2006). *Org. Lett.* 8: 3561–33564.

142 Zhang, Z., Lee, S.D., and Widenhoefer, R.A. (2009). *J. Am. Chem. Soc.* 131: 5372–5373.

143 Ickes, A.R., Ensign, S.C., Gupta, A.K., and Hull, K.L. (2014). *J. Am. Chem. Soc.* 136: 11256–11259.

144 Yin, P. and Loh, T.-P. (2009). *Org. Lett.* 11: 3791–3793.

145 Xiong, Y. and Zhang, G. (2019). *Org. Lett.* 21: 7873–7877.

146 Liu, J., Li, H., Spannenberg, A. et al. (2016). *Angew. Chem. Int. Ed.* 55: 13544–13548.

147 Dong, K., Fang, X., Jackstell, R. et al. (2015). *J. Am. Chem. Soc.* 137: 6053–6058.

148 Li, H., Dong, K., Neumann, H., and Beller, M. (2015). *Angew. Chem. Int. Ed.* 54: 10239–10243.

149 Li, H., Fang, X., Jackstell, R. et al. (2016). *Chem. Commun.* 52: 7142–7145.

150 Gao, B., Zhang, G., Zhou, X., and Huang, H. (2018). *Chem. Sci.* 9: 380–386.

151 Xu, T., Sha, F., and Alper, H. (2016). *J. Am. Chem. Soc.* 138: 6629–6635.

152 Zhu, J., Gao, B., and Huang, H. (2017). *Org. Biomol. Chem.* 15: 2910–2913.

153 Qi, C., Hasenmaile, F., Gandon, V., and Leboeuf, D. (2018). *ACS Catal.* 8: 1734–1739.

154 (a) Zhu, Q., Graff, D.E., and Knowles, R.R. (2018). *J. Am. Chem. Soc.* 140: 741–747. (b) Nguyen, S.T., Zhu, Q., and Knowles, R.R. (2019). *ACS Catal.* 9: 4502–4507.

155 Giner, X. and Nájera, C. (2008). *Org. Lett.* 10: 2919–2922.

156 (a) Utsunomiya, M. and Hartwig, J.F. (2004). *J. Am. Chem. Soc.* 126: 2702–2703. (b) Takaya, J. and Hartwig, J.F. (2005). *J. Am. Chem. Soc.* 127: 5756–5757.

157 Selected reports, see: (a) Utsunomiya, M., Kuwano, R., Kawatsura, M., and Hartwig, J.F. (2003). *J. Am. Chem. Soc.* 125: 5608–5609. (b) Ensign, S.C., Vanable, E.P., Kortman, G.D. et al. (2015). *J. Am. Chem. Soc.* 137: 13748–13751. (c) Couce-Rios, A., Lledós, A., and Ujaque, G. (2016). *Chem. Eur. J.* 22: 9311–9320.

158 Musacchio, A.J., Lainhart, B.C., Zhang, X. et al. (2017). *Science* 355: 727–730.

159 Gurak, J.A. Jr., Yang, K.S., Liu, Z., and Engle, K.M. (2016). *J. Am. Chem. Soc.* 138: 5805–5808.

160 Ryu, J.-S., Li, G.Y., and Marks, T.J. (2003). *J. Am. Chem. Soc.* 125: 12584–12605.

161 Ma, W., Zhang, X., Fan, J. et al. (2019). *J. Am. Chem. Soc.* 141: 13506–13515.

162 Brinkmann, C., Barrett, A.G.M., Hill, M.S., and Procopiou, P.A. (2012). *J. Am. Chem. Soc.* 134: 2193–2207.

163 Xu, X., Zhang, X., Wang, Z., and Kong, M. (2015). *RSC Adv.* 5: 40950–40952.

164 Liu, Y., Hua, R., and Qiu, X. (2005). *Synth. Commun.* 35: 1375–1380.

165 Sitha, S. and Jewell, L.L. (2010). *Tetrahedron* 66: 3030–3036.

166 Lee, S.I., Son, S.U., and Chung, Y.K. (2002). *Chem. Commun.* 1310–1311.

167 Horniakova, J., Komura, K., Osaki, H. et al. (2005). *Catal. Lett.* 102: 191–196.

168 (a) Walsh, P.J., Baranger, A.M., and Bergman, R.G. (1992). *J. Am. Chem. Soc.* 114: 1708–1719. (b) Johnson, J.S. and Bergman, R.G. (2001). *J. Am. Chem. Soc.* 123: 2923–2924. (c) Ayinla, R.O. and Schafer, L.L. (2006). *Inorg. Chim. Acta* 359: 3097–3102.

169 (a) Besson, L., Gore, J., and Cazes, B. (1995). *Tetrahedron Lett.* 36: 3857–3860. (b) AI-Masum, M., Meguro, M., and Yamamoto, Y. (1997). *Tetrahedron Lett.* 38: 6071–6074.

170 (a) Nishina, N. and Yamamoto, Y. (2007). *Synlett* 1767–1770. (b) Zeng, X., Soleilhavoup, M., and Bertrand, G. (2009). *Org. Lett.* 11: 3166–3169. (c) Duncan, A.N. and Widenhoefer, R.A. (2010). *Synlett* 419–422.

171 Blieck, R., Bahri, J., Taillefer, M., and Monnier, F. (2016). *Org. Lett.* 18: 1482–1485.

172 Selected reports, see: (a) LaLonde, R.L., Sherry, B.D., Kang, E.J., and Toste, F.D. (2007). *J. Am. Chem. Soc.* 129: 2452–2453. (b) Tsuhako, A., Oikawa, D.,

Sakai, K., and Okamoto, S. (2008). *Tetrahedron Lett.* 49: 6529–6532. (c) Li, H., Lee, S.D., and Widenhoefer, R.A. (2011). *J. Organomet. Chem.* 696: 316–320. (d) Casavant, B.J., Khoder, Z.M., Berhane, I.A., and Chemler, S.R. (2015). *Org. Lett.* 17: 5958–5961. (e) Gazzola, S., Beccalli, E.M., Bernasconi, A. et al. (2016). *Eur. J. Org. Chem.* 4534–4544.

173 Perego, L.A., Blieck, R., Michel, J. et al. (2017). *Adv. Synth. Catal.* 359: 4388–4392.

174 Grigg, R., Monteith, M., Sridharan, V., and Terrier, C. (1998). *Tetrahedron* 54: 3885–3894.

175 Selected reports for 1,2-Markovnikov hydroamination of 1,3-dienes, see: (a) Minami, T., Okamoto, H., Ikeda, S. et al. (2001). *Angew. Chem. Int. Ed.* 40: 4501–4503. (b) Brouwer, C. and Chuan He, C. (2006). *Angew. Chem. Int. Ed.* 45: 1744–1747. (c) Goldfogel, M.J., Roberts, C.C., and Meek, S.J. (2014). *J. Am. Chem. Soc.* 136: 6227–6230. (d) Yang, X.-H. and Dong, V.M. (2017). *J. Am. Chem. Soc.* 139: 1774–1777. (e) Adamson, N.J., Hull, E., and Malcolmson, S. (2017). *J. Am. Chem. Soc.* 139: 7180–7183.

176 (a) Löber, O., Kawatsura, M., and Hartwig, J.F. (2001). *J. Am. Chem. Soc.* 123: 4366–4367. (b) Banerjee, D., Junge, K., and Beller, M. (2014). *Angew. Chem. Int. Ed.* 53: 1630–1635.

177 Dzhemilev, U.M., Tolstikov, G.A., and Khusnutdinov, R.I. (2009). *Russ. J. Org. Chem.* 45: 957–987.

178 Yang, X.-H., Lu, A., and Dong, V.M. (2017). *J. Am. Chem. Soc.* 139: 14049–14052.

179 Pawlas, J., Nakao, Y., Kawatsura, M., and Hartwig, J.F. (2002). *J. Am. Chem. Soc.* 124: 3669–3679.

180 Li, C., Kähny, M., and Breit, B. (2014). *Angew. Chem. Int. Ed.* 53: 13780–13784.

181 Fang, Y.-Q., Tadross, P.M., and Jacobsen, E.N. (2014). *J. Am. Chem. Soc.* 136: 17966–17968.

182 Nishina, N. and Yamamoto, Y. (2006). *Angew. Chem. Int. Ed.* 45: 3314–3317.

183 Banerjee, D., Junge, K., and Beller, M. (2014). *Org. Chem. Front.* 1: 368–372.

184 Fang, X., Li, H., Jackstell, R., and Beller, M. (2014). *J. Am. Chem. Soc.* 136: 16039–16043.

185 Qin, H., Yamagiwa, N., Matsunaga, S., and Shibasaki, M. (2006). *J. Am. Chem. Soc.* 128: 1611–1614.

186 (a) Hong, S. and Marks, T.J. (2002). *J. Am. Chem. Soc.* 124: 7886–7887. (b) Hong, S., Kawaoka, A.M., and Marks, T.J. (2003). *J. Am. Chem. Soc.* 125: 15878–15892. (c) Tobisch, S. (2005). *J. Am. Chem. Soc.* 127: 11979–11988.

187 Shapiro, N.D., Rauniyar, V., Hamilton, G.L. et al. (2011). *Nature* 470: 245–249.

188 Dion, I. and Beauchemin, A.M. (2011). *Angew. Chem. Int. Ed.* 50: 8233–8235.

189 Selected reviews, see: (a) Roesky, P.W. and Müller, T.E. (2003). *Angew. Chem. Int. Ed.* 42: 2708–2710. (b) Hultzsch, K.C. (2005). *Adv. Synth. Catal.* 347: 367–391.

190 Kawatsura, M. and Hartwig, J.F. (2000). *J. Am. Chem. Soc.* 122: 9546–9547.

191 Sevov, C.S., Zhou, J., and Hartwig, J.F. (2014). *J. Am. Chem. Soc.* 136: 3200–3207.

192 Yang, Y., Shi, S.-L., Niu, D. et al. (2015). *Science* 349: 62–66.

193 Shen, X. and Buchwald, S.L. (2010). *Angew. Chem. Int. Ed.* 49: 564–567.

194 Vanable, E.P., Kennemur, J.L., Joyce, L.A. et al. (2019). *J. Am. Chem. Soc.* 141: 739–742.

195 Lin, J.-S., Yu, P., Huang, L. et al. (2015). *Angew. Chem. Int. Ed.* 54: 7847–7851.

196 Selected reports, see: (a) Reznichenko, A.L., Nguyen, H.N., and Hultzsch, K.C. (2010). *Angew. Chem. Int. Ed.* 49: 8984–8987. (b) Teng, H.-L., Luo, Y., Wang, B. et al. (2016). *Angew. Chem. Int. Ed.* 55: 15406–15410. (c) Nguyen, H.N., Lee, H., Audörsch, S. et al. (2018). *Organometallics* 37: 4358–4379. and references therein.

197 Adamson, N.J., Jeddi, H., and Malcolmson, S.J. (2019). *J. Am. Chem. Soc.* 141: 8574–8583.

198 (a) Shapiro, N.D., Rauniyar, V., Hamilton, G.L. et al. (2011). *Nature* 470: 245–248. (b) Tran, G., Shao, W., and Mazet, C. (2019). *J. Am. Chem. Soc.* 141: 14814–14822.

199 Michael, F.E. and Cochran, B.M. (2006). *J. Am. Chem. Soc.* 128: 4246–4247.

200 Munro-Leighton, C., Blue, E.D., and Gunnoe, T.B. (2006). *J. Am. Chem. Soc.* 128: 1446–1447.

201 (a) Ackermann, L. and Bergman, R.G. (2002). *Org. Lett.* 4: 1475–1478. (b) Rohjans, S.H., Ross, J.H., Lühning, L.H. et al. (2018). *Organometallics* 37: 4350–4357.

202 Lepori, C., Gómez-Orellana, P., Ouharzoune, A. et al. (2018). *ACS Catal.* 8: 4446–4451.

203 Dochnahl, M., Löhnwitz, K., Lühl, A. et al. (2010). *Organometallics* 29: 2637–2645.

204 Liu, X.-Y., Li, C.-H., and Che, C.-M. (2006). *Org. Lett.* 8: 2707–2710.

205 Bender, C.F. and Widenhoefer, R.A. (2008). *Chem. Commun.* 2741–2743.

206 Liu, G.-Q., Li, W., Wang, Y.-M. et al. (2012). *Tetrahedron Lett.* 53: 4393–4396.

207 Spallek, T. and Anwander, R. (2016). *Dalton Trans.* 45: 16393–16403.

208 Crimmin, M.R., Casely, I.J., and Hill, M.S. (2005). *J. Am. Chem. Soc.* 127: 2042–2043.

209 Bender, C.F. and Widenhoefer, R.A. (2006). *Chem. Commun.* 4143–4144.

210 Hesp, K.D., Tobisch, S., and Stradiotto, M. (2010). *J. Am. Chem. Soc.* 132: 413–426.

211 (a) Müller, C., Loos, C., Schulenberg, N., and Doye, S. (2006). *Eur. J. Org. Chem.* 2499–2503. (b) Müller, C., Koch, R., and Doye, S. (2008). *Chem. Eur. J.* 14: 10430–10436. (c) Müller, C., Saak, W., and Doye, S. (2008). *Eur. J. Org. Chem.* 2731–2739.

212 Youn, S.W., Jang, S.S., and Lee, S.R. (2016). *Tetrahedron* 72: 4902–4909.

213 Takemiya, A. and Hartwig, J.F. (2006). *J. Am. Chem. Soc.* 128: 6042–6043.

214 Nguyen, T.M. and Nicewicz, D.A. (2013). *J. Am. Chem. Soc.* 135: 9588–9591.

215 Hu, Y., Shen, Z., and Huang, H. (2016). *ACS Catal.* 6: 6785–6789.

216 Selected reviews, see: (a) Tanaka, M. (2004). *Top. Curr. Chem.* 232: 25–54. (b) Delacroix, O. and Gaumont, A.-C. (2005). *Curr. Org. Chem.* 9: 1851–1882. (c) Glueck, D.S. (2010). *Top Organomet. Chem.* 31: 65–100. (d) Xu, Q. and

Han, L.-B. (2011). *J. Organomet. Chem.* 696: 130–140. (e) Tanaka, M. (2013). *Top. Organomet. Chem.* 43: 167–202. (f) Koshti, V., Gaikwad, S., and Chikkali, S.H. (2014). *Coord. Chem. Rev.* 265: 52–73. (g) Wauters, I., Debrouwer, W., and Stevens, C.V. (2014). *Beilstein J. Org. Chem.* 10: 1064–1096. (h) Bange, C.A. and Waterman, R. (2016). *Chem. Eur. J.* 22: 12598–12605.

217 Shulyupin, M.O., Kazankova, M.A., and Beletskaya, I.P. (2002). *Org. Lett.* 4: 761–763.

218 Webster, R.L. (2018). *Inorganics* 6: 120.

219 Leyva-Pérez, A., Vidal-Moya, J.A., Cabrero-Antonino, J.R. et al. (2011). *J. Organomet. Chem.* 696: 362–367.

220 Liu, Q., Wang, C., Zhang, X. et al. (2016). *New J. Chem.* 40: 10447–10454.

221 Chadwick, A.C., Heckenast, M.A., Race, J.J. et al. (2019). *Organometallics* 38: 3871–3879.

222 (a) Gallagher, K.J. and Webster, R.L. (2014). *Chem. Commun.* 50: 12109–12111. (b) Espinal-Viguri, M., Mahon, M.F., Tyler, S.N.G., and Webster, R.L. (2017). *Tetrahedron* 73: 64–69.

223 Ganushevich, Y.S., Miluykov, V.A., Polyancev, F.M. et al. (2013). *Organometallics* 32: 3914–3919.

224 Yuan, J., Hu, H., and Cui, C. (2016). *Chem. Eur. J.* 22: 5778–5785.

225 Belli, R.G., Burton, K.M.E., Rufh, S.A. et al. (2015). *Organometallics* 34: 5637–5646.

226 Stelmach, J.P.W., Bange, C.A., and Waterman, R. (2016). *Dalton Trans.* 45: 6204–6209.

227 Crimmin, M.R., Barrett, A.G.M., Hill, M.S. et al. (2007). *Organometallics* 26: 2953–2956.

228 (a) Ghebreab, M.B., Bange, C.A., and Waterman, R. (2014). *J. Am. Chem. Soc.* 136: 9240–9243. (b) Bange, C.A., Mucha, N.T., Cousins, M.E. et al. (2016). *Inorganics* 4: 26.

229 Bange, C.A., Ghebreab, M.B., Ficks, A. et al. (2016). *Dalton Trans.* 45: 1863–1867.

230 (a) Zhang, Y., Wang, X., Wang, Y. et al. (2018). *Dalton Trans.* 47: 9090–9095. (b) Zhang, Y., Qu, L., Wang, Y. et al. (2018). *Inorg. Chem.* 57: 139–149.

231 (a) Alonso, F., Moglie, Y., Radivoy, G., and Yus, M. (2012). *Green Chem.* 14: 2699–2702. (b) Moglie, Y., González-Soria, M.J., Martín-García, I. et al. (2016). *Green Chem.* 18: 4896–4907. (c) Bissessar, D., Egly, J., Achard, T. et al. (2019). *RSC Adv.* 9: 27250–27256.

232 Routaboul, L., Toulgoat, F., Gatignol, J. et al. (2013). *Chem. Eur. J.* 19: 8760–8764.

233 (a) Mimeau, D., Delacroix, O., and Gaumont, A.-C. (2003). *Chem. Commun.* 2928–2929. (b) Mimeau, D., Delacroix, O., Join, B., and Gaumont, A.-C. (2004). *C. R. Chim.* 7: 845–854.

234 (a) Dondoni, A., Staderini, S., and Marra, A. (2013). *Eur. J. Org. Chem.* 5370–5375. (b) Pagano, J.K., Bange, C.A., Farmiloe, S.E., and Waterman, R. (2017). *Organometallics* 36: 3891–3895. (c) Dannenberg, S.G. and Waterman, R. (2020). *Chem. Commun.* 56: 14219–14222.

235 Perrier, A., Comte, V., Moïse, C., and Gendre, P.L. (2010). *Chem. Eur. J.* 16: 64–67.

236 (a) Deprèle, S. and Montchamp, J.-L. (2002). *J. Am. Chem. Soc.* 124: 9386–9387. (b) Candy, M., Rousseaux, S.A.L., Román, A.C.S. et al. (2014). *Adv. Synth. Catal.* 356: 2703–2708.

237 Tayama, O., Nakano, A., Iwahama, T. et al. (2004). *J. Org. Chem.* 69: 5494–5496.

238 Ortial, S., Fisher, H.C., and Montchamp, J.-L. (2013). *J. Org. Chem.* 78: 6599–6608.

239 Yoo, W.-J. and Kobayashi, S. (2013). *Green Chem.* 15: 1844–1848.

240 Zhang, G.-Y., Li, C.-K., Li, D.-P. et al. (2016). *Tetrahedron* 72: 2972–2978.

241 Nan, G. and Yue, H. (2018). *Tetrahedron Lett.* 59: 2071–2074.

242 (a) Huang, T.-Z., Chen, T., Saga, Y., and Han, L.-B. (2017). *Tetrahedron* 73: 7085–7093. (b) Salin, A.V., Il'in, A.V., Faskhutdinov, R.I. et al. (2018). *Tetrahedron Lett.* 59: 1630–1634.

243 Shi, Y., Chen, R., Guo, K. et al. (2018). *Tetrahedron Lett.* 59: 2062–2065.

244 Han, L.-B. and Zhao, C.-Q. (2005). *J. Org. Chem.* 70: 10121–10123.

245 Saga, Y., Han, D., Kawaguchi, S.-i. et al. (2015). *Tetrahedron Lett.* 56: 5303–5305.

246 Saga, Y., Mino, Y., Kawaguchi, S.-i. et al. (2017). *Tetrahedron: Asymmetry* 28: 84–89.

247 Hirai, T. and Han, L.-B. (2007). *Org. Lett.* 9: 53–55.

248 Selected reports, see: (a) Midrier, C., Lantsoght, M., Volle, J.-N. et al. (2011). *Tetrahedron Lett.* 52: 6693–6696. (b) Li, Z., Fan, F., Zhang, Z. et al. (2015). *RSC Adv.* 5: 27853–27856. (c) Geant, P.-Y., Mohamed, B.S., Périgaud, C. et al. (2016). *New J. Chem.* 40: 5318–5324.

249 (a) Bunlaksananusorn, T. and Knochel, P. (2004). *J. Org. Chem.* 69: 4595–4601. (b) Martínez-Castro, E., López, Ó., Maya, I. et al. (2010). *Green Chem.* 12: 1171–1174.

250 Lenker, H.K., Richard, M.E., Reese, K.P. et al. (2012). *J. Org. Chem.* 77: 1378–1385.

251 Geng, Z.-C., Zhang, J.-X., Li, N. et al. (2014). *Tetrahedron* 70: 417–426.

252 Bravo-Altamirano, K., Abrunhosa-Thomas, I., and Montchamp, J.-L. (2008). *J. Org. Chem.* 73: 2292–2301.

253 Kovacik, I., Wicht, D.K., Grewal, N.S. et al. (2000). *Organometallics* 19: 950–953.

254 (a) Sadow, A.D., Haller, I., Fadini, L., and Togni, A. (2004). *J. Am. Chem. Soc.* 126: 14704–14705. (b) Sadow, A.D. and Togni, A. (2005). *J. Am. Chem. Soc.* 127: 17012–17024.

255 Selected reports, see: (a) Feng, J.-J., Chen, X.-F., Shi, M., and Duan, W.-L. (2010). *J. Am. Chem. Soc.* 132: 5562–5563. (b) Huang, Y., Pullarkat, S.A., Li, Y., and Leung, P.-H. (2010). *Chem. Commun.* 46: 6950–6952. (c) Chen, Y.-R. and Duan, W.-L. (2011). *Org. Lett.* 13: 5824–5826. (d) Huang, Y., Chew, R.J., Li, Y. et al. (2011). *Org. Lett.* 13: 5862–5865. (e) Du, D. and Duan, W.-L. (2011). *Chem. Commun.* 47: 11101–11103. (f) Huang, Y., Pullarkat, S.A., Li, Y., and Leung, P.-H. (2012). *Inorg. Chem.* 51: 2533–2540. (g) Xu, C., Kennard,

G.J.H., Hennersdorf, F. et al. (2012). *Organometallics* 31: 3022–3026. (h) Du, D., Lin, Z.-Q., Lu, J.-Z. et al. (2013). *Asian J. Org. Chem.* 2: 392–394. (i) Li, C., Bian, Q.-L., Xu, S., and Duan, W.-L. (2014). *Org. Chem. Front.* 1: 541–545. (j) Hao, X.-Q., Huang, J.-J., Wang, T. et al. (2014). *J. Org. Chem.* 79: 9512–9530. (k) Hao, X.-Q., Zhao, Y.-W., Yang, J.-J. et al. (2014). *Organometallics* 33: 1801–1811. (l) Chew, R.J., Lu, Y., Jia, Y.-X. et al. (2014). *Chem. Eur. J.* 20: 14514–14517. (m) Chew, R.J., Teo, K.Y., Huang, Y. et al. (2014). *Chem. Commun.* 50: 8768–8770. (n) Chew, R.J., Li, X.-R., Li, Y. et al. (2015). *Chem. Eur. J.* 21: 4800–4804. (o) Yang, X.-Y., Tay, W.S., Li, Y. et al. (2015). *Organometallics* 34: 5196–5201. (p) Xu, Y., Yang, Z., Ding, B. et al. (2015). *Tetrahedron* 71: 6832–6839. (q) Chew, R.J., Sepp, K., Li, B.-B. et al. (2015). *Adv. Synth. Catal.* 357: 3297–3302. (r) Yang, X.-Y., Gan, J.H., Li, Y. et al. (2015). *Dalton Trans.* 44: 1258–1263. (s) Song, Y.-C., Dai, G.-F., Xiao, F., and Duan, W.-L. (2016). *Tetrahedron Lett.* 57: 2990–2993. (t) Yang, X.-Y., Jia, Y.-X., Tay, W.S. et al. (2016). *Dalton Trans.* 45: 13449–13455.

256 (a) Carlone, A., Bartoli, G., Bosco, M. et al. (2007). *Angew. Chem. Int. Ed.* 46: 4504–4506. (b) Ibrahem, I., Rios, R., Vesely, J. et al. (2007). *Angew. Chem. Int. Ed.* 46: 4507–4510. (c) Ibrahem, I., Hammar, P., Vesely, J. et al. (2008). *Adv. Synth. Catal.* 350: 1875–1884.

257 Huang, Y., Pullarkat, S.A., Teong, S. et al. (2012). *Organometallics* 31: 4871–4875.

258 (a) Feng, J.-J., Huang, M., Lin, Z.-Q., and Duan, W.-L. (2012). *Adv. Synth. Catal.* 354: 3122–3126. (b) Ding, B., Zhang, Z., Xu, Y. et al. (2013). *Org. Lett.* 15: 5476–5479.

259 Lu, J., Ye, J., and Duan, W.-L. (2013). *Org. Lett.* 15: 5016–5019.

260 Chew, R.J., Huang, Y., Li, Y. et al. (2013). *Adv. Synth. Catal.* 355: 1403–1408.

261 Lu, J., Ye, J., and Duan, W.-L. (2014). *Chem. Commun.* 50: 698–700.

262 Huang, J., Zhao, M.-X., and Duan, W.-L. (2014). *Tetrahedron Lett.* 55: 629–631.

263 (a) Sadeer, A., Ong, Y.J., Kojima, T. et al. (2018). *Chem. Asian J.* 13: 2829–2833. (b) Lu, Z., Zhang, H., Yang, Z. et al. (2019). *ACS Catal.* 9: 1457–1463.

264 Huang, Y., Pullarkat, S.A., Yuan, M. et al. (2010). *Organometallics* 29: 536–542.

265 Xu, Q. and Han, L.-B. (2006). *Org. Lett.* 8: 2099–2101.

266 Selected reviews, see: (a) Kondo, T. and Mitsudo, T.-a. (2000). *Chem. Rev.* 100: 3205–3220. (b) Ogawa, A. (2000). *J. Organomet. Chem.* 611: 463–474. (c) Wang, L., He, W., and Yu, Z. (2013). *Chem. Soc. Rev.* 42: 599–621. (d) Beletskaya, I.P. and Ananikov, V.P. (2011). *Chem. Rev.* 111: 1596–1636. (e) Giuseppe, A.D., Castarlenas, R., and Oro, L.A. (2018). *Top. Organomet. Chem.* 61: 31–68.

267 Tamai, T. and Ogawa, A. (2014). *J. Org. Chem.* 79: 5028–5035.

268 Xi, H., Ma, E., and Li, Z. (2016). *Tetrahedron* 72: 4111–4116.

269 Ma, H., Ren, X., Zhou, X. et al. (2015). *Tetrahedron Lett.* 56: 6022–6029.

270 Taniguchi, N. (2020). *J. Org. Chem.* 85: 6528–6534.

271 Kanagasabapathy, S., Sudalai, A., and Benicewicz, B.C. (2001). *Tetrahedron Lett.* 42: 3791–3794.

272 Savolainen, M.A. and Wu, J. (2013). *Org. Lett.* 15: 3802–3804.

273 Sinha, A.K. and Equbal, D. (2019). *Asian J. Org. Chem.* 8: 32–47. and references therein.
274 Upadhyay, R., Rana, R., Sood, A., and Maurya, S.K. (2019). *ACS Omega* 4: 15101–15106.
275 Chun, S., Chung, J., Park, J.E., and Chung, Y.K. (2016). *ChemCatChem* 8: 2476–2481.
276 Biermann, U. and Metzger, J.O. (2018). *Eur. J. Org. Chem.* 730–734.
277 Tamai, T., Fujiwara, K., Higashimae, S. et al. (2016). *Org. Lett.* 18: 2114–2117.
278 Kristensen, S.K., Laursen, S.L.R., Taarning, E., and Skrydstrup, T. (2018). *Angew. Chem. Int. Ed.* 57: 13887–13891.
279 Li, Y., Cai, J., Hao, M., and Li, Z. (2019). *Green Chem.* 21: 2345–2351.
280 Ogawa, A., Kawakami, J.-i., Sonoda, N., and Hirao, T. (1996). *J. Org. Chem.* 61: 4161–4163.
281 Pritzius, A.B. and Breit, B. (2015). *Angew. Chem. Int. Ed.* 54: 3121–3125.
282 Kumar, G., Qu, Z.-W., Ghosh, S. et al. (2019). *ACS Catal.* 9: 11627–11633.
283 Yang, X.-H., Davison, R.T., and Dong, V.M. (2018). *J. Am. Chem. Soc.* 140: 10443–10446.
284 Yang, X.-H., Davison, R.T., Nie, S.-Z. et al. (2019). *J. Am. Chem. Soc.* 141: 3006–3013.
285 (a) Xiao, W.-J., Vasapollo, G., and Alper, H. (1998). *J. Org. Chem.* 63: 2609–2612. (b) Xiao, W.-J. and Alper, H. (1998). *J. Org. Chem.* 63: 7939–7944. (c) Xiao, W.-J., Vasapollo, G., and Alper, H. (2000). *J. Org. Chem.* 65: 4138–4144. (d) Li, C.-F., Xiao, W.-J., and Alper, H. (2009). *J. Org. Chem.* 74: 888–890.
286 Selected reports, see: (a) Kajitani, M., Kamiya, I., Nomoto, A. et al. (2006). *Tetrahedron* 62: 6355–6360. (b) Hirschbeck, V., Gehrtz, P.H., and Fleischer, I. (2016). *J. Am. Chem. Soc.* 138: 16794–16799. (c) Yan, D.-M., Crudden, C.M., Chen, J.-R., and Xiao, W.-J. (2019). *ACS Catal.* 9: 6467–6483.
287 Xiao, W.-J. and Alper, H. (2001). *J. Org. Chem.* 66: 6229–6233.
288 Nie, S., Lu, A., Kuker, E.L., and Dong, V.M. (2021). *J. Am. Chem. Soc.* 143: 6176–6184.
289 Ogawa, A., Kudo, A., and Hirao, T. (1998). *Tetrahedron Lett.* 39: 5213–5216.
290 Tamai, T., Yoshikawa, M., Higashimae, S. et al. (2016). *J. Org. Chem.* 81: 324–329. and references cited therein.
291 (a) Izumi, Y. (1997). *Catal. Today* 33: 371–409. (b) Bruneau, C. (2013). *Top. Organomet. Chem.* 43: 203–230.
292 Wang, X., Sui, D., Huang, M., and Jiang, Y. (2006). *Polym. Adv. Technol.* 17: 163–167.
293 (a) Dong, G., Teo, P., Wickens, Z.K., and Grubbs, R.H. (2011). *Science* 333: 1609–1612. (b) Hintermann, L. (2012). *ChemCatChem* 4: 321–322.
294 Hu, X., Guoting Zhang, G., Bu, F., and Lei, A. (2017). *ACS Catal.* 7: 1432–1437.
295 Ceylan, Y.S. and Cundari, T.R. (2019). *J. Phys. Chem. A* 123: 958–965.
296 Zhang, Z., Lee, S.D., Fisher, A.S., and Widenhoefer, R.A. (2009). *Tetrahedron* 65: 1794–1798.
297 Saito, S., Dobashi, N., and Wakatsuki, Y. (2005). *Chem. Lett.* 34: 504–505.

298 Hirai, T., Hamasaki, A., Nakamura, A., and Tokunaga, M. (2009). *Org. Lett.* 11: 5510–5513.

299 Ke, F., Li, Z., Xiang, H., and Zhou, X. (2011). *Tetrahedron Lett.* 52: 318–320.

300 Yang, C.-G. and He, C. (2005). *J. Am. Chem. Soc.* 127: 6966–6967.

301 (a) Kovács, G., Lledós, A., and Ujaque, G. (2010). *Organometallics* 29: 3252–3260. (b) Kovács, G., Lledós, A., and Ujaque, G. (2010). *Organometallics* 29: 5919–5926.

302 Sevov, C.S. and Hartwig, J.F. (2013). *J. Am. Chem. Soc.* 135: 9303–9306.

303 Dzudza, A. and Marks, T.J. (2010). *Chem. Eur. J.* 16: 3403–3422.

304 Choi, J.-C., Kohno, K., Masuda, D. et al. (2008). *Chem. Commun.* 777–779.

305 Oe, Y., Ohta, T., and Ito, Y. (2010). *Tetrahedron Lett.* 51: 2806–2809.

306 Taylor, J.G., Whittall, N., and Hii, K.K.M. (2005). *Chem. Commun.* 5103–5105.

307 Chandrasekhar, B. and Ryu, J.-S. (2012). *Tetrahedron* 68: 4805–4812.

5

Double Functionalization of Alkynes and Alkenes by Addition of Element–Element Bonds

Transition metal-catalyzed addition reaction of element–element bonds (E—E, E = heteroatom or functional group) to unsaturated hydrocarbons has been well developed for direct approach to double functionalized compounds, and various inter-element linkages from the same groups or different groups have been investigated for this transformation [1]. Particularly, the addition reactions of alkynes are most abstracted to prepare the doubly functionalized alkenes, which are useful synthetic intermediates.

In this chapter, the addition reactions of E—E bonds from the same groups, involving the inter-element linkages of each group from 13 to 16, are described first, and then the inter-element linkages from different groups are summarized in Section 5.5.

The mechanism for most addition reactions of E—E bonds to unsaturated hydrocarbons catalyzed by transition metal complexes is similar to that of the addition of H—E bonds, including oxidative addition of E—E to transition metal, insertion reaction of unsaturated hydrocarbons, and final reductive elimination as depicted in Chapter 3, Scheme 3.2.

5.1 Addition Reaction of Group 13 Element–Element Bonds

The reported procedures for the activation and addition reactions of main group 13 element–element bonds to unsaturated hydrocarbons focus on B—B bond, which provide an efficient and atom economic method for the synthesis of organoboron compounds, and the addition reactions could be achieved in the presence of transition metals or transition metal-free conditions [2]. In particular, vicinal diborated alkenes are the versatile intermediates in organic synthesis via activation of C—B bond and its coupling reactions [3].

5.1.1 *cis*-Addition Reactions to Alkynes

The first transition metal-catalyzed diboration of terminal and internal alkynes was reported by Suzuki and Miyaura's group in 1993 with the use of $Pt(PPh_3)_4$ as catalyst (Scheme 5.1) [4]. The $Pd(PPh_3)_4$-catalyzed cross-coupling reaction of

Addition Reactions with Unsaturated Hydrocarbons, First Edition. Ruimao Hua.

diboration adduct with iodobenzene gives (*Z*)-1,2-diphenyl-1-decene as the sole product to confirm the diboration of alkynes to be *cis*-addition manner. A variety of platinum catalyst systems were then reported for *cis*-diboration of alkynes [5].

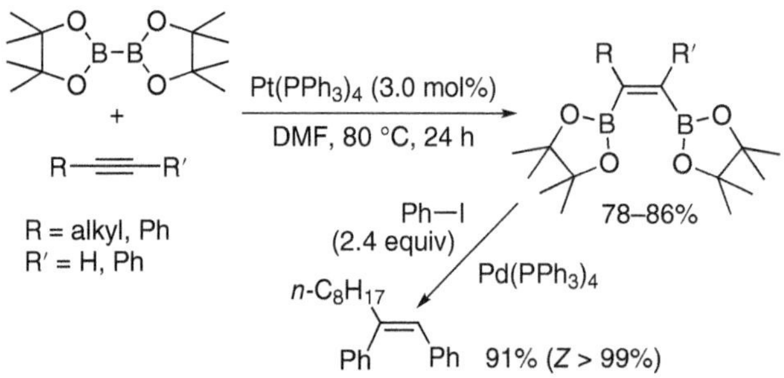

Scheme 5.1 Platinum-catalyzed diboration of alkyne and transformation of adduct.

A proposed mechanism for the diboration of alkynes is shown in Scheme 5.2. It involves the first oxidative addition of B—B bond to the platinum (0) complex, the stereospecific insertion of alkyne to the B—Pt bond, and finally the reductive elimination of the bis(boryl)alkene. The oxidative addition of B—B bond to a variety of transition metal complexes has been well studied, and the complexes having B—M bond have been isolated and characterized [6].

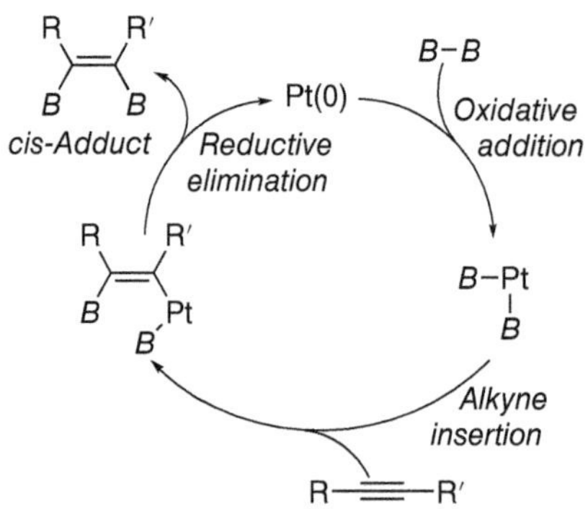

Scheme 5.2 Proposed mechanism for diboration of alkyne.

Copper complexes not only are cheap catalysts but also show interesting catalytic activity in the *cis*-diboration of alkynes. The *cis*-diboration of alkynes and styrenes with bis(catecholato)diboron catalyzed by copper complex was first reported by Fernández and coworker in 2007 [7], and then a few copper catalysis have been developed [8].

Interestingly, in the presence of $Cu(OAc)_2$/phosphines, the addition reaction of internal alkynes with B_2pin_2 occurs smoothly affording diborylated products in a *cis* fashion; arynes also show reactivity to give *vic*-diborylarenes [9]. The most striking feature of this copper catalysis is to undergo the substitution of MeO groups by boryl groups. Therefore, as shown in Scheme 5.3, the reaction of 1,4-dimethoxy-2-butyne with $(pin)_2B_2$ produces a tetraborylated product of 1,2,3,4-tetraboryl-2-butene, and all four C—B bonds form in one pot. The results are completely different from that obtained by using the established platinum catalyst. The simple $FeBr_2$/LiOMe was also used in the *cis*-diboration of alkynes [10].

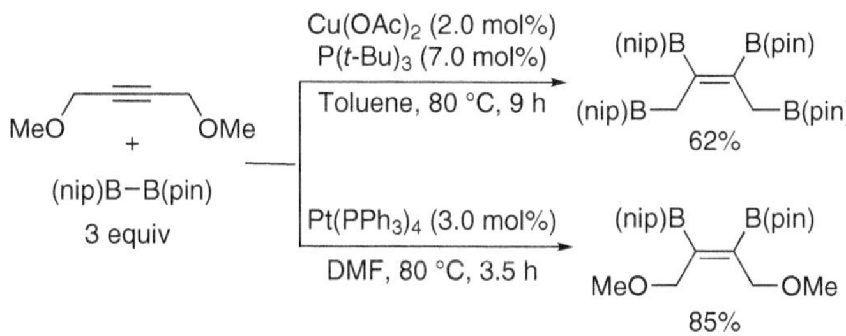

Scheme 5.3 Copper-catalyzed diboration of propargyl ether.

The single-atomic-site catalysts of $Pt_1/Ni(OH)_x$ have been recently prepared by Li's group and show high catalytic efficiency in diboration of a variety of alkynes and alkenes with B_2pin_2 [11]. In the cases of terminal alkynes used, the addition reactions give *cis*-adducts, and the overall turnover frequency is ~3000 hours^{-1} in the diboration of phenylacetylene, which is much higher than other reported heterogeneous catalysts.

5.1.2 *trans*-Addition Reactions to Alkynes

Compared with the *cis*-diboration of alkynes, the catalyst systems for the synthesis of vicinal *trans*-diborated alkenes via a *trans*-diboration of alkynes are relatively rare.

The first *trans*-diboration reaction was reported by Hirano and Uchiyama's group in the reaction of propargyl alcohols with $B_2(pin)_2$ in the presence of BuLi, and the reaction afforded functionalized 4-borylated 1,2-oxaborol-2(5*H*)-oles (vinyldiboronates) [12]. A detailed computational analysis discloses that the directing alkoxide functionality resulted from the reaction of propargyl alcohols with BuLi is the key fact to lower the activation energy of B—C bond formation and to control the stereoselectivity.

The diboration of alkynes can be also achieved under transition metal-free conditions. For example, Ohmiya and Sawamura group have reported a PBu_3-catalyzed *trans*-diboration of alkynoates to produce α,β-diboryl acrylates (Scheme 5.4) [13].

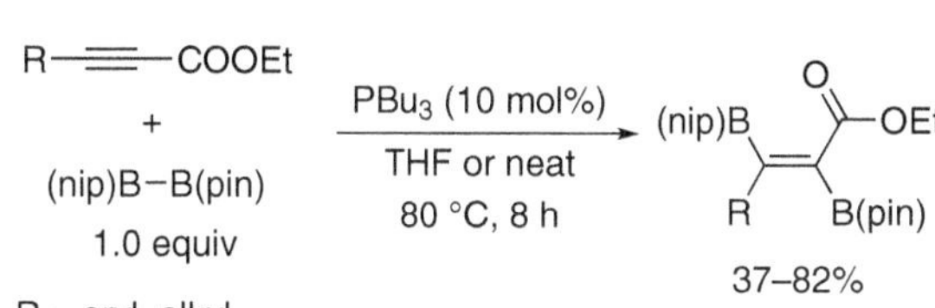

Scheme 5.4 *Anti*-diboration of alkynoates catalyzed by trialkylphosphines.

The *trans*-addition stereoselectivity is complete and robust, and also the ester group is the key functionality to achieve the *trans*-addition via the coordination between carbonyl oxygen with the boron atom.

In addition, a substrate-assisted, transition metal-free, and regio- and stereoselective diboration of alkynamides with mixed diboron to *trans*-1,2-vinyldiboronates has been also developed [14].

Very recently, the inorganic base-catalyzed diborylation of alkynes approach to *cis*-1,2-bis(boryl)alkenes [15], inorganic base-catalyzed diborylation of diaryl alkynes with B_2pin_2 to *trans*-bis(boryl)alkenes [16], and *cis*- or *trans*-1,2-diborylalkenes forming through the addition reactions between unsymmetrical diboranes and a variety of alkynes catalyzed by base [17] have been reported.

A hydroxyl-directed, alkoxide-catalyzed diboration of alkenyl alcohols has been used in the synthesis of dihydroxylation of alkenes via a subsequent oxidation [18].

The metal/metal oxide nanoparticles have been extensively used in the borylation reactions of alkynes [19]. For example, gold(0) nanoparticles [20] and iron oxide nanoparticles supported on magnesia (FeO/MgO) [21] can catalyze direct diboration of alkynes giving vicinal *cis*-diborated alkenes.

5.1.3 Addition Reactions to Alkenes

The vicinal diborated alkanes are also interesting diboron compounds, and like vicinal diborated alkenes, they can be directly prepared by the diboration of alkenes. The organocatalytic diborations of alkenes are also interesting topics, and a variety of organic compounds have been confirmed to be the efficient catalysts. Recently, the cheap and simple amines and pyridines have been reported to activate the diboron reagents and 2-MeTHF and be used in the diboration of alkenes (Scheme 5.5) [22]. The catalytic enantioselective diboration of alkenes has been accomplished with the use of readily available carbohydrate-derived catalysts [23].

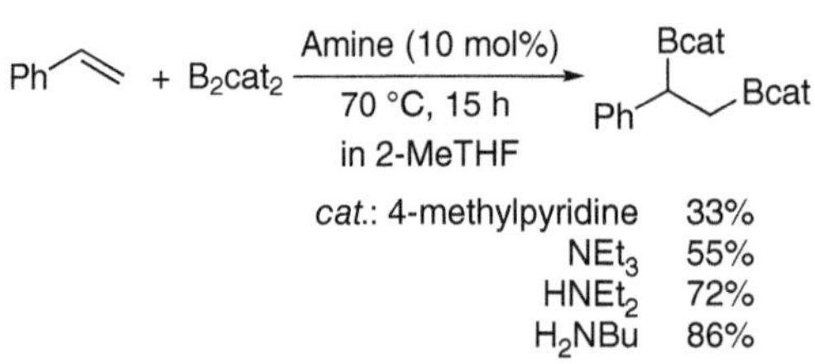

Scheme 5.5 Amine-catalyzed diboration of alkenes.

It has been found that alkoxides are the efficient promoters to catalyze the diboration of alkenes [24], and B_2pin_2 can be activated by alkoxide formed in situ from MeOH and a base through the formation of Lewis acid–base adduct $[Hbase]^+$ $[MeO\text{-}B_2pin_2]^-$. Therefore, under the MeOH/base conditions, 1,2,3-triborated compounds can be prepared by the reaction of allylic alcohols [25] and 1,3-dienes with B_2pin_2 [26].

In addition, vicinal diboronates in high enantiomeric purity could be obtained through tandem site-selective Cu(I)/NHC-catalyzed diboration of terminal alkynes with B_2pin_2 [27].

Recently, under light condition, organosulfide- [28] and organophosphine-catalyzed [29] diborations of alkynes have also been developed.

On the other hand, a few comprehensive reviews on the diboration of alkenes [30] and an elegant review on the transition metal-catalyzed borylation of α,β-unsaturated carbonyl compounds via B—B bond activation giving β-borated carbonyl compounds [31] have appeared.

In addition, in the presence of Au/TiO_2 nanoparticles, terminal allenes underwent diboration with B_2pin_2 exclusively on the terminal double bond in high yields and stereoselectivity (Scheme 5.6) [32]. Under the similar conditions, silaboration reaction also occurred with high regioselectivity to attach boron moiety (Bpin) on the terminal carbon and the silyl group on the sp-carbon. In addition, the catalyst could be recyclable and reusable.

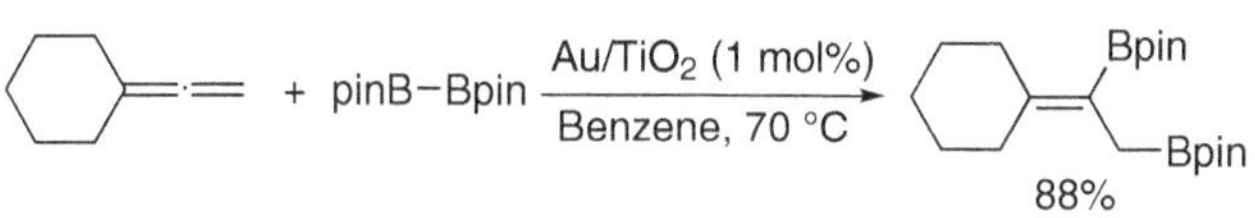

Scheme 5.6 Regioselective diboration of terminal allenes catalyzed by Au/TiO_2 nanoparticles.

Gold(0) nanoparticles stabilized with BINAP ligand show the catalytic diboration of styrene resulting in complete formation of the bis(boronate)esters [33].

In addition, in the presence of a chiral phosphoramidite ligand, Pd_2dba_3-catalyzed enantioselective diboration of allenes with $B_2(pin)_2$ has also been reported [34].

Moreover, a catalytic stereoselective 1,4-diboration of 1,3-dienes with $B_2(pin)_2$ was also reported with the use of $Ni(cod)_2$ and PCy_3 as the catalyst systems, and the intermediate allylboronate could be oxidized to the stereodefined allylic 1,4-diol [35].

Interestingly, palladium-catalyzed reactions of alkenes with $B_2(pin)_2$ could give various diborylalkenes such as 1,1-, *trans*-1,2-, and cyclic 1,2-diborylalkenes via the selective dehydrogenative borylation [36].

5.1.4 Synthesis of 1,1-diborylalkanes/Alkenes via Addition of B—B Bond

1,1-Diboryl compounds (*gem*-diboryl compounds), such as 1,1-diborylalkanes and 1,1-diborylalkenes, are interesting bifunctional reagents in organic synthesis, particularly in the formation of C—C bond; several comprehensive reviews on the recent advances in their preparation, transformation, and application have been recently reported [37].

1,1-Diborylalkanes can be prepared from the regioselective either diboration of terminal alkynes (Scheme 5.7) or double hydroborylation, which was summarized in Section 3.1.

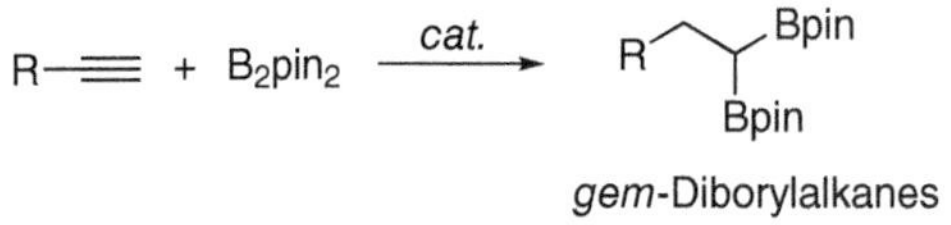

Scheme 5.7 *gem*-Diboryl alkanes from alkynes.

Song and coworkers first reported the procedure for the synthesis of 1,1-diborylalkanes by the addition of B_2pin_2 to terminal alkynes in the presence of a base [38].

Very recently, in the presence of LiOMe/NEt_3, $Ni(cod)_2$-catalyzed 1,1-diboration of terminal alkenes with B_2pin_2 to 1,1-diborylalkanes [39] and quadruple borylation of terminal alkynes with B—B bond under photoinduced and thermal conditions access to 1,1,2,2-tetrakis(boronate)s [40] have also been developed.

Marder and coworker have developed the efficient synthesis of 1,1,2-triborylalkene from the addition reactions of alkyne with B_2pin_2. As shown in Scheme 5.8, 1,1,2-triborylalkene could be obtained from either the desilylative borylation of bis(trimethylsilyl)acetylene, and subsequent diboration in the presence of platinum complex [41], or a $Cu(OAc)_2$-catalyzed triboration of terminal alkynes with B_2pin_2 [42]. In the latter case, aliphatic terminal alkynes and electron-rich and electron-poor aromatic terminal alkynes could undergo the addition reactions to give the corresponding 1,1,2-triborylalkenes in good yields.

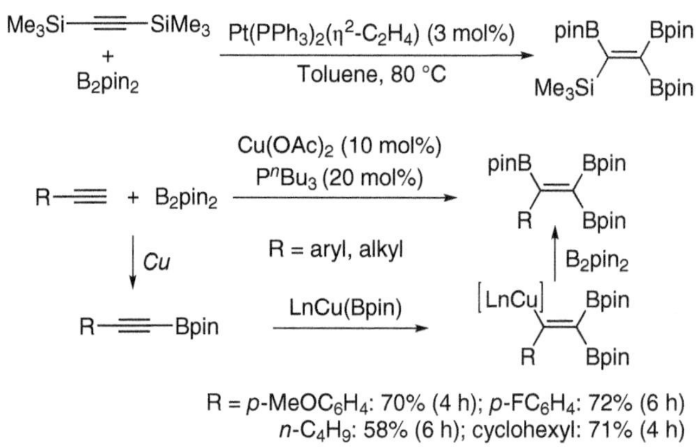

Scheme 5.8 1,1,2-Triborylalkene synthesis from borylation of alkynes with B_2pin_2.

The synthesis of 1,1-diborylalkenes through a Brønsted base, LiO^tBu-catalyzed reactions between terminal alkynes and B_2pin_2 [43], and NaO^tBu-catalyzed reactions between terminal alkynyl esters and amides and unsymmetrical diboron reagents [44] has also been developed.

In addition, very recently, Cu(I)-catalyzed borylation of *gem*-difluoroalkenes with B_2pin_2 via dual C—F bond activation to afford *multi*-borylated compounds, such as 1,2-alkyldiboronates, 1,1,2-alkyltriboronates, and 1,1,1,2-alkyltetraboronates, has also been developed [45].

5.2 Addition Reaction of Group 14 Element–Element Bonds

The known procedures focus on the activation and their addition reactions of Si—Si, Sn—Sn, and Si—Sn bonds (Scheme 5.9) [46].

Scheme 5.9 Addition reaction of group 14 element–element bond to unsaturated hydrocarbons.

The pioneering work on the *cis*-disilylation of alkynes with disilanes was reported by Kumada and coworker in the presence of $PdCl_2(PEt_3)_2$ in refluxed benzene [47]. The addition reactions of 1,1,2,2-tetramethyldisilane to dimethyl acetylenedicarboxylate and phenylacetylene produced (*Z*)-adducts in mild to low yields, and hexamethyldisilane failed to undergo the similar addition reaction under the conditions employed.

At the same time, Sakurai and coworker reported the disilylation of alkynes with 1,1,2,2-tetramethyl-1,2-disilacyclopentane catalyzed by $PdCl_2(PPh_3)_2$, and when dimethyl acetylenedicarboxylate was used, dimethyl 1,1,4,4-tetramethyl-1,4-disilacyclohept-2-ene-2.3-dicarboxylate was obtained in 83% yield (Scheme 5.10) [48]. They then developed the disilylation of alkynes with fluorodisilanes in the presence of $Pd(PPh)_4$ to give (*Z*)-disilylalkenes in 60–95% [49]. Afterward, palladium- and platinum- [50]catalyzed disilylations of alkyne with different disilanes were extensively studied.

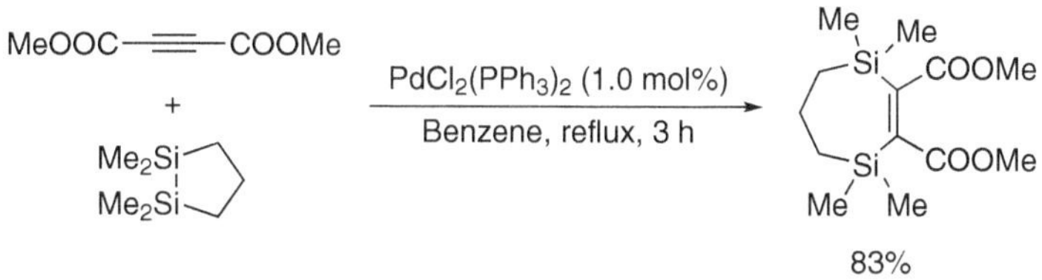

Scheme 5.10 Palladium-catalyzed disilylation of alkyne giving 1,4-disilacyclohept-2-ene.

In addition, palladium-catalyzed insertion of alkynes into Si—Si bond in polymers was also reported by Tanaka's group [51].

Recently, $Pd(acac)_2/^tBuNC$-catalyzed *cis*-disilylation of terminal alkynes with unsymmetrical disilanes with high regioselectivity has also been reported by Song and coworker [52].

Stratakis and coworker reported that the supported gold nanoparticle (Au/TiO_2) showed high catalytic activity in the disilylation of terminal alkynes with hexa-substituted disilanes to give *cis*-adducts in good to high yields with high regio- and stereoselectivity [53]. They then developed the synthesis of substituted 1,4-disila-2,5-cyclohexadienes in moderate to good yields through a tandem Si—Si and Si—H bonds activation of 1,1,2,2-tetramethyldisilane and its reaction with alkynes catalyzed by the same nanoparticles (Scheme 5.11) [54]. The cycloaddition reactions proceed via initial Si—Si activation and addition to alkynes forming the isolable *cis*-1,2-bis(dimethylsilyl)ethenes, which further undergo dehydrogenative cycloaddition to a second alkyne molecule affording the final cycloadducts. In addition, the gold nanoparticles were also used in the regioselective hydrosilylation of allenes [55] and dehydrogenative *cis*-1,2-disilylation of terminal alkynes with dihydrosilanes [56].

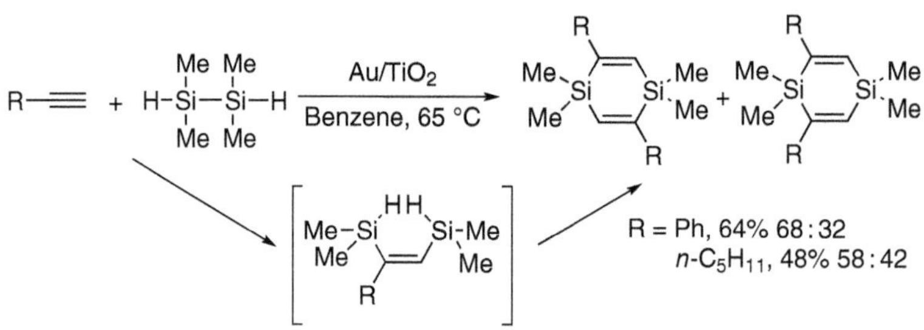

Scheme 5.11 Au/TiO_2-catalyzed cycloaddition of tetramethyldisilane with terminal alkynes.

Substituted 1,4-disila-2,5-cyclohexadienes could also be obtained by the reaction between cyclic hexasilane Si_6Me_{12} and terminal/internal alkynes via successive Si—Si bond activation in the presence of palladium/isocyanide catalysts [57].

Mitchell and coworker reported the formation of (*Z*)-1,2-bis(trimethylstannyl)-1-alkenes by $Pd(PPh_3)_4$-catalyzed addition of Me_3Sn—$SnMe_3$ to terminal alkynes [58], and the transformation of C—Sn bonds was also studied [59].

$Pd(PPh_3)_4$ was also used as the catalyst in the *cis*-distannation of α,β-acetylenic esters and *N*,*N*-dimethyl α,β-acetylenic amides with Me_3Sn—$SnMe_3$ to give the corresponding (*Z*)-2,3-bis(trimethylstannyl)alk-2-enoates and (*Z*)-2,3-bis(trimethyl-stannyl)alk-2-enamides, respectively [60].

The catalytic *cis*-distannylation of terminal and internal alkynes with hex-aalkylditins (alkyl = Me, *n*-Bu) could be achieved in the presence of Pd(II) [61], Cu(I) [62] complexes, and tungsten isonitrile complex [63]. A cationic gold(I) complex bearing a phosphite ligand also catalyzes the *cis*-bis(stannylation) of propiolates with hexabutylditin under mild conditions [64].

In addition, the distannylation of strained carbon–carbon triple bonds such as arynes or cyclohexynes catalyzed by Pd(II) complex has also been developed [65].

On the other hand, the examples of intermolecular *trans*-disilylation of alkynes are few. Recently, Zhang and He's group has reported the intermolecular *trans*-disilylation of alkynones with disilanes catalyzed by $Rh(nbd)_2BF_4$ (Scheme 5.12) [66].

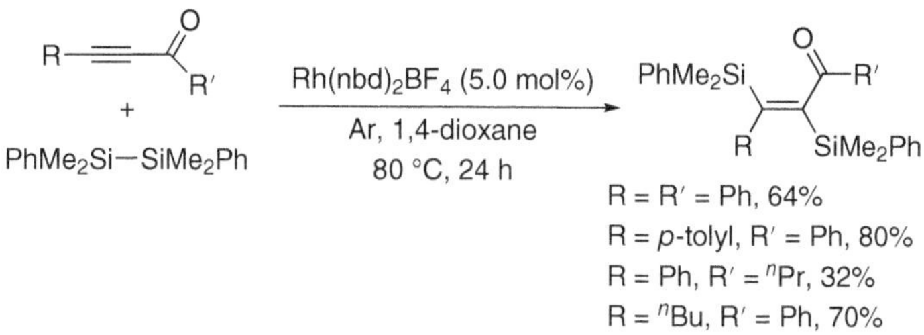

Scheme 5.12 Rhodium-catalyzed intermolecular *trans*-disilylation of alkynones.

Mitchell and coworker have studied the addition reaction of addition of hexaalkylditins R_6Sn_2 (R = Me, Et, Bu) to alkynes in the presence of $Pd(PPh_3)_4$ and found that the addition reaction occurs to give *cis*-adducts, and the (*Z*)-distannylalkenes can be partially or completely isomerized to the (*E*)-alkenes by ultraviolet (UV) irradiation [58].

Note that some catalyst systems previously mentioned are also efficient for the disilylation of alkenes. In addition, $Pt(PPh_3)_4$ shows high catalytic activity for disilylation of terminal and/or internal alkenes [67], and, very recently, a transition metal-free, silylium-ion-promoted disilylation of unactivated terminal and internal alkenes with $Me_3Si—SiMe_3$ has been developed [68].

1,4-Bis(organosilyl)but-2-enes could be prepared by disilylation of 1,3-butadienes with various substituted hexaorganodisilanes (Scheme 5.13).

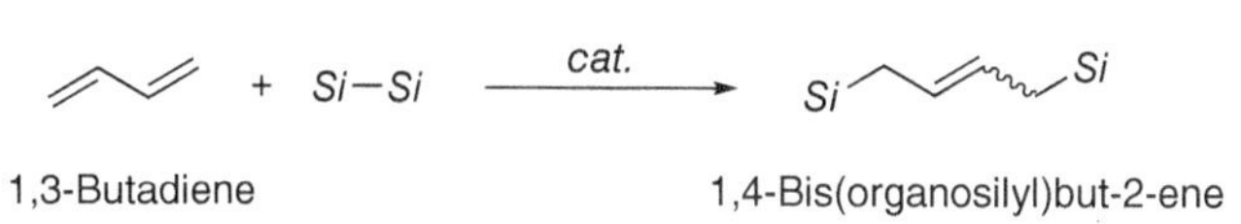

Scheme 5.13 Synthesis of 1,4-bis(organosilyl)but-2-enes by disilylation of 1,3-butadienes.

The pioneering work includes the $NiCl_2(PEt_3)_2$-catalyzed disilylation with 1,2-dihydro-1,1,2,2-tetramethyldisilane to give (*Z*)-1,4-bis(dimethylsilyl)but-2-enes [69]; $PdCl_2L_2$ (L = PPh_3, PEt_3, PhCN)-catalyzed 2 : 1 cycloaddition with strained disilacycloalkanes, hexamethyldisilane, or 1,2-dichloro-1,1,2,2-tetramethyldisilane [70]; $PdCl_2(PPh_3)_2$- and $Pd(PPh_3)_4$-catalyzed disilylation with 1,2-difluorodisilanes giving both 1 : 1 and 1 : 2 adducts [71]; and $Pd(PPh_3)_4$-catalyzed disilylation with chlorodisilanes of type $Cl_{3-n}Me_nSi—SiMe_nCl_{3-n}$ to afford (*Z*)-1,4-bis(chlorosilyl)but-2-enes [72]. Platinum complexes were also reported to be the efficient catalysts for 1,4-disilylation of 1,3-dienes [73].

The first disilylation of allene and 1,2-butadiene with various chloromethyl- and methoxymethyldisilanes, $X_3\_mMe_mSi—SiMe_nX_{3-n}$ (X = Cl, OMe; m, n = 0–2), catalyzed by $Pd(PPh_3)_4$ was reported by Nagai and coworker [74]. The addition reactions took place regioselectively to give the corresponding 1 : 1 adducts of

2,3-bis(organosilyl)prop-1-enes and 2,3-bis(organosilyl)but-1-enes, respectively, in reasonable to good yields.

Interestingly, the highly regio- and stereoselective dimerization–disilylation [75] and dimerization–distannylation [76] of 1,3-dienes were reported in the presence of a catalytic amount of $Pd(dba)_2$.

$Pd(PPh_3)_4$ was used as the efficient catalyst in the distannylation of allenes [77] and distannylation of cyclo-1,2-dienes (C_9–C_{13}) with $Me_3Sn—SnMe_3$ providing 2,3-bis-(trimethylstannyl)cycloalk-1-enes [78]. As shown in Scheme 5.14, under solvent-free conditions, the addition reaction of cyclonona-1,2-diene gave (*E*)-distannane in excellent yield with excellent stereoselectivity, and in the case of cyclodeca-1,2-diene used, both *E*- and *Z*-adducts in a ratio of 1 : 2 were obtained after purification.

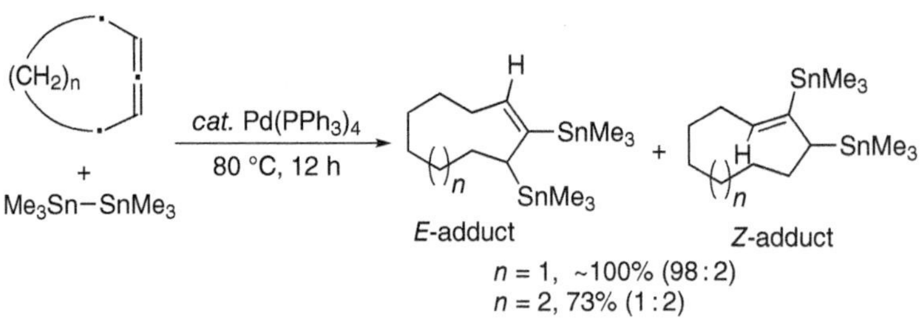

Scheme 5.14 $Pd(PPh_3)_4$-catalyzed distannylation of cyclo-1,2-dienes.

For Si—Sn bond addition reactions, $Pd(PPh_3)_4$ was also found to be of high catalytic activity to catalyze the addition of silylstannanes $R_3Si—SnR_3$ (R = alkyl) to terminal alkynes with high regio- and stereoselectivity to give *cis*-adducts with tin always adding to the internal position [79]. Note that the regio- and stereoselectivity might change depending on the substituents of alkynes and Si—Sn reagents.

However, when the same catalyst was used in the addition reactions of $Bu_3Sn—GeMe_3$ to α,β-acetylenic esters, the addition reactions afforded predominantly the *trans*-adducts of (*E*)-2-tri-*n*-butylstannyl-3-trimethylgermylalk-2-enoates, with $GeMe_3$ situated adjacent to the CO_2Me function [80].

In addition, a palladium-catalyzed selective silylstannylation–cyclization of 1,2-dien-7-ynes and 1,2-dien-8-ynes has also been developed to provide the simple and efficient synthesis of highly functionalized carbocyclic and heterocyclic compounds with silicon and tin substituents on the double bonds [81].

5.3 Addition Reaction of Group 15 Element–Element Bond

The known procedures for activation and addition reactions of group 15 element–element bonds to unsaturated hydrocarbons are only P(III/V)—P(III/V) bond, which are the abstractive and efficient synthesis of diphosphinoalkenes or diphosphinated alkenes [82].

The addition reactions of tetraorganobiphosphine and -arsine ($R_2E—ER_2$, E = P, As; R = Me, Et, Ph) to phenylacetylene under UV irradiation or catalyzed by azobis(isobutyronitrile) giving 1,2-bisphosphino- and 1,2-bisarsinoalkenes were reported in 1971 [83]. Afterwards, the photoinduced addition of tetrafluorodiphosphine to alkynes has also been developed [84].

In 2005, Oshima and coworker reported a V-40-initiated radical diphosphination of terminal alkynes with $Ph_2P—PPh_2$ (formed in situ from Ph_2PH and Ph_2PCl) to provide *trans*-diphosphanylethene as major isomers in good yields after treatment with elemental sulfur (Scheme 5.15) [85].

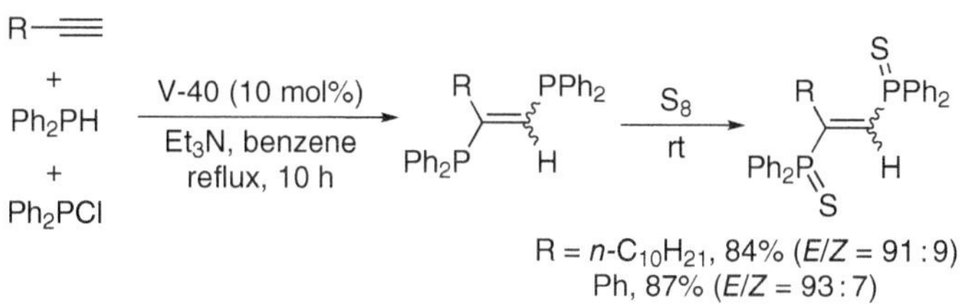

Scheme 5.15 Radical addition of tetraphenyldiphosphine to terminal alkynes.

Under N_2, the photoinduced addition of $Ph_2P—PPh_2$ to terminal alkynes generates a mixture of E/Z-diphosphinated alkenes as air-sensitive compounds, which can be isolated by treatment with elemental sulfur (Scheme 5.16) [86].

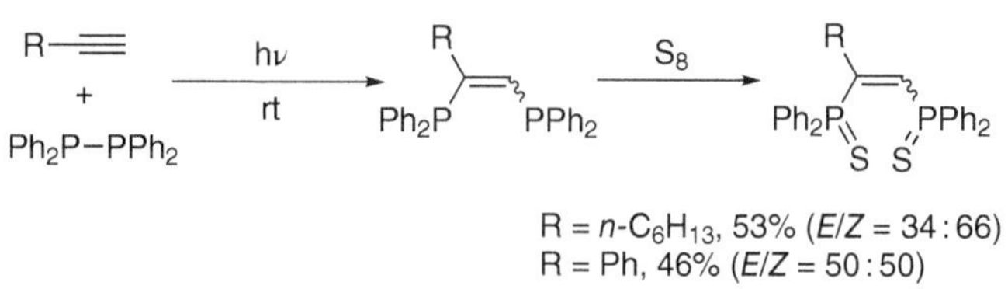

Scheme 5.16 Photoinduced diphosphination of terminal alkynes.

Recently, Hirano and Miura group reported a Brønsted base-mediated stereoselective diphosphination of terminal alkynes with diphosphanes to give (*E*)-1,2-diphosphinoalkenes in good yields. The reaction of aromatic terminal alkynes occurs efficiently in the presence of LiO^tBu, while $MN(TMS)_2$ (M = Li or Na) shows better catalytic activity in the case of aliphatic terminal alkynes used (Scheme 5.17) [87].

In the presence of fluorine- or carbonate-based activators, the same group developed the diphosphination of arynes with tetraaryldiphosphines by using the stable aryne precursors, 2-(trimethylsilyl)aryl triflates [88].

In addition, when activated alkynes were used, such as dimethylacetylene dicarboxylate and methyl propiolate, the highly stereoselective *cis*-diphosphination reactions can occur at ambient temperatures without any initiator and additive [89].

Very recently, an unusual KO^tBu-promoted 1,1-addition of α-C2-bridged biphospholes to terminal alkynes has been reported [90]. The addition reactions occurred

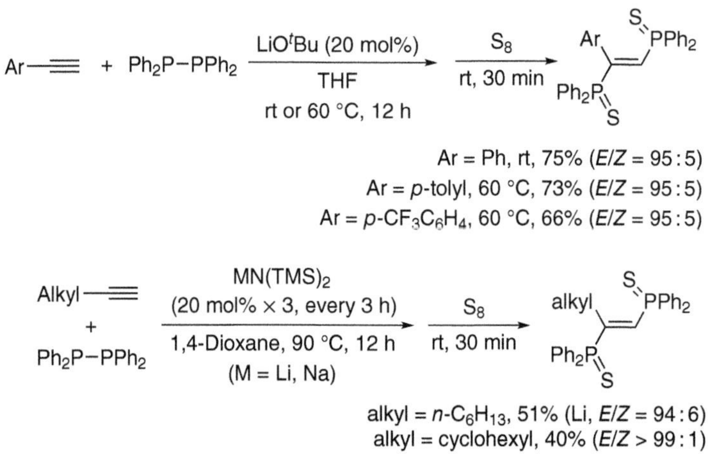

Scheme 5.17 Base-mediated stereoselective diphosphination of terminal alkynes with diphosphanes.

with excellent chemoselectivity to provide simple access to the 1,3-diphosphepines, which have potential applications in the coordination and catalyst chemistry (Scheme 5.18).

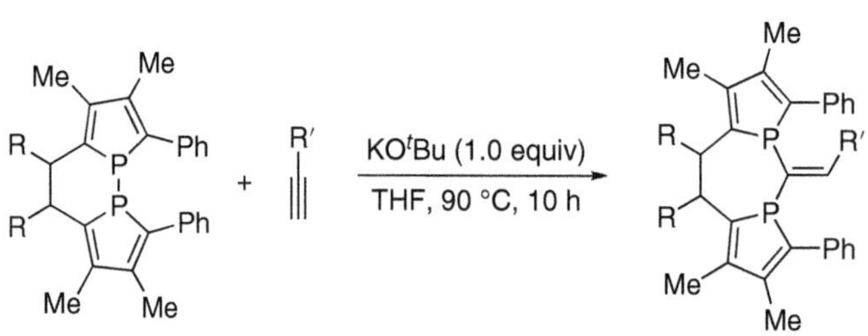

Scheme 5.18 Base-promoted 1,1-addition of α-C2-bridged biphospholes to terminal alkynes.

On the other hand, a pioneering work on the addition reactions of tetramethylbiphosphine to alkenes to give 1,2-bis(dimethylphosphino)ethanes under UV irradiation was reported in 1971 [91].

Interestingly, Ogawa and coworker developed a highly regioselective addition of tetraphenyldiphosphine monoxide [$Ph_2P(O)PPh_2$] to terminal alkenes affording the

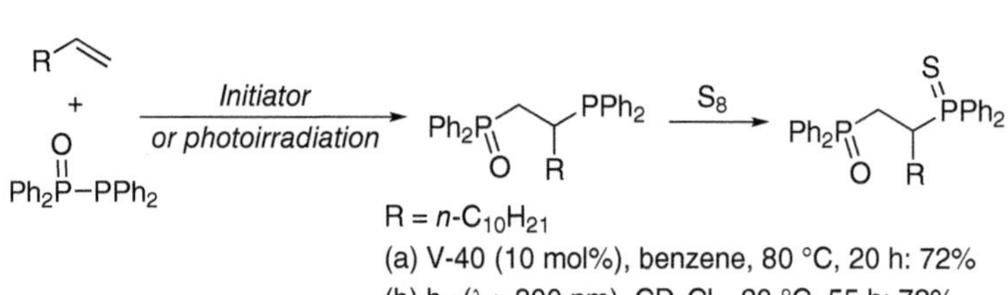

Scheme 5.19 Radical addition of $P^V(O)–P^{III}$ to terminal alkenes.

corresponding 1-phosphinyl-2-phosphinoalkanes (Scheme 5.19) [92]. The addition reactions proceed by the hemolytic cleavage of the $P^V(O)—P^{III}$, followed by selective attack of the phosphinyl radical at the terminal position of the alkenes and selective trapping of the resulting carbon radical by the phosphino group.

The same group then investigated and compared the reactivity of several $P^V(X)—P^{III}$ (X = O, S, Se) compounds in the addition reaction to terminal alkenes under irradiation conditions with a xenon lamp (500 W) through Pyrex without any catalyst, base, or additive [93]. As shown in Scheme 5.20, only diphosphane monosulfide shows high reactivity to give the corresponding 1-thiophosphoryl-2-phosphanylalkane in high yield with excellent regioselectivity, in which diphenylthiophosphoryl and diphenylphosphanyl groups are attached to the terminal and inner carbon, respectively. In addition, the addition reactions of diphosphane monosulfide to internal alkenes, such as maleic ester, and five-membered cyclic alkenes were also successful. Moreover, under the similar conditions, the addition reactions of tetraphenyldiphosphine disulfide to a variety of alkenes, such as terminal, cyclic, internal, and branched alkenes, 1,3-dienes, and terminal alkynes were also developed to afford bis(thiophosphinyl)alkanes and bis(thiophosphinyl)alkenes [94].

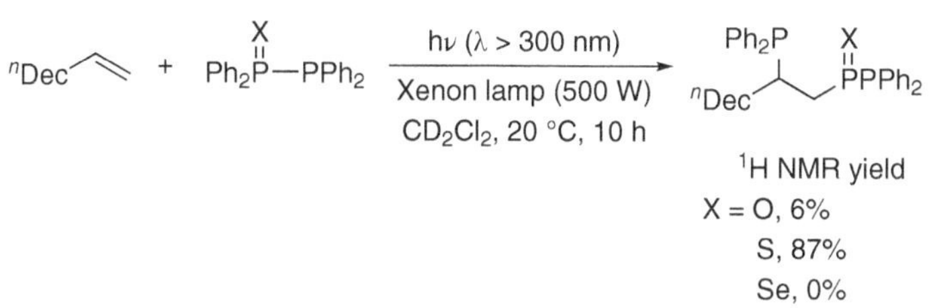

Scheme 5.20 Comparative reactivity of 1,2-addition of diphosphane analogs to alkenes.

The radical 1,4-addition of tetramethyldiphosphine to butadiene that occurs slowly above 100 °C catalyzed by azobis(isobutyronitrile) was reported in 1970 [95]. The formation of *cis/trans*-1,4-bis(dimethylphosphino)but-2-ene was discussed.

Recently, Miura and coworker have developed a 1,4-diphosphination of 1,3-dienes with tetraphenyldiphosphines under $Ir(ppy)_3$-promoted photoredox catalysis to form the corresponding 1,4-diphosphino-2-butenes in good yields with good regioselectivity, and the key to success is the addition of a Br^+ additive (Scheme 5.21) [96]. The similar catalyst system was then applied in the ring-opening diphosphination of methylene- and vinylcyclopropanes [97].

5.4 Addition Reactions of Group 16 Element–Element Bond

The reported procedures for the addition of group 16 element–element bonds (organic dichalcogenides) to unsaturated hydrocarbons have been achieved in

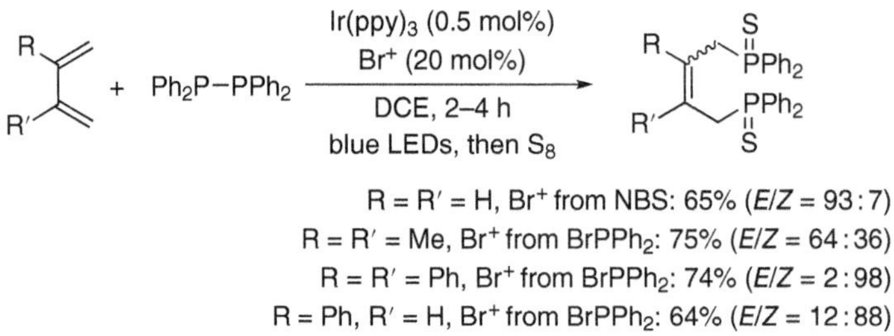

Scheme 5.21 Diphosphination of 1,3-dienes under visible light-promoted photoredox catalysis.

the presence of transition metal complexes or radical initiators (Scheme 5.22) [98]. Since the formed C(sp^2)—S/Se/Te bonds have synthetic utility in organic chemistry [99], the dichalcogenation of alkynes has been extensively investigated [100]. On the basis of the known procedures, transition metal-catalyzed additions of organic dichalcogenides to alkynes usually gave *cis*-adducts (*Z*-isomers), and the free-radical process mainly yielded *trans*-adducts (*E*-isomers).

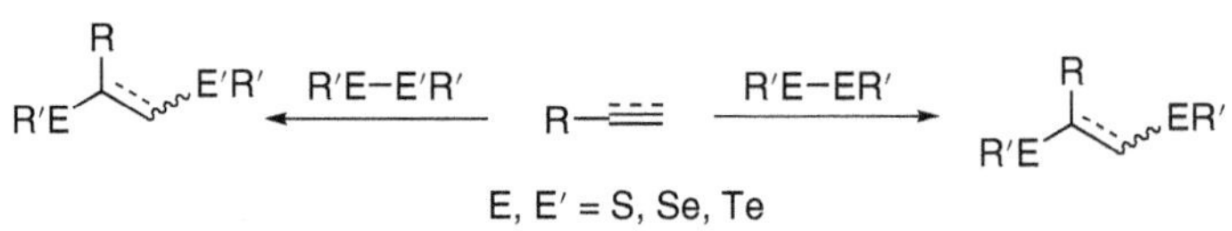

Scheme 5.22 Additions of organic dichalcogenides to alkynes and alkenes.

5.4.1 *cis*-Addition Reactions to Alkynes

The first *cis*-stereoselective disulfidation and diselenation of terminal alkynes with diaryl disulfides and diselenides affording (*Z*)-1,2-bis(arylthio)-alkenes and (*Z*)-1,2-bis(arylseleno)-alkenes were reported by Sonoda and Ogawa's group in 1991 [101]. As shown in Scheme 5.23, in the presence of $Pd(PPh_3)_4$, the addition reactions of 1-octyne or phenyl alkyne occurred with high regio- and stereoselectivities. In addition, when these reactions were performed in the presence of CO, the three-component carbonylative addition reactions took place to give the

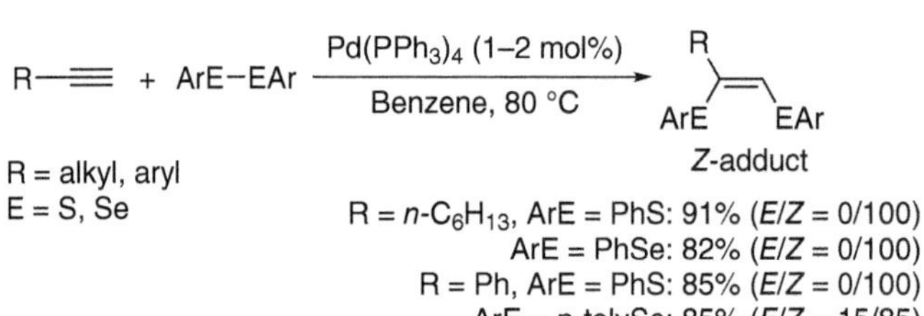

Scheme 5.23 Palladium-catalyzed stereoselective addition of diaryl disulfides/diselenide to terminal alkynes.

(*Z*)-1,3-bis(arylthio)-2-alken-1-ones and (*Z*)-1,3-bis(arylseleno)-2-alken-1-ones, respectively (Scheme 5.24). The carbonylative addition reactions were also completely regioselective and highly stereoselective.

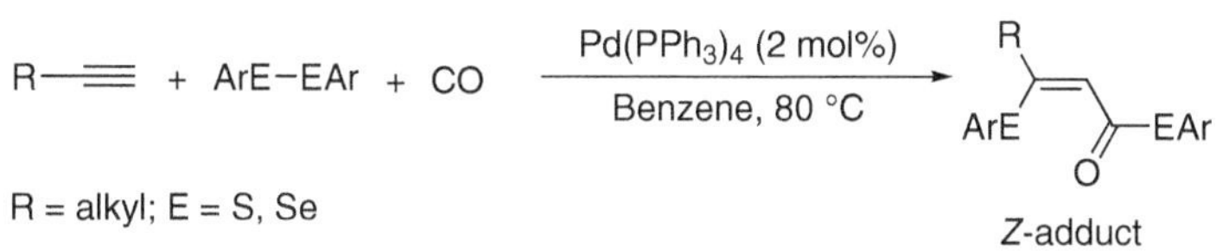

Scheme 5.24 Palladium-catalyzed carbonylative addition of diaryl disulfides/diselenide to terminal alkynes.

In addition, they then studied the addition reactions of propargylic alcohols with diaryl disulfides and CO with the use of same catalyst to develop a novel thiolative lactonization affording β-arylthio- and β-arylseleno-α,β-unsaturated lactones in moderate to good yields [102]. $Co_2(CO)_8$ was also applied in the highly selective carbonylative bisthiolation of internal alkynes with organic disulfides to give (*Z*)-adducts by same group [103].

The *cis*-addition reactions of S—S and Se—Se bonds to alkynes and their application in organic synthesis were then studied extensively in the presence of various transition metal complexes under different conditions [104].

Recent studies have found that in the absence of transition metal complexes, *cis*-disulfidation of terminal alkynes can be also achieved by using CsOH as catalyst [105].

In addition, recently, (*Z*)-1,2-bis(arylthio)alkene derivatives could be obtained by a three-component cascade disulfidation of terminal alkynes, K_2S, and diaryliodonium salts catalyzed by Pd(II)—NHC complexes [106].

Under microwave irradiation conditions, $Pd(OAc)_2$/phosphine-catalyzed addition reactions of Ar_2E_2 (E = S, Se) to terminal alkynes affording *cis*-adducts have been also developed [107].

Note that benzynes can undergo insertion into ArE—EAr (E = S, Se, Te) bonds to afford the corresponding *ortho*-bis(arylthio)-, *ortho*-bis(arylseleno)-, and *ortho*-bis(aryltelluro)benzenes under transition metal-free and radical initiators-free conditions [108].

5.4.2 *trans*-Addition Reactions to Alkynes

As shown in Scheme 5.25, the stereoselective *trans*-disulfidation of terminal alkynes affording (*E*)-1,2-dithio-1-alkenes as the major product can be achieved by free-radical processes [109] and in the presence of $GaCl_3$ [110].

Recently, Xu and coworker have developed the efficient intermolecular atom transfer addition reactions of alkynes with $PhSO_2$—SCF_3, $PhSO_2$—SCF_2H, and $PhSO_2$—SR via the combination of visible light photoredox catalysis and gold

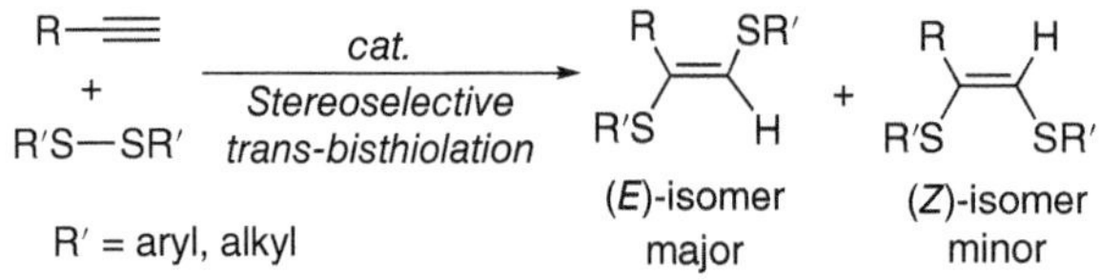

Scheme 5.25 *trans*-Bisthiolation of terminal alkynes.

catalysis [111]. As shown in Scheme 5.26, the thiosulfonylation of terminal alkynes occurred with high regio- and stereoselectivity to afford thio-functionalized vinylsulfones in good yields. The thiosulfonylation of enynes could also occur for constructing functionalized carbo- and heterocycles through a radical cascade cyclization process. In addition, with internal alkynes, the addition reactions also proceeded smoothly; however, the stereoselectivity was not as good as with terminal alkynes.

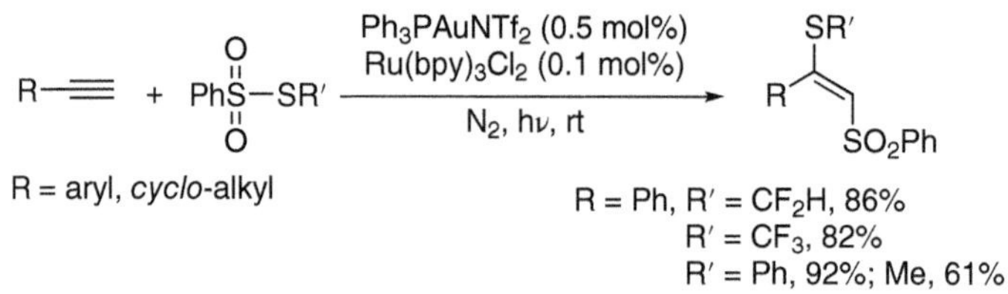

Scheme 5.26 Gold(I)/photoredox-cocatalyzed atom transfer thiosulfonylation of alkynes.

In addition, photoinduced disulfidation [112], diselenation [113], and ditelluration [114] of alkynes with organic disulfides, diselenides, and ditellurides via radical addition and their applications have been well investigated (Scheme 5.27).

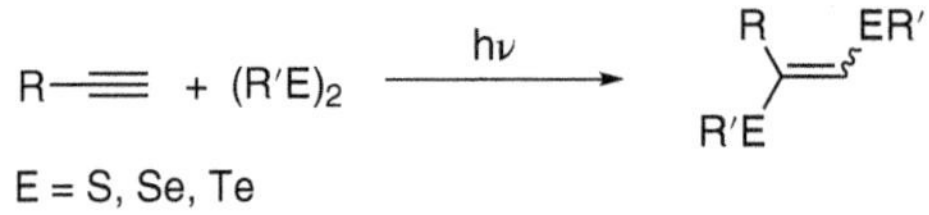

Scheme 5.27 Photoinduced radical additions of organic dichalcogenides to alkynes.

For example, diphenyl ditelluride (PhTeTePh) can undergo the addition reaction to a variety of alkynes upon irradiation with visible light under solvent-free conditions affording 1,2-bis(phenyltelluro)alkenes in good yields (Scheme 5.28) [115]. The reaction was proposed to proceed by a radical chain mechanism initiating the addition of phenyltelluro radical generated in situ by photolysis of diphenyl ditelluride to alkynes. In addition, in the cases of aliphatic alkynes such as 1-octyne, the addition reaction proceeds stereoselectively to provide only (*E*)-1,2-bis(phenyltelluro)alkenes, and aromatic terminal alkynes like phenylacetylene produces a mixture of *E*- and *Z*-isomers of adducts. Interestingly, the addition reaction was found to be reversal; irradiation with visible light (>400 nm) under high concentrations of the substrates induces the addition of $(PhTe)_2$ to alkynes, whereas irradiation with near-UV (>300 nm) under dilution condition causes the reverse reaction.

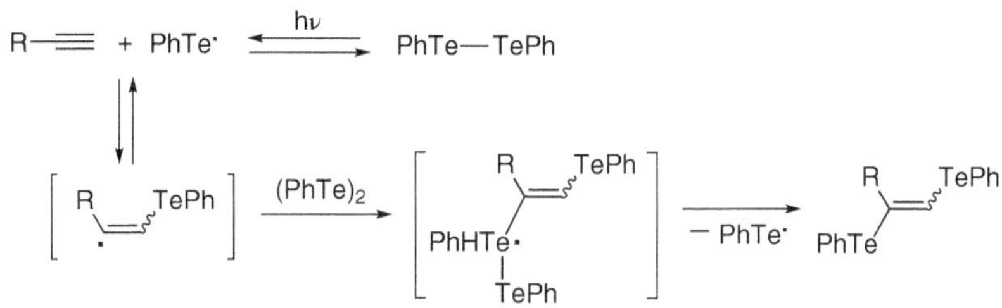

Scheme 5.28 Proposed mechanism for reverse photoinduced ditelluration of alkynes.

5.4.3 Different Heteroatom Bond Addition Reactions to Alkynes

The group 16 heteroatom (E—E′) addition reactions to alkynes can be achieved by using either RE—E'R or the combination of $(RE)_2$ and $(RE')_2$.

The thioselenation of terminal alkynes with diaryl disulfides and diaryl diselenides has been developed in the presence of rhodium complexes [116] or under photoirradiation conditions [117].

For example, CsOH shows highly chemo-, regio-, and stereoselective catalytic activity for the synthesis of (*Z*)-vinylic selenosulfides and (*Z*)-vinylic tellurosulfides in a one-pot reaction of terminal alkynes, diaryl disulfides, and diaryl diselenides or diaryl ditellurides (Scheme 5.29) [118]. The adducts have been also applied in the stereoselective synthesis of tetrasubstituted alkenes due to the different activities of C—S, C—Se, and C—Te bonds and the retention of configuration feature of the chalcogen atoms (S, Se, Te).

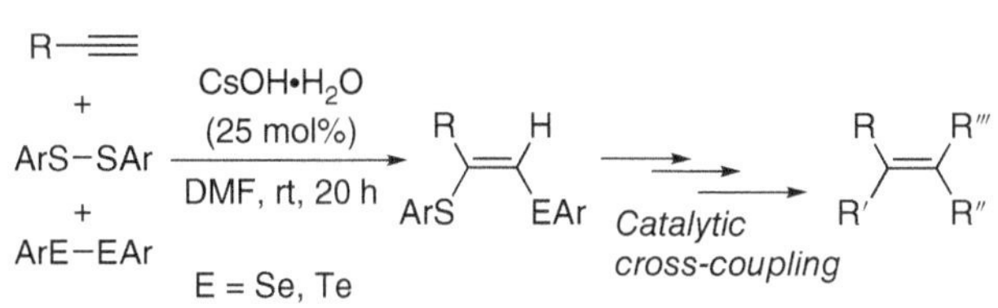

Scheme 5.29 CsOH-catalyzed reaction of terminal alkynes with disulfides, diselenides, and ditellurides.

5.4.4 Addition Reactions to Alkenes

The first transition metal-catalyzed addition of organic disulfides to alkenes affording *vicinal* dithioethers was reported by Mitsudo's group in 1999 [119], although a few metal-free catalytic additions of disulfides to alkenes were reported before [120]. As shown in Scheme 5.30, in the presence of Cp*RuCl(cod), the addition reactions of diphenyl disulfide to ethylene, electron-rich, and electron-poor terminal alkenes proceeded smoothly, affording the dithioether adducts in good yields. In the case of 2-norbornene used, the corresponding adducts were obtained in high yields with diaryl and dialkyl disulfides.

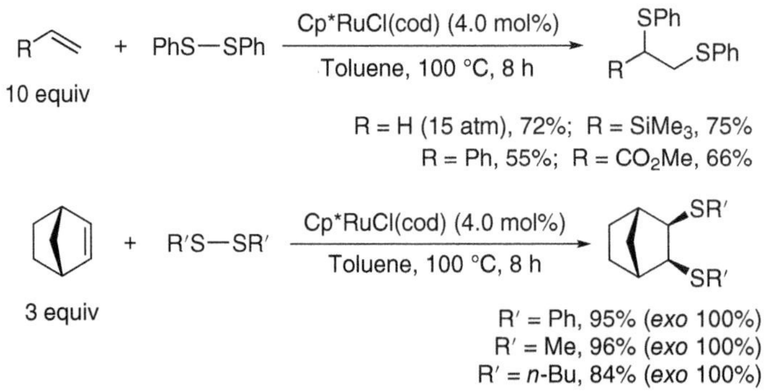

Scheme 5.30 Ruthenium-catalyzed addition of disulfides to alkenes.

The disulfidation of aromatic alkenes also occurred in the presence of iodine in H_2O and 1,2-dichloroethane (DCE) [121], and the diselenation of alkenes could be achieved in the presence of $SnCl_4$ [122].

In addition, a catalytic amount of ArS^+-initiated disulfidation of dienes was also reported [123].

Diselenation of allenes catalyzed by Pd(0) complex [124] and the mechanism studies by using density functional methods [125] have been reported.

For the activation and addition of two different heteroatom bonds of group 16 with alkenes, the thioselenation of alkenes under radical conditions has also been developed. For example, the addition reactions of *S*-benzoyl phenylselenosulfide to alkenes with azobis(isobutyronitrile) (AIBN) afforded selenothiocarboxylated products with high regiospecificity [126].

Under irradiation through Pyrex with a tungsten lamp, with using a disulfide–diselenide mixed system, a highly regioselective thioselenation of terminal alkenes took place to provide 1-(phenylthio)-2-(phenylseleno) alkanes as a sole product in good yields. In the cases of cyclic alkenes, they also underwent thioselenation to give corresponding *E*-isomers stereoselectively. The addition reactions of 1,3-dienes afforded 1,4-thioselenation adduct in good yield [127]. Interestingly, under irradiation with near-UV light, the same group then developed a diphenyl diselenide-assisted disulfidation of 1,3-dienes with diphenyl disulfide to give the corresponding 1,4-adducts selectively in good yields [128].

5.5 Addition Reactions of Element–Element Bonds from Different Group Heteroatoms

Direct activation of E—E′ bonds and their addition to unsaturated hydrocarbons has become an important and efficient synthetic route for the synthesis of difunctionalized alkenes and alkanes via the formation of C—E and C—E′ bonds simultaneously [129].

5.5.1 *cis*-Addition Reactions to Alkynes

Much effort has been devoted to the development of transition metal-catalyzed regio- and stereoselective *cis*-addition reactions of terminal alkynes to give difunctionalized alkenes.

Due to the versatile application of sp^2 C—B bond in the formation of C—C and C–heteroatom bonds, the additions of B—E (E = Sn, Si, N, P, S, Cl, etc.) to alkynes in the presence of transition metal complexes have played a pivotal role in organic synthesis [130]. An early work reported in 1993 by Suzuki's group developed the highly regio- and stereoselective addition reactions of 9-(alkylthio)-9-borabicyclo[3.3.1]nonanes, (9-[RS]-9-BBN) to terminal alkynes in the presence of $Pd(PPh_3)_4$ to produce 9-[(*Z*)-2-(alkylthio)-1-alkenyl]-9-BBN derivatives in high yields [131]. Recently a review has appeared on the intramolecular boron–heteroatom addition reactions, such as oxyboration, aminoboration, and thioboration, to construct heterocyclic compounds [132].

As a pioneering work, Tanaka's group studied the oxidative addition of borylstannanes to palladium complexes to afford *cis*-boryl-(stannyl)palladium species [133]. As shown in Scheme 5.31, the reaction of $Me_3SnB—[NMe(CH_2CH_2)NMe]$ with $Me_2Pd[PMe_2(CH_2CH_2)PMe_2]$ gave palladium complex having Pd—B and Pd—Sn bonds via cleavage of B—Sn bond, which shows catalytic activity for the *cis*-borylstannylation of alkynes. Therefore, a highly stereo- and regioselective palladium-catalyzed borylstannylation of alkynes affording (*Z*)-(β-stannylalkenyl)boranes was developed, and the proposed mechanism involves the oxidative addition of B—Sn bond to palladium as the key step.

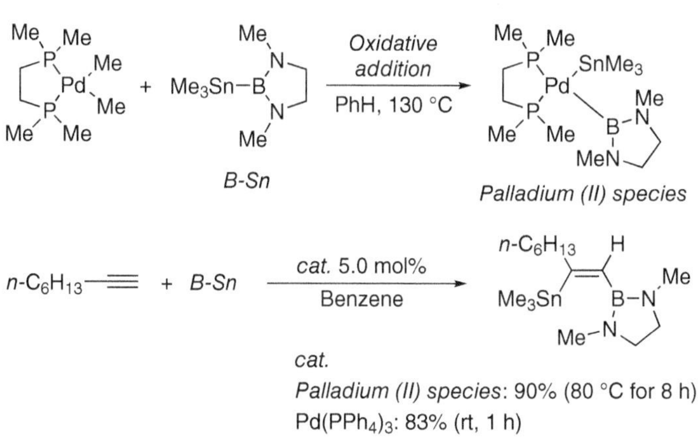

Scheme 5.31 Oxidative addition of borylstannane to palladium complex and palladium-catalyzed borylstannylation of alkynes.

At the same time, the same group also developed a Pd(0)-catalyzed *cis*-selenophosphorylation of terminal alkynes, and the Pd(II) and Pt(II) complexes were isolated and confirmed by the oxidative addition of the phosphorous–selenium bond to Pd(0) and Pt(0) complexes [134].

$PdCl_2(PPh_3)_2$ was then applied in the borylstannylative carbocyclization of diynes with borylstannanes at room temperature and developed a highly regio- and stereoselective formation of 1-(borylmethylidene)-2-(stannylmethylidene) cycloalkane derivatives in high yields by the same group [135].

$PdCl_2(PPh_3)_2$ and $PdCl_2(MeCN)_2/P(t\text{-}Bu)_3$ were also used as the catalysts in the borylstannylation of alkynes and enynes reported by RajanBabu's group [136].

In addition, Tanaka [137] and other groups [138] then developed the palladium-catalyzed borylsilylation of alkynes and borylsilylative carbocyclization of diynes and enynes with borylsilanes.

At the same time, Ito and coworker also reported the addition of $PhMe_2Si$—Bpin bond to alkynes catalyzed by a palladium(0)-*t*-alkyl isocyanide complex or $Pt(PPh_3)_4$ to give (*Z*)-1-boryl-2-silyl alkenes with high regio- and stereoselectivity (Scheme 5.32) [139]. The borylsilylation could also be applicable for internal alkynes, and in addition to the pinacol derivative, diamino derivative of $PhMe_2Si—B(NEt_2)_2$ also underwent the addition reactions to give the corresponding adducts in good yield with high regio- and stereoselectivity catalyzed by palladium-isonitrile catalyst.

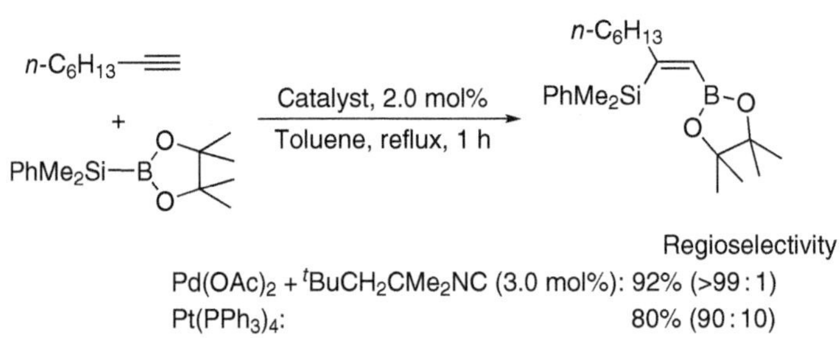

Scheme 5.32 Regio- and stereoselective silylboration of terminal alkynes catalyzed by palladium and platinum complexes.

It was also reported that the regioselectivity of palladium-catalyzed borylsilylation of terminal alkynes could be switched by ligand-dependent control of reductive elimination [140].

Interestingly, with the use of supported gold nanoparticles as the catalysts, the borylsilylation of terminal alkynes with $PhMe_2Si$—Bpin occurs at ambient conditions, and the regioselectivity is opposite to that observed in the presence of Pd or Pt catalysts [141].

When Ni(0) complexes were used as the catalysts, Ito's group then developed a dimerizative borylsilylation of terminal alkynes to give *cis,cis*-1-silyl-4-boryl-1,3-butadiene derivatives in a regio- and stereoselective manner [142].

Very recently, a review paper has appeared on the activation of B—Si bonds and their catalysis [143].

3-Borylated benzofurans and indoles could be prepared by the gold-catalyzed intramolecular alkoxyboration+/+* and aminoboration of *N*/*O*-heterocyclic boronic acid derivatives of *ortho*-alkynylphenols and *ortho*-alkynylanilines, respectively, reported by Blum's group [144].

Tanaka and Han designed the strategy for introducing two different heteroatoms into carbon–carbon triple bonds with the use of disulfides and disilanes catalyzed by $Pt(CH_2=CH_2)(PPh_3)_2$ [145]. This thiosilylation of terminal alkynes occurred with excellent regio- and stereoselectivity to the corresponding (*Z*)-1-silyl-2-thio-1-alkenes (Scheme 5.33). The mechanistic investigation revealed that platinum complex could catalyze the disproportionation of disulfide $(ArS)_2$ with $(SiCl_3)_2$ to give $ArSSiCl_3$ rapidly at the very early stage of the addition reaction, and, in fact, the thiosilylation reactions proceeded smoothly using $PhSSiCl_3$ as the substrate in place of the mixture of $(PhS)_2$ and $(SiCl_3)_2$ (Scheme 5.34).

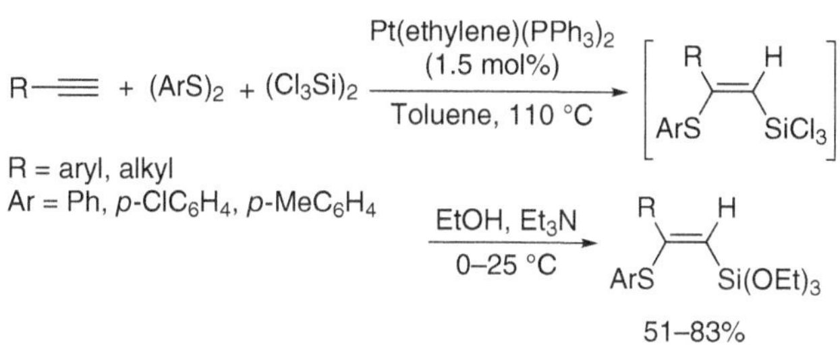

Scheme 5.33 Platinum-catalyzed thiosilylation of terminal alkynes using disulfides and disilanes.

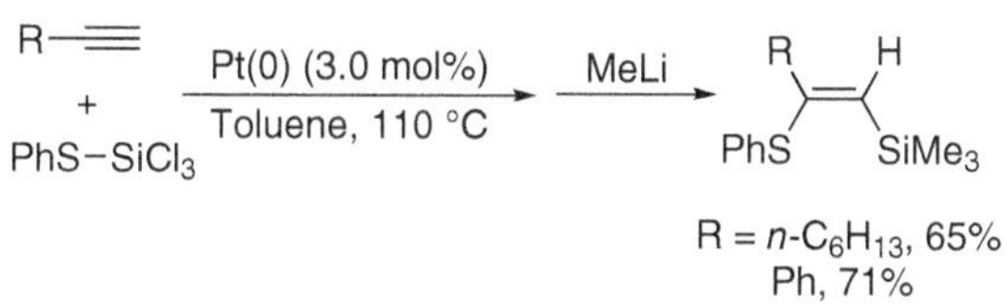

Scheme 5.34 Platinum-catalyzed thiosilylation of terminal alkynes with S−Si bond.

Because of the highly potential applications of alkenyl sulfides as starting materials in organic synthesis and bioactive compounds, various addition reactions of S—E (E = Cl, N, P, etc.) bonds to alkynes have been studied.

The regio- and stereoselectivity of *cis*-chlorothiolation of terminal alkynes with sulfenyl chlorides depends on the reaction conditions (Scheme 5.35).

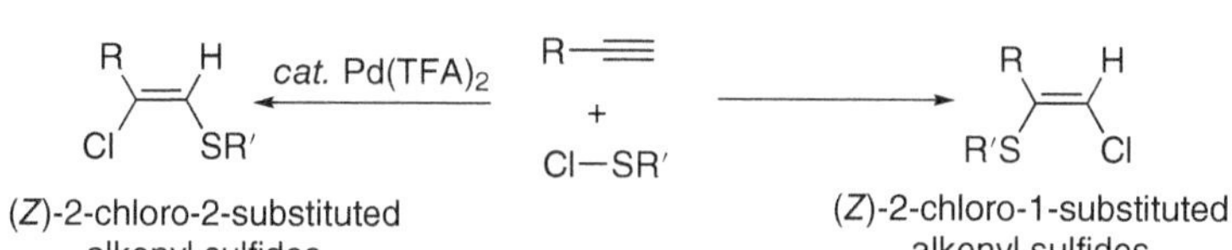

Scheme 5.35 *cis*-Chlorothiolation of alkynes.

In the presence of $Pd(TFA)_2$, a highly regio- and stereoselective chlorothiolation of terminal alkynes with arylsulfenyl chlorides in toluene at room temperature occurred to give (*Z*)-2-chloro-2-substituted-alkenyl aryl sulfide as major adducts [146].

On the other hand, the regio- and stereoselective *cis*-addition of sulfonyl chlorides to terminal alkynes affording (*Z*)-β-chlorovinyl sulfones could be realized with the use of CuCl as the catalysts [147].

In addition, it was found that the stereoselectivity of phosphinylthiolation of terminal alkynes with $Ph_2P(O)SBu$ catalyzed by CpPd(π-allyl)/PEt_3 was highly dependent on the nature of solvents; the addition reactions in *n*-hexanol (or *t*-amyl alcohol) and ethylbenzene afforded *E*- and *Z*-adducts, respectively [148].

The *cis*-addition of other E—E′ bonds to terminal alkynes with high regio- and stereoselectivity to give (*Z*)-adducts as the major adducts is shown in Scheme 5.36. It includes ruthenium-catalyzed aminothiolation with sulfonamides [149], palladium-catalyzed thiocarbamoylation with sulfenamides, and carbon monoxide [150].

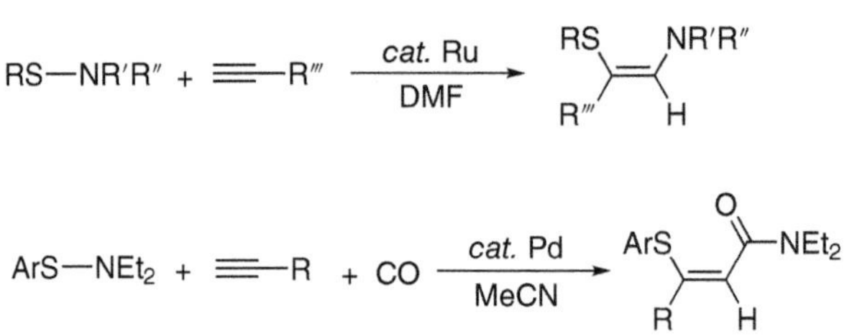

Scheme 5.36 Synthesis of (*Z*)-difunctionalized alkenes via *cis*-addition of activated E—E′ bonds.

The insertion of arynes into E—E′ bonds is the formal *cis*-addition reaction of internal alkynes, which has been developed under different reaction conditions (Scheme 5.37).

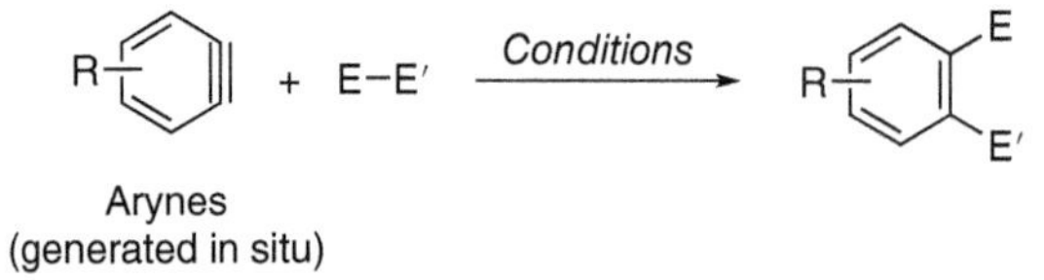

Scheme 5.37 Insertion of arynes into E—E′ bonds for synthesis of *ortho*-difunctionalized benzenes.

For example, Daugulis and coworker have reported a general and novel hindered lithium diadamantylamide (LDAM) base-promoted insertion of arynes into Si—P, Si—S, and Si—N bonds, and the arynes are generated from easily available aryl triflates and halides under the reaction conditions [151]. A few examples are shown in Scheme 5.38; in the presence of LDAM, the reaction of chlorobenzene with Me_3Si—PPh_2 gives the product in 84%. When 2-methoxy-1-chlorobenzene and 2,5-dimethoxy-1-chlorobenzene are used as the aryne precursors, the reactions with Me_3Si—SEt or Me_3Si—NMe_2 afford the corresponding products in good to excellent yields with high regioselectivities. The studies of substrate scope have disclosed that cyano, aryl, alkyl, trifluoromethyl, vinyl, methoxy, chloro, fluoro, and formyl moieties are compatible with the reaction conditions.

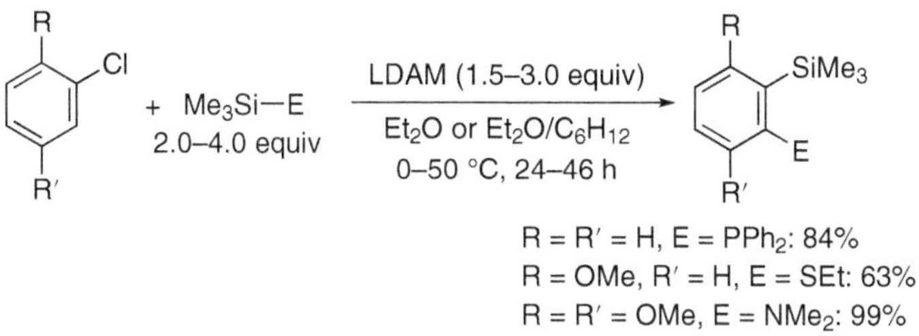

Scheme 5.38 Examples of the LDAM-promoted insertion of arynes into Si–P, Si–S, and Si–N bonds.

Under transition metal-free conditions, the insertion of arynes into S—N bond has also been reported [152].

5.5.2 *trans*-Addition Reactions to Alkynes

The stereoselective *trans*-addition of E—E′ bonds to alkynes can be achieved by using special substrates or bearing directing groups or by radical processes.

Although a palladium-catalyzed *trans*-silylboration of terminal alkynes with the use of an excess amount of Si—B reagent [153], the complete *anti*-selective vicinal silylboration depends on the functional group directed of alkynes. For example, the reaction of ethyl 3-phenylpropiolate with $PhMe_2Si$—B(pin) in the presence of PBu_3 without a solvent at 80 °C for eight hours afforded the corresponding β-boryl-α-silyl acrylate in 84% yield, and the Si—B bond addition is completely regioselective and *anti*-stereoselective (Scheme 5.39) [13]. The ^{11}B-NMR spectra of adducts indicate the coordination between carbonyl oxygen with the boron atom.

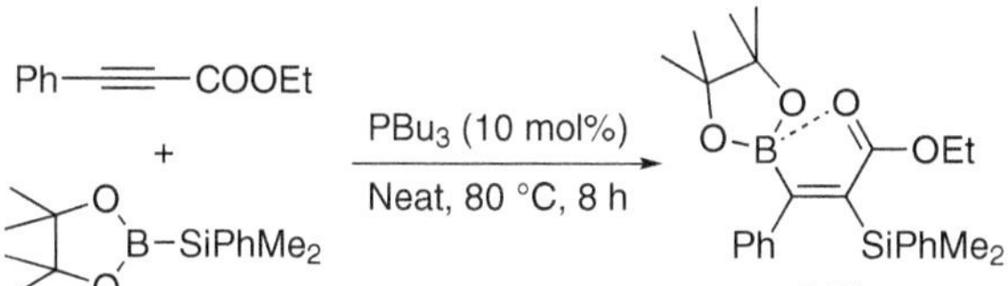

Scheme 5.39 *Anti*-silaboration of alkynoates catalyzed by trialkylphosphines.

Similarly, in the presence of catalytic amounts of PCy_3, the addition of selenoboranes (Se—B bond) to α,β-acetylenic esters and amides occurs stereoselectively to afford *anti*-adducts, and a detailed density functional theory (DFT) study has been performed to understand the mechanism and role of the heteroelement nature [154].

Hayashi and coworker have tried to optimize the catalyst systems for the addition of silylphosphines to alkynes via the activation of Si—P bond catalyzed by transition metal complexes and found that cationic rhodium catalysts, generated by adding AgOTf to chlororhodium complexes, can work as catalyst to catalyze the addition reaction to give protodesilylated products [155]. When $[Rh(cod)Cl]_2$/AgOTf was used, with addition of a small amount of methanol, *cis*-hydrophosphination proceeded to give the *E* isomer with high regio- and stereoselectivities (Scheme 5.40). Under the similar conditions, the activated alkenes, such as ethyl acrylate and acrylonitrile, can also undergo the hydrophosphination to afford the corresponding *anti*-Markovnikov adducts in good yields.

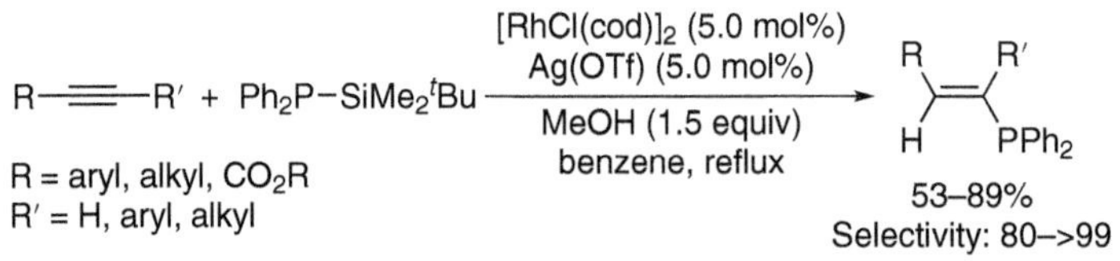

Scheme 5.40 Regio- and stereoselective synthesis of alkenylphosphines via rhodium-catalyzed reaction of alkynes with silylphosphine.

The *trans*-chlorothiolation of alkynes through S—Cl bond cleavage of sulfenyl chlorides can be achieved via the formation of thiirenium intermediate or by a radical pathway (Scheme 5.41).

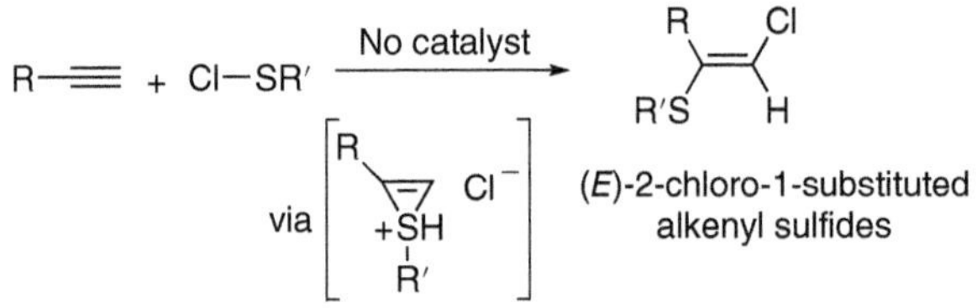

Scheme 5.41 *trans*-Chlorothiolation of alkynes.

Without any catalysts, *trans*-chlorothiolation proposed via the formation of thiirenium intermediate can produce (*E*)-2-chloro-1-substituted alkenyl sulfides as the major adducts [156].

The radical *trans*-addition of aromatic sulfonyl chloride to terminal alkyne with high regioselectivity to produce (*E*)-β-chlorovinylsulfones has also been well-known (Scheme 5.42).

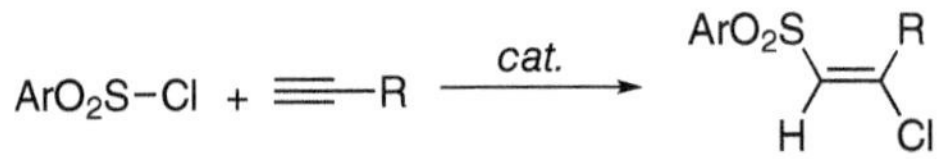

Scheme 5.42 Catalytic regio- and stereoselective chlorosulfonylation of terminal alkynes.

Cu(II)/Cu(I) [157] and Fe(II) complexes [158] have been used as the efficient catalysts in this type of addition reactions.

Addition reactions of sulfonyl iodides [159] and sulfonyl bromides [160] to alkynes were also reported.

On the other hand, $FeCl_2$-catalyzed *trans*-chlorothiolation of aliphatic/aromatic terminal alkynes with sulfenyl chlorides in toluene at room temperature has been reported to give (*E*)-2-chloro-2-substituted alkenyl sulfides, and internal alkynes also undergo the addition reaction with same stereoselectivity [161].

Interestingly, a CuCl-catalyzed *trans*-chlorothiolation of terminal and internal alkynes with sulfonyl chlorides (RSO_2Cl, R = aryl, alkyl) has also been developed with the use of PPh_3 as reductant to reduce sulfonyl chlorides to sulfenyl chlorides first, and the mechanistic investigations revealed a plausible radical process involving a sulfur-centered radical intermediate via copper-mediated homolysis of the S—Cl bond [162].

Alkenyl aryl tetrafluoro-λ^6-sulfanes could be prepared by the addition reactions of substituted chlorotetrafluorosulfanyl arenes to phenyl acetylene via the cleavage of S—Cl bond promoted by Et_3B (Scheme 5.43) [163].

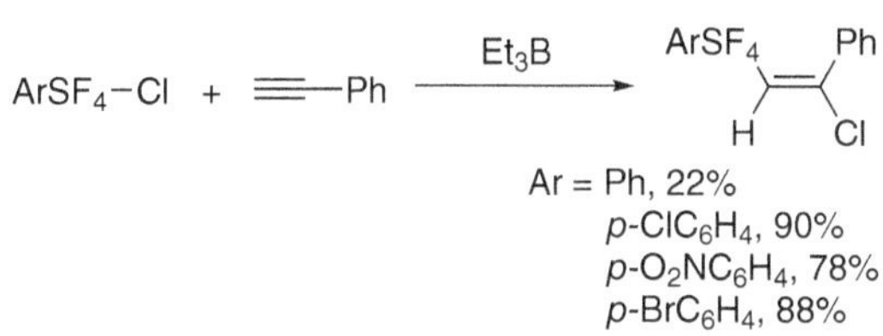

Scheme 5.43 Et_3B-catalyzed addition of chlorotetrafluorosulfanyl arenes to alkynes.

Yorimitsu and Oshima reported the radical addition reactions of alkylthio- and arylthiodiphenylphosphines to terminal alkynes to afford (*E*)-1-thio-2-phosphino-1-alkenes, and this radical thiophosphination usually proceeds in a *trans*-addition to yield *E*-adducts with the sulfur attached to the terminal carbon of alkynes (Scheme 5.44) [164]. With the use of $(Ph_2P)_2/(PhS)_2$ binary system, the photoinduced thiophosphination of terminal alkynes also took place with the same regio- and stereoselectivity [165].

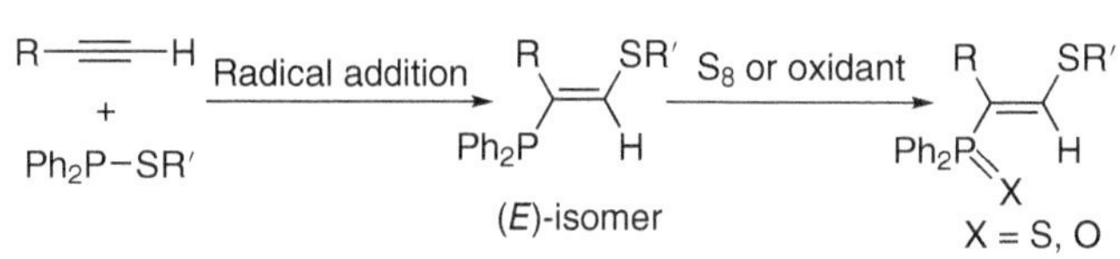

Scheme 5.44 Radical *trans*-thiophosphination of terminal alkynes.

However, Yorimitsu and Oshima's group also disclosed that the radical additions of *S*-thiophosphinyl dithiocarbonate to aromatic terminal alkynes in the presence of a catalytic amount of dilauroyl peroxide (DLP) afforded (*E*)-1-aryl-1-thio-2-thiophosphinylethene derivatives regio- and stereoselectively in high yields (Scheme 5.45) [166]. In this case, the sulfur atom is attached to the internal carbon of terminal alkynes with an opposite regioselectivity.

In addition, other E—E′ bonds activation and their *trans*-addition to terminal alkynes with high regio- and stereoselectivity to give (*E*)-adducts have also been

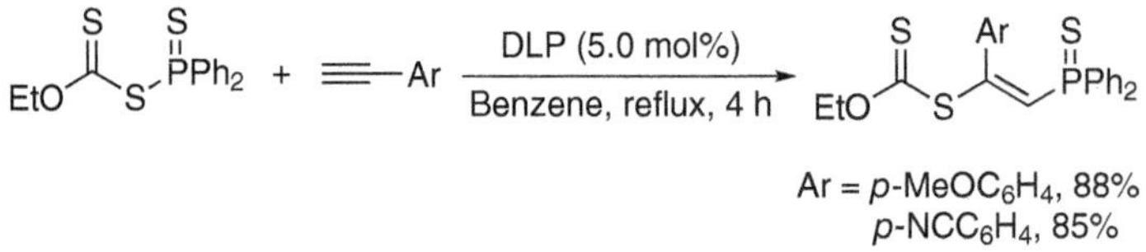

Scheme 5.45 Radical addition of *S*-thiophosphinyl dithiocarbonate to terminal aromatic alkynes.

reported (Scheme 5.46). It includes the selenoarsenation in benzene under reflux conditions [167] and Cu(acac)$_2$-catalyzed chloroamination [168].

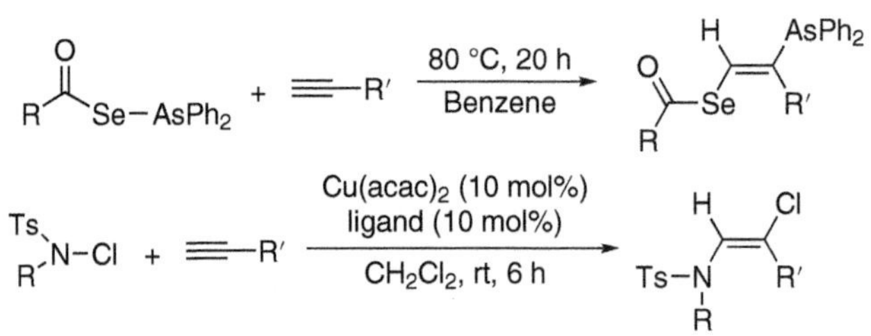

Scheme 5.46 Synthesis of (*E*)-difunctionalized alkenes via *trans*-addition of activated E–E′ bonds.

5.5.3 Addition Reactions to Alkenes

The addition of E—E′ bonds to alkenes are well-known under different conditions. In this section, only a few typical works are briefly described.

For example, the addition of sulfonyl chlorides (S—Cl bond)/bromides (S—Br bond) to alkenes was reported in 1946 [169]. The radical addition of S—Se bond of *S*-benzoyl phenylselenosulfide to alkenes in the presence of AIBN was reported in 1985 [126]. As shown in Scheme 5.47, the selenothiocarboxylation occurs with high regioselectivity in the cases of terminal alkenes, and thiobenzoyloxy and phenylseleno groups attached to the terminal carbon and internal carbon, respectively. In addition, cycloalkenes also undergo the addition reaction with low stereoselectivities.

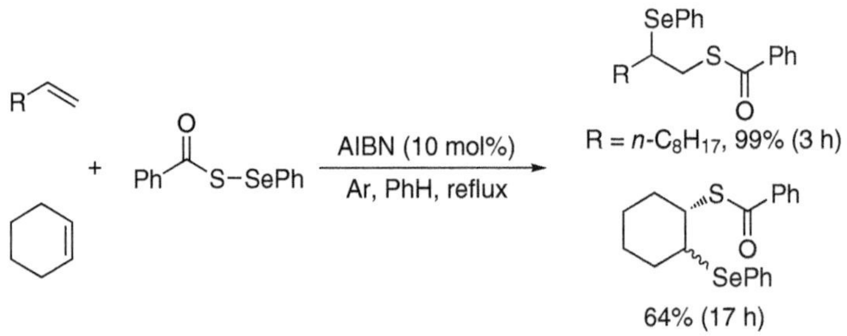

Scheme 5.47 Selenothiocarboxylation of alkenes in the presence of AIBN.

The aminohalogenation of alkenes has been developed for the efficient synthesis of vicinal haloamine derivatives in one-pot manner.

For example, using $AgBF_4$ or $InBr_3/AgBF_4$ (1,3) as catalysts and *N*-halophthalimide as both nitrogen and halogen source, the addition reactions of terminal and internal alkenes with *N*-halophthalimides occurred to give the adducts in good yields with high regioselectivity and diastereoselectivity [170]. In the cases of terminal alkenes used, both styrenes and aliphatic alkenes underwent the Markovnikov-type addition reactions (Scheme 5.48).

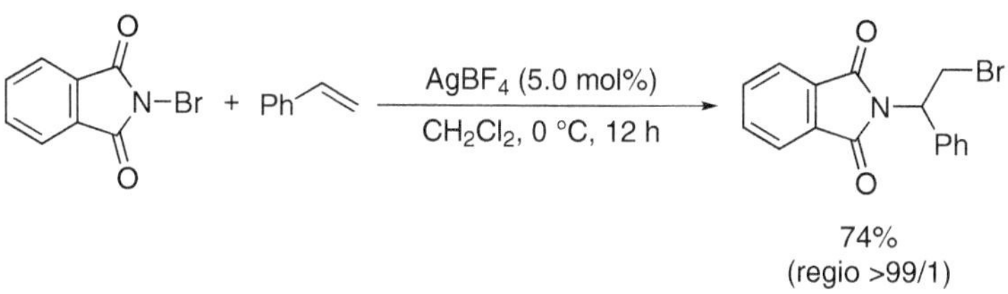

Scheme 5.48 Markovnikov addition of *N*-bromophthalimide to terminal alkenes.

The aminobromination of alkenes with Br—$N(CO_2Me)_2$ in the presence of catalytic $BF_3{\cdot}OEt_2$ also took place with Markovnikov addition [171].

In addition, in MeCN, with the use of $ZnCl_2$ or $Cu(OTf)_2$ as catalysts, the aminochlorination of cinnamic esters with *N*,*N*-dichloro-*p*-toluenesulfonamide as chlorine/nitrogen source occurs with regio- and stereoselectivities to afford the adducts in good to excellent yields (Scheme 5.49) [172]. A metal-free aminobromination of alkenes using $TsNBr_2$ was then also reported [173].

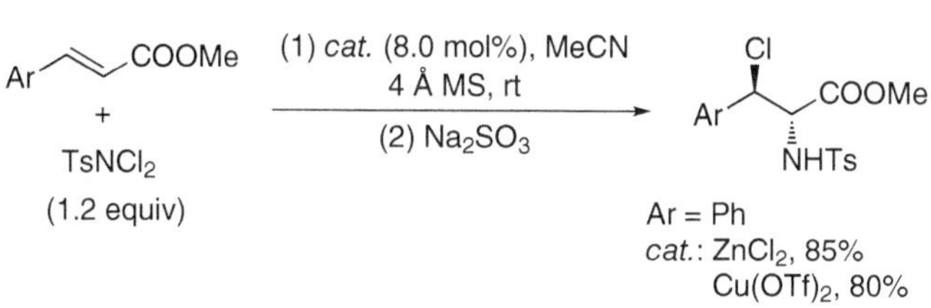

Scheme 5.49 Aminochlorination of methyl *trans*-cinnamates.

The CO_2-promoted aminochlorination of alkenes with chloramine-T (TsNNaCl) without any catalyst [174] and the aminoiodination of alkenes with chloramine-T/I_2 in aqueous media [175] have also been reported.

In the cases of terminal 1,2-allene used, the regioselectivity greatly depends on the substituents of allenes and the use of catalysts.

For example, the borylsilylation of terminal allenes can occur in the presence of either palladium or platinum complexes, but the addition reactions gave either terminal or internal alkenes bearing C—B and C—Si bonds [176].

Recently, in the presence of the commercially available supported Au nanoparticles on TiO_2, terminal allenes undergo silylboration with $PhMe_2Si$—B(pin) exclusively on the terminal double bond in high yields and stereoselectivity with Bpin group attached on the terminal carbon and the silyl group on the sp-carbon [32].

In addition, the addition reactions of sulfonyl halides to terminal allenes with high regioselectivity to give (*E*)-α-halomethyl vinylsulfones have also been developed under different reaction conditions (Scheme 5.50) [177].

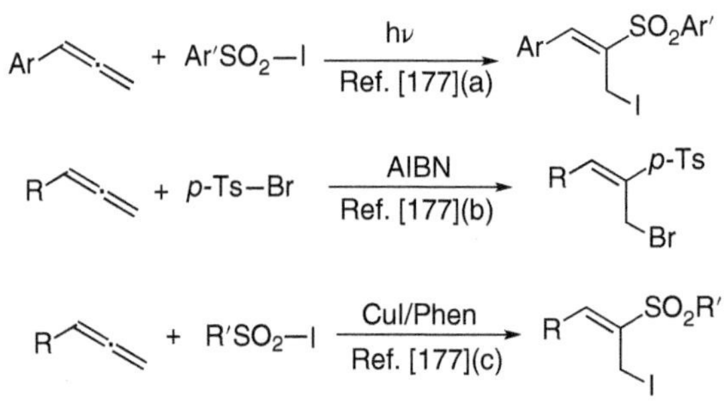

Scheme 5.50 Difunctionalization of allenes with sulfonyl halides leading to (*E*)-α-halomethyl vinylsulfones.

References

1 Selected reviews, see: (a) Han, L.-B. and Tanaka, M. (1999). *Chem. Commun.* 395–402. (b) Beletskaya, I. and Moberg, C. (1999). *Chem. Rev.* 99: 3435–3461. (c) Beletskaya, I. and Moberg, C. (2006). *Chem. Rev.* 106: 2320–2354. (d) Beletskaya, I.P. and Ananikov, V.P. (2011). *Chem. Rev.* 111: 1596–1636. (e) Orlov, N.V. (2015). *ChemistryOpen* 4: 682–697. (f) Yoshida, H. (2016). *ACS Catal.* 6: 1799–1811. (g) Ansell, M.B., Navarro, O., and Spencer, J. (2017). *Coord. Chem. Rev.* 336: 54–77.

2 Selected reviews, see: (a) Marder, T.B. and Norman, N.C. (1998). *Top. Catal.* 5: 63–73. (b) Ishiyama, T. and Miyaura, N. (2000). *J. Organomet. Chem.* 611: 392–402. (c) Ishiyama, T. and Miyaura, N. (2004). *Chem. Rec.* 3: 271–280. (d) Dang, L., Lin, Z., and Marder, T.B. (2009). *Chem. Commun.* 3987–3995. (e) Takaya, J. and Iwasawa, N. (2012). *ACS Catal.* 2: 1993–2006. (f) Yun, J. (2013). *Asian J. Org. Chem.* 2: 1016–1025. (g) Barbeyron, R., Benedetti, E., Cossy, J. et al. (2014). *Tetrahedron* 70: 8431–8452. (h) Yoshida, H. (2016). *ACS Catal.* 6: 1799–1811. (i) Cuenca, A.B., Shishido, R., Ito, H., and Fernández, E. (2017). *Chem. Soc. Rev.* 46: 415–430. (j) Zhao, F., Jia, X., Li, P. et al. (2017). *Org. Chem. Front.* 4: 2235–2255. (k) Yan, G., Huang, D., and Wu, X. (2018). *Adv. Synth. Catal.* 360: 1040–1053. (l) Verma, P.K., Shegavi, M.L., Bose, S.K., and Geetharani, K. (2018). *Org. Biomol. Chem.* 16: 857–873.

3 Selected reviews, see: (a) Suzuki, A. (1982). *Acc. Chem. Res.* 15: 178–184. (b) Miyaura, N. and Suzuki, A. (1995). *Chem. Rev.* 95: 2457–2483. (c) Suzuki, A. (1999). *J. Organomet. Chem.* 576: 147–168. (d) Miyaura, N. (2002). *J. Organomet. Chem.* 653: 54–57. (e) Miyaura, N. (2008). *Bull. Chem. Soc. Jpn.* 81: 1535–1553. (f) Pubill-Ulldemolins, C., Bonet, A., Bo, C. et al. (2010). *Org. Biomol. Chem.* 8: 2667–2682.

4 (a) Ishiyama, T., Matsuda, N., Miyaura, N., and Suzuki, A. (1993). *J. Am. Chem. Soc.* 115: 11018–11019. (b) Ishiyama, T., Matsuda, N., Murata, M. et al. (1996). *Organometallics* 15: 713–720.

5 (a) Iverson, C.N. and Smith, M.R. III, (1995). *J. Am. Chem. Soc.* 117: 4403–4404. (b) Ishiyama, T., Matsuda, N., Murata, M. et al. (1996). *Organometallics* 15: 713–720. (c) Thomas, R.L.I., Souza, F.E.S., and Marder, T.B. (2001). *J. Chem. Soc. Dalton Trans.* 1650–1656. (d) Ali, H.A., El Aziz Al Quntar, A., Goldberg, I., and Srebnik, M. (2002). *Organometallics* 21: 4533–4539. (e) Yoshida, H., Okada, K., Kawashima, S. et al. (2010). *Chem. Commun.* 46: 1763–1765.

6 Selected reports, see: (a) Nguyen, P., Lesley, G., Taylor, N.J., and Marder, T.B. (1994). *Inorg. Chem.* 33: 4623–4624. (b) Dai, C., Stringer, G., Marder, T.B. et al. (1996). *Can. J. Chem.* 74: 2026–2031. (c) Onozawa, S.-y., Hatanaka, Y., Sakakura, T. et al. (1996). *Organometallics* 15: 5450–5452. (d) Crossley, I.R., Hill, A.F., and Willis, A.C. (2008). *Organometallics* 27: 312–315.

7 Lillo, V., Fructos, M.R., Ramírez, J. et al. (2007). *Chem. Eur. J.* 13: 2614–2621.

8 Semba, K., Fujihara, T., Terao, J., and Tsuji, Y. (2015). *Tetrahedron* 71: 2183–2197.

9 Yoshida, H., Kawashima, S., Takemoto, Y. et al. (2012). *Angew. Chem. Int. Ed.* 51: 235–238.

10 Nakagawa, N., Hatakeyama, T., and Nakamura, M. (2015). *Chem. Eur. J.* 21: 4257–4261.

11 Zhang, J., Wu, X., Cheong, W.-C. et al. (2018). *Nat. Commun.* 9: 1002.

12 Nagashima, Y., Hirano, K., Takita, R., and Uchiyama, M. (2014). *J. Am. Chem. Soc.* 136: 8532–8535.

13 Nagao, K., Ohmiya, H., and Sawamura, M. (2015). *Org. Lett.* 17: 1304–1307.

14 Verma, A., Snead, R.F., Dai, Y. et al. (2017). *Angew. Chem. Int. Ed.* 56: 5111–5115.

15 (a) Peng, S., Liu, G., and Huang, Z. (2018). *Org. Lett.* 20: 7363–7366. (b) Gao, G., Kuang, Z., and Song, Q. (2018). *Org. Chem. Front.* 5: 2249–2253. (c) Kuang, Z., Gao, G., and Song, Q. (2019). *Sci. Chin. Chem.* 62: 62–66.

16 Kuang, Z., Mai, S., Yang, K., and Son, Q. (2019). *Sci. Bull.* 64: 1685–1690.

17 Kojima, C., Lee, K.-H., Lin, Z., and Yamashita, M. (2016). *J. Am. Chem. Soc.* 138: 6662–6669.

18 Blaisdell, T.P., Caya, T.C., Zhang, L. et al. (2014). *J. Am. Chem. Soc.* 136: 9264–9267.

19 A review, see:Verma, P.K., Shegavi, M.L., Bose, S.K., and Geetharani, K. (2018). *Org. Biomol. Chem.* 16: 857–873.

20 (a) Chen, Q., Zhao, J., Ishikawa, Y. et al. (2013). *Org. Lett.* 15: 5766–5769. (b) Kidonakis, M. and Stratakis, M. (2017). *Eur. J. Org. Chem.* 2017: 4265–4271.

21 Khan, A., Asiri, A.M., Kosa, S.A. et al. (2015). *J. Catal.* 329: 401–412.
22 Farre, A., Soares, K., Briggs, R.A. et al. (2016). *Chem. Eur. J.* 22: 17552–17556.
23 Fang, L., Yan, L., Haeffner, F., and Morken, J.P. (2016). *J. Am. Chem. Soc.* 138: 2508–2511.
24 Selected reports, see: (a) Bonet, A., Gulyás, H., and Fernández, E. (2010). *Angew. Chem. Int. Ed.* 49: 5130–5134. (b) Bonet, A., Pubill-Ulldemolins, C., Bo, C. et al. (2011). *Angew. Chem. Int. Ed.* 50: 7158–7161. (c) Pubill-Ulldemolins, C., Bonet, A., Bo, C. et al. (2012). *Chem. Eur. J.* 18: 1121–1126.
25 Miralles, N., Alam, R., Szabó, K.J., and Fernández, E. (2016). *Angew. Chem. Int. Ed.* 55: 4303–4307.
26 Davenport, E. and Fernandez, E. (2018). *Chem. Commun.* 54: 10104–10107.
27 Lee, Y., Jang, H., and Hoveyda, A.H. (2009). *J. Am. Chem. Soc.* 131: 18234–18235.
28 Yoshimura, A., Takamachi, Y., Han, L.-B., and Ogawa, A. (2015). *Chem. Eur. J.* 21: 13930–13933.
29 Yoshimura, A., Takamachi, Y., Mihara, K. et al. (2016). *Tetrahedron* 72: 7832–7838.
30 Wen, Y., Deng, C., Xie, J., and Kang, X. (2019). *Molecules* 24: 101.
31 Schiffner, J.A., Müther, K., and Oestreich, M. (2010). *Angew. Chem. Int. Ed.* 49: 1194–1196.
32 Kidonakis, M. and Stratakis, M. (2018). *ACS Catal.* 8: 1227–1230.
33 Ramírez, J., Sanaú, M., and Fernández, E. (2008). *Angew. Chem. Int. Ed.* 47: 5194–5197.
34 Burks, H.E., Liu, S., and Morken, J.P. (2007). *J. Am. Chem. Soc.* 129: 8766–8773.
35 Ely, R.J. and Morken, J.P. (2010). *Org Lett.* 12: 4348–4351.
36 Takaya, J., Kirai, N., and Iwasawa, N. (2011). *J. Am. Chem. Soc.* 133: 12980–12983.
37 (a) Nallagonda, R., Padala, K., and Masarwa, A. (2018). *Org. Biomol. Chem.* 16: 1050–1064. (b) Royes, J., Cuenca, A.B., and Fernandez, E. (2018). *Eur. J. Org. Chem.* 2018: 2728–2739. (c) Wu, C. and Wang, J. (2018). *Tetrahedron Lett.* 59: 2128–2140. (d) Babu, K.N., Massarwe, F., Reddy, R.R. et al. (2020). *Molecules* 25: 959.
38 Gao, G., Yan, J., Yang, K. et al. (2017). *Green Chem.* 19: 3997–4001.
39 Li, L., Gong, T., Lu, X. et al. (2017). *Nat. Commun.* 8: 345.
40 Krautwald, S., Bezdek, M.J., and Chirik, P.J. (2017). *J. Am. Chem. Soc.* 139: 3868–3875.
41 Lesley, G., Nguyen, P., Taylor, N.J. et al. (1996). *Organometallics* 15: 5137–5154.
42 Liu, X., Ming, W., Friedrich, A. et al. (2020). *Angew. Chem. Int. Ed.* 59: 304–309.
43 Morinaga, A., Nagao, K., Ohmiya, H., and Sawamura, M. (2015). *Angew. Chem. Int. Ed.* 54: 15859–15862.
44 Liu, X., Ming, W., Luo, X. et al. (2020). *Eur. J. Org. Chem.* 2020: 1941–1946.
45 Hu, J., Zhao, Y., and Shi, Z. (2018). *Nat. Catal.* 1: 860–869.
46 Xiao, P., Gao, L., and Song, Z. (2019). *Chem. Eur. J.* 25: 2407–2422.
47 Okinoshima, H., Yamamoto, K., and Kumada, M. (1975). *J. Organomet. Chem.* 96: C27–C30.

48 Sakurai, H., Kamiyama, Y., and Nakadaira, Y. (1975). *J. Am. Chem. Soc.* 97: 931–932.
49 Tamao, K., Hayashi, T., and Kumada, M. (1976). *J. Organomet. Chem.* 114: C19–C21.
50 Selected reports, see: (a) Matsumoto, H., Matsubara, I., Takayuki, K. et al. (1980). *Organomet. Chem.* 199: 43–47. (b) Watanabe, H., Kobayashi, M., Higuchi, K., and Nagai, Y. (1980). *J. Organomet. Chem.* 186: 51–62. (c) Watanabe, H., Kobayashi, M., Saito, M., and Nagai, Y. (1981). *J. Organomet. Chem.* 216: 149–157. (d) Ozawa, F., Sugawara, M., and Hayashi, T. (1994). *Organometallics* 13: 3237–3243. (e) Kusukawa, T., Kabe, Y., Nestler, B., and Ando, W. (1995). *Organometallics* 14: 2556–2564. (f) Yoshida, H., Ikadai, J., Shudo, M. et al. (2005). *Organometallics* 24: 156–162.
51 (a) Yamashita, H., Catellani, M., and Tanaka, M. (1991). *Chem. Lett.* 241–244. (b) Yamashita, H. and Tanaka, M. (1992). *Chem. Lett.* 1547–1550.
52 Xiao, P., Cao, Y., Gui, Y. et al. (2018). *Angew. Chem. Int. Ed.* 57: 4769–4773.
53 Gryparis, C., Kidonakis, M., and Stratakis, M. (2013). *Org. Lett.* 15: 6038–6041.
54 Titilas, I., Kidonakis, M., Gryparis, C., and Stratakis, M. (2015). *Organometallics* 34: 1597–1600.
55 Kidonakis, M. and Stratakis, M. (2015). *Org. Lett.* 17: 4538–4541.
56 Saridakis, I., Kidonakis, M., and Stratakis, M. (2018). *ChemCatChem* 10: 980–983.
57 Tahara, A., Nagino, S., Sunada, Y. et al. (2018). *Organometallics* 37: 2531–2543.
58 Mitchell, T.N., Amaria, A., Killing, H., and Rutschow, D. (1986). *J. Organomet. Chem.* 304: 257–265.
59 Mitchell, T.N., Kwetkat, K., Rutschow, D., and Schneider, U. (1989). *Tetrahedron* 45: 969–978.
60 Piers, E. and Skerlj, R.T. (1986). *J. Chem. Soc. Chem. Commun.* 626–627.
61 Mancuso, J. and Lautens, M. (2003). *Org. Lett.* 5: 1653–1655.
62 Yoshida, H., Shinke, A., and Takaki, K. (2013). *Chem. Commun.* 49: 11671–11673.
63 Braune, S. and Kazmaier, U. (2003). *Angew. Chem. Int. Ed.* 42: 306–308.
64 Chahdoura, F., Lassauque, N., Bourissou, D., and Amgoune, A. (2016). *Org. Chem. Front.* 3: 856–860.
65 (a) Yoshida, H., Tanino, K., Ohshita, J., and Kunai, A. (2004). *Angew. Chem. Int. Ed.* 43: 5052–5055. (b) Yoshida, H., Tanino, K., Ohshita, J., and Kunai, A. (2005). *Chem. Commun.* 5678–5680.
66 He, T., Liu, L.-C., Guo, L. et al. (2018). *Angew. Chem. Int. Ed.* 57: 10868–10872.
67 Hayashi, T., Kobayashi, T.-a., Kawamoto, A.M. et al. (1990). *Organometallics* 9: 280–281.
68 Wu, Q., Roy, A., Irran, E. et al. (2019). *Angew. Chem. Int. Ed.* 58: 17307–17311.
69 Okinoshima, H., Yamamoto, K., and Kumada, M. (1972). *J. Am. Chem. Soc.* 94: 9263–9264.
70 (a) Sakurai, H., Kamiyama, Y., and Nakadaira, Y. (1975). *Chem. Lett.* 887–890. (b) Sakurai, H., Eriyama, Y., Kamiyama, Y., and Nakadaira, Y. (1984). *J. Organomet. Chem.* 264: 229–237.

71 Tamao, K., Okazaki, S., and Kumada, M. (1978). *J. Organomet. Chem.* 146: 87–93.

72 Matsumoto, H., Shono, K., Wada, A. et al. (1980). *J. Organomet. Chem.* 199: 185–193.

73 Tsuji, Y., Lago, R.M., Tomohiro, S., and Tsuneishi, H. (1992). *Organometallics* 11: 2353–2355.

74 (a) Watanabe, H., Saito, M., Sutou, N., and Nagai, Y. (1981). *J. Chem. Soc., Chem Commun.* 617–618. (b) Watanabe, H., Saito, M., Sutou, N. et al. (1982). *J. Organomet. Chem.* 225: 343–356.

75 Obora, Y., Tsuji, Y., and Kawamura, T. (1993). *Organometallics* 12: 2853–2856.

76 Tsuji, Y. and Kakehi, T. (1992). *J. Chem. Soc., Chem. Commun.* 1000–1001.

77 (a) Killing, H. and Mitchell, T.N. (1984). *Organometallics* 3: 1318–1320. (b) Schneider, U. and Mitchell, T.N. (1991). *J. Organomet. Chem.* 407: 319–327.

78 Kwetkat, K., Riches, B.H., Rosset, J.-M. et al. (1996). *Chem. Commun.* 773–774.

79 (a) Mitchell, T.N., Killing, H., Dicke, R., and Wickenkamp, R. (1985). *J. Chem. Soc., Chem. Commun.* 354–355. (b) Chenard, B.L. and van Zyl, C.M. (1986). *J. Org. Chem.* 51: 3561–3566. (c) Mitchell, T.N., Wickenkamp, R., Amamria, A. et al. (1987). *J. Org. Chem.* 52: 4868–4874. (d) Murakami, M., Morita, Y., and Ito, Y. (1990). *J. Chem. Soc., Chem. Commun.* 428–429.

80 Piers, E. and Skerlj, R.T. (1987). *J. Chem. Soc., Chem. Commun.* 1025–1026.

81 Kumareswaran, R., Shin, S., Gallou, I., and RajanBabu, T.V. (2004). *J. Org. Chem.* 69: 7157–7170.

82 (a) Hirano, K. and Miura, M. (2017). *Tetrahedron Lett.* 58: 4317–4322. (b) Kawaguchi, S.-i. and Ogawa, A. (2019). *Asian J. Org. Chem.* 8: 1164–1173.

83 Tzschach, V.A. and Baensch, S. (1971). *J. Prakt. Chem.* 313: 254–258.

84 Morse, J.G. and Mielcarek, J.J. (1988). *J. Fluorine Chem.* 40: 41–49.

85 Sato, A., Yorimitsu, H., and Oshima, K. (2005). *Angew. Chem. Int. Ed.* 44: 1694–1696.

86 Kawaguchi, S.-i., Nagata, S., Shirai, T. et al. (2006). *Tetrahedron Lett.* 47: 3919–3922.

87 Okugawa, Y., Hirano, K., and Miura, M. (2017). *Org. Lett.* 19: 2973–2976.

88 Okugawa, Y., Hayashi, Y., Kawauchi, S. et al. (2018). *Org. Lett.* 20: 3670–3673.

89 Dodds, D.L., Haddow, M.F., Orpen, A.G. et al. (2006). *Organometallics* 25: 5937–5945.

90 Liu, Y., Zhang, K., Tian, R. et al. (2020). *Org. Lett.* 22: 6972–6976.

91 Cooper, P., Fields, R., and Haszeldine, R.N.J. (1971). *Chem. Soc. (C)* 3031–3035.

92 Sato, Y., Kawaguchi, S.-i., Nomoto, A., and Ogawa, A. (2016). *Angew. Chem. Int. Ed.* 55: 9700–9703.

93 Sato, Y., Kawaguchi, S.-i., Nomoto, A., and Ogawa, A. (2019). *Chem. Eur. J.* 25: 2295–2302.

94 Sato, Y., Nishimura, M., Kawaguchi, S.-i. et al. (2019). *Chem. Eur. J.* 25: 6797–6806.

95 Hewertson, W. and Taylor, I.C. (1970). *J. Chem. Soc. (C)* 1990–1992.

96 Otomura, N., Hirano, K., and Miura, M. (2018). *Org. Lett.* 20: 7965–7968.

97 Kato, Y., Otomura, N., Hirano, K., and Miura, M. (2020). *J. Org. Chem.* 85: 5981–5994.

98 Selected reports on the activation of E—E bonds (E = S, Se, Te) via oxidative addition to transition-metal complexes, see: (a) Aye, K.-T., Vittal, J.J., and Puddephatt, R.J. *J. Chem. Soc. Dalton Trans.* 1993: 1835–1839. (b) Ogawa, A. (2000). *J. Organomet. Chem.* 611: 463–474. (c) Ananikov, V.P., Kabeshov, M.A., Beletskaya, I.P. et al. (2003). *J. Organomet. Chem.* 687: 451–461. (d) Sugoh, K., Kuniyasu, H., and Kurosawa, H. (2002). *Chem. Lett.* 106–107. (e) Gonzales, J.M., Musaev, D.G., and Morokuma, K. (2005). *Organometallics* 24: 4908–4914. (f) Dutta, P.K., Asatkar, A.K., Zade, S.S., and Panda, S. (2014). *Dalton Trans.* 43: 1736–1743. (g) Karjalainen, M.M., Wiegand, T., Rautiainen, J.M. et al. (2017). *J. Organomet. Chem.* 836 (837): 17–25.

99 Selected reviews, see: (a) Comasseto, J.V., Ling, L.W., Petragnani, N., and Stefani, H.A. (1997). *Synthesis* 1997: 373–403. (b) Chen, J., Chen, S., Xu, X. et al. (2016). *J. Org. Chem.* 81: 3246–3255. (c) Otsuka, S., Nogi, K., and Yorimitsu, H. (2018). *Top. Curr. Chem. (Z)* 376: 13. (d) Lou, J., Wang, Q., Wu, P. et al. (2020). *Chem. Soc. Rev.* 49: 4307–4359.

100 Selected reviews, see: (a) Beletskaya, I.P. and Ananikov, V.P. (2007). *Eur. J. Org. Chem.* 2007: 3431–3444. (b) Beletskaya, I.P. and Ananikov, V.P. (2011). *Chem. Rev.* 111: 1596–1636. (c) Giuseppe, A.D., Castarlenas, R., and Oro, L.A. (2018). *Top. Organomet. Chem.* 61: 31–68. (d) Orlov, N.V. (2015). *ChemistryOpen* 4: 682–697.

101 Kuniyasu, H., Ogawa, A., Miyazaki, S. et al. (1991). *J. Am. Chem. Soc.* 113: 9796–9803.

102 Ogawa, A., Kuniyasu, H., Sonoda, N., and Hirao, T. (1997). *J. Org. Chem.* 62: 8361–8365.

103 Higuchi, Y., Higashimae, S., Tamai, T. et al. (2013). *Chem. Lett.* 42: 1303–1304.

104 Selected reports, see: (a) Arisawa, M. and Yamaguchi, M. (2001). *Org. Lett.* 3: 763–764. (b) Gareau, Y., Tremblay, M., Gauvreau, D., and Juteau, H. (2001). *Tetrahedron* 57: 5739–5750. (c) Ananikov, V.P. and Beletskaya, I.P. (2004). *Org. Biomol. Chem.* 2: 284–287. (d) Ananikov, V.P., Kabeshov, M.A., Beletskaya, I.P. et al. (2005). *Organometallics* 24: 1275–1283. (e) Ananikov, V.P., Kabeshov, M.A., and Beletskaya, I.P. (2005). *Synlett* 1015–1017. (f) Beletskaya, I.P. and Ananikov, V.P. (2007). *Pure Appl. Chem.* 79: 1041–1056. (g) Cai, M., Wang, Y., and Hao, W. (2007). *Green Chem.* 9: 1180–1184. (h) Ananikov, V.P., Gayduk, K.A., Beletskaya, I.P. et al. (2008). *Chem. Eur. J.* 14: 2420–2434. (i) Ananikov, V.P., Gayduk, K.A., Orlov, N.V. et al. (2010). *Chem. Eur. J.* 16: 2063–2071.

105 Zou, K.B., Yin, X.H., Liu, W.Q. et al. (2009). *Synth. Commun.* 39: 2464–2471.

106 Li, J., Li, C., Ouyang, L. et al. (2018). *Adv. Synth. Catal.* 360: 1138–1150.

107 Ananikov, V.P., Orlov, N.V., and Beletskaya, I.P. (2005). *Russ. Chem. Bull.* 54: 576–587.

108 (a) Nakayama, J., Tajiri, T., and Hoshino, M. (1986). *Bull. Chem. Soc. Jpn.* 59: 2907–2908. (b) Petragnani, N. and Toscano, V.G. (1970). *Chem. Ber.* 103: 1652–1653.

109 Heiba, E.I. and Dessau, R.M. (1967). *J. Org. Chem.* 32: 3837–3840.

110 Usugi, S.-i., Yorimitsu, H., Shinokubo, H., and Oshima, K. (2004). *Org. Lett.* 6: 601–603.

111 (a) Li, H., Cheng, Z., Tung, C.-H., and Xu, Z. (2018). *ACS Catal.* 8: 8237–8243. (b) Song, T., Li, H., Wei, F. et al. (2019). *Tetrahedron Lett.* 60: 916–919.

112 Mitamura, T., Iwata, K., and Ogawa, A. (2011). *J. Org. Chem.* 76: 3880–3887.

113 Selected reports, see: (a) Back, T.G. and Krishna, M.V. (1988). *J. Org. Chem.* 53: 2533–2536. (b) Ogawa, A., Ogawa, I., and Sonoda, N. (2000). *J. Org. Chem.* 65: 7682–7685. (c) Tsuchii, K., Doi, M., Ogawa, I. et al. (2005). *Bull. Chem. Soc. Jpn.* 78: 1534–1548. and reference cited therein.

114 Ogawa, A., Yokoyama, K., Yokoyama, H. et al. (1991). *J. Chem. Soc. Chem. Commun.* 1748–1750.

115 Ogawa, A., Yokoyama, K., Obayashi, R. et al. (1993). *Tetrahedron* 49: 1177–1188.

116 Arisawa, M., Kozuki, Y., and Yamaguchi, M. (2003). *J. Org. Chem.* 68: 8964–8967.

117 Ogawa, A., Obayashi, R., Ine, H. et al. (1998). *J. Org. Chem.* 63: 881–884.

118 Chen, J., Tang, Z., Qiu, R. et al. (2015). *Org. Lett.* 17: 2162–2165.

119 (a) Kondo, T., Uenoyama, S.-y., Fujita, K.-i., and Mitsudo, T.-a. (1999). *J. Am. Chem. Soc.* 121: 482–483. (b) Matsumoto, K., Sanada, T., Shimazaki, H. et al. (2013). *Asian J. Org. Chem.* 2: 325–329. and references cited therein.

120 (a) Caserío, M.C., Fisher, C.L., and Kim, J.K. (1985). *J. Org. Chem.* 50: 4390–4393. (b) Kitamura, T., Matsuyuki, J.-I., and Taniguchi, H. (1991). *J. Chem. Soc. Perkin Trans. 1* 1607–1608.

121 Wang, X.-R. and Chen, F. (2011). *Tetrahedron* 67: 4547–4551.

122 Hermans, B., Nicole Colard, N., and Hevesi, L. (1992). *Tetrahedron Lett.* 33: 4629–4632.

123 Matsumoto, K., Fujie, S., Suga, S. et al. (2009). *Chem. Commun.* 5448–5450.

124 Kamiya, I., Nishinaka, E., and Ogawa, A. (2005). *Tetrahedron Lett.* 46: 3649–3652.

125 Wang, M., Cheng, L., Hong, B., and Wu, Z. (2009). *Organometallics* 28: 1506–1513.

126 Toru, A., Seko, T., and Maekawa, E. (1985). *Tetrahedron Lett.* 26: 3263–3266.

127 Ogawa, A., Tanaka, H., Yokoyama, H. et al. (1992). *J. Org. Chem.* 57: 111–115.

128 Ogawa, A., Obayashi, R., Sonoda, N., and Hirao, T. (1998). *Tetrahedron Lett.* 39: 1577–1578.

129 Selected reviews, see: (a) Ogawa, A., Kuniyasu, H., Takeba, M. et al. (1998). *J. Organomet. Chem.* 564: 1–4. (b) Yorimitsu, H. (2013). *Beilstein J. Org. Chem.* 9: 1269–1277. (c) Kawaguchi, S.-i. and Ogawa, A. (2013). *Synlett* 2199–2215. (d) Iwasaki, M. and Nishihara, Y. (2016). *Chem. Rec.* 16: 2031–2045.

130 Yoshida, H. (2016). *ACS Catal.* 6: 1799–1811.

131 Ishiyama, T., Nishijima, K., Miyaura, N., and Suzuki, A. (1993). *J. Am. Chem. Soc.* 115: 7219–7225.

132 Issaian, A., Tu, K.N., and Blum, S.A. (2017). *Acc. Chem. Res.* 50: 2598–2609.

133 Onozawa, S.-y., Hatanaka, Y., Sakakura, T. et al. (1996). *Organometallics* 15: 5450–5452.

134 Han, L.-B., Choi, N., and Tanaka, M. (1996). *J. Am. Chem. Soc.* 118: 7000–7001.

135 Onozawa, S.-y., Hatanaka, Y., Choi, N., and Tanaka, M. (1997). *Organometallics* 16: 5389–5391.

136 Singidi, R.R. and RajanBabu, T.V. (2010). *Org. Lett.* 12: 2622–2625.

137 Onozawa, S.-y., Hatanaka, Y., and Tanaka, M. (1997). *Chem. Commun.* 1229–1230.

138 (a) Xiao, Y.-C. and Moberg, C. (2016). *Org. Lett.* 18: 308–311. (b) Zhang, Q., Liang, Q.-J., Xu, J.-L. et al. (2018). *Chem. Commun.* 54: 2357–2360.

139 (a) Suginome, M., Nakamura, H., and Ito, Y. (1996). *Chem. Commun.* 2777–2778. (b) Suginome, M., Nakamura, H., and Ito, Y. (1997). *Angew. Chem. Int. Ed. Engl.* 36: 2516–2518.

140 Ohmura, T., Oshima, K., Taniguchi, H., and Suginome, M. (2010). *J. Am. Chem. Soc.* 132: 12194–12196.

141 Gryparis, C. and Stratakis, M. (2014). *Org. Lett.* 16: 1430–1433.

142 Suginome, M., Matsuda, T., and Ito, Y. (1998). *Organometallics* 17: 5233–5235.

143 Feng, J.-J., Mao, W., Zhang, L., and Oestreich, M. (2021). *Chem. Soc. Rev.* 50: 2010–2073.

144 (a) Hirner, J.J., Faizi, D.J., and Blum, S.A. (2014). *J. Am. Chem. Soc.* 136: 4740–4745. (b) Chong, E. and Blum, S.A. (2015). *J. Am. Chem. Soc.* 137: 10144–10147.

145 Han, L.-B. and Tanaka, M. (1998). *J. Am. Chem. Soc.* 120: 8249–8250.

146 Iwasaki, M., Fujii, T., Yamamoto, A. et al. (2014). *Chem. Asian J.* 9: 58–62.

147 Liu, X., Duan, X., Pan, Z. et al. (2005). *Synlett* 1752–1754.

148 Hoshi, N., Kashiwabara, T., and Tanaka, M. (2012). *Tetrahedron Lett.* 53: 2078–2081.

149 Kondo, T., Baba, A., Nishi, Y., and Mitsudo, T.-A. (2004). *Tetrahedron Lett.* 45: 1469–1471.

150 Kuniyasu, H., Kato, T., Asano, S. et al. (2006). *Tetrahedron Lett.* 47: 1141–1144.

151 Mesgar, M., Nguyen-Le, J., and Daugulis, O. (2018). *J. Am. Chem. Soc.* 140: 13703–13710.

152 Gaykar, R.N., Bhattacharjee, S., and Biju, A.T. (2019). *Org. Lett.* 21: 737–740.

153 Ohmura, T., Oshima, K., and Suginome, M. (2008). *Chem. Commun.* 1416–1418.

154 García-López, D., Civit, M.G., Vogels, C.M. et al. (2018). *Catal. Sci. Technol.* 8: 3617–3628.

155 Hayashi, M., Matsuura, Y., and Watanabe, Y. (2006). *J. Org. Chem.* 71: 9248–9251.

156 Selected reports, see: (a) Caló, V., Scorrano, G., and Modena, G. (1969). *J. Org. Chem.* 34: 2020–2022. (b) Schmid, G.H., Modro, A., Lenz, F. et al. (1976). *J. Org. Chem.* 41: 2331–2336. (c) Schmid, G.H., Modro, A., Garratt, D.G., and Yates, K. (1976). *Can. J. Chem.* 54: 3045–3049.

157 (a) Amiel, Y. (1971). *J. Org. Chem.* 36: 3691–3696. (b) Amiel, Y. (1971). *J. Org. Chem.* 36: 3697–3702. (c) Amiel, Y. (1971). *Tetrahedron Lett.* 12: 661–663.

158 Zeng, X., Ilies, L., and Nakamura, E. (2012). *Org. Lett.* 14: 954–956.

159 Truce, W.E. and Wolf, G.C. (1971). *J. Org. Chem.* 36: 1727–1732.

160 Amiel, Y. (1974). *J. Org. Chem.* 39: 3867–3870.

161 Iwasaki, M., Fujii, T., Nakajima, K., and Nishihara, Y. (2014). *Angew. Chem. Int. Ed.* 53: 13880–13884.
162 Liang, S., Jiang, L., Yi, W.-b., and Wei, J. (2018). *Org. Lett.* 20: 7024–7028.
163 Zhong, L., Savoie, P.R., Filatov, A.S., and Welch, J.T. (2014). *Angew. Chem. Int. Ed.* 53: 526–529.
164 Wada, T., Kondoh, A., Yorimitsu, H., and Oshima, K. (2008). *Org. Lett.* 10: 1155–1157.
165 Shirai, T., Kawaguchi, S., Nomoto, A., and Ogawa, A. (2008). *Tetrahedron Lett.* 49: 4043–4046.
166 Sato, A., Yorimitsu, H., and Oshima, K. (2009). *Tetrahedron* 65: 1553–1558.
167 Kanda, T., Koike, T., Kagohashi, S. et al. (1995). *Organometallics* 14: 4975–4976.
168 Liu, X.-Y., Gao, P., Shen, Y.-W., and Liang, Y.-M. (2011). *Adv. Synth. Catal.* 353: 3157–3160.
169 Fuson, R.C., Price, C.C., Bauman, R.A. et al. (1946). *J. Org. Chem.* 11: 469–474.
170 Song, L., Luo, S., and Cheng, J.-P. (2013). *Org. Lett.* 15: 5702–5705.
171 Kuszpit, M.R., Giletto, M.B., Jones, C.L. et al. (2015). *J. Org. Chem.* 80: 1440–1445.
172 Guigen Li, G., Wei, H.-X., Kim, S.H., and Neighbors, M. (1999). *Org. Lett.* 1: 395–397.
173 Rajbongshi, K.K., Saikia, I., Chanu, L.D. et al. (2016). *J. Org. Chem.* 81: 5423–5432.
174 Minakata, S., Yoneda, Y., Oderaotoshi, Y., and Komatsu, M. (2006). *Org. Lett.* 8: 967–969.
175 Minakata, S. and Hayakawa, J. (2011). *Chem. Commun.* 47: 1905–1907.
176 Suginome, M., Ohmori, Y., and Ito, Y. (2000). *J. Organomet. Chem.* 611: 403–413.
177 (a) Truce, W.E., Heuring, D.L., and Wolf, G.C. (1974). *J. Org. Chem.* 39: 238–244. (b) Kang, S.-K., Seo, H.-W., and Ha, Y.-H. (2001). *Synthesis* 1321–1326. (c) Lu, N., Zhang, Z., Ma, N. et al. (2018). *Org. Lett.* 20: 4318–4322.

6

Double Functionalization of Alkynes by Addition of Carbon–Element Bonds

Development of efficient transition metal-catalyzed carbon–heteroatom bond-forming reactions has become one of the most important subjects in synthetic chemistry. A simple and efficient protocol is the addition reaction of carbon–heteroatom bonds across C—C unsaturated bonds to form new carbon–heteroatom and carbon–carbon bonds in one step. A wide range of substrates having C—B, C—Si, C—Sn, C—N, C—S, C—Se, and C–halogen bonds have been employed for this transformation.

6.1 Addition Reactions of Carbon–Group 13 Bonds

Two major ways have been developed for the catalytic carboboration of alkynes: direct carboboration via catalytic activation of B—C bond, and indirect carboboration via transmetalative carboborations, in which alkyl or aryl groups are usually introduced by cross-coupling of B—Cl bonds with organometallic reagents. Recently, three reviews have summarized both ways of carboboration of unsaturated hydrocarbon [1].

The first transition metal-catalyzed carboboration of alkynes was intramolecular addition reactions and was reported by Suginome and Murakami's group in 2003 [2]. As shown in Scheme 6.1, in the presence of palladium or nickel complexes, the homopropargylic cyanoboryl ether, which is readily prepared by the reaction of homopropargylic tertiary alcohol with diaminocyanoborane, undergoes the intramolecular carboboration giving five-membered cyclic boryl ethers. The reaction occurs via the *cis*-addition of B—CN bond in a high regioselective 5-*exo* fashion. Although the experimental results confirmed that the intermolecular reactions of bis(dialkylamino)cyanoborane with phenylacetylene in the presence of palladium complexes failed, the same group then successfully developed a [CpPd(η^3-C_3H_5)]/phosphine-catalyzed the intermolecular *cis*-addition of B—CN to internal alkynes with excellent stereoselectivity with the use of cyclic cyanoborane derivatives as carboboration reagents, and in the cases of unsymmetrical alkynes used, the reactions proceed with high regioselectivity to give α,β-unsaturated β-boryl nitriles in good to high yields [3].

Addition Reactions with Unsaturated Hydrocarbons, First Edition. Ruimao Hua.

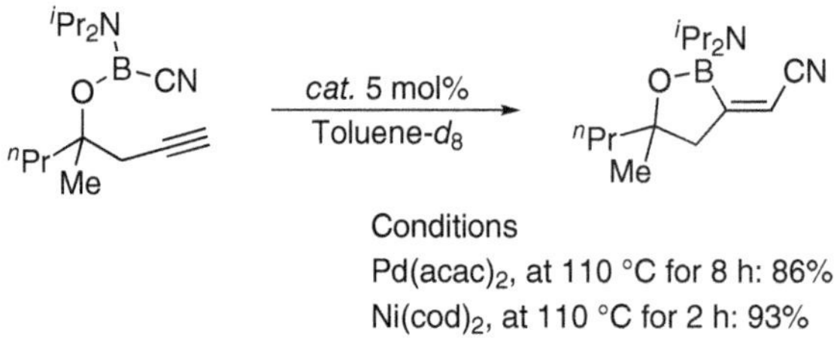

Scheme 6.1 Transition metal-catalyzed intramolecular cyanoboration of alkynes.

The same group also developed a nickel-catalyzed addition of alkynylboranes to internal alkynes with high stereoselectivity (Scheme 6.2) [4]. The addition reactions of aryl and alkyl substituted unsymmetrical alkynes occur with high regioselectivity, and the boryl group is attached selectively to the less sterically hindered position.

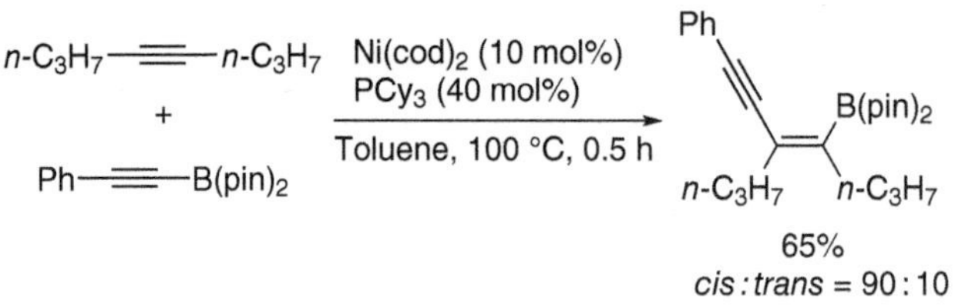

Scheme 6.2 Nickel-catalyzed addition of alkynylborane to 4-octyne.

The *anti*-carboboration of alkynes can be achieved via an oxygen-containing group-assisted. For example, Ohmiya and Sawamura's group developed an *anti*-carboboration of alkynoates with alkyl-, alkenyl-, and arylboranes in the presence of trialkylphosphine to form β-boryl acrylates, and the regioselectivity and the *anti*-addition are both complete and robust (Scheme 6.3) [5].

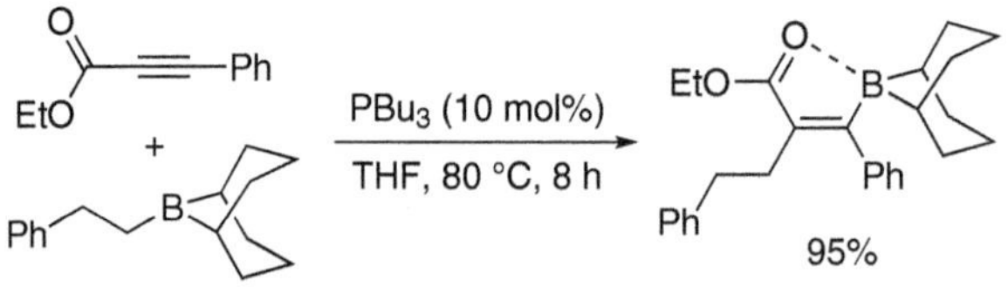

Scheme 6.3 Phosphine-catalyzed *anti*-carboboration of alkynoates.

The hydroxyl group-directing *anti*-carboboration of 3-heteroaryl propargyl alcohols with boronic acids with the use of tartaric acid as promoter to afford boroxole frameworks has been reported by Csákÿ's group [6]. The intermolecular addition reactions proceed with excellent regioselectivity and *anti*-stereoselectivity. Fürstner's group has also developed a regioselective indirect *trans*-carboboration of propargyl alcohols via a subsequent base-promoted *trans*-diboration of propargyl alcohols with $B_2(pin)_2$ and palladium-catalyzed cross-coupling of one of the C—B bond with organic halides; the carbon substituents are invariably placed distal to the —OH group (Scheme 6.4) [7].

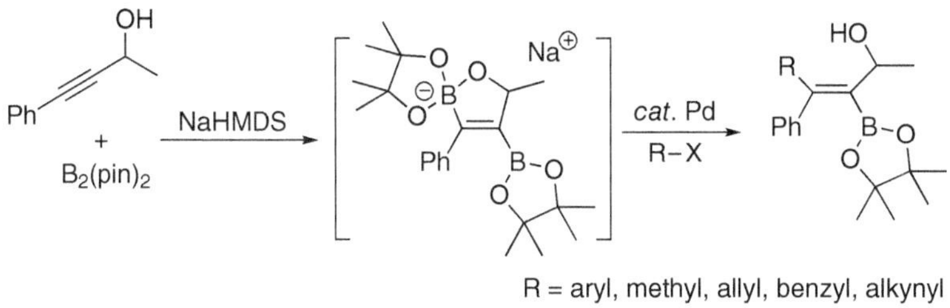

Scheme 6.4 Regioselective indirect *trans*-carboboration of propargyl alcohols.

Hirano and Uchiyama's group then investigated the alkynylboration of propargylic alcohols to develop a *trans*-selective alkynylboration under metal-free conditions affording 4-alkynyl-1,2-oxaborol-2(5*H*)-ols [8].

Recently, the catalytic systems for the indirect alkynylboration [9], vinyl- and arylboration [10] of alkynes, have been extensively studied.

In addition, the copper-catalyzed methylboration [11], allylboration [12], alkylboration [13], and cyanoboration [14] of alkenes have also been reported.

On the other hand, the products from the copper-catalyzed methylboration of terminal alkynes have been applied in the total synthesis of isohericerin, isohericenone, and erinacerin A [15].

In addition, the addition reactions of Al—C bonds to alkynes [16] and to alkenes [17] have also been studied, and the carboalumination is usually applied for further transformation via activation of the formed C—Al bonds.

6.2 Addition Reactions of Carbon–Group 14 Bonds

The activation of C—E bonds (E = Si, Sn) by oxidative addition to transition metal complexes is well-known [18], and the carbosilylation [19] and carbostannylation [20] of unsaturated hydrocarbons via direct activation of C—Si, C—Ge, and C—Sn bonds have been investigated extensively.

6.2.1 Addition Reactions of Carbon–Silicon Bonds

The first example of carbon–silicon activation and its addition to alkynes was reported by Sakurai and Imai in 1975 [21]. As shown in Scheme 6.5, in the presence of $PdCl_2(PPh_3)_2$, 1,1-dimethyl-1-silacyclobutane reacts smoothly with dimethyl acetylenedicarboxylate to give 1,1-dimethyl-2,3-bis(methoxycarbonyl)-1-silacyclohex-2-ene in 95% yield. Other silacyclobutanes react similarly with various alkynes to give the corresponding silacyclohexene derivatives [22].

Cyano group is one of the most important functionalities in organic synthesis due to its easy conversion to carbonyl, carboxyl, amino, and hydroxymethyl groups. The transition metal-catalyzed activation of element–cyano bonds and their addition to unsaturated hydrocarbons have been one of the interesting topics in synthetic chemistry [23]. The first addition reaction of Si—CN to alkynes was reported by Chatani's

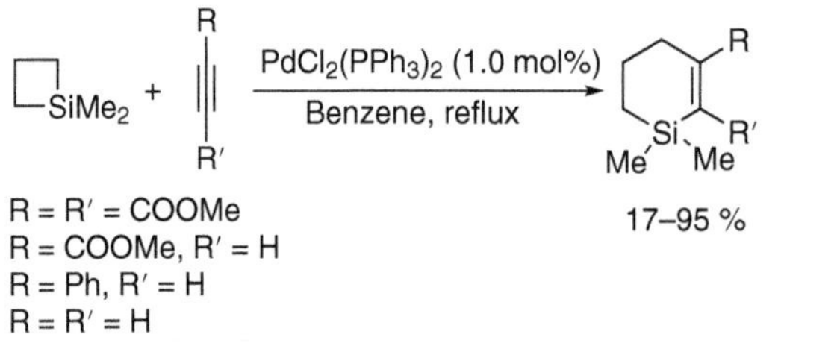

Scheme 6.5 $PdCl_2(PPh_3)_2$-catalyzed cycloaddition of silacyclobutanes with alkynes.

group in 1985, and they found that the reaction of aromatic terminal alkynes with trimethylsilyl cyanide in the presence of $PdCl_2$/pyridine under nitrogen resulted in the formation of β-cyano-β-arylalkenylsilanes with high regio- and stereoselectivity, affording *cis*-adducts (*Z*-isomers) as the major products (Scheme 6.6) [24].

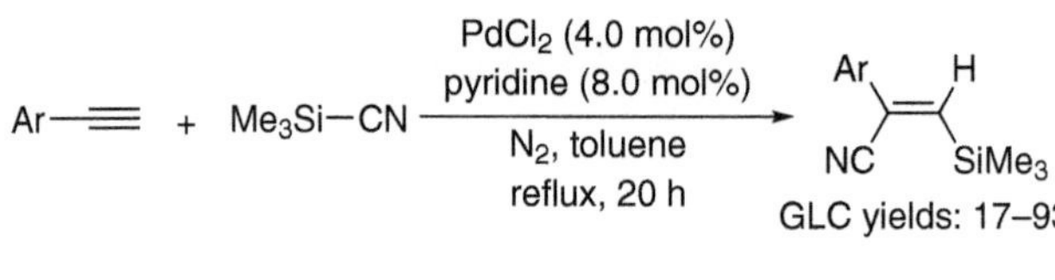

Scheme 6.6 $PdCl_2$-catalyzed addition of trimethylsilyl cyanide to aromatic terminal alkynes.

Under the reaction conditions, the use of 1.6-enyne and 1-phenylpenta-1,4-diyne results in addition of trimethylsilyl cyanide to the terminal carbon–carbon triple bonds, indicating that carbon–carbon double bond and internal carbon–carbon triple bond remained intact (Scheme 6.7).

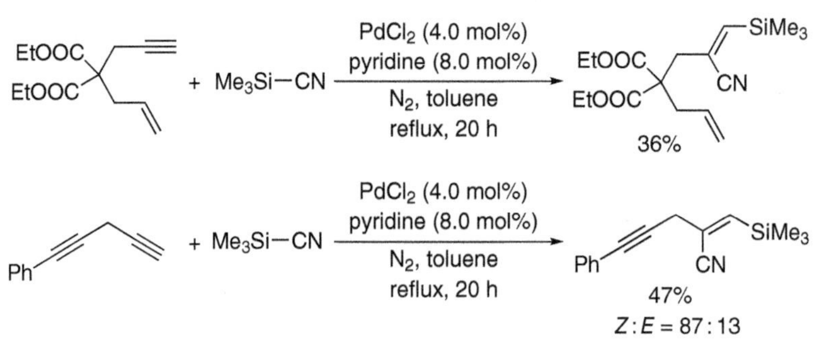

Scheme 6.7 $PdCl_2$-catalyzed addition of trimethylsilyl cyanide to 1,6-enyne and 1,4-diyne.

Interestingly, if $PdCl_2$-catalyzed addition reaction of trimethylsilyl cyanide to terminal alkynes was performed under aerobic conditions, the 1,2-dicyanation of terminal alkynes occurs to give *cis*- and *trans*-1,2-dicyanoalkenes, and *cis*-adducts are major products (Scheme 6.8) [25]. With the use of additive, the internal alkynes also undergo the 1,2-dicyanation reaction but less reactive in the presence of $Pd(CN)_2$.

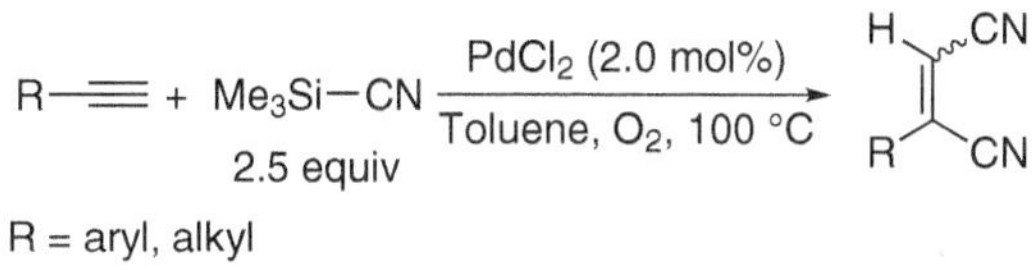

Scheme 6.8 $PdCl_2$-catalyzed 1,2-dicyanation of terminal alkynes under aerobic conditions with trimethylsilyl cyanide.

The intermolecular allylsilylation of simple unactivated alkynes was first reported by Yamamoto and coworker in 1996 [26]. The addition reactions occur smoothly promoted by Lewis acids, such as $EtAlCl_2$ or $AlCl_3$, and in the presence of trimethylsilyl chloride (TMSCl), with high regio- and stereoselectivity. The silyl group is attached to the terminal carbon of terminal alkynes and less hindrance carbon of internal alkynes with *trans*-addition manner (Scheme 6.9).

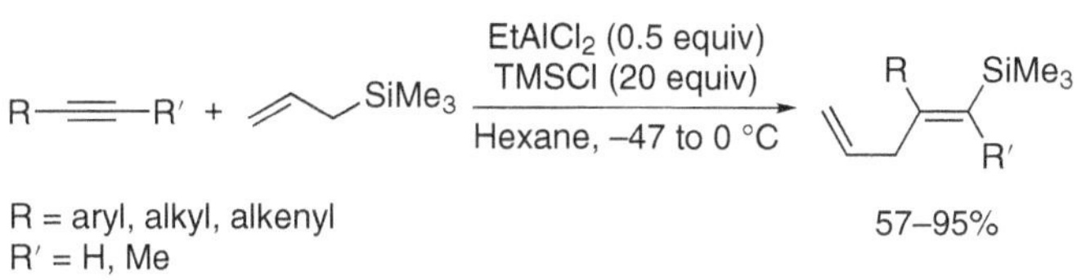

Scheme 6.9 Lewis acid-catalyzed intermolecular allylsilylation of alkynes with allyltrimethylsilane.

Yamamoto's group then investigated several types of intramolecular carbosilylation of alkynes including allylsilylation, vinylsilylation, and arylsilylation of alkynes promoted by Lewis acids, such as $EtAlCl_2$, $AlCl_3$, $AlBr_3$, and $HfCl_4$ [27]. Scheme 6.10 shows the first examples for the *trans*-vinylsilylation of the carbon-tethered alkynyl vinylsilanes to give (*E*)-cyclic dienylsilanes in good to high yields, providing a novel carbocyclization for the preparation of the six- and seven-membered cyclic dienylsilanes. They have further developed the stereoselective intramolecular *trans*-vinylsilylation of the silicon-tethered alkynyl vinylsilanes in the presence of a catalytic amount of $EtAlCl_2$ to give the corresponding six-membered silacycles

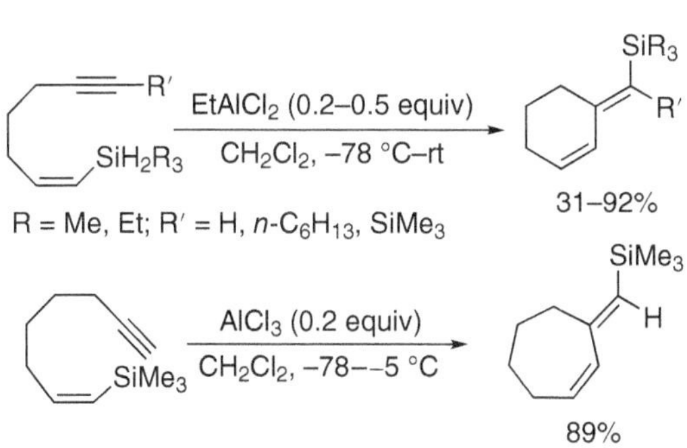

Scheme 6.10 Lewis acid-catalyzed stereoselective intramolecular carbosilylation of the carbon-tethered alkynyl vinylsilanes affording (*E*)-cyclic dienylsilanes.

in high yields (Scheme 6.11). In addition, with the use of $HfCl_4$ and $EtAlCl_2$ as the Lewis acid catalyst, alkynylarylsilanes underwent the *trans*-arylsilylations to afford six- and seven-membered cyclic (*E*)-vinylsilanes (Scheme 6.12). The intramolecular cyclizations proceeded via an *exo*-mode fashion exclusively.

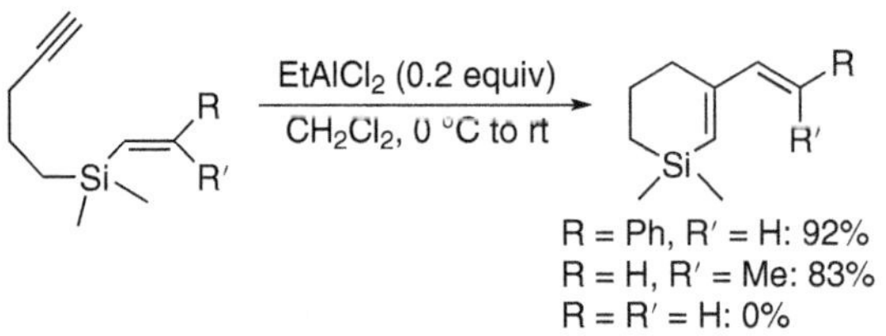

Scheme 6.11 Lewis acid-catalyzed stereoselective intramolecular carbosilylation of the silicon-tethered alkynyl vinylsilanes affording silacycles.

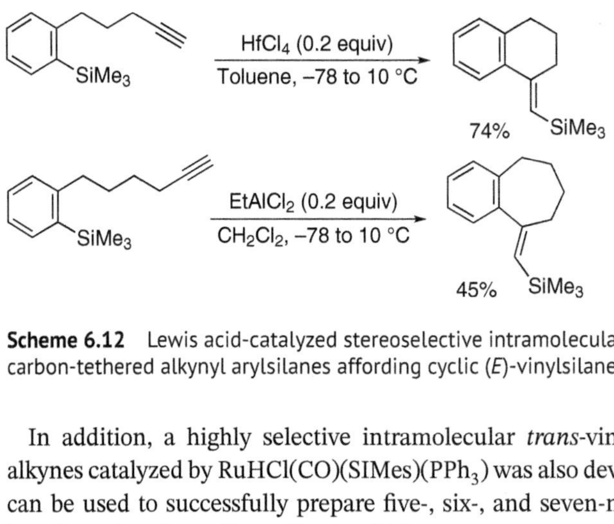

Scheme 6.12 Lewis acid-catalyzed stereoselective intramolecular carbosilylation of the carbon-tethered alkynyl arylsilanes affording cyclic (*E*)-vinylsilanes.

In addition, a highly selective intramolecular *trans*-vinylsilylation of internal alkynes catalyzed by $RuHCl(CO)(SIMes)(PPh_3)$ was also developed, and the process can be used to successfully prepare five-, six-, and seven-membered oxasilacycles by a formal *anti*-*exo*-dig cyclization [28].

6.2.2 Addition Reactions of Carbon–Germanium Bonds

The first example for the transition metal-catalyzed addition of C—Ge bond to unsaturated hydrocarbons was reported by Chatani's group in 1990, and developed a regio- and stereoselective addition of trimethylgermyl cyanide, Me_3GeCN, to terminal alkynes producing β-cyano alkenylgermanes in high yields [29]. For example, the addition reaction of phenylacetylene with Me_3Ge—CN in the presence of $PdCl_2$ at reflux for two hours affords 2-phenyl-3-(trimethylgermyl)-2-propenenitrile in 99% isolated yield with a *Z*/*E* ratio of 97 : 3. While $NiCl_2/{}^iBu_2AlH$ was used, the stereoselectivity is inversed to give *E*-isomer as the major product (Scheme 6.13).

$PdCl_2$-catalyzed reaction of Me_3Ge—CN with 1,6-diynes results in unusual cyclization to give a germole as the main product [30].

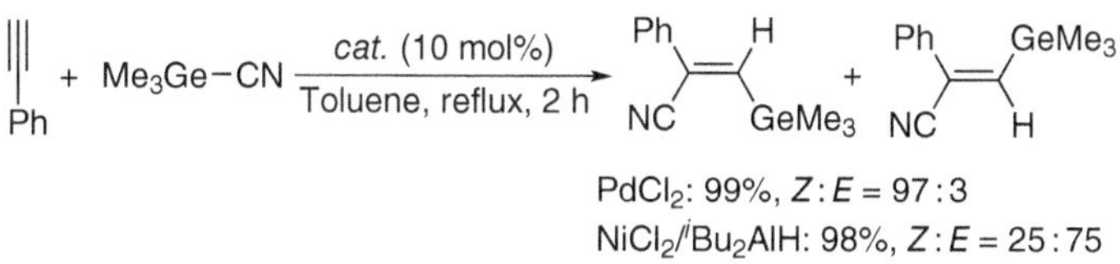

Scheme 6.13 Catalytic addition of trimethylgermyl cyanide to phenylacetylene.

6.2.3 Addition Reactions of Carbon–Tin Bonds

As described in Chapter 5, alkenylstannanes are useful intermediates for the synthesis of various functionalized alkenes via activation of Sn—C bond and its cross-coupling reactions [31]. Although the hydrostannation of alkynes has been one of the simple and efficient reactions for approach to alkenylstannanes, the addition reactions of Sn—C bonds to alkynes (carbostannylation) forming new Sn—C and C—C bonds simultaneously is a much more attractive way in organic synthesis.

The pioneering work on the carbostannylation of alkynes was reported by Yamamoto and Hosomi's groups in 1996 independently. Yamamoto's group developed the $ZrCl_4$-catalyzed *trans*-allylstannylation of alkynes [32], and Hosomi's group developed a radical allylstannylation of activated alkynes such as ethyl propiolate and internal alkynes conjugated with an ester group, to give *trans*-adduct as the major products [33]. Afterward, the allylstannylation of alkynes has also been achieved in the presence of $Pd_2(dba)_3$ [34].

Shirakawa and Hiyama's group studied the oxidative addition of Sn—C bond of tributyl(phenylethynyl)tin to a palladium(0) complex [35], and they then developed the first $[PdCl(\eta^3\text{-}C_3H_5)]_2$/iminophosphine-catalyzed carbostannylation of alkynes with alkynylstannanes to give conjugated (stannyl)enynes in a stereoselective manner with *cis*-adducts only (Scheme 6.14) [36]. However, the regioselectivity depends greatly on the nature of alkynes. With the use of same ligand, they then

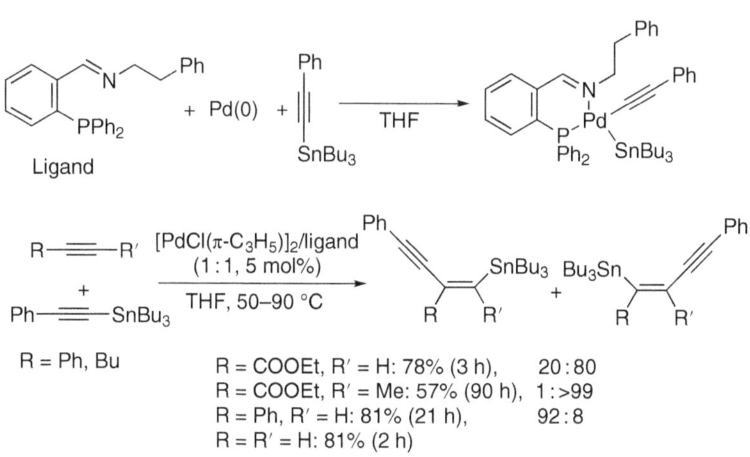

Scheme 6.14 Palladium-catalyzed carbostannylation of alkynes with alkynylstannanes.

developed a $Ni(cod)_2$-catalyzed carbostannylation of terminal and internal alkynes with allylstannylation and acylacylstannylation, providing the stereoselective synthesis of stannylated 1,4-dienes and (*Z*)-β-stannyl-α,β-unsaturated carbonyl compounds, respectively [37]. In addition, the same group has also developed the $Ni(cod)_2$-catalyzed acylstannylation of 1,3-dienes approach to ε-oxoallylstannanes with high stereoselectivity [38].

The application of $[PdCl(\eta^3\text{-}C_3H_5)]_2$/iminophosphine catalyst system in the carbostannylation of arynes with alkynyl- and vinylstannanes affording *ortho*-substituted arylstannanes was also reported by the same group [39].

In addition, when diimine was used as the ligand to replace iminophosphine, the dimerization-carbostannylation of electron-poor alkynes, such as ethyl propiolate and dimethyl acetylenedicarboxylate, took place with alkynyl-, alkenyl-, and allylstannanes to produce highly π-conjugated alkenylstannanes in one-pot [40]. Several examples are shown in Scheme 6.15. The synthesized new stannanes could be transformed to various unsaturated compounds with diverse functionalized structures through palladium-catalyzed cross-coupling reactions with organic electrophiles [41].

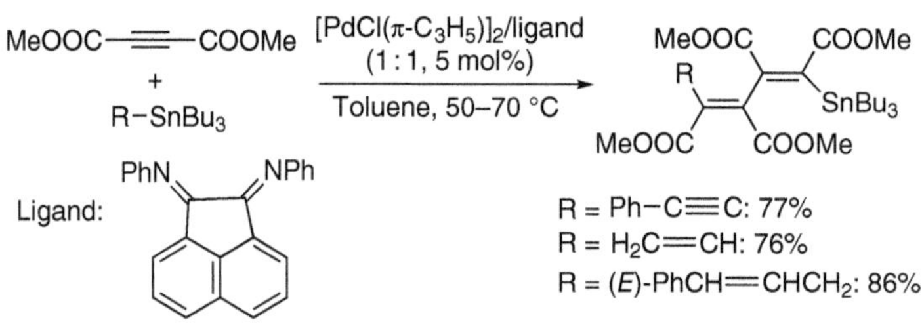

Scheme 6.15 Palladium-catalyzed dimerization-carbostannylation.

A similar gold- and palladium-cocatalyzed intermolecular carbostannylation of propiolates and dimethyl acetylenedicarboxylate with tri-*n*-butyl(vinyl)tins, tri-*n*-butyl(2-furyl)tin, and tri-*n*-butyl(1-propynyl)tin have also been achieved [42].

Tanaka's group also studied the carbamoylstannation of terminal alkynes with carbamoylstannanes to develop a catalyst-dependent regiochemical reversal addition reaction. As shown in Scheme 6.16, when $Rh(acac)(CO)_2$ was used as the catalyst, the carbamoylstannation with (*N*,*N*-diisopropylcarbamoyl)trimethylstannane affords (*Z*)-β-stannyl-α,β-unsaturated amides with terminal attachment of the amide group in excellent stereo- and regioselectivity. However, when $Ni(cod)_2$ was used as catalyst without the use of ligand, it resulted in the reversal of the regioselectivity of the addition reactions to give the adduct with terminal attachment of the stannyl group [43].

In addition, the transition metal-catalyzed addition reactions of tributyltin cyanide (Bu_3Sn—CN) to terminal alkynes [44] or arynes [45] were also developed.

On the other hand, the carbostannylation of 1,2-dienes with acyl- and alkynylstannanes has also been achieved to give α-acylmethyl(vinyl)stannanes and α-alkynylmethyl(vinyl)stannanes by $Ni(cod)_2$ in the presence or absence of ligand [46].

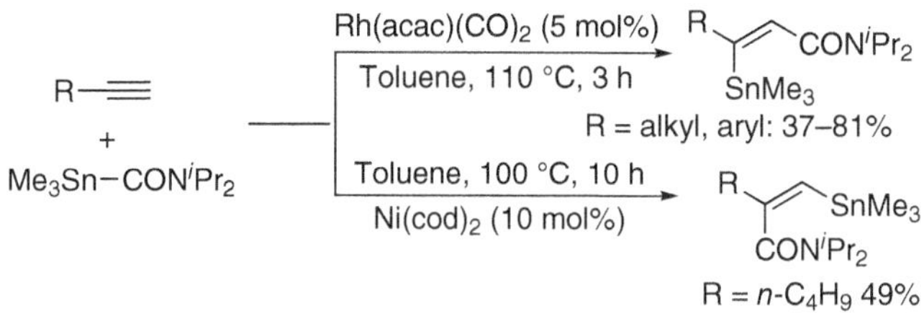

Scheme 6.16 Catalytic carbamoylstannation of terminal alkynes.

On the other hand, noted that, with the use of Pd/C as catalyst, the addition reactions of acylstannanes to propargyl esters occur accompanied with the decarbonylation to give alkenylstannanes having different substituents [47].

6.3 Addition Reactions of Carbon–Group 15 Bonds

The reported addition reactions of carbon–group 15 bonds focus on the activation of C—N bonds of amides and their addition to alkynes (Scheme 6.17) [48]. Recently, two comprehensive reviews on the transition metal-catalyzed aminoacylation of alkynes via activation of C—N single bonds have appeared [49].

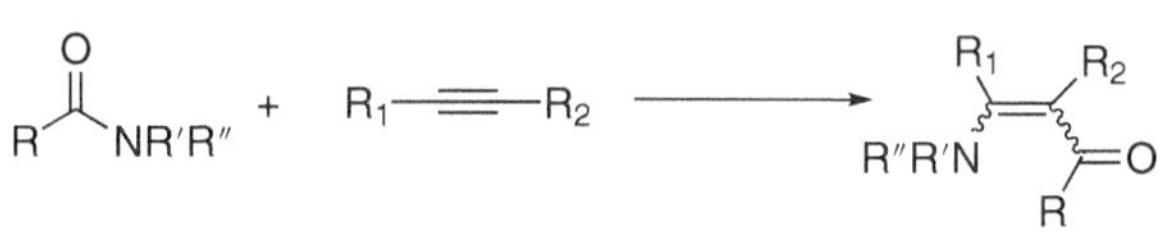

Scheme 6.17 Aminoacylation of alkynes.

Shirakawa and Hiyama's group discovered that when 2-(trimethylsilyl)phenyltriflate was treated with CsF in 1,3-dimethyl-2-imidazolidinone at 20 °C for two hours, 1,4-dimethyl-2,3,4,5-tetrahydro-(1*H*)-1,4-benzodiazepin-5-one could be obtained in 62% yield (Scheme 6.18) [50]. The product is apparent from the addition

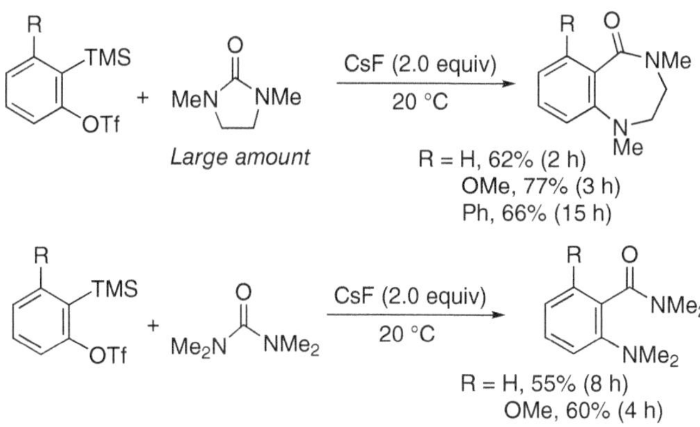

Scheme 6.18 Addition of ureas to arynes.

of C—N bond to aryne, thus they developed a general and efficient approach to 1,4-benzodiazepines, 1,5-benzodiazocines, and 2-aminobenzamides via the reactions of arynes with cycloureas and ureas. More interestingly, when substituted benzynes were used, the reactions occurred with excellent regioselectivity.

In the presence of CsF (2.0 equiv), Larock's group then investigated the reaction of *N*-aryltrifluoroacetamides with 2-(trimethylsilyl)phenyl triflate in MeCN at room temperature, and the products of 2,2,2-trifluoro-1-[2-(arylamino)phenyl]ethanones resulted from C—N insertion into arynes were obtained in good to high yields (Scheme 6.19) [51]. The insertion of C—N bond of ArNH—CN into arynes (amino-cyanation of aryne) was also reported in the presence of CsF in THF at 70 °C [52].

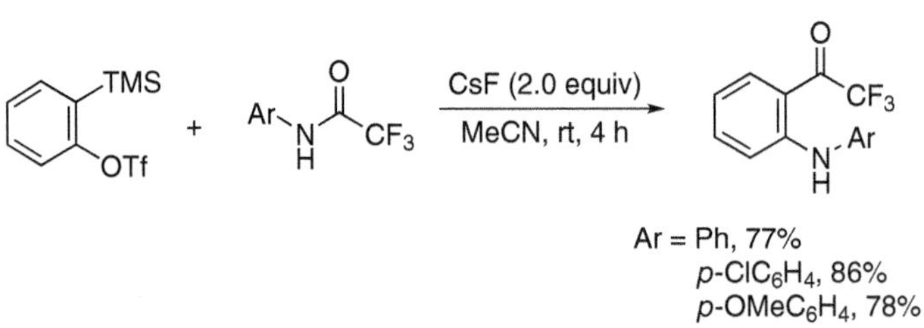

Scheme 6.19 C–N bond addition of *N*-aryltrifluoroacetamides to arynes.

In anisole, $PtCl_2$ could efficiently catalyze the intramolecular aminoacylation of 2-alkynyl amides to give 3-acylindoles as the major adducts, along with deacylated indoles by cleavage of C—N bond of amides (Scheme 6.20) [53]. In the cases of 4/5-methoxy-2-alkynyl benzamides, the reactions selectively produce the acylindoles as the exclusive adduct with excellent yields.

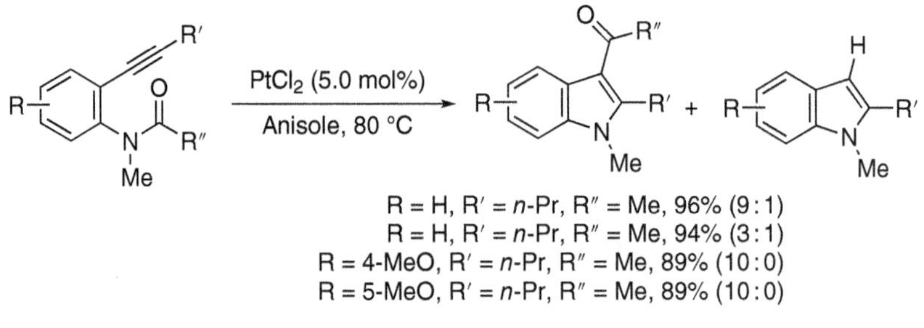

Scheme 6.20 Platinum-catalyzed intramolecular aminoacylation of alkynes.

In the presence of ruthenium complexes, a similar intramolecular annulation of *N*-methyl-*N*-(2-(alkynyl)aryl)formamides or *N*-methyl-*N*-(2-(phenylethynyl) phenyl)-acetamide in 1,2-dichloroethane (DCE) under an argon atmosphere also occurs to give 3-formyl and 3-acetyl indoles [54]. In addition, $PdCl_2(MeCN)_2$ was then used as the catalyst in the similar annulation by cleavage of C—N and S—N bonds to afford indoles with a wide range of functional groups at the C-3 position, including acyl, pyruvoyl, amide, and sulfonyl groups [55].

In addition, in order to get an insight into the mechanism in Lewis acid, $B(C_6F_5)_3$-catalyzed intramolecular aminocyanation and oxycyanation of alkenes, the detailed density functional theory (DFT) calculations have been performed [56].

In the presence of $Ni(cod)_2$ and PMe_3, the decarbonylative addition reactions of *N*-arylphthalimide to internal alkynes occur to form isoquinolones via cleavage of C—N bond of amide group [57]. For example, the reaction of *N*-perfluorophenylphthalimide with 4-octyne in toluene at 110 °C for seven hours led to the corresponding isoquinolone in 84% isolated yield (Scheme 6.21). The mechanism and the role of nickel(0) were then studied by DFT calculations [58].

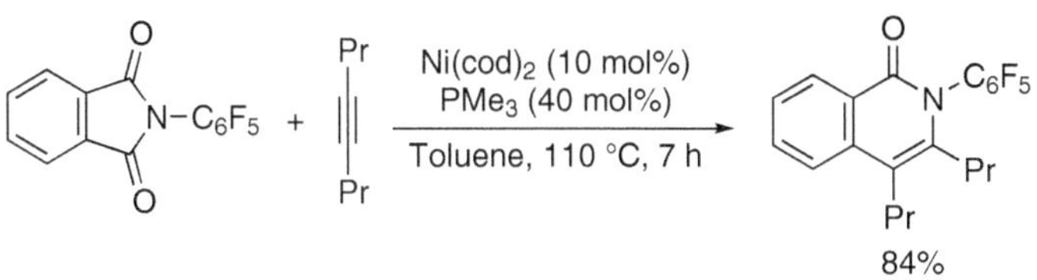

Scheme 6.21 Nickel-catalyzed decarbonylative addition of *N*-arylphthalimides to 4-octyne.

Another two examples on the applications of alkyne aminoacylation in the synthesis of nitrogen-heterocyclic compounds are given here. Liu and coworker developed an efficient route for the formation of pyrazoles via rhodium(I)-catalyzed addition-cyclization of *N*-aryl-*N′*-acylhydrazines with activated internal alkynes [59]. As shown in Scheme 6.22, in the presence of NaOAc, $[Cp^*RhCl_2]_2$ catalyzes the *cis*-addition of *N*-aryl-*N′*-acylhydrazines to alkynes by selective cleavage of C—N bond affording the adducts of *N*-aryl-*N′*-(2-acylvinyl)hydrazines, which undergo the intramolecular nucleophilic addition and dehydration reaction to form pyrazole ring.

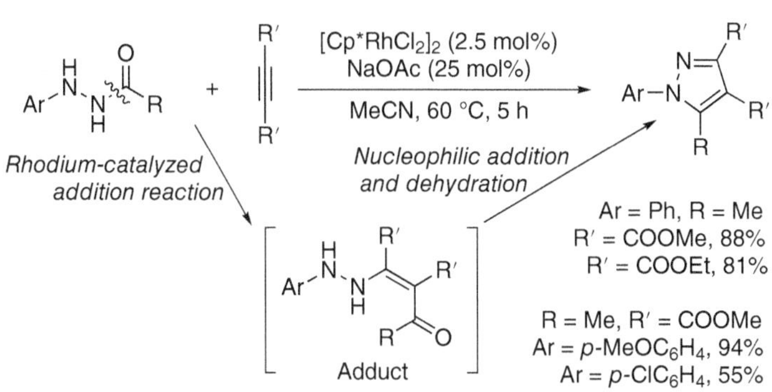

Scheme 6.22 Pyrazole formation via C—N bond cleavage and alkyne annulation.

Li and coworker developed a transition metal-free, Cs_2CO_3-promoted intermolecular aminoacylation of ynones with amides via cleavage of C—N bond in the presence of 1,10-phenanthroline hydrate ($Phen{\cdot}H_2O$) as ligand to give functionalized enaminones or to afford 4-quinolinones through sequential intramolecular nucleophilic substitution of C—Br by N—H, in the case of ynones with an *ortho*-bromo substituent (Scheme 6.23) [60].

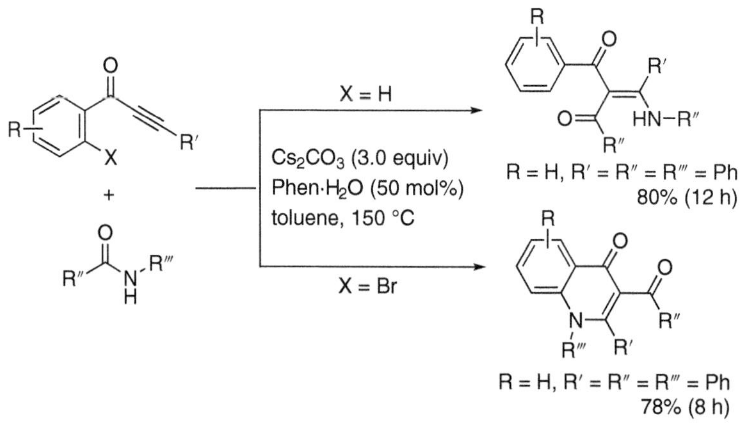

Scheme 6.23 Base-promoted aminoacylation of ynones with amides affording enaminones and 4-quinolinones.

On the other hand, Nakao and coworker reported an intramolecular aminocyanation of alkenes through N—CN bond activation by cooperative palladium/boron catalysis (Scheme 6.24) [61]. The optimizing reaction conditions found that the use of triorganoboron additive and Xantphos ligand is crucial to effectively catalyze the cycloaddition reactions with high chemo- and regioselectivity. A range of substituted indolines and pyrrolidines with both tetra- or trisubstituted carbon and cyano functionalities were readily furnished by this transformation. In addition, a preliminary example of enantioselective aminocyanation was also reported by using chiral ligand.

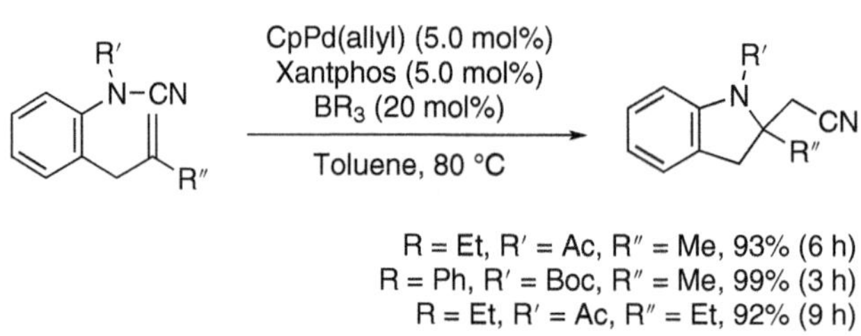

Scheme 6.24 Intramolecular aminocyanation of alkenes catalyzed palladium/BR_3.

At the same time, Douglas's group also developed a metal-free, Lewis acid $B(C_6F_5)_3$-promoted intramolecular aminocyanation of *N*-[2-(2-methylallyl)aryl]-*N*-tosylcyanamides by cleavage of the N—CN bond in conjunction with vicinal addition of sulfonamide and nitrile groups across an alkene [62]. They then developed a palladium-catalyzed activation of C—CN bonds and their enantioselective intramolecular cyanoamidation of alkenes [63]. Using a similar catalyst system of Nakao's group, by designing the substrates, pyrrolidones, piperidinones, isoindolinones, and sultams could be obtained by the intramolecular aminocyanation of alkenes [64].

6.4 Addition Reactions of Carbon–Group 16 Bonds

The preparation, reactivity, and applications of vinylic chalcogenides have been extensively studied [65], and the transition metal-catalyzed carbochalcogenation of alkynes has become one of the direct and attractive synthetic methods with high economy [66].

6.4.1 Addition Reactions of Carbon–Oxygen Bonds

Transition metal-catalyzed activation of carbon–oxygen bonds and their additions to unsaturated hydrocarbons should be one of the most simple and useful reactions for the formation of new carbon–carbon and carbon–oxygen bonds simultaneously.

Kurahashi and Matsubara's group first studied the addition reactions of anhydrides to alkynes in the presence of nickel complex via the activation of acyl C—O bond [67]. Isocoumarins were obtained by the decarbonylative addition of phthalic anhydride to alkyne catalyzed by $Ni(cod)_2/PMe_3$ with the use of Lewis acid as a cocatalyst (Scheme 6.25). A modified catalyst system, $Ni(cod)_2$/phosphine/pyridine, showed high catalytic activity for the cycloaddition of salicylic acid ketal to alkynes via elimination of ketones to give chromones [68]. Thiochromones could be then prepared by $Ni(cod)_2/PPh_3$-catalyzed decarbonylative cycloaddition of readily available thioisatins with alkynes [69].

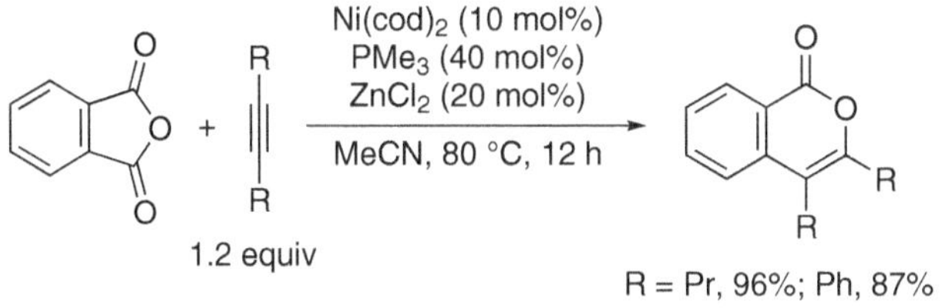

Scheme 6.25 Nickel-catalyzed decarbonylative addition of phthalic anhydride to alkynes.

In addition, the same group also developed a regioselective decarbonylative cycloaddition of phthalic/thiophthalic anhydrides with allenes through cleavage of C—O bond affording δ-lactones and δ-thiolactones catalyzed by $Ni(cod)_2$/phosphine (Scheme 6.26) [70]. When chiral phosphine ligands were used, the asymmetric reactions were also successful.

With the use of readily available benzoxazinones, they studied the similar catalyst system of $Ni(cod)_2$/phosphines-catalyzed cycloaddition with alkynes by tuning a substituent on C_2 of benzoxazinone in favor of the formation of quinolines

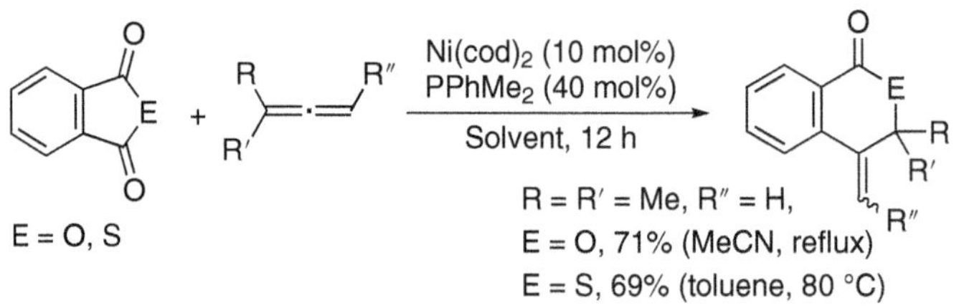

Scheme 6.26 Nickel-catalyzed decarbonylative cycloadditions of (thio)phthalic anhydride with allene.

or quinolones with high chemoselectivity [71]. As depicted in Scheme 6.27, the formation of quinolones results from the reaction between 2-ethoxybenzoxazinones and internal alkynes by ethoxy-directing oxidative addition of Ni(0) to C—O bond of N=C—O moiety and decarboxylation. On the other hand, quinolones form from the cycloaddition of 2-aminobenzoxazinones with internal alkynes through oxidative addition of Ni(0) to C(O)—O bond and acyl migration.

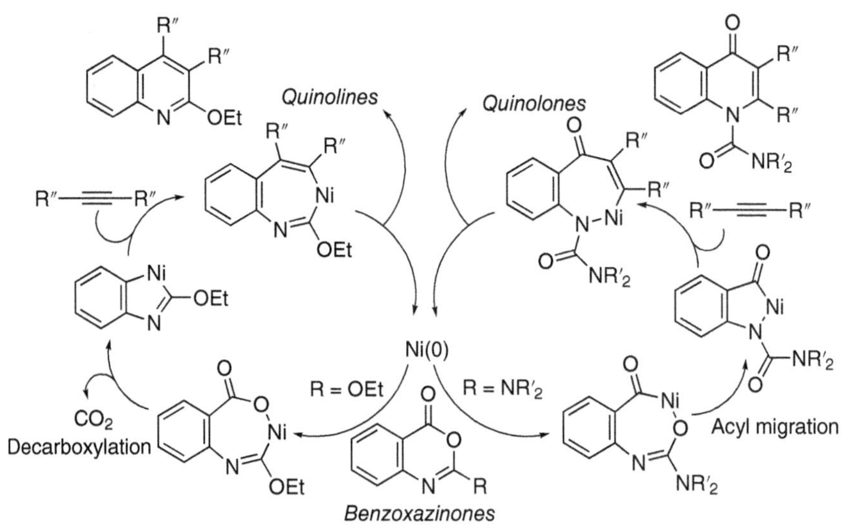

Scheme 6.27 Plausible mechanisms for formation of quinolines and quinolones via nickel (0)-catalyzed cycloadditions of benzoxazinones with internal alkynes.

Arisawa and Yamaguchi first reported the intermolecular addition of aroyl/heteroaroyl acid anhydrides to norbornenes via cleavage of C—O bond in the presence of $Pd_2(dba)_3$ and dppp, and the addition reactions occurred with high stereoselectivity to give *trans*-adducts with the *endo*-benzoyl group and *exo*-benzoyloxy group [72]. As shown in Scheme 6.28, when 4.0 equivalents of norbornadiene and benzoic acid anhydride were refluxed in THF for six hours, 2-benzoyl-3-benzoyloxy-bicyclo[2.2.1]hept-5-ene was obtained in 79% yield. Norbornene and benzonorbornadiene also react with benzoic acid anhydride to give the corresponding adducts in 30 and 67%, respectively.

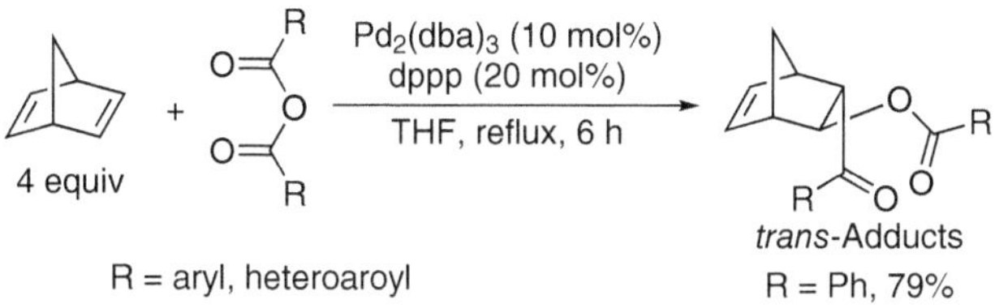

Scheme 6.28 Palladium-catalyzed addition reactions of aroyl/heteroaroyl acid anhydrides to norbornadiene.

In the presence of $Pd(dba)_2/P(OPh)_3$, the first example of intermolecular allylaryloxylation of internal alkyne with allyl aryl ethers was reported to afford *cis*-adducts in moderate yields (Scheme 6.29) [73].

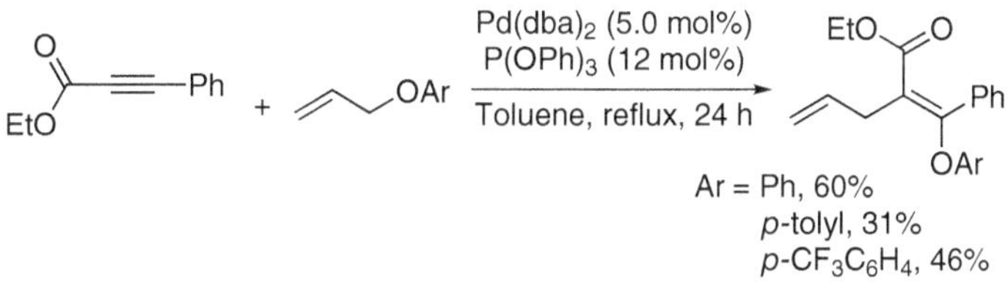

Scheme 6.29 Palladium-catalyzed allylaryloxylation of alkyne with allyl aryl ethers.

Douglas and coworker reported a $[Rh(cod)_2]BF_4$/dppp-catalyzed intramolecular alkoxyacylation of acylated 2-allylphenol derivatives, and chromans can be prepared in good to high yields (Scheme 6.30) [74].

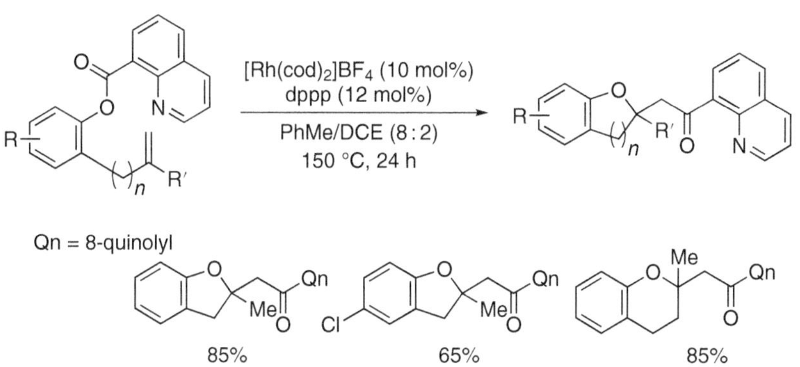

Scheme 6.30 Rhodium-catalyzed intramolecular oxyacylation of alkenes.

In addition, in the presence of $Pd_2(dba)_3$/Xantphos/BPh_3, an intramolecular alkoxycyanation of cyanated 2-allylphenol derivatives through the cleavage of O—CN bonds giving variously substituted dihydrobenzofurans with both a tetrasubstituted carbon and cyano functionality has also been reported (Scheme 6.31) [75].

The mechanism studies on the intramolecular alkene alkoxyfunctionalizations by DFT calculations were then investigated in detail [76].

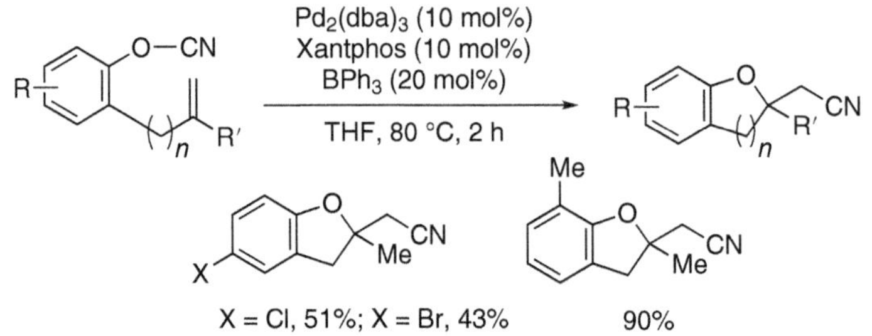

Scheme 6.31 Palladium/BPh_3-catalyzed intramolecular oxycyanation of alkenes.

6.4.2 Addition Reaction of Carbon–Sulfur Bonds

Carbon–sulfur (C—S) bond activation and its transformation have been widely applied in organic synthesis; C—S bond activation and its addition reactions to unsaturated hydrocarbons is an attractive and efficient approach to the new C—S bonds [77]. The early regio- and stereoselective addition reactions of Ts—CN (toluenesulfonyl cyanide) to alkenes, dienes, and 1-hexyne under the free radical conditions initiated by azobis(isobutyronitrile) (AIBN) were reported in 1987 [78], and the first transition metal-catalyzed addition of C—S bond to alkynes were reported in 1991; insertion of dimethyl acetylenedicarboxylate into the S—C bond of allene episulfides afforded sulfur-containing cyclic products in low yields in the presence of $Pd(PPh_3)_4$ [79].

β-Phenylthio acrylates are the versatile and important intermediates for the synthesis of biological molecules [80].

Tanaka and Hua studied the oxidative addition of *O*-methyl *S*-phenyl thiocarbonate (PhS—COOMe) to $Pd(PCy_3)_2$ at room temperature in hexane to afford *trans*-Pd(SPh)(COOMe)$(PCy_3)_2$ and disclosed that C—S bond can be easily cleaved by palladium(0) complex (Scheme 6.32). Therefore, they performed the addition reactions of PhS—COOMe with terminal alkynes in the presence of $Pd(PCy_3)_2$, and successfully developed the thioesterification of terminal alkynes for the synthesis of β-phenylthio acrylates with excellent regio- and stereoselectivity (Scheme 6.33) [81]. The addition reactions occur with simultaneous and selective introduction of thio and ester groups to the internal and terminal carbons of alkynes.

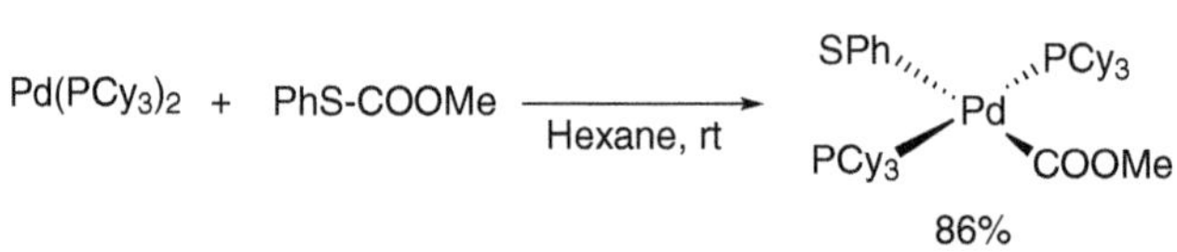

Scheme 6.32 Oxidative addition of *O*-methyl *S*-phenyl thiocarbonate to $Pd(PCy_3)_2$.

They also developed $Ni(cod)_2$- and $Ni(cod)_2$/dppb-catalyzed thioallylation of terminal and internal alkynes with allyl phenyl sulfides affording efficiently thio-1,4-dienes with high regio- and stereoselectivities (Scheme 6.34) [82]. On the basis of isolation of π-allyl nickel complex by oxidative addition of C(sp^3)—S bond to nickel

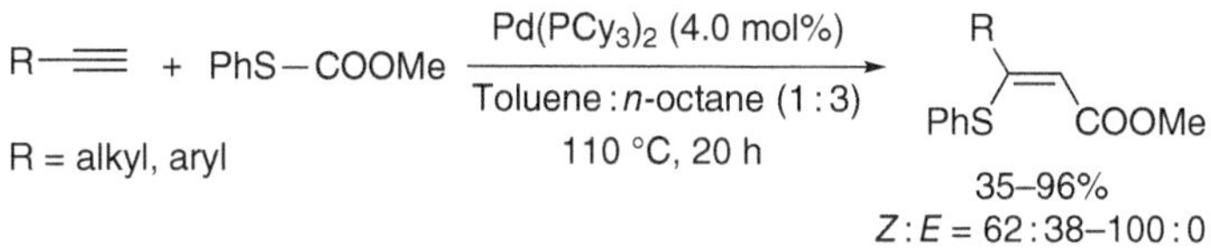

Scheme 6.33 $Pd(PCy_3)_2$-catalyzed thioesterification of terminal alkynes affording β-phenylthio acrylates.

(0), as reported by other transition metal complexes [83], a mechanism involving a π-allyl nickel intermediate as the key intermediate is proposed. In addition, thioallylation of alkynes under possible gold redox catalysis was then reported [84].

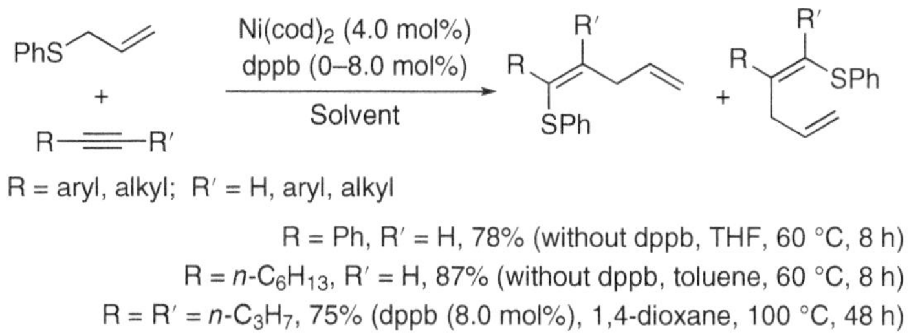

Scheme 6.34 Nickel-catalyzed thioallylation of alkynes with allyl phenyl sulfides.

The activation and addition of another C(sp^3)—S bond to alkynes was reported by Kuninobu and Takai's group [85]. As shown in Scheme 6.35, rhenium-catalyzed the reactions of α-thioketones and terminal alkynes afford γ-thio-α,β-unsaturated ketones in excellent yields with high regio- and stereoselectivity. The formation of adducts results from the insertion of terminal alkynes into the C—S bond of α-thioketones and subsequent isomerization of a double bond.

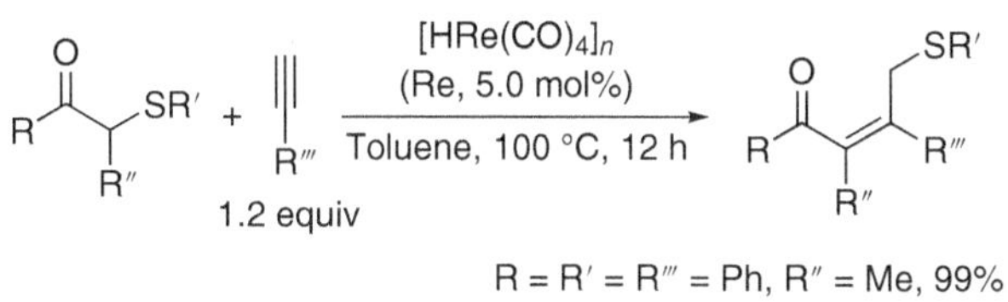

Scheme 6.35 Rhenium-catalyzed regio- and stereoselective synthesis of γ-thio-α,β-unsaturated ketones via insertion of terminal alkynes into the C—S bond.

$Pd(PPh_3)_4$-catalyzed cyanothiolation of terminal alkynes with thiocyanates in benzene or solvent-free conditions to afford β-phenylthio acrylonitriles with high regio- and stereoselectivity was developed in 2006 by Nomoto and Ogawa's group (Scheme 6.36) [86]. The addition reactions show the similar regioselectivity with the introduction of thio and cyano groups to the internal and terminal positions of alkynes, respectively. The mechanism was then theoretically studied with the aid of DFT calculations at the B3LYP level [87].

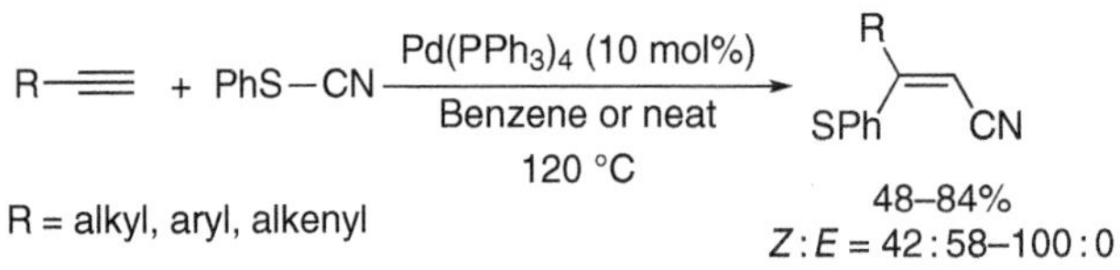

Scheme 6.36 Pd(PPh_3)$_4$-catalyzed cyanothiolation of terminal alkynes with thiocyanates.

In addition, microwave irradiation can dramatically shorten the reaction time with good to excellent isolated yields in the presence of the same catalyst [88].

With the use of same catalyst, the cyanothiolation of internal alkynes can be achieved using organic disulfides and *t*-butyl isocyanide [89].

A Pd(OAc)$_2$/Xantphos-catalyzed activation of carbon–sulfur bonds allow aryne insertion into aryl thiocyanates to generate new C—SAr and C—CN bonds in one step providing an efficient method for the synthesis of a variety of 1,2-thiobenzonitriles under an oxygen atmosphere (Scheme 6.37) [90]. Very recently, the same group has developed a $PdCl_2(PhCN)_2$/Xantphos-catalyzed intramolecular cyanothiolation of aromatic thiocyanate unit across the C—C triple bond to afford sulfur-containing heterocyclic compounds [91].

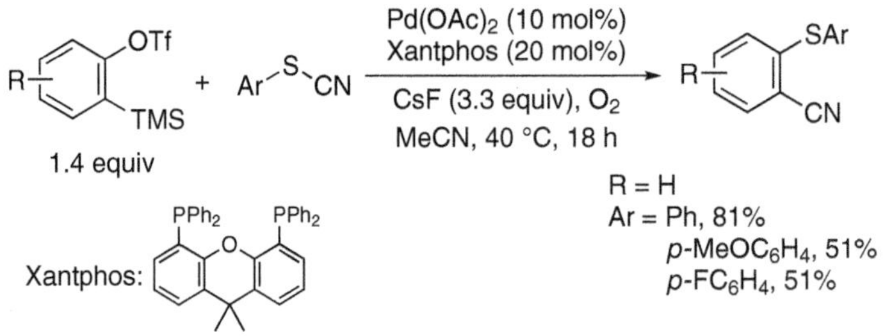

Scheme 6.37 Palladium-catalyzed cyanothiolation of arynes giving 1,2-thiobenzonitriles.

In the presence of Pt(PPh_3)$_4$, the decarbonylative arylthiolation of unsymmetrical internal alkynes bearing oxygen-containing group with thioesters (ArS—C(O)R) take place, the regio- and stereoselective insertion of alkyne into an S—Pt bond is confirmed by reaction with a platinum complex with an S—Pt framework (Scheme 6.38) [92]. The same catalyst also catalyzes the decarbonylative *cis*-thienylthiolation of terminal alkynes with thienylthioesters [(2-thienyl) C(O)SPh] with high regio- and stereoselectivity to afford vinylsulfides in moderate to good yields with thienyl group at terminal carbon of alkynes [93].

The C—S bond activation of thioesters via the oxidative addition reactions with transition metal complexes has been well investigated [94].

In order to undergo the CO-retained carbothiolation of alkynes with thioesters, the choice of catalyst system (Pd(dba)$_2$/dppe) and the introduction of a CF_3 group into the thioesters are the key [95].

The *cis*-thiocarbamoylation of internal alkynes can also occur in the presence of Ni(cod)$_2$/PPr_3 to give tetrasubstituted β-aminocarbonyl vinyl sulfides (Scheme 6.39) [96].

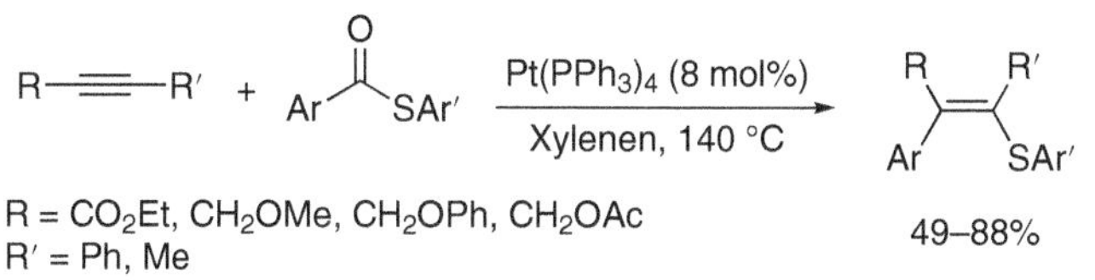

Scheme 6.38 Platinum-catalyzed decarbonylative arylthiolation of internal alkynes.

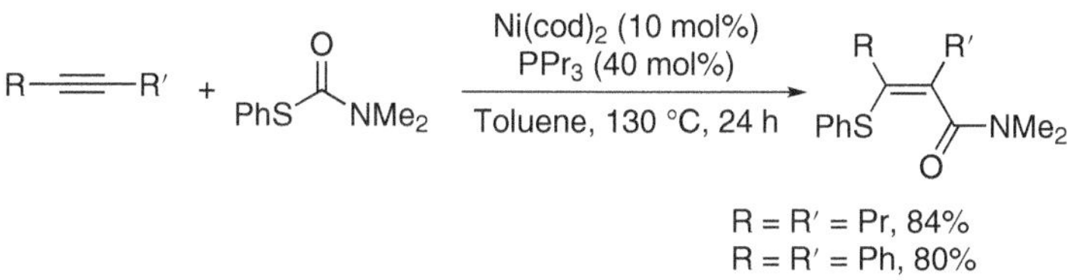

Scheme 6.39 Nickel-catalyzed thiocarbamoylation of internal alkynes.

Interestingly, the decarbonylative cycloadditions of thiophthalic anhydride with internal alkynes affording sulfur-containing heterocyclic compounds were then successfully developed by using Ni(0) complexes as the catalysts, and the reaction selectively produce thioisocoumarins, benzothiophenes, and thiochromones depending on the reaction conditions employed [97].

C(sp)—S bond can also be activated in the presence of rhodium complexes, the carbothiolation of 1,4-diaryl-1,3-butadiynes with 1-alkylthio-1-alkynes catalyzed by $RhH(PPh_3)_4$/phosphine afforded (*Z*)-4-alkylthio-4-aryl-3-arylethynyl-3-buten-1-ynes (Scheme 6.40) [98]. In the cases of terminal alkynes used, such as 1-decyne and (*t*-butylthio)acetylene, the carbothiolation proceeded via *cis*-addition with C—C bond formation at the less hindered alkyne carbon.

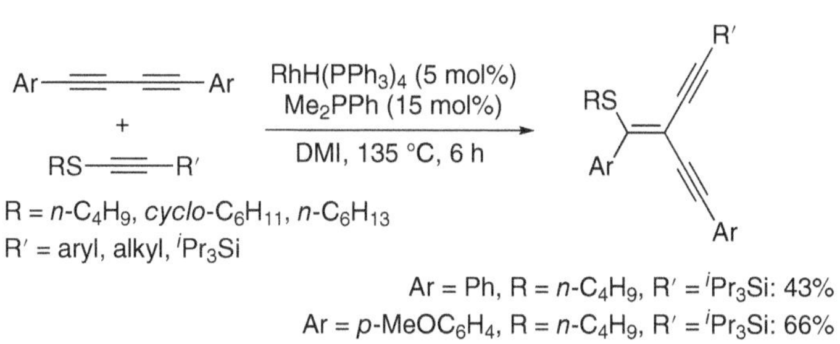

Scheme 6.40 Rhodium-catalyzed carbothiolation of 1,3-butadiynes with 1-alkylthio-1-alkynes.

In addition, for activation of $C(sp^2)$—S bonds, rhodium-catalyzed carbothiolation of alkynes with aryl methyl sulfides (Ar—SMe) [99]; palladium-catalyzed carbothiolation of alkynes with azolyl sulfides [100], with heteroaryl sulfides [101]; and the reaction of in situ generated arynes with vinyl sulfides giving *ortho*-arythiostyrenes [102] have also been reported.

6.4.3 Addition Reactions of Carbon–Selenium Bonds

Under nitrogen, in the presence of CuX (X = Cl, I) and $Et_3N{\cdot}HCl$, the reaction of arylselenoesters with terminal alkynes in dimethylfomamide (DMF) affording (*Z*)-β-arylseleno-α,β-unsaturated ketones was reported by Meng and coworker in 1998 (Scheme 6.41) [103]. The mechanism studies disclosed that it is a cascade reaction including the formation of α,β-unsaturated ketones and arylselenols as the intermediates, which undergo the addition reaction with high regio- and stereoselectivity. Similarly, CuI can catalyze the telluroacylation of terminal alkynes with telluroesters (ArTe—C(O)R) [104].

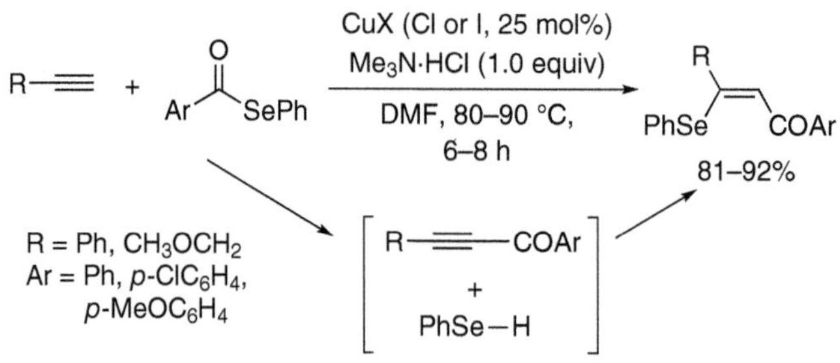

Scheme 6.41 Copper-catalyzed synthesis of (*Z*)-β-arylseleno-α,β-unsaturated ketones by addition of selenoesters to terminal alkynes.

Kambe's group also examined the reactions of terminal alkynes with selenoesters in the presence of $Pt(PPh_3)_4$, and the reactions occur with regio- and stereoselectivity to give vinylselenides in moderate yields with decarbonylation reaction (Scheme 6.42) [105].

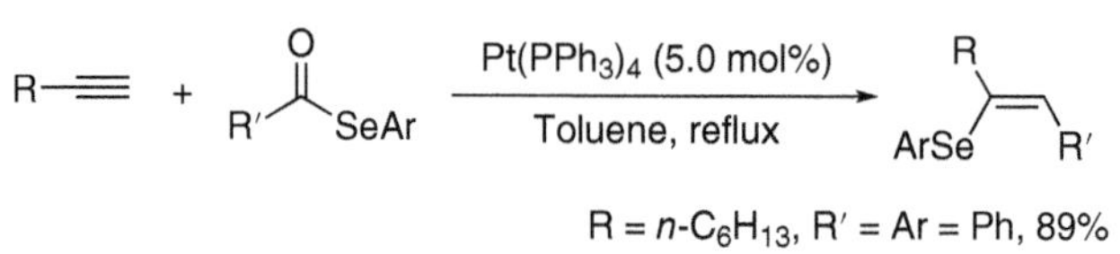

Scheme 6.42 Platinum-catalyzed decarbonylative addition of selenoesters to terminal alkynes.

The same group then investigated the selenocarbamoylation of terminal alkynes by reacting a carbamoselenoate, $PhSe{-}C(O)NMe_2$, with 1-octyne in the presence of $Pd(PPh_3)_4$ [106]. Although the selenocarbamoylation product, β-selenoacrylamide, was obtained in 40% yield, the addition reaction proceeds with high regio- and stereoselectivity (Scheme 6.43). When *Se*-phenyl *N*-methyl-*N*-prop-2-ynyl carbamoselenoate was used, the intramolecular selenocarbamoylation of alkynes occurs smoothly to give the corresponding α-alkylidene-β-lactam in 74% yield within one hour with excellent regio- and stereoselectivity (Scheme 6.44). This reaction could be applied successfully in the synthesis of the α-alkylidene-δ-lactam and -ε-lactam frameworks and applied to the thiocarbamoylation.

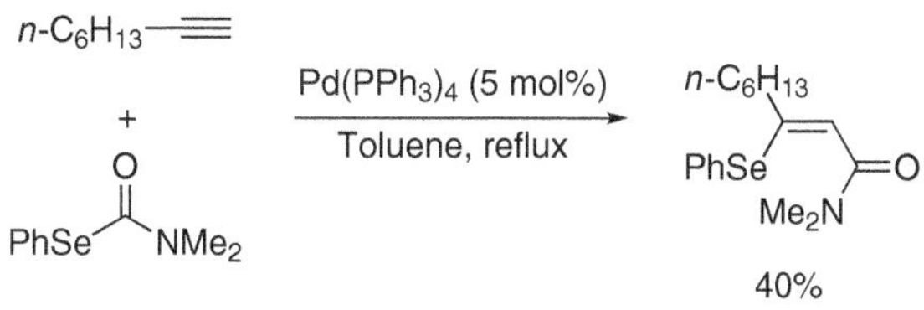

Scheme 6.43 Palladium-catalyzed selenocarbamoylation of alkynes.

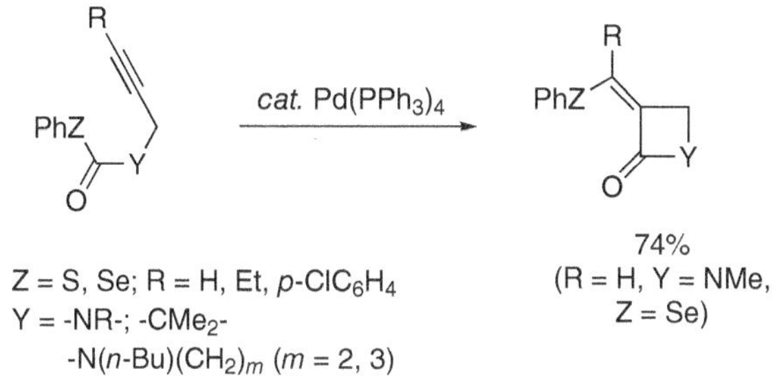

Scheme 6.44 Palladium-catalyzed intramolecular seleno(thio)carbamolyation of alkynes.

An anion stabilizing group on the β-position of acyclic vinyl sulfides and selenides was found to enhance the oxidative addition of $C(sp^2)$—S/Se bonds to Pt(0) complexes [107], and the same group then developed the $Pt(PPh_3)_4$-catalyzed inter- and intramolecular vinylchalcogenation of alkynes with β-phenylchalcogeno-conjugated amides [108]. Scheme 6.45 shows the construction of six-membered lactam framework via an intramolecular *cis*-vinylselenation and -selenation of alkynes.

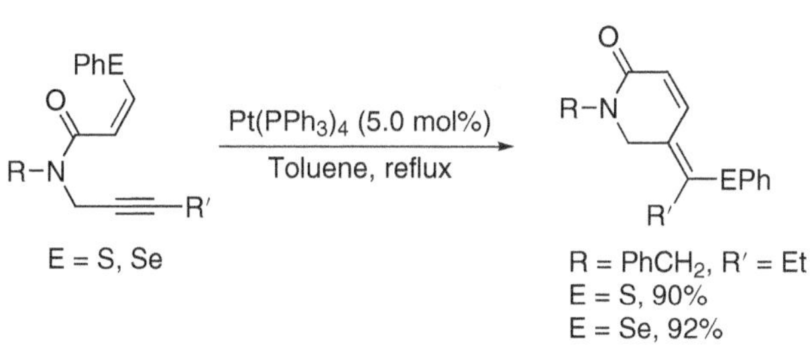

Scheme 6.45 Platinum-catalyzed intramolecular vinylchalcogenation of alkynes affording six-membered lactams.

When $[Co_2(CO)_8]$ was used as the catalyst, the cyanoselenation of 1-octyne with phenyl selenocyanate could occur in the presence of CO with excellent regio- and stereoselectivities, accompanied by the formation of diphenyl diselenide as a by-product, and the adduct was obtained in low to fair yields (Scheme 6.46) [66(c)].

Very recently, Werz and coworker have developed the intramolecular cyanoselenylation of aliphatic selenocyanates catalyzed by $Pd_2(dba)_2$/Xantphos and aromatic selenocyanates catalyzed by $Pd(PPh_3)_4$/SPhos to afford selenium-substituted heterocyclic acrylonitrile derivatives in good to excellent yields [109].

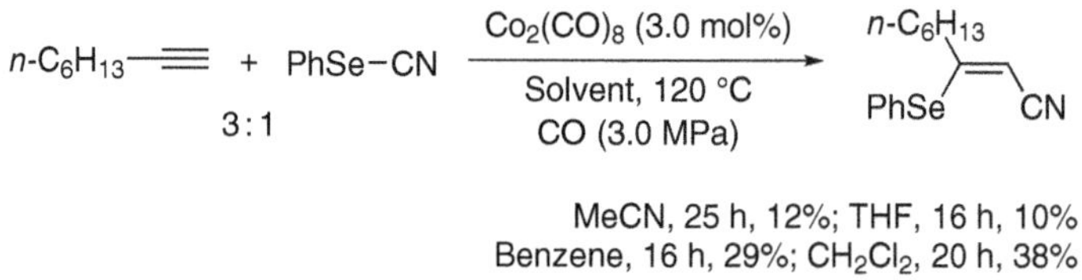

Scheme 6.46 Cobalt-catalyzed addition of PhSeCN to 1-octyne in the presence of CO.

With the use of $Ni(cod)_2$/dppb as catalyst system, the addition reactions of allyl phenyl selenide to terminal alkynes regioselectively afford 2-phenylseleno-1-allyl-1-alkenes in good to excellent yields [110]. A proposed mechanism involving the formation of η^3-allyl–nickel complex is proposed on the basis of the isolation, crystal structure determination, and reactivity study of this complex.

Ogawa's group also developed a $Pd(OAc)_2/P(o\text{-tol})_3$-catalyzed alkynylselenation of acetylenedicarboxylates leading to enyne selenides (Scheme 6.47) [111].

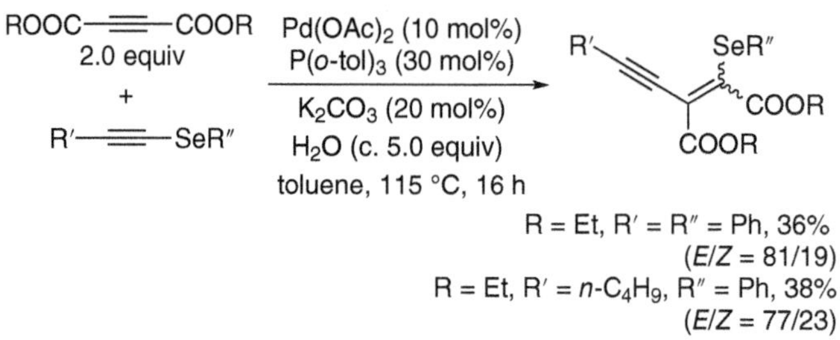

Scheme 6.47 Palladium-catalyzed alkynylselenation of alkynes leading to enyne selenides.

In addition, palladium-catalyzed regio- and stereoselective selenoacylation of allenes with selenoesters producing functionalized allyl selenides [112] and vinylselenation of allenes affording 2-selenomethyl-1,3-dienes [113] were also reported by Kambe group. In both addition reactions, the selenium moiety is introduced at the terminal carbon of allenes.

Moreover, Beletskaya and coworker reported a palladium-catalyzed successful activation of E—E and C—E bonds in diaryl dichalcogenides (E = S, Se) under microwave irradiation conditions [114].

6.5 Addition Reactions of Carbon–Halogen Bonds to Alkynes

Carbon–halogen bonds (C—X) are the fundamental chemical bonds in organic compounds, and the formation and transformation of C—X are the important and interesting research topics in the organic synthesis in lab and industrial scale [115]. The oxidative addition of C—X to low-valent transition metal complexes forming the corresponding C—M—X complexes is well-known in organometallic chemistry [116], which plays an important step to furnish the addition reactions of C—X to unsaturated hydrocarbons. This section focuses on summarizing the recent development of the direct activation and addition reactions of C—X to unsaturated hydrocarbons in the synthesis of functionalized alkenyl halides.

6.5.1 C(sp^3)−X Activation and Its Addition Reactions

The most reported activation of C(sp^3)—X bonds and their addition reactions to unsaturated hydrocarbons include allyl halides, benzyl halides, halocyanation, and polyhalogenated alkanes.

It is well-known that C(sp^3)—X bond can be easily cleaved by the oxidative addition of allyl halides to transition metal compounds [117]. The first addition reaction of C(sp^3)—X to unsaturated hydrocarbons was reported in 1974 by Kaneda and coworker with the use of alkynes and allyl chlorides as the starting materials catalyzed by $PdCl_2(PhCN)_2$ in $MeNO_2$ at room temperature [118]. This chloroallylation provides an atom economic approach to 1-chloro-1,4-dienes (Scheme 6.48). The same group then investigated the similar addition reactions in detail with the use of terminal and internal alkynes in the presence of different palladium complexes [119].

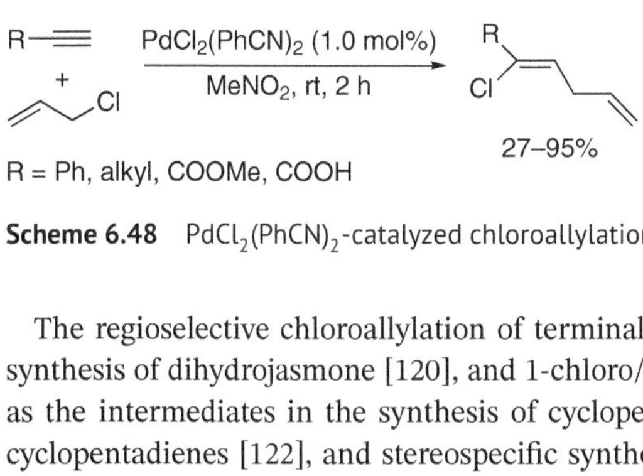

Scheme 6.48 $PdCl_2(PhCN)_2$-catalyzed chloroallylation of alkynes.

The regioselective chloroallylation of terminal alkynes has been applied in the synthesis of dihydrojasmone [120], and 1-chloro/bromo-1,4-dienes have been used as the intermediates in the synthesis of cyclopentenones [121], 1,2-disubstituted cyclopentadienes [122], and stereospecific synthesis of highly substituted skipped 1,4-dienes and enynes [123].

With the use of ynamides, Zhu and coworker developed a $PdCl_2$-catalyzed chloroallylation of ynamides at room temperature approach to highly substituted enamides (Scheme 6.49) [124]. The addition reactions occur with high regio- and stereoselectivity. $PdCl_2$ can also catalyze the haloallylation of aromatic ynolethers with allyl chlorides to give (1*E*)-α-chloroenol ethers [125] and catalyze the bromoallylation of alkynes by using $CuBr_2$ and allylic alcohol as original starting materials [126].

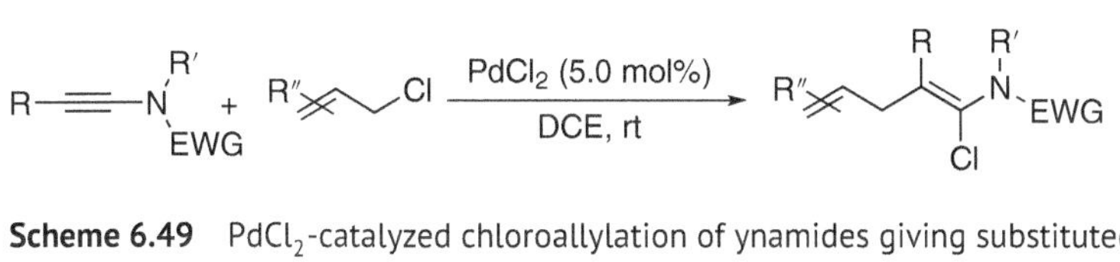

Scheme 6.49 $PdCl_2$-catalyzed chloroallylation of ynamides giving substituted enamides.

In addition, the palladium-catalyzed consecutive one-pot reaction of alkynes with allyl bromide and then the cross-coupling with organotin compounds were also reported [127].

Alkynyl ketones/sulfones were also used as the substrates for the palladium-catalyzed haloallylation for approach to synthesize highly functionalized 1,4-dienes [128].

The radical bromoallylation of alkynes leading to 1-bromo-1,4-dienes [129] and regioselective radical bromoallylation of allenes affording 2-bromo-substituted 1,5-dienes [130] were also developed by Ryu's group. The $C(sp^2)$—Br bond in the bromide dienes could be further functionalized via cross-coupling and carbonylation reactions. In addition, $InCl_3$ has also been found to be an efficient catalyst for the intramolecular bromoallylation of 1,6-enynes with *trans*- or *cis*-bromoally group (Scheme 6.50) [131]. The mechanistic evidence supports a cationic reaction pathway with Lewis acid activation of the allyl halogen, and the incorporated halogen into products is from the substrates or solvent, since the use of CH_2Cl_2 as the solvent results in the formation of the adducts having chloro group to replace bromo group as the major products.

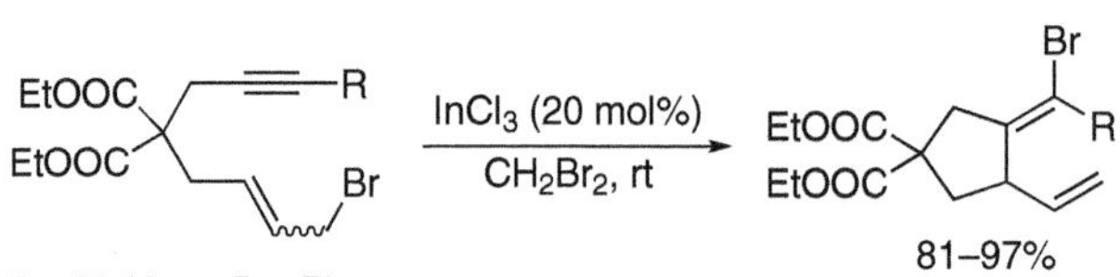

R = H, Me, *n*-Bu, Ph

Scheme 6.50 $InCl_3$-catalyzed intramolecular haloallylation of 1,6-enynes.

$C(sp^3)$—X bond in benzyl halides is also easily cleaved by its oxidative addition to transition metal complexes [132]. $FeCl_3$, as Lewis acid catalyst, was found to be the efficient catalyst to catalyze the direct addition of benzyl halides to arylalkynes affording alkenyl halides [133].

The addition of polyhalogenated alkanes bearing at least a strong electron-withdrawing group (EWG) to alkynes via cleavage of C—X bond has also been developed. As shown in Scheme 6.51, the $Cu(OTf)_2$-catalyzed carboiodination of terminal alkynes with functionalized alkyl iodides, generated in situ by treating alkyl bromides with KI to give β,γ-unsaturated compounds with excellent regioselectivity and good stereoselectivity [134].

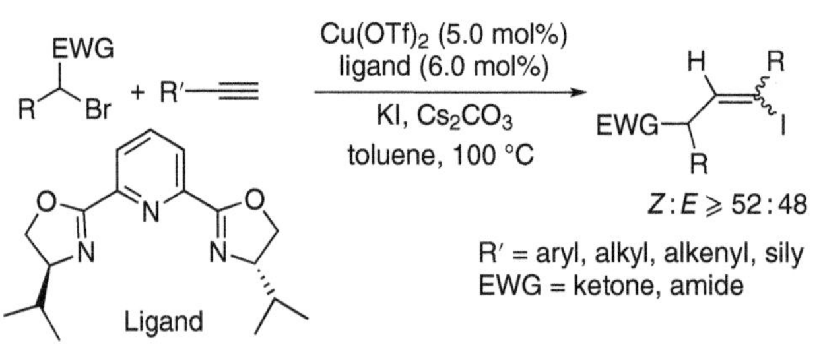

Scheme 6.51 $Cu(OTf)_2$-catalyzed carboiodination of terminal alkynes with alkyl bromides.

A highly *trans*-selective carbohalogenation of terminal alkynes with functionalized tertiary alkyl halides has been developed in the presence of CuBr/ligand (Scheme 6.52). The catalyst system is efficient for carboiodination, carbobromination, and carbochlorination with tertiary alkyl halides activated by ester, ketone, nitro, or nitrile groups to give quaternary carbon-containing alkenyl halides in good yields with excellent regio- and stereoselectivity [135].

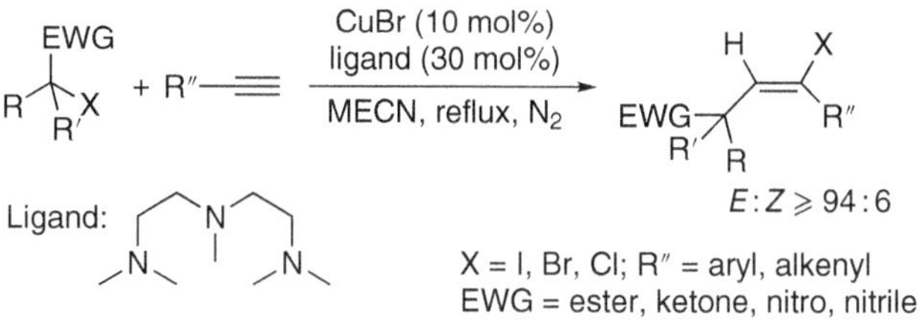

Scheme 6.52 CuBr-catalyzed *trans*-carbohalogenation of terminal alkynes.

Wang's group also developed a visible-light-initiated Na_2-Eosin Y catalyzed highly regio- and stereoselective difunctionalization of terminal alkynes via *trans*-addition of dicyanoalkyl bromides to aryl alkynes under ambient and metal-free conditions (Scheme 6.53) [136].

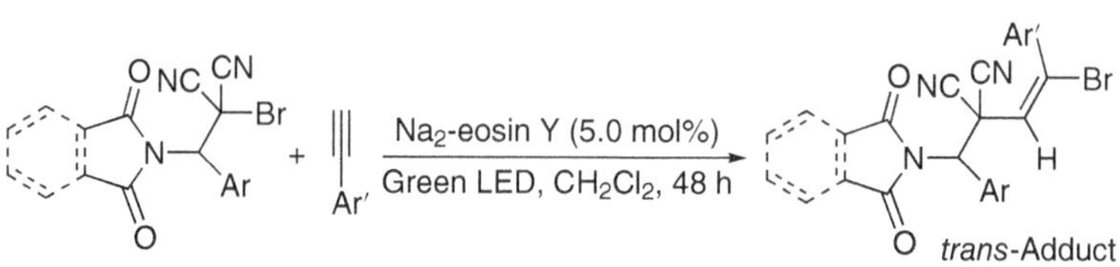

Scheme 6.53 Visible-light-initiated Na_2-Eosin Y catalyzed addition of dicyanoalkyl bromides to aryl alkynes.

The polar halo-fluoroalkanes of R_F—I and R_F—Br bonds are activated C—X bonds, which have been well-applied in the synthesis of fluoroalkylalkenes via their addition reactions to alkynes (Scheme 6.54). For example, Hu and coworker developed a $FeBr_2$-catalyzed addition of perfluoroalkyl iodide to terminal alkynes with moderate to good *E*/*Z*-selectivities in the presence of Cs_2CO_3 [137]. Besset and coworker established the $Cu(OTf)_2$-mediated synthesis of difluoromethyl alkenes from the addition of Br—CF_2CO_2Et to alkynes with low stereoselectivity [138]. Wang and coworkers have also reported a Cu_2O-catalyzed decarboxylative

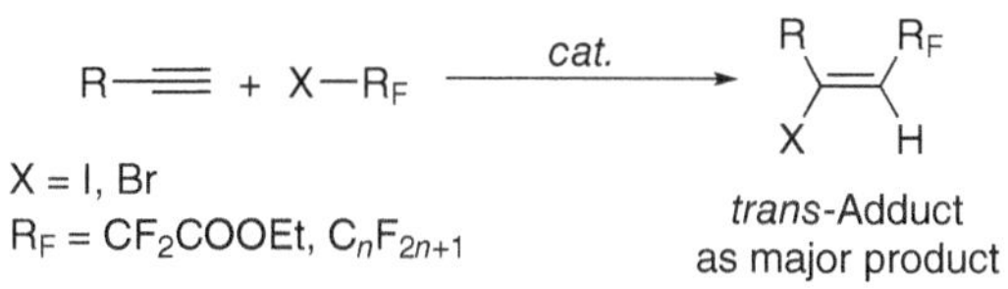

Scheme 6.54 Catalytic halo-fluoroalkylations of alkynes affording fluoroalkylalkenes.

atom transfer radical addition reaction between I—CF_2CO_2Et and substituted propiolic acids [139]. Recently, von Wangelin and coworker have also developed a $CoBr_2$/dppbz-catalyzed bromo- and iodofluoroalkylation of terminal and internal alkynes with high regio- and stereoselectivity in the presence of zinc under mild conditions, and alkenes can also undergo this addition reaction under similar conditions [140].

In addition, the addition of polyhalogenated alkanes bearing at least a strong EWG to alkenes via cleavage of C—X bond is known as the Kharasch reaction [141], and ruthenium [142], rhodium [143] and copper complexes [144] have been well used to catalyze this type of reactions.

Severin and coworker developed the atom transfer radical addition reactions of polychlorinated compounds and sulfonyl chlorides to alkenes in the presence of $RuCp^*Cl_2PPh_3$/AIBN [145]. Two years later, Uozumi's group prepared the amphiphilic polystyrene-polyethylene glycol resin-supported ruthenium complex, PS-PEG-$NHCOC_6H_4PPh_2$-$RuCp^*Cl_2$, which was found to be the efficient catalyst for the atom transfer radical addition of halogenated compounds to alkenes in water with excellent regioselectivity under heterogeneous as well as AIBN-free conditions (Scheme 6.55) [146]. Under similar conditions, the internal alkene of norbornene also reacted with CCl_4, and the addition of C—Br to styrene occurred smoothly.

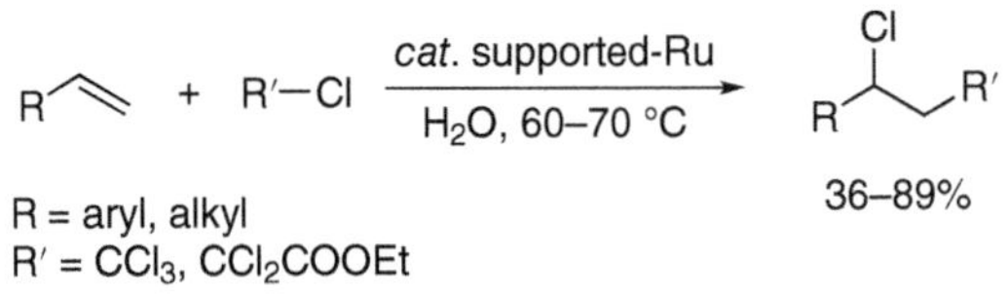

Scheme 6.55 Supported-ruthenium-catalyzed addition of halogenated compounds to alkenes.

6.5.2 C(sp^2)—X Activation and Its Addition Reactions

The reported procedures on the C(sp^2)—X bond activation and their addition reactions to unsaturated hydrocarbons focus on the use of acyl chlorides, chloroformates, ethoxalyl chloride, and α-keto acid chlorides [147].

The C—Cl bond of acyl chlorides is easily cleaved by its oxidative addition to low-valent transition metal complexes [148]. Nomura and coworker first reported the rhodium(I)-catalyzed reactions of aroyl chlorides with alkynes in 1996 [149]. As shown in Scheme 6.56, in the presence of $[RhCl(cod)]_2/PPh_3$, aroyl chlorides react with terminal alkynes accompanied by decarbonylation to afford the (*Z*)-alkenyl chloride derivatives via a *cis*-addition with excellent regio- and stereoselectivity. The proposed mechanism involves the oxidative addition of chloro–carbon bond to Rh(I), decarbonylation of aroylchlororhodium (III), selective insertion of alkynes to Cl—Rh bond of arylchlororhodium (III), and final reductive elimination of carbon–carbon bond to give adducts. Interestingly, when internal alkynes are used, the addition reactions occur without decarbonylation to produce the cyclic compounds of 2,3-disubstituted-1-indenones.

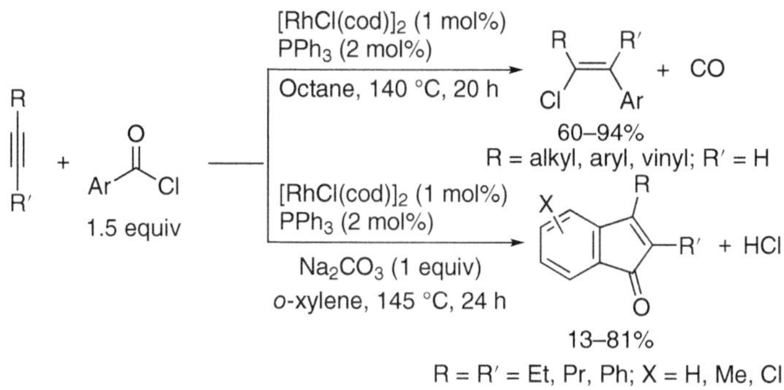

Scheme 6.56 Rhodium-catalyzed reaction of acid chlorides with alkynes.

In addition, the same group reported that in the presence of hexamethyldisilane, aroylarylation of internal alkynes, the 1,2-addition of aroyl and aryl groups to carbon–carbon triple bond occurs to give 1,3-diaryl-2-propen-1-ones in good yields [150].

The later studies in Tsuji's group disclosed that when $[IrCl(cod)]_2$/NHC was used as the catalyst system, the intermolecular additions of acid chlorides to terminal alkynes afforded (*Z*)-β-chloro-α,β-unsaturated ketones with regio- and stereoselectivity with suppression of decarbonylation and β-hydrogen elimination (Scheme 6.57) [151].

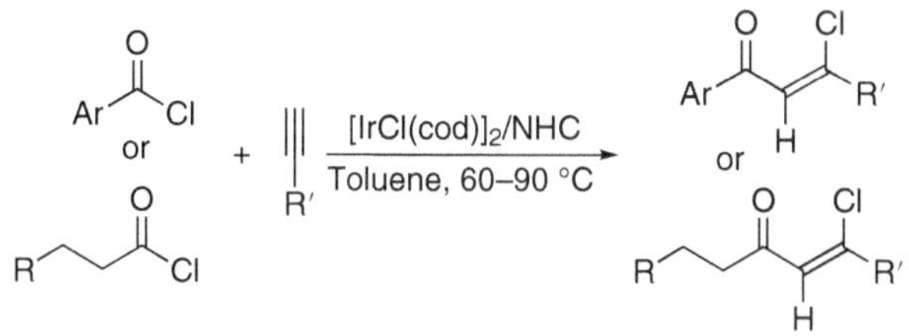

Scheme 6.57 Iridium-catalyzed intermolecular additions of acid chlorides to terminal alkynes without decarbonylation and β-hydrogen elimination.

The reaction of perfluorinated acid chlorides with terminal alkynes was efficiently catalyzed by $Rh(acac)(CO)(PPh_3)$ and proceeded with CO-retentive addition to alkynes to afford (*Z*)-1-perfluoroacyl-2-chloro-1-alkenes selectively in high yields [152]. For instance, hexafluoroglutaroyl dichloride reacts with two equivalents of 1-octyne to afford the corresponding diketone in 62% yield (Scheme 6.58).

$Rh(acac)(CO)(AsPh_3)$ shows high catalytic activity in the addition reaction of chloroacetyl chlorides to terminal alkynes with retention of CO moiety, providing a synthetic method for (*Z*)-1,4-dichloro-3-buten-2-one derivatives [153]. When trichloroacetyl chloride was used, the terminal alkynes underwent a chlorinative dimerization with trichloroacetyl chloride as chlorine donor in the presence of rhodium catalysts to give (*Z*,*Z*)-1,4-dichloro-1,3-butadienes stereoselectively, and the ligand screening revealed that reactions using sterically bulky and

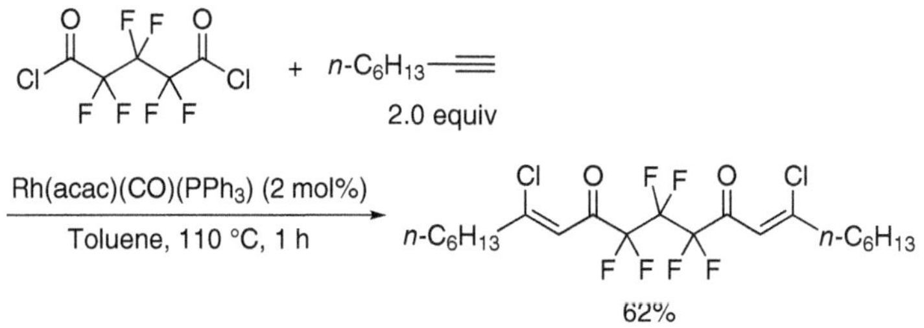

Scheme 6.58 CO-retentive addition of fluorinated acid chloride to alkyne.

electron-donating ligands like trimesitylphosphine [$P(Mes)_3$] were high yielding (Scheme 6.59) [154].

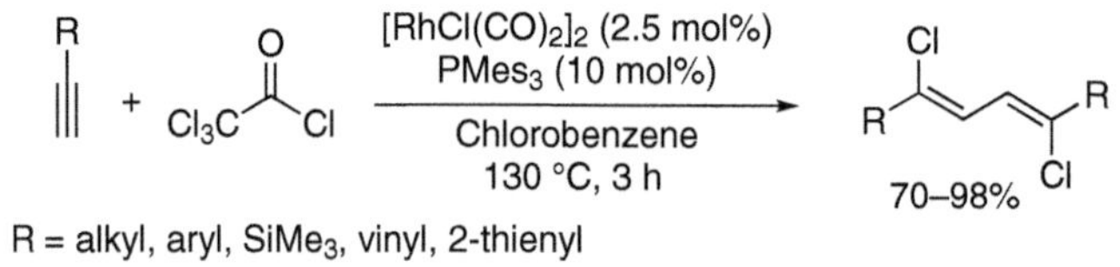

Scheme 6.59 Rhodium-catalyzed chlorinative dimerization of terminal alkynes with trichloroacetyl chloride.

Lewis acids, such as $FeBr_2$ [155] and $FeCl_3$ [156] also show high catalytic activity in the addition reactions of acid chlorides to terminal alkynes yielding (*Z*)-β-chloro-α,β-unsaturated ketones. The selective formation of either β-chloro-α,β-unsaturated ketones (CO-retention) or vinyl chlorides (CO-extrusion) depends on the catalyst system employed, the structures of acid chlorides and alkynes, and the proper choice of ligands [157].

The addition reaction of acid chlorides to alkynes can also be achieved by using metal oxides as catalysts. For example, with the use of ZnO as catalyst, the addition reactions of acid chlorides to terminal alkynes occur under solvent-free conditions at room temperature affording (*Z*)-β-chloro-α,β-unsaturated ketones with high regioselectivity [158]. The nanoparticles of magnetite (Fe_3O_4) also show good catalytic activity for the addition of acid chlorides to internal and terminal alkynes yielding the corresponding chlorovinyl ketones in good yields [159].

Interestingly, chloroaluminate ionic liquids can be used as solvent and Lewis acid catalyst to promote the addition reaction of aromatic and aliphatic acid chlorides to acetylene affording the corresponding β-chlorovinyl ketones in high yields [160].

The iridium-catalyzed annulation of *N*-arylcarbamoyl chlorides with internal alkynes via the key steps of oxidative addition of C—Cl bond, insertion of alkynes, and reductive elimination affording 2-quinolones has been reported by Tsuji's group (Scheme 6.60) [161].

The iridium-catalyzed decarbonylation of aroyl chlorides have been applied in the selective synthesis of phenanthrenes [162]. As shown in Scheme 6.61, 2-arylbenzoyl chlorides effectively undergo annulative coupling with alkynes in the presence of $[IrCl(cod)]_2/P(t\text{-}Bu)_3$ to give phenanthrenes accompanied by elimination of CO and HCl.

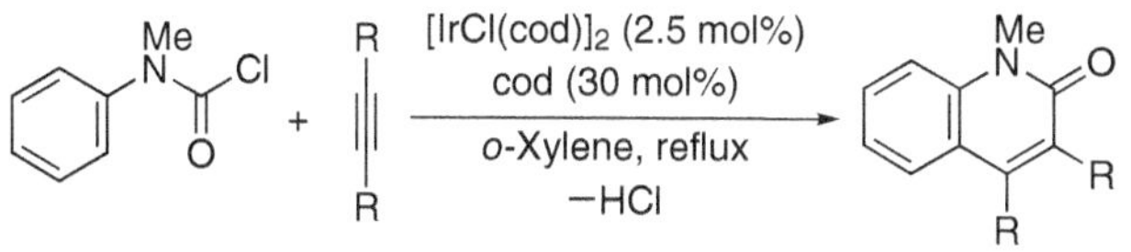

Scheme 6.60 Iridium-catalyzed annulation of *N*-arylcarbamoyl chlorides with internal alkynes affording 2-quinolones.

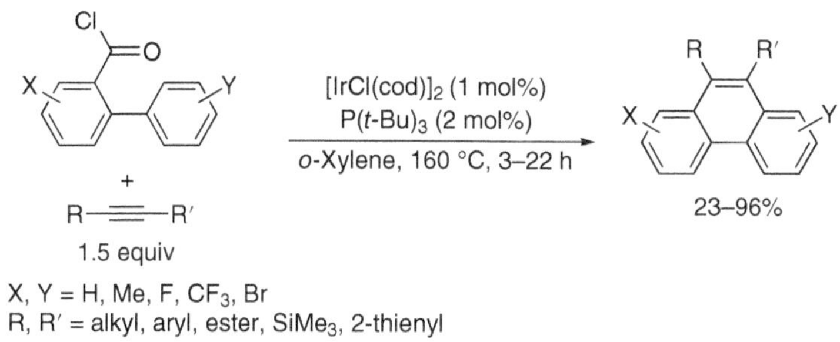

Scheme 6.61 Synthesis of phenanthrenes by iridium-catalyzed annulative coupling of 2-arylbenzoyl chlorides with alkynes.

Schoenebeck and Lautens reported an unexpected intramolecular chlorocarbamoylation of *N*-(*ortho*-alkynylaryl)carbamoyl chlorides in the presence of $Pd_2(dba)_3$/aryl phosphaadamantane [163]. This Pd(0)-catalyzed intramolecular chlorocarbamoylation of alkynes occurs with excellent *trans*-selectivity to provide an access to methylene oxindole scaffolds with $E:Z > 99:1$. Their further studies developed a similar intramolecular annulation of *N*-(*ortho*-alkynylaryl)carbamoyl chlorides with the use of $PdCl_2(PhCN)_2$ as catalyst, and the products of 3-(chloromethylene)oxindoles, however were formed with $Z:E > 95:5$ selectivity in almost all cases (Scheme 6.62) [164].

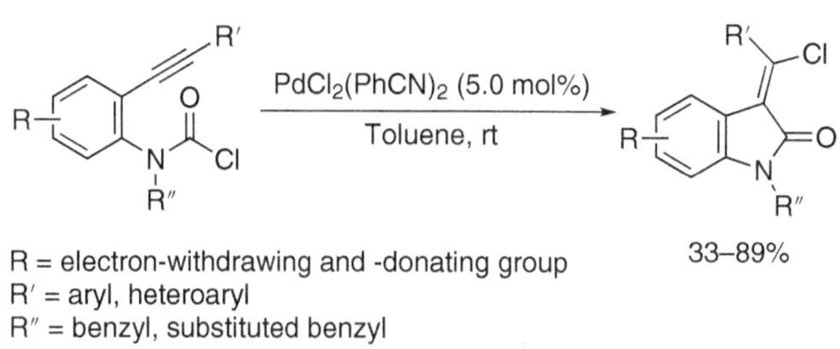

Scheme 6.62 $PdCl_2(PhCN)_2$-catalyzed intramolecular annulation of *N*-(*ortho*-alkynylaryl)-carbamoyl chlorides affording 3-(chloromethylene) oxindoles.

On the other hand, the rhodium-catalyzed nondecarbonylative addition of chloroformates [165], ethoxalyl chloride [166], and α-keto acid chlorides [167] to terminal alkynes are particularly important and useful in the synthesis of (*Z*)-β-chloroacrylates (Scheme 6.63), (*Z*)-γ-chloro-α-oxo-β,γ-alkenoates (Scheme 6.64), and (*Z*)-γ-chloro-α-oxo-β,γ-unsaturated ketones (Scheme 6.65). All the addition reactions occurred with excellent regio- and stereoselectivities to give the *cis*-adducts with internal attachment of the chloro group.

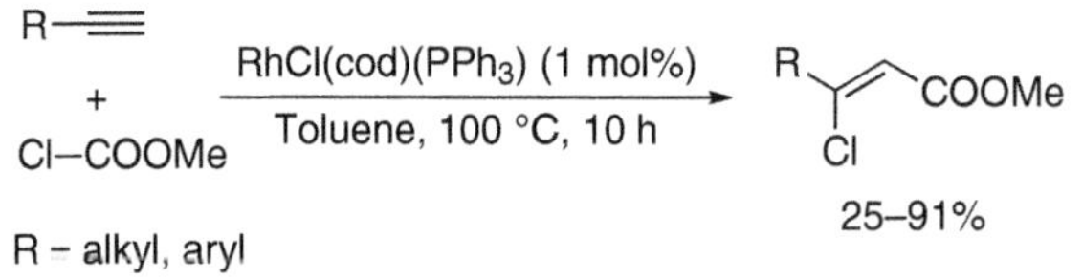

Scheme 6.63 Rhodium-catalyzed chloroesterification of terminal alkynes with chloroformates.

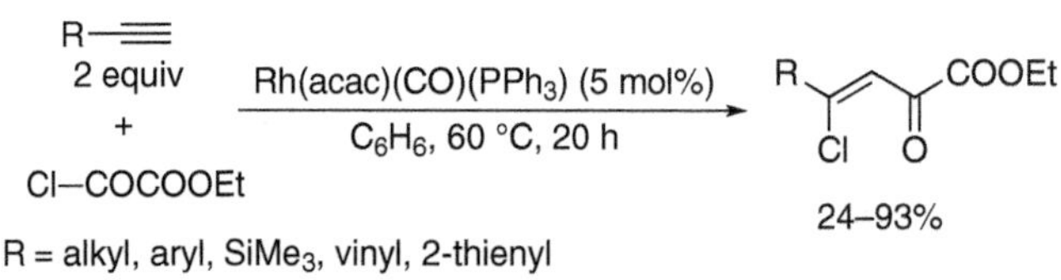

Scheme 6.64 Rhodium-catalyzed addition of ethoxalyl chloride to terminal alkynes.

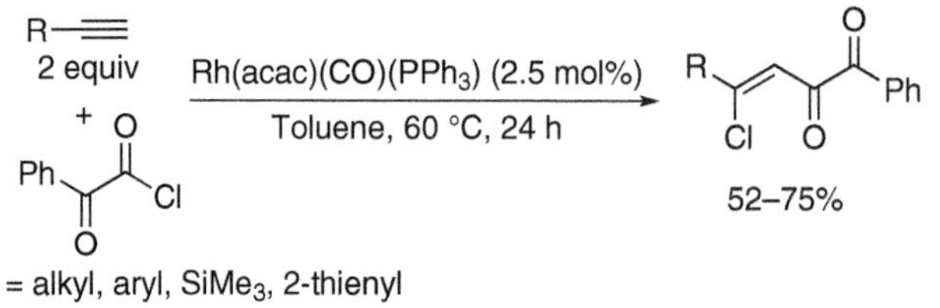

Scheme 6.65 Rhodium-catalyzed addition of α-keto acyl chlorides to terminal alkynes.

In addition, in the presence of rhodium/NHC, the addition of chloroformates to enynes occurs with high regioselectivity to afford 1,3-diene derivatives as the sole adduct, and carbon–carbon double bond is untouched [168]. The rhodium-catalyzed chloroesterification has also been applied in the addition of chloroformates to ribose-, xylose-, and homologated ribose-derived terminal alkynes to give doubly functionalized alkenyl derivatives [169].

β-Chloro-β,γ-unsaturated esters can be regioselectively prepared by the addition reactions of chloroformates to terminal allenes catalyzed by $RhCl(CO)(PPh_3)_2$ (Scheme 6.66) [170].

Uneyama and coworker applied the addition reaction of C(sp^2)—Cl bond of imidoyl chlorides with alkynes in the synthesis of heterocyclic compounds. Scheme 6.67 shows the formation of 2-trifluoromethylated quinolones by an intermolecular addition of *N*-aryl trifluoroacetimidoyl chloride to internal alkyne and subsequent intramolecular cyclization in the presence of base catalyzed by $[RhCl(cod)]_2$/phosphines [171]. The same group then developed a palladium-catalyzed intramolecular chloroimination of alkynes leading to the formation of 4-chloroquinolines (Scheme 6.68) [172].

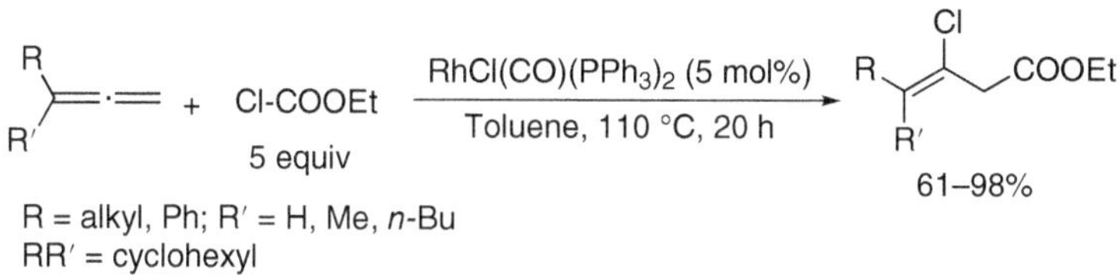

Scheme 6.66 Rhodium-catalyzed chloroesterification of allenes.

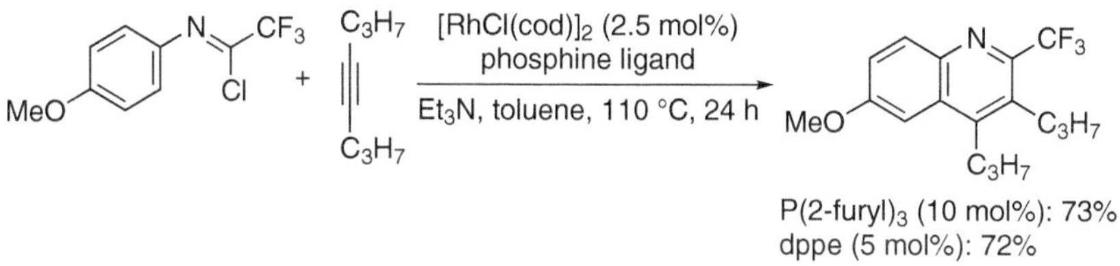

Scheme 6.67 Rhodium-catalyzed coupling cyclization of *N*-aryl fluorinated imidoyl chlorides with 4-octyne.

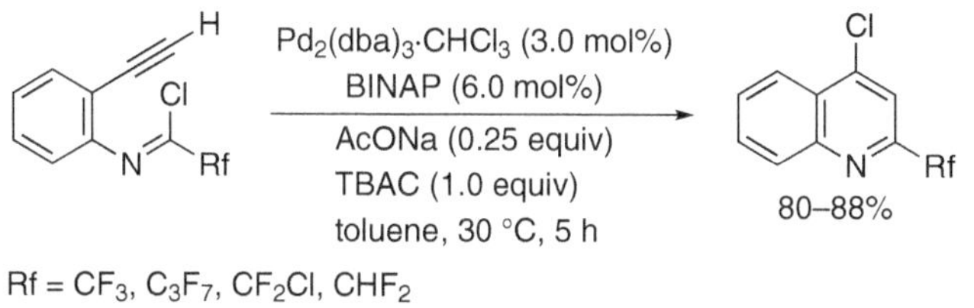

Scheme 6.68 Formation of 4-chloroquinolines via palladium-catalyzed intramolecular chloroimination of alkynes.

Interestingly, the three C—Cl bonds of chlorotriazines have been found to undergo addition reactions to three equivalents of arynes affording triaryl-substituted triazines (Scheme 6.69) [173].

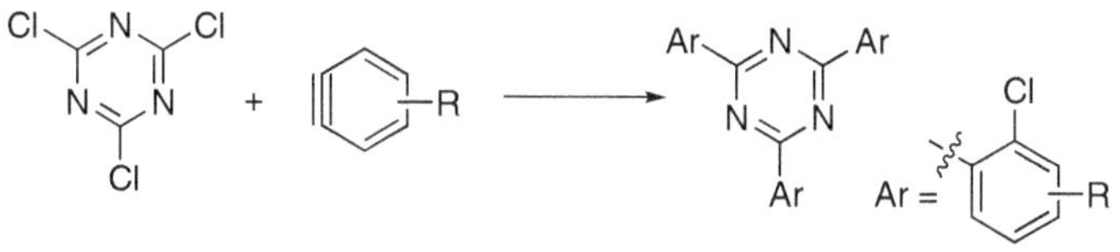

Scheme 6.69 Addition reactions of chlorotriazines to arynes affording triaryl-substituted triazines.

6.5.3 C(sp)–X Activation and Its Addition Reactions

The first example of the addition reaction of C(sp)—X bond of haloalkynes to alkynes was reported by Jiang and coworker in 2010 [174]. In the presence of $Pd(OAc)_2$, the cross-coupling reactions between bromoalkynes and internal alkynes occur smoothly to give conjugated (*Z*)-bromoalkenynes with regio- and stereoselectivity, and terminal alkynes are unsuitable (Scheme 6.70). The preliminary mechanistic experiments have provided evidence in support of a rare Pd(II)/Pd(IV) catalytic cycle for this transformation (Scheme 6.71).

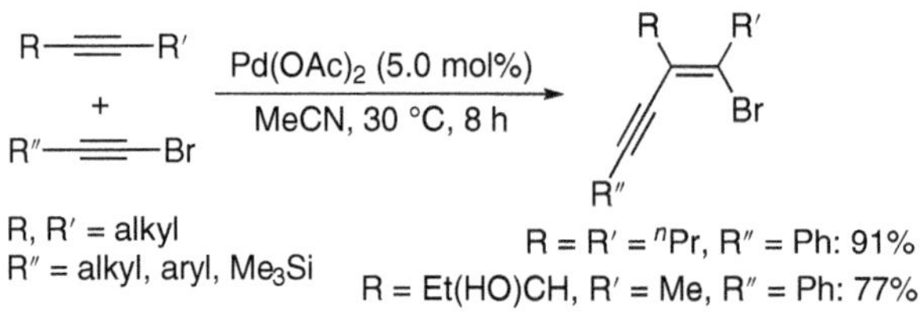

Scheme 6.70 $Pd(OAc)_2$-catalyzed bromoalkynylation of internal alkynes.

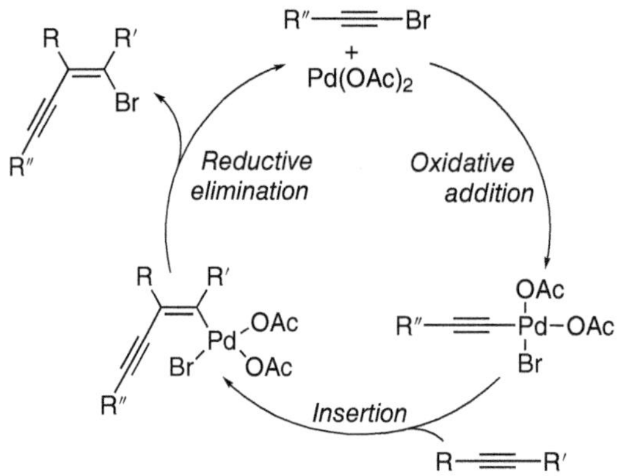

Scheme 6.71 Proposed mechanism for the $Pd(OAc)_2$-catalyzed bromoalkynylation of internal alkynes.

The same group then developed the $Pd(OAc)_2$-catalyzed bromoalkynylation of norbornenes and cyclooctene to provide an unexpected ring structure-dependent synthesis of 7-alkynyl norbornenes and cyclobutenyl bromide under mild conditions [175].

The addition of the C—Cl bond of chloroalkynes to terminal alkynes under palladium catalysis was then reported by Oshima's group [176]. The addition reactions of silyl-substituted chloroalkynes to terminal alkynes in the presence of catalytic amounts of $Pd_2(dba)_3$ and PPh_3 in decalin at 130 °C afford (*Z*)-1-chloro-1,3-enynes with excellent regio- and stereoselectivity, and the studies on the scope of haloalkynes disclosed that the silyl substituent on the haloalkynes is crucial (Scheme 6.72).

Very recently, Au(I)-catalyzed reactions between terminal alkynes and aromatic haloalkynes undergo either *cis*-haloalkynylation of the terminal alkyne or *trans*-hydroalkynylation of the haloalkyne depending on the nature of the catalyst counteranion [177].

Yoshida and coworker also studied a $CuBr_2$-catalyzed bromoalkynylation of arynes formed in situ and found that the insertion of two molar amounts of arynes

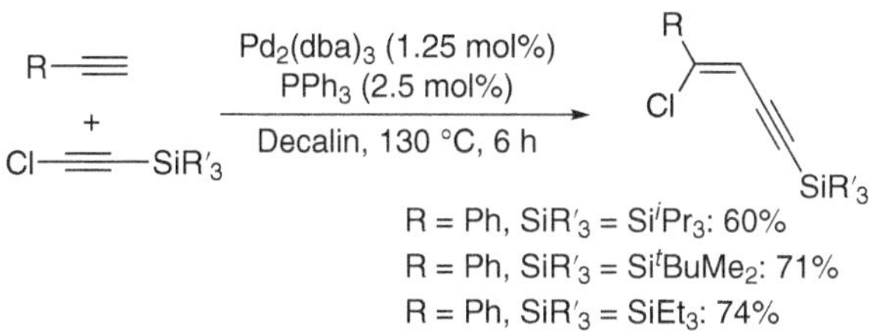

Scheme 6.72 $Pd_2(dba)_3/PPh_3$-catalyzed chloroalkynylation between silyl-substituted chloroalkynes and terminal alkynes.

to C—Br bond takes place predominantly, giving 2-alkynyl-2′-bromo biphenyls as the major products (Scheme 6.73) [178].

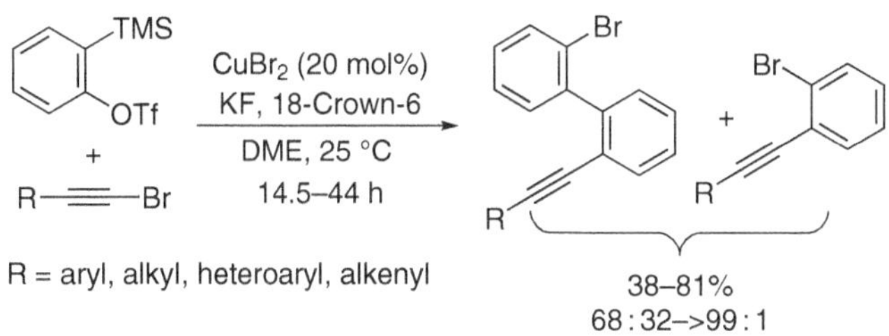

Scheme 6.73 $CuBr_2$-catalyzed bromoalkynylation of arynes.

On the other hand, the halocyanation of alkynes have also been well-studied (Scheme 6.74).

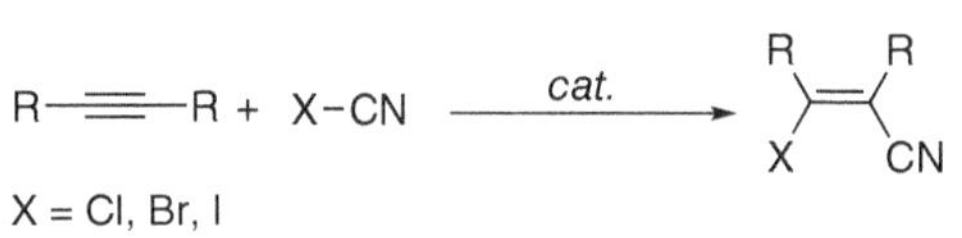

Scheme 6.74 Halocyanation of alkynes.

In 2001, Lukashev and coworker reported the addition reaction of cyanogen bromide to the nucleophilic aminoalkynes without any catalyst in CH_2Cl_2 at room temperature to give (*E*)-β-bromo-α-cyanoenamines by *cis*-addition, which are then finally isomerized to (*Z*)-β-bromo-α-cyanoenamines [23(c)]. $GaCl_3$ was then used as catalyst in the bromocyanation of inactivated aromatic terminal alkynes with cyanogen bromide in CH_2Cl_2 affording (*Z*)-β-bromoacrylonitriles with high regio- and stereoselectivity [179].

$Cu(OAc)_2$ also shows good catalytic activity for the regio- and stereoselective iodocyanation of internal alkynes with cyanogen iodide in MeOH [180].

In addition, although the direct chlorocyanation of alkynes has not been reported, (*Z*)-3-chloroacrylonitriles could be synthesized by the reaction of terminal and internal alkynes with stoichiometric amounts of imidazolium thiocyanates, followed by treatment with BCl_3 [181].

6.6 Addition Reactions of Carbon–Carbon Single Bonds

The transition metal-catalyzed cleavage of carbon–carbon single bonds and their addition reactions to unsaturated hydrocarbons have recently emerged as a useful and important strategy for constructing carbon-increasing molecular structures via simultaneous formation of two new carbon–carbon bonds. The most catalytic procedures are proposed to take place similar to the activation of element–element bonds as described in Chapter 3, including oxidative addition of a C—C bond to transition metal complexes, followed by insertion of an unsaturated bond into the resulting C—M bond, and reductive elimination to complete the catalytic cycle (Scheme 6.75).

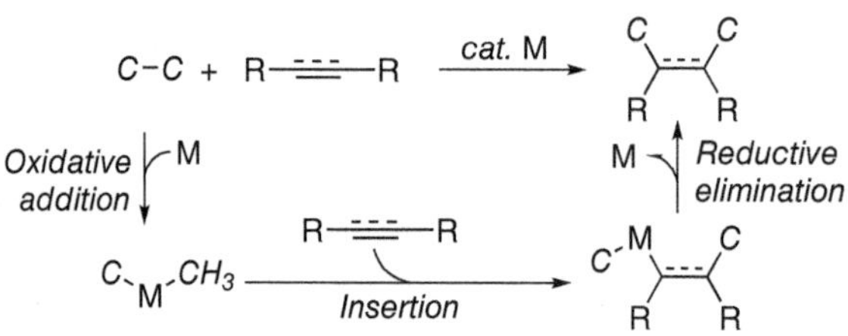

Scheme 6.75 Transition metal-catalyzed addition of carbon–carbon to unsaturated hydrocarbons.

The reported procedures are limited to the activation of strained C—C bonds and C—C bond bearing at least one of the "C" being a functionalized polar carbon group, such as CN, COOR, COR, etc.

6.6.1 Addition Reactions of Strained C–C Bonds

Transition metal-catalyzed activation of C—C bonds of strain rings and their cycloaddition reactions with alkenes is one of the important synthetic methods for the synthesis of five-membered carbocycles. The early work includes the cycloaddition of 1,1-diphenylcyclopropane with tetracyanoethylene [182] and methylenecyclopropanes with activated alkenes [183].

In 1992, Liebeskind and coworker reported the synthesis of substituted phenols by a $Ni(cod)_2$-catalyzed selective ring opening and cycloaddition of cyclobutenones with alkynes (Scheme 6.76) [184]. Their further studies on the reactions of cyclobutenones toward some low-valent transition metal reagents found the formation of η^4-vinylketene cobalt complexes, which could react with alkynes affording substituted phenols [185].

Interestingly, in the presence of $Ni(cod)_2/PPh_3$, the electron-deficient ethyl cyclopropylideneacetate and alkynes could undergo a chemoselective [3 + 2 + 2] cycloaddition reaction to produce cycloheptadienes (Scheme 6.77) [186].

Rhodium(I)-catalyzed intermolecular insertion of norbornene to cyclobutenones [187], intramolecular alkene insertion into cyclobutanones [188], intramolecular alkene insertion into benzocyclobutenones [189], intramolecular allene insertion into cyclobutanones [190], $Ru_3(CO)_{12}/PEt_3$-catalyzed intermolecular insertion

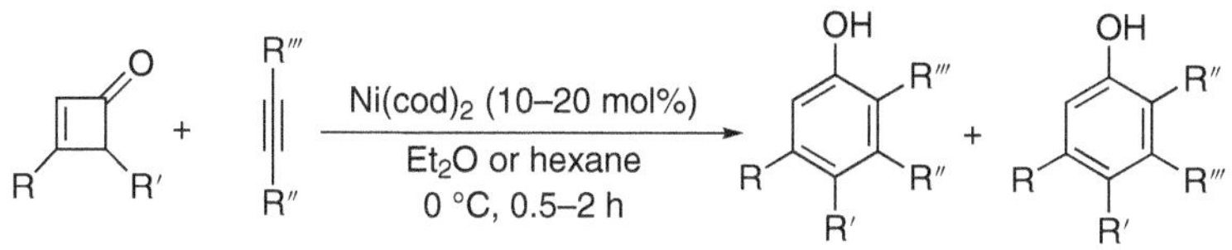

R = Ph, R′ = H, R″ = R‴ = Et, 75%
R″ = Me, R‴ = SiMe3, 58% (45 : 55)
R = Ph, R′ = Me, R″ = R‴ = Et, 60%
R = n-Bu, R′ = H, R″ = R‴ = Et, 76%

Scheme 6.76 $Ni(cod)_2$-catalyzed synthesis of substituted phenols from cyclobutenones and alkynes.

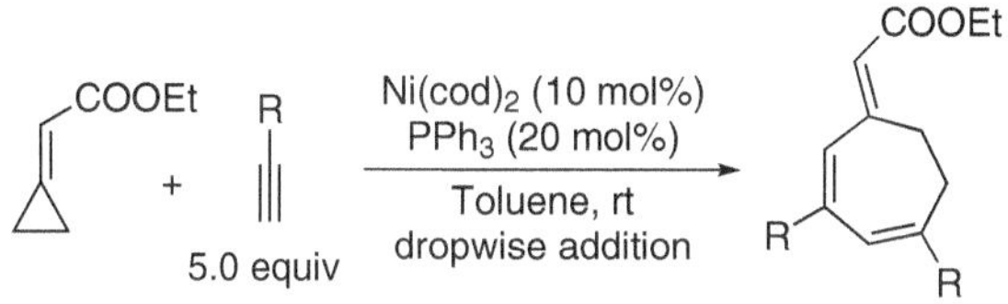

R = Me3Si, 70%
t-Bu, 89
Ph, 74%

Scheme 6.77 Nickel-catalyzed [3 + 2 + 2] cycloaddition of methylenecyclopropanes with two alkynes.

of norbornene to cyclobutenediones [191], and nickel(0) [192] and rhodium(I) complexes [193] catalyzed asymmetric intramolecular alkene insertion into cyclobutanones have become the efficient and atom economic methods for the construction of six-membered rings and cycloketone-fused skeleton.

In the presence of $Ni(cod)_2$ and phosphine ligands, the intermolecular insertion of internal alkynes into C—C single bond between the carbonyl carbon and the α-carbon of cyclobutanone occurred to give α,β-unsaturated cyclohexanone derivatives (Scheme 6.78) [194].

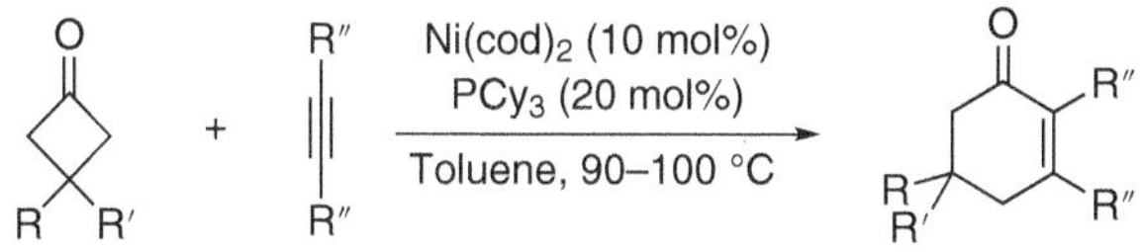

R″ = Pr, R = Me, R′ = Ph, 95%
R = R′ = Ph, 91%

Scheme 6.78 Nickel-catalyzed alkynes insertion to cyclobutanone.

In addition, Matsuda and coworker have recently developed the rhodium(I)-catalyzed intermolecular insertion reactions of alkynes into cyclobutenols [195], (2-pyridylmethylene)cyclobutenes [196], for access to 1,4-cyclohexadienes.

The synthetic applications of the strain rings via the activation of C—C bonds have been reviewed in a few papers, including transition metal-catalyzed carbon–carbon bond activation [197], cyclobutanes in catalysis [198], 3-alkyloxycyclobutanones as useful synthons for heterocycle and carbocycle synthesis [199], and carbon–carbon bond activation of ketones [200].

In addition, the review papers on the transition metal-catalyzed activation of strained rings and their enantioselective reactions have appeared [201].

6.6.2 Addition Reactions of C—CN Bonds

Transition metal-catalyzed addition reactions of R—CN to unsaturated hydrocarbons have been well-investigated, and several reviews have summarized this transformation [202]. Organic nitriles such as acyl cyanides, cyanoformamides, cyanoformates, benzonitriles, alkynyl cyanides, allyl cyanides, and alkyl cyanides have been used as the suitable substrates in the activation of C—CN bonds and their addition to alkynes and/or alkenes.

Although aromatic and heteroaromatic acyl cyanides underwent the decarbonylation catalyzed by palladium(0) complexes to give the corresponding nitriles in excellent yields via the first step of oxidative addition of ArCO—CN bond to palladium(0) [203], Takaya and coworker reported the first example on the cleavage of C—C bonds of aryloyl cyanides and their addition to aromatic terminal alkynes producing (*Z*)-β-cyano-α,β-unsaturated ketones catalyzed by palladium complexes [204]. As shown in Scheme 6.79, the $Pd(OAc)_2$/phosphine-catalyzed intermolecular acylcyanation of the aromatic terminal alkynes occurs with excellent regio- and stereoselectivity, the *cis*-addition resulted in the formation of (*Z*)-isomers as the major products, and acyl group is attached to the terminal carbon of alkynes. The oxidative addition of acyl cyanide to Pd(0) to give acylpalladium cyanide is also proposed to be the first step for this addition reaction.

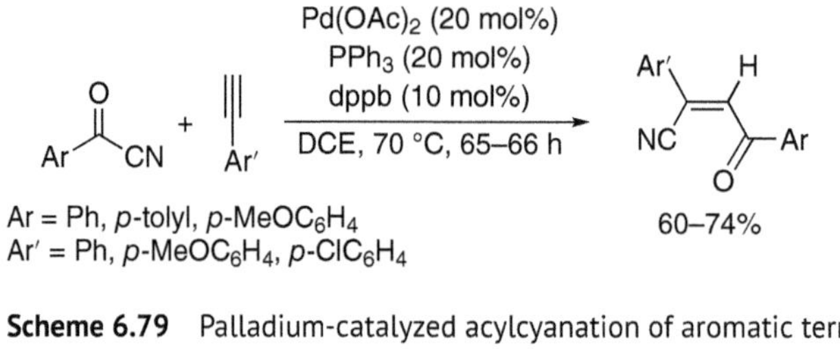

Scheme 6.79 Palladium-catalyzed acylcyanation of aromatic terminal alkynes affording (*Z*)-β-cyano-α,β-unsaturated ketones.

Under transition metal-free conditions, phosphine-catalyzed acylcyanation of the activated alkynoates with acyl cyanides to form acrylonitrile derivatives with complete regioselectivity and high anti-stereoselectivity was then also developed [205].

The intramolecular acylcyanation of alkenes was established in the presence of $Pd(PPh_3)_4$ and $ZnCl_2$, to afford functionalized indanones in moderate to high yields [206].

Takemoto's group has well investigated the palladium-catalyzed intra- and intermolecular amidation of unsaturated hydrocarbons using carbamoyl derivatives

[207]. The intramolecular cyanoamidation of alkenes and alkynes is proposed to start with the oxidative addition of cyanoformamides to palladium(0), the insertion of unsaturated carbon–carbon bond to carbamoyl complex, and the final reductive elimination leading to lactams (Scheme 6.80). Two representative results for the formation of α-alkylidene lactam and 3,3-disubstituted lactam from the intramolecular cyanoamidation of *N*-(*ortho*-alkynyl)- or *N*-(*ortho*-alkenyl)aryl-cyanoformamides are shown in Scheme 6.81.

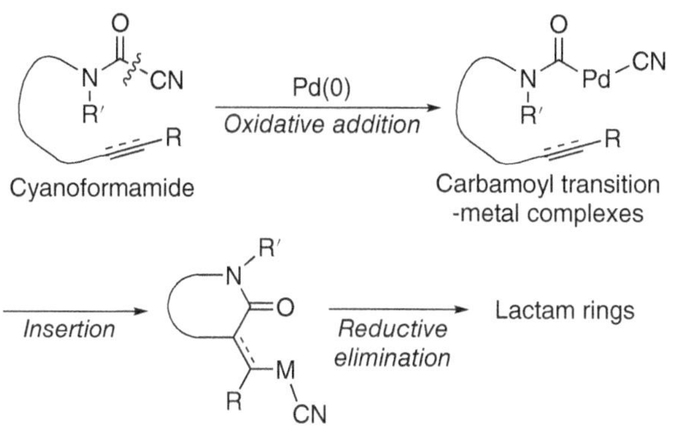

Scheme 6.80 Intramolecular cyanoamidation of unsaturated hydrocarbons via carbamoyl transition metal complexes.

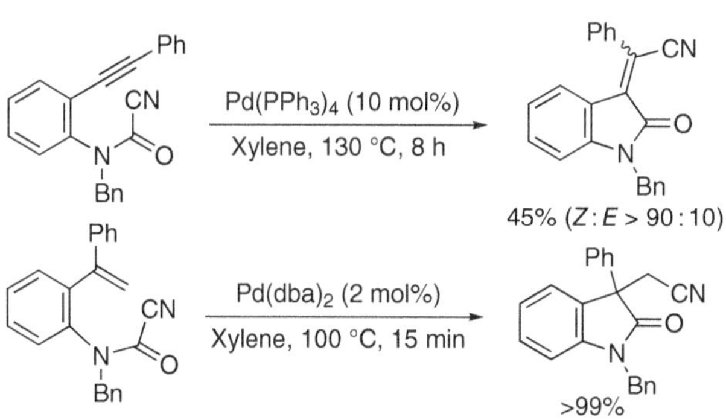

Scheme 6.81 Intramolecular cyanoamidation of unsaturated hydrocarbons affording lactams.

The enantioselective synthesis of 3,3-disubstituted lactam through Pd(dba)$_2$/chiral ligand-catalyzed cyanoamidation was also achieved [208].

Nishihara and coworker first studied the oxidative addition of NC—COOR to Pd(0) to cleave the C—C σ-bond of cyanoformates and developed the direct cyanoesterification of norbornenes (Scheme 6.82) [209].

An intramolecular cyanoesterification of alkynes was then developed in the presence of Pd(PPh$_3$)$_4$, providing an efficient synthetic route for the formation of butenolides in good to excellent yields [210].

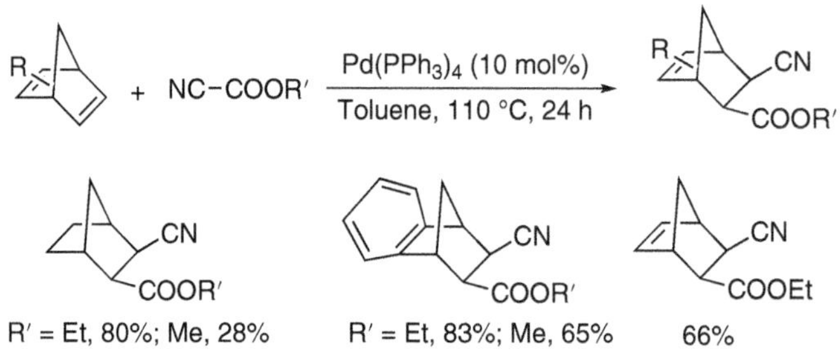

Scheme 6.82 Palladium-catalyzed cyanoesterification of norbornenes with cyanoformates.

Nakao and Hiyama then developed the cyanoesterification of 1,2-dienes affording cyanomethylacrylate esters in the presence of $Ni(cod)_2$ and phosphine ligands [211] and cyanoesterification of alkynes catalyzed by $Ni(cod)_2$/ligand/Lewis acid [212].

Hiyama and coworker also studied the cleavage of aryl–CN bonds and their addition to alkynes [213]. Scheme 6.83 shows the addition of aryl–CN bond to internal alkynes catalyzed by $Ni(cod)_2/PMe_3$, and the stereoselective *cis*-arylcyanation reaction provides an efficient synthetic way for the synthesis of various β-arylsubstituted alkenenitriles [214]. Not only a wide range of functional groups in aryl cyanides tolerate the catalysis, but also the intramolecular arylcyanation can take place to give (*Z*)-2-indene-1-ylidenepropionitrile under similar reaction conditions. When unsymmetrical alkynes were used, the addition reactions gave a mixture of regioisomers, but the isomers having a cyano group at the carbon bearing a larger substituent are the major adducts.

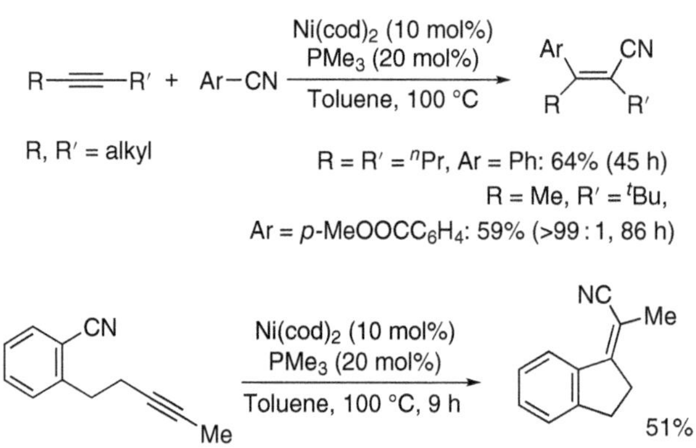

Scheme 6.83 $Ni(cod)_2/PMe_3$-catalyzed arylcyanation of internal alkynes with aryl cyanides.

In addition, with the use of nickel(0)/Lewis acid catalyst system, the selective cleavage of aryl–CN bond in polyfluorobenzonitriles (C_6F_5—CN) and 2,3,5,6-tetrafluorobenzonitrile (C_6HF_4—CN) and their addition of the fluorophenyl and cyano moieties to internal alkynes and alkenes were also achieved, and the reactive C—H and C—F bonds are unaffected [215].

Carbon–carbon bonds in alkyl cyanides can be cleaved by nickel(0) [216] and cobalt(I) complexes [217]. The aforementioned catalyst systems show the catalytic activities in the alkylcyanation of terminal alkynes with acetonitrile and substituted acetonitriles to give a range of variously substituted acrylonitriles [218], and the heteroatom-directed alkylcyanation of alkynes was then developed [219]. The effect of Lewis acid catalysts on nickel-catalyzed arylcyanation of alkynes was also investigated in details [220].

The same group then developed the nickel-catalyzed alkynylcyanation of alkynes and 1,2-dienes via the cleavage and addition of alkynyl–CN to internal alkynes or dienes [221].

Moreover, the catalytic asymmetric intramolecular arylcyanation of unactivated alkenes giving benzo-fused cyclic compounds with quaternary carbon stereogenic centers in good yields and high enantioselectivity was also developed [222].

The intramolecular arylcyanation of alkenes catalyzed by $Ni(cod)_2$/phosphine/$AlMe_2Cl$ has been used in the synthesis of natural product of rhazinilam [223].

On the other hand, the oxidative addition of allyl cyanide, a sp^3 C—CN bond to nickel complexes gives C—CN cleavage products [224], and the allyl cyanation of internal alkynes with allyl cyanides was therefore developed in the presence of $Ni(cod)_2/P(p\text{-}CF_3C_6H_4)_3$ to produce polysubstituted 2,5-hexadienenitriles with high stereo- and regioselectivity. For example, the reaction of 4-octyne with four equivalents of allyl cyanide in acetonitrile at 80 °C gave (*Z*)-2,3-dipropylhexa-2,5-dienenitrile in 78% isolated yield (Scheme 6.84) [225].

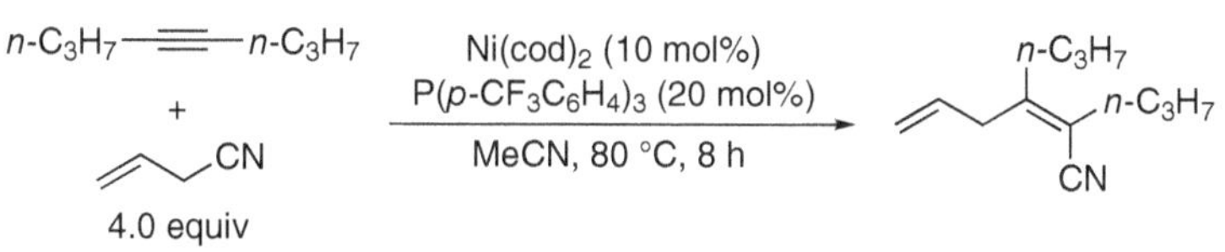

Scheme 6.84 Nickel-catalyzed allylcyanation of 4-octyne.

Inoue and coworker studied the Cu(I)-catalyzed reactions of chlorinated cyanides with electronically non-polarized alkenes to obtain the adducts from the cleavage of either C—Cl or C—CN bond. As shown in Scheme 6.85, the CuCl/dppf-catalyzed atom transfer radical reactions of chlorinated cyanides such as Cl_3CCN, $Cl_2C(R)CN$, and $Cl_2C(CN)_2$ with inherently unreactive terminal alkenes

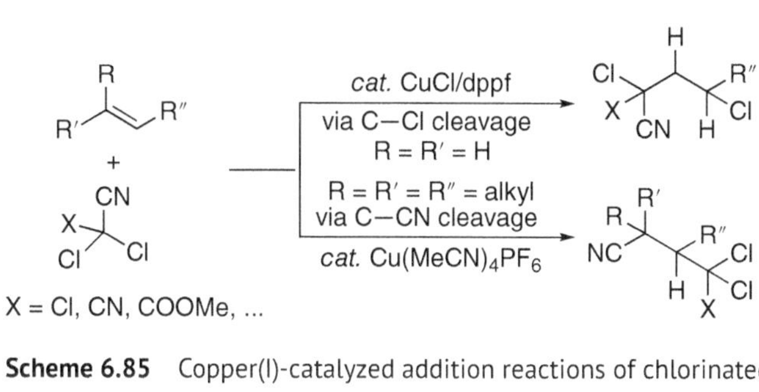

Scheme 6.85 Copper(I)-catalyzed addition reactions of chlorinated cyanides to alkenes.

(R = R′ = H, R″ = alkyl) afford 1,3-dichlorinated adducts via C—Cl bond cleavage [226], and this transformation can also occur without any catalysts under microwave irradiation at 200 °C [227]. Further studies found that $Cu(MeCN)_4PF_6$-catalyzed reactions of trisubstituted alkenes (R = R′ = R″ = alkyl) led to the carbocyanation products through C—CN bond cleavage [228].

6.6.3 Other Carbon–Carbon Bond Cleavage and Their Addition Reactions

In the presence of $[ReBr(CO)_3(thf)]_2$, Kuninobu and Takai's group found that the reactions of β-keto esters with terminal alkynes afforded a mixture of alkene products through the insertion of alkynes into a carbon–carbon single bond of the β-keto ester [229], the insertion reactions to cyclic β-keto esters could be applied in the synthesis of bicyclo[3.3.1]nonenes [230].

The same group then developed an interesting carbon–carbon cleavage of β-keto sulfones and its addition to terminal alkynes catalyzed by $Re_2(CO)_{10}$ (Scheme 6.86) [231]. The insertion reactions of alkynes into carbon–carbon bond occur with high regio- and stereoselectivity to give (*Z*)-unsaturated δ-keto sulfones, and acyl groups are attached to the terminal carbons of alkynes. The proposed mechanism involves a formal [2+2] cycloaddition of enol of the β-keto sulfone with alkyne to give cyclobutene intermediate as the key intermediate, which undergoes the regioselective cleavage of carbon–carbon single bond to give the adduct.

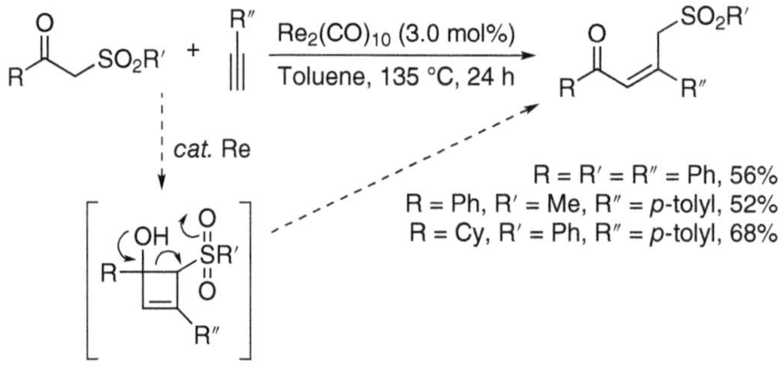

Scheme 6.86 Rhenium-catalyzed regio- and stereoselective addition of β-keto sulfones to terminal alkynes via carbon–carbon bond cleavage.

Aryne insertion to cyclic C—C bond can easily construct benzo-fused carbocycles. As shown in Scheme 6.87, Stoltz and coworker developed a protocol based on aryne insertion into the C—C bond of β-ketoesters and cyclic β-ketoesters to afford *ortho*-disubstituted arenes and benzannulated carbocycles [232]. This transformation was then applied in the synthesis of natural product (+)-amurensinine [233].

Zeng and coworker also reported a similar aryne insertion into the C—C bond of α-benzocyclic ketones and α-aryl cycloketones to give a straightforward formation of medium ring-fused benzocarbocycles, including seven- and eight-membered rings (Scheme 6.88) [234].

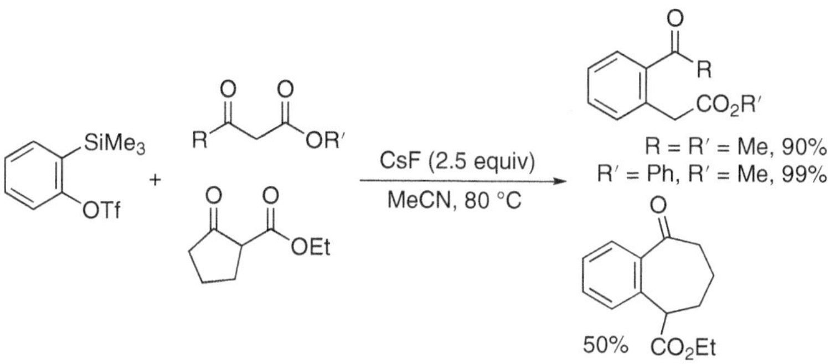

Scheme 6.87 Aryne insertion into the C–C bond of β-ketoesters and cyclic β-ketoesters.

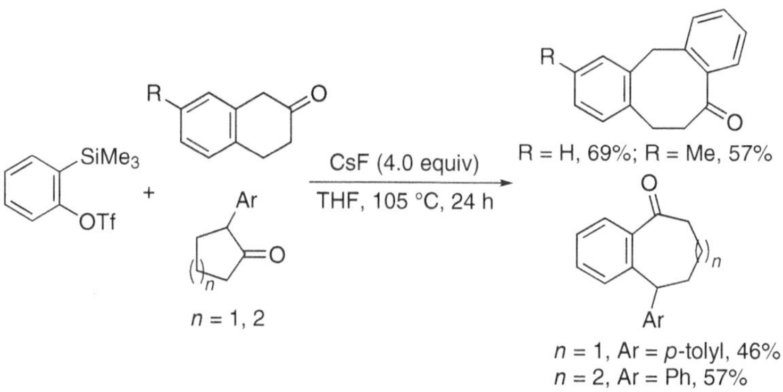

Scheme 6.88 Aryne insertion into the C–C bond of α-benzocyclic ketones and α-aryl cycloketones.

Srihari and Mehta's group has recently also developed the aryne insertion into the C—C bond of cyclic β-diketones to provide a versatile route to benzoannulated medium-ring carbocycles, which have been applied to establish an efficient total synthesis of Radermachol [235]. For example, in the presence of CsF, the reaction of 2-(trimethylsilyl)phenyl trifluoromethanesulfonate (1.25 equiv) with dimedone afforded benzocyclooctanone in 82% yield (Scheme 6.89).

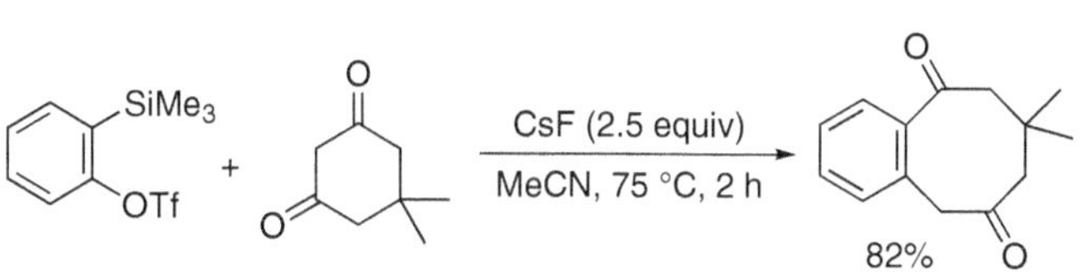

Scheme 6.89 Cyclooctane-1,5-dione from aryne insertion into the C–C bond of cyclic β-diketones.

In addition, Yoshida and Kunai's group studied the aryne insertion into the carbonyl-cyanomethyl σ-bond of α-cyanocarbonyl compounds in the presence of KF/18-crown-6, to directly introduce the carbonyl and cyanomethyl groups into adjacent positions of arenes (Scheme 6.90) [236].

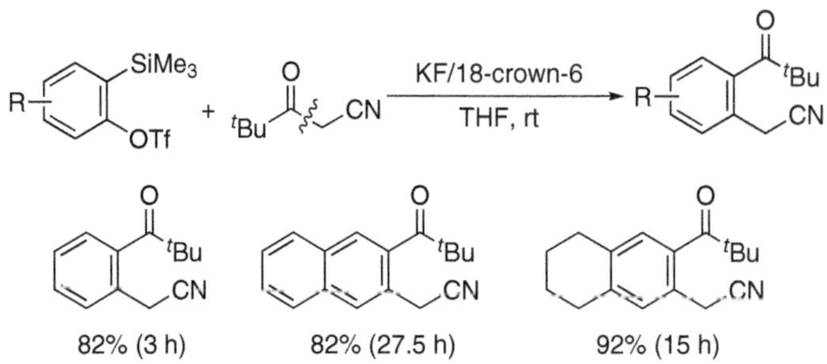

Scheme 6.90 Aryne insertion into α-cyanocarbonyl compounds.

References

1 (a) Suginome, M. (2010). *Chem. Rec.* 10: 348–358. (b) Barbeyron, R., Benedetti, E., Cossy, J. et al. (2014). *Tetrahedron* 70: 8431–8452. (c) Yoshida, H. (2016). *ACS Catal.* 6: 1799–1811.

2 Suginome, M., Yamamoto, A., and Murakami, M. (2003). *J. Am. Chem. Soc.* 125: 6358–6359.

3 Suginome, M., Yamamoto, A., and Murakami, M. (2005). *Angew. Chem. Int. Ed.* 44: 2380–2382.

4 Suginome, M., Shirakura, M., and Yamamoto, A. (2006). *J. Am. Chem. Soc.* 128: 14438–14439.

5 Nagao, K., Ohmiya, H., and Sawamura, M. (2014). *J. Am. Chem. Soc.* 136: 10605–10608.

6 Roscales, S. and Csákÿ, A.G. (2015). *Org. Lett.* 17: 1605–1608.

7 Jin, H. and Fürstner, A. (2019). *Org. Lett.* 21: 3446–3450.

8 Nogami, M., Hirano, K., Kanai, M. et al. (2017). *J. Am. Chem. Soc.* 139: 12358–12361.

9 Selected reports, see: (a) Yamamoto, A. and Suginome, M. (2005). *J. Am. Chem. Soc.* 127: 15706–15707. (b) Han, J.T. and Yun, J. (2018). *Org. Lett.* 20: 2104–2107.

10 Daini, M. and Suginome, M. (2008). *Chem. Commun.* 5224–5226.

11 Chen, B., Cao, P., Liao, Y. et al. (2018). *Org. Lett.* 20: 1346–1349.

12 Kim, N., Han, J.T., Ryu, D.H., and Yun, J. (2017). *Org. Lett.* 19: 6144–6147.

13 (a) Alfaro, R., Parra, A., Alemán, J. et al. (2012). *J. Am. Chem. Soc.* 134: 15165–15168. (b) Kageyuki, I., Osaka, I., Takaki, K., and Yoshida, H. (2017). *Org. Lett.* 19: 830–833.

14 He, S.-J., Wang, B., Lu, X. et al. (2018). *Org. Lett.* 20: 5208–5212.

15 Mun, B., Kim, S., Yoon, H. et al. (2017). *J. Org. Chem.* 82: 6349–6357.

16 Selected reports, see: (a) Ireland, R.E. and Wipf, P. (1990). *J. Org. Chem.* 55: 1425–1426. (b) Negishi, E.-i., Kondakov, D.Y., Choueiry, D. et al. (1996). *J. Am. Chem. Soc.* 118: 9577–9588. (c) Negishi, E.-i., Montehamp, J.-L., Anastasia, L. et al. (1998). *Tetrahedron Lett.* 39: 2503–2506. (d) Lipshutz, B.H., Butler, T., and Lower, A. (2006). *J. Am. Chem. Soc.* 128: 15396–15398. (e) Lipshutz, B.H., Butler, T., Lower, A., and Servesko, J. (2007). *Org. Lett.* 9: 3737–3740.

(f) Lipshutz, B.H. and Amorelli, B. (2009). *Tetrahedron Lett.* 50: 2144–2146. (g) Takimoto, M., Usami, S., and Hou, Z. (2009). *J. Am. Chem. Soc.* 131: 18266–18268.

17 Selected reports, see: (a) Kondakov, D.Y., Wang, S., and Negishi, E.-i. (1996). *Tetrahedron Lett.* 37: 3803–3806. (b) Millward, D.B., Cole, A.P., and Waymouth, R.M. (2000). *Organometallics* 19: 1870–1878. (c) Tan, Z., Liang, B., Huo, S. et al. (2006). *Tetrahedron: Asymmetry* 17: 512–515. (d) Negishi, E.-i. (2011). *ARKIVOC* (viii): 34–53. (e) Sultanov, R.M., Samoilova, E.V., Popod'ko, N.R. et al. (2013). *Tetrahedron Lett.* 54: 6619–6623. (f) Xu, S. and Negishi, E.-i. (2016). *Acc. Chem. Res.* 49: 2158–2168.

18 Schubert, U. (1994). *Angew. Chem. Int. Ed. Engl.* 33: 419–421.

19 Wang, L. and Duan, Z. (2013). *Chin. Sci. Bull.* 58: 307–315.

20 Eiji, S. and Tamejiro, H. (2002). *Bull. Chem. Soc. Jpn.* 75: 1435–1450.

21 Sakurai, H. and Imai, T. (1975). *Chem. Lett.* 4: 891–894.

22 Takeyama, Y., Nozaki, K., Matsumoto, K. et al. (1991). *Bull. Chem. Soc. Jpn.* 64: 1461–1466.

23 (a) Wang, R. and Falck, J.R. (2014). *Catal. Rev.* 56: 288–331. (b) Wang, R. and Falck, J.R. (2014). *RSC Adv.* 4: 1062–1066. Transition-metal-free cyanation of nucleophilic alkynes: (c) Lukashev, N.V., Kazantsev, A.V., Borisenko, A.A., and Beletskaya, I.P. (2001). *Tetrahedron* 57: 10309–10317.

24 (a) Chatani, N. and Hanafusa, T. (1985). *J. Chem. Soc. Chem. Commun.* 838–839. (b) Chatani, N., Takeyasu, T., Horiuchi, N., and Hanafusa, T. (1988). *J. Org. Chem.* 53: 3539–3548.

25 (a) Arai, S., Sato, T., Koike, Y. et al. (2009). *Angew. Chem. Int. Ed.* 48: 4528–4531. (b) Arai, S., Sato, T., and Nishida, A. (2009). *Adv. Synth. Catal.* 351: 1897–1904.

26 Asao, N., Yoshikawa, E., and Yamamoto, Y. (1996). *J. Org. Chem.* 61: 4874–4875.

27 (a) Imamura, K.-i., Yoshikawa, E., Gevorgyan, V., and Yamamoto, Y. (1998). *J. Am. Chem. Soc.* 120: 5339–5340. (b) Asao, N., Shimada, T., and Yamamoto, Y. (1999). *J. Am. Chem. Soc.* 121: 3797–3798. (c) Asao, N., Shimada, T., Shimada, T., and Yamamoto, Y. (2001). *J. Am. Chem. Soc.* 123: 10899–10902.

28 (a) Liu, S., Zhao, J., Kaminsky, L. et al. (2014). *Org. Lett.* 16: 4456–4459. (b) Wilson, R.J., Kaminsky, L., Ahmed, I., and Clark, D.A. (2015). *J. Org. Chem.* 80: 8290–8299.

29 Chatani, N., Horiuchi, N., and Hanafusa, T. (1990). *J. Org. Chem.* 55: 3393–3395.

30 Chatani, N., Morimoto, T., Muto, T., and Murai, S. (1994). *J. Organomet. Chem.* 473: 335–342.

31 Stille, J.K. (1986). *Angew. Chem. Int. Ed.* 25: 508–524. and references therein.

32 (a) Asao, N., Matsukawa, Y., and Yamamoto, Y. (1996). *Chem. Commun.* 1513–1514. (b) Matsukawa, Y., Asao, N., Kitahara, H., and Yamamoto, Y. (1999). *Tetrahedron* 55: 3779–3790.

33 Miura, K., Ltoh, D., Hondo, T. et al. (1996). *Tetrahedron Lett.* 37: 8539–8542.

34 Shirakawa, E., Yoshida, H., Nakao, Y., and Hiyama, T. (2000). *Org. Lett.* 2: 2209–2211.

35 Shirakawa, E., Yoshida, H., and Hiyama, T. (1997). *Tetrahedron Lett.* 38: 5177–5180.
36 Shirakawa, E., Yoshida, H., Kurahashi, T. et al. (1998). *J. Am. Chem. Soc.* 120: 2975–2976.
37 Shirakawa, E., Yamasaki, K., Yoshida, H., and Hiyama, T. (1999). *J. Am. Chem. Soc.* 121: 10221–10222.
38 Shirakawa, E., Nakao, Y., Yoshida, H., and Hiyama, T. (2000). *J. Am. Chem. Soc.* 122: 9030–9031.
39 Yoshida, H., Honda, Y., Shirakawa, E., and Hiyama, T. (2001). *Chem. Commun.* 1880–1881.
40 Shirakawa, E., Yoshida, H., Nakao, Y., and Hiyama, T. (1999). *J. Am. Chem. Soc.* 121: 4290–4292.
41 Shirakawa, E. and Hiyama, T. (2002). *J. Organomet. Chem.* 653: 114–121.
42 Shi, Y., Peterson, S.M., Haberaecker, W.W. III, and Blum, S.A. (2008). *J. Am. Chem. Soc.* 130: 2168–2169.
43 Hua, R., Onozawa, S., and Tanaka, M. (2000). *Organometallics* 19: 3269–3271.
44 Obora, Y., Baleta, A.S., Tokunaga, M., and Tsuji, Y. (2002). *J. Organomet. Chem.* 660: 173–177.
45 Yoshida, H., Kubo, T., Kuriki, H. et al. (2017). *ChemistrySelect* 2: 3212–3215.
46 (a) Shirakawa, E., Nakao, Y., and Hiyama, T. (2001). *Chem. Commun.* 263–264. (b) Nakao, Y., Shirakawa, E., Tsuchimoto, T., and Hiyama, T. (2004). *J. Organomet. Chem.* 689: 3701–3721.
47 Nakao, Y., Satoh, J., Shirakawa, E., and Hiyama, T. (2006). *Angew. Chem. Int. Ed.* 45: 2271–2274.
48 Zhao, F., Li, P., Liu, X. et al. (2019). *Molecules* 24: 164.
49 (a) Ouyang, K., Hao, W., Zhang, W.-X., and Xi, Z. (2015). *Chem. Rev.* 115: 12045–12090. (b) Wang, Q., Su, Y., Li, L., and Huang, H. (2016). *Chem. Soc. Rev.* 45: 1257–1272.
50 Yoshida, H., Shirakawa, E., Honda, Y., and Hiyama, T. (2002). *Angew. Chem. Int. Ed.* 41: 3247–3249.
51 Liu, Z. and Larock, R. (2005). *J. Am. Chem. Soc.* 127: 13112–13113.
52 Rao, B. and Zeng, X. (2014). *Org. Lett.* 16: 314–317.
53 Shimada, T., Nakamura, I., and Yamamoto, Y. (2004). *J. Am. Chem. Soc.* 126: 10546–10547.
54 Wu, C.-Y., Hu, M., Liu, Y. et al. (2012). *Chem. Commun.* 48: 3197–3199.
55 Zhao, F., Zhang, D., Nian, Y. et al. (2014). *Org. Lett.* 16: 5124–5127.
56 Zhao, J., Wang, G., and Li, S. (2015). *Chem. Commun.* 51: 15450–15453.
57 Kajita, Y., Matsubara, S., and Kurahashi, T. (2008). *J. Am. Chem. Soc.* 130: 6058–6059.
58 Poater, A., Vummaleti, S.V.C., and Cavallo, L. (2013). *Organometallics* 32: 6330–6336.
59 Li, D.Y., Mao, X.F., Chen, H.J. et al. (2014). *Org. Lett.* 16: 3476–3479.
60 Zheng, Z., Tao, Q., Ao, Y. et al. (2018). *Org. Lett.* 20: 3907–3910.
61 Miyazaki, Y., Ohta, N., Semba, K., and Nakao, Y. (2014). *J. Am. Chem. Soc.* 136: 3732–3735.

62 Pan, Z., Pound, S.M., Rondla, N.R., and Douglas, C.J. (2014). *Angew. Chem. Int. Ed.* 53: 5170–5174.

63 Frost, G.B., Serratore, N.A., Ogilvie, J.M., and Douglas, C.J. (2017). *J. Org. Chem.* 82: 3721–3726.

64 Pan, Z., Wang, S., Brethorst, J.T., and Douglas, C.J. (2018). *J. Am. Chem. Soc.* 140: 3331–3338.

65 Comasseto, J.V., Ling, L.W., Petragnani, N., and Stefani, H.A. (1997). *Synthesis* 373–403.

66 Selected reports, see: (a) Kuniyasu, H. and Kambe, N. (2006). *Chem. Lett.* 35: 1320–1325. (b) Fujiwara, S., Toyofuku, M., Kuniyasu, H., and Kambe, N. (2010). *Pure Appl. Chem.* 82: 565–575. (c) Ozaki, T., Nomoto, A., Kamiya, I. et al. (2011). *Bull. Chem. Soc. Jpn.* 84: 155–163. (d) Nakao, Y. (2012). *Bull. Chem. Soc. Jpn.* 85: 731–745.

67 Kajita, Y., Kurahashi, T., and Matsubara, S. (2008). *J. Am. Chem. Soc.* 130: 17226–17227.

68 Ooguri, A., Nakai, K., Kurahashi, T., and Matsubara, S. (2009). *J. Am. Chem. Soc.* 131: 13194–13195.

69 Inami, T., Kurahashi, T., and Matsubara, S. (2014). *Org. Lett.* 16: 5660–5662.

70 Ochi, Y., Kurahashi, T., and Matsubara, S. (2011). *Org. Lett.* 13: 1374–1377.

71 Maizuru, N., Inami, T., Kurahashi, T., and Matsubara, S. (2011). *Chem. Lett.* 40: 375–377.

72 Arisawa, M., Tanii, S., and Yamaguchi, M. (2014). *Chem. Commun.* 50: 15267–15270.

73 Kuniyasu, H., Takekawa, K., Sanagawa, A. et al. (2011). *Tetrahedron Lett.* 52: 5501–5503.

74 Hoang, G.T., Reddy, V.J., Nguyen, H.H.K., and Douglas, C.J. (2011). *Angew. Chem. Int. Ed.* 50: 1882–1884.

75 Koester, D.C., Kobayashi, M., Werz, D.B., and Nakao, Y. (2012). *J. Am. Chem. Soc.* 134: 6544–6547.

76 Vummaleti, S.V.C., Al-Ghamdi, M., Poater, A. et al. (2015). *Organometallics* 34: 5549–5554.

77 Selected reviews, see: (a) Luh, T.-Y. (1991). *Acc. Chem. Res.* 24: 257–263. (b) Luh, T.-Y., Leung, M.-K., and Wong, K.-T. (2000). *Chem. Rev.* 100: 3187–3204. (c) Kondo, T. and Mitsudo, T.-a. (2000). *Chem. Rev.* 100: 3205–3220. (d) Otsuka, S., Nogi, K., and Yorimitsu, H. (2018). *Top. Curr. Chem. (Z)* 376: 13.

78 Fang, J.-M. and Chen, M.-Y. (1987). *Tetrahedron Lett.* 28: 2853–2856.

79 Choi, N., Kabe, Y., and Ando, W. (1991). *Tetrahedron Lett.* 32: 4573–4576.

80 Selected reports, see: (a) Schwerdtfeger, A.E. and Chan, T.H. (1993). *J. Org. Chem.* 58: 6513–6516. (b) Pereira, O.Z. and Chan, T.H. (1994). *J. Org. Chem.* 59: 6710–6716. (c) Kinoshita, Y. and Kitahara, T. (1997). *Tetrahedron Lett.* 38: 4993–4996.

81 Hua, R., Takeda, H., Onozawa, S.-y. et al. (2001). *J. Am. Chem. Soc.* 123: 2899–2900.

82 Hua, R., Takeda, H., Onozawa, S.-y. et al. (2007). *Org. Lett.* 9: 263–266.

83 (a) Osakada, K., Ozawa, Y., and Yamamoto, A. (1990). *J. Organomet. Chem.* 399: 341–348. (b) Miyauch, Y., Watanabe, S., Kuniyasu, H., and Kurosawa, H. (1995). *Organometallics* 14: 5450–5453. (c) Planas, J.G., Hirano, M., and Komiya, S. (1998). *Chem. Lett.* 123–124.
84 Wang, J., Zhang, S., Xu, C. et al. (2018). *Angew. Chem. Int. Ed.* 57: 6915–6920.
85 Nishi, M., Kuninobu, Y., and Takai, K. (2012). *Org. Lett.* 14: 6116–6118.
86 Kamiya, I., Kawakami, J.-i., Yano, S. et al. (2006). *Organometallics* 25: 3562–3564.
87 Zheng, W., Ariafard, A., and Lin, Z. (2008). *Organometallics* 27: 246–253.
88 Lee, Y.T., Choi, S.Y., and Chung, Y.K. (2007). *Tetrahedron Lett.* 48: 5673–5677.
89 Higashimae, S., Kurata, D., Kawaguchi, S.-i. et al. (2018). *J. Org. Chem.* 83: 5267–5273.
90 Pawliczek, M., Garve, L.K.B., and Werz, D.B. (2015). *Org. Lett.* 17: 1716–1719.
91 Bürger, M., Loch, M.N., Jones, P.G., and Werz, D.B. (2020). *Chem. Sci.* 11: 1912–1917.
92 (a) Sugoh, K., Kuniyasu, H., Sugae, T. et al. (2001). *J. Am. Chem. Soc.* 123: 5108–5109. (b) Yamashita, F., Kuniyasu, H., Terao, J., and Kambe, N. (2008). *Org. Lett.* 10: 101–104.
93 Hirai, T., Kuniyasu, H., and Kambe, N. (2004). *Chem. Lett.* 33: 1148–1149.
94 Kundu, S., Brennessel, W.W., and Jones, W.D. (2011). *Organometallics* 30: 5147–5154.
95 Minami, Y., Kuniyasu, H., Miyafuji, K., and Kambe, N. (2009). *Chem. Commun.* 3080–3082.
96 Inami, T., Kurahashi, T., and Matsubara, S. (2015). *Chem. Commun.* 51: 1285–1288.
97 Inami, T., Baba, Y., Kurahashi, T., and Matsubara, S. (2011). *Org. Lett.* 13: 1912–1915.
98 Arisawa, M., Igarashi, Y., Tagami, Y. et al. (2011). *Tetrahedron Lett.* 52: 920–922.
99 Hooper, J.F., Chaplin, A.B., González-Rodríguez, C. et al. (2012). *J. Am. Chem. Soc.* 134: 2906–2909.
100 Iwasaki, M., Topolovčan, N., Hu, H. et al. (2016). *Org. Lett.* 18: 1642–1645.
101 Uno, D., Nogi, K., and Yorimitsu, H. (2019). *Org. Lett.* 21: 8295–8299.
102 Li, Y., Mück-Lichtenfeld, C., and Studer, A. (2016). *Angew. Chem. Int. Ed.* 55: 14435–14438.
103 Zhao, C.-Q., Huang, X., and Meng, J.-B. (1998). *Tetrahedron Lett.* 39: 1933–1936.
104 Zhao, C.-Q., Li, J.-L., Meng, J.-B., and Wang, Y.-M. (1998). *J. Org. Chem.* 63: 4170–4171.
105 Hirai, T., Kuniyasu, H., Kato, T. et al. (2003). *Org. Lett.* 5: 3871–3873.
106 Toyofuku, M., Fujiwara, S.-i., Shin-ike, T. et al. (2005). *J. Am. Chem. Soc.* 127: 9706–9707.
107 (a) Kuniyasu, H., Ohtaka, A., Nakazono, T. et al. (2000). *J. Am. Chem. Soc.* 122: 2375–2376. (b) Kuniyasu, H., Kato, T., Inoue, M. et al. (2006). *J. Organomet. Chem.* 691: 1873–1878.
108 Toyofuku, M., Fujiwara, S.-i., Shin-ike, T. et al. (2008). *J. Am. Chem. Soc.* 130: 10504–10505.

109 Bürger, M., Röttger, S.H., Loch, M.N. et al. (2020). *Org. Lett.* 22: 5025–5029.

110 Yamashita, K.-i., Takeda, H., Kashiwabara, T. et al. (2007). *Tetrahedron Lett.* 48: 6655–6659.

111 Mitamura, T. and Ogawa, A. (2010). *Tetrahedron Lett.* 51: 3538–3541.

112 Toyofuku, M., Murase, E., Fujiwara, S.-i. et al. (2008). *Org. Lett.* 10: 3957–3960.

113 Fujiwara, S.-i., Okuyama, M., Tsuda, S. et al. (2012). *Tetrahedron* 68: 10523–10529.

114 Ananikov, V.P., Orlov, N.V., and Beletskaya, I.P. (2005). *Russ. Chem. Bull. Int. Ed.* 54: 576–587.

115 Petrone, D.A., Ye, J., and Lautens, M. (2016). *Chem. Rev.* 116: 8003–8104.

116 Selected reports, see: (a) Halpern, J. (1970). *Acc. Chem. Res.* 3: 386–392. (b) Stille, J.K. and Lau, K.S.Y. (1977). *Acc. Chem. Res.* 10: 434–442. (c) Fahey, D.R. and Mahan, J.E. (1977). *J. Am. Chem. Soc.* 99: 2501–2508. (d) Ebert, G.W. and Rieke, R.D. (1988). *J. Org. Chem.* 53: 4482–4488. (e) van Asselt, R., Vrieze, K., and Elsevier, C.J. (1994). *J. Organomet. Chem.* 480: 27–40. (f) Zeni, G. and Larock, R.C. (2006). *Chem. Rev.* 106: 4644–4680.

117 The representative reports on the oxidative addition of allyl halides to transition-metal compounds, see: (a) Sbrana, G., Braca, G., Piacenti, F., and Pino, P. (1968). *J. Organomet. Chem.* 13: 240–242. (b) Pearson, R.G. and Poulos, A.T. (1979). *Inorg. Chim. Acta* 34: 67–76. (c) Nagashima, H., Mukai, K., Shiota, Y. et al. (1990). *Organometallics* 9: 799–807. (d) Kurosawa, H., Kajimaru, H., Ogoshi, S. et al. (1992). *J. Am. Chem. Soc.* 114: 8417–8424. (e) Matsushima, Y., Onitsuka, K., and Takahashi, S. (2004). *Organometallics* 23: 3763–3765.

118 Kaneda, K., Kawamoto, F., Fujiwara, Y. et al. (1974). *Tetrahedron Lett.* 1067–1070.

119 (a) Kaneda, K., Kobayashi, H., Fujiwara, Y. et al. (1975). *Tetrahedron Lett.* 2833–2836. (b) Kaneda, K., Uchiyama, T., Terasawa, M. et al. (1976). *Chem. Lett.* 449–452. (c) Kaneda, K., Uchiyama, T., Fujiwara, Y. et al. (1979). *J. Org. Chem.* 44: 55–63.

120 Tsuji, J. and Yasuda, H. (1978). *Synth. Commun.* 8: 103–107.

121 (a) Camps, F., Coll, J., Llebaria, A., and Moreto, J.M. (1988). *Tetrahedron Lett.* 29: 5811–5814. (b) Lleberia, A., Camps, F., and Moreto, J.M. (1993). *Tetrahedron* 49: 1283–1296.

122 Topolovčan, N., Panov, I., and Kotora, M. (2016). *Org. Lett.* 18: 3634–3637.

123 (a) Thadani, A.N. and Rawal, V.H. (2002). *Org. Lett.* 4: 4317–4320. (b) Thadani, A.N. and Rawal, V.H. (2002). *Org. Lett.* 4: 4321–4323. (c) Yamaguchi, R., Kawasaki, H., Yoshitome, T., and Kawanishi, M. (1982). *Chem. Lett.* 1485–1486.

124 Lu, Z., Kong, W., Yuan, Z. et al. (2011). *J. Org. Chem.* 76: 8524–8529.

125 Cai, H., Yuan, Z., Zhu, W., and Zhu, G. (2011). *Chem. Commun.* 47: 8682–8684.

126 Wen, Y. and Jiang, H. (2013). *Tetrahedron Lett.* 54: 4034–4037.

127 Masanori, K., Takashi, S., Shinji, O., and Toshihiko, M. (1993). *Bull. Chem. Soc. Jap.* 66: 3058–3061.

128 Xie, M., Wang, J., Fang, K. et al. (2015). *Tetrahedron Lett.* 56: 4388–4391.

129 (a) Kippo, T., Fukuyama, T., and Ryu, I. (2010). *Org. Lett.* 12: 4006–4009. (b) Kippo, T., Hamaoka, K., Ueda, M. et al. (2016). *Tetrahedron* 72: 7866–7874.

130 Kippo, T., Fukuyama, T., and Ryu, I. (2011). *Org. Lett.* 13: 3864–3867.
131 Cook, G.R. and Hayashi, R. (2006). *Org. Lett.* 8: 1045–1048.
132 Selected reports, see: (a) Lau, K.S.Y., Wong, P.K., and Stille, J.K. (1976). *J. Am. Chem. Soc.* 98: 5832–5840. (b) Stille, J.K. and Cowell, A.B. (1977). *J. Organomet. Chem.* 124: 253–261. (c) Roberts, J.S. and Klabunde, K.J. (1977). *J. Am. Chem. Soc.* 99: 2509–2519. (d) Amatore, C., Catellani, M., Deledda, S. et al. (2008). *Organometallics* 27: 4549–4554. (e) Aseman, M.D., Rashidi, M., Nabavizadeh, S.M., and Puddephatt, R.J. (2013). *Organometallics* 32: 2593–2598.
133 Liu, Z., Wang, J., Zhao, Y., and Zhou, B. (2009). *Adv. Synth. Catal.* 351: 371–374.
134 Xu, T. and Hu, X. (2015). *Angew. Chem. Int. Ed.* 54: 1307–1311.
135 Che, C., Zheng, H., and Zhu, G. (2015). *Org. Lett.* 17: 1617–1620.
136 Wang, K., Meng, L.-G., and Wang, L. (2016). *J. Org. Chem.* 81: 7080–7087.
137 Xu, T., Cheung, C., and Hu, X. (2014). *Angew. Chem. Int. Ed.* 53: 4910–4914.
138 Belhomme, M.-C., Dru, D., Xiong, H.-Y. et al. (2014). *Synthesis* 46: 1859–1870.
139 Li, G., Cao, Y.-X., Luo, C.-G. et al. (2016). *Org. Lett.* 18: 4806–4809.
140 Wu, G. and von Wangelin, A. (2018). *J. Chem. Sci.* 9: 1795–1802.
141 Kharasch, M.S., Reinmuth, O., and Urry, W.H. (1947). *J. Am. Chem. Soc.* 69: 1105–1110.
142 (a) Simal, F., Wlodarczak, L., Demonceau, A., and Noels, A.F. (2000). *Tetrahedron Lett.* 41: 6071–6074. (b) Lee, J., Grandner, J.M., Engle, K.M. et al. (2016). *J. Am. Chem. Soc.* 138: 7171–7177.
143 (a) Pereira, S. and Srebnik, M. (1996). *J. Am. Chem. Soc.* 118: 909–910. (b) Chen, B., Fang, C., Liu, P., and Ready, J.M. (2017). *Angew. Chem. Int. Ed.* 56: 8780–8784.
144 (a) Clark, A.J. (2016). *Eur. J. Org. Chem.* 2016: 2231–2243. (b) Pu, W., Sun, D., Fan, W. et al. (2019). *Chem. Commun.* 55: 4821–4824.
145 Quebatte, L., Thommes, K., and Severin, K. (2006). *J. Am. Chem. Soc.* 128: 7440–7441.
146 Oe, Y. and Uozumi, Y. (2008). *Adv. Synth. Catal.* 350: 1771–1775.
147 Tanaka, M. and Hua, R. (2002). *Pure Appl. Chem.* 74: 181–186.
148 (a) Kubota, M. and Blake, D.M. (1971). *J. Am. Chem. Soc.* 93: 1368–1373. (b) Bennett, M.A., Jeffery, J.C., and Robertson, G.B. (1981). *Inorg. Chem.* 20: 323–330.
149 Kokubo, K., Matsumasa, K., Miura, M., and Nomura, M. (1996). *J. Org. Chem.* 61: 6941–6946.
150 Kokubo, K., Matsumasa, K., Miura, M., and Nomura, M. (1998). *J. Organomet. Chem.* 560: 217–222.
151 (a) Iwai, T., Fujihara, T., Terao, J., and Tsuji, Y. (2009). *J. Am. Chem. Soc.* 131: 6668–6669. (b) Iwai, T., Fujihara, T., Terao, J., and Tsuji, Y. (2012). *J. Am. Chem. Soc.* 134: 1268–1274.
152 Kashiwabara, T., Kataoka, K., Hua, R. et al. (2005). *Org. Lett.* 7: 2241–2244.
153 Kashiwabara, T., Fuse, K., Hua, R., and Tanaka, M. (2008). *Org. Lett.* 10: 5469–5472.

154 Kashiwabara, T., Fuse, K., Muramatsu, T., and Tanaka, M. (2009). *J. Org. Chem.* 74: 9433–9439.

155 Wang, B., Wang, S., Li, P., and Wang, L. (2010). *Chem. Commun.* 46: 5891–5893.

156 Gandeepan, P., Parthasarathy, K., Su, T.-H., and Cheng, C.-H. (2012). *Adv. Synth. Catal.* 354: 457–468.

157 Gooßen, L.J., Rodríguez, N., and Gooßen, K. (2009). *Angew. Chem. Int. Ed.* 48: 9592–9594.

158 Hosseini-Sarvari, M. and Mardaneh, Z. (2011). *Bull. Chem. Soc. Jpn.* 84: 778–782.

159 Cano, R., Yus, M., and Ramón, D.J. (2013). *Tetrahedron* 69: 7056–7065.

160 Dennis, J.M., Snelders, D.J.M., and Dyson, P.J. (2011). *Org. Lett.* 13: 4048–4051.

161 Iwai, T., Fujihara, T., Terao, J., and Tsuji, Y. (2010). *J. Am. Chem. Soc.* 132: 9602–9603.

162 Nagata, T., Hirano, K., Satoh, T., and Miura, M. (2014). *J. Org. Chem.* 79: 8960–8967.

163 Le, C.M., Hou, X., Sperger, T. et al. (2015). *Angew. Chem. Int. Ed.* 54: 15897–15900.

164 Le, C.M., Sperger, T., Fu, R. et al. (2016). *J. Am. Chem. Soc.* 138: 14441–14448.

165 Hua, R., Shimada, S., and Tanaka, M. (1998). *J. Am. Chem. Soc.* 120: 12365–12366.

166 Hua, R., Onozawa, S.-y., and Tanaka, M. (2005). *Chem. Eur. J.* 11: 3621–3630.

167 Kashiwabara, T. and Tanaka, M. (2011). *Adv. Synth. Catal.* 353: 1485–1490.

168 Baek, J.Y., Lee, S.I., Sim, S.H., and Chung, Y.K. (2008). *Synlett* 551–554.

169 Stanislaw, F., Wnuk, S.F., Valdez, C.A., and Valdez, N.X. (2001). *J. Carbohydr. Chem.* 20: 71–79.

170 Hua, R. and Tanaka, M. (2004). *Tetrahedron Lett.* 45: 2367–2370.

171 Amii, H., Kishikawa, Y., and Uneyama, K. (2001). *Org. Lett.* 3: 1109–1112.

172 Isobe, A., Takagi, J., Katagiri, T., and Uneyama, K. (2008). *Org. Lett.* 10: 2657–2659.

173 Hiroto, Y., Yasihiro, M., and Joji, O. (2009). *Chem. Lett.* 38: 1132–1133.

174 Li, Y., Liu, X., Jiang, H., and Feng, Z. (2010). *Angew. Chem. Int. Ed.* 49: 3338–3341.

175 Li, Y., Liu, X., Jiang, H. et al. (2011). *Angew. Chem. Int. Ed.* 50: 6341–6345.

176 Wada, T., Iwasaki, M., Kondoh, A. et al. (2010). *Chem. Eur. J.* 16: 10671–10674.

177 García-Fernández, P.D., Iglesias-Sigüenza, J., Rivero-Jerez, P.S. et al. (2020). *J. Am. Chem. Soc.* 142: 16082–16089.

178 Morishita, T., Yoshida, H., and Ohshita, J. (2010). *Chem. Commun.* 46: 640–642.

179 Murai, M., Hatano, R., Kitabata, S., and Ohe, K. (2011). *Chem. Commun.* 47: 2375–2377.

180 Sakata, N., Sasakura, K., Matsushita, G. et al. (2017). *Org. Lett.* 19: 3422–3425.

181 Barrado, A.G., Zieliński, A., Goddard, R., and Alcarazo, M. (2017). *Angew. Chem. Int. Ed.* 56: 13401–13405.

182 Martini, T. and Kampmeier, J.A. (1970). *Angew. Chem.* 82: 216–217.

183 Noyori, R., Odagi, T., and Takaya, H. (1970). *J. Am. Chem. Soc.* 92: 5780–5781.

184 Huffman, M.A. and Liebeskind, L.S. (1991). *J. Am. Chem. Soc.* 113: 2771–2772.

185 Huffman, M.A., Liebeskind, L.S., and Pennington, W.T. (1992). *Organometallics* 11: 255–266.
186 Saito, S., Masuda, M., and Komagawa, S. (2004). *J. Am. Chem. Soc.* 126: 10540–10541.
187 Kondo, T., Taguchi, Y., Kaneko, Y. et al. (2004). *Angew. Chem. Int. Ed.* 43: 5369–5372.
188 (a) Murakami, M., Itahashi, T., and Ito, Y. (2002). *J. Am. Chem. Soc.* 124: 13976–13977. (b) Matsuda, T., Fujimoto, A., Ishibashi, M., and Murakami, M. (2004). *Chem. Lett.* 33: 876–877. (c) Murakami, M. and Ashida, S. (2006). *Chem. Commun.* 4599–4601. (d) Ko, H.M. and Dong, G. (2014). *Nat. Chem.* 6: 739–744.
189 Xu, T. and Dong, G. (2012). *Angew. Chem. Int. Ed.* 51: 7567–7571.
190 Zhou, X. and Dong, G. (2015). *J. Am. Chem. Soc.* 137: 13715–13721.
191 Kondo, T., Nakamura, A., Okada, T. et al. (2000). *J. Am. Chem. Soc.* 122: 6319–6320.
192 Liu, L., Ishida, N., and Murakami, M. (2012). *Angew. Chem. Int. Ed.* 51: 2485–2488.
193 Souillart, L., Parker, E., and Cramer, N. (2014). *Angew. Chem. Int. Ed.* 53: 3001–3005.
194 (a) Murakami, M., Ashida, S., and Matsuda, T. (2005). *J. Am. Chem. Soc.* 127: 6932–6933. (b) Murakami, M., Ashida, S., and Matsuda, T. (2006). *Tetrahedron* 62: 7540–7546.
195 Matsuda, T. and Miura, N. (2013). *Org. Biomol. Chem.* 11: 3424–3427.
196 Matsuda, T. and Matsumoto, T. (2016). *Org. Biomol. Chem.* 14: 5023–5027.
197 (a) Jun, C.-H. (2004). *Chem. Soc. Rev.* 33: 610–618. (b) Murakami, M. and Matsuda, T. (2011). *Chem. Commun.* 47: 1100–1105. (c) Ruhland, K. (2012). *Eur. J. Org. Chem.* 2012: 2683–2706.
198 Seiser, T., Saget, T., Tran, D.N., and Cramer, N. (2011). *Angew. Chem. Int. Ed.* 50: 7740–7752.
199 Chen, S., Shan, G., Nie, P., and Rao, Y. (2015). *Asian J. Org. Chem.* 4: 16–26.
200 (a) Xu, T., Dermenci, A., and Dong, G. (2014). *Top. Curr. Chem.* 346: 233–258. (b) Deng, L. and Dong, G. (2020). *Trends Chem.* 2: 183–198.
201 (a) Seiser, T. and Cramer, N. (2009). *Org. Biomol. Chem.* 7: 2835–2840. (b) Winter, C. and Krause, N. (2009). *Angew. Chem. Int. Ed.* 48: 2460–2462.
202 Selected reviews, see: (a) Tobisu, M. and Chatani, N. (2008). *Chem. Soc. Rev.* 37: 300–307. (b) Nakao, Y. (2012). *Bull. Chem. Soc. Jpn.* 85: 731–745. (c) Wen, Q., Lu, P., and Wang, Y. (2014). *RSC Adv.* 4: 47806–47826. (d) Chen, F., Wang, T., and Jiao, N. (2014). *Chem. Rev.* 114: 8613–8661. (e) Souillart, L. and Cramer, N. (2015). *Chem. Rev.* 115: 9410–9464. (f) Chen, P.-h., Billett, B.A., Tsukamoto, T., and Dong, G. (2017). *ACS Catal.* 7: 1340–1360.
203 Murahashi, S.-I., Naota, T., and Nakajima, N. (1986). *J. Org. Chem.* 51: 898–901.
204 (a) Nozaki, K., Sato, N., and Takaya, H. (1994). *J. Org. Chem.* 59: 2679–2681. (b) Nozaki, K., Sato, N., and Takaya, H. (1996). *Bull. Chem. Soc. Jpn.* 69: 1629–1637.

205 Murayama, H., Nagao, K., Ohmiya, H., and Sawamura, M. (2016). *Org. Lett.* 18: 1706–1709.

206 Rondla, N.R., Ogilvie, J.M., Pan, Z., and Douglas, C.J. (2014). *Chem. Commun.* 50: 8974–8977.

207 (a) Kobayashi, Y., Kamisaki, H., Yanada, R., and Takemoto, Y. (2006). *Org. Lett.* 8: 2711–2713. (b) Kobayashi, Y., Kamisaki, H., Takeda, H. et al. (2007). *Tetrahedron* 63: 2978–2989. (c) Yasui, Y. and Takemoto, Y. (2008). *Chem. Rec.* 8: 386–394. (d) Yasui, Y., Kamisaki, H., Ishida, T., and Takemoto, Y. (2010). *Tetrahedron* 66: 1980–1989.

208 Yasui, Y., Kamisaki, H., and Takemoto, Y. (2008). *Org. Lett.* 10: 3303–3306.

209 (a) Nishihara, Y., Inoue, Y., Itazaki, M., and Takagi, K. (2005). *Org. Lett.* 7 (13): 2639–2641. (b) Nishihara, Y., Inoue, Y., Izawa, S. et al. (2006). *Tetrahedron* 62: 9872–9882.

210 Rondla, N.R., Levi, S.M., Ryss, J.M. et al. (2011). *Org. Lett.* 13: 1940–1943.

211 (a) Nakao, Y., Hirata, Y., and Hiyama, T. (2006). *J. Am. Chem. Soc.* 128: 7420–7421. (b) Hirata, Y., Inui, T., Nakao, Y., and Hiyama, T. (2009). *J. Am. Chem. Soc.* 131: 6624–6631.

212 (a) Hirata, Y., Yada, A., Morita, E. et al. (2010). *J. Am. Chem. Soc.* 132: 10070–10077. (b) Zhu, B., Du, G.-F., Ren, H. et al. (2017). *Organometallics* 36: 4713–4720.

213 Nakao, Y. and Hiyama, T. (2008). *Pure Appl. Chem.* 80: 1097–1107.

214 (a) Nakao, Y., Oda, S., and Hiyama, T. (2004). *J. Am. Chem. Soc.* 126: 13904–13905. (b) Nakao, Y., Oda, S., Yada, A., and Hiyama, T. (2006). *Tetrahedron* 62: 7567–7576.

215 Minami, Y., Yoshiyasu, H., Nakao, Y., and Hiyama, T. (2013). *Angew. Chem. Int. Ed.* 52: 883–887.

216 García, J.J., Arévalo, A., Brunkan, N.M., and Jones, W.D. (2004). *Organometallics* 23: 3997–4002.

217 Xu, H., Williard, P.G., and Bernskoetter, W.H. (2012). *Organometallics* 31: 1588–1590.

218 (a) Yada, A., Yukawa, T., Nakao, Y., and Hiyama, T. (2009). *Chem. Commun.* 3931–3933. (b) Yada, A., Yukawa, T., Idei, H. et al. (2010). *Bull. Chem. Soc. Jpn.* 83: 619–634.

219 Nakao, Y., Yada, A., and Hiyama, T. (2010). *J. Am. Chem. Soc.* 132: 10024–10026.

220 Nakao, Y., Yada, A., Ebata, S., and Hiyama, T. (2007). *J. Am. Chem. Soc.* 129: 2428–2429.

221 (a) Nakao, Y., Hirata, Y., Tanaka, M., and Hiyama, T. (2008). *Angew. Chem. Int. Ed.* 47: 385–387. (b) Hirata, Y., Tanaka, M., Yada, A. et al. (2009). *Tetrahedron* 65: 5037–5050.

222 (a) Watson, M.P. and Jacobsen, E.N. (2008). *J. Am. Chem. Soc.* 130: 12594–12595. (b) Nakao, Y., Ebata, S., Yada, A. et al. (2008). *J. Am. Chem. Soc.* 130: 12874–12875.

223 Yamada, Y., Ebata, S., Hiyama, T., and Nakao, Y. (2015). *Tetrahedron* 71: 4413–4417.

224 Brunkan, N.M., Brestensky, D.M., and Jones, W.D. (2004). *J. Am. Chem. Soc.* 126: 3627–3641.

225 (a) Nakao, Y., Yukawa, T., Hirata, Y. et al. (2006). *J. Am. Chem. Soc.* 128: 7116–7117. (b) Hirata, Y., Yukawa, T., Kashihara, N. et al. (2009). *J. Am. Chem. Soc.* 131: 10964–10973.

226 Kamijo, S., Yokosaka, S., and Inoue, M. (2012). *Tetrahedron* 68: 5290–5296.

227 Kamijo, S., Matsumura, S., and Inoue, M. (2012). *Tetrahedron Lett.* 53: 4368–4371.

228 Kamijo, S., Yokosaka, S., and Inoue, M. (2012). *Tetrahedron Lett.* 53: 4324–4327.

229 Kuninobu, Y., Kawata, A., Nishi, M. et al. (2008). *Chem. Commun.* 6360–6362.

230 Kuninobu, Y., Morita, J., Nishi, M. et al. (2009). *Org. Lett.* 11: 2535–2537.

231 Kuninobu, Y., Matsuzaki, H., Nishi, M., and Takai, K. (2011). *Org. Lett.* 13: 2959–2961.

232 Tambar, U.K. and Stoltz, B.M. (2005). *J. Am. Chem. Soc.* 127: 5340–5341.

233 Tambar, U.K., Ebner, D.C., and Stoltz, B.M. (2006). *J. Am. Chem. Soc.* 128: 11752–11753.

234 Rao, B., Tang, J., Wei, Y., and Zeng, X. (2016). *Chem. Asian J.* 11: 991–995.

235 Samineni, R., Srihari, P., and Mehta, G. (2016). *Org. Lett.* 18: 2832–2835.

236 Yoshida, H., Watanabe, M., Ohshita, J., and Kunai, A. (2005). *Tetrahedron Lett.* 46: 6729–6731.

7

Carbocycles from Annulation of Alkynes and Alkenes

There is a huge literature on the construction of different sizes of carbocycles based on inter- and intramolecular cycloaddition reactions starting from alkynes, alkenes, and enynes, and a few comprehensive reviews are cited here [1]. This chapter highlights the typical and new procedures for the construction of carbocycles using alkyne/alkene as carbon building blocks, which have been developed in recent years.

7.1 Four-Membered Carbocycles

Cyclobutanes and cyclobutenes are particularly interesting substrates in organic synthesis as useful building blocks due to their unique reactivity [2] and also are the key moieties occurring in a range of natural products [3]. The [2 + 2] cycloaddition reactions of either two molecules of alkenes or alkene with alkynes under different conditions are the most straightforward and atom economic approach to prepare these strained carbocyclic compounds (Scheme 7.1) [4].

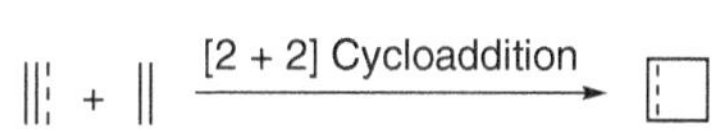

Scheme 7.1 Cyclobutanes and cyclobutenes from [2 + 2] cycloaddition of carbon–carbon unsaturated bonds.

7.1.1 Construction of Cyclobutenes

The [2 + 2] cycloaddition reactions of alkynes bearing different functional groups with strained bicyclic alkenes, such as norbornenes and norbornadienes as the alkene partners, have been extensively studied by Tam's group in the presence of ruthenium complexes, and alkynes usually have electron-withdrawing group, such as ester, phosphonate, and sulfone groups [5] (Scheme 7.2). Other groups have also investigated this transformation using other transition metal complexes, such as rhenium [6] and cobalt complexes [7].

The reactivity of various bicyclic alkenes involving the presence of oxygen and nitrogen in the bridgehead of the bicyclic alkene have also been studied in the presence of ruthenium complexes by Tam's group [8]. Ynamides could be used in

Addition Reactions with Unsaturated Hydrocarbons, First Edition. Ruimao Hua.

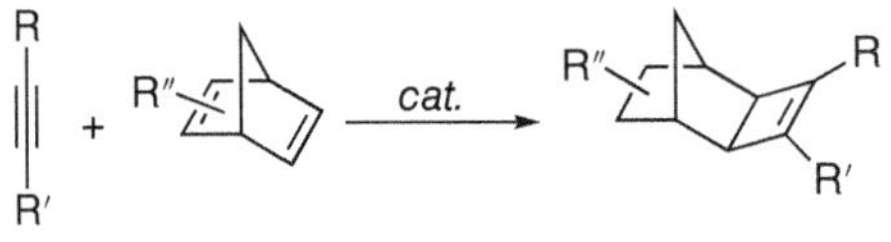

Scheme 7.2 Synthesis of cyclobutenes from cycloaddition of bicyclic alkenes with alkynes.

the ruthenium-catalyzed cycloaddition giving the corresponding amino-substituted cyclobutenes, and diastereoselective cycloaddition utilizing chiral cyclic ynamides can also be achieved with a low to moderate level of asymmetric induction [9].

In addition, rhodium-catalyzed enantioselective [2+2] cycloaddition of alkynyl esters and norbornenes [10], iridium- or nickel-catalyzed asymmetric [2+2] cycloaddition of hetero-bicyclic alkenes and alkynes [11], and ruthenium-catalyzed enantioselective [2+2] cycloadditions of chiral acyl camphorsultam-functionalized alkynes and bicyclic alkenes [12] have also been developed.

Yi and coworker have found that the cationic ruthenium–alkylidene complex $[(PCy_3)_2(CO)(Cl)Ru{=}CHCH{=}C(CH_3)_2]^+BF_4^-$ can efficiently catalyze both hydrovinylation and [2+2] cycloaddition reactions of alkynes with ethylene, and in the case of dimethyl acetylenedicarboxylate (DMAD) used, the reactions with ethylene and norbornene afford the [2+2] adducts in high yields with excellent chemoselectivity (Scheme 7.3) [13]. $[ReBr(CO)_3(thf)]_2$ also shows highly catalytic activity for the [2+2] cycloaddition of norbornenes with electron-poor internal and terminal alkynes giving cyclobutenes in good to excellent yields [6].

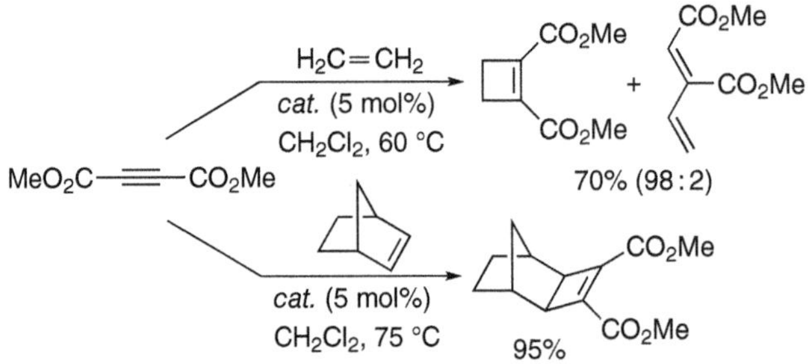

Scheme 7.3 Ruthenium-catalyzed intermolecular [2 + 2] cycloaddition of active alkynes with alkenes.

In the presence of $Ni(PPh_3)_2Cl_2/PPh_3/Zn$, the activated cyclic alkenes, such as oxabenzonorbornadienes and azabenzonorbornadiene, undergo [2+2] cycloaddition with terminal or internal alkynes to afford the corresponding cyclobutenes in fair to excellent yields [14]. The enantioselective cycloaddition reactions with the use of similar substrates have also been achieved in the presence of $Ni(cod)_2$/chiral ligands [15].

Echavarren and coworker investigated the intermolecular [2+2] cycloaddition of terminal alkynes with alkenes in the presence of Au(I) complexes and found that the key for the success of the [2+2] cycloaddition leading to cyclobutenes is the use of gold(I) complexes with bulky ligands, which selectively activate alkynes

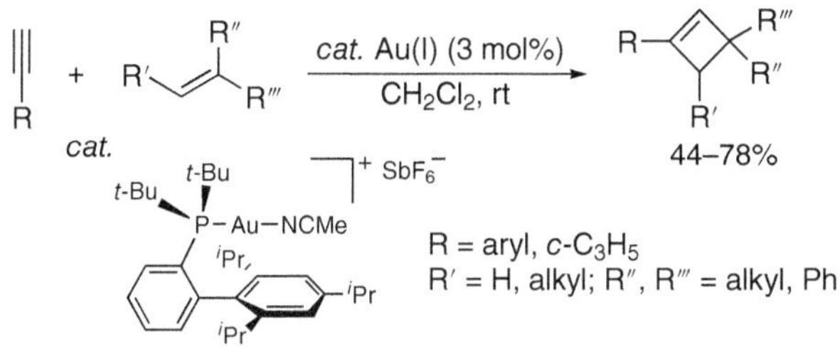

Scheme 7.4 Gold-catalyzed intermolecular [2 + 2] cycloaddition of terminal alkynes with alkenes.

in the presence of alkenes (Scheme 7.4) [16]. The same group then developed an enantioselective intermolecular gold(I)-catalyzed [2 + 2] cycloaddition of terminal alkynes with alkenes, and the approach is applied to the enantioselective total synthesis of rumphellaone A [17].

Cobalt(II)/Zn catalytic system has been confirmed to efficiently catalyze the [2 + 2] cycloaddition reactions of alkenes with alkynes [18]. Under similar cobalt catalysis, 1,3-enynes also facilitate [2 + 2] cycloaddition with alkenes to produce vinylcyclobutenes [19].

In the presence of gold complexes, 1-vinylcyclobutenes, 1-vinyl-3-alkynylcyclobutenes, and 1,3-divinylcyclobutenes can be prepared by the intermolecular [2 + 2] cycloaddition reactions of 1,3-enynes with alkenes, two molecules of 1,3-enynes, or 1,3-enynes with 1,3-dienes, and all the cycloaddition reactions occur with excellent regioselectivity [20].

Ogoshi's group developed a $Ni(cod)_2$-catalyzed intermolecular [2 + 2] cycloaddition of conjugated enynes or alkynes with electron-deficient alkenes [21]. As shown in Scheme 7.5, either 1,3-bis(2,6-diisopropylphenyl)imidazol-2-ylidene (IPr) or PCy_3 is required as ligand to achieve the cycloadditions, depending on the use of the substrates. When [Cp*RuCl(cod)] is used as catalyst, an asymmetric cycloaddition between electronically neutral norbornenes and electron-poor chiral alkynes occurs with high stereo- and regioselectivity affording chiral cyclobutenes in excellent levels of asymmetric induction [22].

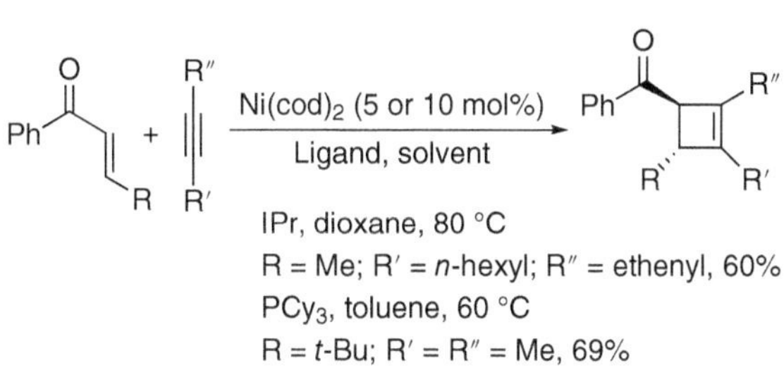

Scheme 7.5 $Ni(cod)_2$-catalyzed intermolecular [2 + 2] cycloaddition of alkynes with alkenes.

Saá and coworker have found that in the presence of $[Cp^*Ru(MeCN)_3]PF_6$, a chemoselective [2 + 2 + 2] dimerization of *o*-alkenylarylalkynes occurs to give dihydrobiphenylenes, and then the [2 + 2 + 2] cycloadditions of *o*-alkenylarylalkynes with terminal alkynes was developed to provide a new synthetic route to dihydrobiphenylenes (Scheme 7.6) [23].

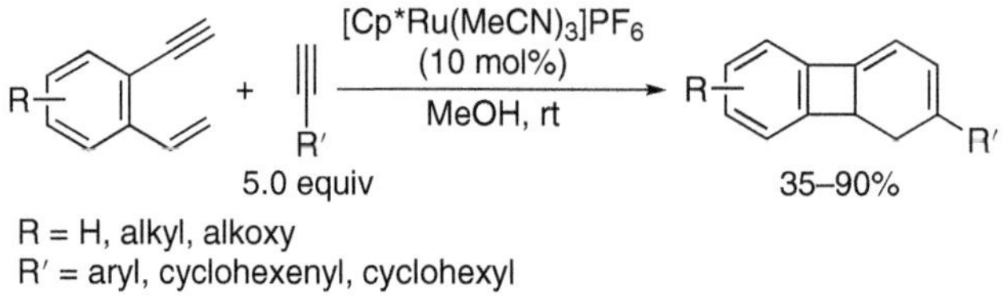

Scheme 7.6 Dihydrobiphenylene synthesis from ruthenium-catalyzed [2 + 2 + 2] cycloaddition of *o*-alkenylarylacetylenes with alkynes.

$InBr_3$-catalyzed [2 + 2] cycloaddition of internal alkynones, electron-deficient alkynes with allylsilanes, serves as an efficient method for the synthesis of cyclobutenones (Scheme 7.7) [24], and a density functional theory (DFT) study on the mechanism has also been reported [25].

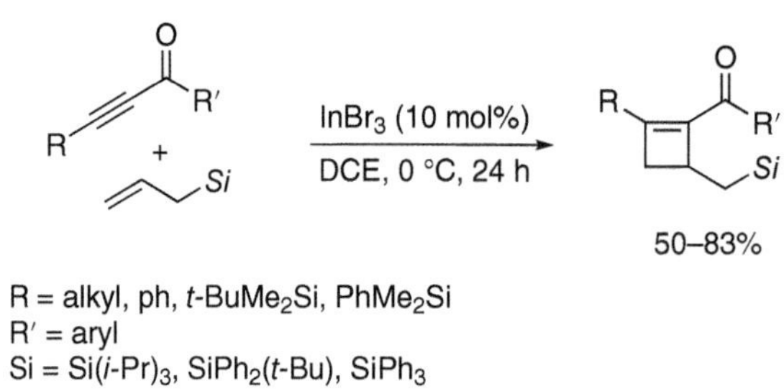

Scheme 7.7 Indium-catalyzed [2 + 2] cycloaddition of allylsilanes to internal alkynones.

$In(OTf)_3$ in $MeNO_2$ shows the catalytic activity in the intramolecular [2 + 2] cycloaddition of ene-allenones to give the strained bicyclo[n.2.0] frameworks in good to excellent yields [26]. This Lewis acid also shows highly catalytic activity for the intermolecular cyclocondensation of enynals with alkynes to construct cyclobutenes and bromoalkynes; terminal alkynes can be used as efficient substrates as well to afford the desired products in modest to high yields (Scheme 7.8) [27].

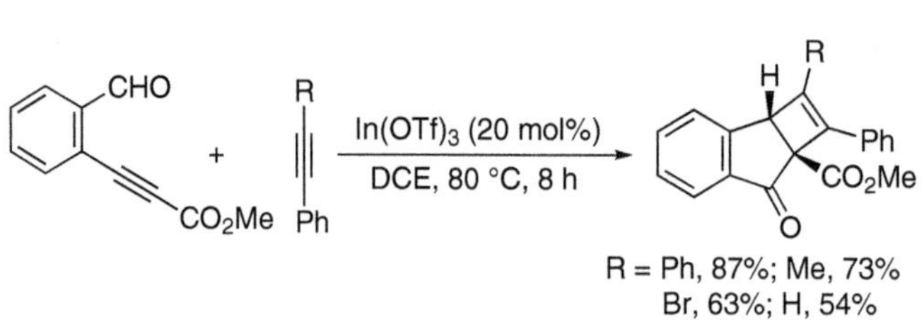

Scheme 7.8 Cyclobutene synthesis from indium-catalyzed cyclocondensation of enynals with alkynes.

In addition, a combination of In(III) salt and trimethylsilyl halide has also been used as the catalyst system in [2 + 2] cycloaddition of aryl alkynes with acrylates [28].

Tetrasubstituted cyclobutenes can also be obtained via a zirconium-induced cyclodimerization of *N*-heteroaryl-substituted alkynes in the presence of H_2O with high regio- and diastereoselectivity [29].

Various functionalized cyclobutenes have been prepared by the [2 + 2] cycloaddition of alkynes with alkenes. As shown in Scheme 7.9, bromo- and iodocyclobutenes can be prepared by palladium-catalyzed intermolecular [2 + 2] cycloaddition of aromatic alkynyl bromides and iodides with cyclooctene [30]. Bromo- and iodocyclobutenes can also be obtained by an intramolecular [2 + 2] cycloisomerization of enynes, which bear a bromide or iodide at their alkyne terminus in the presence of ruthenium complexes [31].

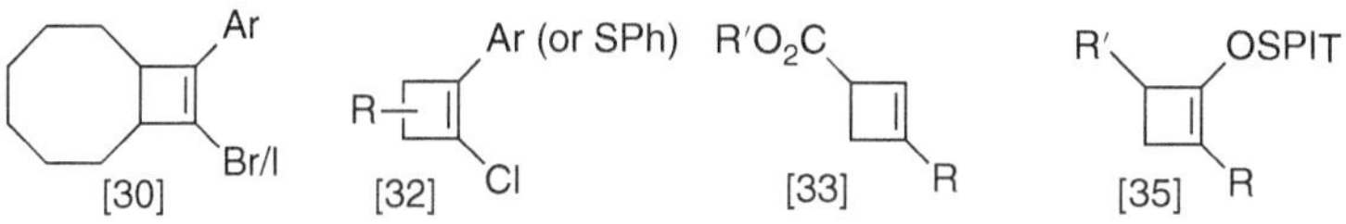

Scheme 7.9 Functionalized cyclobutenes from intermolecular [2 + 2] cycloaddition.

Gold-catalyzed [2 + 2] cycloadditions of (chloroethynyl)arenes or phenyl chloroethynyl sulfide with alkenes give 1-chlorocyclobutenes and 1-chloro-2-phenylthiocyclobutenes, respectively, which can undergo further reactions including various cross-coupling reactions via activation of $C(sp^2)$—Cl and/or $C(sp^2)$—SPh bonds [32].

Rhodium-catalyzed intermolecular [2 + 2] cycloadditions of terminal alkynes with electron-deficient terminal alkenes proceed with complete regioselectivity to form 1,3-disubstituted cyclobutenes [33], and [2 + 2] cycloadditions of ynamides with nitroalkenes provide cyclobutenamides [34].

Silver-catalyzed [2 + 2] cycloadditions of siloxy alkynes with unsaturated ketones, esters, and nitriles afford highly functionalized siloxy cyclobutenes (R′ = acyl, ester, cyano) with excellent regio- and diastereoselectivity [35].

In addition, the highly activated alkyne of ethyl 4-chloro-2-oxobut-3-ynoate shows unusual ability to furnish [2 + 2] cycloaddition with terminal alkenes in the absence of irradiation and catalysts, and in the case of 1,2-disubstituted alkenes used, $SnCl_4$ can effectively catalyze the cycloaddition reactions [36]. A regiospecific [2 + 2] cycloaddition of cyclic isoimidium salts with ynamides can also be realized under catalyst-free conditions [37].

On the other hand, the intermolecular alkyne-allene [2 + 2] cycloaddition has provided a direct synthesis of alkylidenecyclobutenes (Scheme 7.10). When either activated alkynes or activated allenes are used, the cycloaddition reactions occur smoothly under thermal conditions [38]. In the presence of a cobalt/diphosphine catalyst, intermolecular [2 + 2] cycloaddition reactions of unactivated alkynes with unactivated mono- and disubstituted allenes occur smoothly to afford a variety of 3-alkylidenecyclobutenes in good yields with high regioselectivity [39].

Scheme 7.10 Alkylidenecyclobutenes from intermolecular [2 + 2] cycloaddition of alkynes with allenes.

In addition, the enantioselective [2+2] cycloaddition using chiral complexes as catalysts affording chiral cyclobutenes with a variety of functional alkynes and alkenes has also been well developed [40].

7.1.2 Construction of Cyclobutanes

The intermolecular [2+2] homo-cyclodimerization and cross-dimerization of alkenes can construct cyclobutanes. Chirik and coworker developed an iron-catalyzed intermolecular [2+2] cycloaddition of unactivated alkenes and cross-cycloaddition of alkenes and 1,3-dienes as regio- and stereoselective routes for the synthesis of cyclobutanes and vinylcyclobutanes (Scheme 7.11) [41].

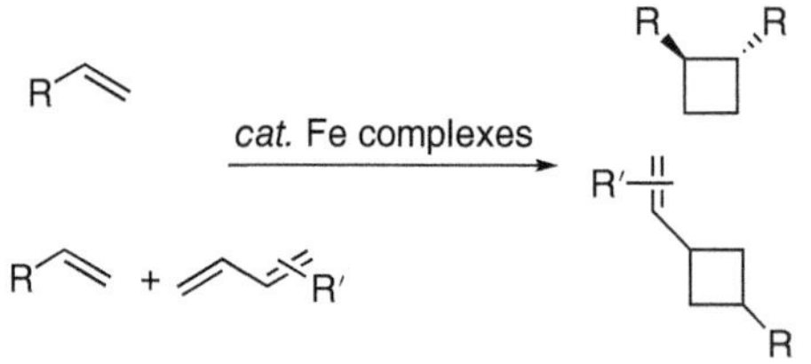

Scheme 7.11 Cyclobutanes from iron-catalyzed intermolecular [2 + 2] cycloaddition.

In the presence of ruthenium(II) [42] and iridium(I) complexes [43] as the photocatalysts, the [2+2] homo- and cross-cyclodimerization and intramolecular of styrenes upon irradiation with visible light were also reported to give substituted cyclobutanes and fused cyclobutanes (Scheme 7.12).

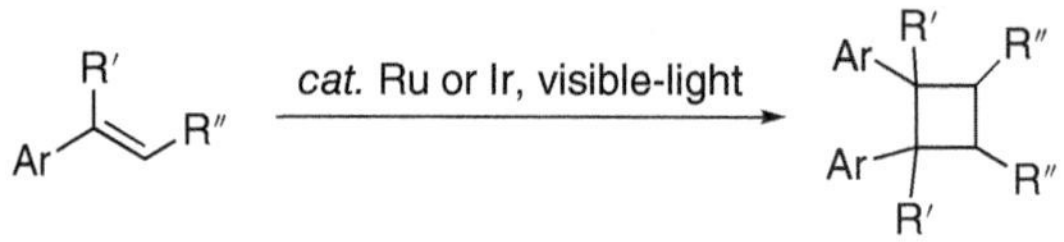

Scheme 7.12 Visible light-mediated [2 + 2] cycloadditions of styrenes.

Interestingly, Munshi and coworker have reported an ambient sunlight assisted regioselective photodimerization of acrylonitrile based electron-push/pull (Z)-alkenes under catalyst/sensitizer-free conditions to give a novel library of densely functionalized cyclobutane derivatives (Scheme 7.13) [44]. It also demonstrated stereoselective heterodimerization of selected substrates to afford highly substituted cyclobutanes.

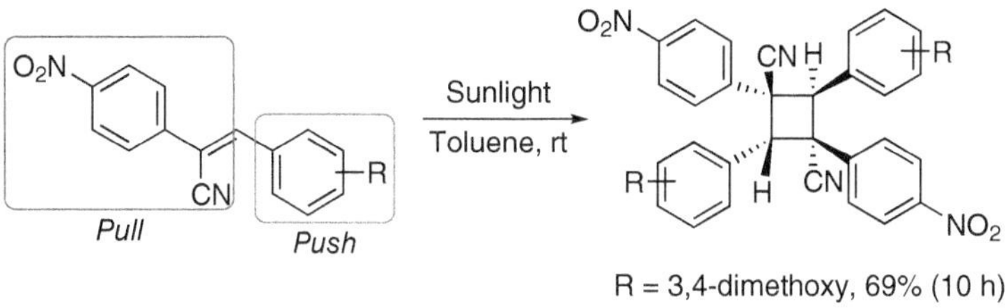

Scheme 7.13 Sunlight-promoted regioselective photodimerization of acrylonitrile.

A DFT study on the mechanism of [2 + 2] cycloadditions induced by visible light photocatalysts has also been performed [45].

Very recently, the visible light-mediated intermolecular [2 + 2] photocycloaddition of 1-aryl-2-nitroethenes with alkenes [46] and regio- and diastereoselective homo- and hetero-intermolecular [2 + 2] photocycloaddition of 4-vinylbenzoic acids photocatalyzed by quantum dots [47] have been reported.

On the other hand, the intermolecular [2 + 2] cycloaddition of alkene and allene has also been studied for the synthesis of alkylidenecyclobutanes (Scheme 7.14).

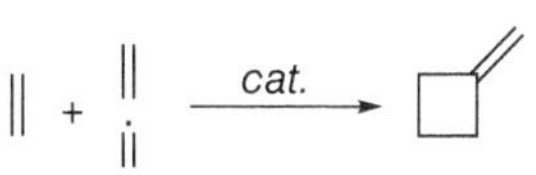

Scheme 7.14 Alkylidenecyclobutanes from intermolecular [2 + 2] cycloaddition of alkenes with allenes.

As shown in Scheme 7.15, in the presence of Lewis acids of $EtAlCl_2$, $AlCl_3$, the alkene–allenoate cycloaddition reactions affording methylene cyclobutanes was reported [48]. Bismethylenecyclobutanes can be prepared by a highly regioselective head-to-head [2 + 2] cyclodimerization of electron-deficient terminal allenes catalyzed by nickel(0) complexes [49]. A highly regio- and stereoselective rhodium-catalyzed intermolecular head-to-head [2 + 2] cycloaddition of allenamides affording *trans*-dimethylenecyclobutane-1,2-diamine derivatives has also been developed [50].

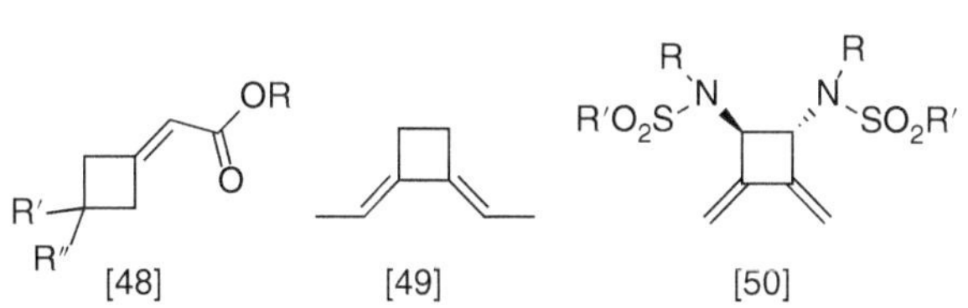

Scheme 7.15 Functionalized cyclobutanes from intermolecular [2 + 2] cycloaddition.

The enantioselective allenoate-alkene [2 + 2] cycloaddition is also developed in the presence of *N*-protonated oxazaborolidine at room temperature in CH_2Cl_2 by Brown's group [51].

$PPh_3AuCl/AgBF_4$-catalyzed intramolecular ene-allene [2 + 2] cycloaddition produces cyclopentane-fused methylenecyclobutanes (Scheme 7.16) [52]. Ruthenium

[53] and nickel complexes [54] have also shown the catalytic activity for this type of cycloaddition.

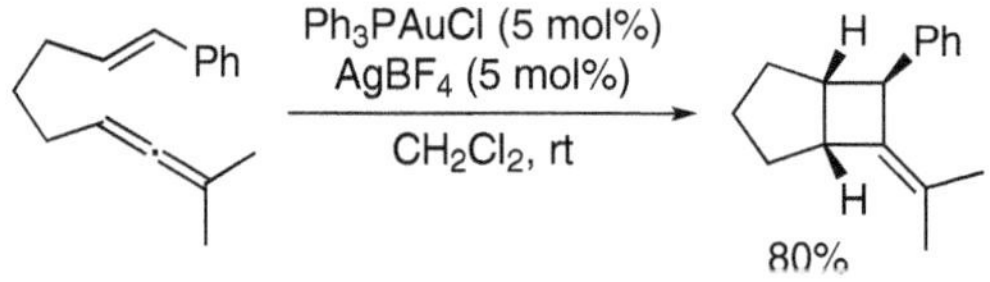

Scheme 7.16 Ene-allene intramolecular [2 + 2] cycloaddition.

The asymmetric [2 + 2] cycloaddition reactions of alkenes catalyzed by chiral titanium complexes [55] to give the corresponding cyclobutanes in high enantioselectivity have also been developed.

In addition, an asymmetric [2 + 2] photocycloaddition of α,β-unsaturated ketones to the corresponding cyclobutane derivatives was developed with the use of $Ru(bpy)_3Cl_2$ as the visible light-absorbing transition metal photocatalyst and $Eu(OTf)_3$/chiral ligand as the stereocontrolling Lewis acid cocatalyst [56]. The enantioselective [2 + 2] cycloaddition of phenyl vinyl sulfide with α-substituted acroleins could be achieved by BBr_3-assisted chiral phosphoric acid catalysts [57].

Moreover, a regioselective and stereoselective [2 + 2] cycloaddition of styrene derivatives using a heterogeneous organic photocatalyst under the irradiation of visible light has also been developed with high conversion and selectivity [58].

7.2 Five-Membered Carbocycles

The known procedures of [2 + 2 + 1] and [3 + 2] cycloaddition reactions of alkynes and/or alkenes for the formation of five-membered carbocycles are relatively few compared to other small-sized carbocycles with the use of unsaturated hydrocarbons. The formation of five-membered carbocycles by activation of C—C bonds of cyclopropanes or cyclopropenes and their cycloaddition reactions with alkynes and alkenes are summarized in Chapter 6, Section 6.6.1. Recently, a review paper has reported the synthesis of indane (benzocyclopentane) and indene (benzocyclopentene) [59].

7.2.1 Five-Membered Carbocycles by [2 + 2 + 1] Cycloaddition

Transition metal-catalyzed [2 + 2 + 2] homo- and cross-cyclotrimerization of alkynes has become one of the most important protocols to construct benzene rings, which will be summarized in Section 7.3. However, only a few examples of the catalytic [2 + 2 + 1] homo-cyclotrimerization of alkynes afford *multi*-substituted fulvenes with high chemoselectivity catalyzed by titanium [60] and palladium(II) complexes [61] (Scheme 7.17). The first example of the catalytic [2 + 2 + 1] cross-cyclotrimerization of two different alkynes to produce substituted silylfulvenes was reported by Tanaka's group in 2011, with the use of triisopropylsilylacetylene and two alkynyl esters (Scheme 7.18) [62]. The use of other bulky silylacetylenes, such as *t*-butyldimethylsilylacetylene, *t*-butyldiphenylsilylacetylene, triethylsilylacetylene, and alkynyl amides, resulted in no reaction or low yields of the desired carbocycles.

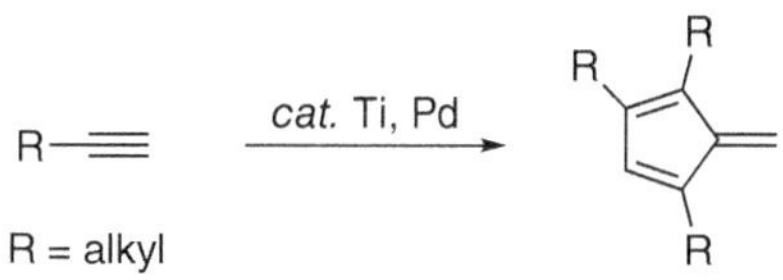

Scheme 7.17 Transition metal-catalyzed homo-cyclotrimerization of alkynes constructing fulvenes.

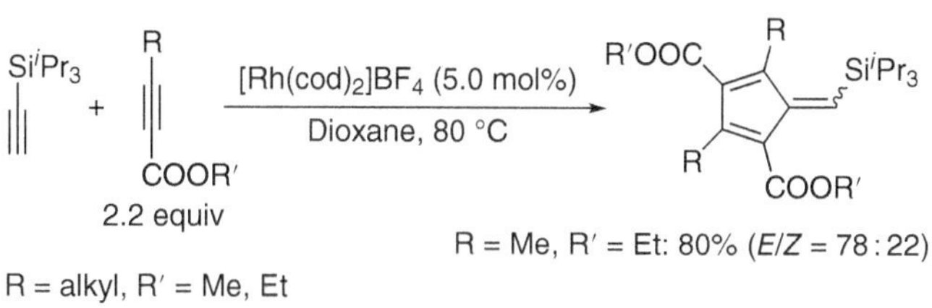

Scheme 7.18 Cross-cyclotrimerization of triisopropylsilylacetylene with alkynyl esters.

Li and coworker demonstrated an efficient $[Cp^*RhCl_2]_2/AgNTf_2$-catalyzed [2 + 2 + 1] cyclo-coupling reactions of quinoline-8-carbaldehydes with two internal alkynes to provide a straightforward access to various pentasubstituted cyclopentadienols (Scheme 7.19) [63]. Mechanistic studies suggest that the formal rearrangement of the hydroxyl group may result from a dehydration–rehydration process with thermodynamics as the driving force.

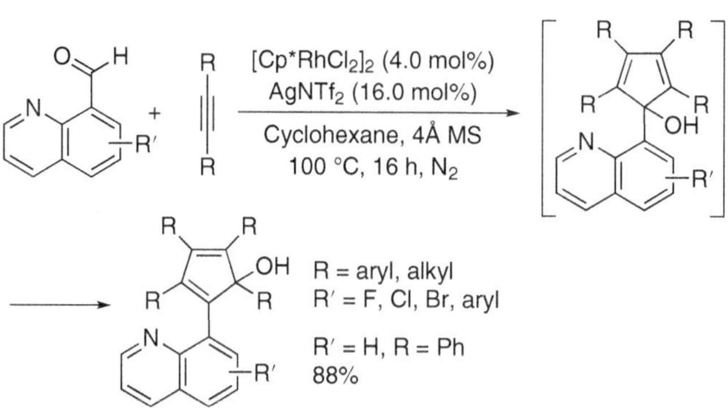

Scheme 7.19 Rhodium-catalyzed synthesis of cyclopentadienols via a [2 + 2 + 1] carbocyclization of 8-formylquinolines with internal alkynes.

7.2.2 Five-Membered Carbocycles by [3 + 2] Cycloaddition

With the use of cationic rhodium(I) complex of $[Rh(cod)_2]SbF_6$, Tanaka and coworker developed a [3 + 2] cycloadditions of alkoxycarbonyl-substituted propargyl esters with electron-deficient alkynes to give fully substituted cyclopentadienes in good to high yields (Scheme 7.20) [64]. With two aromatic terminal alkynes, the cycloaddition reactions led to the formation of substituted dihydropentalenes with elimination of carboxylic acid (Scheme 7.21) [65].

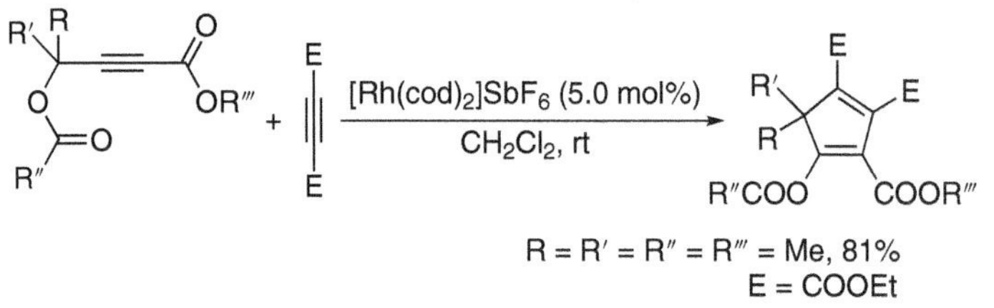

Scheme 7.20 $[Rh(cod)_2]SbF_6$-catalyzed [3 + 2] cycloadditions of alkoxycarbonyl-substituted propargyl esters with electron-deficient alkynes.

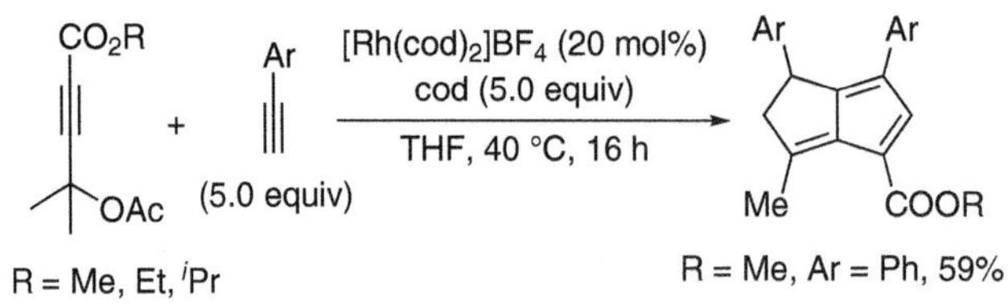

Scheme 7.21 $[Rh(cod)_2]BF_4$-catalyzed cotrimerization of propargyl esters with terminal alkynes affording dihydropentalenes.

In the presence of LiBr, $Pd(OAc)_2$-catalyzed intermolecular [3 + 2] carbocyclization of alkynols with propiolates provides an efficient entry to bromo-substituted cyclopentadienes (Scheme 7.22) [66]. Instead of LiBr, the use of $CuCl_2{\cdot}2H_2O$ resulted in the formation of the expected chlorinated cyclopentadienes.

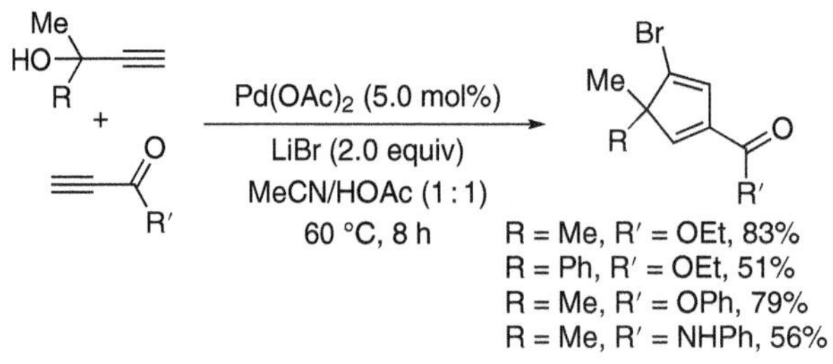

Scheme 7.22 $Pd(OAc)_2$-catalyzed intermolecular [3 + 2] cycloaddition of alkynols with electron-deficient alkynes.

In the presence of $AgSbF_6$, Chen and coworker developed an atom- and step-efficient construction of highly substituted cyclopentenes via intermolecular [3 + 2] cyclization of alkenes with alkynes; alkenes serve as C3 sources (Scheme 7.23) [67].

On the other hand, the procedures for the formation of benzo-fused five-membered carbocycle from the annulation of alkynes have also been extensively studied, including the construction of indenes, indenols, and benzofulvene.

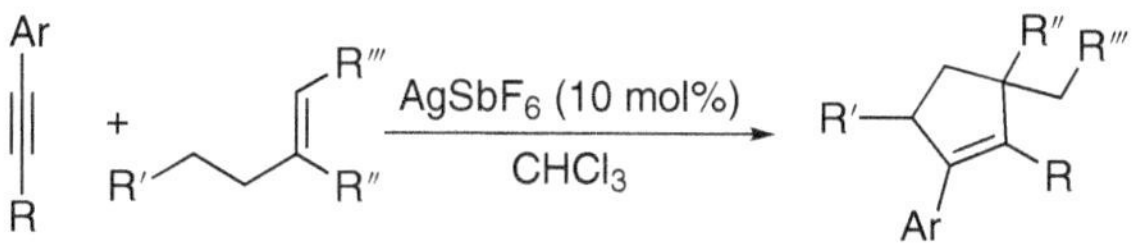

Scheme 7.23 Five-membered carbocycles from cyclization of alkynes and alkenes.

Larock and other groups developed three different synthetic methods for preparing substituted indenes by the transition metal-catalyzed carboannulation of alkynes [68]. As shown in Scheme 7.24, the first method involves $Pd(OAc)_2$-catalyzed carboannulation of aryl halides with internal alkynes with excellent regioselectivity, and is particularly suited for the synthesis of hindered indenes with single regioisomer bearing the bulky group in the second-position of the indene ring. The second synthetic method is the CuI-catalyzed intramolecular cyclization of arylated alkynes to give 2-substituted indenes, and the third synthesis of highly substituted indenes involves the $Pd(OAc)_2$-catalyzed cross-coupling of arylated alkynes with various aryl halides.

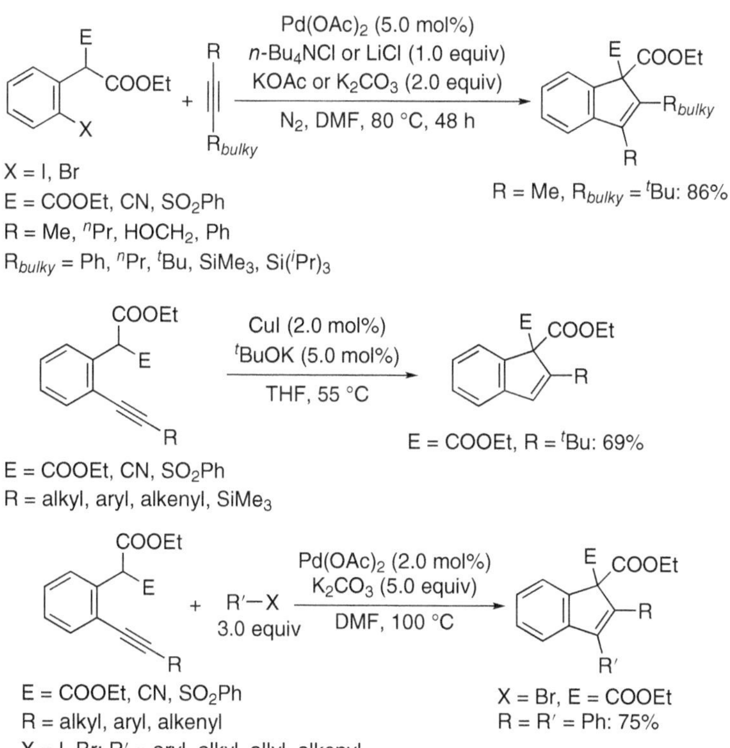

Scheme 7.24 Construction of indenes by carboannulation of alkynes.

3-Iodoindene derivatives can be synthesized by iodonium-promoted 5-*endo*-dig carbocyclization of 2-substituted ethynylmalonates [69]. The substituted indenes could also be prepared from the cyclocondensation of benzylic alcohols with internal alkynes catalyzed by $FeCl_3{\cdot}6H_2O$ [70], *ortho*-bromobenzyl zinc bromide with terminal and internal alkynes catalyzed by $NiI_2(PPh_3)_2$ [71], internal alkynes with hindered Grignard reagents catalyzed by $Pd(OAc)_2$ [72], arylated alkynes with naphthalenesulfonyl chloride catalyzed by CuCl [73], arylated alkynes with diazoester catalyzed by CuI [74], and aldimines with enones catalyzed by rhodium(III) complexes [75].

In addition, substituted indenes can also be obtained via the intramolecular addition strategy; for example, the intramolecular addition of triarylmethanes to alkynes promoted by KO^tBu/dimethylfomamide (DMF) [76], intramolecular Michael reaction of styrenes catalyzed by $FeCl_3{\cdot}6H_2O$ [77], and intramolecular cyclization of *ortho*-vinylcinnamates mediated by $FeCl_3$ [78].

An efficient procedure for the synthesis of optically active indenes has also been developed by the cyclocondensation of *ortho*-boronate-substituted cinnamic ketones with internal alkynes catalyzed by $Pd(OTf)_2{\cdot}2H_2O$/chiral ligands [79].

It has been well-known that the structures of indenols and benzofulvene are found in organic materials and medicinal molecules, thus various synthetic approaches include the transition metal-catalyzed [3 + 2] annulation of aryl ketones with internal alkynes via C—H activation catalyzed by transition metal complexes (Scheme 7.25). Rhodium [80], iridium [81], ruthenium [82], and cobalt [83] have been found to have high catalytic activity for this type of transformation.

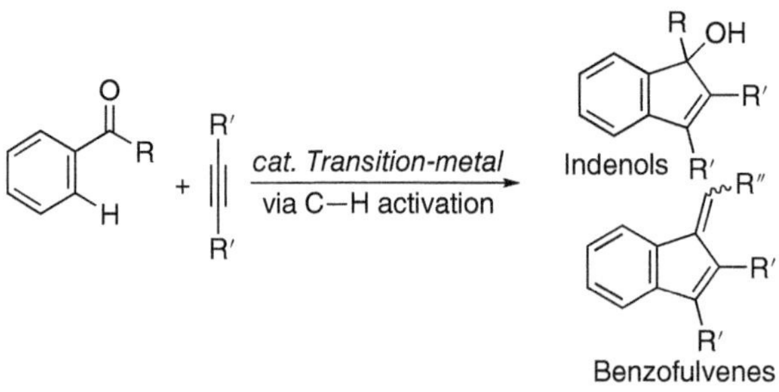

Scheme 7.25 Indenols and benzofulvenes synthesis via [3 + 2] annulation between aromatic ketones and alkynes.

The $[Rh(OH)(cod)]_2$-catalyzed annulations of *o*-carbonylated arylboronic acids with internal alkynes show high regioselectivity to produce indenols bearing bulky and electron-withdrawing groups in second-position (Scheme 7.26) [84]. Combined $[Ru(p\text{-cymene})Cl_2]_2$, $AgSbF_6$, and $TsNH_2$ as catalyst system, the cyclization between *N*-sulfonyl imines of benzaldehydes and internal alkynes produces indenamines (Scheme 7.27) [85]. When aryl- and alkyl-substituted alkynes are used, 2-arylindenamines are the major products. In addition, in the presence of $[IrCl(cod)]_2$, [3 + 2] annulation of ketimines with alkynes also gives indenamines [86].

Recently, Hua and coworker have also established an efficient and simple approach to benzofulvenes by a direct cyclodimerization of diarylacetylenes via

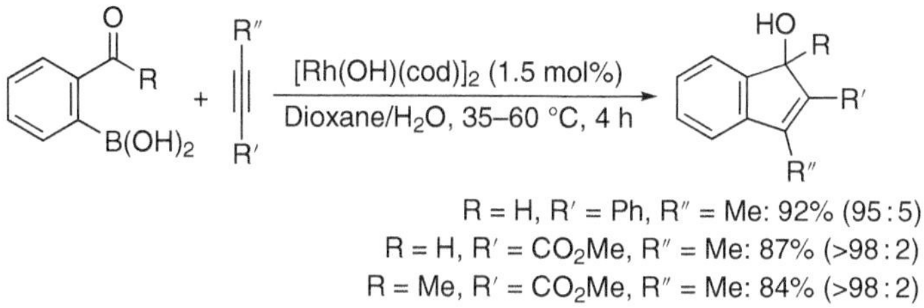

Scheme 7.26 $[Rh(OH)(cod)]_2$-catalyzed regioselective synthesis of indenols.

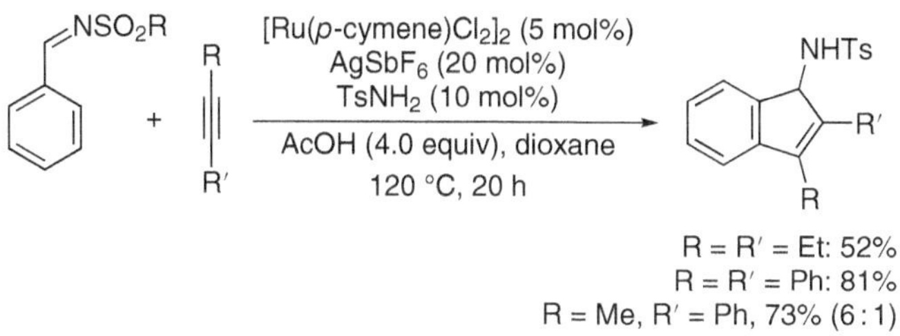

Scheme 7.27 Ruthenium-catalyzed cyclization of *N*-sulfonyl imines of benzaldehydes with internal alkynes.

$Pd(OAc)_2$-catalyzed alkyne-directed *ortho* C—H activation followed by cyclization in DMAc with the use of LiBr, HOPiv, and NaOAc as the necessary additives to achieve good to high yields (Scheme 7.28) [87].

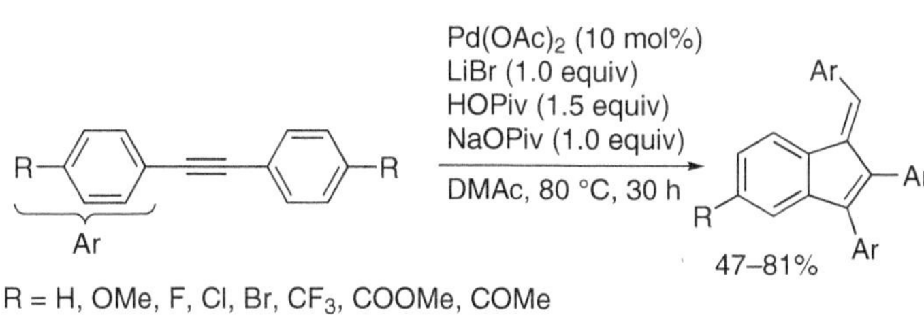

Scheme 7.28 Synthesis of benzofulvenes via $Pd(OAc)_2$-catalyzed cyclodimerization of diarylacetylenes.

$PdCl_2/PPh_3$-catalyzed reactions of (*E*)-3-(2-bromophenyl) acrylaldehyde with internal alkynes are also the efficient catalytic procedures for the synthesis of substituted benzofulvenes [88].

The functionalized indenes and benzofulvenes can also be synthesized through gold-catalyzed cycloisomerizations of *ortho*-(alkynyl)styrenes [89] and iodocyclization of *ortho*-(alkynyl)styrenes [90].

On the other hand, allenes are the suitable C3 sources for the formation of five-membered carbocycles. Phosphine-catalyzed [3 + 2] annulation of allenes with electron-deficient alkenes was first reported by Lu and coworker in 1995 [91], which then has been developed to be a powerful approach for the construction of functionalized cyclopentenes (Scheme 7.29) [92]. In the case of chiral phosphine used, diverse optically active cyclopentenes can be easily obtained [93].

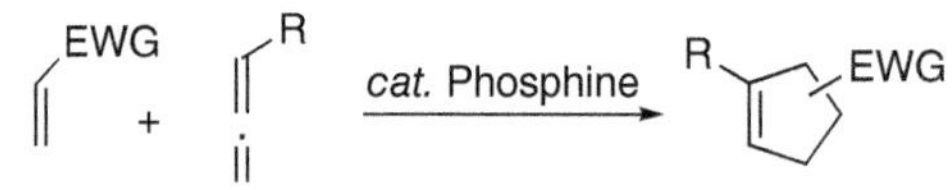

EWG = electron-withdrawing group

Scheme 7.29 Phosphine-catalyzed [2 + 3] annulation of electron-deficient alkenes with allenes affording cyclopentenes.

7.2.3 Intramolecular Cycloaddition of Active sp^3 C—H to Carbon–Carbon Unsaturated Bonds

The active C—H bonds in β-dicarbonyl compounds, such as β-ketoester, β-ketoaldehyde, and β-ketoamide, have been well applied in the construction of five-membered and fused five-membered carbocyclic compounds via either intermolecular or intramolecular cycloaddition of active C—H bond to carbon–carbon unsaturated bond.

For example, Toste and coworker developed an Au(I)-catalyzed 5-*endo*-dig carbocyclization of acetylenic β-ketoester (Scheme 7.30) [94]. Under the optimized reaction conditions, β-diketones also undergo efficient carbocyclization giving cyclopentene derivatives and alkenyl cyclopentanes [95]. The first enantioselective intramolecular carbocyclization of acetylenic β-ketoester in the presence of Pd(II)/Yb(III) complexes was also reported by the same group [96].

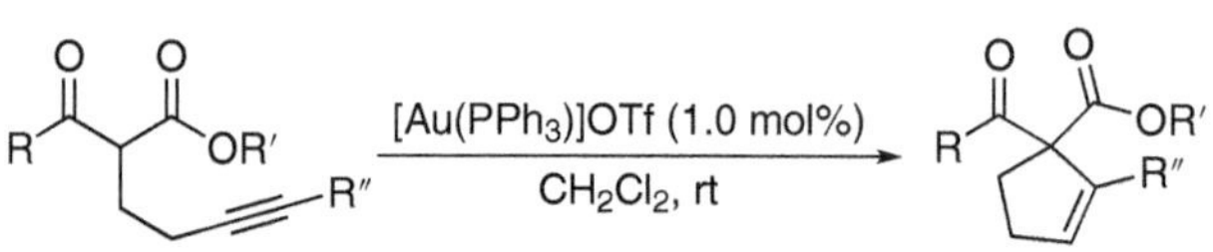

R = alkyl, Ph; R′ = Me, Et
R″ = H, alkyl, Ph, vinyl

74–99%

Scheme 7.30 Gold(I)-catalyzed 5-*endo*-dig carbocyclization of acetylenic β-ketoester.

In addition, $Ni(acac)_2/Yb(OTf)_3$ [97] and $FeCl_3$ [98] show high catalytic activity in the similar carbocyclization of acetylenic β-ketoester giving various mono- and bicyclic alkenyl cyclopentanes in a highly regioselective manner.

Iodocyclopentenes were formed at room temperature by the iodocarbocyclization of δ-alkynyl-β-ketoesters with I_2, and the terminal and substituted alkynes are used (Scheme 7.31) [99]. This report presents the first example of the iodonium-promoted 5-*endo*-dig carbocyclization of active C—H bond to alkynes.

Che and coworker have demonstrated that the gold(I) complex can efficiently catalyze the intramolecular cycloaddition of β-ketoamides to unactivated alkenes to afford highly substituted lactams in high yields with excellent regioselectivity under mild conditions (Scheme 7.32) [100]. The proposed mechanism involves the

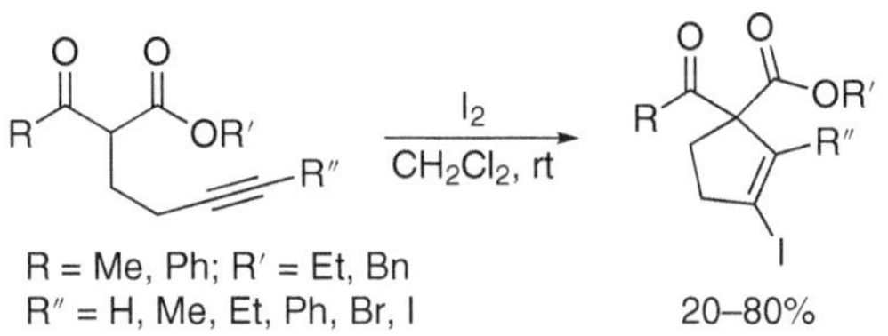

Scheme 7.31 Iodocyclopentene formation via iodocarbocyclization of β-ketoesters and alkynes.

cationic gold(I) coordinating to alkene and the *exo*-trig addition of the enol form of β-ketoamide.

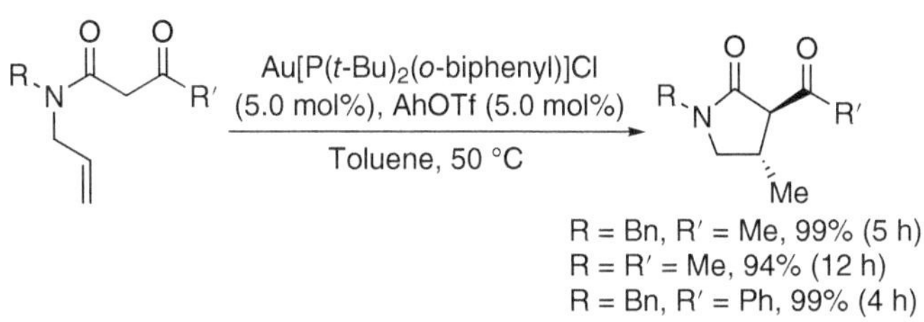

Scheme 7.32 Lactam formation via gold(I)-catalyzed intramolecular cycloaddition of β-ketoamide to alkenes.

Intramolecular 5-*exo*-dig cyclization via hydrocarbonation of alkyne is one of the efficient ways for the construction of five-membered carbocycles [101]. As shown in Scheme 7.33, the direct cyclization of formyl alkynes catalyzed by the combined action of an amine and a cationic gold(I) complex affords either methylene cyclopentane or cyclopentene derivatives, depending on the substituents in the starting material.

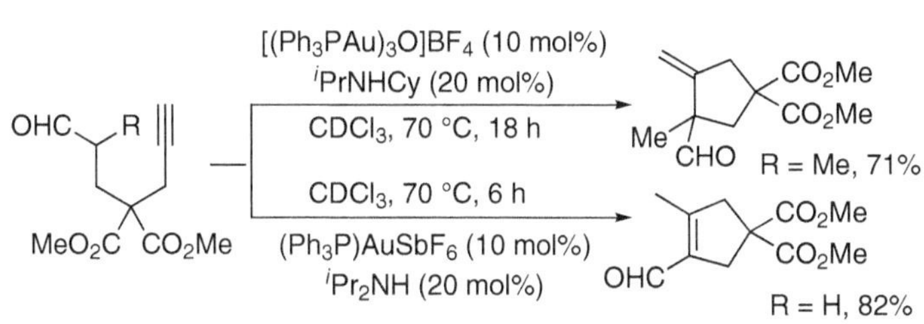

Scheme 7.33 Cyclopentane and cyclopentene formation via intramolecular carbocyclization of formyl alkyne.

Combined use of $In(OTf)_3$ and Tf_2NH effectively promotes the cyclization of alk-5-ynyl ketones to cyclopent-1-enyl ketones via the Markovnikov sp^3 C—H bond addition to alkynes (Scheme 7.34) [102]. In some cases, cyclohept-2-enones were also obtained mainly resulting from the intramolecular *anti*-Markovnikov addition depending on the substrates and reaction conditions.

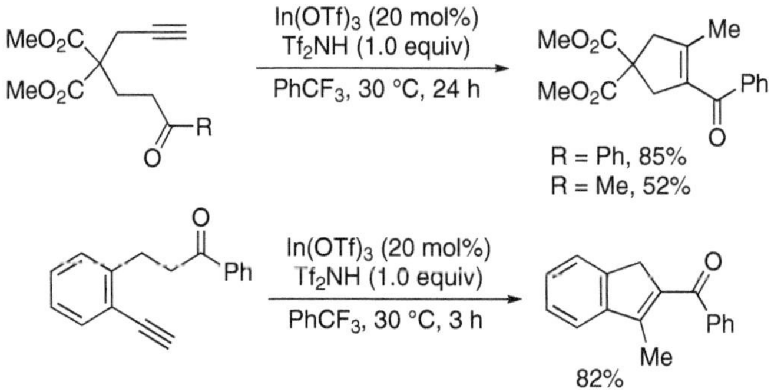

Scheme 7.34 $In(OTf)_3/Tf_2NH$-promoted cyclization of alk-5-ynyl ketones via C—H addition to alkynes.

Rhodium-catalyzed selective dimerization of aromatic terminal alkynes producing 1,3-enynes, and the subsequent cycloaddition of double C—H bonds of malonates forming substituted cyclopentenes were also developed (Scheme 7.35) [103]. The one-pot reaction using palladium-catalyzed alkyne dimerization in conjunction with rhodium-catalyzed addition of malonates was also possible, and the yields of cyclopentenes were slightly improved.

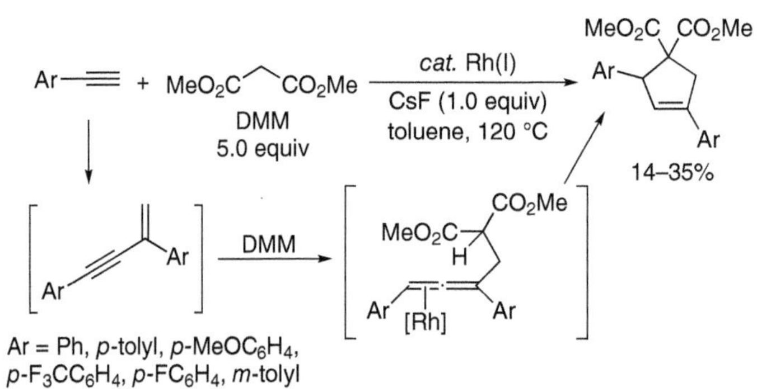

Scheme 7.35 Cyclopentene formation via one-pot sequential rhodium-catalyzed reactions between aromatic terminal alkynes and dimethyl malonate.

The intramolecular hydrocarbonation of allenes bearing active C—H bonds catalyzed by $[(\eta^3\text{-}C_3H_5)PdCl]_2$/dppf is also the efficient way for the construction of five-membered carbocycles [104].

7.2.4 Five-Membered Carbocycles from Intramolecular Cycloaddition of Unsaturated Bonds

Chloro-/bromo-substituted benzofulvenes could be prepared by a PdX_2 (X = Cl, Br)-catalyzed intramolecular cyclization of 1,2-dialkynylbenzenes in the presence of CuX_2 in MeCN (Scheme 7.36) [105] or $Pd(OAc)_2$-catalyzed intramolecular cyclization of *ortho*-alkynylbenzylidene ketones in the presence of CuX_2 (Scheme 7.37) [106] or LiX [107].

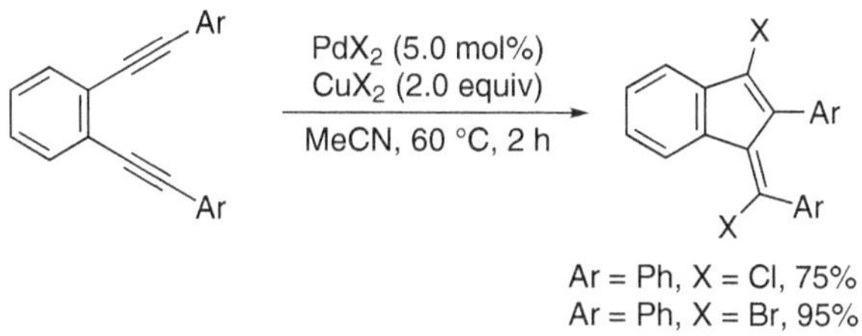

Scheme 7.36 Synthesis of benzofulvenes via palladium-catalyzed cyclization of 1,2-dialkynylbenzenes.

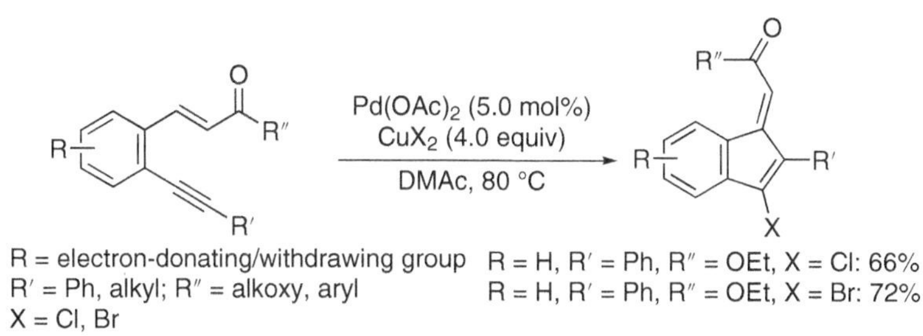

Scheme 7.37 $Pd(OAc)_2$-catalyzed synthesis of 3-halo-benzofulvenes from 2-alkenylphenyl-acetylenes in the presence of copper(II) halide.

In addition, iodo-substituted benzofulvenes were obtained by iodine-mediated intramolecular electrophilic iodocyclization of 1,2-dialkynylbenzenes promoted by K_2CO_3 under transition metal-free conditions [108]. Iodine(III)-induced regioselective intramolecular carbocyclization of 2-(2-ethynylbenzyl)malonates provides a facile approach to 1,1-diiodomethylene-substituted indanes [109].

7.3 Six-Membered Carbocycles

This section is limited to highlighting the recent progress on the construction of benzene, naphthalene, polyaromatic hydrocarbons (PAHs), and 1,3-cyclohexadiene rings via [2 + 2 + 2] and [4 + 2] annulation using alkyne as the reaction partner in one-pot manner.

7.3.1 Benzene Ring Formation

Alkyne usually provides two carbons to undergo the annulations for construction of benzene ring, thus the transition metal-catalyzed [2 + 2 + 2] cyclotrimerization of alkynes is the most extensively investigated for the synthesis of benzene derivatives in one step. In 1948, Reppe and coworkers first reported the example of [2+2+2] cyclotrimerization of alkynes in the presence of homogeneous nickel complexes [110], and then a variety of transition metal complexes-mediated or catalyzed procedures including nickel [111], palladium [112], rhodium [113], cobalt [114], ruthenium [115], iridium [116], iron [117], and others [118] for the synthesis of substituted benzenes and bicyclic benzenes have been developed, and the detailed advances have been summarized in a few review papers [119].

With the use of cocatalysts of Zn and ZnI_2, in CH_3CN, anhydrous $CoBr_2$ can catalyze [2+2+2] cyclotrimerization of phenylacetylene in almost quantitative yield with excellent regioselective formation of asymmetrical 1,2,4-triphenylbenzene (Scheme 7.38) [120]. The addition of dicyclohexyldiimine as ligand could improve the catalytic activity, and similar results could be obtained even with a decrease of reaction time to 15 minutes. Further studies disclosed that the regioselective formation of trisubstituted benzenes was greatly solvent dependent, and the cyclotrimerization of phenylacetylene in $CHCl_2$ to replace CH_3CN afforded the symmetrical 1,3,5-triphenylbenzene as major product (up to 85%) [121].

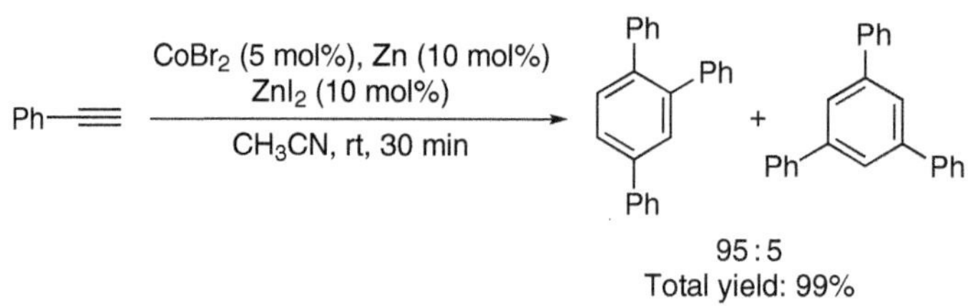

Scheme 7.38 [2 + 2 + 2] Cyclotrimerization of phenylacetylene catalyzed by $CoBr_2$ activated by Zn and ZnI_2.

An alternative cobalt complex catalytic system, $CoCl_2{\cdot}6H_2O/Zn$, with the use of 2-(2,6-diisopropylphenyl) iminomethylpyridine as ligand also shows high catalytic activity in the intermolecular cyclotrimerization of terminal alkynes to regioselectively yield 1,2,4-trisubstituted benzenes as major products. In the case of phenylacetylene used, the regioselectivity of 1,2,4-triphenyl benzene is up to 99% (Scheme 7.39) [122].

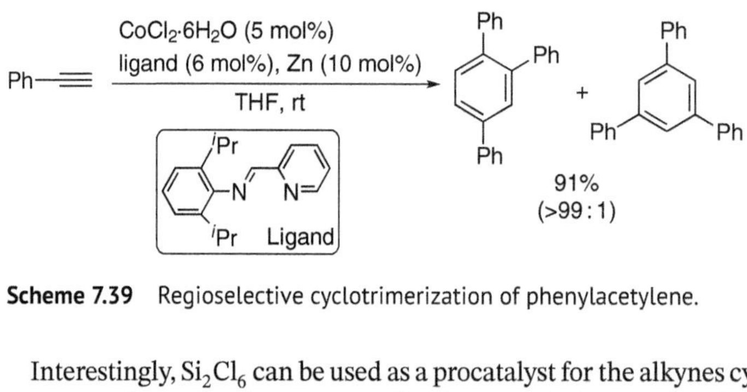

Scheme 7.39 Regioselective cyclotrimerization of phenylacetylene.

Interestingly, Si_2Cl_6 can be used as a procatalyst for the alkynes cyclotrimerization at 170–200 °C to give the corresponding benzene derivatives [123]. $InCl_3$ can catalyze the cyclotrimerization of terminal alkynes in the presence of 2-iodophenol affording 1,3,5-trisubstituted benzenes in excellent yields with complete regioselectivity [124], and $FeCl_2$ ligated with 2-(benzimidazolyl)-6-(1-[arylimino]ethyl)pyridine in the presence of zinc powder and zinc iodide can also effectively catalyze regioselective cyclotrimerization of terminal alkynes to 1,2,4-trisubstituted benzenes [125]. Similar catalyst system of $FeCl_2$/DPPP/Zn shows good catalytic activity in the [2 + 2 + 2] intermolecular cycloaddition of the unsymmetrical 2-trifluoromethyl aromatic alkynes to afford the corresponding trifluoromethyl group substituted benzenes in high yield with excellent selectivity [126].

In addition, under an oxygen atmosphere, $PdCl_2$ [127] and $Pd(OAc)_2/HPMo_8V_4/CeCl_3$ [128] show good catalytic activity for the highly regioselective cyclotrimerization of electron-deficient terminal alkenes to 1,3,5-trisubstituted benzenes, and $PdCl_2(HNMe_2)_2$ can catalyze the cross [2 + 2 + 2] cyclization of alkynoates with alkenes affording pentasubstituted benzenes [129].

Polysubstituted benzenes can be also synthesized from the intermolecular dehydrative cycloaddition reaction of 1,3-dicarbonyl compounds with alkynes. For example, in the presence of dimethyl aminopyridine (DMAP), in 1,2-dimethoxyethane (DME), β-ketoesters and DMAD undergo a sequent tandem addition–cyclization–dehydration reaction leading to polysubstituted benzenes in good yields (Scheme 7.40) [130].

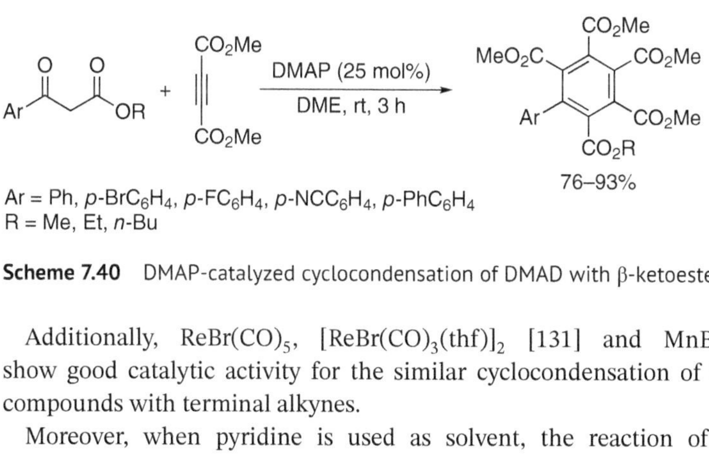

Scheme 7.40 DMAP-catalyzed cyclocondensation of DMAD with β-ketoesters.

Additionally, $ReBr(CO)_5$, $[ReBr(CO)_3(thf)]_2$ [131] and $MnBr(CO)_5$ [132] show good catalytic activity for the similar cyclocondensation of 1,3-dicarbonyl compounds with terminal alkynes.

Moreover, when pyridine is used as solvent, the reaction of DMAD with cyclobutene-1,2-diones affords hexasubstituted benzenes [133], and the cyclocondensation of alkynyl ketones with 1,3-dicarbonyl compounds in the presence of NaOH in DMF produces polysubstituted phenols [134].

In CH_3CN, $Ni(dppe)Br_2$/Zn was found to be an efficient catalyst system for highly regio- and chemoselective [2 + 2 + 2] cycloaddition of two molecules of propiolates with one molecule of allene to provide polysubstituted benzenes in good to excellent yields [135]. The cycloaddition process features the use of allene as equivalent to terminal alkyne, and cyclohexadiene is proposed to be the key intermediate. The further application of this catalyst system is in the cycloaddition of nonconjugated diynes with allenes [136] or 1,4-disubstituted-1,3-diynes [137] to afford bicyclic polysubstituted benzenes (Scheme 7.41) and aryl alkynes (Scheme 7.42). In addition, another cheap catalyst system of $CoI_2(PPh_3)_2$/Zn also shows highly catalytic activity for the [2 + 2 + 2] cycloaddition of 1,6-heptadiynes with allenes [138].

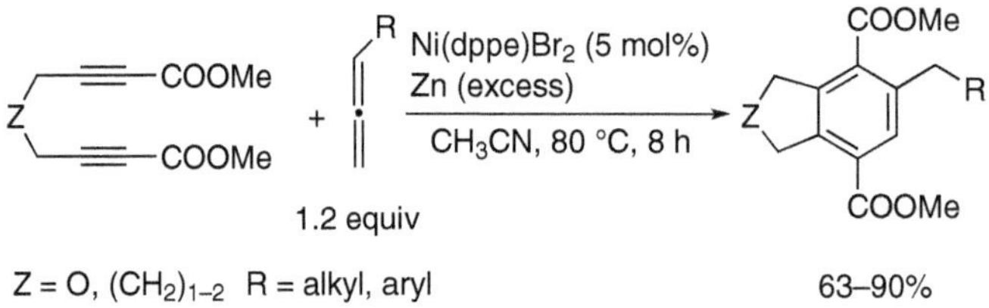

Scheme 7.41 Cycloaddition of propiolates with allene.

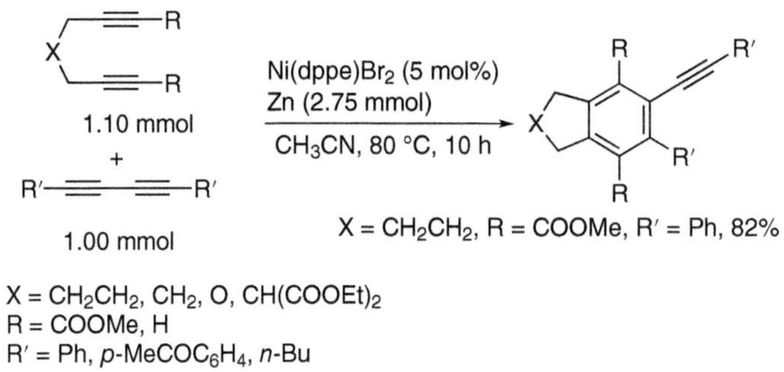

Scheme 7.42 Synthesis of arylalkynes from nonconjugated diynes and 1,3-diynes.

The nickel catalytic system of $Ni(acac)_2/PPh_3$/DIBAL-H can catalyze the [2 + 2 + 2] cocyclization of arylated alkynes with two equivalents of acetylene gas (balloon) or α, ω-diyne having an aryl group at α-position with one equivalent of acetylene gas affording biaryls in good to excellent yields [139].

In addition, $[PhCH_2NRu(PCy_3)_2Cl_2]$ [140], $CoCl_2$/IPr/Zn [141], and $FeCl_3$/Zn combined with imidazol-2-ylidene or 2-iminomethylpyridine as ligand [142] have also been confirmed to be the efficient catalyst systems for the intramolecular cyclotrimerization of triynes to annulated benzenes.

$Pd_2(dba)_3/PPh_3$ catalyst system shows good catalytic activity not only in the intermolecular [2 + 2 + 2] cycloaddition of nonconjugated diyne with DMAD giving highly substituted bicyclic benzenes but also in the intramolecular cyclization of triynes to afford the expected tricycle (Scheme 7.43) [143]. A palladium(0) complex

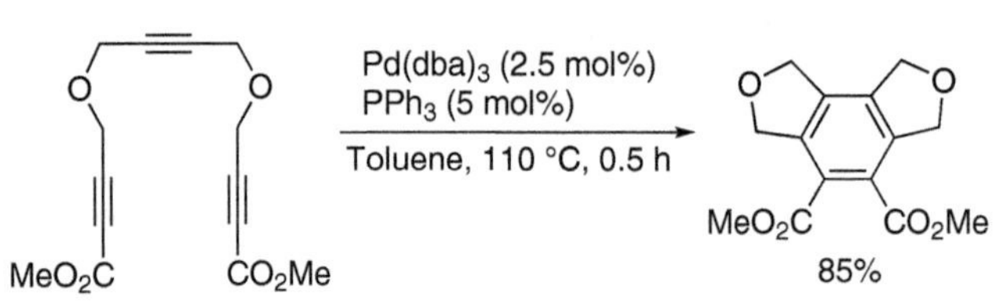

Scheme 7.43 Palladium(0)-catalyzed intramolecular cyclization of triyne.

of a linear 1,6,11-triyne could be prepared by the reaction of $Pd_2(dba)_3 \cdot CHCl_3$ with dimethyl 4,9-dioxatrideca-2,7,12-triyne-1,13-dioate, which can be used as an efficient catalyst precursor for the cyclization of triyne and considered to be the reactive intermediate (Scheme 7.44) [144]. In addition, the chiral tricycle can be prepared by using a chiral iridium catalyst via an enantioselective intramolecular [2 + 2 + 2] cycloaddition of triynes [145].

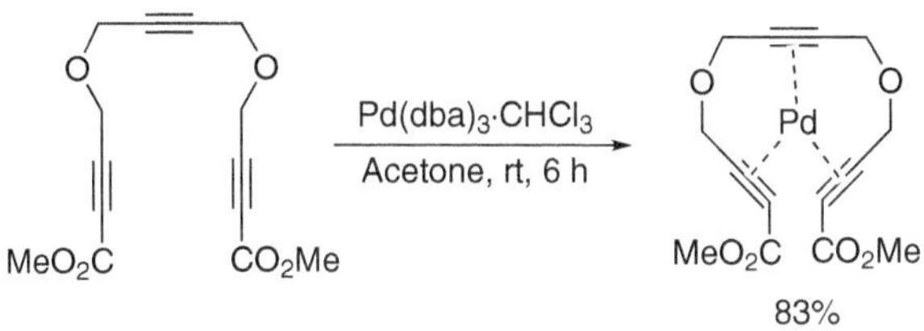

Scheme 7.44 Formation of palladium(0) complex of a linear 1,6,11-triyne.

Although the catalytic activity of $[Ir(cod)Cl]_2$ in the cycloaddition of 1,6-diynes with monoalkynes depends on the ligand employed, the use of dppe resulted in the formation of polysubstituted bicyclic benzenes in 84% yield at room temperature (Scheme 7.45) [146].

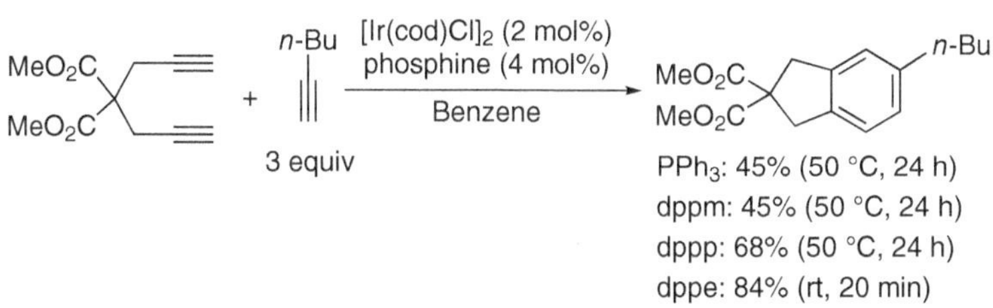

Scheme 7.45 Effect of phosphines on the [2 + 2 + 2] cycloaddition of alkynes.

Cp*RuCl(cod) shows high catalytic activity for the intermolecular [2+2+2] cycloaddition of a variety of alkynes [147]. For instance, the intermolecular cycloaddition of 1,2-bis(propiolyl)benzenes with monoalkynes in 1,2-dichloroethane (DCE) at room temperature to afford substituted anthraquinones in moderate to high yields (Scheme 7.46) [148].

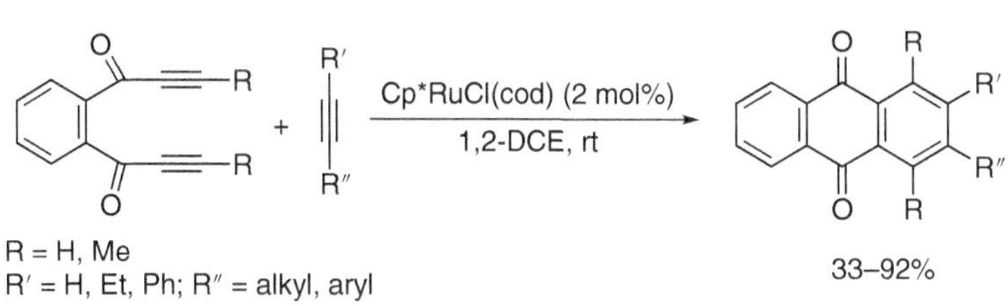

Scheme 7.46 Synthesis of anthraquinones via [2 + 2 + 2] cycloaddition of 1,2-bis(propiolyl)benzenes with monoalkynes.

In the presence of Cp*RuCl(cod), a wide variety of 1,6-diynes can chemoselectively react with monoalkynes at ambient temperature to afford bicyclic benzene derivatives in good yields. Even the unsymmetrical 1,6-diynes undergo the cycloaddition giving *meta*-substituted products with excellent regioselectivity. Completely intramolecular alkyne cyclotrimerization is also accomplished using triyne substrates to obtain tricyclic aromatic compounds fused with five- to seven-membered rings in good yield [149].

Cp*RuCl(cod) can catalyze the cycloaddition of 1,6- and 1,7-diynes with ethynylboronates to give bi- and tricyclic arylboronates [150], and 1,7-diiodo-1,6-diynes with acetylene to produce bicyclic 1,4-diiodobenzenes (Scheme 7.47) [151], or with 2-ethynyl-5,5-dimethyl-1,3,2-dioxaborinane to afford diiodophenylboronates (Scheme 7.48) [152]. The obtained products containing C—I or/and C—B bonds are important intermediates for the synthesis of highly functionalized aromatic molecules by the cross-coupling reactions.

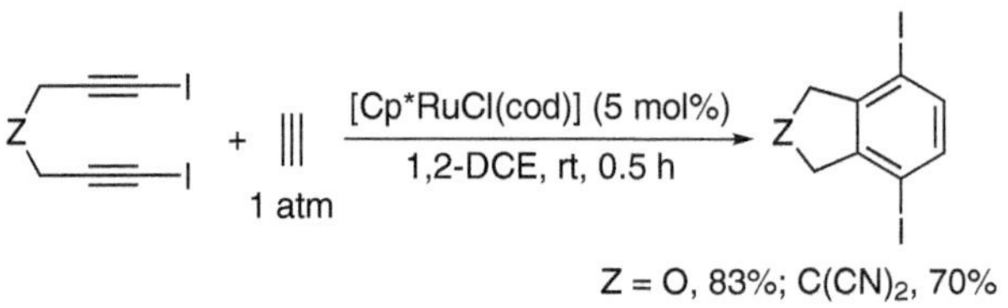

Scheme 7.47 Synthesis of iodobenzenes from diynes with acetylene.

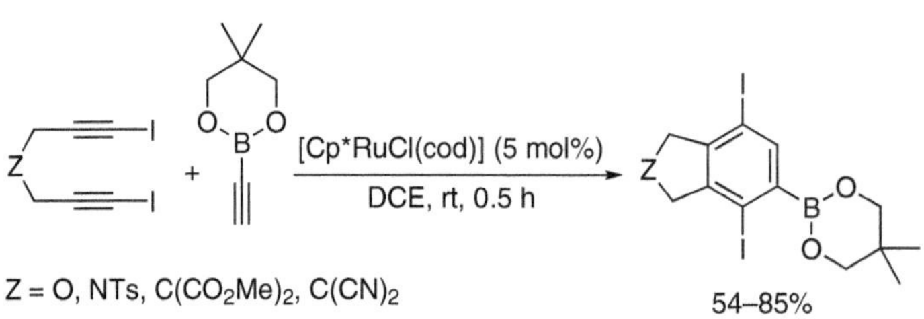

Scheme 7.48 Synthesis of diiodophenylboronates from diiododiynes with ethynylboronate.

The cycloaddition of nonconjugated diynes with 1,3-diynes can provide the efficient ways for the formation of polysubstituted aryl alkynes and biaryls or chiral biaryls in good to excellent yields with high regio- and chemoselectivity depending on the transition metals used and reaction conditions.

Yamamoto and coworker developed a $Pd(PPh_3)_4$-catalyzed dimerizative cyclocondensation of the conjugated 1,3-enynes to obtain the corresponding disubstituted arenes in good to high yields [153]; they then studied the reactions of conjugated 1,3-enynes with 1,3-diynes in the presence of the same catalyst to yield polysubstituted aryl alkynes with excellent regioselectivity via an intermolecular [4 + 2] cyclocondensation (Scheme 7.49) [154], and disclosed the role of Lewis acids and bases [155]. When 2-siloxy- [156] and 2-fluoro- or 2-perfluoroalkyl- [157] substituted enynes were used, polysubstituted phenols and fluoro- or perfluoroalkyl arenes could be efficiently synthesized. Unsymmetrical 1,3-diynes and 1,3,5-triynes are also the suitable substrates for the synthesis of polysubstituted aryl alkynes [158].

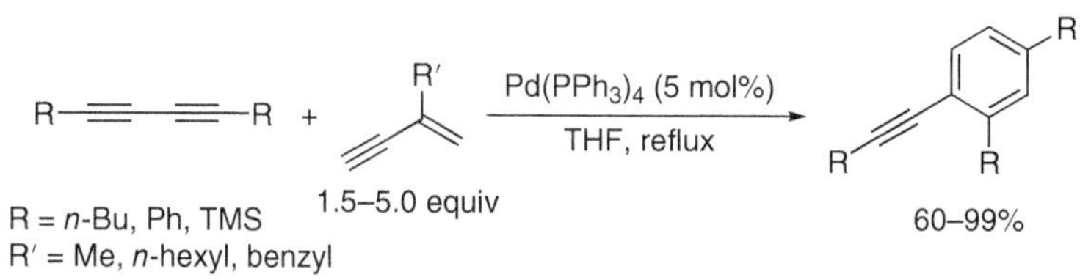

Scheme 7.49 Palladium-catalyzed enyne-diyne cross-benzannulation.

In the presence of $[Rh(cod)_2]BF_4$/chiral ligand, planar-chiral metacyclophanes can be synthesized via cyclotrimerization of 1,9-, 1,11-diynes with DMAD or intramolecular cyclotrimerization of triynes [159], and [RhI(H_8-BINAP)] (Figure 7.1) catalyzes an enantioselective cross alkyne cyclotrimerization of unsymmetrical 1,6-diynes with both terminal and internal alkynes, providing easy access to axially chiral phthalides having one or two oxymethylene functionalities [160].

$[RhCl(cod)]_2$/*rac*-BINAP/$AgBF_4$ shows high catalytic activity for the double [2 + 2 + 2] cycloaddition of 1,4-bis(diphenylphosphinoyl)buta-1,3-diyne with terminal nonconjugated 1,6- or 1,7-diynes to produce achiral biaryl diphosphine oxides in excellent yields (Scheme 7.50) [161]. Although under the same reaction conditions, the internal diyne of 2,8-decadiyne could not undergo the similar cycloaddition reaction, in chlorobenzene at 100 °C, the formation of bicyclic benzenes via a mono[2 + 2 + 2] cycloaddition was achieved. However, when SEGPHOS was used as a chiral ligand (a modified BINAP ligand) (Figure 7.1), the internal 2,7-diynes could undergo the double [2 + 2 + 2] cycloaddition with phosphonate- or ester-substituted 1,3-butadiyne to produce the C_2-symmetric axially chiral biaryl diphosphonates or dicarboxylates in good yields with excellent enantioselectivity (Scheme 7.51) [162].

Substituted 1,3-dienes [163] and 1,4-enynes [164] have also been successfully employed as four-carbon synthons to participate in the [4 + 2] benzannulation to access substituted benzenes and benzene-fused heteroaromatic compounds.

Very recently, Lei's group has reported a visible light-induced [4 + 2] annulation of thiophenes with alkynes driven by in situ generated thiophene radical cation,

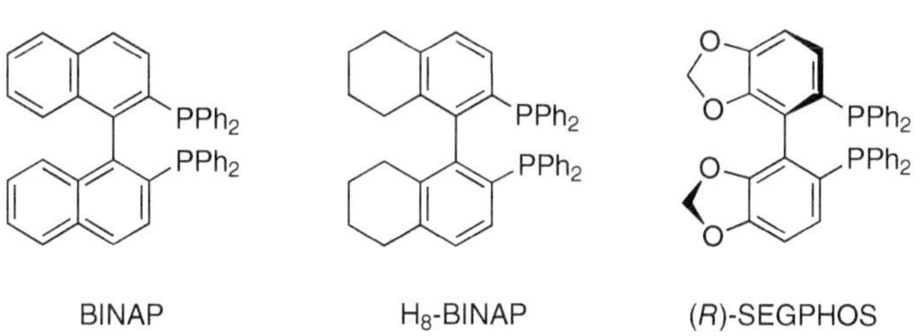

Figure 7.1 Structures of BINAP and modified BINAP-type bisphosphine ligands.

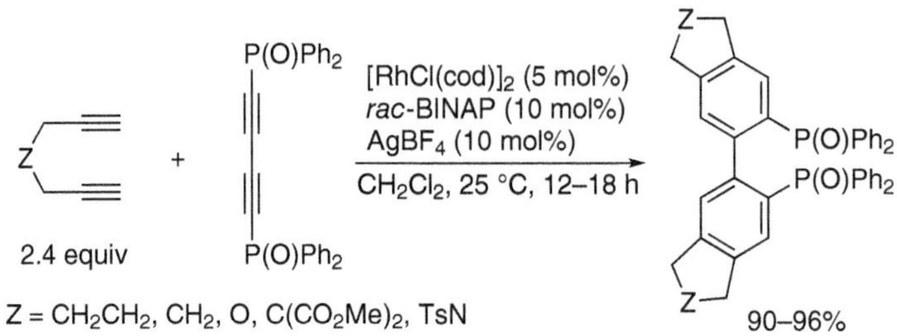

Scheme 7.50 Rhodium-catalyzed double [2 + 2 + 2] cycloaddition of 1,4-bis-(diphenyl-phosphinoyl)buta-1,3-diyne with diynes.

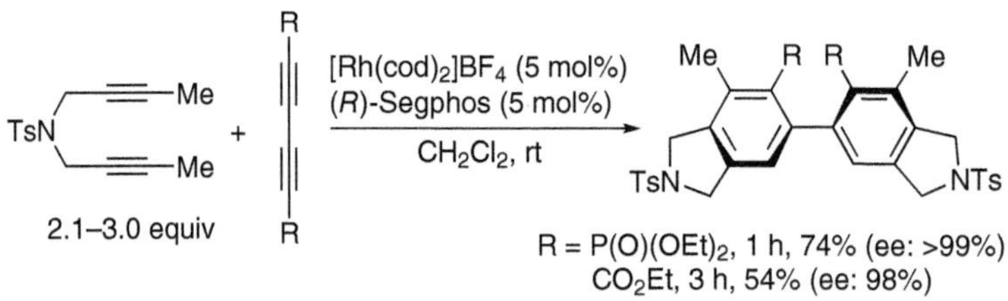

Scheme 7.51 Enantioselective double [2 + 2 + 2] cycloaddition of internal 2,7-diynes with 1,3-diynes.

and thiophene is used as a four-carbon synthon with desulfuration reaction to allow facile synthesis of aromatic rings with a diverse array of substituted alkynes under oxidant- and metal-free conditions [165].

7.3.2 Naphthalene and Polyaromatic Hydrocarbons (PAHs) Ring Formation

Naphthalene is the simplest structure of PAHs, and its derivatives are the fundamental precursors for the synthesis of other PAHs, which have been well studied as organic materials due to their remarkable electronic and optical properties [166].

With the use of alkynes as the cyclocondensation partner, naphthalene ring can be constructed by a $Pd_2(dba)_3$-catalyzed cocyclization of benzyne with two molecules of DMAD [167], $Pd(OAc)_2$-catalyzed annulation of *ortho*-(2-alkenyl)aryl halides with internal alkynes [168], and $CaCl_3$- [169], $AuCl_3/AgSbF_6$- [170], and $HNTf_2$- [171] catalyzed cyclocondensation of phenyl acetaldehydes or phenyl ketones with alkynes with high regioselectivity.

Miura's group developed an $[IrCl(cod)]_2/P(t\text{-}Bu)_3$-catalyzed reaction of aroyl chlorides with two aliphatic alkyne molecules affording naphthalenes (Scheme 7.52) [172] and the directly oxidative coupling of benzoic acids with internal alkynes effectively catalyzed by $[Cp^*IrCl_2]_2$ with the use of Ag_2CO_3 as oxidant via decarboxylation and regioselective C—H bond cleavage [173].

Transition metal-catalyzed oxidative carbocyclization is an atom economic process and has been well studied [174]. For example, Wu and coworker have developed a novel route for the synthesis of highly substituted naphthalenes via a

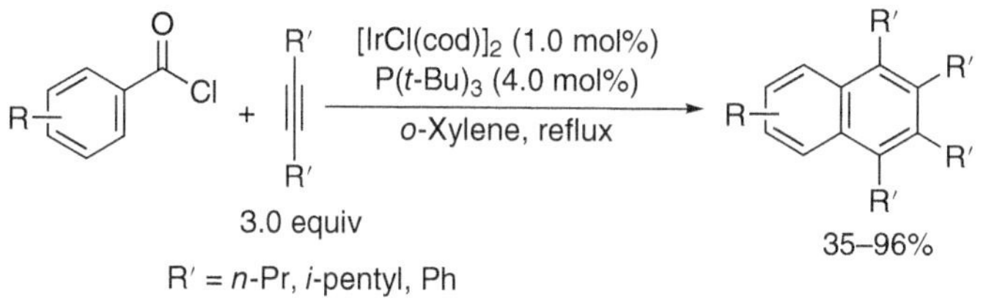

Scheme 7.52 Synthesis of naphthalenes by the reaction of aroyl chlorides with internal alkynes.

one-pot [2 + 2 + 2] benzannulation of arenes with internal alkynes in the presence of $Pd(OAc)_2$ and AgOAc (Scheme 7.53) [175]. In this protocol, an arene provides a benzo source for the construction of a naphthalene core through dual C—H bond activation, and more importantly, when R and R′ groups are not the functional groups, the reactions occur smoothly.

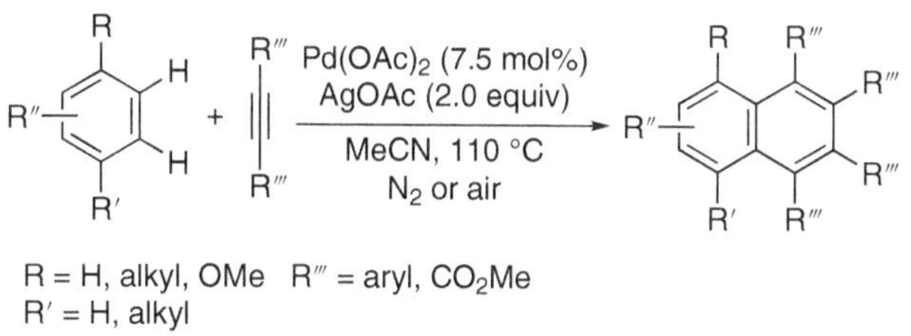

Scheme 7.53 Synthesis of naphthalenes from arenes and internal alkynes via dual C–H bond activation.

When R or R′ group in Scheme 7.53 is a functional group playing a directing group for the activation of C—H bonds, the similar protocols for the synthesis of PAHs have been achieved. For example, using $[Cp^*RhCl_2]_2$ as catalyst and $Cu(OAc)_2$ as the oxidant, the oxidative benzannulation of *N*-adamantyl-1-naphthylamines with two molecules of internal alkynes via dual C—H bond activation to afford highly substituted anthracenes in satisfactory to good yields [176]. A similar catalyst system of $[Cp^*RhCl_2]_2/AgSbF_6/Cu(OAc)_2{\cdot}H_2O$ also shows high catalytic activity in the directly dehydrogenative annulation of 2-phenoxypyridines with internal alkynes by using a pyridyloxy group as an auxiliary, and the pyridyl moiety can be removed to afford α-naphthols [177]. The impact of directing group, solvent , and phosphine ligand on the [2 + 2 + 2] benzannulation of aromatic amides with internal alkynes has also been investigated in the presence of $Ni(acac)_2$ and has been found that the use of a rigid chelating group and a strong aprotic polar solvent results in the favorable formation of 1-naphthamides via a [2 + 2 + 2] benzannulation process [178].

In addition, with the use of O_2 as oxidant, Jiao's group developed a $Pd(OAc)_2$-catalyzed [4 + 2] annulation of 2- or 3-arylindoles and 2,2′-bis(*N*-methylindolyl) with one molecule of internal alkynes through dual activation of C—H bonds to carbazoles and polycyclic carbazoles [179]. Chuang and coworker have also developed $Pd(OAc)_2$-catalyzed selective aryl C—H activation of *N*-acyl-2-aminobiaryls to provide an efficient access to *multi*-aryl-substituted naphthalenes [180].

A metal-free protocol for the synthesis of a series of functionalized 1-amino-2-naphthalenes has also been achieved from the hypervalent iodine(III)-mediated benzannulation of enamines with terminal and internal alkynes with high regioselectivity (Scheme 7.54) [181].

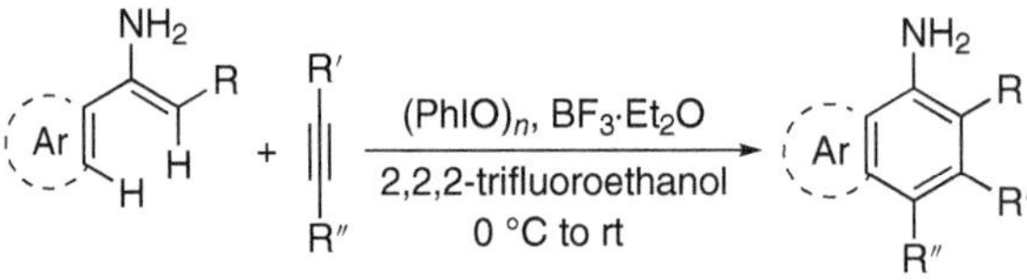

R = COOMe, CN
R′ = H, Ph, Et; R″ = aryl, *n*-Bu

Scheme 7.54 Naphthalene formation via hypervalent iodine(III)-mediated benzannulation of enamines with alkynes.

In the presence of Lewis acids, *ortho*-alkynylbenzaldehydes can be converted into isochromenylium cations, which are a type of useful reactive intermediates in the synthesis of benzo-fused six-membered carbocyclic compounds bearing acyl group via a key step of Diels–Alder [4 + 2] cycloaddition with alkenes [182]. In the case of alkynes used, the [4 + 2] cycloaddition protocol has been applied in the synthesis of acyl-substituted naphthalenes. For example, as shown in Scheme 7.55, in the presence of $AuCl_3$, the reaction of *ortho*-alkynylbenzaldehydes with terminal or internal alkynes undergoes a formal [4 + 2] benzannulation with high regioselectivity to afford α-benzoyl naphthalenes [183]. *ortho*-Alkynylbenzaldehydes and enynals have been well used as four-carbon synthons in [4 + 2] benzannulation to form substituted naphthalenes [184]. A reasonably proposed mechanism including the formation of isochromenylium and its Diels–Alder [4 + 2] cycloaddition with alkyne is depicted in Scheme 7.56.

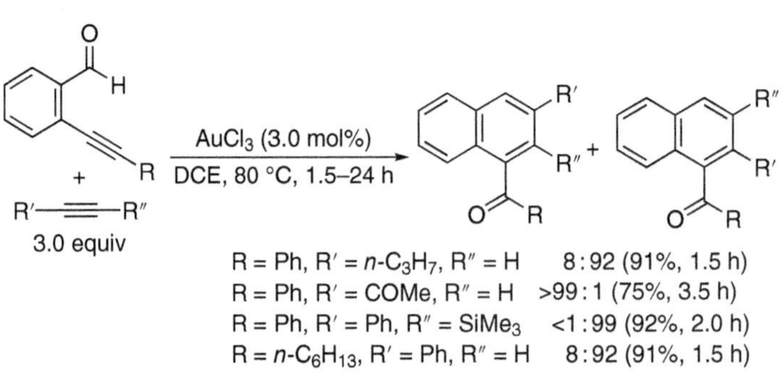

Scheme 7.55 Naphthalene ring formation by $AuCl_3$-catalyzed reaction of *ortho*-alkynylbenzaldehydes with alkynes.

The $CuCl_2$- or $CuBr_2$-mediated [4 + 2] benzannulation of *ortho*-alkynylbenzaldehydes with alkynes gives a variety of haloaromatic compounds including α-halonaphthalenes stereoselectively in good to high yields, and 2,2′-binaphthyl skeletons are also readily prepared by using diynes instead of momoalkynes [185].

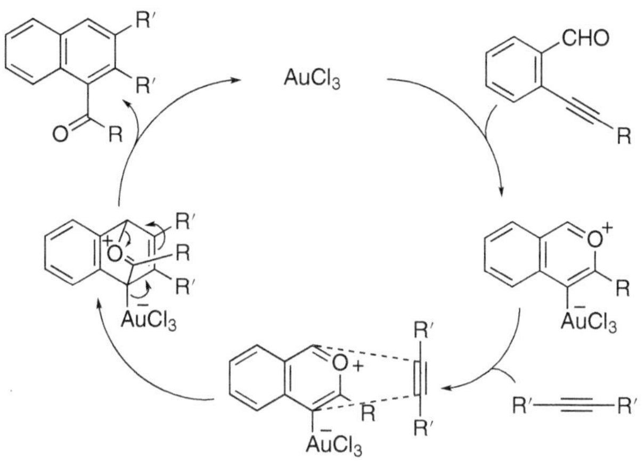

Scheme 7.56 Proposed mechanism for the formal [4 + 2] benzannulation forming naphthalene ring.

On the other hand, the arylated 1,3-butadiynes have been demonstrated to be the useful starting materials for construction of naphthalene ring. For example, Hua and coworker have reported that 1,4-diaryl-1,3-butadiynes react with an excess amount of cyclic amines, such as pyrrolidine, piperidine, and morpholine without solvent to produce diamino-substituted naphthalenes in good to high yields in the presence of CuCl (Scheme 7.57) [186]. In the presence of bromine, 1,4-diaryl-1,3-butadiynes undergo the intramolecular cyclization to afford 1,2,3-tribromonaphthalenes at low temperature (Scheme 7.58) [187]. The obtained products are very useful intermediates for the synthesis of functionalized naphthalenes via the activation

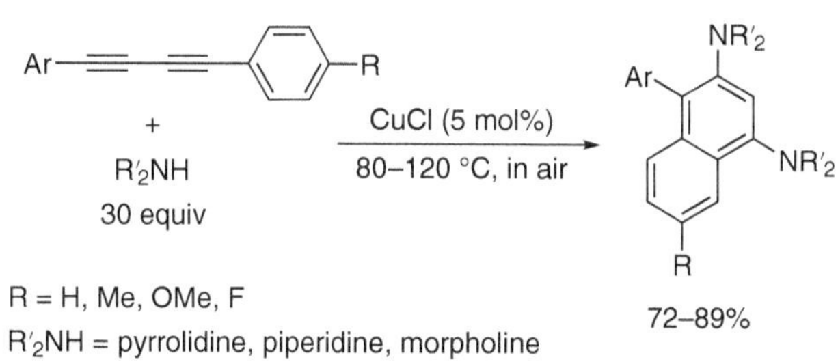

Scheme 7.57 Amino-substituted naphthalene synthesis catalyzed by CuC.

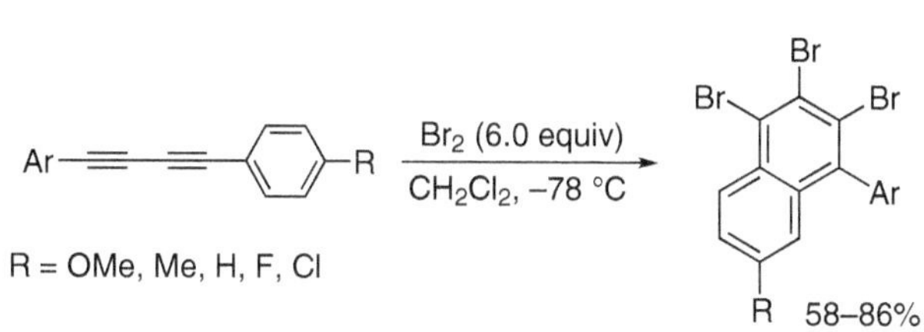

Scheme 7.58 1,2,3-Tribromonaphthalene synthesis from 1,3-butadiynes and bromine.

of C—Br bond and its further cross-coupling reaction in the presence of transition metal catalyst.

In addition, Barluenga and coworker have reported an IPy_2BF_4-promoted regioselective synthesis of substituted naphthalenes at room temperature via the cycloaddition of *ortho*-alkynylbenzaldehydes with either alkynes or alkenes, and in the case of alkynes used, the reaction produces α-iodonaphthalenes, whereas the reaction with alkenes instead of alkynes affords related α-acylnaphthalenes [188]. Loh and coworkers have studied a $PdCl_2$-catalyzed oxidative cyclocondensation of 1-(ethynyl)-2-vinylbenzenes with acrylates or styrene derivatives in DMSO with the use of O_2 as oxidant to afford naphthalenes with high regioselectivity, and the formation of naphthalene ring is considered from a bisalkenylation of C—C triple bonds [189].

Nishiyama's group developed either $Cu(OTf)_2$- or $ReCl(CO)_5$-catalyzed double benzannulation of *ortho*-(phenylethynyl)benzaldehyde with 1,3-butadiynes producing 3,3′-disubstituted 2,2′-binaphthyls in moderate to good yields in the presence of trichloroacetic acid (Scheme 7.59) [190]. The electron-rich aryl group(s) in 1,3-butadiynes greatly accelerate this type of cyclization.

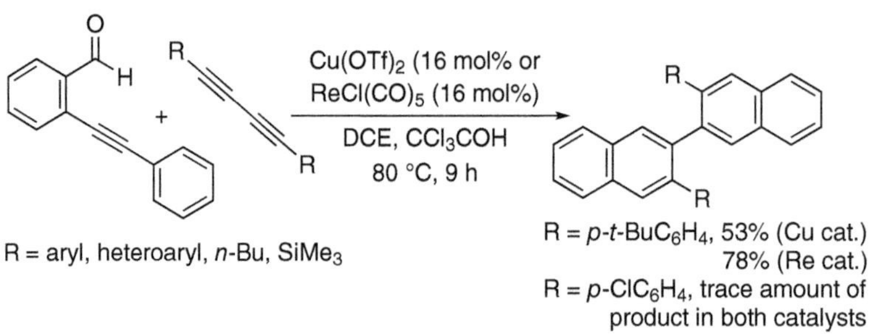

Scheme 7.59 2,2′-Binaphthyls from double benzannulation of *ortho*-(phenylethynyl)-benzaldehyde with 1,3-butadiynes.

Scheme 7.60 shows a representative procedure for the construction of naphthalene ring via a selective 6-*endo* intramolecular hydroarylation of (*E*)-2-ethynyl/alkynyl cinnamates catalyzed by $PtCl_4$ [191]. This procedure provides an efficient synthetic method of functionalized naphthalenes bearing alkyl, alkenyl, aryl, and heteroaryl groups on the 4-position and ethoxycarbonyl group on the 2-position.

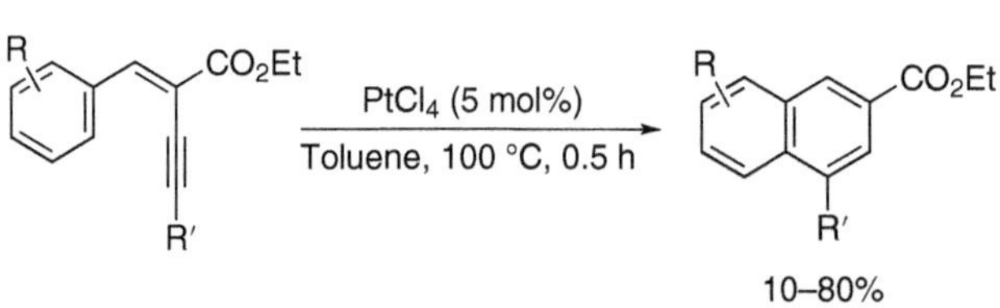

Scheme 7.60 Synthesis of naphthalenes via $PtCl_4$-catalyzed intramolecular hydroarylation of aryl enynes.

Chrysenes, as one important class of PAHs, have attracted growing attention due to their potential applications in the fields of electronic devices such as field-effect transistors (FETs) and organic light-emitting diodes (OLEDs) [192]. Isochromenylium was first reported to undergo the acid-promoted dimerization for the synthesis of chrysene derivatives in 1982 [193], and Liu and coworker have developed a one-pot approach to synthesize chrysenes via $PtCl_2$-catalyzed hydrative dimerization of 2-alkynyl-1-acetylbenzenes under CO atmosphere [194]. Hua and coworker have then also developed an efficient route for the synthesis of highly substituted chrysenes by simple copper-catalyzed one-pot homo-cyclodimerization of 2-alkynyl-1-acetylbenzenes [195] or by HOTf-promoted cross-cyclodimerization of 2-[2-(4-methoxylphenyl)-alkynyl]acetophenones with 2-alkynylbenzaldehydes (Scheme 7.61) [196]. The active intermediate of isochromenylium generated in situ from the reaction of $CuCl_2$ or HOTf with 2-alkynyl-1-acetylbenzenes is proposed to be the key intermediate for the formation of chrysene ring. In addition, in the case of HOTf used, the stable isochromenylium can be isolated.

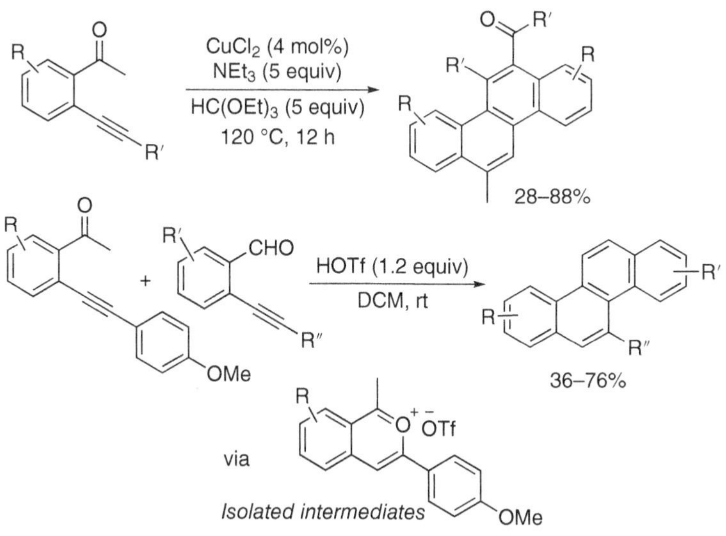

Scheme 7.61 Chrysene synthesis from 2-alkynyl-1-acetylbenzenes.

7.3.3 1,3-Cyclohexadiene Ring Formation Via Cycloaddition of Alkynes

1,3-Cyclohexadienes are important partners in Diels–Alder reaction to construct fused cycles in a stereoselective manner; one of the simple and efficient synthetic methods is the catalytic [2 + 2 + 2] cycloaddition of two alkynes with one alkene.

Low-valent niobium compounds are thermally stable and therefore have been widely used as catalysts in organic transformations. Obora and coworker first reported the synthesis of substituted 1,3-cyclohexadiene from an intermolecular [2 + 2 + 2] cycloaddition of alkynes and alkenes in the presence of $NbCl_3$(DME) [197]. In the case of *tert*-butylacetylene used, the cycloaddition reactions with a variety of alkenes afford trisubstituted 1,3-cyclohexadienes with a mixture of products (Scheme 7.62). Their further studies with the use of $NbCl_5$ as an efficient catalyst for the similar intermolecular [2+2+2] cycloaddition afforded 1,3-cyclohexadienes with high chemo- and regioselectivity [198]. The catalyst system of $NbCl_5/Zn/PCy_3$ shows high catalytic activity in the intramolecular [2+2+2] cycloadditions of diynes and alkenes to form bicyclic cyclohexadienes [199].

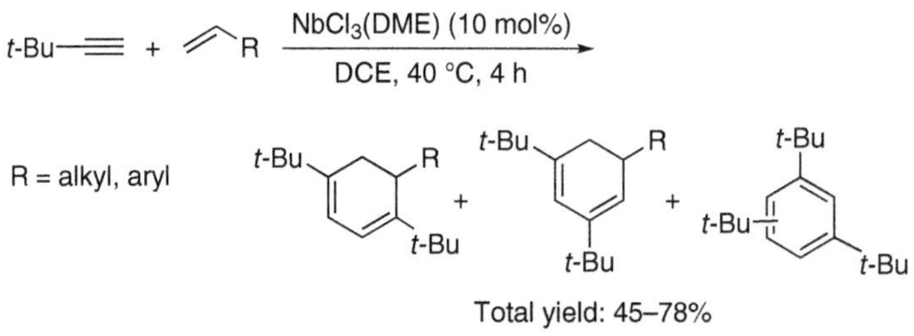

Scheme 7.62 1,3-Cyclohexadiene synthesis from niobium(III)-catalyzed intermolecular [2+2+2] cycloaddition of alkynes with alkenes.

Jiang and coworker have developed an efficient synthetic method of substituted 1,3-cyclohexadienes through metal-catalyzed cycloaddition of enynals with alkenes, and the substituted group can be regioselectively bonded to different position, depending on the use of metal catalyst (Scheme 7.63) [200].

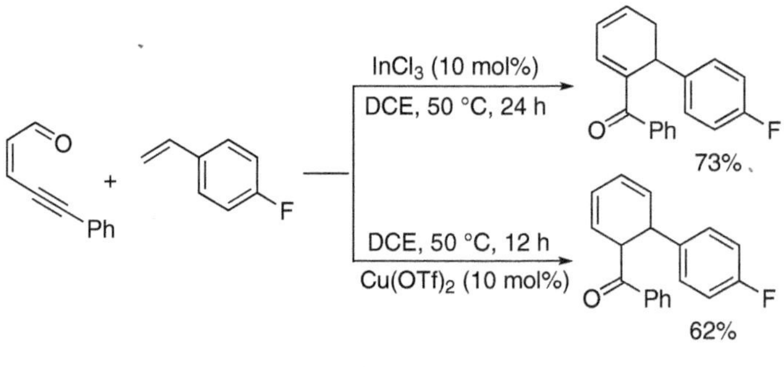

Scheme 7.63 Catalyst-dependent formation of substituted 1,3-cyclohexadienes.

With the use of $Ni(cod)_2$/NHC as catalysts, Louie's group investigated the cycloisomerization of 1,7- and 1,6-enynes, and found that 1,3-bis(2,5-di-*t*-butylphenyl)-imidazol-2-ylidene (IDTB) ligand is the best choice for the cycloisomerization of 1,7- and 1,6-enynes to cyclic 1,3-dienes in high yields (Scheme 7.64) [201].

The synthesis of chiral cyclic compounds possessing 1,3-cyclohexadiene skeleton via the enantioselective [2+2+2] cycloaddition protocols is much more important and interesting research topic, and many procedures have been reported [202]. For example, Shibata's group extensively investigated the enantioselective [2+2+2] cycloaddition of nitrogen-bridged enyne with 1,4-dimethoxybut-2-yne catalyzed by $[Rh(cod)_2]X$/chiral phosphine (X = BF_4, SbF_6, and OTf) [203], and as shown in Scheme 7.65, $[Rh(cod)_2]_2$/tolBINAP can efficiently catalyze the enantioselective cycloaddition affording the chiral bicyclic cyclohexa-1,3-dienes in high yield with

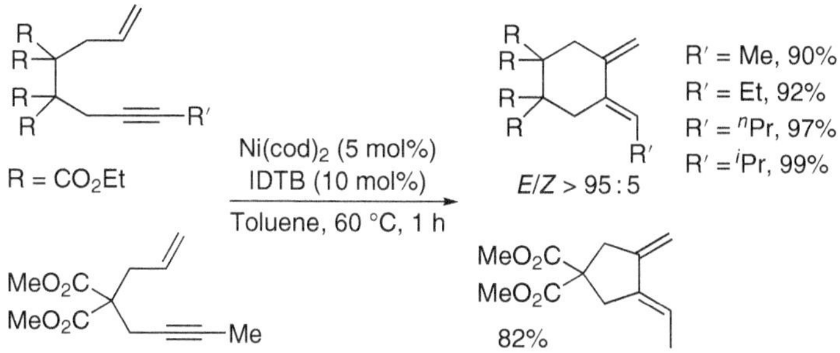

Scheme 7.64 Nickel(0)-catalyzed cycloisomerization of enynes to cyclic 1,3-dienes.

excellent ee value. Evan's group then also reported a $[RhCl(cod)]_2/AgBF_4$/chiral ligand-catalyzed regio- and enantioselective intermolecular [2+2+2] cycloaddition of terminal 1,6-enynes with methyl arylpropiolates to give chiral bicyclic cyclohexa-1,3-dienes [204].

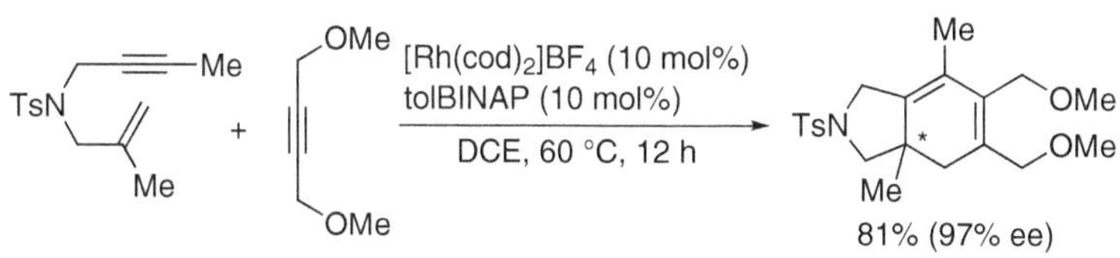

Scheme 7.65 Rhodium-catalyzed enantioselective [2 + 2 + 2] cycloaddition of enyne alkyne affording chiral bicyclic cyclohexa-1,3-dienes.

In addition, the chiral polycyclic fused 1,3-cyclohexadienes have also been obtained from the transition metal-catalyzed enantioselective [2+2+2] cycloaddition of either diynes with alkenes [205] and enynes with alkenes [206] or cycloisomerization of 1,4-diene-ynes [207] and enediynes [208].

In the presence of $Hg(OTf)_2$-$(TMU)_3$, ω-arylalkynes undergo the intramolecular cyclohydroarylation at ambient temperature in acetonitrile leading to the formation of dihydronaphthalenes [209]. One of the examples is shown in Scheme 7.66, and the expected product is obtained in 95% yield. Dihydronaphthalenes can also be prepared by organocatalysis reported by Hua's group [210], and as shown in Scheme 7.67, in the presence of *N*,*N*-dimethylformamide dimethyl acetal (DMF·DMA), the intermolecular cycloaddition of electron-deficient alkynes with excess amounts of vinylarenes afforded substituted dihydronaphthalenes in mild to good yields.

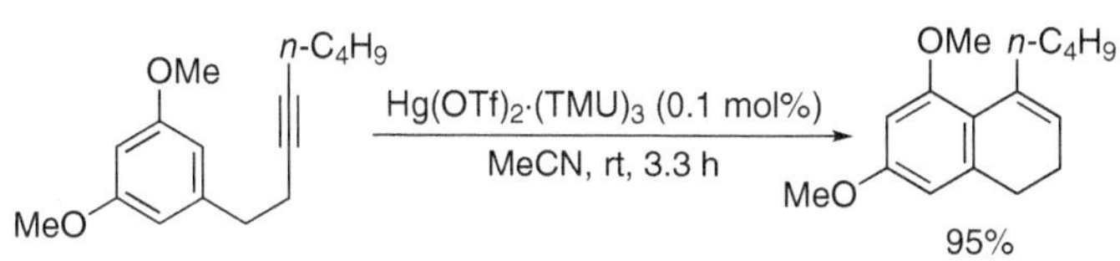

Scheme 7.66 $Hg(OTf)_2$-$(TMU)_3$-catalyzed cyclization of ω-arylalkynes affording dihydronaphthalenes.

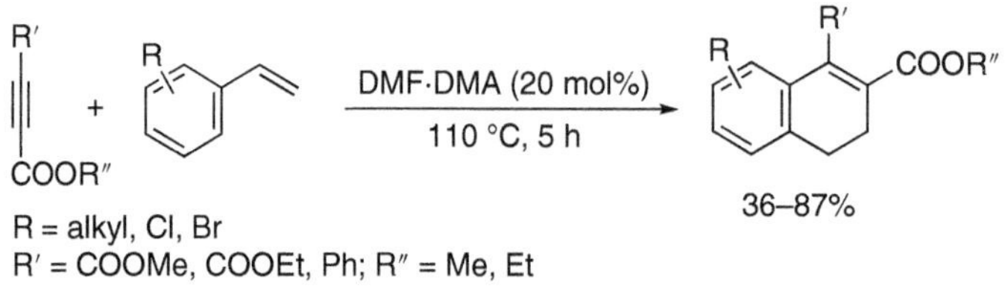

Scheme 7.67 DMF DMA-catalyzed formation of substituted dihydronaphthalenes.

7.4 Seven-Membered Carbocycles

The formation of seven-membered carbocycles starting from the unsaturated hydrocarbons can be finished through [5 + 2], [3 + 2 + 2], [4 + 3], and [6 + 1] annulation reactions, and several reviews are available [211].

Particularly, rhodium-catalyzed inter- and intramolecular [5 + 2] cycloaddition of vinylcyclopropanes (VCPs) and 2π components such as alkyne, alkene, and allene have become an atom- and step-economical way to construct seven-membered carbocycles, which have been extensively investigated (Scheme 7.68) [212].

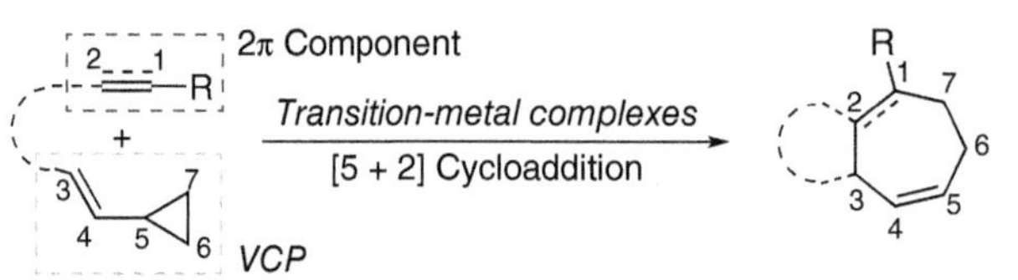

Scheme 7.68 Rhodium-catalyzed intramolecular [5 + 2] cycloaddition of vinylcyclopropanes with alkyne or alkene.

Wender and coworker first reported the intramolecular [5 + 2] cycloaddition of yne-VCPs with alkynes in 1995 in the presence of $RhCl(PPh_3)_3$/AgOTf (Scheme 7.69) [213].

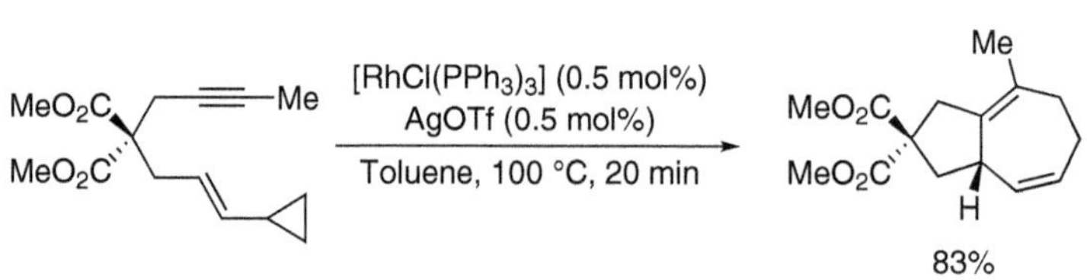

Scheme 7.69 Rhodium-catalyzed intramolecular [5 + 2] cycloaddition of vinylcyclopropanes with alkyne giving seven-membered carbocycles.

In addition, ruthenium [214] and nickel complexes [215] have also been used as the catalysts for this type of intramolecular [5 + 2] cycloaddition of yne-VCPs reported by other groups. The experimental and computational studies have been carried out to gain insights into the mechanisms of [5 + 2] cycloadditions [216].

Wender's group further developed the rhodium-catalyzed intermolecular [5 + 2] cycloadditions of VCPs with alkynes [217], intramolecular [5 + 2] cycloaddition of ene-VCP [218], and regio- and stereoselective intramolecular [5 + 2] cycloadditions [219].

Other groups also reported the iridium-catalyzed intermolecular [5 + 2] cycloadditions of VCPs with alkynes [220], rhodium-catalyzed (asymmetric) intramolecular [5 + 2] cycloaddition of VCPs with alkynes [221], and rhodium(I)-catalyzed inter- and intramolecular [5 + 2] cycloaddition of VCPs with allenes [222].

Evans and coworker also developed a rhodium(I)-catalyzed regio- and diastereoselective intermolecular [5 + 2] carbocyclization of alkenylidenecyclopropanes (ACPs) with activated alkynes for constructing seven-membered carbocycles (Scheme 7.70) [223]. The detailed mechanistic studies for this carbocyclization with the aid of theoretical calculations were then performed by Dang and Fu's group [224].

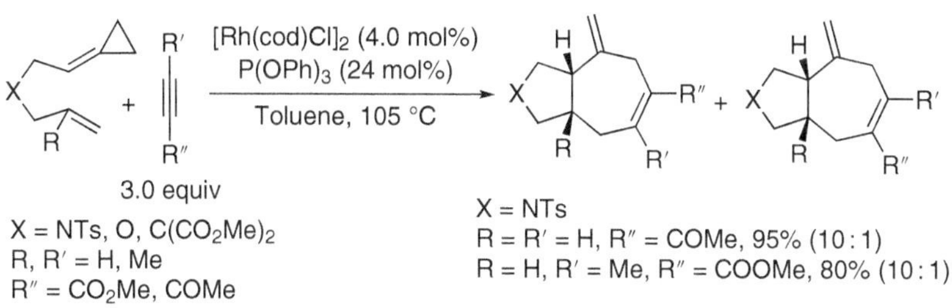

Scheme 7.70 Rhodium(I)-catalyzed intermolecular [5 + 2] carbocyclization of ACPs with activated alkynes giving seven-membered carbocycles.

Evans's then developed the synthesis of methylene-substituted seven-membered carbocycles through rhodium-catalyzed intermolecular [5 + 2] carbocyclization of alkynylidenecyclopropanes with substituted allenes [225].

Rhodium(I)-catalyzed intramolecular [5 + 2] carbocyclization of alkyne-allenylcyclopropanes also provides a facile preparation of the bicyclo[5.4.0]undecatrienes and bicyclo[5.5.0]dodecatrienes (Scheme 7.71) [226]. When the highly strained allenylcyclopropane is replaced by the strained allenylcyclobutane, the intramolecular [6 + 2] carbocyclization of alkyne–allenylcyclobutanes resulted in the formation of eight-membered rings [227].

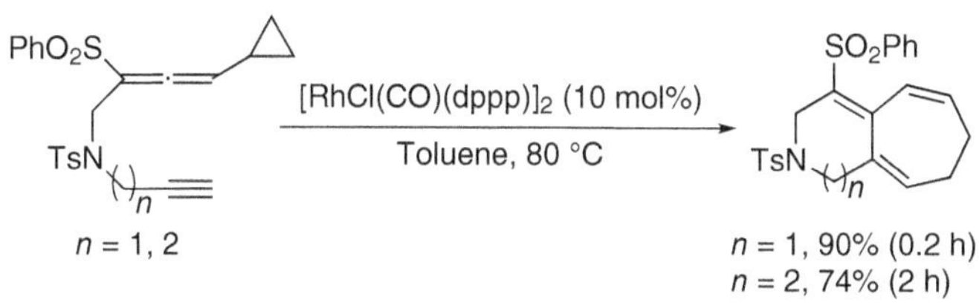

Scheme 7.71 Rhodium(I)-catalyzed intramolecular [5 + 2] carbocyclization of alkyne–allenylcyclopropanes.

The protocols of rhodium(I)-catalyzed intramolecular [5 + 2] cycloaddition of yne-VCPs have been applied in the synthesis of natural products [228].

Very recently, a review on the rhodium-catalyzed [5 + 2] cycloadditions using 1,4-enynes as five-carbon synthons in the synthesis of seven-membered carbocycles has appeared [229].

Nickel(0)/phosphine-catalyzed intermolecular [3 + 2 + 2] cocyclization of ethyl cyclopropylideneacetate with two alkynes gave the multi-substituted cycloheptadienes (Scheme 7.72) [230]. The same group also developed the nickel(0)-catalyzed synthesis of nine-membered carbocycles by the [4 + 3 + 2] cycloaddition reaction of ethyl cyclopropylideneacetate with dienynes, and the literature on the formation of seven-membered carbocycles and cycloketones is cited in detail [231].

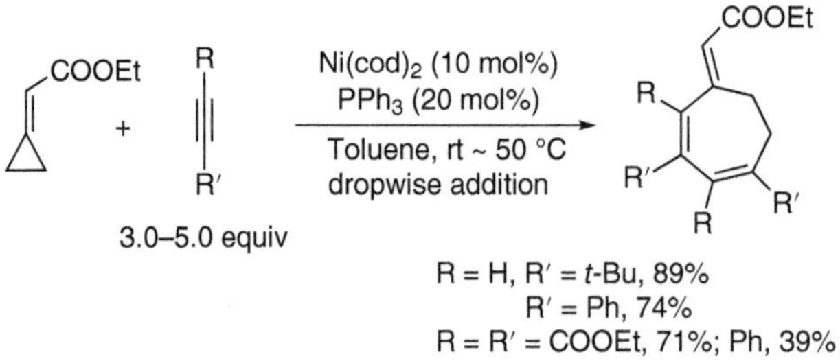

Scheme 7.72 $Ni(cod)_2$-catalyzed intermolecular [3 + 2 + 2] cocyclization of ethyl cyclopropylideneacetate with two alkynes.

In addition, an electrochemical synthesis of functionalized seven-membered carbocycles through a 5-*exo*-trig/7-*endo*-trig radical cyclization of 1,6-dienes has also been developed [232].

7.5 Eight-Membered and Larger Carbocycles

The syntheses of eight-membered and larger carbocycles, particularly medium-sized ring systems (usually 8- to 12-membered rings), are quite challenging in organic synthesis. As other size carbocycles formation, the transition metal-catalyzed intra- and intermolecular cycloadditions of unsaturated carbon–carbon bonds are the most efficient routes, and several reviews have been reported [233]. In this section, only a few typical catalytic procedures reported in the past one decade are highlighted.

Eight-membered carbocycles are found in a wide variety of natural products that exhibit a broad range of biological and medicinal activities [234].

Kuninobu and Takai's group developed an interesting $[ReBr(CO)_3(thf)]_2$-catalyzed insertion of alkyne into the C—C bond of cyclohexanone-2-carboxylic acid ethyl ester under solvent-free conditions to give eight-membered carbocycles in excellent yields (Scheme 7.73) [235]. The insertion reactions can be used in the synthesis of 9- and 10-membered cyclic esters from 7- and 8-membered cyclic 1,3-dicarbonyl compounds. The rhenium and manganese complex-catalyzed insertion of alkynes into cyclic and acyclic 1,3-dicarbonyl compounds to give ring-expanded and carbon-chain extension products was then also reported [236].

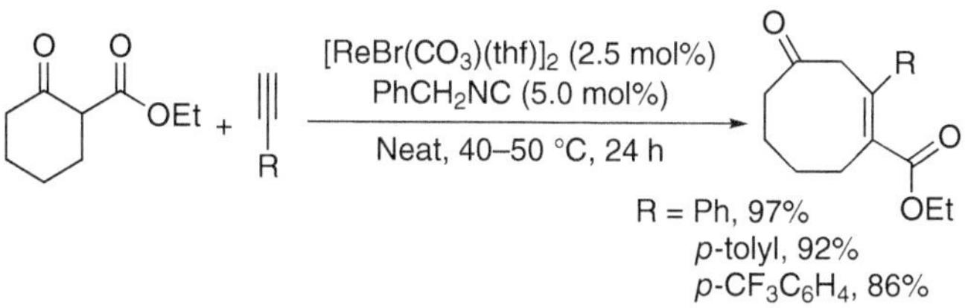

Scheme 7.73 Rhenium-catalyzed insertion of alkyne into the C—C bond of cyclohexanone-2-carboxylic acid ethyl ester.

With the use of $CoBr_2$(*c*-Hex-diimine), Zn and ZnI_2 as the catalyst system, Hilt and coworker examined the cyclocondensation of isoprene with phenylacetylene in CH_2Cl_2 at 20 °C to find that the reaction produced not only 1,4-cyclohexadiene by Diels–Alder reaction but also eight-membered carbocyclic product in 31% yield by a [4 + 2 + 2] cocyclotrimerization of isoprene with two molecules of alkynes (Scheme 7.74) [237]. Then they optimized the reaction conditions to develop a cobalt-catalyzed [4 + 2 + 2] cycloaddition for the synthesis of 1,3,6-cyclooctatrienes (Scheme 7.75) [238]. In addition, the cobalt(II)-catalyzed dimerizative [4 + 4] cycloadditions of 1-alkyl-1,3-dienes was observed by the change of the ligand [239]. Very recently, Chirik and coworker have also developed a regio- and diastereoselective iron-catalyzed homo-/cross-dimerizative [4 + 4] cycloadditions of 1- or 2-substituted 1,3-dienes to construct cyclooctadienes (Scheme 7.76) [240].

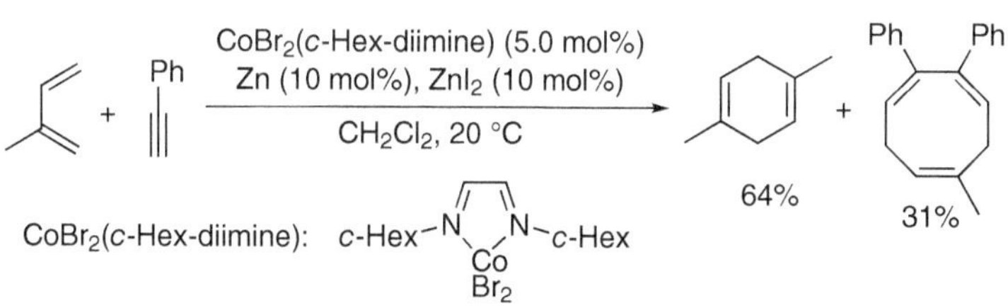

Scheme 7.74 Cobalt(II)-catalyzed [4 + 2 + 2] carbocyclization of isoprene with alkynes.

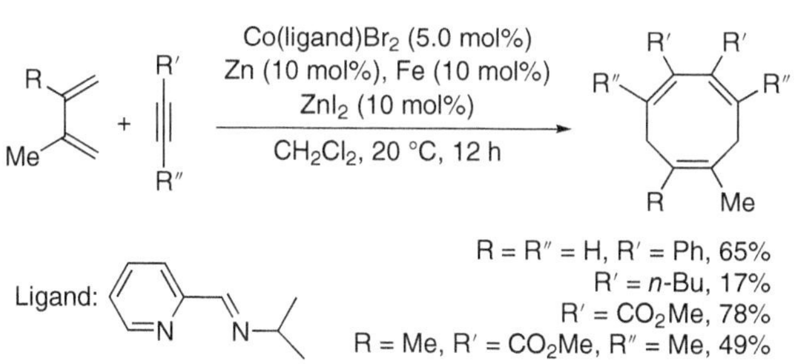

Scheme 7.75 Cobalt(II)-catalyzed intermolecular [4 + 2 + 2] carbocyclization.

Danheiser and coworker reported an intramolecular [4 + 4] annulation of conjugated enyne with cyclobutene for the synthesis of eight-membered carbocycles

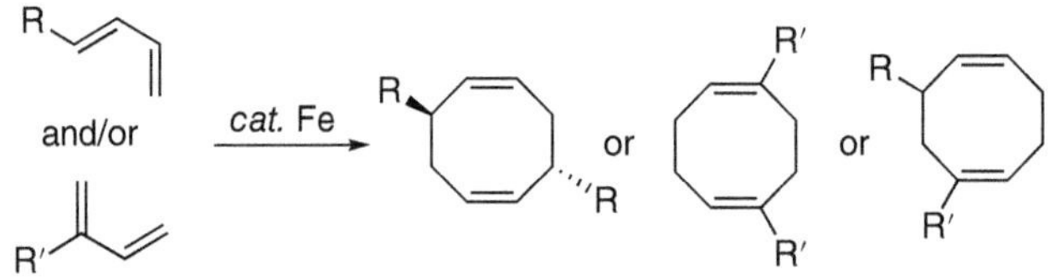

Scheme 7.76 Iron-catalyzed homo-/cross-dimerizative [4 + 4] cycloaddition of 1,3-dienes.

(Scheme 7.77) [241]. This eight-membered ring formation strategy involves a [4 + 2] cycloaddition of enyne with cyclobutene generating a strained six-membered cyclic allene, which isomerizes to the 1,3-cyclohexadiene intermediate, followed by an electrocyclic ring opening constructing eight-membered ring (Scheme 7.78).

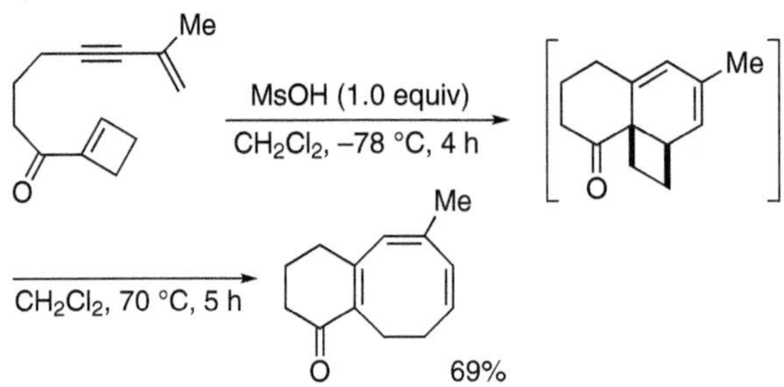

Scheme 7.77 Intramolecular [4 + 4] annulation of conjugated enyne with cyclobutene.

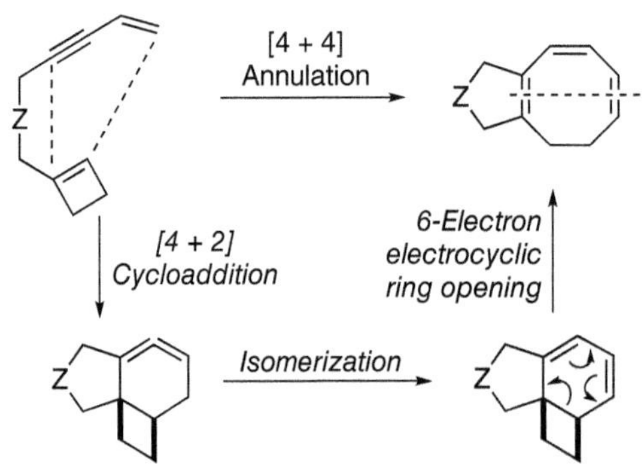

Scheme 7.78 Intramolecular [4 + 4] annulation strategy for the synthesis of eight-membered carbocycle.

Magauer and coworker have recently reported a review on the synthesis of functionalized nine-membered carbocyclic rings highlighting the methods of ring expansion of smaller rings and the direct cyclization reactions [242].

Gilbertson and coworker reported alternative rhodium-catalyzed synthesis of eight-membered carbocycles through cyclodimerization of dienynes or cross-cyclocondensation between a dienyne and another terminal alkyne (Scheme 7.79) [243].

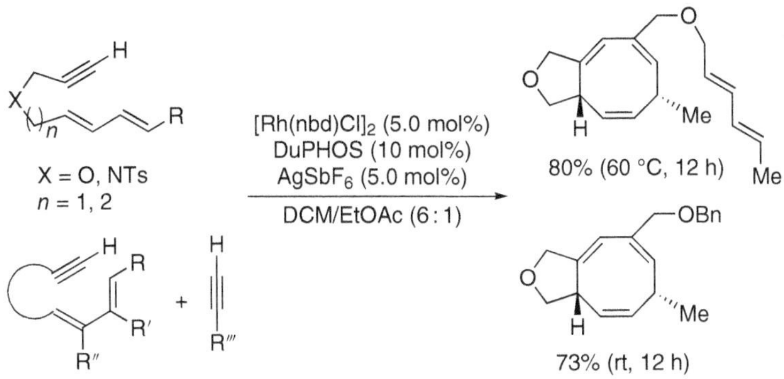

Scheme 7.79 Synthesis of eight-membered carbocycles via rhodium-catalyzed cyclocondensation from dienynes.

The rhodium-catalyzed intramolecular [7 + 2] cyclization of allenylcyclopentane-alkynes affording bicyclo[7.4.0]tridecatrienes, the formation of nine-membered carbocycle ring was developed by Mukai's group (Scheme 7.80) [244], and the studies of mechanism by means of DFT calculations [245] were also reported.

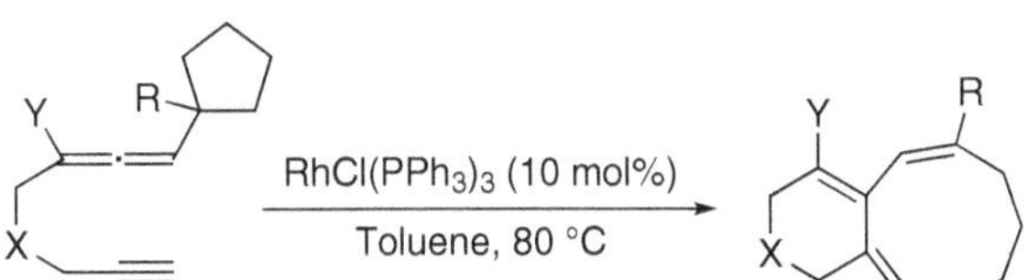

$X = C(SO_2Ph)_2$, $Y = SO_2Ph$, R = Me, 85% (0.2 h)
$X = C(CH_2OH)_2$, $Y = SO_2Ph$, R = Me, 51% (0.75 h)
$X = C(CO_2Me)_2$, Y = *n*-Bu, R = Me, 71% (1 h)

Scheme 7.80 Rhodium(I)-catalyzed cycloaddition of allenylcyclopentane-alkyne.

Nickel(0)-catalyzed cycloaddition reactions have been well applied in the synthesis of different-sized carbocycles; Baran and coworker studied the nickel(0)-catalyzed [4 + 4 + 2] cycloadditions of dienes with alkynes to selectively give 1*Z*,4*Z*,8*E*-cyclodecatriene and the transformation mechanism by DFT calculations to elucidate the reactivity of alkynes and *Z*/*E* selectivity of cyclodecatriene products, as well as the substrate limitation (Scheme 7.81) [246].

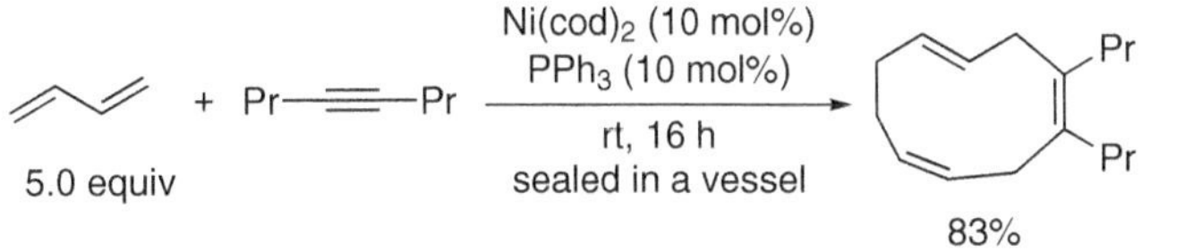

Scheme 7.81 Nickel(0)-catalyzed [4 + 4 + 2] cycloadditions of dienes with alkynes forming 10-membered carbocycles.

The proposed mechanism for the cycloadditions of dienes with alkynes giving 10-membered carbocycles involves the nickel-catalyzed oxidative cyclization of dienes forming the 9-membered ring intermediate, which undergoes alkyne insertion to produce the 11-membered ring intermediate, followed by subsequent reductive elimination generating the desired 10-membered ring product (Scheme 7.82) [247].

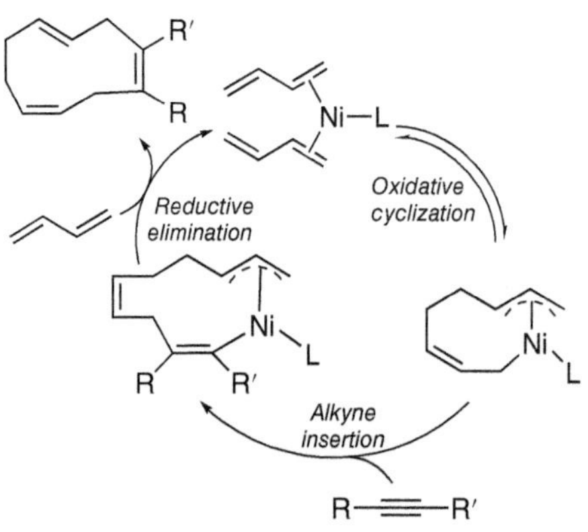

Scheme 7.82 Proposed mechanism for cycloadditions of dienes with alkynes giving 10-membered carbocycles.

Very recently, a review summarizes the synthesis of 10-membered and larger macrocycles, including carbocycles by free radical reactions, and also their applications as the key steps in the synthesis of natural products [248].

References

1 Selected reviews, see: (a) Ojima, I., Tzamarioudaki, M., Li, Z., and Donovan, R.J. (1996). *Chem. Rev.* 96: 635–662. (b) Sohel, S.M.A. and Liu, R.-S. (2009). *Chem. Soc. Rev.* 38: 2269–2281. (c) Gulías, M., López, F., and Mascareñas, J.L. (2011). *Pure Appl. Chem.* 83: 495–506. (d) Goudedranche, S., Raimondi, W., Bugaut, X. et al. (2013). *Synthesis* 45: 1909–1930. (e) Jiao, L. and Yu, Z.-X. (2013). *J. Org. Chem.* 78: 6842–6848. (f) Vladimir, A., D'yakonov, V.A., Trapeznikova, O.A. et al. (2014). *Chem. Rev.* 114: 5775–5814. (g) Chen, J.-R., Hu, X.-Q., Lu, L.-Q., and Xiao, W.-J. (2015). *Chem. Rev.* 115: 5301–5365. (h) Zhang, D., Liu, J., Córdova, A., and Liao, W.-W. (2017). *ACS Catal.* 7: 7051–7063. (i) Trofimov, B.A. and Schmidt, E.Y. (2018). *Acc. Chem. Res.* 51: 1117–1130. (j) Blaszczyk, S.A., Glazier, D.A., and Tang, W. (2020). *Acc. Chem. Res.* 53: 231–243. (k) Wu, L., Yu, B., and Li, E.-Q. (2020). *Adv. Synth. Catal.* 362: 4010–4026.

2 Selected reviews, see: (a) Namyslo, J.C. and Kaufmann, D.E. (2003). *Chem. Rev.* 103: 1485–1537. (b) Lee-Ruff, E. and Mladenova, G. (2003). *Chem. Rev.* 103: 1449–1484. (c) Misale, A., Niyomchon, S., and Maulide, N. (2016). *Acc. Chem. Res.* 49: 2444–2458.

3 (a) Dembitsky, V.M. (2008). *J. Nat. Med.* 62: 1–33. (b) Sergeiko, A., Poroikov, V.V., Hanuš, L.O., and Dembitsky, V.M. (2008). *Open Med. Chem. J.* 2: 26–37.

4 Selected reviews, see: (a) Lautens, M., Klute, W., and Tam, W. (1996). *Chem. Rev.* 96: 49–92. (b) Alcaide, B., Almendros, P., and Aragoncillo, C. (2010). *Chem. Soc. Rev.* 39: 783–816. (c) Secci, F., Frongia, A., and Piras, P.P. (2013). *Molecules* 18: 15541–15572. (d) Xu, Y., Conner, M.L., and Brown, M.K. (2015). *Angew. Chem. Int. Ed.* 54: 11918–11928. (e) Fructos, M.R. and Prieto, A. (2016). *Tetrahedron* 72: 355–369. (f) Poplata, S., Tröster, A., Zou, Y.-Q., and Bach, T. (2016). *Chem. Rev.* 116: 9748–9815.

5 Selected reports, see: (a) Jordan, R.W. and Tam, W. (2000). *Org. Lett.* 2: 3031–3034. (b) Villeneuve, K., Riddell, N., Jordan, R.W. et al. (2004). *Org. Lett.* 6: 4543–4546. (c) Riddell, N., Villeneuve, K., and Tam, W. (2005). *Org. Lett.* 7: 3681–3684. (d) Riddell, N. and Tam, W. (2006). *J. Org. Chem.* 71: 1934–1937. (e) Liu, P., Jordan, R.W., Kibbee, S.P. et al. (2006). *J. Org. Chem.* 71: 3793–3803. (f) Tam, W., Goodreid, J., and Cockburn, N. (2009). *Curr. Org. Synth.* 6: 219–238. g Cockburn, N., Karimi, E., and Tam, W. (2009). *J. Org. Chem.* 74: 5762–5765. h Tsui, G.C., Villeneuve, K., Carlson, E., and Tam, W. (2014). *Organometallics* 33: 3847–3856.

6 Kuninobu, Y., Yu, P., and Takai, K. (2007). *Chem. Lett.* 36: 1162–1163.

7 Chao, K.C., Rayabarapu, D.K., Wang, C.-C., and Cheng, C.-H. (2001). *J. Org. Chem.* 66: 8804–8810.

8 Burton, R.R. and Tam, W. (2007). *J. Org. Chem.* 72: 7333–7336.

9 Villeneuve, K., Riddell, N., and Tam, W. (2006). *Tetrahedron* 62: 3823–3836.

10 Shibata, T., Takami, K., and Kawachi, A. (2006). *Org. Lett.* 8: 1343–1345.

11 (a) Fan, B.-M., Li, X.-J., Peng, F.-Z. et al. (2010). *Org. Lett.* 12: 304–306. (b) Qin, H., Chen, J., Li, K. et al. (2018). *Chem. Asian J.* 13: 2431–2434.

12 Goodreid, J., Villeneuve, K., Carlson, E., and Tam, W. (2014). *J. Org. Chem.* 79: 10002–10012.

13 Yi, C.S., Lee, D.W., and Chen, Y. (1999). *Organometallics* 18: 2043–2045.

14 Huang, D.-J., Rayabarapu, D.K., Li, L.-P. et al. (2000). *Chem. Eur. J.* 6: 3706–3713.

15 Qin, H., Chen, J., Li, K. et al. (2018). *Chem. Asian J.* 13: 2431–2434.

16 (a) López-Carrillo, V. and Echavarren, A.M. (2010). *J. Am. Chem. Soc.* 132: 9292–9294. (b) de Orbe, M.E., Amenós, L., Kirillova, M.S. et al. (2017). *J. Am. Chem. Soc.* 139: 10302–10311.

17 García-Morales, C., Ranieri, B., Escofet, I. et al. (2017). *J. Am. Chem. Soc.* 139: 13628–13631.

18 (a) Treutwein, J. and Hilt, G. (2008). *Angew. Chem. Int. Ed.* 47: 6811–6813. (b) Fujihara, T., Kukita, S., Yagi, Y. et al. (2018). *Asian J. Org. Chem.* 7: 2456–2458.

19 (a) Hilt, G., Paul, A., and Treutwein, J. (2010). *Org. Lett.* 12: 1536–1539. (b) Nishimura, A., Tamai, E., Ohashi, M., and Ogoshi, S. (2014). *Chem. Eur. J.* 20: 6613–6617.

20 de Orbe, M.E. and Echavarren, A.M. (2018). *Eur. J. Org. Chem.* 2018: 2740–2752.

21 (a) Nishimura, A., Ohashi, M., and Ogoshi, S. (2012). *J. Am. Chem. Soc.* 134: 15692–15695. (b) Abulimiti, A., Nishimura, A., Ohashi, M., and Ogoshi, S. (2013). *Chem. Lett.* 42: 904–905.

22 Villeneuve, K. and Tam, W. (2004). *Angew. Chem. Int. Ed.* 43: 610–613.

23 García-Rubín, S., González-Rodríguez, C., García-Yebra, C. et al. (2014). *Angew. Chem. Int. Ed.* 53: 1841–1844.

24 Okamoto, K., Shimbayashi, T., Tamura, E., and Ohe, K. (2015). *Org. Lett.* 17: 5843–5845.

25 Zhang, X.H. (2017). *J. Chem. Sci.* 129: 505–513.

26 Zhao, J.-F. and Loh, T.-P. (2009). *Angew. Chem. Int. Ed.* 48: 7232–7235.

27 Liang, R., Jiang, H., and Zhu, S. (2015). *Chem. Commun.* 51: 5530–5533.

28 Shen, L., Zhao, K., Doitomi, K. et al. (2017). *J. Am. Chem. Soc.* 139: 13570–13578.

29 Liu, Y., Liu, M., and Song, Z. (2005). *J. Am. Chem. Soc.* 127: 3662–3663.

30 Li, Y., Liu, X., Jiang, H. et al. (2011). *Angew. Chem. Int. Ed.* 50: 6341–6345.

31 Fürstner, A., Schlecker, A., and Lehmann, C.W. (2007). *Chem. Commun.* 4277–4279.

32 Bai, Y.-B., Luo, Z., Wang, Y. et al. (2018). *J. Am. Chem. Soc.* 140: 5860–5865.

33 (a) Sakai, K., Kochi, T., and Kakiuchi, F. (2013). *Org. Lett.* 15: 1024–1027. (b) Zhao, L., Zhang, L., and Fang, D.-C. (2016). *Organometallics* 35: 3577–3586.

34 Smith, D.L., Chidipudi, S.R., Goundry, W.R., and Lam, H.W. (2012). *Org. Lett.* 14: 4934–4937.

35 Sweis, R.F., Schramm, M.P., and Kozmin, S.A. (2004). *J. Am. Chem. Soc.* 126: 7442–7443.

36 Koldobskii, A.B., Solodova, E.V., Godovikov, I.A., and Kalinin, V.N. (2008). *Tetrahedron* 64: 9555–9560.

37 Yuan, Y., Bai, L., Nan, J. et al. (2014). *Org. Lett.* 16: 4316–4319.

38 (a) Kimura, M., Horino, Y., Wakamiya, Y. et al. (1997). *J. Am. Chem. Soc.* 119: 10869–10870. (b) Horino, Y., Kimura, M., Tanaka, S. et al. (2003). *Chem. Eur. J.* 9: 2419–2438.

39 Ding, W. and Yoshikai, N. (2019). *Angew. Chem. Int. Ed.* 58: 2500–2504.

40 Selected reports, see: (a) Villeneuve, K. and Tam, W. (2004). *Angew. Chem. Int. Ed.* 43: 610–613. (b) Takenaka, Y., Ito, H., Hasegawa, M., and Iguchi, K. (2006). *Tetrahedron* 62: 3380–3388. (c) Ishihara, K. and Fushimi, M. (2008). *J. Am. Chem. Soc.* 130: 7532–7533. (d) Schotes, C. and Mezzetti, A. (2011). *Angew. Chem. Int. Ed.* 50: 3072–3074. (e) Hu, J., Yang, Q., Yu, L. et al. (2013). *Org. Biomol. Chem.* 11: 2294–2301. (f) Maturi, M.M. and Bach, T. (2014). *Angew. Chem. Int. Ed.* 53: 7661–7664. (g) Enomoto, K., Oyama, H., and Nakada, M. (2015). *Chem. Eur. J.* 21: 2798–2802. (h) Kang, T., Ge, S., Lin, L. et al. (2016). *Angew. Chem. Int. Ed.* 55: 5541–5544. (i) Kossler, D. and Cramer, N. (2017). *Chem. Sci.* 8: 1862–1866. (j) Jiao, Z., Shi, Q., and Zhou, J.S. (2017). *Angew. Chem. Int. Ed.* 56: 14567–14571. (k) Parsutkar, M.M., Pagar, V.V., and RajanBabu, T.V. (2019). *J. Am. Chem. Soc.* 141: 15367–15377. (l) Meng, X., Li, J., Zuo, Y. et al. (2019). *Organometallics* 38: 3111–3123.

41 Hoyt, J.M., Schmidt, V.A., Tondreau, A.M., and Chirik, P.J. (2015). *Science* 349: 960–963.

42 Ischay, M.A., Ament, M.S., and Yoon, T.P. (2012). *Chem. Sci.* 3: 2807–2811.

43 (a) Hurtley, A.E., Lu, Z., and Yoon, T.P. (2014). *Angew. Chem., Int. Ed.* 53: 8991–8994. (b) Wang, C. and Lu, Z. (2017). *Org. Lett.* 19: 5888–5891. (c) Pagire, S.K., Hossain, A., Traub, L. et al. (2017). *Chem. Commun.* 53: 12072–12075. (d) Singh, K., Trinh, W., and Weaver, J.D. III (2019). *Org. Biomol. Chem.* 17: 1854–1861.

44 Jha, K.K., Dutta, S., Sar, S. et al. (2018). *Tetrahedron* 74: 7326–7334.

45 Jiao, M., Ju, Y.-W., and Chen, B.-Z. (2018). *Tetrahedron Lett.* 59: 1651–1660.

46 Mohr, L.-M., Bauer, A., Jandl, C., and Bach, T. (2019). *Org. Biomol. Chem.* 17: 7192–7203.

47 Jiang, Y., Wang, C., Rogers, C.R. et al. (2019). *Nat. Chem.* 11: 1034–1040.

48 (a) Snider, B.B. and Spindell, D.K. (1980). *J. Org. Chem.* 45: 5017–5021. (b) Hoffmann, H.M.R., Ismail, Z.M., and Weber, A. (1981). *Tetrahedron Lett.* 22: 1953–1956. (c) Snider, B.B. and Ron, E. (1986). *J. Org. Chem.* 51: 3643–3652. (d) Conner, M.L. and Brown, M.K. (2016). *J. Org. Chem.* 81: 8050–8060.

49 Saito, S., Hirayama, K., Kabuto, C., and Yamamoto, Y. (2000). *J. Am. Chem. Soc.* 122: 10776–10780.

50 Zheng, W.-F., Bora, P.P., Sun, G.-J., and Kang, Q. (2016). *Org. Lett.* 18: 3694–3697.

51 (a) Conner, M.L., Xu, Y., and Brown, M.K. (2015). *J. Am. Chem. Soc.* 137: 3482–3485. (b) Wiest, J.M., Conner, M.L., and Brown, M.K. (2018). *J. Am. Chem. Soc.* 140: 15943–15949.

52 Luzung, M.R., Mauleýn, P., and Toste, F.D. (2007). *J. Am. Chem. Soc.* 129: 12402–12403.

53 Gulías, M., Collado, A., Trillo, B. et al. (2011). *J. Am. Chem. Soc.* 133: 7660–7663.

54 Noucti, N.N. and Alexanian, E.J. (2015). *Angew. Chem. Int. Ed.* 54: 5447–5450.

55 (a) Yujiro, H. and Koichi, N. (1989). *Chem. Lett.* 18: 793–796. (b) Narasaka, K., Hayashi, Y., Shimadzu, H., and Niihata, S. (1992). *J. Am. Chem. Soc.* 114: 8869–8885.

56 Du, J., Skubi, K.L., Schultz, D.M., and Yoon, T.P. (2014). *Science* 344: 392–396.

57 Sakamoto, T., Mochizuki, T., Goto, Y. et al. (2018). *Chem. Asian J.* 13: 2373–2377.

58 Li, R., Ma, B.C., Huang, W. et al. (2017). *ACS Catal.* 7: 3097–3101.

59 Gabriele, B., Mancuso, R., and Veltri, L. (2016). *Chem. Eur. J.* 22: 5056–5094.

60 Johnson, E.S., Balaich, G.J., Fanwick, P.E., and Rothwell, I.P. (1997). *J. Am. Chem. Soc.* 119: 11086–11087.

61 (a) Radhakrishnan, U., Gevorgyan, V., and Yamamoto, Y. (2000). *Tetrahedron Lett.* 41: 1971–1974.

62 Shibata, Y. and Tanaka, K. (2011). *Angew. Chem. Int. Ed.* 50: 10917–10921.

63 Li, X., Yang, X., and Qi, Z. (2016). *ACS Catal.* 6: 6372–6376.

64 Shibata, Y., Noguchi, K., and Tanaka, K. (2010). *J. Am. Chem. Soc.* 132: 7896–7898.

65 Shibata, Y., Noguchi, K., and Tanaka, K. (2010). *Org. Lett.* 12: 5596–5599.

66 Gao, Y., Wu, W., Huang, H. et al. (2014). *Chem. Commun.* 50: 846–848.

67 Su, X., Chen, B., Wang, S. et al. (2018). *ACS Catal.* 8: 7760–7765.

68 (a) Zhang, D., Yum, E.K., Liu, Z., and Larock, R.C. (2005). *Org. Lett.* 7: 4963–4966. (b) Guo, L.-N., Duan, X.-H., Bi, H.-P. et al. (2006). *J. Org. Chem.* 71: 3325–3327. (c) Duan, X.-H., Guo, L.-N., Bi, H.-P. et al. (2006). *Org. Lett.* 8: 3053–3056. (d) Zhang, D., Liu, Z., Yum, E.K., and Larock, R.C. (2007). *J. Org. Chem.* 72: 251–262. (e) Guo, L.-N., Duan, X.-H., Bi, H.-P. et al. (2007). *J. Org. Chem.* 72: 1538–1540.

69 Khan, Z.A. and Wirth, T. (2009). *Org. Lett.* 11: 229–231.

70 Bu, X., Hong, J., and Zhou, X. (2011). *Adv. Synth. Catal.* 353: 2111–2118.

71 Deng, R., Sun, L., and Li, Z. (2007). *Org. Lett.* 9: 5207–5210.

72 Dong, C.-G., Yeung, P., and Hu, Q.-S. (2007). *Org. Lett.* 9: 363–366.

73 Zeng, X., Ilies, L., and Nakamura, E. (2011). *J. Am. Chem. Soc.* 133: 17638–17640.

74 Kumaraswamy, G. and Balakishan, G. (2015). *Eur. J. Org. Chem.* 2015: 3141–3146.

75 Han, S.H., Kim, S., Mishra, N.K. et al. (2017). *Asian J. Org. Chem.* 6: 1823–1829.

76 Chen, Y.-Y., Chen, Z.-Y., Zhang, N.-N. et al. (2016). *Eur. J. Org. Chem.* 599–606.

77 Dethe, D.H., Murhade, G.M., and Ghosh, S. (2015). *J. Org. Chem.* 80: 8367–8376.

78 Dethe, D.H. and Murhade, G.M. (2013). *Chem. Commun.* 49: 8051–8053.

79 Zhou, F., Yang, M., and Lu, X. (2009). *Org. Lett.* 11: 1405–1408.

80 (a) Muralirajan, K., Parthasarathy, K., and Cheng, C.-H. (2011). *Angew. Chem. Int. Ed.* 50: 4169–4172. (b) Patureau, F.W., Besset, T., Kuhl, N., and Glorius, F. (2011). *J. Am. Chem. Soc.* 133: 2154–2156.

81 Tsuchikama, K., Kasagawa, M., Endo, K., and Shibata, T. (2010). *Synlett* 97–100.

82 (a) Chinnagolla, R.K. and Jeganmohan, M. (2012). *Eur. J. Org. Chem.* 417–423. (b) Zhang, J., Ugrinov, A., and Zhao, P. (2013). *Angew. Chem. Int. Ed.* 52: 6681–6684.

83 Yu, Y., Wu, Q., Liu, D. et al. (2019). *J. Org. Chem.* 84: 7449–7458.

84 Shintani, R., Okamoto, K., and Hayashi, T. (2005). *Chem. Lett.* 34: 1294–1295.

85 Zhao, P., Wang, F., Han, K., and Li, X. (2012). *Org. Lett.* 14: 5506–5509.

86 Nagamoto, M. and Nishimura, T. (2014). *Chem. Commun.* 50: 6274–6277.

87 Guo, B., Zheng, L., Zhang, L., and Hua, R. (2015). *J. Org. Chem.* 80: 8430–8434.

88 Dhanorkar, R.J., Gupta, V.K., and Gocher, C.P. (2019). *Asian J. Org. Chem.* 8: 909–915.

89 Sanjuán, A.M., Rashid, M.A., García-García, P. et al. (2015). *Chem. Eur. J.* 21: 3042–3052. and references cited therein.

90 García-García, P., Sanjuán, A.M., Rashid, M.A. et al. (2017). *J. Org. Chem.* 82: 1155–1165.

91 (a) Zhang, C. and Lu, X. (1995). *J. Org. Chem.* 60: 2906–2908. (b) Lu, X., Zhang, C., and Xu, Z. (2001). *Acc. Chem. Res.* 34: 535–544.

92 (a) Wei, Y. and Shi, M. (2017). *Org. Chem. Front.* 4: 1876–1890. (b) Gao, Y.-W., Niu, H.-Y., Zhang, Q.-Y. et al. (2018). *Adv. Synth. Catal.* 360: 2813–2819. and references cited therein.

93 Selected reports, see: (a) Wilson, J.E. and Fu, G.C. (2006). *Angew. Chem. Int. Ed.* 45: 1426–1429. (b) Pinto, N., Retailleau, P., Voituriez, A., and Marinetti, A. (2011). *Chem. Commun.* 47: 1015–1017. (c) Fujiwara, Y. and Fu, G.C. (2011). *J. Am. Chem. Soc.* 133: 12293–12297. (d) Steurer, M., Jensen, K.L., Worgull, D., and Jørgensen, K.A. (2012). *Chem. Eur. J.* 18: 76–79. (e) Gicquel, M., Zhang, Y., Aillard, P. et al. (2015). *Angew. Chem. Int. Ed.* 54: 5470–5473.

94 Staben, S.T., Kennedy-Smith, J.J., and Toste, F.D. (2004). *Angew. Chem. Int. Ed.* 43: 5350–5352.

95 Kennedy-Smith, J.J., Staben, S.T., and Toste, F.D. (2004). *J. Am. Chem. Soc.* 126: 4526–4527.

96 Corkey, B.K. and Toste, F.D. (2005). *J. Am. Chem. Soc.* 127: 17168–17169.

97 Gao, Q., Zheng, B.-F., Li, J.-H., and Yang, D. (2005). *Org. Lett.* 7: 2185–2188.

98 Chan, L.Y., Lim, S., Park, Y., and Lee, P.H. (2012). *J. Org. Chem.* 77: 5239–5244.

99 Barluenga, J., Palomas, D., Rubio, E., and González, J.M. (2007). *Org. Lett.* 9: 2823–2826.

100 Zhou, C.-Y. and Che, C.-M. (2007). *J. Am. Chem. Soc.* 129: 5828–5829.

101 Binder, J.T., Crone, B., Haug, T.T. et al. (2008). *Org. Lett.* 10: 1025–1028.

102 Kinoshita, H., Miyama, C., and Miura, K. (2016). *Tetrahedron Lett.* 57: 5065–5069.

103 Mochizuki, K., Sakai, K., Kochi, T., and Kakiuchi, F. (2013). *Synthesis* 45: 2088–2092.

104 Meguro, M., Kamijo, S., and Yamamoto, Y. (1996). *Tetrahedron Lett.* 37: 7453–7456.

105 Lee, C.-Y. and Wu, M.-J. (2007). *Eur. J. Org. Chem.* 2007: 3463–3467.

106 Ye, S., Gao, K., Zhou, H. et al. (2009). *Chem. Commun.* 5406–5408.

107 Zhou, F., Han, X., and Lu, X. (2011). *J. Org. Chem.* 76: 1491–1494.

108 Saunthwal, R.K., Danodia, A.K., Patel, M. et al. (2016). *Chem. Asian J.* 11: 3001–3007.

109 Wang, Z., Zhong, J., Zheng, C., and Fan, R. (2017). *Org. Chem. Front.* 4: 1005–1010.

110 Reppe, W., Schlichting, O., Klager, K., and Toepel, T. (1948). *Justus Liebigs Ann. Chem.* 560: 1–92.

111 Selected reports, see: (a) Sato, Y., Nishimata, T., and Mori, M. (1994). *J. Org. Chem.* 59: 6133–6155. (b) Sato, Y., Ohashi, K., and Mori, M. (1999). *Tetrahedron Lett.* 40: 5231–5234. (c) Xi, C., Sun, Z., and Liu, Y. (2013). *Dalton Trans.* 42: 13327–13330. (d) Xue, F., Loh, Y.K., Song, X. et al. (2017). *Chem. Asian J.* 12: 168–173. (e) Shen, L., Zhao, Y., Luo, Q. et al. (2019). *Dalton Trans.* 48: 4643–4649. (f) Orsino, A.F., del Campo, M.G., Lutz, M., and Moret, M.-E. (2019). *ACS Catal.* 9: 2458–2481. (g) Yang, K., Wang, P., Sun, Z.-Y. et al. (2021). *Org. Lett.* 23: 3933–3938.

112 Selected reports, see: (a) Jhingan, A.K. and Maier, W.F. (1987). *J. Org. Chem.* 52: 1161–1165. (b) Yamamoto, Y., Nagata, A., and Itoh, K. (1999). *Tetrahedron*

Lett. 40: 5035–5038. (c) Carvalho, M.F.N.N., Almeida, F.M.T., Galvão, A.M., and Pombeiro, A.J.L. (2003). *J. Organomet. Chem.* 679: 143–147. (d) Cheng, J.-S., Jiang, H.-F., Ouyang, X.-Y., and Zhang, Q.-J. (2003). *Chin. J. Chem.* 21: 361–364. (e) Li, J.-H. and Xie, Y.-X. (2004). *Synth. Commun.* 34: 1737–1737. (f) Li, J.-H. and Xie, Y.-X. (2004). *Chin. J. Chem.* 22: 1421–1424.

113 Selected reports, see: (a) Grigg, R., Scott, R., and Stevenson, P. (1988). *J. Chem. Soc., Perkin Trans. 1* 1357–1364. (b) Witulski, R. and Stengel, T. (1999). *Angew. Chem. Int. Ed.* 38: 2426–2430. (c) Witulski, B. and Alayrac, C. (2002). *Angew. Chem. Int. Ed.* 41: 3281–3284. (d) Tanaka, K. and Shirasaka, K. (2003). *Org. Lett.* 5: 4697–4699. (e) Yoshida, K., Morimoto, I., Mitsudo, K., and Tanaka, T. (2007). *Chem. Lett.* 36: 998–999. (f) Wu, W., Zhang, X.Y., and Kang, S.X. (2010). *Chin. Chem. Lett.* 21: 18–22. (g) Yoshida, K., Morimoto, I., Mitsudo, K., and Tanaka, H. (2008). *Tetrahedron* 64: 5800–5807. (h) Tanaka, K., Sagae, H., Toyoda, K., and Hirano, M. (2008). *Tetrahedron* 64: 831–846. (i) Wang, Y.-H., Huang, S.-H., Lin, T.-C., and Tsai, T.-C. (2010). *Tetrahedron* 66: 7136–7141. (j) Torres, Ò., Fernàndez, M., Díaz-Jiménez, À. et al. (2019). *Organometallics* 38: 2853–2862.

114 Selected reports, see: (a) Vollhardt, K.P.C. (1984). *Angew. Chem. Int. Ed. Engl.* 23: 539–644. (b) Mao, F., Schut, D.M., and Tyler, D.R. (1996). *Organometallics* 15: 4770–4775. (c) Doszczak, L., Fey, P., and Tacke, R. (2007). *Synlett* 753–756. (d) Kumon, T., Yamada, S., Agou, T. et al. (2018). *J. Fluorine Chem.* 213: 11–17.

115 Selected reports, see: (a) Yamamoto, Y., Ogawa, R., and Itoh, K. (2000). *Chem. Commun.* 549–550. (b) Witulski, B., Stengel, T., and Fernández-Hernández, J.M. (2000). *Chem. Commun.* 1965–1966. (c) Yamamoto, Y., Ishii, J.-i., Nishiyama, H., and Itoh, K. (2004). *J. Am. Chem. Soc.* 126: 3712–3713. (d) Yamamoto, Y., Hashimoto, T., Hattori, K. et al. (2006). *Org. Lett.* 8: 3565–3568. (e) Gierada, M., Czelusniak, I., and Handzlik, J. (2019). *Mol. Catal.* 469: 18–26. (f) Silvestri, A.P. and Oakdale, J.S. (2020). *Chem. Commun.* 56: 13417–13420.

116 Selected reports, see: (a) Farnetti, E. and Marsich, N. (2004). *J. Organomet. Chem.* 689: 14–17. (b) Shibata, T., Fujimoto, T., Yokota, K., and Takagi, K. (2004). *J. Am. Chem. Soc.* 126: 8382–8383. (c) Shibata, T. and Tsuchikama, K. (2005). *Chem. Commun.* 6017–6019. (d) Shibata, T., Arai, Y., Takami, K. et al. (2006). *Adv. Synth. Catal.* 348: 2475–2483. (e) Kezuka, S., Tanaka, S., Ohe, T. et al. (2006). *J. Org. Chem.* 71: 543–552. (f) Shibata, T., Yoshida, S., Arai, Y. et al. (2008). *Tetrahedron* 64: 821–830.

117 Selected reports, see: (a) Brenna, D., Villa, M., Gieshoff, T.N. et al. (2017). *Angew. Chem. Int. Ed.* 56: 8451–8454. (b) Gawali, S.S. and Gunanathan, C. (2019). *J. Organomet. Chem.* 881: 139–149. (c) Witzke, R.J., Hait, D., Chakarawet, K. et al. (2020). *ACS Catal.* 10: 7800–7807. (d) Provis-Evans, C.B., Lau, S., Krewald, V., and Webster, R.L. (2020). *ACS Catal.* 10: 10157–10168. (e) Neumeier, M., Chakraborty, U., Schaarschmidt, D. et al. (2020). *Angew. Chem. Int. Ed.* 59: 13473–13478.

118 Kakeya, M., Fujihara, T., Kasaya, T., and Nagasawa, A. (2006). *Organometallics* 25: 4131–4137.

119 (a) Lautens, M., Klute, W., and Tam, W. (1996). *Chem. Rev.* 96: 49–92. (b) Saito, S. and Yamomoto, Y. (2000). *Chem. Rev.* 100: 2901–2915. (c) Chopade, P.R. and

Louie, J. (2006). *Adv. Synth. Catal.* 348: 2307–2327. (d) Domínguez, G. and Pérez-Castells, J. (2011). *Chem. Soc. Rev.* 40: 3430–3444. (e) Chinchilla, R. and Nájera, C. (2014). *Chem. Rev.* 114: 1783–1826. (f) Amatore, M. and Aubert, C. (2015). *Eur. J. Org. Chem.* 2015: 265–286. (g) Kotha, S., Lahiri, K., and Sreevani, G. (2018). *Synlett* 29: 2342–2361.

120 Hilt, G., Vogler, T., Hess, W., and Galbiati, F. (2005). *Chem. Commun.* 1474–1475.

121 Hilt, G., Hengst, C., and Hess, W.S. (2008). *Eur. J. Org. Chem.* 2008: 2293–2297.

122 Saino, N., Amemiya, F., Tanabe, E. et al. (2006). *Org. Lett.* 8: 1439–1442.

123 Yang, J. and Verkade, J.G. (2000). *Organometallics* 19: 893–900.

124 Xu, Y.-L., Pan, Y.-M., Wu, Q. et al. (2011). *J. Org. Chem.* 76: 8472–8476.

125 Liu, Y., Yan, X., Yang, N., and Xi, C. (2011). *Catal. Commun.* 12: 489–492.

126 Minakawa, M., Ishikawa, T., Namioka, J. et al. (2014). *RSC Adv.* 4: 41353–41356.

127 Jiang, H.-F., Shen, Y.-X., and Wang, Z.-Y. (2007). *Tetrahedron Lett.* 48: 7542–7545.

128 Tamaso, K., Hatamoto, Y., Sakaguchi, S. et al. (2007). *J. Org. Chem.* 72: 3603–3605.

129 Shen, Y., Jiang, H., and Chen, Z. (2010). *J. Org. Chem.* 75: 1321–1324.

130 Nair, V., Vidya, N., Biju, A.T. et al. (2006). *Tetrahedron* 62: 10136–10140.

131 Kuninobu, Y., Nishi, M., Kawata, A. et al. (2010). *J. Org. Chem.* 75: 334–341.

132 (a) Kuninobu, Y., Nishi, M., Salprima, Y.S., and Takai, K. (2008). *Org. Lett.* 10: 3009–3011. (b) Tsuji, H., Yamagata, K.-i., Fujimoto, T., and Nakamura, E. (2008). *J. Am. Chem. Soc.* 130: 7792–7793. (c) Yoshikai, N., Zhang, S.-L., Yamagata, K.-i. et al. (2009). *J. Am. Chem. Soc.* 131: 4099–4109.

133 Nair, V., Pillai, A.N., Beneesh, P.B., and Suresh, E. (2005). *Org. Lett.* 7: 4625–4628.

134 Li, J., Hua, R., and Liu, T. (2013). *Curr. Org. Synth.* 10: 486–491.

135 Shanmugasundaram, M., Wu, M.-S., and Cheng, C.-H. (2001). *Org. Lett.* 3: 4233–4236.

136 Shanmugasundaram, M., Wu, M.-S., Jeganmohan, M. et al. (2002). *J. Org. Chem.* 67: 7724–7729.

137 Jeevanandam, A., Korivi, R.P., Huang, I.-w., and Cheng, C.-H. (2002). *Org. Lett.* 4: 807–810.

138 Wu, M.-S., Shanmugasundaram, M., and Cheng, C.-H. (2003). *Chem. Commun.* 718–719.

139 Sato, Y., Ohashi, K., and Mori, M. (1999). *Tetrahedron Lett.* 40: 5231–5234.

140 Peters, J.-U. and Blechert, S. (1997). *Chem. Commun.* 1983–1984.

141 Saino, N., Kogure, D., and Okamoto, S. (2005). *Org. Lett.* 7: 3065–3067.

142 Saino, N., Kogure, D., Kase, K., and Okamoto, S. (2006). *J. Organomet. Chem.* 691: 3129–3136.

143 Yamamoto, Y., Nagata, A., and Itoh, K. (1999). *Tetrahedron Lett.* 40: 5035–5038.

144 Yamamoto, Y., Nagata, A., Arikawa, Y. et al. (2000). *Organometallics* 19: 2403–2404.

145 Shibata, T., Tsuchikama, K., and Otsuka, M. (2006). *Tetrahedron: Asymmetry* 17: 614–619.

146 Takeuchi, R., Tanaka, S., and Nakaya, Y. (2001). *Tetrahedron Lett.* 42: 2991–2994.

147 (a) Yamamoto, Y., Kinpara, K., Saigoku, T. et al. (2004). *Org. Biomol. Chem.* 2: 1287–1294. (b) Ura, Y., Sato, Y., Tsujita, H. et al. (2005). *J. Mol. Catal. A: Chem.* 239: 166–171.

148 Yamamoto, Y., Hata, K., Arakawa, T., and Itoh, K. (2003). *Chem. Commun.* 1290–1291.

149 Yamamoto, Y., Arakawa, T., Ogawa, R., and Itoh, K. (2003). *J. Am. Chem. Soc.* 125: 12143–12160.

150 Yamamoto, Y., Hattori, K., Ishii, J.-i. et al. (2005). *Chem. Commun.* 4438–4440.

151 Yamamoto, Y., Hattori, K., and Nishiyama, H. (2006). *J. Am. Chem. Soc.* 128: 8336–8340.

152 Yamamoto, Y. and Hattori, K. (2008). *Tetrahedron* 64: 847–855.

153 Saito, S., Salter, M.M., Gevorgyan, V. et al. (1996). *J. Am. Chem. Soc.* 118: 3970–3971.

154 Gevorgyan, V., Takeda, A., and Yamamoto, Y. (1997). *J. Am. Chem. Soc.* 119: 11313–11314.

155 Rubina, M., Conley, M., and Gevorgyan, V. (2006). *J. Am. Chem. Soc.* 128: 5818–5827.

156 Gevorgyan, V., Quan, L.G., and Yamamoto, Y. (1998). *J. Org. Chem.* 63: 1244–1247.

157 Zatolochnaya, O.V. and Gevorgyan, V. (2013). *Org. Lett.* 15: 2562–2565.

158 Gevorgyan, V., Takeda, A., Homma, M. et al. (1999). *J. Am. Chem. Soc.* 121: 6391–6402.

159 Tanaka, K., Sagae, H., Toyoda, K. et al. (2007). *J. Am. Chem. Soc.* 129: 1522–1523.

160 Tanaka, K., Nishida, G., Wada, A., and Noguchi, K. (2004). *Angew. Chem. Int. Ed.* 43: 6510–6512.

161 Doherty, S., Knight, J.G., Smyth, C.H. et al. (2007). *Org. Lett.* 9: 4925–4928.

162 Nishida, G., Ogaki, S., Yusa, Y. et al. (2008). *Org. Lett.* 10: 2849–2852.

163 Selected reports, see: (a) Chang, M., Wu, C., Zheng, J., and Huang, Y. (2016). *Chem. Asian J.* 11: 1512–1517. (b) Zhou, C.-J., Gao, H., Huang, S.-L. et al. (2019). *ACS Catal.* 9: 556–564. and references cited therein. (c) Saunthwal, R.K., Saini, K.M., Patel, M., and Verma, A.K. (2017). *Tetrahedron* 73: 2415–2431.

164 Reddy, C.R., Dilipkumar, U., and Reddy, M.D. (2014). *Org. Lett.* 16: 3792–3795.

165 Song, C., Dong, X., Wang, Z. et al. (2019). *Angew. Chem. Int. Ed.* 58: 12206–12210.

166 Selected reviews, see: (a) Watson, M.D., Fechtenkötter, A., and Müllen, K. (2001). *Chem. Rev.* 101: 1267–1300. (b) de Koning, C.B., Rousseau, A.L., and van Otterlo, W.A.L. (2003). *Tetrahedron* 59: 7–36. (c) Coropceamu, V., Cornil, J., Da Silva Filho, D.A. et al. (2007). *Chem. Rev.* 107: 926–952. (d) Sergeyev, S., Pisula, W., and Geerts, Y.H. (2007). *Chem. Soc. Rev.* 36: 1902–1929. (e) Dong, H., Zhu, H., Meng, Q. et al. (2012). *Chem. Soc. Rev.* 41: 1754–1808. (f) Keyte,

I.J., Harrisonz, R.M., and Lammel, G. (2013). *Chem. Soc. Rev.* 42: 9333–9391. (g) Hein, S.J., Lehnherr, D., Arslan, H. et al. (2017). *Acc. Chem. Res.* 50: 2776–2788.

167 Peña, D., Pérez, D., Guitián, E., and Castedo, L. (1999). *J. Am. Chem. Soc.* 121: 5827–5828.

168 Huang, Q. and Larock, R.C. (2002). *Org. Lett.* 4: 2505–2508.

169 Viswanathan, G.S., Wang, M., and Li, C.-J. (2002). *Angew. Chem. Int. Ed.* 41: 2138–2141.

170 Balamurugan, R. and Gudla, V. (2009). *Org. Lett.* 11: 3116–3119.

171 Ponra, S., Vitale, M.R., Michelet, V., and Ratovelomanana-Vidal, V. (2015). *J. Org. Chem.* 80: 3250–3257.

172 Yasukawa, T., Satoh, T., Miura, M., and Nomura, M. (2002). *J. Am. Chem. Soc.* 124: 12680–12681.

173 Ueura, K., Satoh, T., and Miura, M. (2007). *J. Org. Chem.* 72: 5362–5367.

174 Zhang, D., Liu, J., Córdova, A., and Liao, W.-W. (2017). *ACS Catal.* 7: 7051–7063.

175 Wu, Y.-T., Huang, K.-H., Shin, C.-C., and Wu, T.-C. (2008). *Chem. Eur. J.* 14: 6697–6703.

176 Zhang, X., Yu, X., Ji, D. et al. (2016). *Org. Lett.* 18: 4246–4249.

177 Wang, L., Yu, Y., Yang, M. et al. (2017). *Adv. Synth. Catal.* 359: 3818–3825.

178 He, Z. and Huang, Y. (2016). *ACS Catal.* 6: 7814–7823.

179 Shi, Z., Ding, S., Cui, Y., and Jiao, N. (2009). *Angew. Chem. Int. Ed.* 48: 7895–7898.

180 Annamalai, P., Chen, W.-Y., Raju, S. et al. (2016). *Adv. Synth. Catal.* 358: 3642–3648.

181 Gao, P., Liu, J., and Wei, Y. (2013). *Org. Lett.* 15: 2872–2875.

182 Hua, Z.-L., Qian, W.-J., Wang, S. et al. (2009). *J. Org. Chem.* 74: 8787–8793. and references cited therein.

183 Asao, N., Takahashi, K., Lee, S. et al. (2002). *J. Am. Chem. Soc.* 124: 12650–12651.

184 (a) Asao, N., Aikawa, H., and Yamamoto, Y. (2004). *J. Am. Chem. Soc.* 126: 7458–7459. (b) Yue, D., Cá, N.D., and Larock, R.C. (2006). *J. Org. Chem.* 71: 3381–3388. (c) Asao, N. and Aikawa, H. (2006). *J. Org. Chem.* 71: 5249–5253.

185 Isogai, Y., Menggenbateer, Khan, F.N., and Asao, N. (2009). *Tetrahedron* 65: 9575–9582.

186 Sun, H., Wu, X., and Hua, R. (2011). *Tetrahedron Lett.* 52: 4408–4411.

187 Singha, R., Nandi, S., and Ray, J.K. (2012). *Tetrahedron Lett.* 53: 6531–6534.

188 Barluenga, J., Vázquez-Villa, H., Ballesteros, A., and González, J.M. (2003). *Org. Lett.* 5: 4121–4123.

189 Feng, C. and Loh, T.-P. (2010). *J. Am. Chem. Soc.* 132: 17710–17712.

190 Umeda, R., Tabata, H., Tobe, Y., and Nishiyama, Y. (2014). *Chem. Lett.* 43: 883–884.

191 Kang, D., Kim, J., Oh, S., and Lee, P.H. (2012). *Org. Lett.* 14: 5636–5639.

192 Selected reports, see: (a) Okamoto, H., Kawasaki, N., Kaji, Y. et al. (2008). *J. Am. Chem. Soc.* 130: 10470–10471. (b) Ionkin, A.S., Marshall, W.J., Fish, B.M. et al. (2008). *Chem. Commun.* 2319–2321. (c) Shinamura, S., Osaka, I.,

Miyazaki, E. et al. (2011). *J. Am. Chem. Soc.* 133: 5024–5035. (d) Wu, T.-L., Chou, H.-H., Huang, P.-Y. et al. (2014). *J. Org. Chem.* 79: 267–274.

193 Kuznetsov, E.V., Shcherbakova, I.V., and Balaban, A.T. (1990). *Adv. Heterocycl. Chem.* 50: 157–254.

194 Das, A., Liao, H.-H., and Liu, R.-S. (2007). *J. Org. Chem.* 72: 9214–9218.

195 Guo, B., Zheng, L., Yang, L., and Hua, R. (2014). *J. Org. Chem.* 79: 4352–4357.

196 Guo, B., Zhou, Y., Zhang, L., and Hua, R. (2015). *J. Org. Chem.* 80: 7635–7641.

197 Obora, Y., Takeshita, K., and Ishii, Y. (2009). *Org. Biomol. Chem.* 7: 428–431.

198 Kamei, M., Watanabe, K., Fuji, M., and Obora, Y. (2016). *Chem. Lett.* 45: 943–945.

199 Watanabe, K., Satoh, Y., Kamei, M. et al. (2017). *Org. Lett.* 19: 5398–5401.

200 Zhu, S., Huang, X., Zhao, T.-Q. et al. (2015). *Org. Biomol. Chem.* 13: 1225–1233.

201 Tekavec, T.N. and Louie, J. (2008). *Tetrahedron* 64: 6870–6875.

202 Selected review papers, see: (a) Shibata, T. and Tsuchikama, K. (2008). *Org. Biomol. Chem.* 6: 1317–1323. (b) Amatore, M. and Aubert, C. (2015). *Eur. J. Org. Chem.* 2015: 265–286.

203 Shibata, T., Arai, Y., and Tahara, Y.-k. (2005). Enantioselective construction of quaternary carbon centers by catalytic [2 + 2 + 2] cycloaddition of 1,6-enynes and alkynes. *Org. Lett.* 7: 4955–4957.

204 Evans, P.A., Lai, K.W., and Sawyer, J.R. (2005). *J. Am. Chem. Soc.* 127: 12466–12467.

205 (a) Shibata, T., Kawachi, A., Ogawa, M. et al. (2007). *Tetrahedron* 63: 12853–12859. (b) Tanaka, K., Takahashi, M., Imase, H. et al. (2008). *Tetrahedron* 64: 6289–6293. (c) Aida, Y., Sugiyama, H., Uekusa, H. et al. (2016). *Org. Lett.* 18: 2672–2675.

206 (a) Ueda, H., Masutomi, K., Shibata, Y., and Tanaka, K. (2017). *Org. Lett.* 19: 2913–2916.

207 Selected reports, see: Shibata, T.; Tahara, Y.-k. *J. Am. Chem. Soc.* 2006, 128, 11766–11767.

208 (a) Shibata, T., Kurokawa, H., and Kanda, K. (2007). *J. Org. Chem.* 72: 6521–6525. (b) Geny, A., Gaudrel, S., Slowinski, F. et al. (2009). *Adv. Synth. Catal.* 351: 271–275.

209 Nishizawa, M., Takao, H., Yadav, V.K. et al. (2003). *Org. Lett.* 5: 4563–4565.

210 Jiang, J.-L., Ju, J., and Hua, R. (2007). *Org. Biomol. Chem.* 5: 1854–1857.

211 Selected reviews, see: (a) Battiste, M.A., Pelphrey, P.M., and Wright, D.L. (2006). *Chem. Eur. J.* 12: 3438–3447. (b) Butenschön, H. (2008). *Angew. Chem. Int. Ed.* 47: 5287–5290. (c) Pellissier, H. (2011). *Adv. Synth. Catal.* 353: 189–218.

212 Selected reviews, see: (a) Pellissier, H. (2011). *Adv. Synth. Catal.* 353: 189–218. (b) Ylijoki, K.E.O. and Stryker, J.M. (2013). *Chem. Rev.* 113: 2244–2266. (c) Jiao, L. and Yu, Z.-X. (2013). *J. Org. Chem.* 78: 6842–6848. (d) Gao, K., Zhang, Y.-G., Wang, Z., and Ding, H. (2019). *Chem. Commun.* 55: 1859–1878. (e) Wang, J., Blaszczyk, S.A., Li, X., and Tang, W. (2021). *Chem. Rev.* 121: 110–139.

213 (a) Wender, P.A., Takahashi, H., and Witulski, B. (1995). *J. Am. Chem. Soc.* 117: 4720–4721. (b) Wender, P.W. and Sperandio, D. (1998). *J. Org. Chem.* 63: 4164–4165.

214 Selected Trost's group work, see: (a) Trost, B.M. and Shen, H.C. (2000). *Org. Lett.* 2: 2523–2525. (b) Trost, B.M., Toste, F.D., and Shen, H. (2000). *J. Am. Chem. Soc.* 122: 2379–2380. (c) Trost, B.M. and Shen, H.C. (2001). *Angew. Chem. Int. Ed.* 40: 2313–2316. (d) Trost, B.M., Shen, H.C., Schulz, T. et al. (2003). *Org. Lett.* 5: 4149–4151. (e) Trost, B.M., Shen, H.C., Horne, D.B. et al. (2005). *Chem. Eur. J.* 11: 2577–2590.

215 Zuo, G. and Louie, J. (2005). *J. Am. Chem. Soc.* 127: 5798–5799.

216 (a) Liu, P., Sirois, L.E., Cheong, P.H.-Y. et al. (2010). *J. Am. Chem. Soc.* 132: 10127–10135. (b) Zhou, H., Wu, B., Ma, J.-A., and Dang, Y. (2018). *Org. Biomol. Chem.* 16: 4295–4303.

217 (a) Wender, P.A., Rieck, H., and Fuji, M. (1998). *J. Am. Chem. Soc.* 120: 10976–10977. (b) Wender, P.A., Dyckman, A.J., Husfeld, C.O., and Scanio, M.J.C. (2000). *Org. Lett.* 2: 1609–1611. (c) Wender, P.A., Barzilay, C.M., and Dyckman, A.J. (2001). *J. Am. Chem. Soc.* 123: 179–180.

218 (a) Wender, P.A., Husfeld, C.O., Langkopf, E. et al. (1998). *J. Am. Chem. Soc.* 120: 1940–1941. (b) Wender, P.A., Husfeld, C.O., Langkopf, E. et al. (1998). *Tetrahedron* 54: 7203–7220. (c) Wender, P.A. and Williams, T.J. (2002). *Angew. Chem. Int. Ed.* 41: 4550–4553.

219 (a) Wender, P.A., Dyckman, A.J., Husfeld, C.O. et al. (1999). *J. Am. Chem. Soc.* 121: 10442–10443. (b) Wender, P.A., Haustedt, L.O., Lim, J. et al. (2006). *J. Am. Chem. Soc.* 128: 6302–6303.

220 Melcher, M.-C., von Wachenfeldt, H., Sundin, A., and Strand, D. (2015). *Chem. Eur. J.* 21: 531–535.

221 (a) Wang, B., Cao, P., and Zhang, X. (2000). *Tetrahedron Lett.* 41: 8041–8044. (b) Shintani, R., Nakatsu, H., Takatsu, K., and Hayashi, T. (2009). *Chem. Eur. J.* 15: 8692–8694.

222 (a) Hong, X., Stevens, M.C., Liu, P. et al. (2014). *J. Am. Chem. Soc.* 136: 17273–17283. (b) Liu, C.-H. and Yu, Z.-X. (2017). *Angew. Chem. Int. Ed.* 56: 8667–8671.

223 Evans, P.A. and Inglesby, P.A. (2008). *J. Am. Chem. Soc.* 130: 12838–12839.

224 Yu, H., Lu, Q., Dang, Z., and Fu, Y. (2013). *Chem. Asian J.* 8: 2262–2273.

225 Evans, P.A., Negru, D.E., and Shang, D. (2015). *Angew. Chem. Int. Ed.* 54: 4768–4772.

226 Inagaki, F., Sugikubo, K., Miyashita, Y., and Mukai, C. (2010). *Angew. Chem. Int. Ed.* 49: 2206–2210.

227 Inagaki, F., Sugikubo, K., Oura, Y., and Mukai, C. (2011). *Chem. Eur. J.* 17: 9062–9065.

228 Wender, P.A., Bi, F.C., Brodney, M.A., and Gosselin, F. (2001). *Org. Lett.* 3: 2105–2108.

229 Blaszczyk, S.A., Glazier, D.A., and Tang, W. (2020). *Acc. Chem. Res.* 53: 231–243.

230 (a) Komagawa, S. and Saito, S. (2006). *Angew. Chem. Int. Ed.* 45: 2446–2449. (b) Saito, S., Komagawa, S., Azumaya, I., and Masuda, M. (2007). *J. Org. Chem.* 72: 9114–9120.

231 (a) Saito, S., Maeda, K., Yamasaki, R. et al. (2010). *Angew. Chem. Int. Ed.* 49: 1830–1833. and references cited therein. (b) Yamasaki, R., Ohashi, M., Maeda, K. et al. (2013). *Chem. Eur. J.* 19: 3415–3425.

232 Long, H., Song, J., and Xu, H.-C. (2018). *Org. Chem. Front.* 5: 3129–3132.

233 (a) Petasis, N.A. and Patane, M.A. (1992). *Tetrahedron* 48: 5757–5821. (b) Yu, Z.-X., Wang, Y., and Wang, Y. (2010). *Chem. Asian J.* 5: 1072–1088. (c) Anand, A., Singh, P., Kumar, V., and Bhargava, G. (2019). *RSC Adv.* 9: 25554–25568.

234 Hu, Y.-J., Li, L.-X., Han, J.-C. et al. (2020). *Chem. Rev.* 120: 5910–5953. and cited references therein.

235 Kuninobu, Y., Kawata, A., and Takai, K. (2006). *J. Am. Chem. Soc.* 128: 11368–11369.

236 Kuninobu, Y., Kawata, A., Nishi, M. et al. (2009). *Chem. Asian J.* 4: 1424–1433.

237 Hilt, G., Hess, W., Vogler, T., and Hengst, C. (2005). *J. Organomet. Chem.* 690: 5170–5181.

238 Hilt, G. and Janikowski, J. (2008). *Angew. Chem. Int. Ed.* 47: 5243–5245.

239 Bohn, M.A., Schmidt, A., Hilt, G. et al. (2011). *Angew. Chem. Int. Ed.* 50: 9689–9693.

240 Kennedy, C.R., Zhong, H., Macaulay, R.L., and Chirik, P.J. (2019). *J. Am. Chem. Soc.* 141: 8557–8573.

241 Robinson, J.M., Tlais, S.F., Fong, J., and Danheiser, R.L. (2011). *Tetrahedron* 67: 9890–9898.

242 Huber, T., Wildermuth, R.E., and Magauer, T. (2018). *Chem. Eur. J.* 24: 12107–12120.

243 DeBoef, B., Counts, W.R., and Gilbertson, S.R. (2007). *J. Org. Chem.* 72: 799–804.

244 Mukai, C., Ohta, Y., Oura, Y. et al. (2012). *J. Am. Chem. Soc.* 134: 19580–19583.

245 Huang, G. (2015). *J. Org. Chem.* 80: 7564–7571.

246 (a) Holte, D., Götz, D.C.G., and Baran, P.S. (2012). *J. Org. Chem.* 77: 825–842. (b) Hong, X., Holte, D., Götz, D.C.G. et al. (2014). *J. Org. Chem.* 79: 12177–12184.

247 Benn, R., Büssemeier, B., Holle, S. et al. (1985). *J. Organomet. Chem.* 279: 63–86.

248 López-Valdez, L.G., Zuleta-Prada, H., Reyes-Yrejo, B., and Cuevas-Yañez, E. (2018). *Tetrahedron* 74: 1581–1612.

8

Heterocycles from Cycloaddition of Alkynes

Different sized heterocycles containing one or more heteroatoms not only display an intrinsic application as versatile building blocks in organic synthesis and organic materials [1] but also occur in a vast number of biologically active compounds and drugs [2]. Therefore, the development of efficient preparative protocols for heterocyclic compounds is an important and popular research area in modern synthetic organic chemistry [3]. The synthetic methods for construction of heterocycles with the use of unsaturated hydrocarbons of alkynes, enynes, 1,3-butadynes, alkenes, and allenes show generally atom economical, which have been well established to develop efficient approaches to functionalized and *multi*-substituted heterocycles [4].

This chapter focuses on summarizing the synthetic methodologies for the formation of four-, five-, and six-membered heterocyclic compounds via the cyclization process with the use of alkynes.

8.1 Four-membered Heterocycles

The strained four-membered heterocycles are very important and are interesting compounds due to their special structural property, and they possess considerable synthetic/medicinal application [5], but there are very few examples for the synthesis of four-membered heterocycles from alkyne annulation.

Four-membered oxetane ring has been increasingly exploited in its synthesis, its physicochemical properties as a stable motif in medicinal chemistry, and its propensity to undergo ring-opening reactions as a synthetic intermediate [6].

Zhang and coworker developed an efficient synthetic method for the formation of oxetan-3-ones by an Au(I)-catalyzed oxidative intramolecular cycloisomerization of propargylic alcohols in the presence of pyridine *N*-oxide derivatives as oxidants (Scheme 8.1) [7]. The formation of α-oxo gold carbenes is considered to be the key intermediate for the formation of oxetane ring. By suitable selection of ligand for catalyst and the oxidants, both terminal and internal propargylic alcohols bearing electron-withdrawing carboxylate group showed good reactivity to afford substituted oxetan-3-one derivatives in good yields.

Addition Reactions with Unsaturated Hydrocarbons, First Edition. Ruimao Hua.

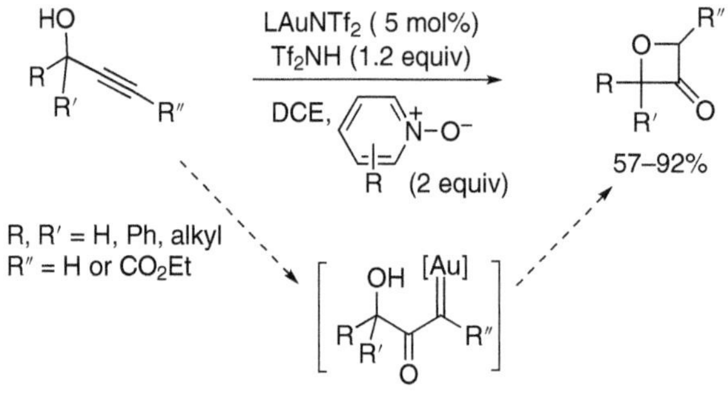

Scheme 8.1 Gold-catalyzed synthesis of oxetan-3-ones from propargylic alcohols.

Very interestingly, Rosenberg's group reported a [2 + 2] cycloaddition reaction of Ru=P double bond with alkenes [8] and alkynes [9]. As shown in Scheme 8.2, the five-coordinate half-sandwich complexes [Ru(η^5-indenyl)(PR_2)(PPh_3)] (R = Cy, ^{i}Pr) react rapidly with acrylonitrile to give two four-membered ruthenium- and phosphorus-heterocyclic complexes, and ethylene (1 atm) also undergoes [2 + 2] cycloaddition reactions rapidly and quantitatively, but the cycloaddition reaction of 1-hexene and the strained internal alkene of norbornene require longer reaction times. Similarly, phenyl acetylene also undergoes the [2 + 2] cycloaddition reaction rapidly to give four-membered phosphametallacyclobutene complexes regioselectively.

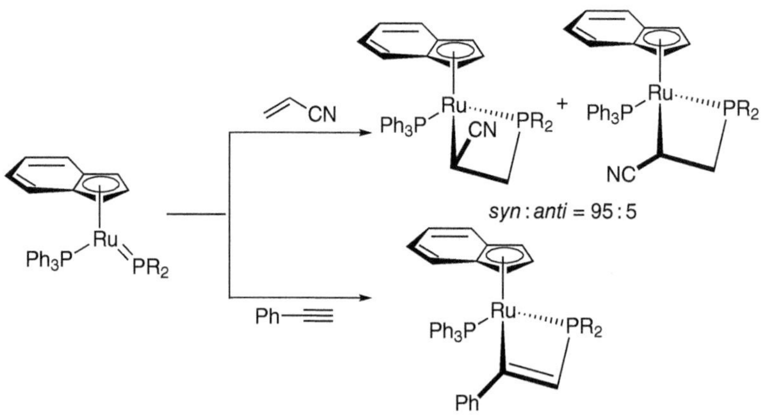

Scheme 8.2 [2 + 2] Cycloaddition reactions of Ru=P with alkenes and alkynes.

On the other hand, the synthesis of four-membered heterocycles from the electrophilic heteroatom cyclization of alkenes has been extensively studied, and a comprehensive review paper has been published by Rousseau and coworker [10].

8.2 Five-membered Heterocycles

Five-membered heterocycles, such as pyrroles, furans, thiophenes, pyrazoles, imidazoles, isoxazoles, oxazoles, etc. are commonly found in many important natural products, pharmaceuticals, and polymers, and they are also the versatile intermediates approach to other heterocyclic compounds (Scheme 8.3) [11]. Therefore, the syntheses of these fundamental structural building blocks have received significant attention, and a wide variety of the synthetic methods have been developed [12]. In this section, a number of essential processes for the formation of five-membered heterocycles and their benzo-fused derivatives from alkynes are introduced.

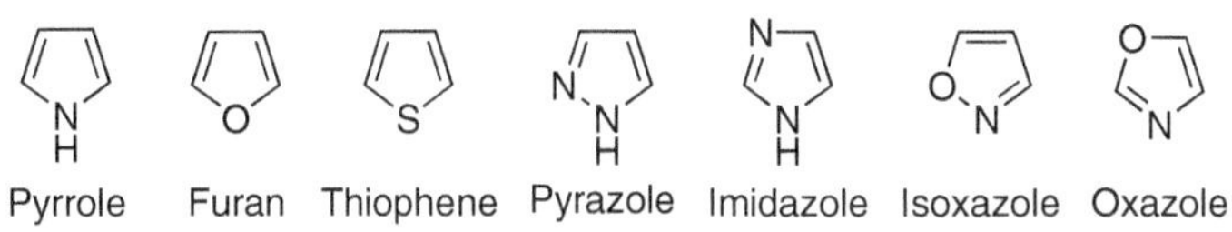

Scheme 8.3 Representative five-membered heterocycles summarized in this section.

8.2.1 Pyrroles, Furans, and Thiophenes Synthesis

One-heteroatom five-membered heterocycles such as pyrroles, furans, thiophenes, and their benzo-fused compounds are very important compounds in organic chemistry and medicinal chemistry. Development of the synthetic methods is one of interesting research topics in organic chemistry.

Pyrroles are not only important building blocks in organic synthesis [13] but also widely used for biological studies [14] and material molecules [15]. Therefore, numerous processes have been developed for the construction of pyrrole ring; in particular, the cycloaddition of alkynes and vinyl-containing compounds with nitrogen sources to develop the straightforward, convenient, and efficient routes to pyrrole derivatives has been well studied [16].

Scheme 8.4 shows the transition metal-catalyzed [2 + 2 + 1] cycloaddition of two alkynes with primary amines giving pentasubstituted pyrroles, including $AgBF_4$ [17] and $Cu(OAc)_2$ [18]-catalyzed oxidative coupling of electron-deficient internal alkynes with primary amines in the presence of 1.2 equivalents of $PhI(OAc)_2$ and 1.0 equivalent of Cs_2CO_3 under O_2 atmosphere, respectively. The oxidative coupling of diarylalkynes and primary amines catalyzed by $PdCl_2$ combined with the use of 4.0 equivalents of CuCl also affords the pentasubstituted pyrroles [19].

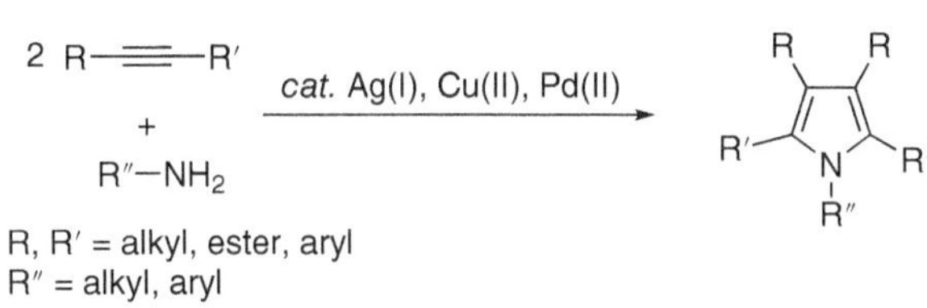

Scheme 8.4 [2 + 2 + 1] Cycloaddition of two alkynes with primary amines affording pentasubstituted pyrroles.

Tonks and coworker developed another type of [2 + 2 + 1] cyclocoupling of two alkynes with azobenzenes affording tetra- and pentasubstituted pyrroles in the presence of titanium [20] or vanadium complexes (Scheme 8.5) [21].

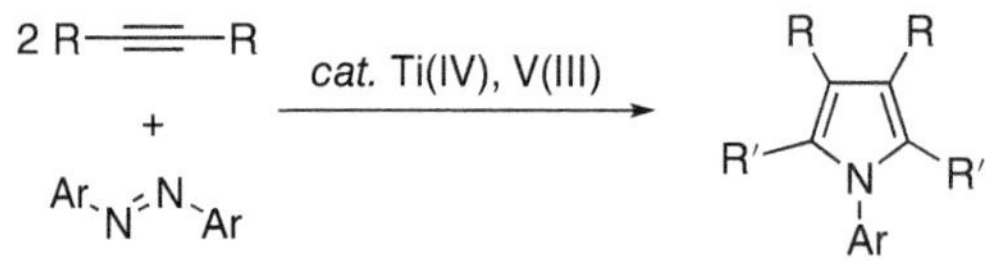

Scheme 8.5 [2 + 2 + 1] Cyclocoupling of two alkynes with azobenzenes affording pentasubstituted pyrroles.

Camp and coworker studied a $PPh_3AuCl/AgOTf$-catalyzed condensation of oximes with electron-deficient internal alkynes to afford pyrroles with high regioselectivity (Scheme 8.6) [22]. The proposed mechanism involves the initial gold-promoted addition of the oxime oxygen to the activated internal alkyne giving the *O*-vinyloxime in situ and subsequently transforming into the pyrrole ring via gold-catalyzed tautomerization, [3.3]-sigmatropic rearrangement, and final cyclodehydration process.

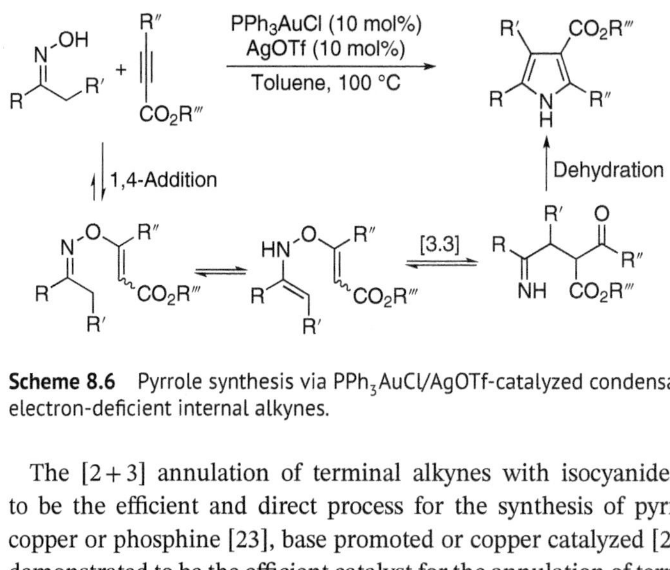

Scheme 8.6 Pyrrole synthesis via $PPh_3AuCl/AgOTf$-catalyzed condensation of oximes with electron-deficient internal alkynes.

The [2 + 3] annulation of terminal alkynes with isocyanides has been found to be the efficient and direct process for the synthesis of pyrroles catalyzed by copper or phosphine [23], base promoted or copper catalyzed [24]. Ag_2CO_3 is also demonstrated to be the efficient catalyst for the annulation of terminal alkynes with isocyanides to afford 2,3-disubstituted pyrroles in good yields with high selectivity (Scheme 8.7) [25].

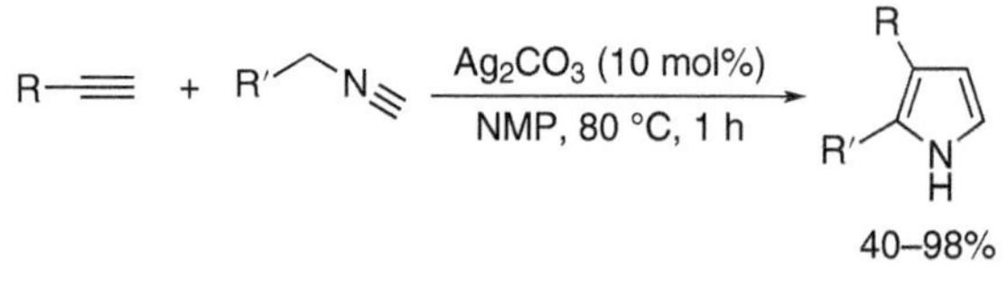

Scheme 8.7 Pyrrole synthesis by silver-catalyzed cycloaddition of alkynes with isocyanides.

Enamines are of great importance in the synthesis of nitrogen-containing heterocyclic compounds [26]. Glorius and coworker studied a $[Cp^*RhCl_2]_2/AgSbF_6$-

catalyzed reaction of enamines with internal alkynes to disclose a substituted group-controlled formation of pyrrole ring via activation of either vinylic sp^2 C—H bond or allylic sp^3 C—H bond and the subsequent coupling with alkynes yielding pyrroles (Scheme 8.8) [27].

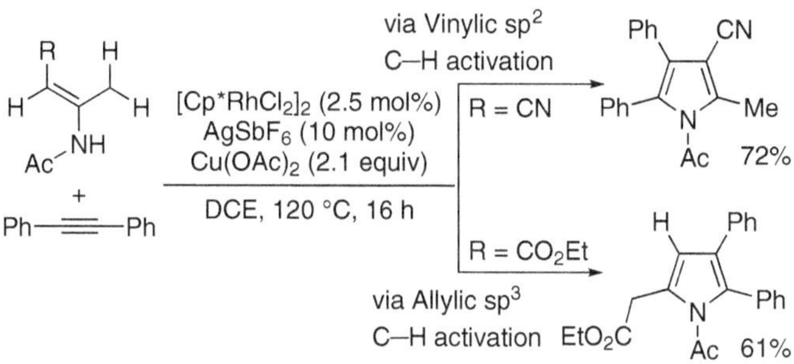

Scheme 8.8 Rhodium-catalyzed and substituted group-controlled formation of pyrroles.

Liang and coworker developed an oxidative annulation of β-enamino ketones or esters with electron-deficient internal alkynes catalyzed by CuI in the presence of O_2 to provide a straightforward method for the synthesis of pentasubstituted pyrroles (Scheme 8.9) [28]. A similar procedure of enamides with internal alkynes via the cleavage of C(sp^2)—H/N—H bonds in the presence of $[RuCl_2(p\text{-cymene})]_2$ as the catalyst and $Cu(OAc)_2 \cdot H_2O$ as the oxidant afforded *N*-acetyl pyrroles, and with the addition of $AgSbF_6$ and MeOH to the previously mentioned reaction system, the reaction gave unprotected pyrroles (Scheme 8.10) [29].

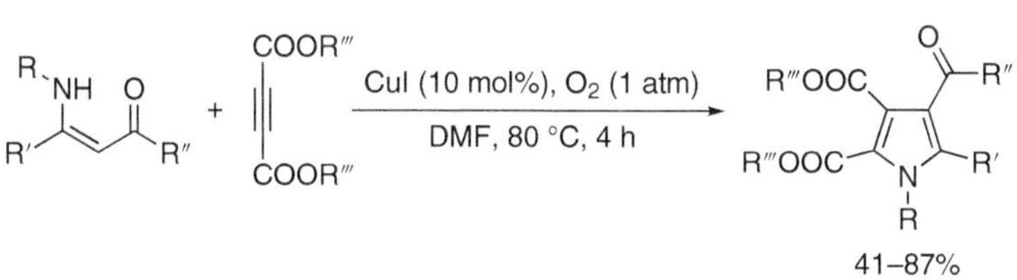

Scheme 8.9 CuI-catalyzed pentasubstituted pyrrole synthesis via oxidative annulation of β-enamides and internal alkynes.

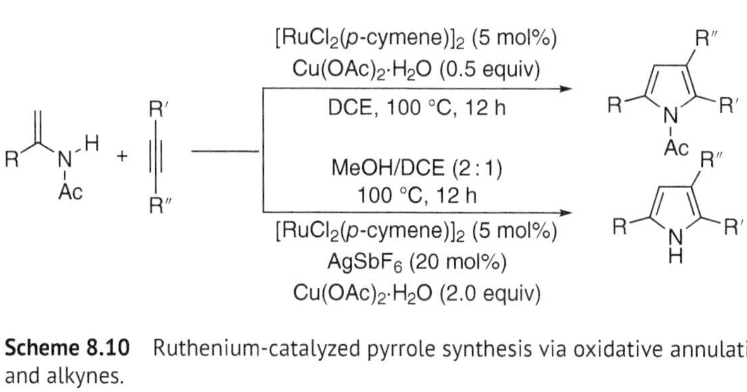

Scheme 8.10 Ruthenium-catalyzed pyrrole synthesis via oxidative annulation of enamides and alkynes.

In addition, *multi*-substituted pyrroles can be achieved from enamines and allenes with high regioselectivity catalyzed by iodine molecule [30].

Moreover, as shown in Scheme 8.11, the CuI-catalyzed domino C—N coupling of haloenynes with primary amide (H_2NBoc) and subsequent intramolecular hydroamidation provides an alternatively efficient route for the synthesis of substituted pyrroles [31].

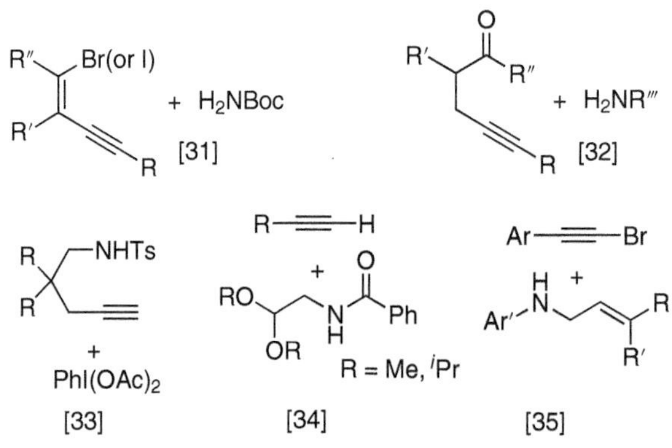

Scheme 8.11 Substrates for the synthesis of pyrroles.

1,2,3,5-Tetrasubstituted pyrroles can be obtained via the $FeCl_3$-catalyzed [4C + 1 N] cyclization of 4-acetylenic ketones with primary amines, involving the formation of enamines by the condensation of carbonyl group with primary amine with the release of one H_2O, the regioselective 5-*exo*-dig intramolecular cyclohydroamination of alkynes, and isomerization to give pyrroles [32].

$PhI(OAc)_2$-mediated cascade reactions of 3-alkynyl amines afford 1,2,3,4-tetrasubstituted pyrroles under metal-free conditions, and $PhI(OAc)_2$ is used as oxidant and incorporating reagent [33].

$(PPh_3)Au(NTf)$ catalyzes the addition–cyclization sequence of *N*-(2,2-dialkoxyethyl)-benzamides and terminal alkynes to develop the efficient synthetic method approach to *multi*-substituted pyrroles; this sequence is promoted by autotandem gold catalysis, and gold catalyst is proposed to serve a dual role in the activation of both nucleophilicity and electrophilicity of an alkyne by forming gold acetylide and by π-coordination, respectively [34].

With the use of benzoquinone (BQ) as oxidant, $PdCl_2$-catalyzed intermolecular oxidative annulation between bromoalkynes and *N*-allylamines provided 3-bromopyrroles in moderate to excellent yields [35].

1,3-Butadiynes are the easily available and important starting materials for the synthesis of carbo- and heterocyclic compounds [36]. Hua and coworker studied the cycloaddition of 1,4-diaryl-1,3-butadiynes with excess amount of primary amines (10 equiv.) in the presence of CuCl to produce 1,2,5-trisubsituted pyrroles (Scheme 8.12) [37]. The formation of pyrrole ring is resulted from the intermolecular and intramolecular double hydroamination of carbon–carbon triple bond with N—H bond.

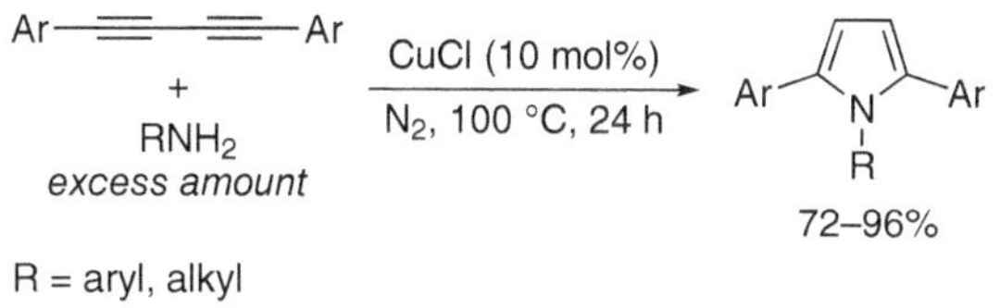

Scheme 8.12 CuCl-catalyzed 1,2,5-trisubstituted pyrrole synthesis from the cycloaddition of 1,3-butadiynes with primary amines.

The same group has also designed the annulation of terminal propargylamines with electron-deficient alkynes in the presence of silver salts (Scheme 8.13) [38]. It has been found that with the use of AgOAc as catalyst and 1,8-diazabicyclop [5.4.0]undec-7-ene (DBU) as base, *N*-substituted propargylamines undergo the chemoselective [3 + 2] cycloaddition with electron-deficient alkynes to give pyrroles, whereas AgOTf catalyzes a [4 + 2] cycloaddition of unprotected propargylamine with electron-deficient alkynes in the presence of $K_2S_2O_8$ as oxidant to afford pyridines. Both pyrroles and pyridines have electron-withdrawing groups at three positions.

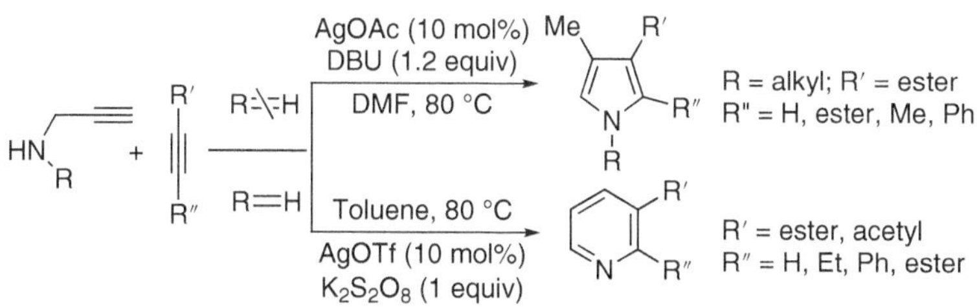

Scheme 8.13 Silver-catalyzed chemoselective pyrroles and pyridines synthesis.

A proposed mechanism for the chemoselective [3 + 2] and [4 + 2] annulation of propargylamines with alkynes affording pyrroles and pyridines selectively is depicted in Scheme 8.14. It involves the first key step of regioselective Michael addition of propargylamine with alkyne to yield *N*-propargylic β-enamino esters and β-enaminones and the carbocyclization forming five-membered and six-membered nitrogen-heterocyclic intermediates selectively to construct pyrrole and pyridine ring via further aromatization under different conditions.

Multi-substituted furans are ubiquitous in biologically active molecules and have also been applied as building blocks for the synthesis of various functionalized compounds [39]. Therefore, a tremendous number of synthetic methods to approach substituted furans have been developed from alkynes and alkenes [40].

Dixneuf and coworker established a one-pot facile synthesis of 2,5-disubstituted furans from a Cu(II)-catalyzed intramolecular cyclization of 1,3-dienyl alkyl ethers, which are resulted from a ruthenium(II)-catalyzed dimerization of terminal alkynes in the presence of alcohols with high stereoselectivity. Therefore, 2,5-disubstituted furans can be directly obtained by a successive ruthenium(II)- and copper(II)-catalyzed process from terminal alkynes (Scheme 8.15) [41].

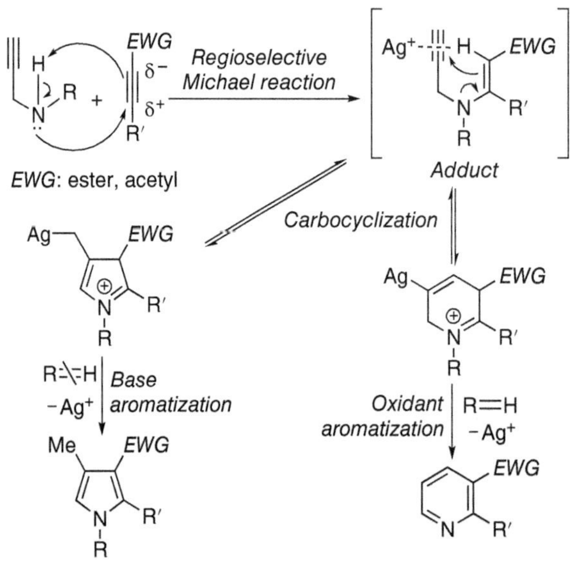

Scheme 8.14 Proposed mechanism for silver-catalyzed formation of pyrroles and pyridines.

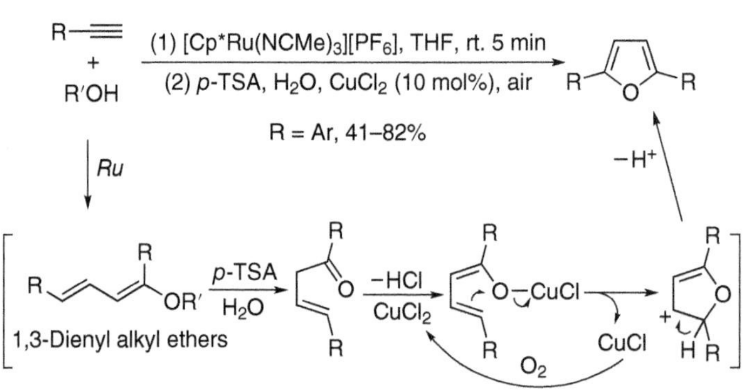

Scheme 8.15 One-pot synthesis of 2,5-disubstituted furans directly from terminal alkynes.

Skrydstrup and coworker reported a gold(I)-catalyzed reaction for the formation of 2,4-disubstituted furans from terminal alkynes with sulfur ylides involving the step of gold carbine (Scheme 8.16) [42]. The procedure represents a formal [2 + 3] cycloaddition of alkynes with sulfur ylides.

Willis and coworker designed an approach route to furans via an acid-promoted dehydrative cyclization of γ-hydroxy-α,β-enones, which are formed in situ by the intermolecular hydroacylation of an aldehyde bearing *S*-chelating alkyl group with propargylic alcohols catalyzed by Rh (I) complex (Scheme 8.17) [43].

2,3,5-Trisubstituted furans, diethyl 5-formylfuran-2,3-dicarboxylate, or diethyl 5-methylfuran-2,3-dicarboxylate can be synthesized from a one-pot, CuI-, or $AgBF_4$-catalyzed cyclocondensation of electron-deficient alkynes with 2-yn-1-ols

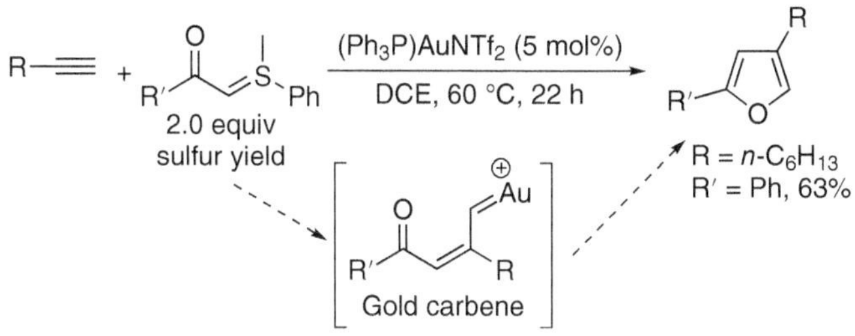

Scheme 8.16 2,4-Disubstituted furan synthesis via gold carbene formation.

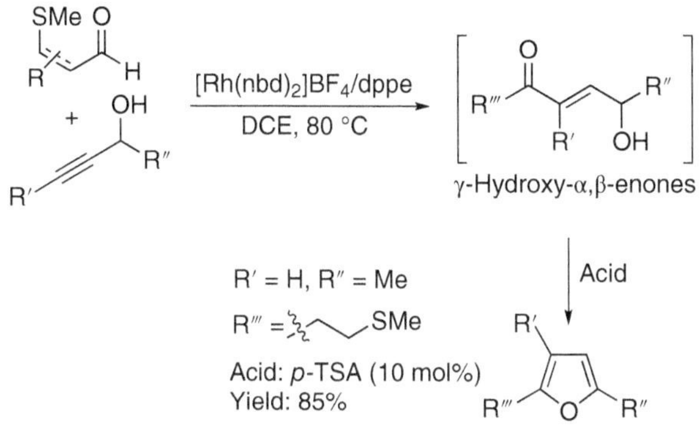

Scheme 8.17 Furan formation via an alkyne hydroacylation route.

under different conditions reported by Jiang's group (Scheme 8.18) [44]. The same group also reported a 1,4-diazabicyclo[2.2.2]octane (DABCO)- or PBu_3-promoted nucleophilic addition of propargylic alcohol to electron-deficient alkynes to give 1,5-enynes, which undergo a nano-Cu_2O-catalyzed sequent cyclization/ rearrangement/ oxidation to afford α-carbonyl furans [45].

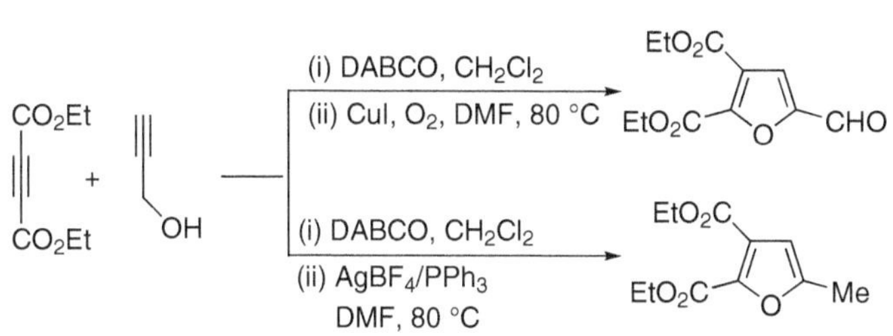

Scheme 8.18 Trisubstituted furans from electron-deficient alkynes with propargylic alcohol.

In the presence of DABCO, tetrasubstituted furans having trifluoromethyl group can be prepared from the intermolecular cyclization of aryl-substituted propargyl alcohols with methyl 2-perfluoroalkynoate (Scheme 8.19) [46]. The established allene enol and control experiments indicate that the reaction proceeds via a Michael addition, Claisen rearrangement, and cyclization process.

Scheme 8.19 DABCO-catalyzed synthesis of tetrasubstituted furans.

Gold/copper-catalyzed intermolecular cascade reaction of propargylic alcohol with terminal alkynes also provides the way access to di-, tri-, and tetrasubstituted furans in good to excellent yields [47].

Lei and coworker reported a silver-mediated highly selective oxidative coupling of C(sp^3)—H with C(sp)—H of 1,3-dicarbonyl compounds and terminal alkynes giving 2-alkynyl-1,3-dicarbonyl intermediates for the formation of polysubstituted furans (Scheme 8.20) [48].

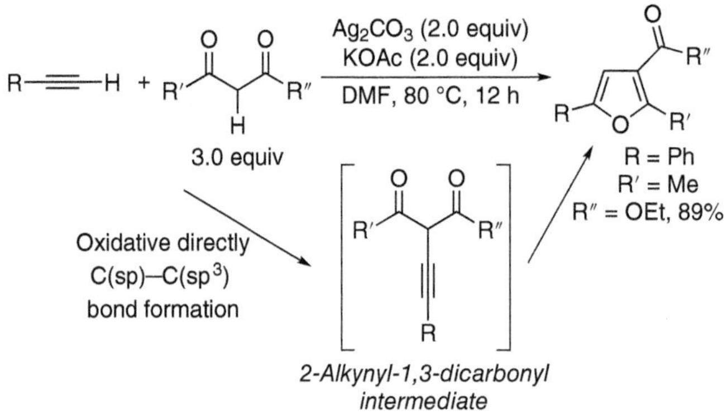

Scheme 8.20 Silver-mediated oxidative C—H/C—H functionalization route to polysubstituted furans.

In the presence of $Pd(OAc)_2$ and $Zn(OTf)_2$, aromatic internal alkynes undergo oxidation and cyclization to give tetrasubstituted furans under oxygen atmosphere (Scheme 8.21) [49].

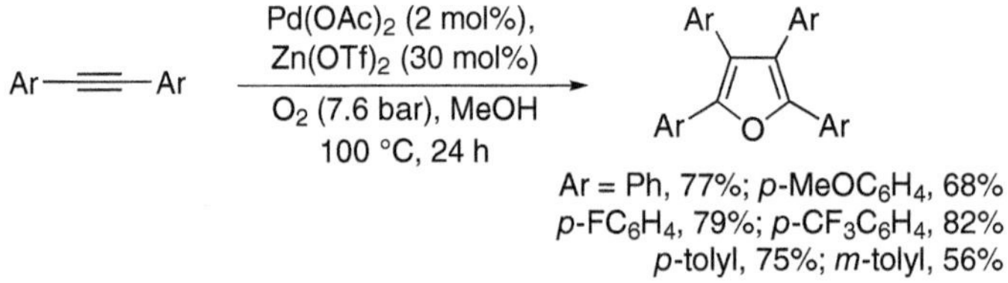

Scheme 8.21 Catalytic oxidation-cyclization of alkyne to tetrasubstituted furans.

Oh and coworker developed a one-pot synthesis of 2,5-disubstituted furans using β-halovinyl ketones as starting materials (Scheme 8.22) [50]. The formation of furan ring is verified from the CuCl-catalyzed intramolecular cyclization of allenyl ketones and unconjugated alkynyl ketones, which are formed from the elimination reaction of β-chlorovinyl ketones.

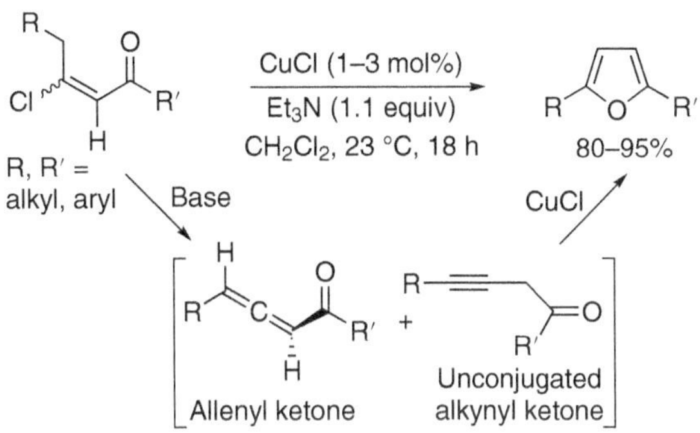

Scheme 8.22 One-pot approach to 2,5-disubstituted furans from β-chlorovinyl ketones.

1,4-Alkynediols serve as starting materials for isomerization to 1,4-diketones catalyzed by $RuH_2(PPh_3)_3(CO)$/xantphos, which can be further converted into 2,5-disubstituted furans by acid-catalyzed dehydration [51]. Under the similar reaction conditions, various 1,2,5-substituted pyrroles have been synthesized [52].

Using the combination of $(Ph_3P)AuCl$ with either $AgNTf_2$ or AgOTf as catalyst systems, the intramolecular cyclization of 3-alkyne-1,2-diols affords a variety of substituted furans in excellent yields (Scheme 8.23) [53]. The similar transformation with the use of 1-amino-3-alkyn-2-ols produces the substituted pyrroles.

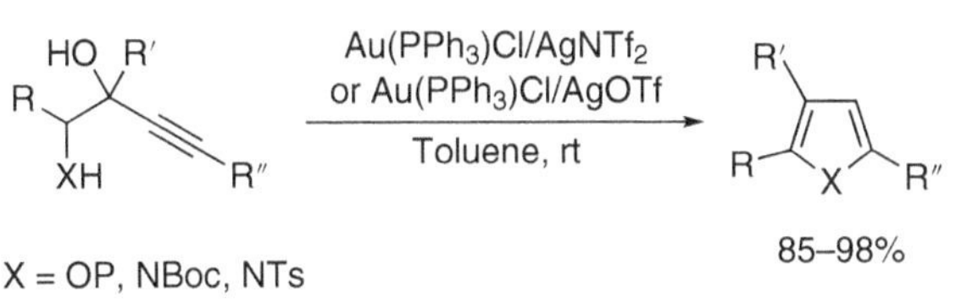

Scheme 8.23 Synthesis of furans and pyrroles from 3-alkyne-1,2-diols.

Ring expansion of hetero three-membered rings is one of the important transformations for the synthesis of other heterocyclic compounds [54]. 2,4-Disubstituted furans can be prepared by $AuCl_3$-catalyzed the isomerization of alkynyl epoxides at room temperature, and the additional functional groups like hydroxyl groups or arylbromides did not need to be protected (Scheme 8.24) [55]. The reactions also occur in the presence of AgOTf and *p*-TSA [56]. $PtCl_2$ shows good catalytic activity for the similar transformation and also for the synthesis of pyrroles by using alkynyl aziridines [57].

In the presence of a copper(I) catalyst and a pyridine oxide, alkynyl epoxides can be converted into functionalized five-membered α,β-unsaturated lactones [58]. The cycloisomerization of alkynyl ketones in the presence of CuI provides an efficient

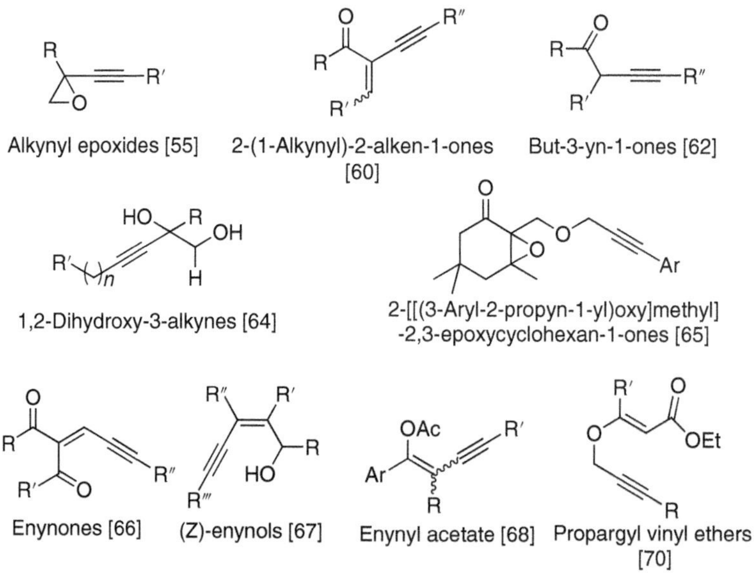

Scheme 8.24 Precursors for the synthesis of furans via intramolecular cyclization.

synthesis of 2-mono- and 2,5-disubstituted furans [59]. $AuCl_3$ also catalyzes the cyclization of 2-(1-alkynyl)-2-alken-1-ones with various nucleophiles under mild conditions to give highly substituted furans [60]. In the presence of palladium(II) complexes, but-3-yn-1-ones have been applied as the easily available starting materials for the synthesis of furans [61]. Similarly, in the presence of a catalytic amount of $ZnCl_2$, 1,4-di- and 1,2,4-trisubstituted but-3-yn-1-ones undergo a 5-*endo*-dig cycloisomerization at room temperature to provide 2,5-di- and 2,3,5-trisubstituted furans in high yields [62].

2-Methylfuran can be prepared by an effective cyclization of 1-alkyn-5-ones in the presence of $Hg(OTf)_2$ [63]. With the use of same catalyst, 3,5-disubstituted furans formed through a 5-*endo*-dig cyclization of 1,2-dihydroxy-3-alkyne [64]. 3,4-Disubstituted furans can be available from a $BF_3{\cdot}OEt_2$-promoted cyclization of 3,5,5-trimethyl-2,3-epoxycyclohexan-1-ones with a tethered arylpropargyl methyl ether at the C-2 position under an atmosphere of oxygen [65].

2,4,5-Trisubstituted furans with 2-acyl group were obtained by a cyclization of enynones catalyzed by $AgBF_4$ under an atmosphere of oxygen in a fluorous biphasic system of perfluorodecalin and dimethylformamide (DMF) [66]. The fully substituted furans can be prepared by a cycloisomerization of (*Z*)-enynols at ambient conditions in the presence of $AuCl_3$ or $(PPh_3)AuCl/AgOTf$ [67] and have also been readily prepared from (*Z*)- or (*E*)-conjugated enynyl acetates with *N*-bromosuccinimide (NBS) or *N*-iodosuccinimide (NIS) under metal-free conditions at room temperature via a haloallenyl ketone intermediates, and the formation of furan ring was accelerated by electron-rich aryl group [68]. The application of (*Z*)-conjugated enynyl acetates formed in situ from terminal alkynes in the synthesis of 2,5-disubstituted furans has also been reported [69].

Tri- and tetrasubstituted furans can also be obtained by a cascade reaction of propargyl Claisen rearrangement and heterocyclization process in the presence of $(PPh_3)AuCl/AgBF_4$ [70].

1,4-Disubstituted-1,3-butadiynes are also the good precursors to develop the simple and valuable approach to symmetrical and nonsymmetrical 2,5-disubstituted furans via their reaction with another reactants having heteroatom. For example, both symmetrical and nonsymmetrical 2,5-disubstituted furans were synthesized by a gold(I)-catalyzed hydration of 1,4-diaryl/alkyl 1,3-butadiynes [71]. In the presence of $Pd(PPh_3)_4$/2,5-norbornadiene/KOH, the cycloaddition of 1,4-diaryl-1,3-butadiynes with H_2O gave 2,5-diarylfurans in good yields, and both electron-rich and electron-poor aryl were suitable groups for the formation of 2,5-diarylfurans [72].

DBU-promoted intramolecular cycloaddition of diynyl-1,6-diols provides an efficient method to construct 2,5-disubstituted furans through a cascade 1,3-H shift/cyclization process (Scheme 8.25) [73].

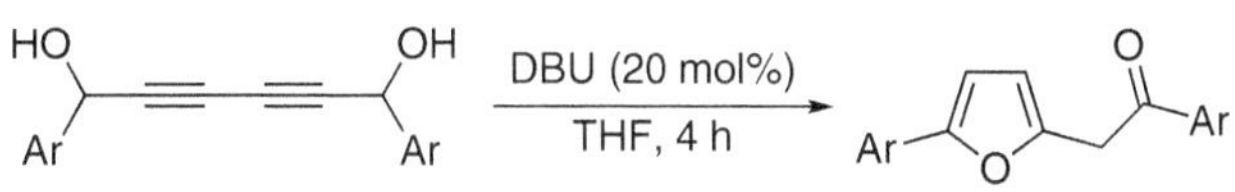

Scheme 8.25 DBU-promoted synthesis of 2,5-disubstituted furans.

Hua and coworker developed a general synthetic method for the formation of 2,5-diarylpyrroles, 2,5-diarylfurans, and 2,5-diarylthiophenes by the cyclocondensation of 1,4-diaryl-1,3-butadiynes with primary amines, water, and $Na_2S·9H_2O$, respectively, in the presence of KOH in DMSO (Scheme 8.26) [74]. The procedure provides a promising protocol using simple and cheap catalytic system for the construction of five-membered heterocycles bearing a variety of electron-rich and electron-deficient aryl groups in good to high yields. In addition, the five-membered heterocycles can also be prepared by a one-pot, two-step procedure from terminal alkynes in DMSO first catalyzed by CuCl and then via addition of KOH to promote the cyclocondensation of 1,3-butadiynes generated in situ with nucleophiles

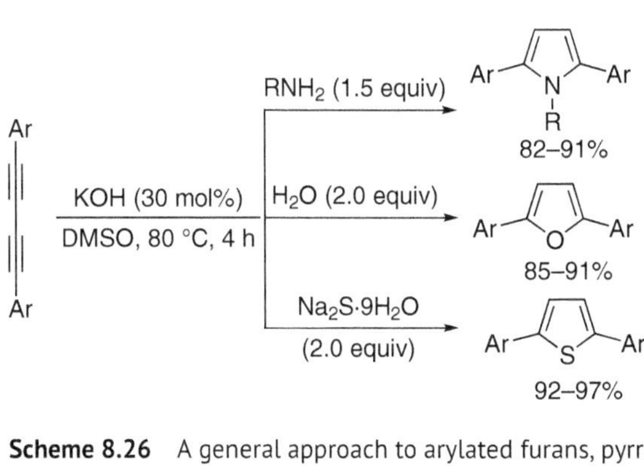

Scheme 8.26 A general approach to arylated furans, pyrroles, and thiophenes.

(Scheme 8.27). Moreover, the procedure was then applied in the synthesis of RITA and its analogs [75].

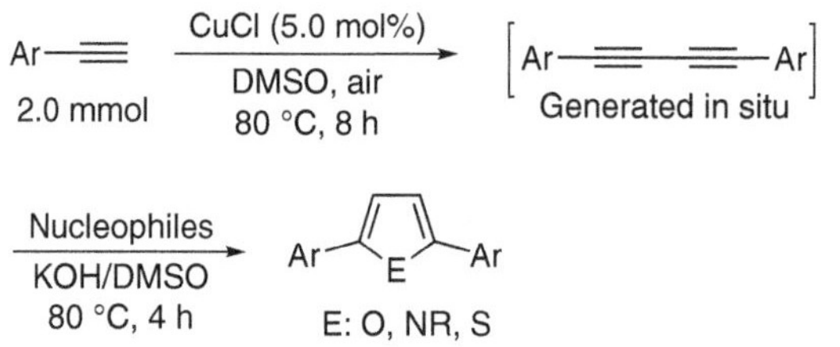

Scheme 8.27 One-pot synthesis of arylated furans, pyrroles, and thiophenes from terminal alkynes.

2-(Tosylamido)- and 2,5-bis(tosylamido)thiophenes have been reported to be synthesized from the direct reaction of ynamide-derived buta-1,3-diynes with $Na_2S{\cdot}9H_2O$ [76].

Nolan and coworker have shown [Au(IPr)OH]/$HNTf_2$ to be a very convenient catalyst system for the formation of furans and pyrroles from the hydration or hydroamination of 1,3-butadiynes [77]. In the case of pyrrole synthesis, the reactions were performed under microwave irradiation due to a lower reactivity under the optimized reaction conditions for the formation of furans. Also a detailed computational study has been made to support the proposed mechanism.

Jiang and coworker also reported the synthesis of 2,5-disubstituted furans using CuI/1,10-phenanthroline catalyst system from bromoalkyne or iodoalkyne in a one-pot procedure under a superbase of KOH/DMSO conditions [78]. The formation of furans proceeds through the hydration reaction of 1,3-butadiynes, which is readily formed from the coupling reaction of haloalkynes in the presence of CuI. The procedure could also be used for the facile synthesis of 2,5-disubstituted thiophenes.

In the presence of a KOH/DMSO, Trofimov and coworker studied the reaction of ketones bearing bulky aromatic, heteroaromatic, and ferrocene substituents with acetylene to give polysubstituted furans in good yields, and the procedure involves a domino sequence in which one molecule of ketone reacts with two molecules of acetylene [79].

In addition, in the presence of base, the cycloisomerization of 4-oxahepta-1,6-diynes and 4-azahepta-1,6-diynes affords *multi*-substituted furans and pyrroles, respectively [80].

Alkenes have been applied as the precursors for the intermolecular condensation in the synthesis of furan ring. For example, Ishii and coworker studied the cyclocondensation of acrylates with aldehydes in the presence of catalytic amounts of $Pd(OAc)_2$, $H_4PMo_{11}VO_{40}{\cdot}28H_2O$, and Lewis acid of $CeCl_3{\cdot}7H_2O$ to provide a direct route to 5-substituted 3-furoates under atmospheric dioxygen (Scheme 8.28) [81]. The reaction is found to proceed through the palladium-catalyzed acetalization of acrylates with methanol followed by the reaction of acetals with aldehydes. With the use of acetyl acetones, 2,5-disubstituted 3-furoate was also obtained.

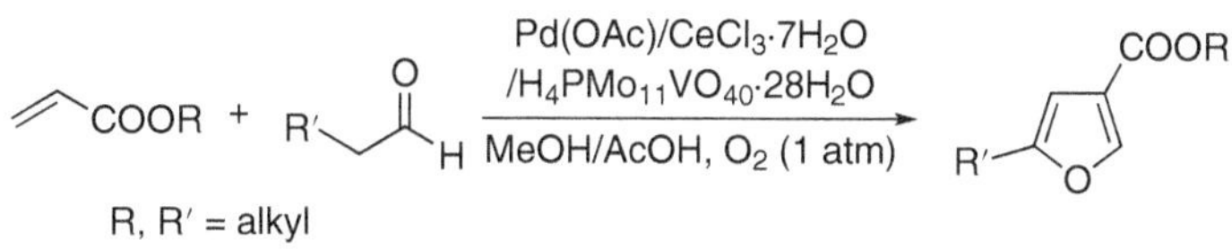

Scheme 8.28 Synthesis of 5-substituted 3-furoates from acrylates and aldehydes.

As previously described, some synthetic methods for the formation of pyrroles and furans are also used for the synthesis of thiophenes. The interesting and efficient alternative routes for the synthesis of substituted thiophenes involve the intramolecular thioannulation of alkynyl thiols and alkynyl thioacetates [82]. As shown in Scheme 8.29, in the presence of 18-crown-6 and KO*t*Bu in *t*BuOH, β-alkynyl thiols undergo the cycloisomerization to afford 2-substituted thiophenes [83]; PdI_2/KI-catalyzed cycloisomerization of (*Z*)-2-en-4-yne-1-thiols gave di- or trisubstituted thiophenes [84]. With the use of same catalyst systems, a variety of readily available 1-mercapto-3-yn-2-ols were converted into the corresponding thiophenes in good to high yields in MeOH [85], and 1-mercapto-3-yn-2-ols undergo the iodocyclization with molecular iodine affording 3-iodothiophenes in good yields in the presence of $NaHCO_3$ at room temperature using MeCN as the solvent [86].

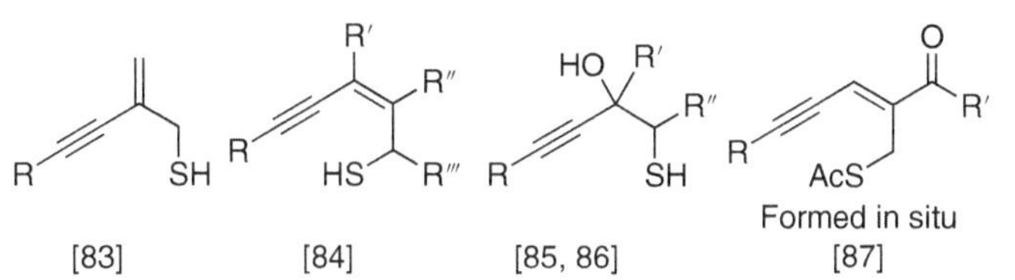

Scheme 8.29 Cycloisomerization of alkynyl thiols to thiophenes.

Reddy and coworker studied the reaction of 3-acetate-1-penten-4-yn-3-ol with KSAc to provide allylic thioacetate, which underwent K_2CO_3-mediated tandem allylic substitution, and deacetylative 5-*exo*-dig-thiocycloisomerization to produce 2,4-disubstituted thiophenes [87]. Punniyamurthy's group also developed a DABCO-mediated domino reaction of 1,3-enynes with mercaptoacetaldehyde to assemble tetrasubstituted thiophenes at room temperature via a Michael addition, 5-*exo*-dig carboannulation, and oxidation sequence under air [88]

2,3,5-Trisubstituted thiophene derivatives could be prepared from the reaction of aryl-substituted 1,3-butadiynes with equivalent of arylmethanethiols in DMSO promoted by KOH (Scheme 8.30) [89].

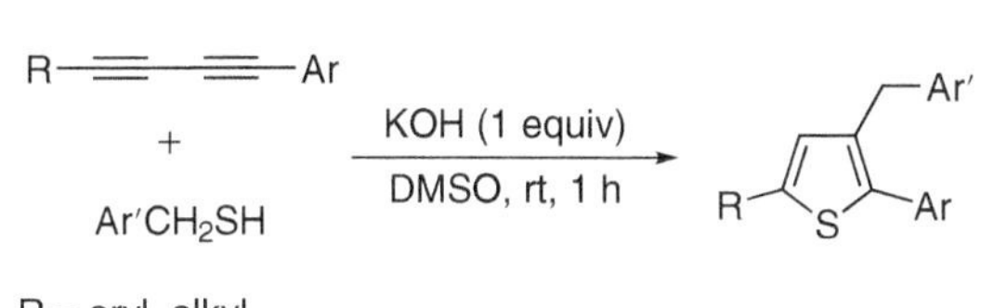

Scheme 8.30 Thiophene formation from reaction of 1,3-butadiynes with arylmethanethiols.

8.2.2 Indoles, Benzo[*b*]Furans, Benzo[*b*]Thiophenes, and Benzo[*b*]Selenophenes

The representative examples of benzo-fused five-membered heterocyclic compounds of indoles (benzo[*b*]pyrroles) [90], benzo[*b*]furans [91], benzo[*b*]thiophenes [92], and benzo[*b*]selenophenes [93] are commonly found in various pharmaceuticals and natural products having a wide range of biological and medicinal properties.

Indoles are one of the most important classes of *N*-heterocycles, which have been well applied in organic transformation [94]. Although Fischer indole synthesis via the condensation of a substituted phenylhydrazine with a carbonyl compound under acidic conditions is one of the most important methods [95], many other synthetic methods have been developed for the construction of indole rings. The most attractive synthetic methods are the intramolecular cycloisomerization of 2-alkynylanilines and the intermolecular cyclocondensation of alkynes with *N*-aromatic precursors, which have become one of the hot research topics in recent years in organic synthetic chemistry [96].

Scheme 8.31 shows the intramolecular cycloisomerization (hydroamination) of 2-alkynylanilines to be an efficient method for the synthesis of 2-substituted indoles. Knochel and coworker developed a general preparation of 2-substituted indoles from 2-alkynylanilines in the presence of inorganic bases, and the base-mediated reaction has also been used to synthesize benzofurans and isoindolones [97]. Also a variety of metal compounds and complexes such as $Cu(OTf)_2$ [98], $Hg(OTf)_2$ [99], iridium(III) hydrides [100], $AuCl_3$ [101], AuCl [102], and carbon-supported gold nanoparticles [103] have been used as the efficient catalysts for the transformation to afford the unprotected indoles.

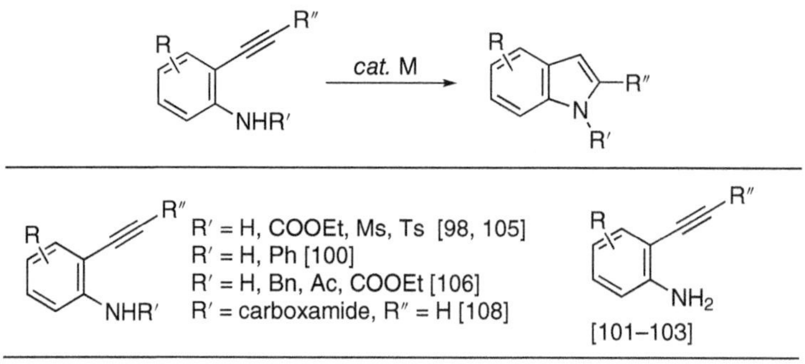

Scheme 8.31 2-Substituted indole from cycloisomerization of 2-alkylaniline.

Hiroya and coworker found that in 1,2-dichloroethane, a few Cu(II) salts could efficiently catalyze the formation of indole ring from 2-alkynylanilines [104]. Their further studied disclosed that $Cu(OCOCF_3)_2{\cdot}xH_2O$ showed high catalytic activity for the cycloisomerization of 2-alkynylanilines in a mixture of H_2O and MeOH in the presence of 1-ethylpiperidine at room temperature to give *N*-Ms or Ts-substituted indoles [105].

Sakai and coworker also studied the Lewis acid-promoted cycloisomerization of 2-alkynylanilines and found that $InBr_3$ could catalyze the intramolecular cyclization efficiently to produce the corresponding indoles with high chemoselectivity, when R″ is an alkyl or aryl group [106]. $InBr_3$ also catalyzed the similar cycloisomerization to provide an efficient synthesis of β-(*N*-Indolyl)-α,β-unsaturated esters [107].

$[Au(PPh_3)]Cl/Ag_2CO_3$ catalyzed the cycloisomerization of *N*′-substituted *N*-(2-alkynylphenyl)ureas in water under microwave irradiation affording indole-1-carboxamides [108]. Under the reaction conditions, no expected 2-substituted indole-1-carboxamides could be obtained, when R″ is alkyl and phenyl groups.

As previously mentioned, iridium hydride complexes have been demonstrated to be the efficient catalysts in the cycloisomerization of 2-alkynylanilines to give 2-substituted indoles; interestingly, Leong and coworker also developed a $[Cp^*IrCl_2]_2$-catalyzed a novel formation of 2,2′-biindoles from unprotected 2-ethynylanilines (Scheme 8.32) [109]. The proposed pathway involves the formation of a vinylidene intermediate, intramolecular hydroamination, and a subsequent insertion reaction.

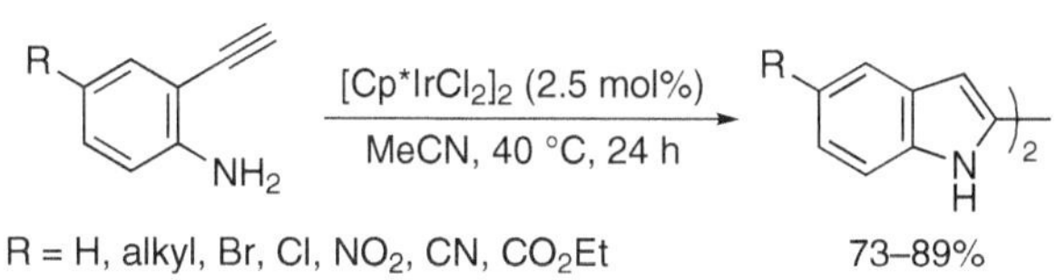

Scheme 8.32 Iridium-catalyzed formation of 2,2′-biindoles from 2-alkynylanilines.

However, in the presence of $NaBF_4$, $[Cp^*IrCl_2]_2$ catalyzed the cycloisomerization of 2-alkynylanilines (internal alkynes) under similar reaction conditions to give 2-substituted indoles, and when 1,4-bis(2′-aminopheny)-1,3-butadiyne, 1,4-bis[2-(2′-aminophenyl)ethynyl]benzene, and 1,3,5-tris[2-(2′-aminophenyl)ethynyl] benzene were used, 2,2′-biindole, 1,4-diindolylbenzene, and 1,3,5-triindolylbenzene could be obtained, respectively, in good yields in one-pot manner [110].

Saito and coworker developed a mild and facile preparation of 2-methylindoles catalyzed by $RhH(CO)(PPh_3)_3$ in hexafluoroisopropanol (HFIP) from the cycloisomerization of *N*-propargylanilines (Scheme 8.33) [111]. The formation of indole ring

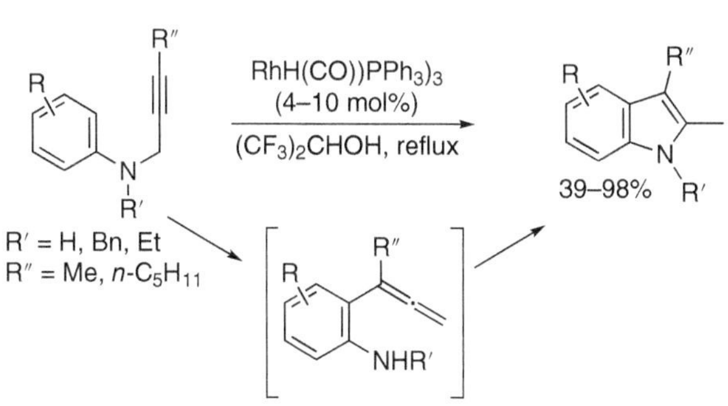

Scheme 8.33 Rhodium(I)-catalyzed synthesis of 2-methylindoles via amino-Claisen rearrangement of *N*-propargylanilines.

is proposed to be derived from an *ortho*-allenylaniline intermediate generated by the Rh(I)-catalyzed amino-Claisen rearrangement of *N*-propargylanilines.

More interestingly, in the presence of [CpRuCl(dppe)] and $NaBAr^F{}_4$ (Ar^F = 3,5-$(CF_3)_2(C_6H_3)$), Saito and coworker reported a cycloisomerization of 2-alkynylanilines to give 3-substituted indoles in high yields via the disubstituted vinylidene ruthenium complex that was formed by a 1,2-carbon migration (Scheme 8.34) [112].

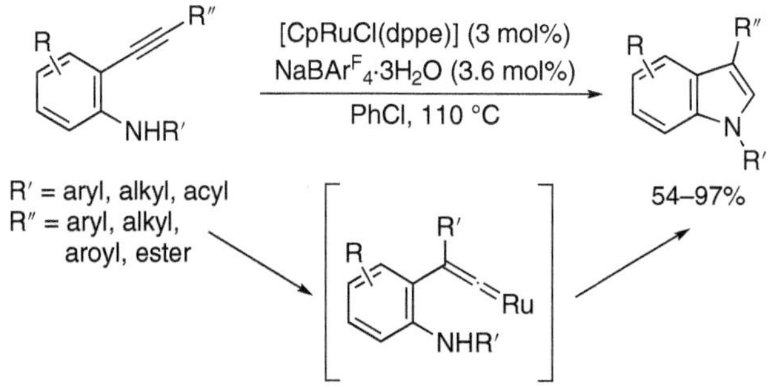

Scheme 8.34 Ruthenium-catalyzed synthesis of 3-substituted indoles from cycloisomerization of 2-alkynylaniline via 1,2-carbon migration.

In addition, if another reaction partner is present, 2,3-disubstituted indoles could be synthesized. For example, the reaction of 2-alkynylanilines with α,β-enones with the use of simple and stable $NaAuCl_4·2H_2O$ [113] and $PdCl_2$ [114] as catalysts could give 3-alkylindoles. 2-Alkynylanilines reacted with ethyl propiolate or dimethyl acetylenedicarboxylate catalyzed by $PtCl_2$ to generate the corresponding 2,3-disubstituted indoles via a sequential cyclization to give indole ring and the intermolecular addition of C—H bond at C-3 position of indole to electron-poor alkynes [115]. In the presence of Ph_3PAuCl/AgOTf, the reaction of 2-alkynylanilines with alkynols has realized a formal carboamination of alkynes affording C-3-substituted indoles [116].

The direct *ortho*-C—H activation of anilines and the cyclocondensation with alkynes has the significant advantages of not only higher atom economic efficiency but also without need for prefunctionalization of anilines, and a variety of functional groups have been found to be suitable directing group (DG) for such type of transformation.

As shown in Scheme 8.35, Fagnou and coworker reported an NHCOR (R = Me, NMe_2)-directed dehydrogenative cyclization between acetoanilines or *N*-aryl ureas and internal alkynes catalyzed by rhodium(III) complex to afford *N*-COMe and *N*-$CONMe_2$ indoles [117]. With the use of Cu(II) salts as oxidants, $Pd(OAc)_2$ also showed catalytic activity for the C—H activation of *N*-aryl amides and subsequent coupling with internal alkynes to give *N*-acyl-2,3-disubstituted indoles in fair to good yields [118].

Li and coworker demonstrated a simple $PdCl_2(MeCN)_2$-catalyzed oxidative coupling between *N*-aryl-2-pyridines with internal alkynes using pyridinyl group as DG for the synthesis of *N*-(2-pyridyl)indoles [119].

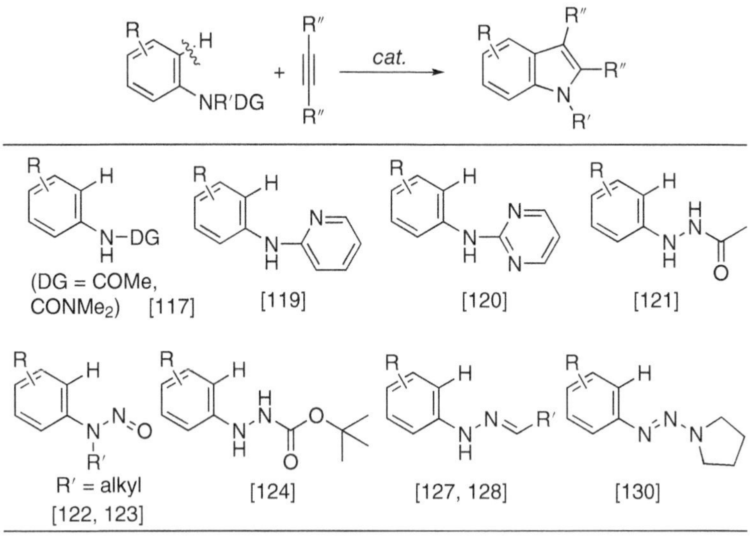

Scheme 8.35 Indole synthesis via directed C—H bond activation.

Ackermann and coworker also developed the dehydrogenative cyclization of *N*-2-pyrimidyl-substituted anilines with internal alkynes in the presence of $[RuCl_2(p\text{-cymene})]_2$ with the use of $Cu(OAc)_2 \cdot H_2O$ as oxidant and KPF_6 as additive in water, and the active catalyst is considered to be the cationic ruthenium(II) complexes formed in situ [120].

Glorius and coworker developed an efficient rhodium(III)-catalyzed redox-neutral C—H activation/cyclization of 2-acetyl-1-arylhydrazines with internal alkynes to give unprotected indoles, and the reaction did not require any external oxidant due to the cleavage of the N—N bond [121]. At the same time, Zhu [122] and Huang [123] also developed an essentially similar rhodium(III)-catalyzed cyclization of *N*-nitrosoanilines with internal alkynes for the synthesis of indoles using N—N bond as an internal oxidant.

Glorius group's further studies disclosed that $[Cp^*Co(CO)I_2]$ could catalyze the redox-neutral synthesis of unprotected indoles with N—N bond cleavage using Boc-arylhydrazines as starting materials [124]. Using the same Co(III) catalyst, *N*-arylureas also underwent the annulation with internal alkynes to give *N*-carbamoyl indoles [125]. Arylhydrazines are the good candidates for the synthesis of indoles using the same cobalt(III) complexes [126].

Hua and coworker have developed an efficient and external oxidant-free rhodium(III)-catalyzed indole synthesis from readily available aryl hydrazines and internal alkynes via hydrazone (formed in situ) as a DG and with the use of N—N bond as an internal oxidant [127]. The similar strategy and catalytic system have been then reported by Matsuda's group [128].

However, with the use of 1,3-dinitrobenzene as an external oxidant, rhodium(III)-catalyzed oxidative annulation of 2-acyl-1-arylhydrazines with internal alkynes afforded 1-aminoindole derivatives in good to high yields [129].

Triazenes have been used as an efficient and removable DG for the synthesis of unprotected indoles in the presence of rhodium (III) complex by the reaction of triazenyl aromatics with internal alkynes reported by Huang's group [130]. They then surveyed various reaction parameters for the direct synthesis of indolo[2,1-*a*]isoquinolines via combining the indole synthesis and the second N—H/C—H annulation with another equivalent of the internal alkyne [131].

In addition, cross-dehydrogenative coupling (CDC) reactions have been also well explored as an alternative way for the synthesis of indoles (Scheme 8.36).

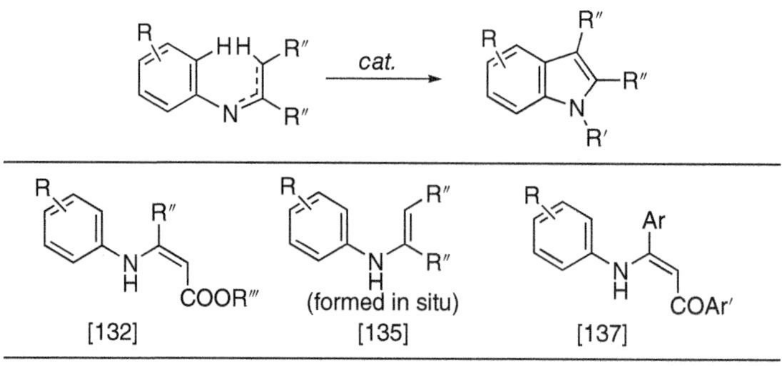

Scheme 8.36 Indole synthesis via intramolecular cross-dehydrogenative coupling reaction.

With the use of $Cu(OAc)_2$ as the oxidant, $Pd(OAc)_2$ catalyzed the intramolecular oxidative coupling of *N*-aryl enaminoesters to give the corresponding indoles in good yields [132]; $FeCl_3/Cu(OAc)_2 \cdot CuCl_2$ also showed the high activity to realize this type of C—H activation [133]. The indole synthesis via radical cyclization of *N*-aryl enaminoesters in air could be also realized via photoredox catalytic reaction by using a catalytic amounts of iridium(III) photosensitizer under visible light irradiation [134].

$Pd(OAc)_2$ also catalyzed the intermolecular cyclization reaction of *N*-nonsubstituted and *N*-alkyl monosubstituted anilines with electron-deficient internal alkynes with the use of O_2 (1 atm) as the oxidant; enamine generated in situ from the hydroamination of alkyne with aniline is considered to be the key intermediate in the formation of indoles [135]. With the use of similar starting materials, unprotected 2,3-disubstituted indoles bearing electron-withdrawing group at three position could be prepared by $PhI(OAc)_2$ or $PhI(CF_3COO)_2$-mediated oxidative carbon-carbon bond formation without transition metals [136]. Under an atmosphere of air, *N*-aryl enaminones underwent the CDC reaction catalyzed by CuI/1,10-phenanthroline to give 2-aryl-3-aroylindoles, and under the similar reaction conditions, the indoles could be also obtained from anilines and α,β-ynones omitting the isolation of enaminone intermediates [137].

In addition, with the use of $PdBr_2$/CuI as cocatalyst system and ^{t}BuOOH as an oxidant and a reactant, 3-acylindoles could be synthesized through the oxidative cyclization reaction of 2-alkynyl-*N*,*N*-dimethylanilines [138].

In the presence of transition metal catalysts, as shown in Scheme 8.37, the cycloisomerization of 2-alkynylanilines [139–141], 2-alkynylphenyl ethers [141–143], 2-alkynylphenyl sulfides [144], and 2-alkynyl-*N*-sulfonylanilines [145] with a migrating group (E) on the heteroatom (Y) gave the corresponding 2,3-disubstituted indoles, benzofurans, and benzothiophenes, respectively. Allyl, propargyl, acyl, alkoxyalkyl, and sulfonyl groups could be employed as migrating groups.

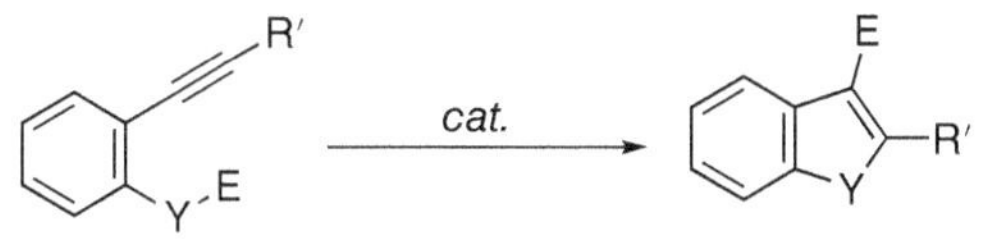

Y	E	Catalyst	Y in product	E in product	Ref.
$NCOCF_3$	Allyl	$Pd(PPh_3)_4$ or Pd_2dba_3/ttmpp	NH	Allyl	[139]
NMe	COMe, $COCF_3$, COH, COBn COPh	$PtCl_2$	NMe	Same as starting E	[140]
NMs	Allyl	$PtCl_2$	NMs		[141]
O	Propargyl	$Pd(PPh_3)_4$	O	Allenyl or propargyl	[142]
O	Allyl, benzyl, CH_2OR	$PtCl_2$	O	Same as starting E	[141]
O	Alkoxyalkyl	$PtCl_2$/COD	O	α-Alkoxyalkyl	[143]

Scheme 8.37 Cycloisomerization with a migrating group E.

Transition metal-catalyzed intermolecular cyclocondensation of 2-haloanilines with alkynes is another efficient and practical synthetic method for the formation of indole ring, and Scheme 8.38 summarizes the representative starting materials.

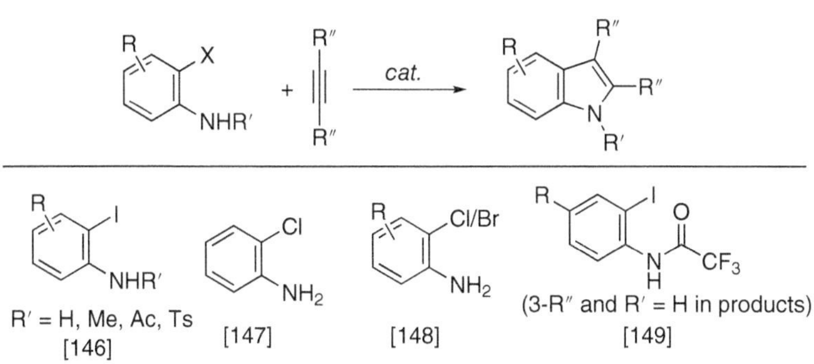

Scheme 8.38 Indole synthesis via *o*-haloaniline and alkyne.

Larock and coworker first reported the synthesis of 2,3-disubstituted indoles in good to excellent yields from the cyclocondensation of 2-iodoaniline and the corresponding *N*-methyl, *N*-acetyl, and *N*-tosyl derivatives with internal alkynes catalyzed by $Pd(OAc)_2$ [146]. Unprotected 2,3-disubstituted indoles could be prepared by a highly regioselective $TiCl_4$-catalyzed hydroamination and a 5-*endo* Heck cyclization starting from 2-chloroaniline and internal alkynes [147] or by first activation of aromatic C—X (X = Cl, Br) bond via oxidative addition to palladium [148].

Muthusubramanian and coworker have studied the cuprous halides/amino acid-catalyzed intermolecular decarboxylative cross-coupling reactions of arylpropiolic acids with 2-iodo-*N*-CF_3CO-anilines and then subsequent cycloisomerization under basic conditions; unprotected 2-arylindoles were formed in good to high yields with the use of catalytic amount of CuBr and L-proline [149].

As shown in Scheme 8.39, Cacchi and coworker have reported $Pd(PPh_3)_4$-catalyzed synthesis of 3-arylated indoles from ArN_2BF_4 (arenediazonium) and 2-alkynyltrifluoro-acetanilides via an alkynyl-coordinated arylpalladium intermediate [150]. A similar procedure can be also promoted by dual gold/photoredox catalysis under ambient conditions [151]. Yanada and coworker have developed an alternative method for the synthesis of 3-arylated indoles by $Pd(OAc)_2$-catalyzed a regioselective C-arylative 5-*endo*-dig ring-closing reaction of 2-alkynylphenylcarbamates with the use of Ar_2IOTf as the arylating reagent. It is thought that $Pd(OAc)_2$ reacts with diaryliodonium salts to form intermediate that act as both ring-closing catalysts and active arylating reagents [152].

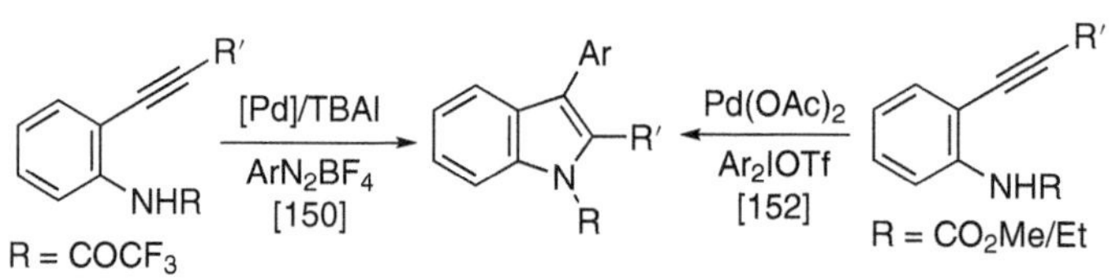

Scheme 8.39 3-Arylated indoles from nucleophilic aminoarylation of alkyne.

A straightforward route for the preparation of unprotected 2-substituted 3-alkynylindoles could be realized by a $Pd(PPh_3)_4$-catalyzed reaction of 2-alkynyltrifluoroacetanilides with 1-bromoalkynes, as an electrophilic coupling partner in good to high yield [153]. A similar synthetic way by $Pd(OAc)_2$-catalyzed oxidative cyclization of 2-alkynylanilines with terminal alkynes under mild aerobic conditions has also been reported by Zhu and coworker [154].

The benziodoxolone derivatives of 1-[(triisopropylsilyl)ethynyl]-1,2-benziodoxol-3(1*H*)-one (TIPS-EBX) has been found to be efficient reagents for the direct C_3-selective alkynylation of indoles [155]. Waser and coworker then developed the one-pot combination of the cyclization of *o*-alkynylanilines using $NaAuCl_4$ as catalyst, followed by C_3-alkynylation with AuCl and TIPS-EBX, providing an operationally simple access to 3-silylalkynyl indoles [156].

In addition, 3-substituted indoles and 3-substituted benzofurans could be synthesized via electrophilic cyclization of 2-alkynylanilines and 2-alkynylphenols, respectively [157]. For example, 3-iodoindoles could be prepared via electrophilic cyclization of 2-alkynylanilines with the use of electrophilic halogenation reagents

[158]; 3-cyanoindoles were obtained by using CuCN [159]; 3-iodobenzo[*b*]furans were synthesized by iodocyclization of 2-alkynylanisoles with I_2, ICl [160]; and 2-alkynyl-1-(1-ethoxyethoxy)benzenes were prepared with the use of bis(2,4,6-collidine)iodonium hexafluorophosphate and $BF_3{\cdot}OEt_2$ combination [161].

Unprotected 2-substituted indoles could be obtained through a rhodium-catalyzed vinyl C—H bond amination of α-azide (—N_3) styrenes [162]. 2,3-Disubstituted indoles were synthesized from the reaction of *ortho*-azidoarylalkyne with nucleophiles catalyzed by Au(I) complex via a gold carbene intermediate having nucleophilic group at C-3 [163] and the coupling of aryl bromides with 2-alkynyl arylazides via a key intermediate of iminophosphorane generated in situ in the presence of Pd_2dba_3/phosphine [164].

In addition, unprotected 2,3-disubstituted indoles were produced regioselectively in moderate yields from the reductive annulation of nitroaromatics with terminal and internal alkynes catalyzed by $[Cp^*Ru(CO)_2]_2$ with the use of CO as reductant [165].

Moreover, cascade reactions of nitrones with allenes have been known to facilitate the efficient synthesis of substitute indoles [166].

As described previously, one of the simple and efficient methods for the formation of indole ring is the electrophilic cyclization of *ortho*-alkynylanilines, which bears an electrophilic group. The similar protocol has also been applied in the synthesis of benzofurans, benzothiophenes, and benzoselenophenes (Scheme 8.40) [167].

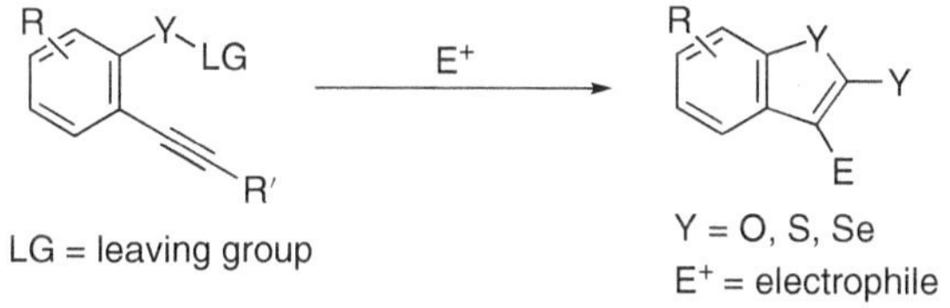

Scheme 8.40 Synthesis of benzo-fused five-membered heterocycles via electrophilic cyclization.

For example, Larock and coworker studied and developed the efficient route approach to 2,3-disubstituted benzo[*b*]furans in high yields by the electrophilic cyclization of *ortho*-alkynylanisoles (Y = O, LG = Me) with a variety of electrophiles such as I_2, PhSe—Cl, and *p*-$O_2NC_6H_4S$—Cl [168], and 2,3-disubstituted benzo[*b*]selenophenes were also obtained via the similar electrophilic cyclization process [169].

The iodocyclization of *ortho*-[(2-methylthio)ethynyl)]thioanisoles gave 3-iodo-2-thiomethyl heterocycles, setting up the synthesis of thieno-fused compounds via a subsequent iteration of alkyne coupling and iodocylization [170]. Similarly, the reaction of *ortho*-alknylthioanisole with alcohols and I_2 in MeCN at room temperature afforded 3-iodo-substituted benzo[*b*]thiophene in good to high yields via an iodocyclization/etherification reaction sequence [171].

In EtOH, at room temperature, $FeCl_3{\cdot}6H_2O$ and NaI were used to generate iodine for the iodocyclization of 2-alkynylthioanisoles to yield the

3-iodobenzo[*b*]thiophenes in high yields [172]. Also, in the presence of $CuSO_4 \cdot 5H_2O$, halogenated thiophenes, selenophenes, and benzo[*b*]selenophenes could be prepared by an electrophilic cyclization process using NaX (X = I, Br, Cl) as the source of electrophilic halogens in EtOH at room temperature [173].

The Larock's group [174] and other groups [175] have also studied the synthesis of benzothiophenes via the similar transformation by using *ortho*-(1-alkynyl)thioanisoles and 1-(2-benzyl-thioxyaryl)prop-2-yn-ol as starting materials, respectively.

In EtOH, in the presence of *p*-toluenesulfonic acid under microwave irradiation, 2-aryl substituted benzo[*b*]furans and benzo[*b*]thiophenes have been readily prepared from the cyclization of *ortho*-(2-arylethynyl)anisoles or *ortho*-(2-arylethynyl)thioanisoles [176].

In addition, Ingleson and coworker studied a BCl_3-induced borylative cyclization of aromatic alkynes possessing *ortho*-EMe (E = S, O) groups with pinacol to develop a simple and metal-free method for the synthesis of C3-borylated benzothiophenes and benzofurans [177].

Another representative methods for construction of benzo[*b*]furans involve the cyclization of *ortho*-alkynylphenols (or generated in situ via Sonogashira cross-coupling reaction of 2-halophenols with terminal alkynes) (Scheme 8.41) [178] or transition metal-catalyzed oxidative annulation of phenols with unactivated internal alkynes [179].

Scheme 8.41 Construction of benzo[*b*]furans from cyclization of *ortho*-alkynylphenols or phenols with alkynes.

Similarly, benzo[*b*]thiophenes could be easily synthesized from the reaction of *ortho*-haloaromatic alkynes with different sulfur sources as thiol precursors in the presence or absence of transition metal complexes (Scheme 8.42) [180].

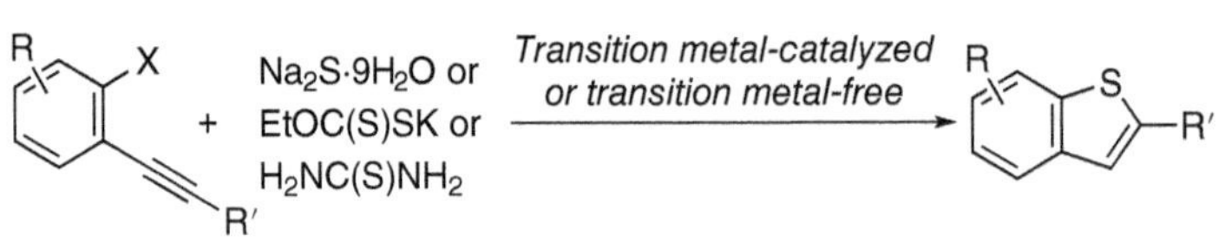

X = F, Cl, Br, I

Scheme 8.42 Synthesis of benzo[*b*]thiophenes from *ortho*-haloaromatic alkynes.

8.2.3 Five-membered Rings with Two Heteroatoms

Five-membered heterocyclic compounds having two heteroatoms show important and interesting biological and pharmacological activities, which have also been found in numerous natural products. Therefore, their synthetic methods have been well studied [181]. In this section, the representatively synthetic routes for the construction of 1,2- and 1,3-diheteroatom five-membered rings, such as pyrazoles, imidazoles, isoxazoles, oxazoles, and their derivatives from the cyclization reactions of alkynes and alkenes, are summarized.

Pyrazoles and its derivatives have been recognized as an important framework in pharmaceutical science [182]; various efficient approaches on the basis of alkyne transformation have been developed for the construction of such type nitrogen-containing heterocyclic skeleton.

Zora and coworker reported the synthesis of α,β-alkynic hydrazones and their intramolecular electrophilic cyclization into 4-iodopyrazoles by using molecular iodine in the presence of $NaHCO_3$ or pyrazoles by CuI-mediated conditions (Scheme 8.43) [183].

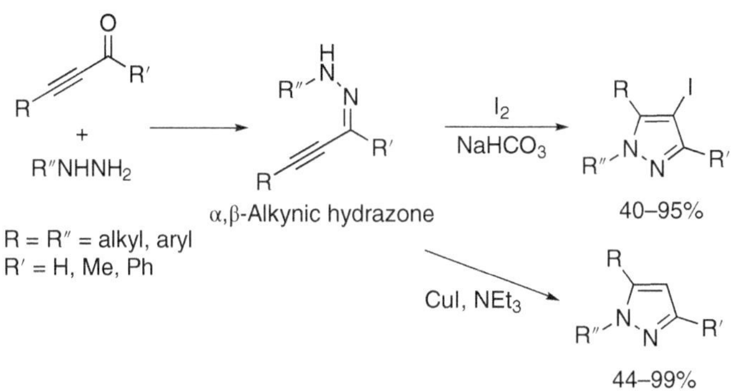

Scheme 8.43 Synthesis of pyrazoles via intramolecular electrophilic cyclization of α,β-alkynic hydrazones.

Oxone®-mediated electrophilic cyclization of α,β-alkynic hydrazones with diorganyl diselenides affording 4-organoselanylpyrazoles has also been recently reported (Scheme 8.44) [184].

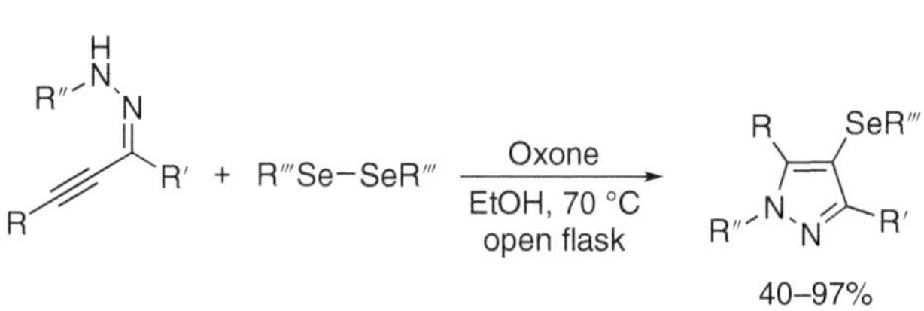

Scheme 8.44 Synthesis of 4-arylselenylpyrazoles from α,β-alkynic hydrazones.

Yoshimatsu and coworker studied the hydrazinationin of propargyl alcohols in the presence of $Sc(OTf)_3$ or $La(OTf)_3$ to afford propargyl hydrazides and which underwent an intramolecular hydroamination under different conditions giving *N*-tosylpyrazoles and unprotected N—H pyrazoles (Scheme 8.45) [185].

Unprotected N—H 3,5-disubstituted pyrazoles could also be obtained in good to excellent yields via a cope-type hydroamination of 1,4-disubstituted-1,3-butadiynes

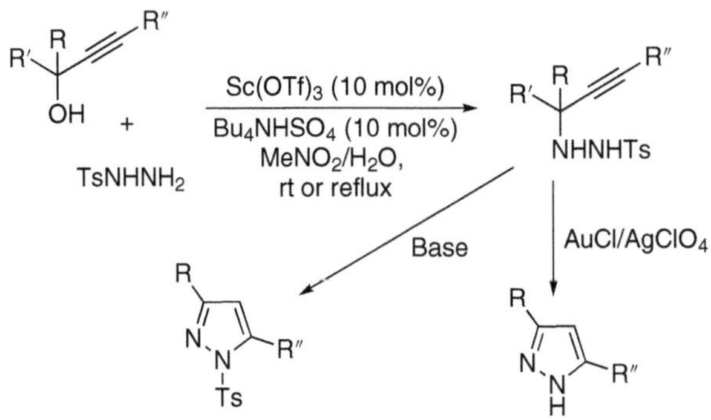

Scheme 8.45 Synthesis of pyrazoles via intramolecular hydroamination of propargyl hydrazides.

with aqueous hydrazine solution in DMSO without the use of catalyst (Scheme 8.46) [186].

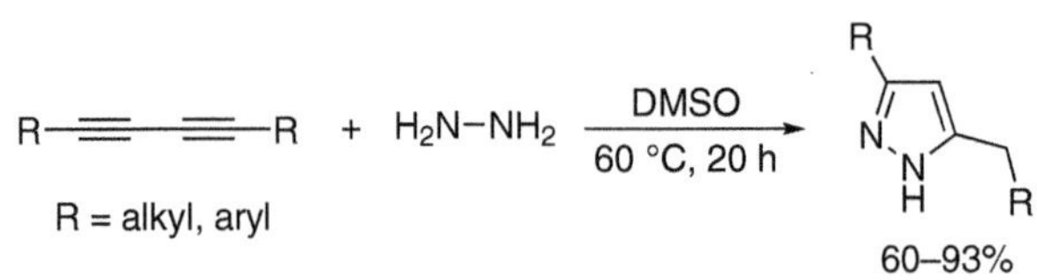

Scheme 8.46 3,5-Disubstituted pyrazoles from hydroamination of 1,3-butadiynes with aqueous hydrazine solution.

Zhan and coworker developed an alternative route for the synthesis of a variety of highly substituted pyrazoles via a $PtCl_4$-catalyzed [3.3] sigmatropic rearrangement/cyclization cascade of *N*-propargylhydrazones (Scheme 8.47) [187]. The proposed mechanism for this process involves the formation of key intermediate of diazoallene.

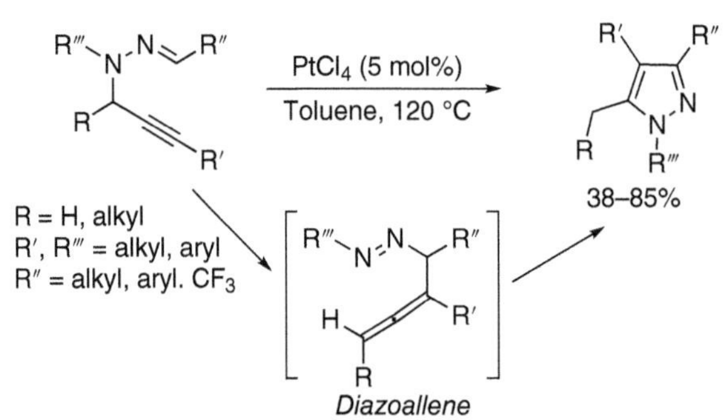

Scheme 8.47 Synthesis of polysubstituted pyrazoles from *N*-propargylhydrazones.

The iodocyclization of propargyl hydrazide with the use of bis(2,4,6-collidine)-iodonium(I) hexafluorophosphate $[I(coll)_2PF_6]$ as the iodinating reagent at room temperature in CH_2Cl_2 afforded 2,5-dihydropyrazole (Scheme 8.48) [188].

Interestingly, the presence of $BF_3{\cdot}OEt_2$ and using NIS as I^+ source instead of $I(coll)_2PF_6$ resulted in the formation of 4-iodopyrazoles, which would be formed by oxidative aromatization of dihydropyrazole, developing a novel reagent-controlled oxidative aromatization in iodocyclization to afford dihydropyrazoles and pyrazoles with high selectivity.

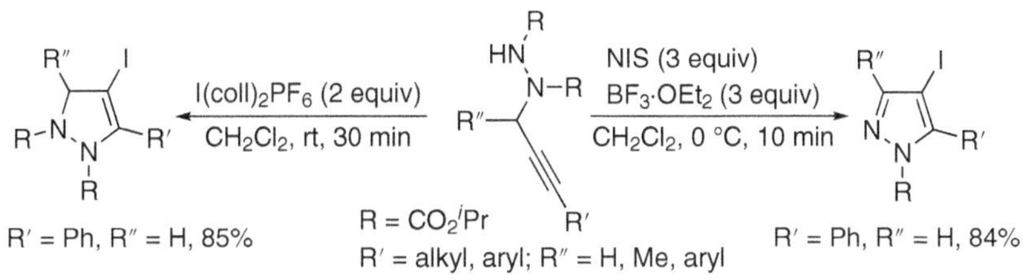

Scheme 8.48 Iodocyclization of propargylic hydrazide.

In addition, Au(I)/Ag(I)-catalyzed 5-*endo*-dig intramolecular hydroamination of propargyl hydrazides, which were formed in situ via A^3-coupling of alkynes, aldehydes/ketones, and hydrazines [189], also afforded 2,3-dihydropyrazoles [190].

It was found that $BF_3{\cdot}OEt_2$ could also promote the reaction of propargyl alcohols with *N*-sulfonylhydrazones for straightforward synthesis of 4-methylene-1-(phenylsulfonyl)-4,5-dihydropyrazole, and *N*-sulfonyl allenamide is postulated to be the key intermediate (Scheme 8.49) [191].

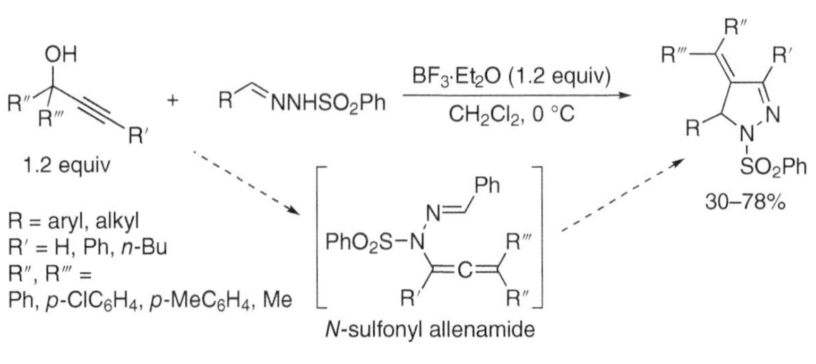

Scheme 8.49 Synthesis of 4-methylene-1-(phenylsulfonyl)-4,5-dihydropyrazole.

The reactions of propargyl alcohols with hydrazines using *t*-BuOK as a catalyst produced 4,5-dihydropyrazole derivatives (Scheme 8.50) [192]. The reaction proceeds through the base-induced isomerization of propargyl alcohols followed by cyclization of α,β-unsaturated hydrazones.

A catalytic asymmetric 1,3-dipolar cycloaddition of terminal alkynes with acyclic azomethine imines, which were generated in situ from the corresponding aldehydes and hydrazides, in the presence of CuOAc/pybox and axially chiral dicarboxylic acid cocatalysts afforded a variety of *multi*-substituted 2,3-dihydropyrazoles in high yields with excellent enantioselectivity [193].

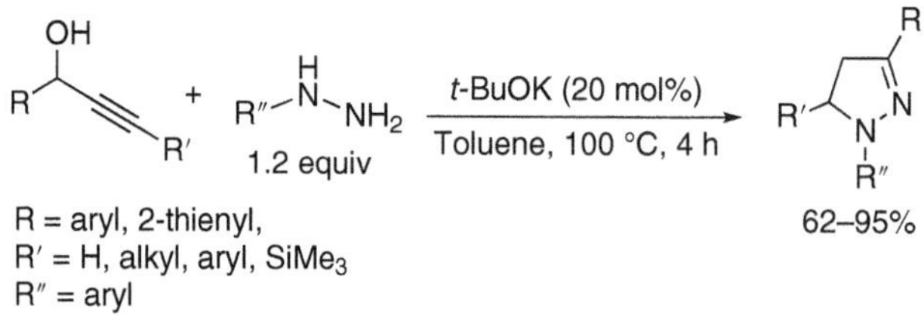

Scheme 8.50 Synthesis of 4,5-dihydropyrazoles by the reaction of propargyl alcohols with hydrazines.

Imidazole is a 1,3-dinitrogen five-membered heterocyclic compounds. Imidazoles have an important feature of a variety of biological activities and useful applications in organic synthesis [194]. The formation of imidazoles using alkyne as one of the reactants via a formal [3 + 2] cyclocondensation has been well developed with high atom utilization. For example, the cyclocondensation of amidines with terminal alkynes catalyzed by $CuCl_2{\cdot}2H_2O$ in pyridine in the presence of Na_2CO_3 under atmospheric oxygen afforded 1,2,4-trisubstituted imidazoles (Scheme 8.51) [195]. Amidine hydrochlorides reacted with bromoacetylenes promoted by K_2CO_3 under air in the presence of 2,2′-bipyridine and water giving 2,5-disubstituted imidazoles (Scheme 8.52) [196]. Under transition metal-free and ligand-free conditions, Cs_2CO_3-promoted annulation of amidoximes with terminal alkynes in DMSO produced 2,4-disubstituted imidazoles, and internal alkynes could also undergo the similar reaction to give 2,4,5-trisubstituted imidazoles (Scheme 8.53) [197].

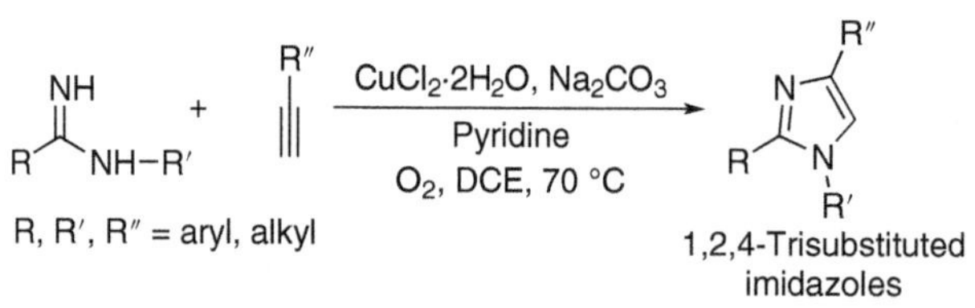

Scheme 8.51 Formation of 1,2,4-trisubstituted imidazoles from cyclocondensation of amidines with terminal alkynes.

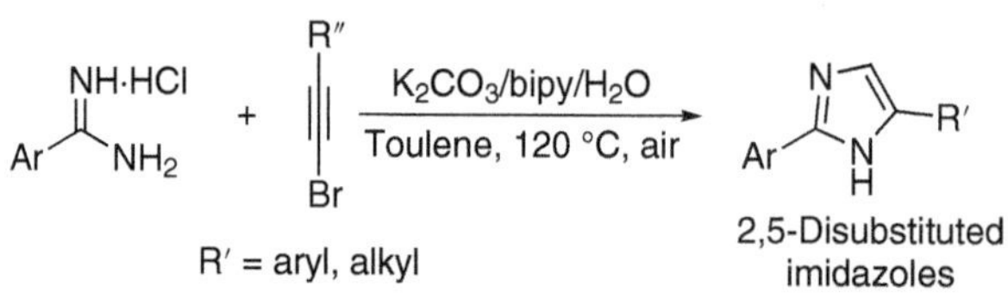

Scheme 8.52 Formation of 2,5-disubstituted imidazoles from annulation of amidine hydrochlorides with bromoacetylenes.

In addition, the formation of 1,2,4-triaryl imidazole-5-carbaldehydes by the cyclocondensation of amidines with ynals and 1,2,5-trisubstituted imidazoles by a three-component reaction of amidines, ynals, alcohols, phenols, or water has also been developed with the use of Ag(I) salts as catalysts [198].

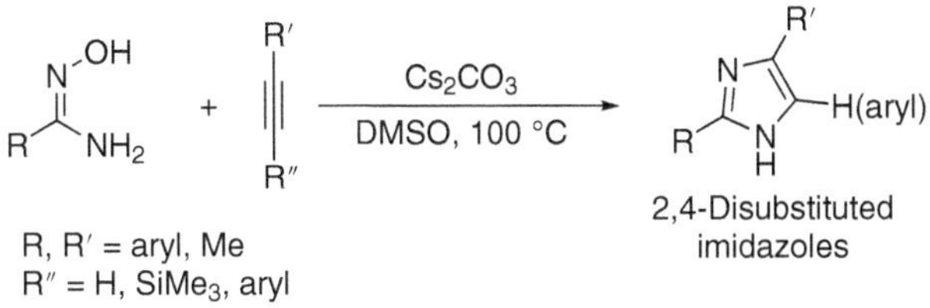

Scheme 8.53 Cs_2CO_3-promoted annulation of amidoximes with terminal alkynes affording 2,4-disubstituted imidazoles.

Isoxazole derivatives are very important due to their wide applications in organic chemistry and pharmaceutical chemistry [199]. The 1,3-dipolar cycloaddition of nitrile oxides with alkynes is probably the most simple and direct route to approach to isoxazoles.

As shown in Scheme 8.54, in the presence of Et_3N (for generating nitrile oxides in situ from hydroximoyl chlorides), the cycloaddition of phenylacetylene with 4-chloro-*N*-hydroxybenzimidoyl chloride in 1,2-dichloroethane (DCE) at room temperature gave exclusively 3,5-disubstituted isoxazole in 72% GC yield [200]. The formation of regioisomers greatly depended on the use of ruthenium complexes and the catalyst loadings. When 5 mol% of [CpRuCl(PPh_3)$_2$] was used, 3,5-disubstituted regioisomers were also the major product, but the use of 2 mol% of [Cp*RuCl(cod)] resulted in the preferential formation of 3,4-disubstituted isoxazole, and increasing amount of catalyst afforded 3,4-disubstituted isomer as major product. In addition, 3,4,5-trisubstituted isoxazoles could be produced by using internal alkynes catalyzed by [Cp*RuCl(cod)].

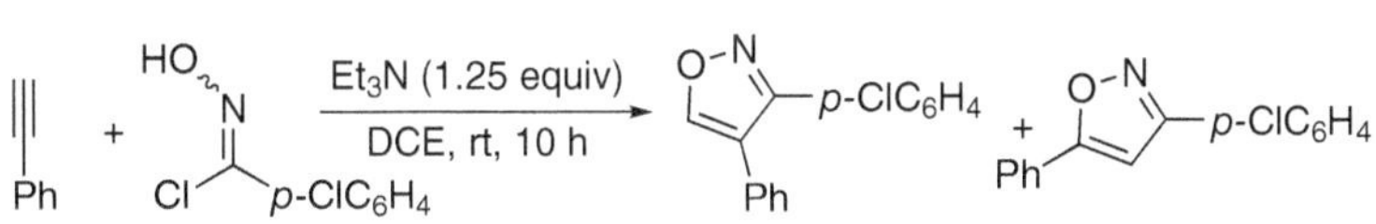

No catalyst:	0 : 72
[CpRuCl(PPh_3)$_2$] (5 mol%):	1 : 74
[Cp*RuCl(cod)] (2 mol%):	53 : 30
[Cp*RuCl(cod)] (5 mol%):	95 : 3

Scheme 8.54 Formation of 3,4- and 3,5-disubstituted isoxazoles.

The regioselective synthesis of 3,5-disubstituted isoxazoles could also be achieved through the cycloaddition of 1,1-disubstituted bromoalkenes with hydroximoyl chlorides in good to high yields (Scheme 8.55) [201]. The formation of isoxazoles was resulted from the elimination of HBr from 5,5-disubstituted bromoisoxazoline intermediates.

3,5-Disubstituted isoxazoles could be also synthesized by the reaction of aryl-substituted propargyl alcohols with *t*-BuONO in the presence of Fe(OTf)$_3$ under aerobic oxidative reaction conditions (Scheme 8.56) [202]. The reaction is highly solvent dependent with optimal isolated yields in acetonitrile, and both

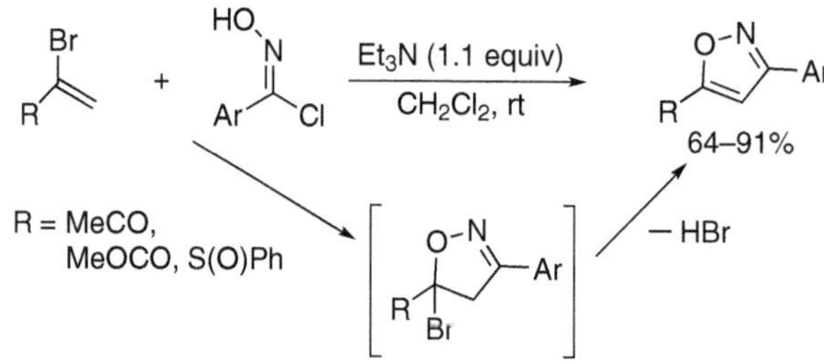

Scheme 8.55 Isoxazoles from cycloaddition of bromoalkenes with hydroximoyl chlorides.

electron-rich and electron-deficient aryl groups show high reactivity to afford isoxazoles in good yields. In addition, the mechanism studies have revealed that both iron catalyst and H_2O are essential for the transformation.

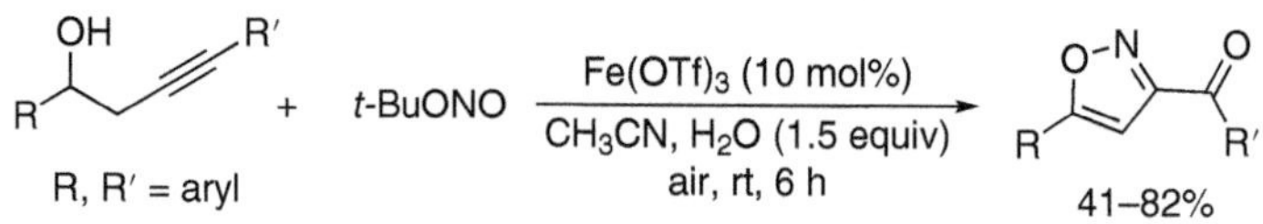

Scheme 8.56 Formation of 3,5-disubstituted isoxazoles from homopropargylic alcohol, *t*-BuONO, and H_2O.

The cycloaddition of 1,3-butadiynes with hydroxylamine hydrochloride (2.5 equiv) in the presence of Et_3N in DMSO also afforded 3,5-disubstituted isoxazoles in satisfactory to excellent yields (Scheme 8.57) [203]. The intermolecular Cope-type hydroamination of 1,3-butadiynes with $NH_2OH{\cdot}HCl$ and the intramolecular electrophilic addition are involved in the proposed mechanism.

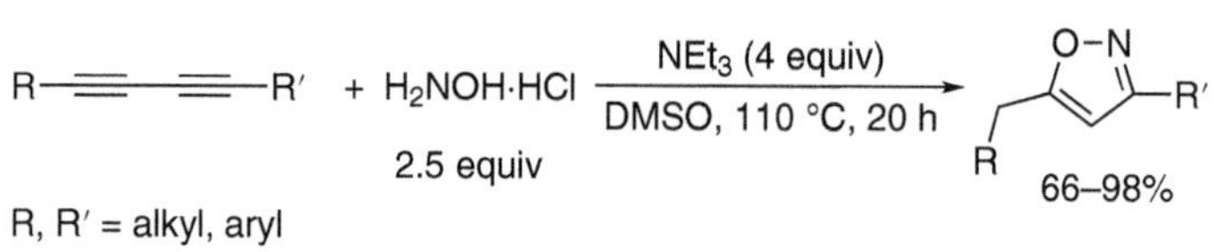

Scheme 8.57 Isoxazoles from Cope-type hydroamination of 1,3-butadiynes.

3,5-Disubstituted isoxazoles were synthesized by the reaction of terminal alkynes with nitric acid in the presence of [TBA][$AuCl_4$] [204], $CuSO_4{\cdot}5H_2O$-catalyzed cycloaddition between terminal alkynes and in situ generated nitrile oxides from oximes [205], CuI-catalyzed cycloaddition of aromatic terminal alkynes with hydroximinoyl chlorides [206], cyclization of alkynylnitrones [207], $AgBF_4$-catalyzed cyclization of alkynyl oxime ethers with the use of phenol as proton source [208].

With the use of $KF{\cdot}2H_2O$ as a hydrohalide scavenger, the reactions of dibromoformaldehyde oxime (a precursor for generation of bromonitrile oxide in situ) with nonactivated 1,2-disubstituted alkynes in a refluxed ethylene glycol dimethyl ether (EGDE) afforded the bromo-bearing 3,4,5-trisubstituted isoxazoles in fair to mild yields (Scheme 8.58) [209]. The obtained 3-bromo-4-hydroxymethyl-5-methylisoxazole was successfully applied in developing an alternative and more efficient synthesis of 2-amino-3-(3-hydroxy-5-methyl-4-isoxazolyl)propionic acid (AMPA), which is well-known as receptor antagonist [210].

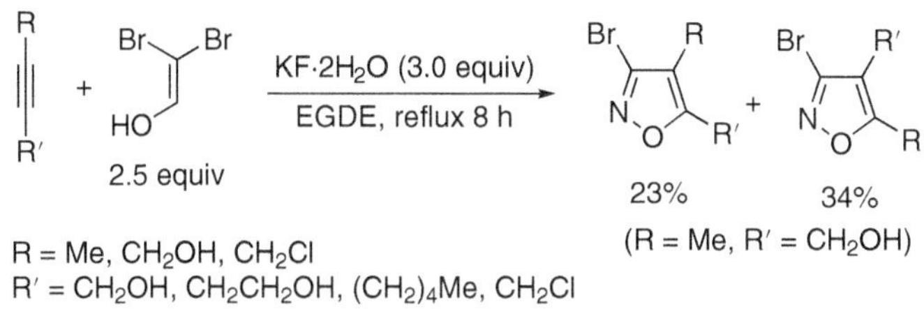

Scheme 8.58 Isoxazoles from cycloaddition of dibromoformaldehyde oxime with alkynes.

Larock and coworker have developed the synthesis of numerous highly substituted isoxazoles via the cyclization of various 2-alkyn-1-one *O*-methyl oximes by a variety of electrophiles (E–X) (Scheme 8.59) [211]. When E group is the halogen atom, the corresponding 4-haloisoxazoles can undergo the various palladium-catalyzed cross-coupling reaction via the activation of C—E bond.

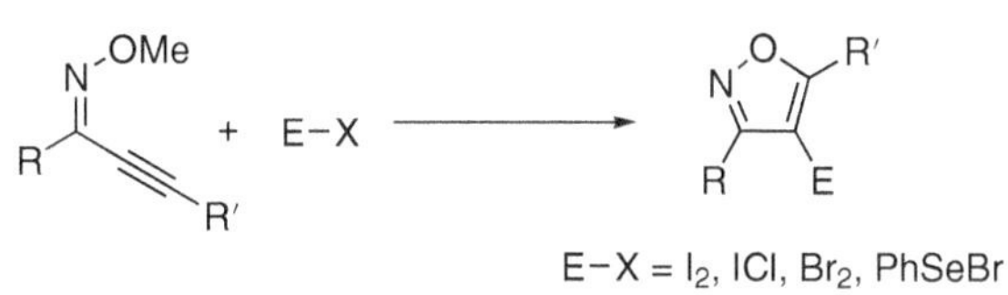

Scheme 8.59 3,4,5-Trisubstituted isoxazoles synthesis from the cyclization of 2-alkyn-1-one *O*-methyl oximes by electrophiles.

In the presence of $AuCl_3$, alkynyl oxime ether underwent a domino reaction involving cyclization and subsequent Claisen-type rearrangement to afford trisubstituted isoxazoles in a regioselective manner (Scheme 8.60) [212]. The synthesized trisubstituted isoxazoles bear allyl groups, which are used as the versatile building blocks for further transformation into a variety of different heterocycles.

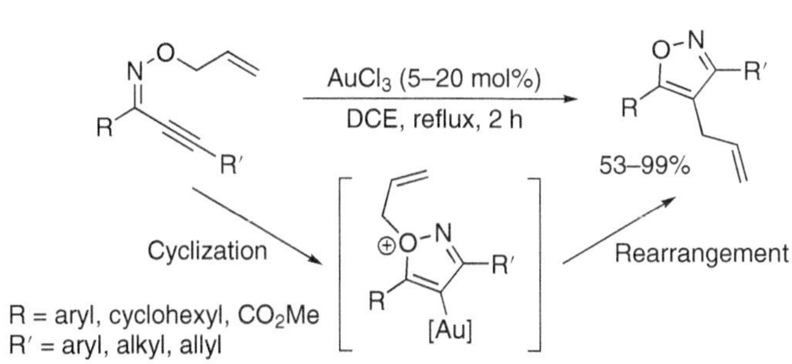

Scheme 8.60 Isoxazoles from Au-catalyzed cyclization-rearrangement reaction.

3,4,5-Trisubstituted isoxazoles could also be obtained from [3 + 2] cycloaddition reactions of alkynyldimethylsilyl ethers with ethyl nitrile oxide (generated in situ from 1-nitropropane and phenyl isocyanate) or with benzonitrile oxide (generated in situ from chlorooxime and $KHCO_3$) [213]. The further palladium-catalyzed cross-coupling reactions of C—Si bonds with aryl iodides can afford a variety of 3,4,5-trisubstituted isoxazoles.

On the basis of 1,3-dipolar cycloaddition reaction, isoxazoles can be prepared by soluble polymer-supported methodology [214] and solid-phase methodology [215].

In addition, *N*-heterocyclic carbine (NHC), as an efficient organocatalyst, could also catalyze the 1,3-dipolar cycloaddition reaction of alkynes with nitrile oxides to give 3,5-di- and 3,4,5-trisubstituted isoxazoles [216].

Moreover, it has reported that the activated nitroalkanes are the good precursors [217], and cycloaddition of aldoximes with alkynes by using a stoichiometric amount of hypervalent iodine as oxidant [218] and by hypervalent iodine-catalyzed oxidation using oxone as oxidant [219] for generating nitrile oxides in situ to react with alkynes to give substituted isoxazoles.

As another typical five-membered heterocyclic compounds containing 1,3-nitrogen/oxygen heteroatoms, 1,3-oxazole rings serve as important sources of valuable biological and physiological activities, as well as the building block in organic synthesis [220].

Alkynes have been well used as a C2 synthon for the construction of 1,3-oxazole ring. For example, with the use of $BF_3{\cdot}OEt_2$ as the additive, Jiang's group developed a $Cu(OAc)_2$-catalyzed oxidative [2 + 2 + 1] cyclocondensation of internal alkynes, nitriles, and H_2O (providing oxygen atom) to give 2,4,5-trisubstituted oxazoles with high regioselectivity (Scheme 8.61) [221].

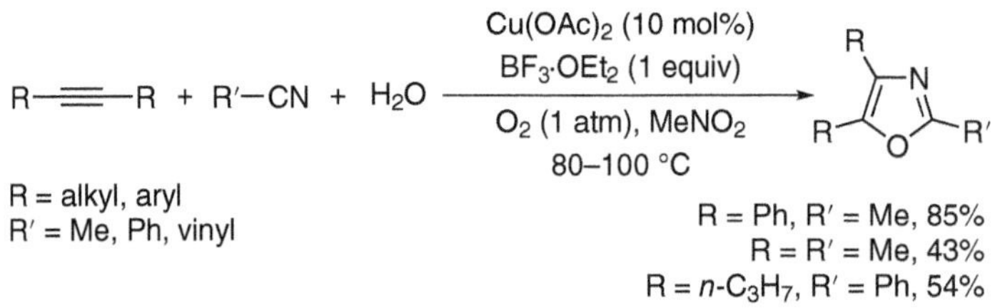

Scheme 8.61 $Cu(OAc)_2$-catalyzed oxidative [2 + 2 + 1] cyclocondensation of internal alkynes, nitriles, and H_2O affording 2,4,5-trisubstituted oxazoles.

Terminal alkynes also underwent a [2 + 2 + 1] annulation with nitrile and an oxygen atom (from pyridine/quinoline *N*-oxides) in the presence of Au(I) complexes, offering a generally efficient synthesis of 2,5-disubstituted oxazoles with broad substrate scope, and nitrile was used as both the reactant and solvent (Scheme 8.62) [222]. The key intermediate is considered to be α-oxo gold carbenes generated via alkyne oxidation. The similar Au(I)-catalyzed oxidative cyclocondensation of propynals with amides affording 2,5-disubstituted oxazoles bearing one acyl substituent [223] and internal alkynes with nitriles giving 2,4,5-trisubstituted oxazoles [224] were then also developed.

On the other hand, the use of hypervalent iodine(III) compound of PhI(OH)X generated in situ to provide oxygen atom [225] and the iodine(III) catalyst, generated in

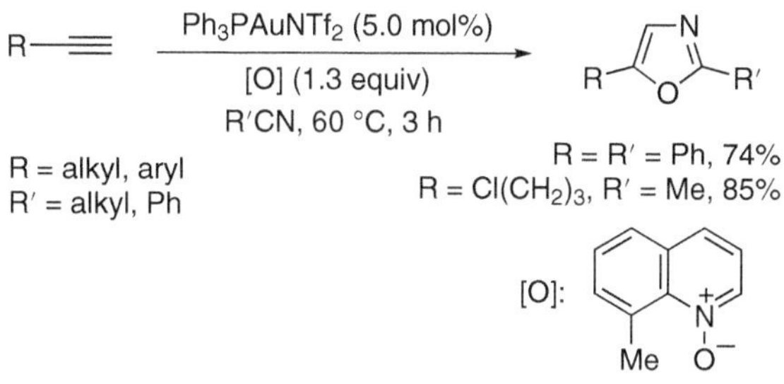

Scheme 8.62 2,5-Disubstituted oxazoles via Au(I)-catalyzed [2 + 2 + 1] annulation of terminal alkynes with nitrile and an oxygen atom.

situ from iodoarene as a precatalyst with *m*-CPBA and Tf_2NH [226], realized the metal-free [2 + 2 + 1] cyclocondensations of alkynes, nitriles, and oxygen atom to produce 2,4-disubstituted and 2,4,5-trisubstituted oxazoles.

Molecular oxygen has also been used as the green oxidant in the synthesis of 2,4,5-trisubstituted oxazoles via a $CuBr_2$-catalyzed aerobic oxidative dehydrogenative annulation of amines with electron-deficient internal alkyne (Scheme 8.63) [227].

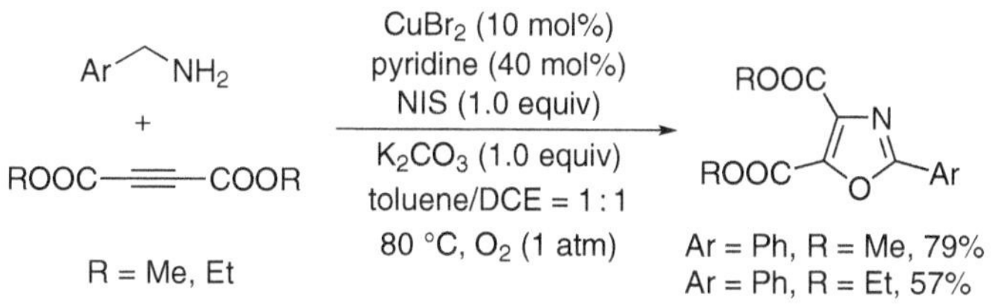

Scheme 8.63 $CuBr_2$-catalyzed annulation of amines, internal alkyne, and oxygen affording 2,4,5-trisubstituted oxazoles

It is well-known that the copper-catalyzed azide-alkyne cycloaddition (CuAAC) reaction has become one of the most efficient routes for the synthesis of 1,2,3-triazoles [228]. Interestingly, in the presence of $[Tpm^{*,Br}Cu(NCMe)]BF_4$, the reaction of terminal alkynes with aryl carbonyl azides provided 2,5-oxazoles in moderate to high yields (Scheme 8.64) [229].

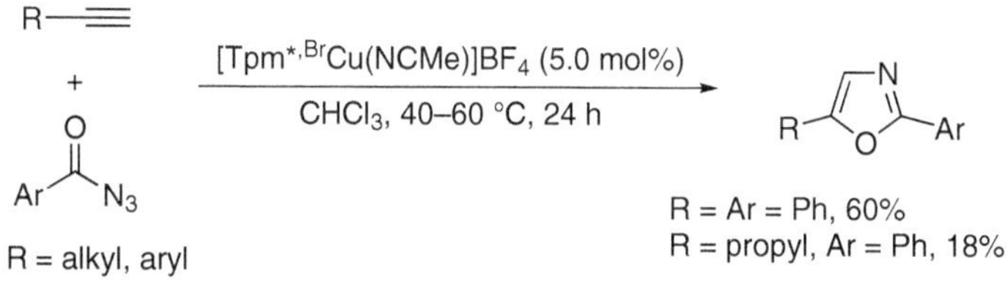

Scheme 8.64 Copper(I)-catalyzed cyclocondensation of terminal alkynes with aryl carbonyl azides affording 2,5-oxazoles

2,5-Disubstituted oxazoles could also be obtained via $[Cp^*Co(MeCN)_3(SbF_6)_2$-catalyzed [3 + 2] cyclocoupling reactions of *N*-pivaloyloxyamides with alkynes with a broad scope of substrates (Scheme 8.65) [230].

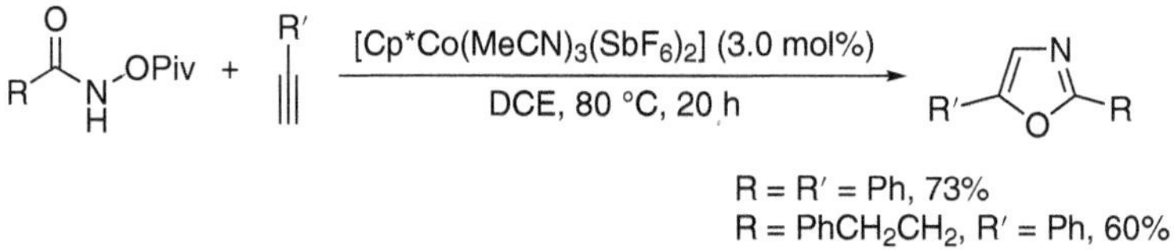

Scheme 8.65 2,5-Disubstituted oxazoles via Co(III)-catalyzed cyclocoupling of *N*-pivaloyloxyamides with alkynes.

In addition, 4,5-disubstituted oxazoles could be synthesized through a metal-free [3 + 2] cyclization of siloxyalkynes with electron-deficient isocyanides in the presence of tetrabutylammonium fluoride (TBAF) as the activator, and siloxyalkynes contribute a C—O unit to the cyclization [231].

In DMSO, Cs_2CO_3-promoted cycloisomerization of propargylamides could also afford 2,4-disubstituted and 2,4,5-trisubstituted oxazoles in good yields (Scheme 8.66) [232]. DFT study on a model substrate revealed that the cycloisomerization involves the formation of allene, intermolecular nucleophilic addition of oxygen anion to allenyl group, and final double-bond isomerization.

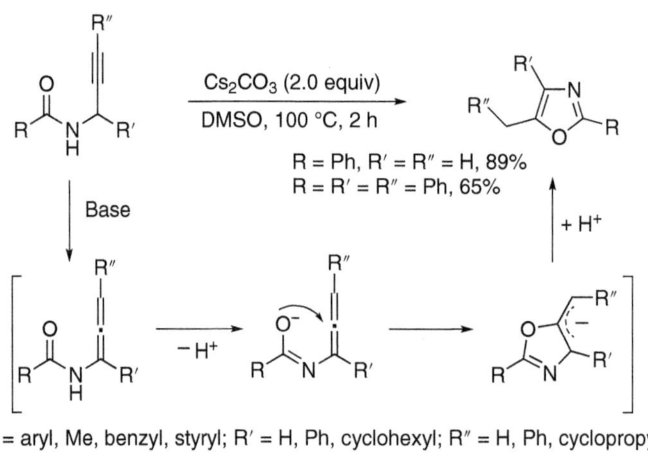

Scheme 8.66 Base-promoted cycloisomerization of propargylamides forming oxazoles.

The cycloisomerization of propargylamides has been well applied in the synthesis of oxazoles [233]. 2,5-Disubstituted selenazoles could be prepared from the similar starting materials via the cycloisomerization of propargyl selenoamides generated in situ using oxygen–selenium exchange reaction (Scheme 8.67) [234]. The formation of 1,3-selenazoles is similar to 1,3-oxazole synthesis; once the selenoamide is formed, the Se atom promotes a 5-*exo*-dig cyclization spontaneously.

Hua and coworker reported a K_2CO_3-promoted [3 + 2] annulation reaction employing 2-alkynyl pyrroles as [3C]-synthon and isothiocyanates as [2C]-synthon for the regioselective construction of the fused heterocycle of pyrrolo[1,2-*c*]thiazol-3-imines, having a 1,3-nitrogen/sulfur five-membered ring

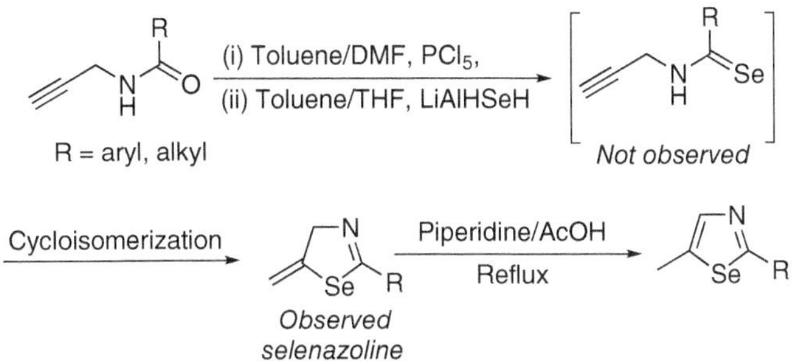

Scheme 8.67 Synthesis of selenazoles from propargylamides.

(Scheme 8.68) [235]. In the proposed mechanism, the interaction between sulfur atoms of isothiocyanate with alkynyl group to enhance the electrophilicity of isothiocyanate is considered to be the key factor to promote the annulation (Scheme 8.69).

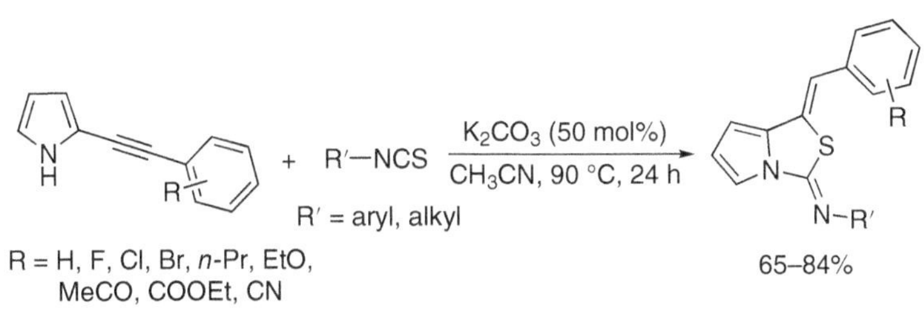

Scheme 8.68 Formation of pyrrolo[1,2-*c*]thiazol-3-imines via [3 + 2] annulation.

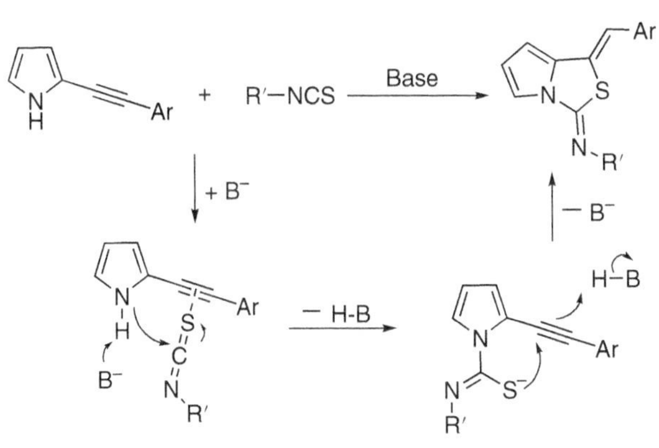

Scheme 8.69 Proposed mechanism for the formation of pyrrolo[1,2-*c*]thiazol-3-imines.

8.3 Six-membered Heterocycles

Cycloaddition reactions of alkynes, enynes, and 1,3-butadiynes are the elegant, atom-efficient transformations for the synthesis of six-membered heterocycles, mostly heteroaromatic compounds. In this section, the interesting and useful protocols for the construction of six-membered heterocyclic rings such as pyridine, quinoline, isoquinoline, 2-pyridone, quinolinone, isoquinolinone, 2-pyrone, coumarin, isocoumarin, chromone, etc. are summarized [236].

8.3.1 Pyridine Derivatives via Cycloaddition of Alkynes with Nitriles

Substituted pyridines and pyridine-fused cyclic compounds are important nitrogen-containing heterocycles; the chemoselective [2 + 2 + 2] cycloaddition of two alkynes with one nitrile is the most efficient and atom-economic process (Scheme 8.70) [237]. After the first example of cobalt-catalyzed [2 + 2 + 2] cycloaddition was reported by Yamazaki and coworkers in 1973 [238], the transition metal-catalyzed or -mediated such type of reactions have been well developed.

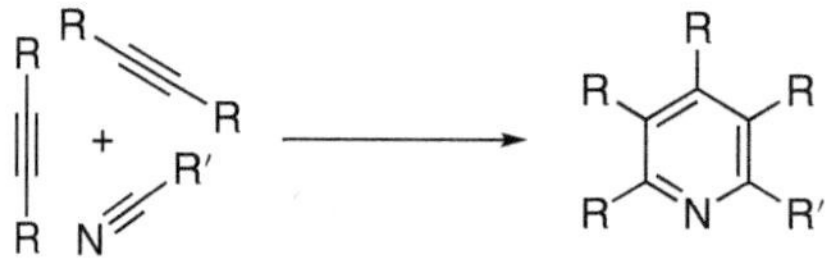

Scheme 8.70 Construction of pyridine ring by [2 + 2 + 2] cycloaddition.

2-Substituted pyridines could be prepared by the [2 + 2 + 2] photocycloaddition of a variety of applicable nitriles with two equivalents of acetylene gas in the presence of [CpCo(cod)] [239]. As shown in Scheme 8.71, irradiation (two 460 W lamps, ~420 nm) of a mixture of 3,3-diethoxypropionitrile and CpCo(cod) (0.5 mol%) in toluene with a bubbling of acetylene for four hours resulted in the formation of 2-(2,2-diethoxy-ethyl)pyridine in 87% yield. The functionalized aliphatic and aromatic nitriles were found to be the suitable substrates for the formation of 2-substituted pyridines. In addition, the same group then developed an asymmetric [2 + 2 + 2] cycloaddition of alkynes with nitriles to yield the axially chiral 2-arylpyridines in the presence of chiral cobalt (I) complexes as catalysts [240].

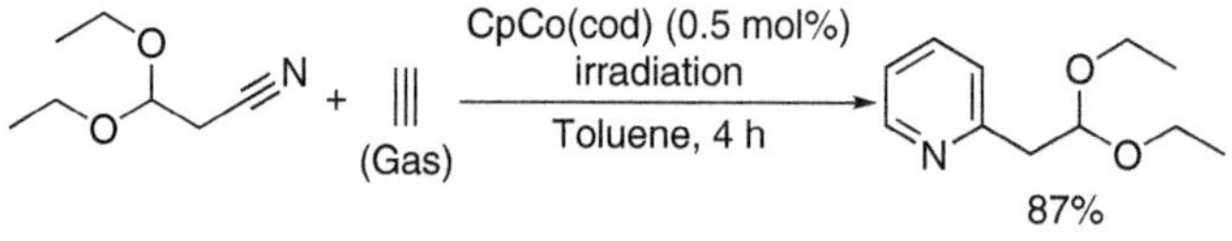

Scheme 8.71 CpCo(cod)-catalyzed [2 + 2 + 2] photocycloaddition of nitriles with acetylene.

$CpCo(CO)_2$ complex was also found to be the cheap and efficient catalyst in [2 + 2 + 2] cycloaddition of diynes with nitriles to afford bicyclic pyridines. As shown in Scheme 8.72, it catalyzed the cycloaddition reaction of dimethyl 2,2-diprop-2-ynylmalonate and *N*-cyanopyrrolidine in 1,4-dioxane to yield 2-aminopyridines in 88% yield [241].

Under microwave irradiation, $CpCo(CO)_2$ also showed high catalytic activity for the intra- and intermolecular [2 + 2 + 2] cycloaddition [242]. For example, tetrahydro-1,6-naphthyridine could be prepared in 80% yield (Scheme 8.73).

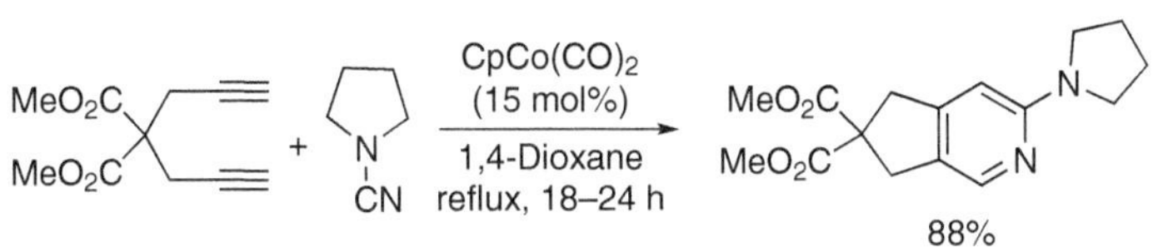

Scheme 8.72 $CpCo(CO)_2$-catalyzed [2 + 2 + 2] cycloaddition of dines with nitriles affording bicyclic pyridines.

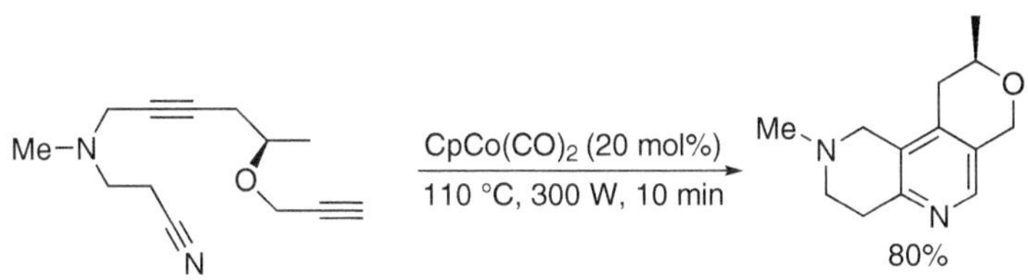

Scheme 8.73 $CpCo(CO)_2$-catalyzed intramolecular cyclization with microwave promotion.

The solid-supported formation of fused pyridine rings could be obtained under microwave irradiation by [$CpCo(CO)_2$]-catalyzed [2 + 2 + 2] cocyclotrimerizations of either nonconjugated alkyne nitriles with alkynes or nonconjugated diynes with nitriles [243].

In MeCN, CoI_2(dppe)/Zn catalyst system showed good catalytic activity in an intramolecular [2 + 2 + 2] cocyclotrimerization of nitrile diynes affording tetra- and pentacyclic pyridine derivatives in good to excellent yields [244].

On the other hand, the simple cobalt salts combined with phosphine ligands have also been found to be the effective catalysts for the intermolecular [2 + 2 + 2] cycloaddition of diynes with nitriles to construct bicyclic pyridine ring. For example, with the use of $CoCl_2{\cdot}6H_2O$/phosphine/Zn as catalyst system, *multi*-substituted pyridines could be formed by the reaction of 1,6-diynes with nitriles, and inactivated nitriles such as acetonitrile and benzonitrile also showed high reactivity when dppe was used as ligand (Scheme 8.74) [245].

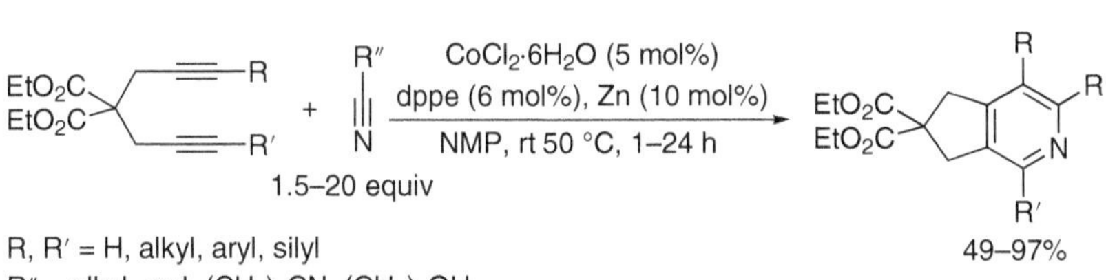

Scheme 8.74 Synthesis of bicyclic pyridine by $CoCl_2{\cdot}6H_2O$/Zn/dppe catalyst system.

The same catalyst system was then used in the cycloaddition of either 1,6-diynes with 2-cyanopyridine or (2-pyridyl)-1,6-diynes with nitriles to regioselectively afford 2,2′-bipyridine derivatives, which are an important structure occurring in biologically active compounds and particularly as a ligand in transition metal complexes [246].

In addition, other cobalt complexes have also been confirmed to be the efficient catalysts in the cycloaddition of alkynes with nitriles to construct pyridine ring [247].

The combination of $Ni(cod)_2$ and sterically hindered, electron-donating *N*-heterocyclic carbene (NHC) ligands could catalyze the cycloaddition of alkynes with nitriles at ambient temperature to afford pyridines in excellent yields [248]. As shown in Schemes 8.75 and 8.76, in the presence of $Ni(cod)_2$/SIPr, the cycloaddition of 3-hexyne or dimethyl 2,2-di-but-2-ynylmalonate with benzonitrile afforded 2,3,4,5-tetraethyl- 6-phenylpyridine or bicylic pyridine derivative in 82 and 86% yields, respectively. The same group further explored the use of $Ni(acac)_2$ and NHC salt as the air-stable, readily available precursors that produce the active Ni(0)/NHC catalyst in situ for the synthesis of bicylic pyridine derivatives [249].

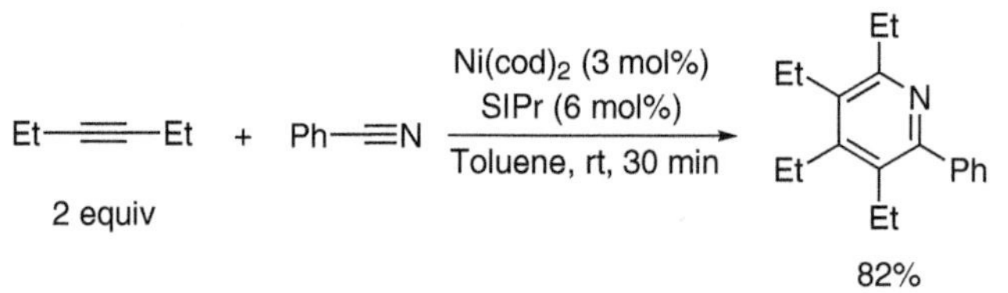

Scheme 8.75 Nickel-catalyzed cycloaddition of 3-hexyne with benzonitrile.

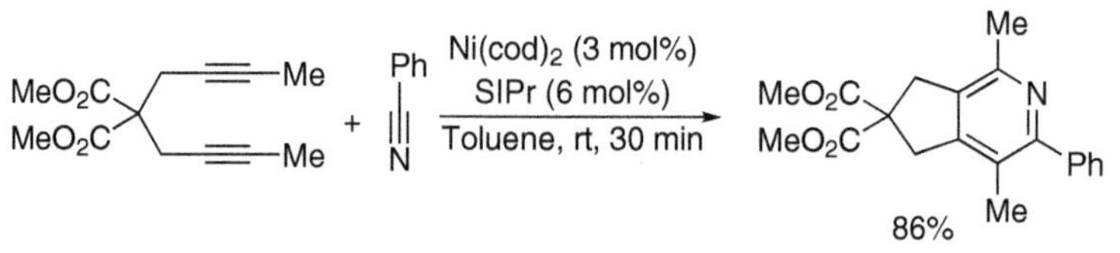

Scheme 8.76 Nickel-catalyzed formation of bicylic pyridines from diyne with benzonitrile.

Moreover, the same group has also developed the general iron (II)/ligand(amines, phosphines, and NHCs)-catalyzed method for pyridine synthesis [250].

Rhodium complexes were also employed as the catalysts in the formation of pyridine derivatives. For example, in the presence of $[Rh(cod)_2]BF_4$/BINAP, monoalkynes or 1,6-, 1,7-, 1,8-diynes and nitriles smoothly underwent the [2 + 2 + 2] cycloaddition affording a broad variety of highly functionalized pyridines (Schemes 8.77 and 8.78) [251].

In addition, ruthenium [252], cobalt complexes [253], $[Ir(cod)Cl]_2$/phosphine [254], and iron (II)/(III) halides/phosphines [255] have also been applied in the catalysis of [2 + 2 + 2] cycloaddition of alkynes with nitriles affording fused pyridine derivatives.

The synthesis of pyridines via the cycloaddition of alkynes could also be achieved by using oximes (or generated in situ from the reaction of hydroxylamine with carbonyl compounds) as the alternative nitrogen source. As shown in Scheme 8.79, in HFIP, $[RhCl(cod)]_2/AgSbF_6$ could efficiently catalyze the formation of bicyclic pyridine derivatives by an intramolecular hetero-[4 + 2] cycloadditions of ω-alkynyl-vinyl oximes [256].

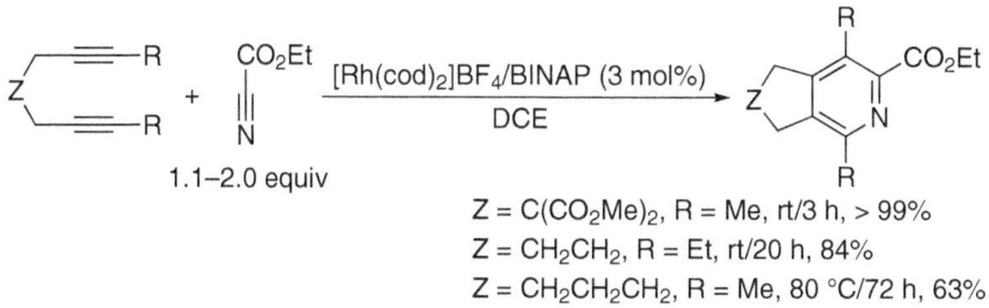

Scheme 8.77 Rhodium-catalyzed [2 + 2 + 2] cycloaddition of diynes with nitriles.

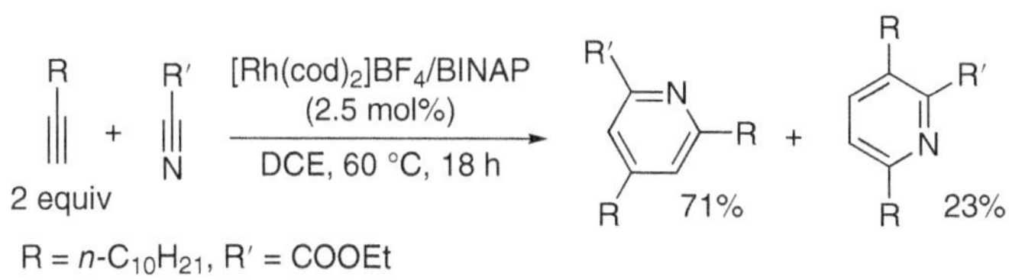

Scheme 8.78 [2 + 2 + 2] Cycloaddition of 1-dodecyne with ethyl cyanformate.

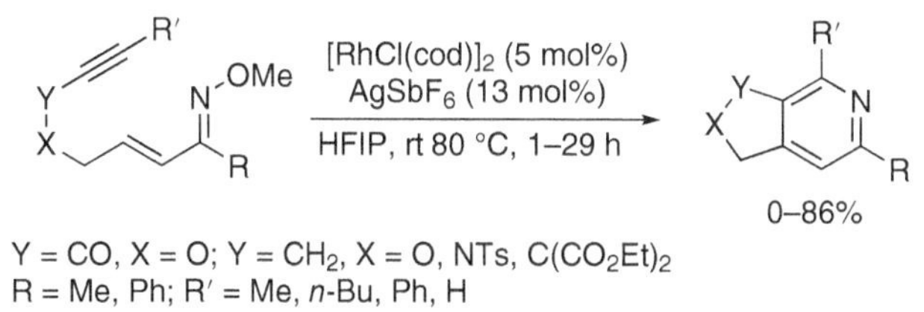

Scheme 8.79 Synthesis of pyridines via intramolecular hetero-[4 + 2] cycloadditions of ω-alkynyl-vinyl oximes.

$RhCl(PPh_3)_3$ [257], $[RhCp^RCl_2]_2$ (Cp^R = Cp*, Cp^t, Cp^{CF3}) [258], $[RhCl(coe)_2]_2$ [259], and $[Cp^*RhCl_2]_2$ [260] also showed high catalytic activity in the cycloaddition of α,β-unsaturated oximes with alkynes affording highly substituted pyridines, and $[Rh(cod)_2]BF_4$/phosphine could efficiently catalyze [2+2+2] cycloaddition of oximes and diynes giving bicyclic pyridines [261]. In addition, $Ni(cod)_2$/IPr catalyzed the cycloaddition of either α,β-unsaturated oximes or β,γ-unsaturated oximes with alkynes to furnish 2,3,4,6-tetrasubstituted pyridines [262].

The rhodium-catalyzed cyclocondensation of α,β-unsaturated oxime esters with alkenes in the presence of oxidants has also been developed to synthesize the pyridines [263].

The highly substituted pyridines could also be synthesized from α,β-unsaturated *N*-benzyl aldimines and ketimines with alkynes through dihydropyridine intermediates in the presence of $[RhCl(coe)_2]_2$/phosphine [264]. For example, the reaction of *N*-(2-methyl-2-propen-1-ylidene)benzenemethanamine with 3-hexyne yielded 2,3-diethyl-5-methylpyridine in 80% overall yield (Scheme 8.80).

The polysubstituted pyridines could be obtained on the basis of the one-pot reaction of *in situ* generated α,β-unsaturated imines with CH nucleophiles

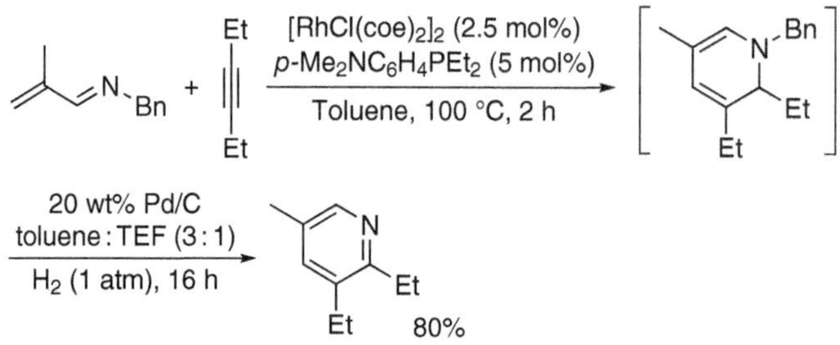

Scheme 8.80 One-pot synthesis of pyridines from α,β-unsaturated imines and alkynes via dihydropyridine intermediates.

in the presence of oxidant [265]. $CoBr_2$/triarylphosphine/^{i}PrMgBr catalyzed the annulation of α,β-unsaturated imines with internal alkynes to afford polysubstituted dihydropyridines [266].

[(*p*-cymene)$RuCl_2]_2$-catalyzed formal dehydrative [4 + 2] cycloaddition of enamides and alkynes is one of another efficient synthetic methods for the formation of highly substituted pyridines (Scheme 8.81) [267].

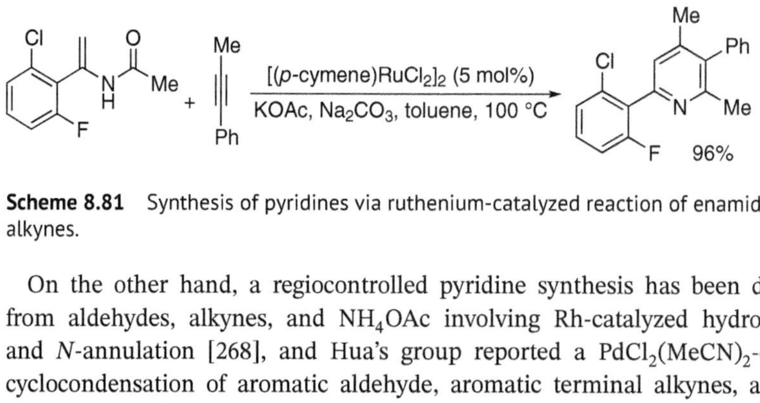

Scheme 8.81 Synthesis of pyridines via ruthenium-catalyzed reaction of enamides with alkynes.

On the other hand, a regiocontrolled pyridine synthesis has been developed from aldehydes, alkynes, and NH_4OAc involving Rh-catalyzed hydroacylation and *N*-annulation [268], and Hua's group reported a $PdCl_2(MeCN)_2$-catalyzed cyclocondensation of aromatic aldehyde, aromatic terminal alkynes, and using ammonium bifluoride as nitrogen source in the presence of *t*-butyl peroxide (TBP), providing an alternative efficient synthesis of unsymmetrical 2,4,6-triaryl pyridines; the strategy is an attractive and practical method due to its versatility of starting materials and high regioselectivity with good yields, as well as compatible with various functional groups (Scheme 8.82) [269].

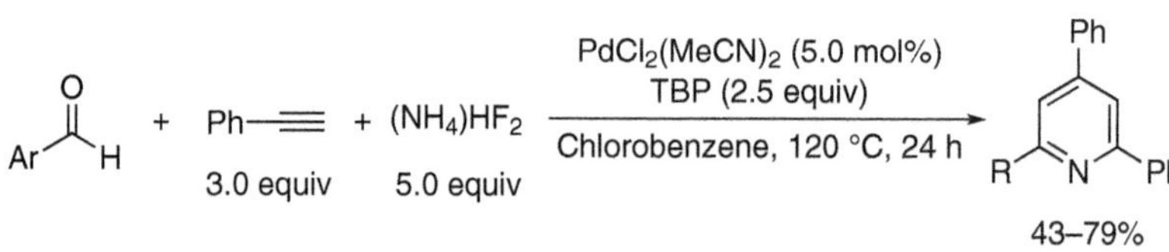

Scheme 8.82 Palladium-catalyzed formation of 2,4,6-triarylpyridines.

Moreover, highly functionalized pyridines could be synthesized under metal-free conditions. For example, a variety of tetrasubstituted pyridines were obtained via a Me_3SiCl-promoted [3+2+1] intermolecular three-component coupling cyclization of a functionalized enamine, *N*,*N*-dimethylformamide diethyl acetal, and acetylenedicarboxylates in good to excellent yields (Scheme 8.83) [270].

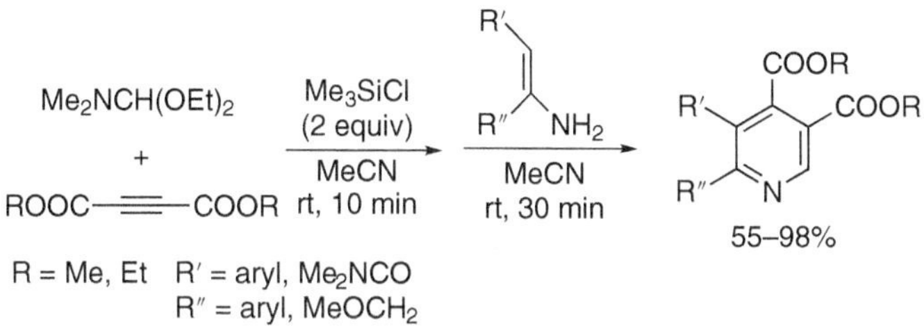

Scheme 8.83 Synthesis of pyridines by Me_3SiCl-promoted [3 + 2 + 1] cyclization.

Propargyl amines as versatile building blocks have been well applied in the construction of pyridine ring [271], and the substituted pyridines from the intramolecular cyclization of *N*-propargylic β-enaminones have been well documented [272].

Bohlmann–Rahtz pyridine synthesis is one of the most important procedures with the use of ethynylketones, including the condensation with enamines leading to an aminodiene intermediate, and then subsequent E/Z isomerization and cyclodehydration [273].

Organobase-catalyzed cyclocondensation of 2,3-butadienoate with *N*-sulfonyl-1-aza-1,3-dienes [274], Brønsted acid-promoted intermolecular cycloaddition of alkynylnitrile or nitriles with alkynes [275], thermal [4+2] cycloaddition of 1,3-dienes with nitriles [276], Ni(cod)/phosphine-catalyzed dehydrogenative [4+2] cycloaddition of 1,3-dienes with nitriles [277], Cs_2CO_3-promoted annulation of alkynylimines with benzylamines [278], $Re_2(CO)_{10}$-catalyzed regioselective annulation of β-enamino ketones and alkynes via C—C bond cleavage [279], $[Cp^*RhCl_2]_2$-catalyzed *N*-annulation of α,β-unsaturated imine (generated in situ via a $Cu(OAc)_2$-promoted dehydrogenation of the allylamine) with alkyne [280], two-step procedure involving the formation of allenyl imines in situ from amino allenes and aldehydes, a subsequent $Pd(PPh_3)_4$-catalyzed cyclization with aryl iodides [281], $[Cp^*RhCl_2]_2/AgBF_4$-catalyzed annulation of *N*-sulfonyl ketimines with alkynes using N—S bond as an internal oxidant [282], and $AgSbF_6$- and I_2-mediated aza-annulation of 2-en-4-ynyl azides [283] have also provided the alternative ways approach to *multi*-substituted pyridine derivatives.

In addition, polysubstituted pyridines could also be synthesized by a multicomponent reaction of arynes, isocyanides, and terminal alkynes [284]; condensation of a β-ketoester, ammonia, and alkynone under acidic conditions [285]; ruthenium-catalyzed cycloisomerization of 3-azadienynes [286]; base-promoted cyclocondensation of 1-arylethylamines and ynones [287]; lithiation/isomerization/intramolecular carbolithiation of *N*-allyl-ynamides [288]; and three-component heteroannulation reaction of alkynone, 1,3-dicarbonyl compound, and ammonium acetate [289].

8.3.2 Benzopyridine Derivatives (Quinolines and Isoquinolines)

The representative structures of benzopyridine are quinolines and isoquinolines, which are very important types of *N*-heterocyclic compounds, and the development of their synthetic methods is the interesting research topics in organic synthetic chemistry [290].

Cheng and coworker reported an efficient and convenient $NiBr_2$(dppe)-catalyzed cyclization of 2-iodoanilines with alkynyl aryl ketones to give 2,4-disubstituted quinolines (Scheme 8.84) [291]. On the basis of the regiochemistry of the products, the proposed mechanism for the formation of quinoline ring involves the oxidative addition of C—I bond to Ni(0) species and the insertion of alkynes to Ni—C bond with the formation of the intermediate of amino chalcone, which undergoes the intramolecular nucleophilic addition of amino group to ketone to construct quinoline ring with dehydration.

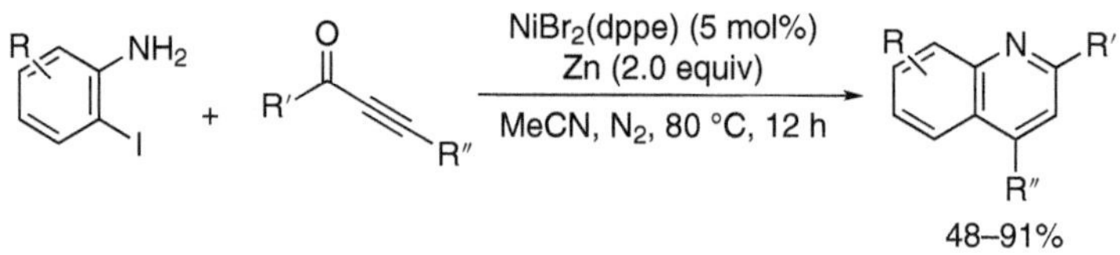

Scheme 8.84 Nickel-catalyzed synthesis of quinolines from 2-iodoaniline and aroylalkyne.

There have two important and efficient catalyst systems to approach to substituted quinolines by the annulation of anilines with alkynes and C1 source. One is from Balaraman's group; they developed an $[Rh(cod)Cl]_2$/dppm-catalyzed C—H activation of unprotected anilines and its addition reaction with electron-deficient alkynes, and subsequent annulation with CO gas or paraformaldehyde to construct quinoline ring [292]. The other is from Yi's group; in the presence of $[Cp^*Co(CO)I_2]/AgNTf_2$, using DMSO as both the solvent and the C1, the cyclization of anilines with alkynes could directly give the quinolines with high regioselectivity (Scheme 8.85) [293]. A similar procedure without the use of transition metal catalyst approach for the synthesis of 4-arylquinoline from aromatic terminal alkynes, anilines, and DMSO has also been developed [294].

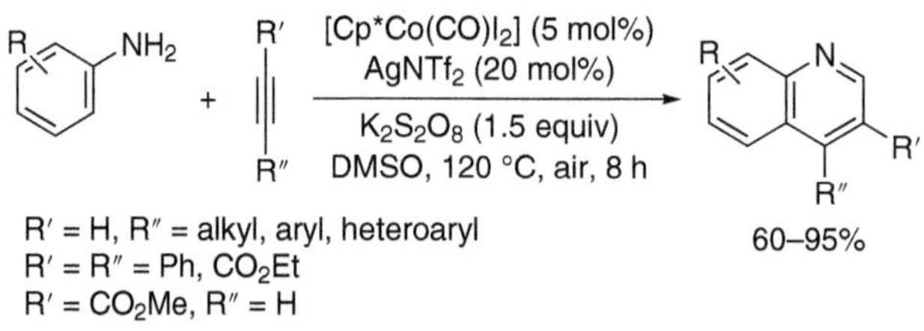

Scheme 8.85 Quinoline ring formation from intermolecular cyclization of aniline, alkyne, and DMSO.

$[Cp^*CoCl_2]/AgNTf_2$ was applied in the synthesis of quinolines via a redox-neutral annulative coupling of arylamides with alkynes by Li's group (Scheme 8.86) [295].

In this protocol, the acetamino group is used as the DG to assist the activation of *ortho*-C—H bond of anilines, and the reaction affords quinolines with water as the sole by-product.

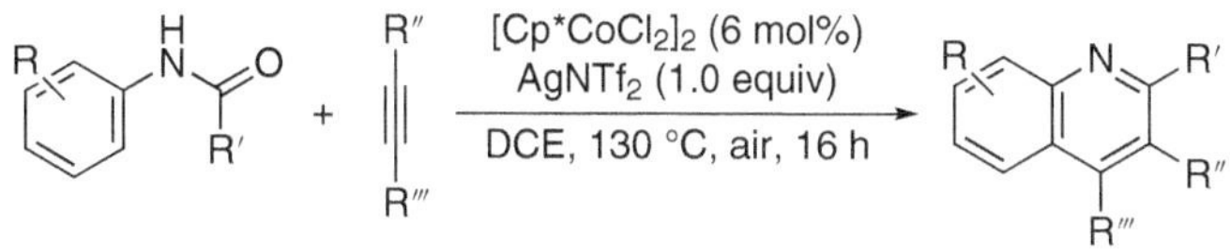

Scheme 8.86 Quinoline ring formation from intermolecular cyclization of arylamides with alkyne.

In addition, the substituted quinolines could also be synthesized by the reductive cyclization of *ortho*-nitrophenyl propargyl alcohols with the use of Zn or Fe/HCl or AcOH [296], by an electrophilic intramolecular cyclization of 1-azido-2-(2-propynyl)benzene with the use of a variety of electrophiles [297], and by superbase KOH/DMSO-promoted regioselective [4+2] cycloaddition of azadienes (generated in situ from *ortho*-aminobenzyl alcohol) with internal alkynes [298]; *multi*-substituted 2-aminoquinolines have been achieved from the annulation of 1-aryl tetrazoles with internal alkynes via double C—H activation and denitrogenation [299].

Moreover, other unsaturated hydrocarbons such as allenes have also been used in the synthesis of quinolines [300].

Similarly, isoquinoline ring can be constructed from either the intramolecular cyclization of alkynylarene having *N*-source group or the intermolecular cyclization of aromatic alkynes with nitrogen-containing compounds.

As shown in Scheme 8.87, the intramolecular cyclization of alkynylarene having *N*-source group involves iodine-mediated electrophilic cyclization [301], $AgSbF_6$ [302], $AuCl_3/AgSbF_6$ [303]-catalyzed cyclization of 2-alkynyl benzylazides, and AgOTf-catalyzed cyclization of 2-alkynyl benzaldoximes [304, 305]. In addition, under different reaction conditions, 2-alkynylbenzaldoximes were found to be the versatile building blocks for the construction of *N*-heterocycles [306].

A number of substituted isoquinolines have been prepared by the electrophilic cyclocoupling of 2-(1-alkynyl)arylaldimines with a variety of electrophiles or organic halides catalyzed by transition metal complexes or metal-free conditions (Scheme 8.88), for example, electrophiles of I_2, ICl, PhSCl, PhSeCl, and *p*-$O_2NC_6H_4SCl$ [307], with organic halides of aryl, allylic, and alkynyl halides [308] and with aryl halides in the presence of CO [309].

Scheme 8.89 summarizes the synthesis of isoquinolines by either intermolecular cyclization of aromatic alkynes with nitrogen-containing compounds or intermolecular cyclization of aryl aldimines with alkynes, including the thermal imination/annulation of 2-ethynylbenzaldehydes with ammonia to give 3-substituted isoquinoline [310], or under microwave irradiation [311], Pd(II)/CuI or Pd(II)-catalyzed cyclocoupling of 2-halobenzaldimines with terminal and internal alkynes [312], nickel(II)/Zn-catalyzed annulation of 2-iodobenzaldimines with

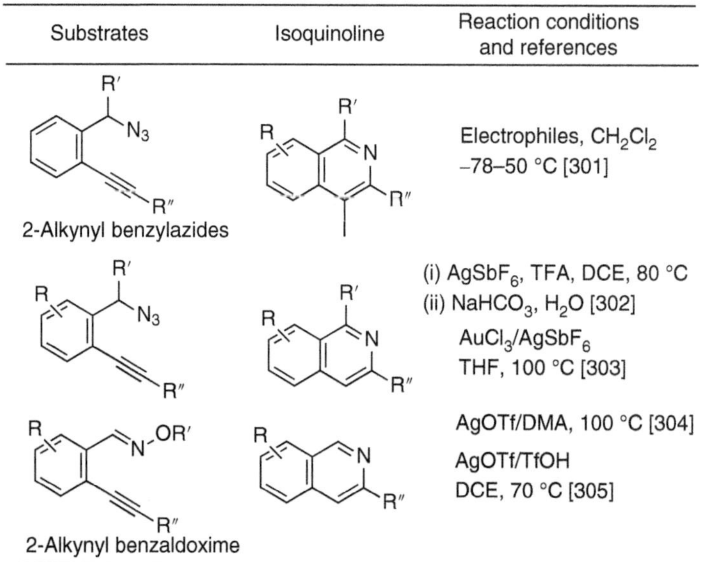

Scheme 8.87 Isoquinoline ring formation from intramolecular cyclization of alkynylarene having *N*-source group.

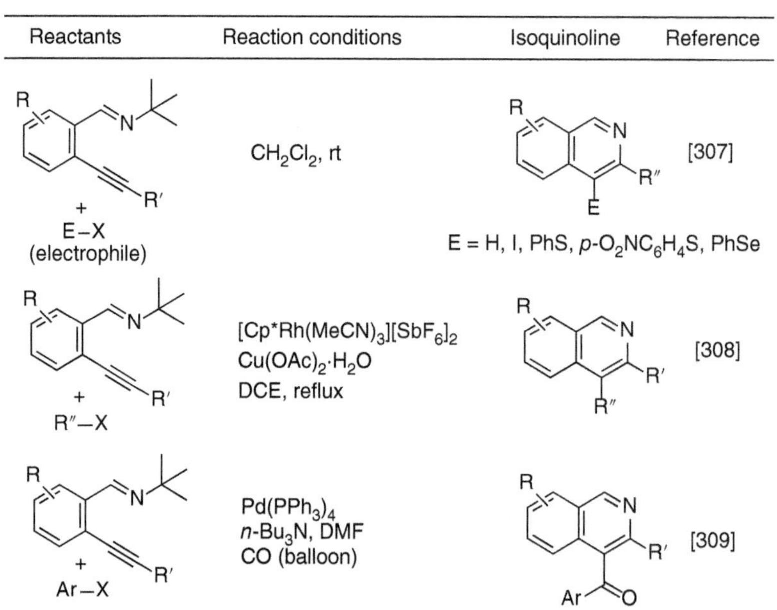

Scheme 8.88 Isoquinoline ring formation from intermolecular cyclocoupling of 2-(1-alkynyl)arylaldimines with electrophiles.

alkynes [313], rhodium(I)-catalyzed oxidative cross-coupling/cyclization of aryl aldimines with internal alkynes with the use of $Cu(OAc)_2 \cdot H_2O$ as oxidant [314], rhodium(I) [315] and cationic ruthenium(II) [316]-catalyzed cyclocondensation of aromatic ketoximes with alkynes, rhodium(III)-catalyzed cyclocondensation of aromatic ketoximes with alkynes [317], or annulation of aryl ketone *O*-acyloximes with internal alkynes [318].

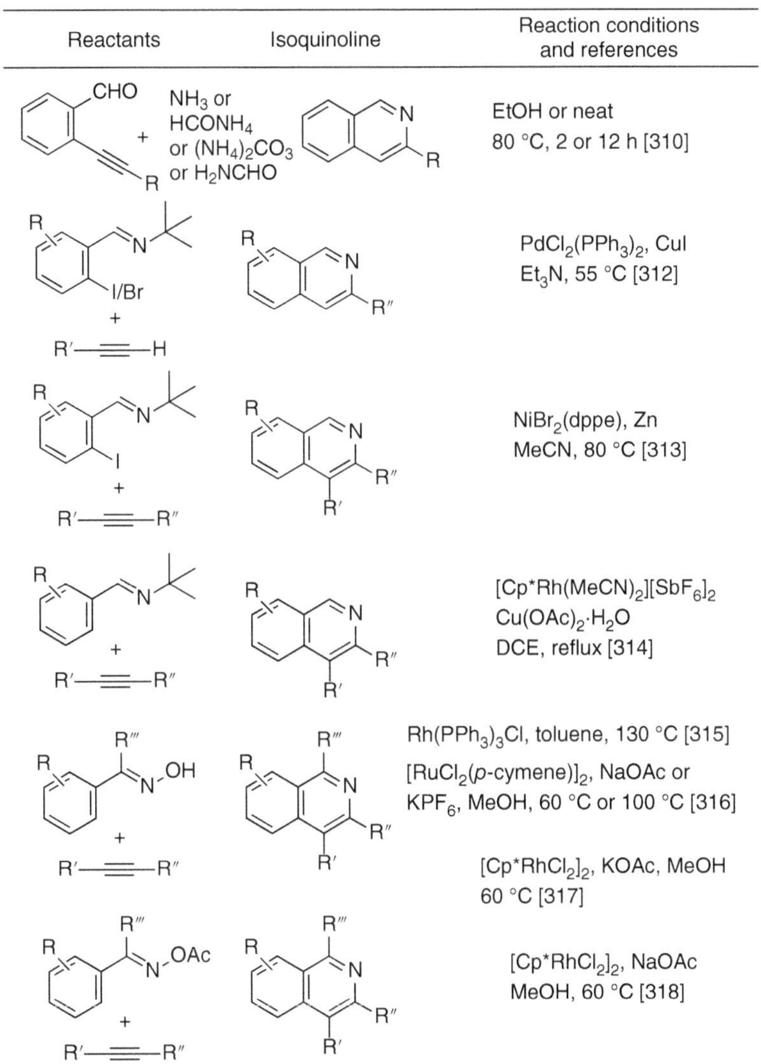

Scheme 8.89 Isoquinoline ring formation from intermolecular cyclization of aryl aldimines with alkynes.

Scheme 8.90 shows the protocols for the formation of isoquinoline ring by intermolecular cyclization of aryl aldimines with alkynes in the presence of transition metal catalyst via the activation of carbon–hydrogen bond.

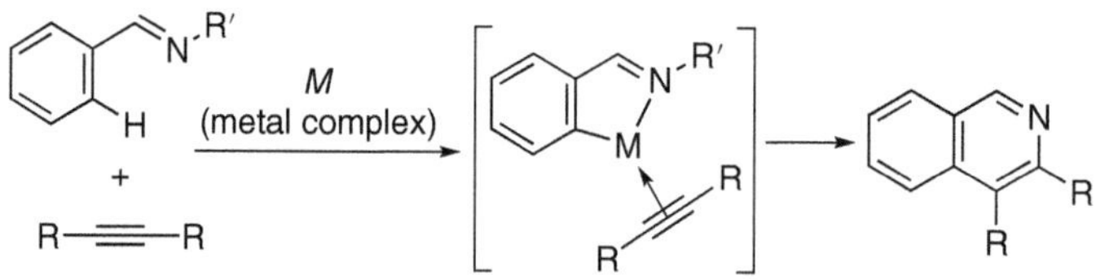

Scheme 8.90 Protocol for the synthesis of isoquinolines from the annulation of aldimines with alkynes catalyzed by transition metal complexes.

Hua and coworker also developed a cyclocondensation of *ortho*-alkynyl aromatic aldehydes/ketones with urea catalyzed by $CuSO_4 \cdot 5H_2O$ affording isoquinolines and isoquinoline-fused heterocyclic compounds, and the procedure has the advantages of cheap catalyst, urea as nitrogen source, wide substrate scope, and high chemoselectivity (Scheme 8.91) [319]. Interestingly, a three-component reaction of *ortho*-bromobenzaldehyde, phenylacetylene, and urea in the presence of CuI could also afford the desired isoquinolines in good yield.

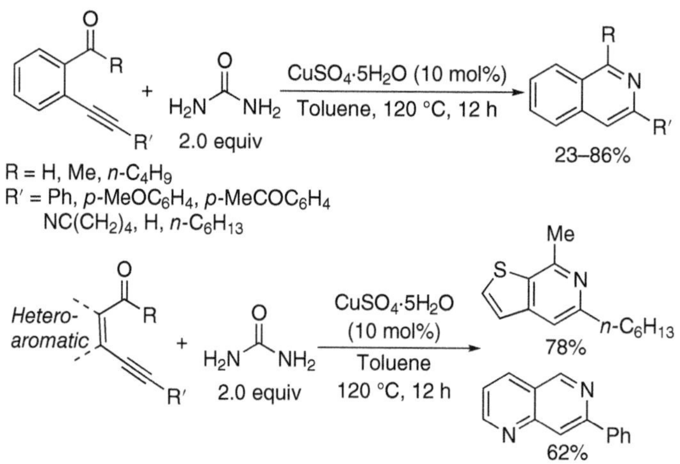

Scheme 8.91 Formation of isoquinoline ring by cyclocondensation.

In addition, *multi*-substituted isoquinolines could be achieved from $PdBr_2$-catalyzed cyclization of 2-(1-alkynyl)arylaldimines with various alkenes [320], AgOTf/CuI-catalyzed cyclization of 2-alkynylbenzaldoximes with aldehydes or alcohols [321], and $[Cp^*RhCl_2]_2/AgSbF_6$-catalyzed regioselective annulation of aryloxime esters with 1,3-dienes under redox-neutral conditions [322]. The 1,2-dihydroisoquinolines were prepared from the transition metal-catalyzed direct addition of nucleophiles to *ortho*-alkynylarylaldimines (or formed in situ) [323].

8.3.3 2-Pyridone Derivatives and Their Benzo-derivatives (Quinolinones and Isoquinolonones)

2-Pyridone (2-pyridinone) moiety is present in many natural substances with various biological activities; 2-pyridone derivatives are interesting and useful intermediates in the syntheses of other *N*-heterocyclic compounds [324].

As previously described, the [2 + 2 + 2] cycloaddition of two alkynes with nitrile has become an efficient way to construct pyridine ring in an atom-economic manner. Similarly, 2-pyridones can be also obtained by the [2 + 2 + 2] cycloaddition of two alkynes with isocyanate.

As shown in Scheme 8.92, Louie and coworker reported an $Ni(cod)_2$/SIPr-catalyzed cycloaddition of diynes with isocyanates for the preparation of a fused five-membered pyridones in excellent yields under mild conditions [325]. The fused six- and seven-membered pyridines could be also obtained depending on the structures of diynes. The three-component cycloaddition of two alkynes and isocyanate was also investigated by the same group, and it has been found that the cycloaddition of asymmetrical alkynes with various isocyanates catalyzed by $Ni(cod)_2/PEt_3$ occurs with high regioselectivity (Scheme 8.93) [326].

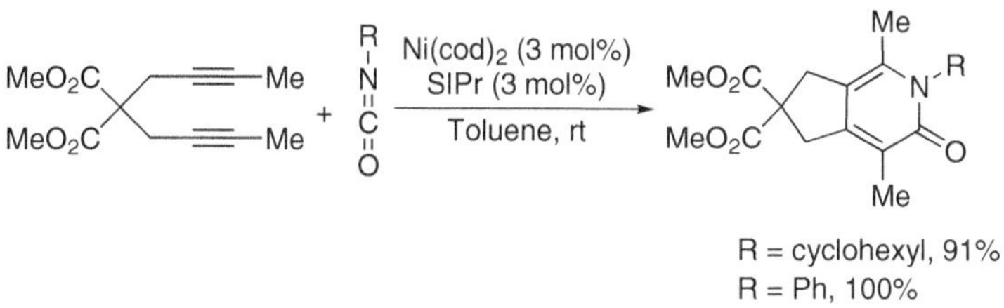

Scheme 8.92 Synthesis of 2-pyridones via the [2 + 2 + 2] cycloaddition of alkynes with isocyanate.

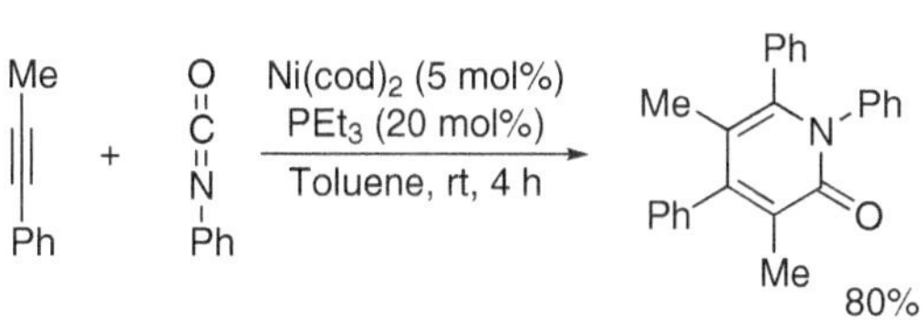

Scheme 8.93 $Ni(cod)_2$-catalyzed regioselective cycloaddition of two alkynes and isocyanate.

Kondo and coworker also studied the cyclocotrimerization of internal alkynes with isocyanates using $[RhCl(C_2H_4)_2]_2/PPh_3$ as a catalyst system and found that the selective formation of 2-pyridones and pyrimidine-2,4-diones could be controlled by changing the molar ratio of alkynes and isocyanates [327]. As indicated in Scheme 8.94, with the use of an excess amount of 3-hexyne, the reaction afforded the corresponding 2-pyridones as major product, whereas in the case of an excess amount of *n*-hexyl isocyanate (20 equiv), pyrimidine-2,4-dione was obtained in high yield.

The application of $[RhCl(C_2H_4)_2]_2$ combined with ligands in the cycloaddition of terminal alkynes with isocyanates to produce 2,4-disubstituted 2-pyridones has also been reported by Rovis's group [328].

The enantioselective [2 + 2 + 2] cycloaddition of diynes with isocyanates to synthesize axially chiral 2-pyridones was developed by Tanaka's group with the use of cationic rhodium(I)/modified-BINAP catalyst systems [329].

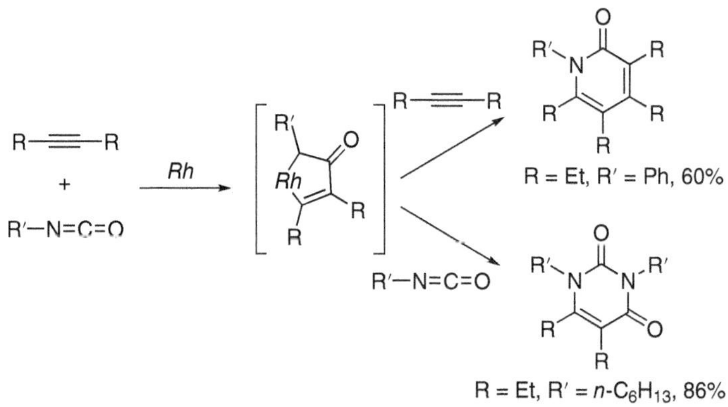

Scheme 8.94 Selective synthesis of 2-pyridones and pyrimidine-2,4-diones by cyclocotrimerization of alkynes with isocyanates.

Other examples for the formation of 2-pyridone ring from the annulation reactions of alkynes or alkenes as one of the starting materials are summarized in Scheme 8.95.

The Diels–Alder cycloadditions of 2-azadienes with the electron-deficient alkynes, such as dimethyl acetylenedicarboxylate and methyl propiolate, gave a primary six-membered cyclo-adduct, which spontaneously lost a molecule of alcohol or silanol by hydrolysis to afford *multi*-substituted pyridines [330].

$[RhCp^*Cl_2]_2$-catalyzed oxidative cyclocoupling reaction of acrylamides with aromatic internal alkynes with the use of $Cu(OAc)_2$ as oxidant produced 2-pyridones in good to high yields [331]. The catalyst systems of $Cp^*Rh(MeCN)_3(SbF_6)_2/Cu(OAc)_2{\cdot}H_2O$ [332], $[RuCl_2(p\text{-cymene})]_2/Cu(OAc)_2{\cdot}H_2O$ [333], and $Pd(OAc)_2/CuBr_2/O_2$ [334] have also shown the highly catalytic activity for the similar transformation.

In addition, 2-pyridones could be also prepared with the use of 1,3-dicarbonyl compounds as starting materials in the presence of $In(OTf)_3$ [335] or under microwave irradiation [336].

Substituted quinolinones and heterocycle-fused quinolinones exhibit various biological activities [337]. One of their efficient syntheses is the intramolecular hydroarylation of C—C triple bond of *N*-arylpropiolamides (Scheme 8.96) catalyzed by $Pd(OAc)_2$/TFA [338] and $Pd(OAc)_2$/AgOAc [339].

Li and coworker developed an alternative transition metal-catalyzed intramolecular carbocyclization of alkynes for the synthesis of quinolinones via the activation of $C(sp^3)$—H bond adjacent to carbonyl group (Scheme 8.97). In the presence of $RuCl_3/CuCl_2$, with the use of O_2 as the terminal oxidant, 2-acetamido-*N*-(2-ethynyl)-aryl)acetamides underwent an intramolecular oxidative hydration–deprotonation–cyclization reaction affording 3-acetamido substituted quinolinones. It has been confirmed that oxygen atom in the newly formed carbonyl group (at four-position) is from water via hydration [340]. A further study of the catalyst system using $Cu(acac)_2$ as only metal catalyst has realized the similar transformation producing the substituted quinolinones via carbocyclization–oxygenation procedure, and, in this case, the oxygen of carbonyl group is from O_2 [341]. In both

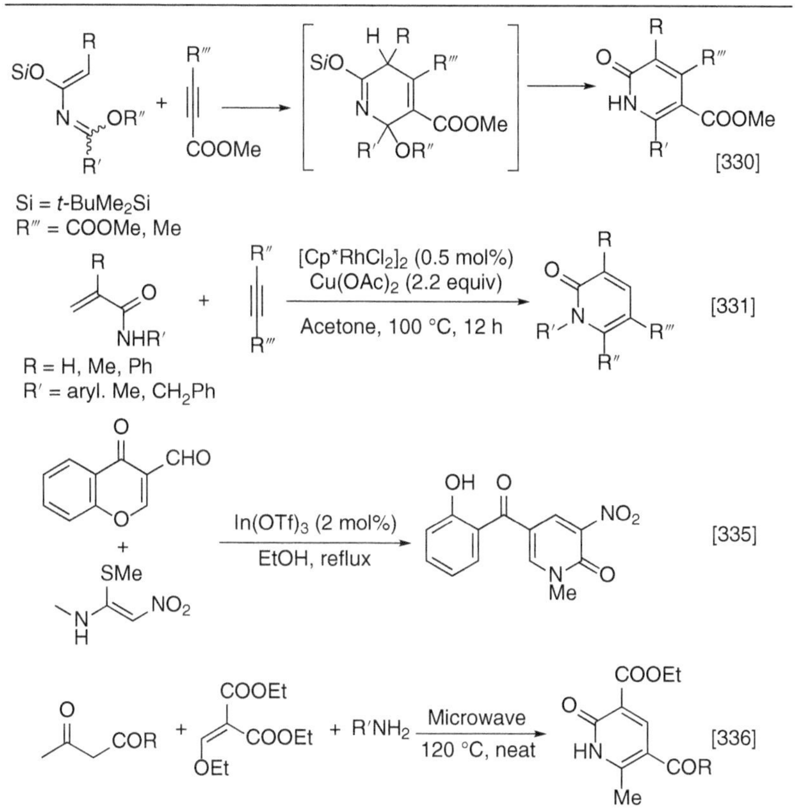

Scheme 8.95 Pyridone ring formation from unsaturated hydrocarbons.

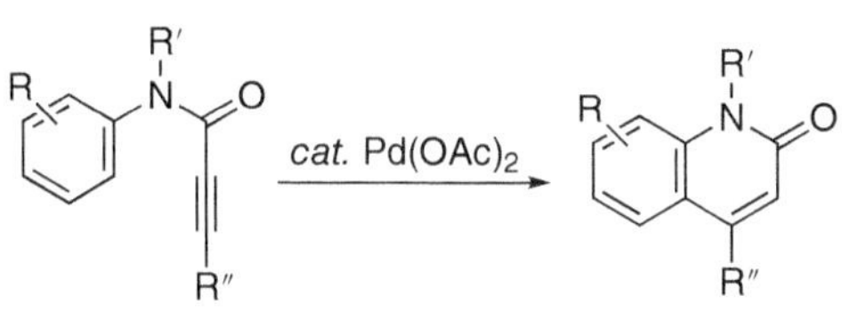

Scheme 8.96 Synthesis of quinolinone from intramolecular hydroarylation of *N*-arylpropiolamides.

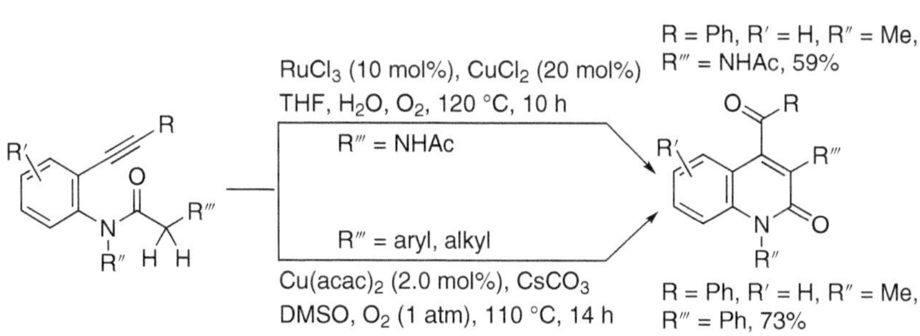

Scheme 8.97 Synthesis of quinolinones via C(sp^3)–H bond functionalization.

transformations, the key steps are the C—H bond cleavage by either oxidants or under basic conditions.

In addition, under an O_2 atmosphere, $PdCl_2/Cu(OAc)_2$-catalyzed intramolecular dehydrogenative coupling of C—H with N—H of 3-arylacrylamides could efficiently produce quinolinones [342]. The $Pd(OAc)_2$-intermolecular cyclocarbonylation of 2-iodoanilines with alkynes and CO [343], the oxidative cyclocarbonylation of 2-vinyl anilines with CO [344], $[RuCl_2(p\text{-cymene})]_2/AgSbF_6$-catalyzed cyclization of anilides with propiolates or acrylates [345], and $Pd(OAc)_2$-catalyzed oxidative annulation between acrylamides and arynes (generated in situ) [346] have also provided the useful routes for the synthesis of quinolinones.

The synthetic method for the formation of isoquinolinone ring has also been developed with the use of unsaturated hydrogen carbons [347]. As the examples of the high atom utilization procedures, the oxidative dehydrogenative cyclo-coupling reaction of benzamides with alkynes has become one of the interesting methods for the formation of isoquinolinones via C—H activation. As shown in Scheme 8.98, with the use of $Cu(OAc)_2 \cdot H_2O$ as oxidant, $[Cp^*RhCl_2]_2$ catalyzed the oxidative annulation of *N*-alkyl- and *N*-arylbenzamides with internal alkynes [348], and Ag_2CO_3 is also an efficient oxidant [349] to afford the corresponding isoquinolinones efficiently.

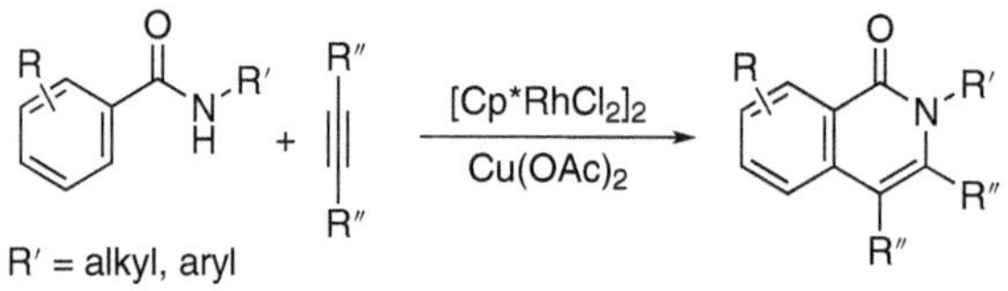

Scheme 8.98 Formation of isoquinolinones from cyclocoupling reactions of benzamides with alkynes.

With the use of the same rhodium catalyst, Guimond and coworker developed an external oxidant-free protocol to construct isoquinolinone ring using benzhydroxamic acids as starting materials [350], and with the similar starting materials, $[RuCl_2(p\text{-cymene})]_2$ shows high catalytic activity to catalyze the transformation at room temperature without use of oxidant [351].

In addition, the intermolecular aerobic cyclocondensation of 2-(1-alkynyl)benzaldehydes with primary amines under different conditions has also become one of the efficient routes to construct isoquinolinone ring (Scheme 8.99). In the presence of an excess amount of $CuBr \cdot SMe_2$, the cyclocondensation reactions gave 4-bromo-2,3-disubstituted isoquinolinones [352], and 2,3-diaryl isoquinolinones were obtained through aerobic cyclocondensation of 2-(1-alkynyl)-benzaldehydes with arylamines catalyzed by $Cu(OAc)^2$ [353], or $Zn(OTf)_2$ [354].

8.3.4 Six-membered *N*-heterocycles Having Two Nitrogen Atoms

Two examples for the synthesis of six-membered *N*-heterocycles having two nitrogen atoms are shown in Schemes 8.100 and 8.101.

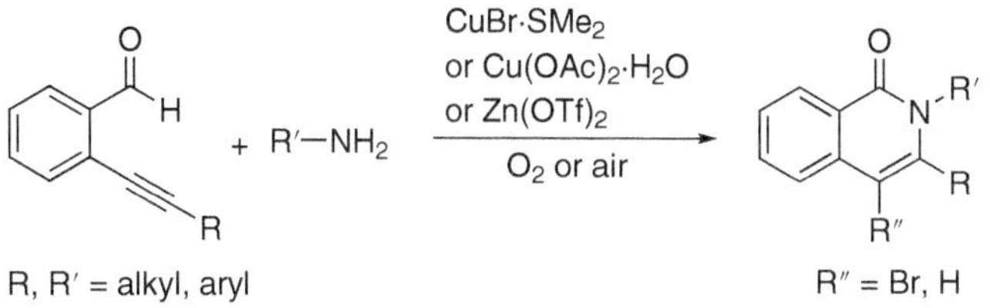

Scheme 8.99 Formation of isoquinolinones from intermolecular aerobic cyclocondensation of 2-(1-alkynyl)benzaldehydes with primary amines.

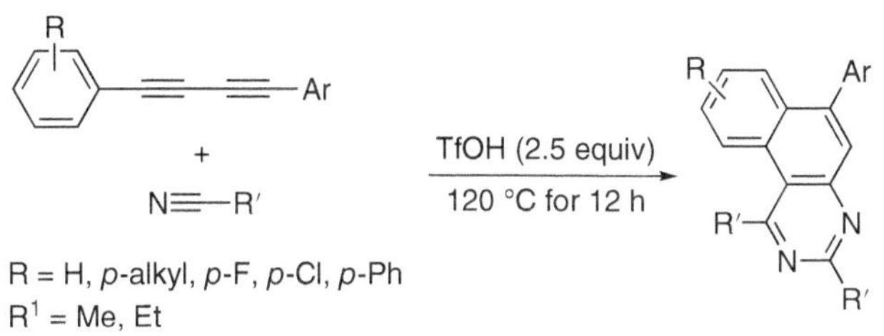

Scheme 8.100 Acid-promoted [4 + 2 + 2] cyclocondensation of 1,3-butadiynes with alkylnitriles.

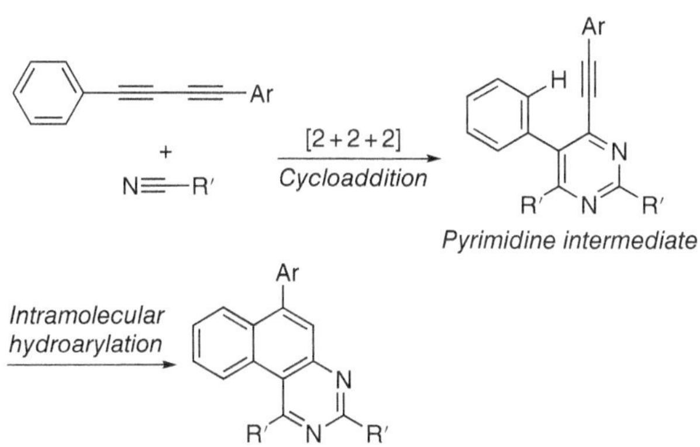

Scheme 8.101 Proposed route for the formation of benzoquinazoline ring via [4 + 2 + 2] cyclocondensation of 1,3-butadiynes with alkylnitriles.

Aryl-substituted 1,3-butadiynes can undergo the [4 + 2 + 2] cyclocondensation with alkylnitriles promoted by TfOH affording benzo[*f*]quinazolines in mild yields (Scheme 8.100) [355]. The formation of benzo[*f*]quinazoline ring is proposed to involve a chemoselective [2 + 2 + 2] cycloaddition of aryl-substituted 1,3-butadiyne with two molecules of nitrile to form pyrimidine ring as intermediate and subsequent intramolecular hydroarylation of other C—C triple bond to achieve the [4 + 2 + 2] cycloaddition as shown in Scheme 8.101.

The same group has also developed a one-pot synthesis of 3-arylmethyl-2,5-dimethylpyrazines in good yields by an Au(PPh_2Cy)Cl-catalyzed condensation of aryl aldehydes with an excess amount of propargyl amine (Scheme 8.102) [356].

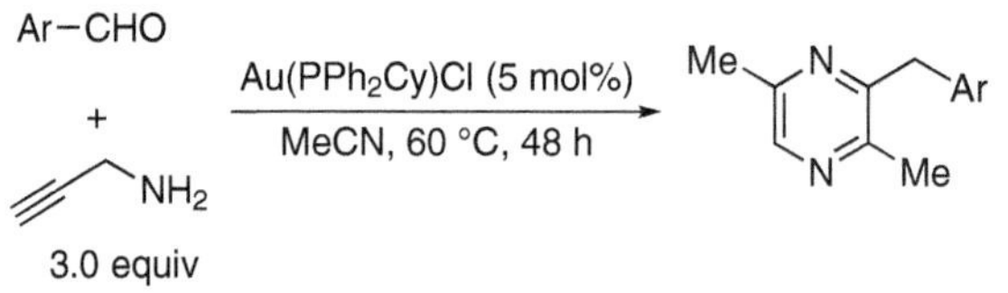

Scheme 8.102 Pyrazine synthesis by Au(I)-catalyzed condensation of aryl aldehydes with propargyl amine.

8.3.5 2-Pyrone, Coumarin, Isocoumarin, and Chromone Derivatives

2-Pyrones (α-2*H*-pyran-2-ones, α-pyrones,) [357], coumarins [358], isocoumarins [359], and chromones [360] are representative six-membered oxygen-heterocyclic compounds (Scheme 8.103), which have been found in a wide range of natural products with interesting biological and physiological activity; thus the synthesis and their application have been well investigated, and the development of synthetic methods from the annulation of alkynes has been greatly achieved [361].

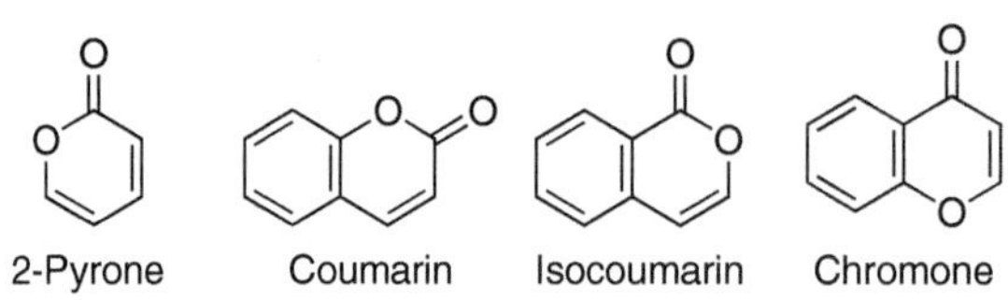

Scheme 8.103 Six-membered oxygen-heterocycles of 2-pyrone, coumarin, isocoumarin, and chromone.

It has been well documented that the C—H bond of 1,3-dicarbonyl compounds can catalytically add to alkynes to produce 2-alkenylated 1,3-dicarbonyl compounds; some of them can isomerize to enolic 2-alkenyl derivatives depending on the structures of 1,3-dicarbonyl compounds [362], and this property can be applied in the construction of 2-pyrone ring.

As shown in Scheme 8.104, in the presence of $ReBr(CO)_5$, the reaction of 1,3-dicarbonyl compounds with internal electron-deficient alkynes affords enolic

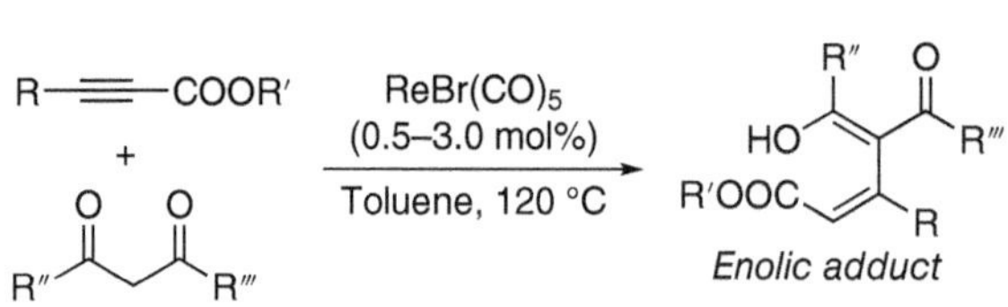

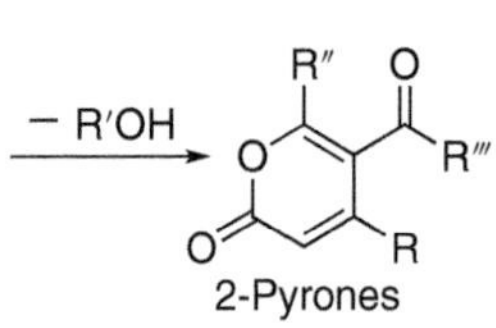

R	R′	R″	R‴	Yield (%) Enolic adduct	Yield (%) 2-Pyrone
Ph	Et	Me	OEt	87	Trace
Ph	Me	Me	Me	47	32
Me	Me	Me	OMe	0	94

Scheme 8.104 $ReBr(CO)_5$-catalyzed addition of propiolates with 1,3-dicarbonyl compounds.

adducts and/or 4,5,6-trisubstituted 2-pyrones [363]. The selective formation of enolic 2-alkenyl intermediate and 2-pyrones depends on the structure and substituents of both reactants, and 2-pyrones are resulted from the dealcoholic cyclization of enolic 2-alkenyl intermediate. The tetrasubstituted 2-pyrones could also be synthesized under the similar conditions via the reaction of methyl phenylpropiolate with methyl 2-benzylacetoacetate (Scheme 8.105).

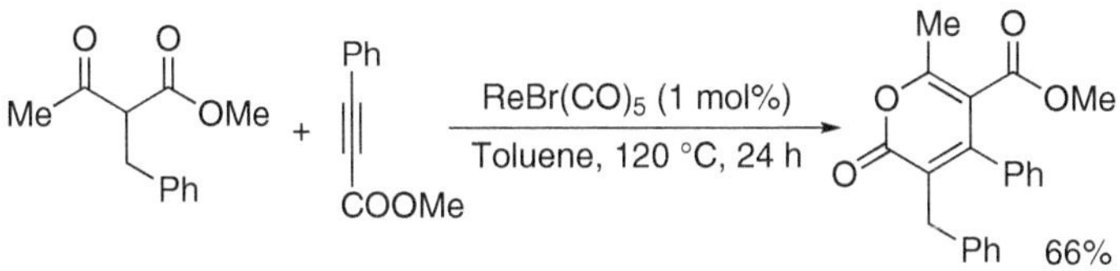

Scheme 8.105 Synthesis of tetrasubstituted 2-pyrones.

3,5,6-Tri- and 3,4,5,6-tetrasubstituted 2-pyrones can be obtained in modest to good yields via a $Pd(OAc)_2$-catalyzed reaction of β-iodoacrylate and cyclic vinylic bromides and triflates bearing β-ester functionality with internal alkynes, and selected examples are shown in Scheme 8.106 [364]. 3,4-Disubstituted isocoumarins were prepared under the similar conditions, when halogen- or triflate-containing aromatic esters are employed [365].

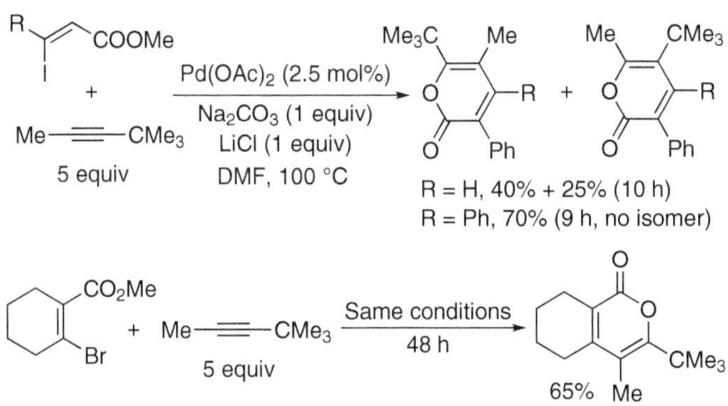

Scheme 8.106 Palladium-catalyzed construction of 2-pyrones ring.

In addition, tetrasubstituted 2-pyrones could be prepared via $Ru_3(CO)_{12}$-catalyzed carbonylative [3+2+1] cycloaddition of silylacetylenes, α,β-unsaturated ketones, and CO [366].

A $PdCl_2(PPh_3)_2$-catalyzed cyclocondensation of β-chloroacrylates with internal alkynes in the presence of Et_3N also constructs 2-pyrone ring bearing 4,5,6-trisubstituents (Scheme 8.107) [367]. However, when unsymmetrical alkynes were used, the reactions afforded a mixture of two regioisomers. A similar synthesis of 3,4-difluoro-6-substituted-2-pyrones via $PdCl_2(PPh_3)_2$/CuI-catalyzed cyclocondensation of (2*E*)-2,3-difluoro-3-iodoacrylic acid with a variety of terminal alkynes has been also reported [368].

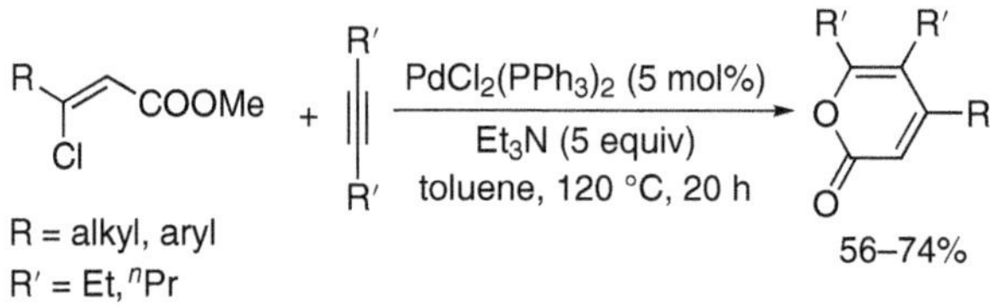

Scheme 8.107 Palladium-catalyzed annulation of β-chloroacrylates with internal alkynes.

On the other hand, aromatic acids, acrylic acids, and propiolic acids can also be used as the starting materials in the construction of 2-pyrone ring. As one example shown in Scheme 8.108, in the presence of $[Cp^*RhCl_2]_2/Cu(OAc)_2{\cdot}H_2O$, 3,4-diphenylisocoumarin was prepared in 93% yield by an oxidative coupling of benzoic acid with diphenylacetylene via regioselective *ortho*-C—H bond cleavage [369], and the catalyst system is efficient for *ortho*-substituted benzoic acids [370]. When Ag_2CO_3 was used as oxidant to replace $Cu(OAc)_2{\cdot}H_2O$, the reaction of variously substituted acrylic acids with alkynes gave the corresponding 2-pyrones in high yields [371].

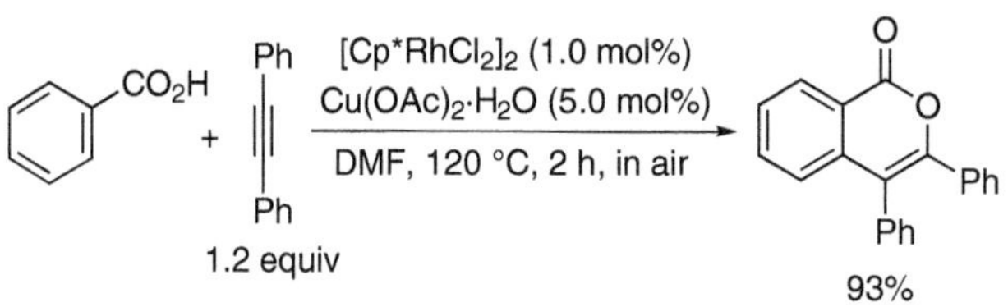

Scheme 8.108 Isocoumarin from the annulation of benzoic acid with diphenylacetylene.

Without the use of external oxidant, by the cleavage of an oxidizing O—O bond as an internal oxidant, a redox economic strategy for the synthesis of 2-pyrones has been developed via $[Cp^*RhCl_2]_2$-catalyzed direct coupling of *t*-butyl peroxybenzoate with internal alkynes [372].

In addition, the reaction of propiolic acids with terminal alkynes catalyzed by$(Ph_3P)AuCl$ is also an efficient procedure for the synthesis of 4,6-disubstituted 2-pyrones (Scheme 8.109) [373].

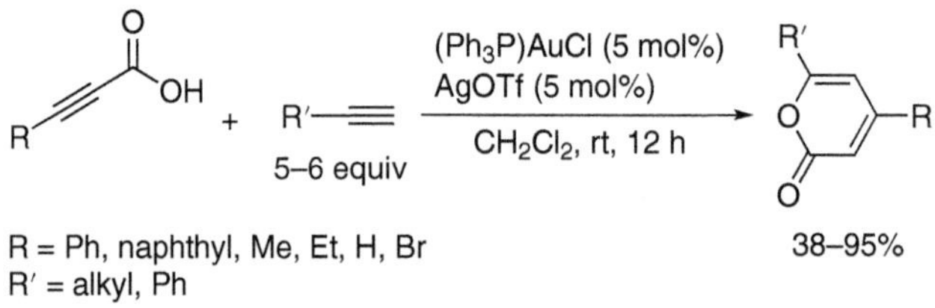

Scheme 8.109 Gold(I)-catalyzed synthesis of 2-pyrones from propiolic acids and alkynes.

Interestingly, as shown in Scheme 8.110, bis-(2-pyrones) were synthesized from the rhodium(I)-catalyzed reaction of 1,3-butadiynes with maleic acids via by decarboxylative and dehydrogenative cyclocondensation [374].

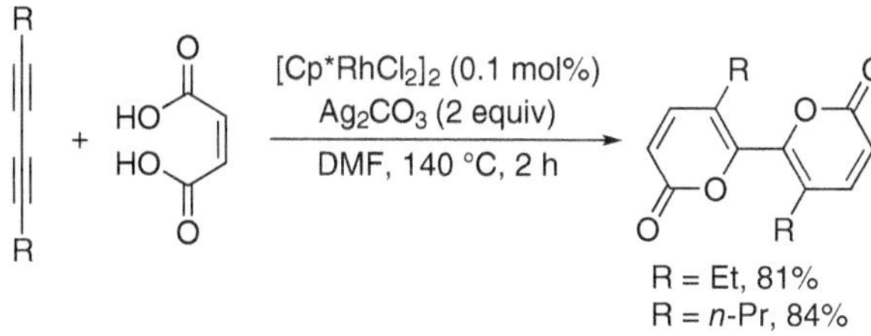

Scheme 8.110 Bis-(α-2*H*-pyran-2-ones) from rhodium(I)-catalyzed reaction of 1,3-butadiynes with maleic acids.

One of the most interesting chemistries of CO_2 is to develop the catalytic systems to incorporate CO_2 into heterocycles [375]. Inoue and coworker first reported the catalytic atom-economic route to prepare 2-pyrones via [2 + 2 + 2] cycloaddition of two alkynes with CO_2 catalyzed by Ni(cod)$_2$/phosphine in low yield together with the formation of trimers and higher oligomers of 1-hexyne (Scheme 8.111) [376]. However, in the cases of 1,6-, 1,7-, and 1,8-diynes used, Ni(cod)$_2$/phosphines can efficiently catalyze the cyclization with CO_2 to afford bicyclic 2-pyrones fused with five- and six-membered carbo- or heterocycles in good yields [377]. Also Inoue's group has found that Ni(cod)$_2$/IPr shows high catalytic activity for the similar transformation to give the bicyclic pyrones under atmospheric pressure of CO_2 in good to excellent yields (Scheme 8.112) [378]. Although symmetrical diynes containing bulky groups such as $SiMe_3$ and *t*-butyl at the terminal positions cannot undergo a similar cyclization, the use of asymmetrical diynes bearing relatively bulky $SiMe_3$ and a relatively unhindered methyl group results in the regioselective formation of fused pyrones with the bulky groups in the three position (Scheme 8.113) [379].

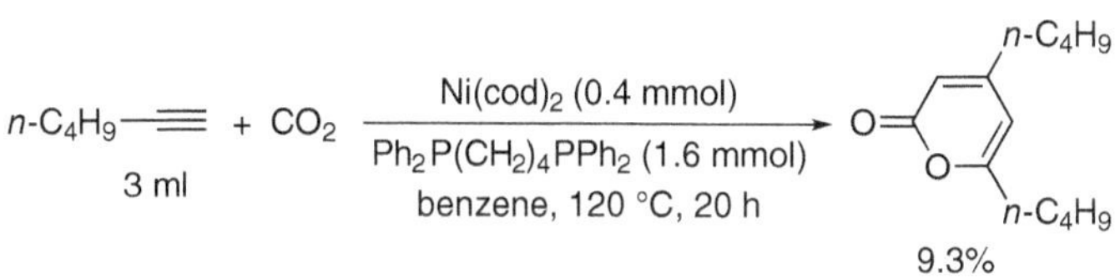

Scheme 8.111 Formation of substituted 2-pyrones from 1-hexyne and CO_2.

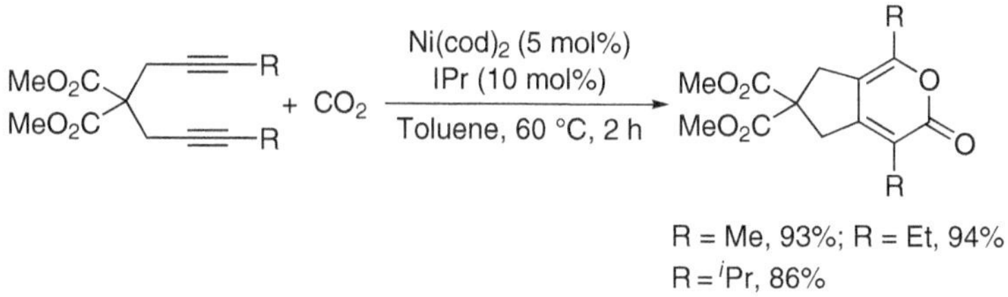

Scheme 8.112 Ni(0)-catalyzed cycloaddition of diynes and CO_2.

In addition, this protocol has also been well investigated by other groups [380], and the poly-(2-pyrones) were prepared via the copolymerization of acyclic and cyclic diynes with CO_2 catalyzed by Ni(cod)$_2$/phosphines [381].

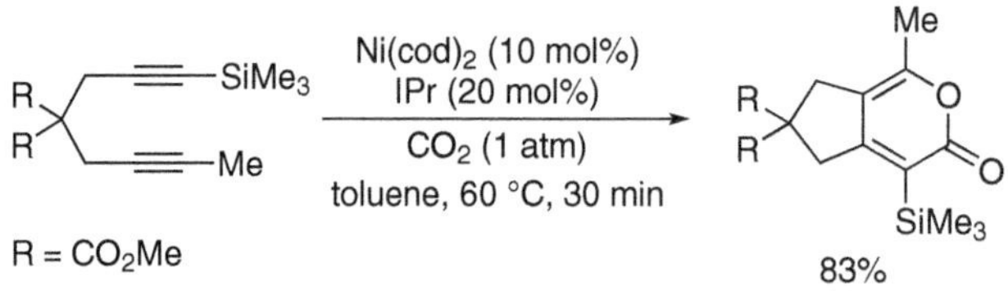

Scheme 8.113 Nickel(0)-catalyzed regioselective cycloaddition of diyne with CO_2.

Coumarins and isocoumarins are the benzo derivatives of 2-pyrone; most of their synthetic methods are resulted from the inter- or intramolecular cyclization of alkynes- or alkynyl-bearing aromatic compounds (Scheme 8.114).

Coumarins	Isocoumarins
R″ = I [382]	R′, R″ = alkyl, Ph [388, 389]
R″ = CF_2CO_2Et [383]	R′ = alkyl, phenyl, $SiMe_3$
R′, R″ = alkyl, aryl [384]	R″ = alkyl, phenyl [390]
R′ = aryl, R″ = acyl [385, 386]	R′ = SeR‴, SR‴, TeR‴
R″ = CH_2CN, CH_2COMe [387]	R″ = alkyl, aryl [391]

Scheme 8.114 *O*-heterocyclic compounds from alkynes.

As shown in Scheme 8.114, 4-substituted 3-iodocoumarins could be obtained via an exclusive 6-*endo*-dig iodocyclization of 3-ethoxy-1-(2-alkoxyaryl)-2-yn-1-ols at room temperature, and oxygens in OMe and OMOM groups are used as efficient nucleophiles for this intramolecular cyclization [382].

With the use of *fac*-Ir(ppy)$_3$ as the photocatalyst, 3-difluoroacetylated coumarins could be synthesized through a visible light-promoted aryldifluoroacetylation of alkynes with ethyl bromodifluoroacetate via a photoredox catalysis; the reaction allows the direct formation of C(sp^2)—CF_2CO_2Et and C—C bonds by a tandem radical cyclization process [383].

With the use of $Cu(OAc)_2$ as oxidant, $[Cp^*RhCl_2]_2$/AgOTf also shows high catalytic activity for the formation of 3,4-disubstituted coumarins via an oxidative annulation of aryl thiocarbamates with internal alkynes by C—H bond activation [384].

Aryl alkynoates have been confirmed to be the useful precursors for the synthesis of coumarins under different conditions. 3-acyl-4-arylcoumarins can be prepared through the tetrabutylammonium bromide (TBAB)-mediated metal-free oxidative tandem coupling of alkynoates with aldehydes by the addition of acyl radical to alkynes and a C—H bond functionalization to form two new C—C bonds simultaneously, and $K_2S_2O_8$ is the best oxidant [385]. Using the same oxidant, the similar products could also be obtained by an $AgNO_3$-mediated radical cyclization of aryl alkynoates with α-keto acids [386].

A tandem oxidative cyclization of aryl alkynoates with acetonitrile under transition metal-free conditions affords 3-cyanomethylated coumarins promoted by peroxide of TBPB, and under the same reaction conditions, 3-(2-oxopropyl) substituted coumarins can be also obtained in satisfactory yields by using acetone instead of acetonitrile [387].

Isocoumarins (isochromenones) have been efficiently synthesized through an oxidative cyclocoupling of benzoic acids with internal alkynes catalyzed by $[Cp^*RhCl_2]_2$ [388] and $[RuCl_2(p\text{-cymene})]_2/AgSbF_6$ [389] with the use of $Cu(OAc)_2{\cdot}H_2O$ as oxidant. The $Ni(cod)_2/PMe_3/ZnCl_2$-catalyzed decarbonylative addition of anhydrides to internal alkynes was also demonstrated to be the efficient method for the formation of isocoumarins [390]. A series of 4-Se-(Te, S)-isochromenones and 3-substituted isocoumarins were synthesized in good yields at room temperature via $FeCl_3$-mediated cyclization of alkynylaryl esters with different diorganyl dichalcogenides [391].

Very recently, a regioselective synthesis of isocoumarins in good yields in water from intramolecular addition of 2-alkynylbenzoic acid via a metal-free radical pathway [392] and a ruthenium(II)-catalyzed electrooxidative [4 + 2] annulation of benzylic alcohols with internal alkynes in water [393] have been reported.

In addition, chromones (γ-benzopyrone)-containing natural products have been widely found exhibiting a wide variety of biological activities [394].

Larock and coworker reported the synthesis of substituted 3-iodochromones via an ICl-induced iodocyclization of 2-methoxyaryl-containing alkynones, and thiochromenones and quinolinones could also be readily prepared by similar ICl-induced cyclizations [395]. Substituted chromones were also reported to be prepared by the oxidative cyclocoupling of salicylaldehydes with internal alkynes in the presence of $[Cp^*RhCl_2]_2$ with the use of $Cu(OAc)_2{\cdot}H_2O$ as oxidant [396], the regioselective intramolecular cyclization of *ortho*-alkynoylphenols promoted by TfOH via a 6-*endo* cyclization manner [397], and $Cu(OAc)_2$-mediated intermolecular [4 + 2] annulation of sulfonylacetylenes with salicylic acids [398].

8.4 Other Heterocycles

In this section, several recently reported procedures access to five- to eight-membered heterocyclic compounds are given to show the important applications of alkynes, alkenes, and allenes in the synthesis of different heterocycles.

In the presence of $[Cp^*Ru(MeCN)_3]PF_6$, 1,3-butadiynes undergo the double hydrosilylation with 9-silafluorene providing a novel synthetic route access to *spiro*-type 2,5-diarylsiloles (Scheme 8.115) [399]. It has been found that the double hydrosilylation depends greatly on the structures of dihydrosilanes; Ph_2SiH_2 and $PhSiClH_2$ show mild reactivity, but Et_2SiH_2 and $PhMeSiH_2$ show no reactivity.

The synthesis of metallacyclopentadienes can also be from the cycloaddition reaction of 1,3-butadiynes with metal-hydrogen bonds. With the use of same ruthenium catalyst, the same group has developed the double *trans*-hydrogermylation of

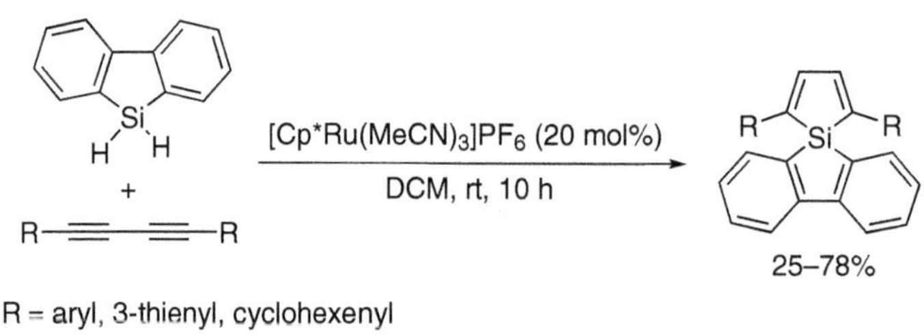

Scheme 8.115 Ruthenium-catalyzed double hydrosilylation of 1,3-butadiynes.

1,3-butadiynes with dihydrogermane at room temperature to afford 2,5-disubstituted germoles in good to high yields (Scheme 8.116) [400].

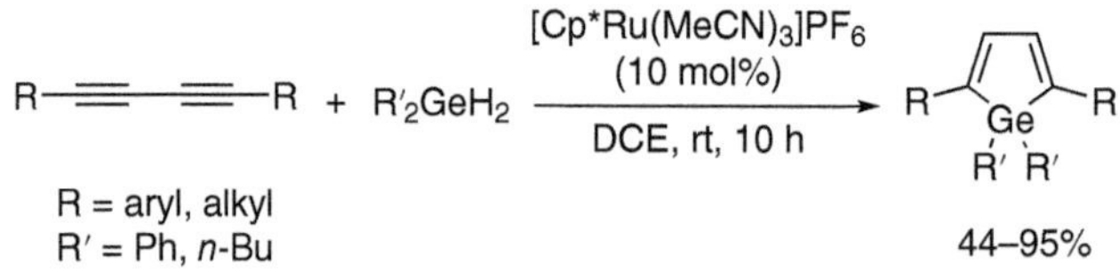

Scheme 8.116 Ruthenium-catalyzed double *trans*-hydrogermylation of 1,3-butadiynes.

Synthesis of organoselenium compounds is an active research area due to their distinct chemical, physical, and biological properties [401]. The electrophilic cyclization of (*Z*)-selenoenynes with different electrophiles such as I_2, ICl, PhSeBr, and PhSeCl is one of the efficient synthetic methods for the synthesis of 3-substituted selenophenes (Scheme 8.117) [402]. The obtained 3-iodoselenophenes are useful intermediates for the further transformation via transition metal-catalyzed cross-coupling reaction of C—I bond.

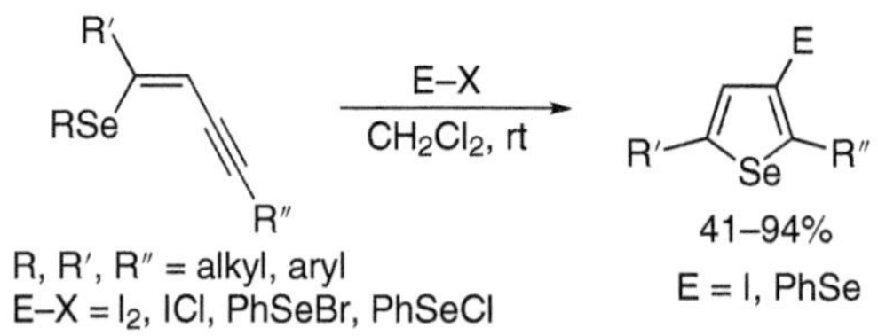

Scheme 8.117 Selenophene synthesis from the electrophilic cyclization of (*Z*)-selenoenynes.

2*H*-chromene (2*H*-1-benzopyran) structures can be found as parent molecules in a wide variety of important natural products and have been also well applied as the versatile intermediates in the synthesis of numerous pharmaceutical and biologically active compounds [403].

A simple and direct one-pot route for the formation of 2,2-dimethyl-2*H*-chromenes by $Re(CO)_5Cl$-catalyzed cyclocondensation of phenols with 2-methyl-3-butyn-2-ol has been reported by Hua's group (Scheme 8.118) [404]. Since the condensation proceeds with high chemoselectivity and regioselectivity including the activation of C—H bond at the *ortho*-position of the phenols, a reasonable route including the first dehydration and then hydroarylation is proposed. The similar protocol has been used to provide an efficient strategy for the synthesis of 1,2-dihydroquinolines via a Cu(I)/Cu(II)-catalyzed reaction of anilines with propargyl alcohols [405].

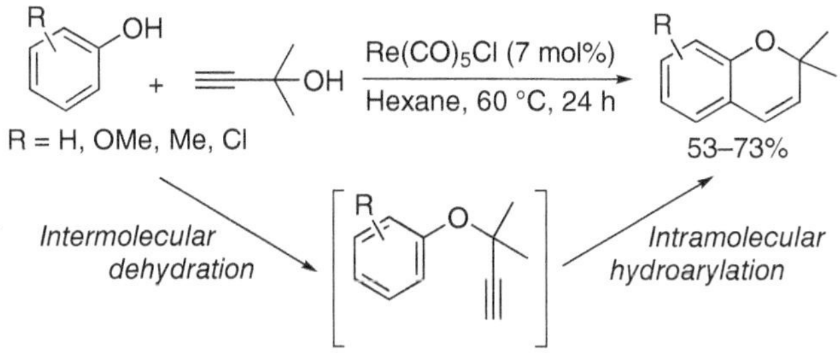

Scheme 8.118 Synthesis of 2,2-dimethyl-2*H*-chromenes by cyclocondensation of phenols with 2-methyl-3-butyn-2-ol.

Very recently, Rovis's group has developed a novel and interesting formation of chromene ring via a carbocarbation of triple bonds proceeding by undirected, sequential activation of arene C(sp^2)—H and methoxy C(sp^3)—bonds of anisoles, generating reactive metalacycles that insert difluoroalkynes under oxygen atmosphere with the use of iridium(III) complex bearing the electron-deficient cyclopentadienyl ligand as catalyst and cesium adamantane-1-carboxylate (1-$AdCO_2Cs$) as base, as well as dimethyl diacetoxysilane as additive (Scheme 8.119) [406].

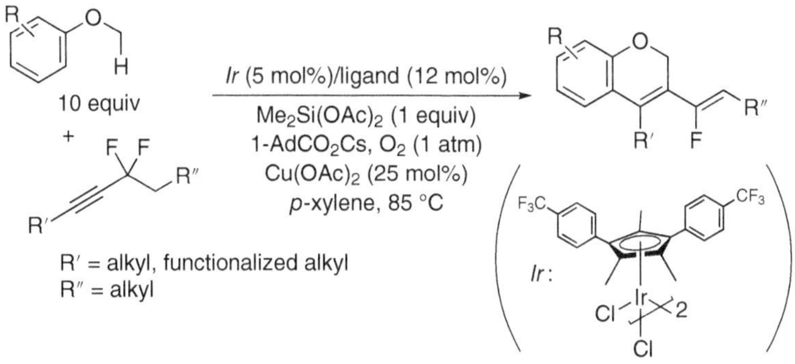

Scheme 8.119 Chromene formation via iridium-catalyzed carbocarbation of triple bond involving activation of C(sp^2)−H and C(sp^3)−H bonds.

Kwon and coworker have reported an efficient synthesis of *multi*-substituted tetrahydropyridines via a phosphine-catalyzed [4 + 2] annulation of allene ester with *N*-tosylimines under mild conditions (Scheme 8.120) [407].

Scheme 8.121 shows three examples for the formation of seven-membered nitrogen- and oxygen-heterocyclic compounds via the [5 + 2] cycloaddition reaction with alkynes. With the use of cyclopropyl imines as five-atom components and an alkyne as a two-carbon component, $[Rh(CO)_2Cl]_2$-catalyzed hetero-[5 + 2] cycloaddition of cyclopropyl imines and alkynes affords the dihydroazepine derivatives in

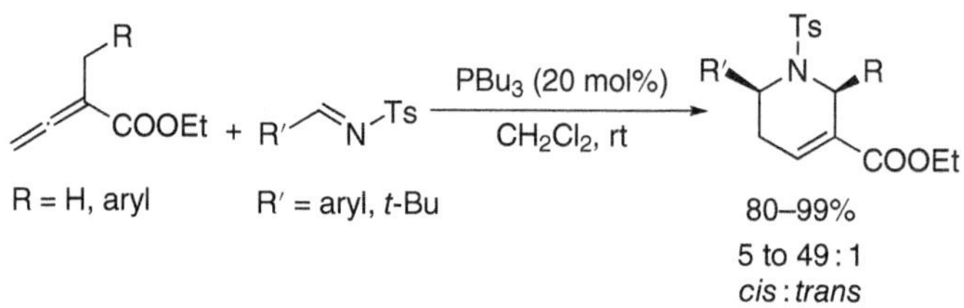

Scheme 8.120 Tetrahydropyridine formation via a phosphine-catalyzed [4 + 2] annulation of alléné with aldimines.

good to high yields with excellent regioselectivity [408]. The dihydroazepine ring can also be formed by $AgSbF_6$-catalyzed [5 + 2] cycloaddition of γ-amino ketones with alkynes [409].

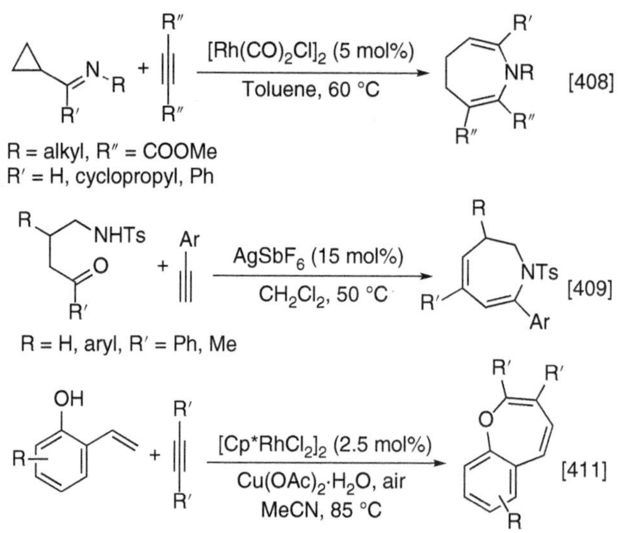

Scheme 8.121 Formation of seven-membered heterocycles via heter-[5 + 2] cycloaddition with alkynes.

In the presence of $[Cp^*RhCl_2]_2$ and $Cu(OAc)_2$, *N*-vinyl or *N*-arylpyridin-2(1*H*)-ones undergo the oxidative cyclization with internal alkynes to afford 4*H*-quinolizin-4-ones and 4*H*-benzoquinolizin-4-ones via double C—H activation (Scheme 8.122) [410].

In the presence of $[Cp^*RhCl_2]_2$ and $Cu(OAc)_2$, under air atmosphere, *ortho*-vinylphenols undergo a formal [5 + 2] cycloaddition to alkynes to generate benzoxepine skeletons in a practical and atom-economical manner [411].

In addition, 1,4-diazepine derivatives can be prepared from a three-component reaction of pyridines, 1-sulfonyl-1,2,3-triazoles, and alkynes through an air-stable azomethine ylide intermediate, which is resulted from the rhodium-catalyzed reaction between pyridines and 1-sulfonyl-1,2,3-triazoles, and to then undergo the [5 + 2] cycloaddition reaction with alkynes [412].

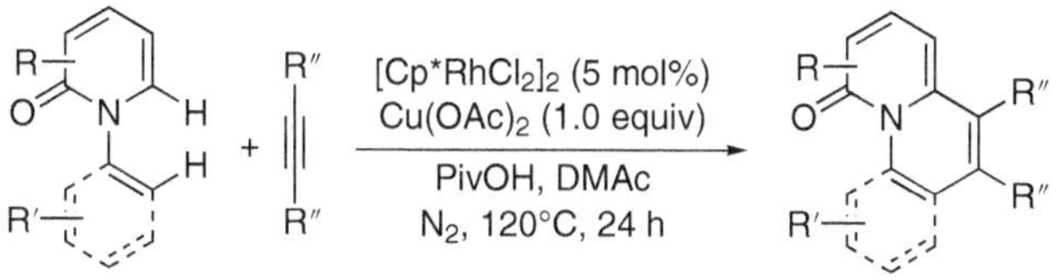

Scheme 8.122 Rh(III)-catalyzed oxidative annulation of *N*-vinyl/*N*-arylpyridin-2(1*H*)-ones with alkynes via double C—H activation.

Hetero-[4 + 4] annulation is one of the simplest and most practical reactions for the construction of eight-membered heterocyclic compound. As shown in Scheme 8.123, DABCO-mediated [4 + 4] domino annulation reactions of ynones and α-cyano-α, β-unsaturated ketones afford an efficient route to eight-membered cyclic ethers in good yields under mild conditions [413].

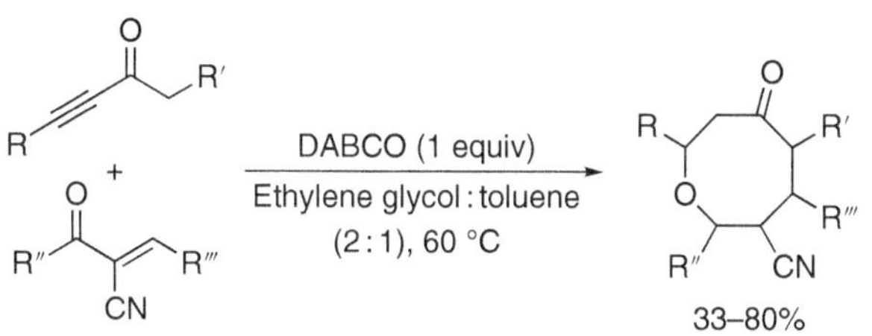

R, R″, R‴ = aryl; R′ = H, Me, di-Me

Scheme 8.123 DABCO-mediated [4 + 4] annulation of ynones and α-cyano-α,β-unsaturated ketones.

With the use of amino acid-derived phosphine as organocatalyst, the formal [4+4] annulations of benzofuran/indole-derived α,β-unsaturated imines and allene ketones construct the eight-membered nitrogen-heteroring to furnish a wide range of benzofuran- or indole-fused azocines in high yields with excellent enantioselectivity; the examples are shown in Scheme 8.124 [414].

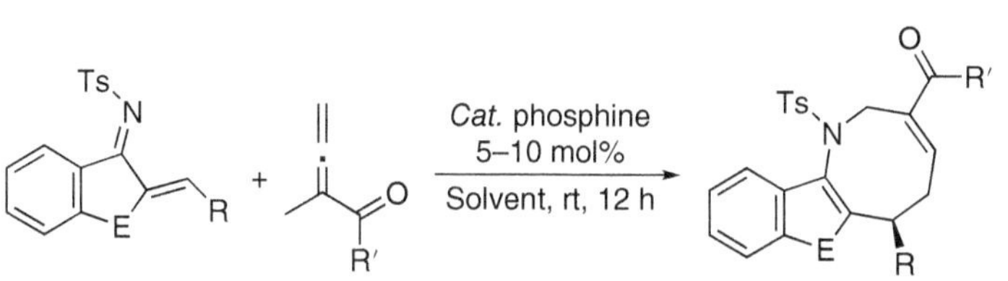

E = O, NAc

Cat.: **A**; E = O, R = R′ = Ph, CH_2Cl_2; 94% yield, 98% ee
Cat.: **B**; E = NAc, R = R′ = Ph, toluene; 85% yield, 95% ee

Scheme 8.124 Formai [4 + 4] annulation of α,β-unsaturated imine with allene ketones.

Recently, in the presence of KOH, a facile and efficient solvent-controlled chemo- and regioselective synthesis of 1,4-benzoxazepin-5(4*H*)-ones and 1,3-benzoxazin-4(4*H*)-ones via intermolecular cyclization of *ortho*-fluorobenzamides with

2-propyn-1-ol was reported (Scheme 8.125) [415]. The cyclization reaction is proposed to involve the S_NAr reaction of C—F bond with 2-propyn-1-ol to give *ortho*-[(2-propynyl)oxy]benzamide intermediates and subsequent intramolecular of either a 7-*exo*-dig cyclization in a superbase medium of KOH/DMSO or a Michael addition of N—H to allenyl intermediate in KOH/MeCN medium to form different size benzo-fused cyclic compounds.

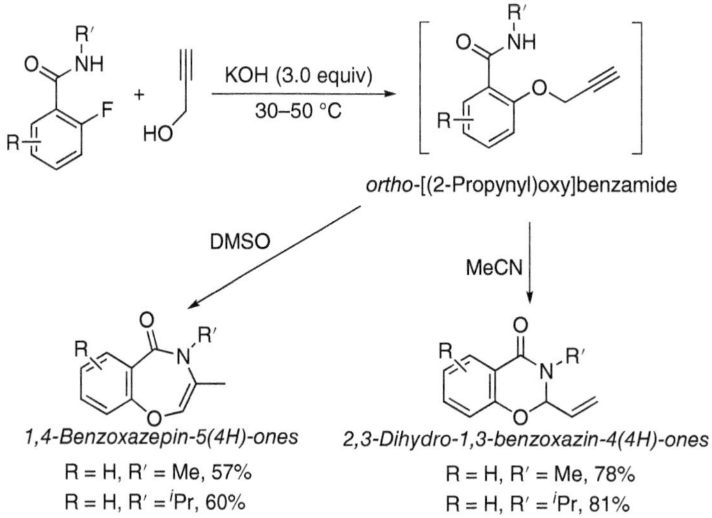

Scheme 8.125 Base-promoted chemodivergent formation of 1,4-benzoxazepin-5(4*H*)-ones and 1,3-benzoxazin-4(4*H*)-ones switched by solvents.

References

1 Selected reviews on the application of heterocycles in organic synthesis and material synthesis, see: (a) Lipshutz, B.H. (1986). *Chem. Rev.* 86: 795–819. (b) Shipman, M. (1995). *Contemp. Org. Synth.* 2: 1–17. (c) Chinchilla, R., Nájera, C., and Yus, M. *ARKIVOC* 2007 (x): 152–231. (d) Merino, P., Tejero, T., Delso, J.I., and Matute, R. (2007). *Curr. Org. Chem.* 11: 1076–1091. (e) Isambert, N. and Lavilla, R. (2008). *Chem. Eur. J.* 14: 8444–8454. (f) Gao, H., Zhang, Q., and Shreeve, J.M. (2020). *J. Mater. Chem. A* 8: 4193–4216.

2 Selected reviews, see: (a) Nefzi, A., Ostresh, J.M., and Houghten, R.A. (1997). *Chem. Rev.* 97: 449–472. (b) Dua, R., Shrivastava, S., Sonwane, S.K., and Srivastava, S.K. (2011). *Adv. Bio. Res.* 5: 120–144. (c) Gandhi, D., Kalal, P., and Agarwal, S. (2017). *Chem. Bio. Interface* 7: 79–101.

3 Selected reviews, see: (a) Padwa, A. (2003). *Pure Appl. Chem.* 75: 47–62. (b) Krause, N. and Winter, C. (2011). *Chem. Rev.* 111: 1994–2009. (c) Wu, X.-F., Neumann, H., and Beller, M. (2013). *Chem. Rev.* 113: 1–35. (d) Chen, J.-R., Hu, X.-Q., Lu, L.-Q., and Xiao, W.-J. (2015). *Chem. Rev.* 115: 5301–5365. e Trofimov, B.A. and Schmidt, E.Y. (2018). *Acc. Chem. Res.* 51: 1117–1130. f Feng, J.-J. and

Zhang, J. (2016). *ACS Catal.* 6: 6651–6661. (g) Brandão, P., Pineiro, M., and Pinho e Melo, T.M.V.D. (2019). *Eur. J. Org. Chem.* 7188–7217. (h) Pathan, S.I., Chundawat, N.S., Chauhan, N.P.S., and Singh, G.P. (2020). *Synth. Commun.* 50: 1251–1285.

4 The literatures on the heterocyclic synthesis from unsaturated hydrocarbons, refers to the cited review papers in preface.

5 Selected reviews, see: (a) Dejaegher, Y., Kuźmenok, N.M., Zvonok, A.M., and De Kimpe, N. (2002). *Chem. Rev.* 102: 29–60. (b) Carreira, E.M. and Fessard, T.C. (2014). *Chem. Rev.* 114: 8257–8322. (c) Anaya, J. and Sánchez, R.M. (2016). *Prog. Heterocycl. Chem.* 28: 95–120.

6 Bull, J.A., Croft, R.A., Davis, O.A. et al. (2016). *Chem. Rev.* 116: 12150–12233.

7 Ye, L., He, W., and Zhang, L. (2010). *J. Am. Chem. Soc.* 132: 8550–8551.

8 Derrah, E.J., Pantazis, D.A., McDonald, R., and Rosenberg, L. (2010). *Angew. Chem. Int. Ed.* 49: 3367–3370.

9 Derrah, E.J., McDonald, R., and Rosenberg, L. (2010). *Chem. Commun.* 46: 4592–4594.

10 Robin, S. and Rousseau, G. (2002). *Eur. J. Org. Chem.* 3099–3114.

11 (a) Lipshutz, B.H. (1986). *Chem. Rev.* 86: 795–819. (b) Shipman, M. (1995). *Contemp. Org. Synth.* 2: 1–17. (c) Baumann, M., Baxendale, I.R., Ley, S.V., and Nikbin, N. (2011). *Beilstein J. Org. Chem.* 7: 442–495. (d) Kaur, N., Grewal, P., Bhardwaj, P. et al. (2019). *Synth. Commun.* 49: 3058–3100.

12 Selected recent reviews, See: (a) Pineiro, M. and Pinho e Melo, T.M.V.D. (2009). *Eur. J. Org. Chem.* 5287–5307. (b) Kaur, N., Dwivedi, J., and Kishore, D. (2014). *Synth. Commun.* 44: 1671–1729. (c) Varvounis, G., Gerontitis, I.E., and Gkalpinos, V. (2018). *Chem. Heterocyl. Compd.* 54: 249–268. (d) Kaur, N., Ahlawat, N., Verma, Y. et al. (2019). *Curr. Org. Chem.* 23: 1901–1944. (e) Kaur, N., Grewal, P., Bhardwaj, P. et al. (2019). *Synth. Commun.* 49: 3058–3100.

13 Selected reviews, see: (a) Trofimov, B.A., Sobenina, L.N., Demenev, A.P., and Mikhaleva, A.I. (2004). *Chem. Rev.* 104: 2481–2506. (b) Donohoe, T.J. and Thomas, R.E. (2007). *Chem. Record.* 7: 180–190.

14 Selected reviews, see: (a) Bando, T. and Sugiyama, H. (2006). *Acc. Chem. Res.* 39: 935–944. (b) Walsh, C.T., Garneau-Tsodikova, S., and Howard-Jones, A.R. (2006). *Nat. Prod. Rep.* 23: 517–531. (c) Fan, H., Peng, J., Hamann, M.T., and Hu, J.-F. (2008). *Chem. Rev.* 108: 264–287.

15 Selected reviews, see: (a) Deronzier, A. and Moutet, J.-C. (1989). *Acc. Chem. Res.* 22: 249–255. (b) Deronzier, A. and Moutet, J.-C. (1996). *Coord. Chem. Rev.* 147: 339–371. (c) Anzenbacher, P. Jr., Nishiyabu, R., and Palacios, M.A. (2006). *Coord. Chem. Rev.* 250: 2929–2938. (d) Misra, R. and Chandrashekar, T.K. (2008). *Acc. Chem. Res.* 41: 265–279. (e) Kim, D.S. and Sessler, J.L. (2015). *Chem. Soc. Rev.* 44: 532–546. (f) Anguera, G. and Sánchez-García, D. (2017). *Chem. Rev.* 117: 2481–2516.

16 Selected reviews, see: (a) Ferreira, V.F., de Souza, M.C.B.V., Cunha, A.C. et al. (2001). *Org. Prep. Proced. Int.* 33: 411–454. (b) Balme, G. (2004). *Angew. Chem. Int. Ed.* 43: 6238–6241. (c) Schmuck, C. and Rupprecht, D. (2007). *Synthesis*

3095–3110. (d) Russel, J.S., Pelkey, E.T., and Yoon-Miller, S.J.P. (2009). *Prog. Heterocycl. Chem.* 21: 145–178. (e) Estévez, V., Villacampa, M., and Menéndez, J.C. (2010). *Chem. Soc. Rev.* 39: 4402–4421. (f) Estévez, V., Villacampa, M., and Menéndez, J.C. (2014). *Chem. Soc. Rev.* 43: 4633–4657. (g) Lopchuk, J.M. (2016). *Prog. Heterocycl. Chem.* 28: 165–218. (h) Gao, M., Zhao, W., Zhao, H. et al. (2018). *Beilstein J. Org. Chem.* 14: 884–890.

17 Liu, W., Jiang, H., and Huang, L. (2010). *Org. Lett.* 12: 312–315.

18 Zhang, L., Wang, X., Li, S., and Wu, J. (2013). *Tetrahedron* 69: 3805–3809.

19 Chen, X., Li, X., Wang, N. et al. (2012). *Eur. J. Org. Chem.* 4380–4386.

20 (a) Davis-Gilbert, Z.W., Hue, R.J., and Tonks, I.A. (2016). *Nat. Chem.* 8: 63–68. (b) Guo, J., Deng, X., Song, C. et al. (2017). *Chem. Sci.* 8: 2413–2425. (c) Chiu, H.-C. and Tonks, I.A. (2018). *Angew. Chem. Int. Ed.* 57: 6090–6094. (d) Davis-Gilbert, Z.W., Wen, X., Goodpaster, J.D., and Tonks, I.A. (2018). *J. Am. Chem. Soc.* 140: 7267–7281. (e) Chiu, H.-C., See, X.Y., and Tonks, I.A. (2019). *ACS Catal.* 9: 216–223.

21 Kawakita, K., Beaumier, E.P., Kakiuchi, Y. et al. (2019). *J. Am. Chem. Soc.* 141: 4194–4198. and references therein.

22 Ngwerume, S., Lewis, W., and Camp, J.E. (2013). *J. Org. Chem.* 78: 920–934.

23 Kamijo, S., Kanazawa, C., and Yamamoto, Y. (2005). *J. Am. Chem. Soc.* 127: 9260–9266.

24 (a) Larionov, O.V. and de Meijere, A. (2005). *Angew. Chem. Int. Ed.* 44: 5664–5667. (b) Lygin, A.V., Larionov, O.V., Korotkov, V.S., and de Meijere, A. (2009). *Chem. Eur. J.* 15: 227–236.

25 Gao, M., He, C., Chen, H. et al. (2013). *Angew. Chem. Int. Ed.* 52: 6958–6961.

26 Selected reviews, see: (a) Granik, V.G. (1984). *Russ. Chem. Rev.* 53: 383–400. (b) Erian, A.W. (1993). *Chem. Rev.* 93: 1991–2005. (c) Granik, V.G., Makarov, V.A., and Párkányi, C. (1998). *Adv. Heterocycl. Chem.* 72: 283–359. (d) Adams, J.P. (2000). *J. Chem. Soc., Perkin Trans. 1* 125–139. (e) Cheng, Y., Huang, Z.-T., and Wang, M.-X. (2004). *Curr. Org. Chem.* 8: 325–351.

27 Rakshit, S., Patureau, F.W., and Glorius, F. (2010). *J. Am. Chem. Soc.* 132: 9585–9587.

28 Yan, R.-L., Luo, J., Wang, C.-X. et al. (2010). *J. Org. Chem.* 75: 5395–5397.

29 Li, B., Wang, N., Liang, Y. et al. (2013). *Org. Lett.* 15: 136–139.

30 Wang, Y., Jiang, C.-M., Li, H.-L. et al. (2016). *J. Org. Chem.* 81: 8653–8658.

31 Martín, R., Rivero, M.R., and Buchwald, S.L. (2006). *Angew. Chem. Int. Ed.* 45: 7079–7082.

32 Wang, Y., Bi, X., Li, D. et al. (2011). *Chem. Commun.* 47: 809–811.

33 Mo, D.-L., Ding, C.-H., Dai, L.-X., and Hou, X.-L. (2011). *Chem. Asian J.* 6: 3200–3204.

34 Ueda, H., Yamaguchi, M., Kameya, H. et al. (2014). *Org. Lett.* 16: 4948–4951.

35 Zheng, J., Huang, L., Li, Z. et al. (2015). *Chem. Commun.* 51: 5894–5897.

36 Selected reviews, see: (a) Shi, W. and Lei, A. (2014). *Tetrahedron Lett.* 55: 2763–2772. (b) Nizami, T.A. and Hua, R. (2014). *Molecules* 19: 13788–13802. (c) Asiri, A.M. and Hashmi, A.S.K. (2016). *Chem. Soc. Rev.* 45: 4471–4503.

37 Zheng, Q. and Hua, R. (2010). *Tetrahedron Lett.* 51: 4512–4514.

38 Nizami, T.A. and Hua, R. (2017). *Tetrahedron* 73: 6080–6084.

39 Selected reviews, see: (a) Kappe, C.O., Murphree, S.S., and Padwa, A. (1997). *Tetrahedron* 53: 14179–14233. (b) Wong, H.N.C., Yu, P., and Yick, C.-Y. (1999). *Pure Appl. Chem.* 71: 1041–1044. (c) Lee, H.-K., Chan, K.-F., Hui, C.-W. et al. (2005). *Pure Appl. Chem.* 77: 139–143.

40 Selected reviews, see: (a) Hou, X.L., Cheung, H.Y., Hon, T.Y. et al. (1998). *Tetrahedron* 54: 1955–2020. (b) Keay, B.A. (1999). *Chem. Soc. Rev.* 28: 209–215. (c) Brown, R.C.D. (2005). *Angew. Chem. Int. Ed.* 44: 850–852. (d) Kirsch, S.F. (2006). *Org. Biomol. Chem.* 4: 2076–2080. (e) Moran, W.J. and Rodríguez, A. (2012). *Org. Prep. Proced. Int.* 44: 103–130. (f) Blanc, A., Bénéteau, V., Weibel, J.-M., and Pale, P. (2016). *Org. Biomol. Chem.* 14: 9184–9205.

41 Zhang, M., Jiang, H.-F., Neumann, H. et al. (2009). *Angew. Chem. Int. Ed.* 48: 1681–1684.

42 Kramer, S. and Skrydstrup, T. (2012). *Angew. Chem. Int. Ed.* 51: 4681–4684.

43 Lenden, P., Entwistle, D.A., and Willis, M.C. (2011). *Angew. Chem. Int. Ed.* 50: 10657–10660.

44 (a) Cao, H., Jiang, H., Yao, W., and Liu, X. (2009). *Org. Lett.* 11: 1931–1933. (b) Cao, H., Jiang, H., Mai, R. et al. (2010). *Adv. Synth. Catal.* 352: 143–152.

45 Cao, H., Jiang, H., Yuan, G. et al. (2010). *Chem. Eur. J.* 16: 10553–10559.

46 Chong, Q., Xin, X., Wang, C. et al. (2014). *J. Org. Chem.* 79: 2105–2110.

47 Hosseyni, S., Su, Y., and Shi, X. (2015). *Org. Lett.* 17: 6010–6013.

48 He, C., Guo, S., Ke, J. et al. (2012). *J. Am. Chem. Soc.* 134: 5766–5769.

49 Wang, A., Jiang, H., and Xu, Q. (2009). *Synlett* 929–932.

50 Kim, H.Y., Li, J.-Y., and Oh, K. (2012). *J. Org. Chem.* 77: 11132–11145.

51 Pridmore, S.J., Slatford, P.A., and Williams, J.M.J. (2007). *Tetrahedron Lett.* 48: 5111–5114.

52 Pridmore, S.J., Slatford, P.A., Daniel, A. et al. (2007). *Tetrahedron Lett.* 48: 5115–5120.

53 Egi, M., Azechi, K., and Akai, S. (2009). *Org. Lett.* 11: 5002–5005.

54 (a) Mack, D.J. and Njardarson, J.T. (2013). *ACS Catal.* 3: 272–286. (b) Huang, C.-Y. and Doyle, A.G. (2014). *Chem. Rev.* 114: 8153–8198.

55 Hashmi, A.S.K. and Ainha, P. (2004). *Adv. Synth. Catal.* 346: 432–438.

56 Blanc, A., Tenbrink, K., Weibel, J.-M., and Pale, P. (2009). *J. Org. Chem.* 74: 4360–4363.

57 Yoshida, M., Al-Amin, M., and Shishido, K. (2009). *Synthesis* 2454–2466.

58 Gronnier, C., Kramer, S., Odabachian, Y., and Gagosz, F. (2012). *J. Am. Chem. Soc.* 134: 828–831.

59 Kel'in, A.V. and Gevorgyan, V. (2002). *J. Org. Chem.* 67: 95–98.

60 Yao, T., Zhang, X., and Larock, R.C. (2004). *J. Am. Chem. Soc.* 126: 11164–11165.

61 (a) Utimoto, K. (1983). *Pure Appl. Chem.* 55: 1845–1852. (b) Fukuda, Y., Shiragami, H., Utimota, K., and Nozaki, H. (1991). *J. Org. Chem.* 56: 5816–5819.

62 (a) Sniady, A., Durham, A., Morreale, M.S. et al. (2007). *Org. Lett.* 9: 1175–1178. (b) Sniady, A., Durham, A., Morreale, M.S. et al. (2008). *J. Org. Chem.* 73: 5881–5889.

63 Imagawa, H., Kurisaki, T., and Nishizawa, M. (2004). *Org. Lett.* 6: 3679–3681.
64 Ravindar, K., Reddy, M.S., and Deslongchamps, P. (2011). *Org. Lett.* 13: 3178–3181.
65 Yeh, M.-C.P., Lin, M.-N., Hsu, C.-H., and Liang, C.-J. (2013). *J. Org. Chem.* 78: 12381–12396.
66 Chen, Z.-W., Luo, M.-T., Ye, D.-N. et al. (2014). *Synth. Commun.* 44: 1825–1831.
67 Liu, Y., Song, F., Song, Z. et al. (2005). *Org. Lett.* 7: 5409–5412.
68 (a) Chen, Z., Huang, G., Jiang, H. et al. (2011). *J. Org. Chem.* 76: 1134–1139. (b) Okamoto, N. and Yanada, R. (2012). *J. Org. Chem.* 77: 3944–3951.
69 Lee, H., Yi, Y., and Jun, C.-H. (2015). *Adv. Synth. Catal.* 357: 3485–3490.
70 Suhre, M.H., Reif, M., and Kirsch, S.F. (2005). *Org. Lett.* 7: 3925–3927.
71 Kramer, S., Madsen, J.L.H., Rottländer, M., and Skrydstrup, T. (2010). *Org. Lett.* 12: 2758–2761.
72 Zheng, Q., Hua, R., and Yin, T. (2013). *Curr. Org. Synth.* 10: 161–164.
73 Wang, Y.-H., Liu, H., Zhu, L.-L. et al. (2011). *Adv. Synth. Catal.* 353: 707–712.
74 Zheng, Q., Hua, R., Jiang, J., and Zhang, L. (2014). *Tetrahedron* 70: 8252–8256.
75 Jiang, J., Ding, C., Li, L. et al. (2014). *Tetrahedron Lett.* 55: 6635–6638.
76 Talbi, I., Alayrac, C., Lohier, J.-F. et al. (2016). *Org. Lett.* 18: 2656–2659.
77 Nun, P., Dupuy, S., Gaillard, S. et al. (2011). *Catal. Sci. Technol.* 1: 58–61.
78 Jiang, H., Zeng, W., Li, Y. et al. (2012). *J. Org. Chem.* 77: 5179–5183.
79 Schmidt, E.Y., Bidusenko, I.A., Cherimichkina, N.A. et al. (2015). *Chem. Eur. J.* 21: 15944–15946.
80 Yoshimatsu, M., Watanabe, H., and Koketsu, E. (2010). *Org. Lett.* 12: 4192–4194.
81 Tamaso, K.-i., Hatamoto, Y., Obora, Y. et al. (2007). *J. Org. Chem.* 72: 8820–8823.
82 Mancuso, R. and Gabriele, B. (2014). *Molecules* 19: 15687–15719.
83 Marshall, J.A. and DuBay, W.J. (1993). *Synlett* 209.
84 Gabriele, B., Salerno, G., and Fazio, A. (2000). *Org. Lett.* 2: 351–352.
85 Gabriele, B., Mancuso, R., Veltri, L. et al. (2012). *J. Org. Chem.* 77: 9905–9909.
86 Gabriele, B., Mancuso, R., Salerno, G., and Larock, R.C. (2012). *J. Org. Chem.* 77: 7640–7645.
87 Reddy, C.R., Valleti, R.R., and Reddy, M.D. (2013). *J. Org. Chem.* 78: 6495–6502.
88 Bharathiraja, G., Sathishkannan, G., and Punniyamurthy, T. (2016). *J. Org. Chem.* 81: 2670–2674.
89 Freeman, F., Lu, H., Zeng, Q., and Rodriguez, E. (1994). *J. Org. Chem.* 59: 4350–4354.
90 Selected reviews, see: (a) Huffman, J.W. and Padgett, L.W. (2005). *Curr. Med. Chem.* 12: 1395–1411. (b) Brancale, A. and Silvestri, R. (2007). *Med. Res. Rev.* 27: 209–238. (c) Sharma, V., Kumar, P., and Pathak, D. (2010). *J. Heterocycl. Chem.* 47: 491–502. (d) Patel, H., Darji, N., Pillai, J., and Patel, B. (2012). *Int. J. Drug Res. Tech.* 2: 225–230. (e) Zhang, M.-Z., Chen, Q., and Yang, G.-F. (2015). *Eur. J. Med. Chem.* 89: 421–441. (f) Sravanthi, T.V. and Manju, S.L. (2016). *Eur. J. Pharm. Sci.* 91: 1–10. (g) Chadha, N. and Silakari, O. (2017). *Eur. J. Med. Chem.* 134: 159–184. (h) Suzen, S. (2017). *Curr. Org. Chem.* 21: 2068–2076. (i) Singh, T.P. and Singh, O.M. (2018). *Mini-Rev. Med. Chem.* 18: 9–25. (j) Hendrikx, T. and Schnabl, B.J. (2019). *Internal Med.* 286: 32–40.

91 Selected reviews, see: (a) Galal, S.A., El-All, A.S.A., Abdallah, M.M., and El-Diwani, H.I. (2009). *Bioorg. Med. Chem. Lett.* 19: 2420–2428. (b) Chand, K., Hiremathad, A., Singh, M. et al. (2017). *Pharmacol. Rep.* 69: 281–295. (c) Radadiya, A. and Shah, A. (2015). *Eur. J. Med. Chem.* 97: 356–376. (d) Khanam, H. and Shamsuzzaman (2015). *Eur. J. Med. Chem.* 97: 483–504. (e) Nevagi, R.J., Dighe, S.N., and Dighe, S.N. (2015). *Eur. J. Med. Chem.* 97: 561–581.

92 Selected reports, see: (a) Horton, D.A., Bourne, G.T., and Smythe, M.L. (2003). *Chem. Rev.* 103: 893–930. (b) Qin, Z., Kastrati, I., Chandrasena, R.E.P. et al. (2007). *J. Med. Chem.* 50: 2682–2692.

93 Mahmoud, A., Ahmed, B., Kirsch, G., and Peagle, E. (2017). *Curr. Org. Synth.* 14: 1091–1101.

94 Selected reports, see: (a) Gribble, G.W. (2000). *J. Chem. Soc. Perkin Trans. 1* 1045–1075. (b) Shiri, M. (2012). *Chem. Rev.* 112: 3508–3549. (c) Dalpozzo, R. (2015). *Chem. Soc. Rev.* 44: 742–778. (d) Sandtorv, A.H. (2015). *Adv. Synth. Catal.* 357: 2403–2435. (e) Chen, J.-B. and Jia, Y.-X. (2017). *Org. Biomol. Chem.* 15: 3550–3567. (f) Chen, S., Li, Y., Ni, P. et al. (2016). *Org. Lett.* 18: 5384–5387. (g) Huang, L., Dai, L.-X., and You, S.-L. (2016). *J. Am. Chem. Soc.* 138: 5793–5796. (h) Haak, E. (2019). *Synlett* 30: 245–251.

95 (a) Robinson, B. (1963). *Chem. Rev.* 63: 373–401. (b) Robinson, B. (1969). *Chem. Rev.* 69: 227–250. (c) Hughes, D.L. (1993). *Org. Prep. Proc. Int.* 25: 607–632. (d) Heravi, M.M., Rohani, S., Zadsirjan, V., and Zahedi, N. (2017). *RSC Adv.* 7: 52852–52887.

96 Selected reviews, see: (a) Gribble, G.W. (2000). *J. Chem. Soc., Perkin Trans. 1* 1045–1075. (b) Cacchi, S. and Fabrizi, G. (2005). *Chem. Rev.* 105: 2873–2920. (c) Humphrey, G.R. and Kuethe, J.T. (2006). *Chem. Rev.* 106: 2875–2911. (d) Krüger, K., Tillack, A., and Beller, M. (2008). *Adv. Synth. Catal.* 350: 2153–2167. (e) Vicente, R. (2011). *Org. Biomol. Chem.* 9: 6469–6480. (f) Taber, D.F. and Tirunahari, P.K. (2011). *Tetrahedron* 67: 7195–7210. (g) Inman, M. and Moody, J.C. (2013). *Chem. Sci.* 4: 29–41. (h) Guo, T., Huang, F., Yu, L., and Yu, Z. (2015). *Tetrahedron Lett.* 56: 296–302. (i) Zhang, B. and Studer, A. (2015). *Chem. Soc. Rev.* 44: 3505–3521. (j) Anderson, L.L., Kroc, M.A., Reidl, T.W., and Son, J. (2016). *J. Org. Chem.* 81: 9521–9529. (k) Chelucci, G. (2017). *Coord. Chem. Rev.* 331: 37–53. (l) Hou, Z.-W., Mao, Z.-Y., and Xu, H.-C. (2017). *Synlett* 28: 1867–1872. (m) Nosova, E.V., Lipunova, G.N., Charushin, V.N., and Chupakhin, O.N. (2018). *J. Fluor. Chem.* 212: 51–106. (n) Haak, E. (2019). *Synlett* 30: 245–251. (o) Neto, J.S.S. and Zeni, G. (2020). *Org. Chem. Front.* 7: 155–210. (p) Colella, M., Degennaro, L., and Luisi, R. (2020). *Molecules* 25: 3242.

97 Koradin, C., Dohle, W., Rodriguez, A.L. et al. (2003). *Tetrahedron* 59: 1571–1587.

98 Hiroya, K., Itoh, S., Ozawa, M. et al. (2002). *Tetrahedron Lett.* 43: 1277–1280.

99 Kurisaki, T., Naniwa, T., Yamamoto, H. et al. (2007). *Tetrahedron Lett.* 48: 1871–1874.

100 Li, X., Chianese, A.R., Vogel, T., and Crabtree, R.H. (2005). *Org. Lett.* 7: 5437–5440.

101 Majumdar, K.C., Samanta, S., and Chattopadhyay, B. (2008). *Tetrahedron Lett.* 49: 7213–7216.

102 La-Venia, A., Testero, S.A., Mischne, M.P., and Mata, E.G. (2012). *Org. Biomol. Chem.* 10: 2514–2517.

103 Perea-Buceta, J.-E., Wirtanen, T., Laukkanen, O.-V. et al. (2013). *Angew. Chem. Int. Ed.* 52: 11835–11839.

104 Hiroya, K., Itoh, S., and Sakamoto, T. (2004). *J. Org. Chem.* 69: 1126–1136.

105 Hiroya, K., Itoh, S., and Sakamoto, T. (2005). *Tetrahedron* 61: 10958–10964.

106 Sakai, N., Annaka, K., Fujita, A. et al. (2008). *J. Org. Chem.* 73: 4160–4165.

107 Murai, K., Hayashi, S., Takaichi, N. et al. (2009). *J. Org. Chem.* 74: 1418–1421.

108 Ye, D., Wang, J., Zhang, X. et al. (2009). *Green Chem.* 11: 1201–1208.

109 Kumaran, E., Fan, W.Y., and Leong, W.K. (2014). *Org. Lett.* 16: 1342–1345.

110 Kumaran, E. and Leong, W.K. (2014). *Tetrahedron Lett.* 55: 5495–5498.

111 (a) Saito, A., Kanno, A., and Hanzawa, Y. (2007). *Angew. Chem. Int. Ed.* 46: 3931–3933. (b) Saito, A., Oda, S., Fukaya, H., and Hanzawa, Y. (2009). *J. Org. Chem.* 74: 1517–1524.

112 Watanabe, T., Mutoh, Y., and Saito, S. (2017). *J. Am. Chem. Soc.* 139: 7749–7752.

113 Alfonsi, M., Arcadi, A., Aschi, M. et al. (2005). *J. Org. Chem.* 70: 2265–2273.

114 Janreddy, D., Kavala, V., Kuo, C.-W. et al. (2013). *Tetrahedron* 69: 3323–3330.

115 Li, X., Wang, J.-Y., Yu, W., and Wu, L.-M. (2009). *Tetrahedron* 65: 1140–1146.

116 Patil, N.T., Singh, V., Konala, A., and Mutyala, A.K. (2010). *Tetrahedron Lett.* 51: 1493–1496.

117 (a) Stuart, D.R., Bertrand-Laperle, M., Burgess, K.M.N., and Fagnou, K. (2008). *J. Am. Chem. Soc.* 130: 16474–16475. (b) Stuart, D.R., Alsabeh, P., Kuhn, M., and Fagnou, K. (2010). *J. Am. Chem. Soc.* 132: 18326–18339. (c) Huestis, M.P., Chan, L.N., Stuart, D.R., and Fagnou, K. (2011). *Angew. Chem. Int. Ed.* 50: 1338–1341.

118 Zhou, F., Han, X., and Lu, X. (2011). *Tetrahedron Lett.* 52: 4681–4685.

119 Chen, J., Pang, G., Sun, Y., and Li, X. (2011). *J. Org. Chem.* 76: 3523–3526.

120 Ackermann, L. and Lygin, A.V. (2012). *Org. Lett.* 14: 764–767.

121 Zhao, D., Shi, Z., and Glorius, F. (2013). *Angew. Chem. Int. Ed.* 52: 12426–12429.

122 Liu, B., Song, C., Sun, C. et al. (2013). *J. Am. Chem. Soc.* 135: 16625–16631.

123 Wang, C. and Huang, Y. (2013). *Org. Lett.* 15: 5294–5297.

124 Lerchen, A., Vásquez-Céspedes, S., and Glorius, F. (2016). *Angew. Chem. Int. Ed.* 55: 3208–3211.

125 Zhang, Z.-Z., Liu, B., Xu, J.-W. et al. (2016). *Org. Lett.* 18: 1776–1779.

126 Zhou, S., Wang, J., Wang, L. et al. (2016). *Org. Lett.* 18: 3806–3809.

127 Zheng, L. and Hua, R. (2014). *Chem. Eur. J.* 20: 2352–2356.

128 Matsuda, T. and Tomaru, Y. (2014). *Tetrahedron Lett.* 55: 3302–3304.

129 Li, D.Y., Chen, H.J., and Liu, P.N. (2014). *Org. Lett.* 16: 6176–6179.

130 Wang, C., Sun, H., Fang, Y., and Huang, Y. (2013). *Angew. Chem. Int. Ed.* 52: 5795–5798.

131 Sun, H., Wang, C., Yang, Y.-F. et al. (2014). *J. Org. Chem.* 79: 11863–11872.

132 (a) Würtz, S., Rakshit, S., Neumann, J.J. et al. (2008). *Angew. Chem. Int. Ed.* 47: 7230–7233. (b) Neumann, J.J., Rakshit, S., Dröge, T. et al. (2011). *Chem. Eur. J.* 17: 7298–7303.

133 Guan, Z.-H., Yan, Z.-Y., Ren, Z.-H. et al. (2010). *Chem. Commun.* 46: 2823–2825.

134 Liu, W.-Q., Lei, T., Song, Z.-Q. et al. (2017). *Org. Lett.* 19: 3251–3254.

135 Shi, Z., Zhang, C., Li, S. et al. (2009). *Angew. Chem. Int. Ed.* 48: 4572–4576.

136 Yu, W., Du, Y., and Zhao, K. (2009). *Org. Lett.* 11: 2417–2420.

137 Bernini, R., Fabrizi, G., Sferrazza, A., and Cacchi, S. (2009). *Angew. Chem. Int. Ed.* 48: 8078–8081.

138 Xia, X.-F., Zhang, L.-L., Song, X.-R. et al. (2013). *Chem. Commun.* 49: 1410–1412.

139 (a) Cacchi, S., Fabrizi, G., and Pace, P. (1998). *J. Org. Chem.* 63: 1001–1011.

140 Shimada, T., Nakamura, I., and Yamamoto, Y. (2004). *J. Am. Chem. Soc.* 126: 10546–10547.

141 Fürstner, A. and Davies, P.W. (2005). *J. Am. Chem. Soc.* 127: 15024–15025.

142 Cacchi, S., Fabrizi, G., and Moro, L. (1998). *Tetrahedron Lett.* 39: 5101–5104.

143 Nakamura, I., Mizushima, Y., and Yamamoto, Y. (2005). *J. Am. Chem. Soc.* 127: 15022–15023.

144 Nakamura, I., Sato, T., and Yamamoto, Y. (2006). *Angew. Chem. Int. Ed.* 45: 4473–4475.

145 Nakamura, I., Yamagishi, U., Song, D. et al. (2007). *Angew. Chem. Int. Ed.* 46: 2284–2287.

146 (a) Larock, R.C. and Yum, E.K. (1991). *J. Am. Chem. Soc.* 113: 6689–6690. (b) Larock, R.C., Yum, E.K., and Refvik, M.D. (1998). *J. Org. Chem.* 63: 7652–7662.

147 Ackermann, L., Kaspar, L.T., and Gschrei, C.J. (2004). *Chem. Commun.* 2824–2825.

148 Shen, M., Li, G., Lu, B.Z. et al. (2004). *Org. Lett.* 6: 4129–4132.

149 Ponpandian, T. and Muthusubramanian, S. (2012). *Tetrahedron Lett.* 53: 4248–4252.

150 Cacchi, S., Fabrizi, G., Goggiamani, A. et al. (2010). *Org. Lett.* 12: 3279–3281.

151 Qu, C., Zhang, S., Du, H., and Zhu, C. (2016). *Chem. Commun.* 52: 14400–14403.

152 Minami, H., Kanayama, T., Tanaka, R. et al. (2016). *Eur. J. Org. Chem.* 5990–6000.

153 Arcadi, A., Cacchi, S., Fabrizi, G. et al. (2005). *J. Org. Chem.* 70: 6213–6217.

154 Yao, B., Wang, Q., and Zhu, J. (2012). *Angew. Chem. Int. Ed.* 51: 12311–12315.

155 Brand, J.P., González, D.F., Nicolai, S., and Waser, J. (2011). *Chem. Commun.* 47: 102–115.

156 Brand, J.P., Chevalley, C., and Waser, J. (2011). *Beilstein J. Org. Chem.* 7: 565–569.

157 Godoi, B., Schumacher, R.F., and Zeni, G. (2011). *Chem. Rev.* 111: 2937–2980.

158 (a) Barluenga, J., Trincado, M., Rubino, E., and González, J.M. (2003). *Angew. Chem. Int. Ed.* 42: 2406–2409. (b) Yue, D., Yao, T., and Larock, R.C. (2006). *J. Org. Chem.* 71: 62–69.

159 Swamy, N.K., Yazici, A., and Pyne, S.G. (2010). *J. Org. Chem.* 75: 3412–3419.
160 Yue, D., Yao, T., and Larock, R.C. (2005). *J. Org. Chem.* 70: 10292–10296.
161 Okitsu, T., Nakazawa, D., Taniguchi, R., and Wada, A. (2008). *Org. Lett.* 10: 4967–4970.
162 Shen, M., Leslie, B.E., and Driver, T.G. (2008). *Angew. Chem. Int. Ed.* 47: 5056–5059.
163 Lu, B., Luo, Y., Liu, L. et al. (2011). *Angew. Chem. Int. Ed.* 50: 8358–8362.
164 Zhou, Q., Zhang, Z., Zhou, Y. et al. (2017). *J. Org. Chem.* 82: 48–56.
165 Penoni, A. and Nicholas, K.M. (2002). *Chem. Commun.* 484–485.
166 Anderson, L.L., Kroc, M.A., Reidl, T.W., and Son, J. (2016). *J. Org. Chem.* 81: 9521–9529.
167 Mehta, S., Waldo, J.P., and Larock, R.C. (2009). *J. Org. Chem.* 74: 1141–1147.
168 (a) Yue, D., Yao, T., and Larock, R.C. (2005). *J. Org. Chem.* 70: 10292–10296. (b) Manarin, F., Roehrs, J.A., Gay, R.M. et al. (2009). *J. Org. Chem.* 74: 2153–2162.
169 Kesharwani, T., Worlikar, S.A., and Larock, R.C. (2006). *J. Org. Chem.* 71: 2307–2312.
170 Aurelio, L., Volpe, R., Halim, R. et al. (2014). *Adv. Synth. Catal.* 356: 1974–1978.
171 Kesharwani, T., Craig, J., Rosario, C.D. et al. (2014). *Tetrahedron Lett.* 55: 6812–6816.
172 Kesharwani, T., Kornman, C.T., Tonnaer, A.L., and Royappa, A.D. (2016). *Tetrahedron Lett.* 57: 411–414.
173 Kesharwani, T., Giraudy, K.A., Morgan, J.L. et al. (2017). *Tetrahedron Lett.* 58: 638–641.
174 Yue, D. and Larock, R.C. (2002). *J. Org. Chem.* 67: 1905–1909.
175 Hessian, K.O. and Flynn, B.L. (2003). *Org. Lett.* 5: 4377–4380.
176 Jacubert, M., Hamze, A., Provot, O. et al. (2009). *Tetrahedron Lett.* 50: 3588–3592.
177 Warner, A.J., Churn, A., McGough, J.S., and Ingleson, M.J. (2017). *Angew. Chem. Int. Ed.* 56: 354–358.
178 (a) Zhang, Y., Xin, Z.-J., Xue, J.-J., and Li, Y. (2008). *Chin. J. Chem.* 26: 1461–1464. (b) Arcadi, A., Blesi, F., Cacchi, S. et al. (2013). *Tetrahedron* 69: 1857–1871. (c) Markina, N.A., Chen, Y., and Larock, R.C. (2013). *Tetrahedron* 69: 2701–2713.
179 (a) Kuram, M.R., Bhanuchandra, M., and Sahoo, A.K. (2013). *Angew. Chem. Int. Ed.* 52: 4607–4612. (b) Zhu, R., Wei, J., and Shi, Z. (2013). *Chem. Sci.* 4: 3706–3711.
180 (a) Kashiki, T., Shinamura, S., Kohara, M. et al. (2009). *Org. Lett.* 11: 2473–2475. (b) Kuhn, M., Falk, F.C., and Paradies, J. (2011). *Org. Lett.* 13: 4100–4103. (c) Guilarte, V., Fernández-Rodríguez, M.A., García-García, P. et al. (2011). *Org. Lett.* 13: 5100–5103. (d) Prasad, D.J.C. and Sekar, G. (2013). *Org. Biomol. Chem.* 11: 1659–1665. (e) Xiong, X., Liu, Q., Zhang, J. et al. (2013). *RSC Adv.* 3: 17707–17711. (f) Li, Y., Cheng, L., Li, B. et al. (2016). *ChemistrySelect* 5: 1092–1095.

181 Selected reviews, see: (a) Verrier, C., Lassalas, P., Théveau, L. et al. (2011). *Beilstein J. Org. Chem.* 7: 1584–1601. (b) Bresciani, S. and Tomkinson, N.C.O. (2014). *Heterocycles* 89: 2479–2543. (c) Cicchi, S., Cordero, F.M., and Giomi, D. (2012). *Prog. Hetero. Chem.* 24: 317–342. (d) Zheng, X., Liu, W., and Zhang, D. (2020). *Molecules* 25: 1594.

182 (a) Fustero, S., Simón-Fuentes, A., and Sanz-Cervera, J.F. (2009). *Org. Prep. Proced Int.* 41: 253–290. (b) Korablina, D.D., Vorozhtsov, N.I., Sviridova, L.A. et al. (2016). *Pharm. Chem. J.* 50: 281–295.

183 (a) Zora, M., Kivrak, A., and Yazici, C. (2011). *J. Org. Chem.* 76: 6726–6742. (b) Zora, M. and Kivrak, A. (2011). *J. Org. Chem.* 76: 9379–9390.

184 Perin, G., Nobre, P.C., Mailahn, D.H. et al. (2019). *Synthesis* 51: 2293–2305.

185 Yoshimatsu, M., Ohta, K., and Takahashi, N. (2012). *Chem. Eur. J.* 18: 15602–15606.

186 Wang, L., Yu, X., Feng, X., and Bao, M. (2013). *J. Org. Chem.* 78: 1693–1698.

187 Wen, J.-J., Tang, H.-T., Xiong, K. et al. (2014). Synthesis of polysubstituted pyrazoles by a platinum-catalyzed sigmatropic rearrangement/cyclization cascade. *Org. Lett.* 16: 5940–5943.

188 Okitsu, T., Sato, K., and Wada, A. (2010). *Org. Lett.* 12: 3506–3509.

189 For reviews, see: (a) Wei, C., Li, Z., and Li, C.-J. (2004). *Synlett* 1472–1483. (b) Li, C.-J. (2010). *Acc. Chem. Res.* 43: 581–590.

190 Suzuki, Y., Naoe, S., Oishi, S. et al. (2012). *Org. Lett.* 14: 326–329.

191 Zhu, Y., Wen, S., Yin, G. et al. (2011). *Org. Lett.* 13: 3553–3555.

192 Wang, X., Pan, Y.-m., Huang, X.-c. et al. (2014). A novel methodology for synthesis of dihydropyrazole derivatives as potential anticancer agents. *Org. Biomol. Chem.* 12: 2028–2032.

193 Hashimoto, T., Takiguchi, Y., and Maruoka, K. (2013). *J. Am. Chem. Soc.* 135: 11473–11476.

194 Selected reviews, see: (a) Narasimhan, B., Sharma, D., and Kumar, P. (2011). *Med. Chem. Res.* 20: 1119–1140. (b) Rani, N., Sharma, A., and Singh, R. (2013). *Mini-Rev. Med. Chem.* 13: 1812–1835. (c) Zhang, L., Peng, X.-M., Damu, G.L.V. et al. (2014). *Med. Res. Rev.* 34: 340–437. (d) Zhang, L., Peng, X.-M., Damu, G.L.V. et al. (2014). *Med. Res. Rev.* 34: 340–437. (e) Ali, I., Lonea, M.N., and Aboul-Enein, H.Y. (2017). *Med. Chem. Commun.* 8: 1742–1773.

195 Li, J. and Neuville, L. (2013). *Org. Lett.* 15: 1752–1755.

196 Chen, X.Y., Englert, U., and Bolm, C. (2015). *Chem. Eur. J.* 21: 13221–13224.

197 Mehmood, H., Iqbal, M.A., Lu, L., and Hua, R. (2020). *Molecules* 25: 3621.

198 (a) Wang, C., Wang, E., Chen, W. et al. (2017). *J. Org. Chem.* 82: 9144–9153. (b) Wang, C., Jiang, H., Chen, W. et al. (2017). *Org. Biomol. Chem.* 15: 6463–6466.

199 (a) Baraldi, P.G., Barco, A., Benetti, S. et al. (1987). *Synthesis* 857–869. (b) Lakhvich, F.A., Koroleva, E.V., and Akhrem, A.A. (1989). *Chem. Heterocycl. Compd.* 25: 359–375. (c) Sperry, J. and Wright, D. (2005). *Curr. Opin. Drug Discov. Devel.* 8: 723–740. (d) Barmade, M.A., Murumkar, P.R., Sharma, M.K., and Yadav, M.R. (2016). *Curr. Top. Med. Chem.* 16: 2863–2883. (e) Sysak, A. and Obminska-Mrukowicz, B. (2017). *Eur. J. Med. Chem.* 137: 292–309. (f) Zhu, J.,

Mo, J., Lin, H.Z. et al. (2018). *Bioorg. Med. Chem.* 23: 3065–3075. (g) Zimecki, M., Bąchor, U., and Mączyński, M. (2018). *Molecules* 23: 2724. (h) Morita, T., Yugandar, S., Fuse, S., and Nakamura, H. (2018). *Tetrahedron Lett.* 59: 1159–1171.

200 Grecian, S. and Fokin, V.V. (2008). *Angew. Chem. Int. Ed.* 47: 8285–8287.

201 Dadiboyena, S., Xu, J., and Hamme, A.T. (2007). *Tetrahedron Lett.* 48: 1295–1298.

202 Gao, P., Li, H.-X., Hao, X.-H. et al. (2014). *Org. Lett.* 16: 6298–6301.

203 Wang, L., Yu, X., Feng, X., and Bao, M. (2012). *Org. Lett.* 14: 2418–2421.

204 Gasparrini, F., Giovannoli, M., Misiti, D. et al. (1993). *J. Am. Chem. Soc.* 115: 4401–4402.

205 Hansen, T.V., Wu, P., and Fokin, V.V. (2005). *J. Org. Chem.* 70: 7761–7764.

206 Vieira, A.A., Bryk, F.R., Conte, G. et al. (2009). *Tetrahedron Lett.* 50: 905–908.

207 Murarka, S. and Studer, A. (2011). *Org. Lett.* 13: 2746–2749.

208 Ueda, M., Ikeda, Y., Sato, A. et al. (2011). *Tetrahedron* 67: 4612–4615.

209 Pevarello, P., Amici, R., Colombo, M., and Varasi, M.J. (1993). *J. Chem. Soc., Perkin Trans. 1.* 2151–2152.

210 Bang-Andersen, B., Ahmadian, H., Lenz, S.M. et al. (2000). *J. Med. Chem.* 43: 4910–4918.

211 (a) Waldo, J.P. and Larock, R.C. (2005). *Org. Lett.* 23: 5203–5205. (b) Waldo, J.P. and Larock, R.C. (2007). *J. Org. Chem.* 72: 9643–9647.

212 Ueda, M., Sato, A., Ikeda, Y. et al. (2010). *Org. Lett.* 12: 2594–2597.

213 Denmark, S.E. and Kallemeyn, J.M. (2005). *J. Org. Chem.* 70: 2839–2842.

214 (a) Shang, Y.-J. and Wang, Y.-G. (2002). *Tetrahedron Lett.* 43: 2247–2249. (b) Singh, I., Zarafshani, Z., Lutz, J.-F., and Heaney, F. (2009). *Macromolecules* 42: 5411–5413.

215 (a) Kantorowski, E.J. and Kurth, M.J. (1997). *J. Org. Chem.* 62: 6797–6803. (b) Cereda, E., Ezhaya, A., Quai, M., and Barbaglia, W. (2001). *Tetrahedron Lett.* 42: 4951–4953. (c) De Luca, L., Giacomelli, G., and Riu, A. (2001). *J. Org. Chem.* 66: 6823–6825. (d) Dadiboyena, S. and Nefzi, A. (2012). *Tetrahedron Lett.* 53: 2096–2099.

216 Kankala, S., Vadde, R., and Vasam, C.S. (2011). *Org. Biomol. Chem.* 9: 7869–7876.

217 (a) Giacomelli, G., De Luca, L., and Porcheddu, A. (2003). *Tetrahedron* 59: 5437–5440. (b) Lin, L., Zhang, J., and Wang, R. (2012). *Asian J. Org. Chem.* 1: 222–225.

218 Jawalekar, A.M., Reubsaet, E., Rutjes, F.P.J.T., and van Delft, F.L. (2011). *Chem. Commun.* 47: 3198–3200.

219 Yoshimura, A., Middleton, K.R., Todora, A.D. et al. (2013). *Org. Lett.* 15: 4010–4013.

220 Selected reviews, see: (a) Turchi, I.J. and Dewar, M.J.S. (1975). *Chem. Rev.* 75: 389–437. (b) Yeh, V.S.C. (2004). *Tetrahedron* 60: 11995–12042. (c) Jin, Z. (2009). *Nat. Prod. Rep.* 26: 382–445. (d) Zhang, H.-Z., Zhao, Z.-L., and Zhou, C.-H. (2018). *Eur. J. Med. Chem.* 144: 444–492. (e) Kakkar, S. and Narasimhan, B. (2019). *BMC Chem.* 13: 16.

221 Li, X., Huang, L., Chen, H. et al. (2012). *Chem. Sci.* 3: 3463–3467.

222 He, W., Li, C., and Zhang, L. (2011). *J. Am. Chem. Soc.* 133: 8482–8485.

223 Xu, Y., Wang, Q., Wu, Y. et al. (2019). *Adv. Synth. Catal.* 361: 2309–2314.

224 (a) Dubovtsev, A.Y., Dar'in, D.V., and Kukushkin, V.Y. (2019). *Adv. Synth. Catal.* 361: 2926–2935. (b) Wang, Q., Hoffmann, S., Schießl, J. et al. (2020). *Eur. J. Org. Chem.* 2384–2388.

225 Saito, A., Taniguchi, A., Kambara, Y., and Hanzawa, Y. (2013). *Org. Lett.* 15: 2672–2675.

226 Yagyu, T., Takemoto, Y., Yoshimura, A. et al. (2017). *Org. Lett.* 19: 2506–2509.

227 Pan, J., Li, X., Qiu, X. et al. (2018). *Org. Lett.* 20: 2762–2765.

228 Selected reviews, see: (a) Kolb, H.C., Finn, M.G., and Sharpless, K.B. (2001). *Angew. Chem., Int. Ed.* 40: 2004–2021. (b) Moses, J.E. and Moorhouse, A.D. (2007). *Chem. Soc. Rev.* 36: 1249–1262. (c) Medal, M. and Tornøe, C.W. (2008). *Chem. Rev.* 108: 2952–3015. (d) Cañeque, T., Müller, S., and Rodriguez, R. (2018). *Nat. Rev. Chem.* 2: 202–215.

229 (a) Cano, I., Álvarez, E., Nicasio, M.C., and Pérez, P.J. (2011). *J. Am. Chem. Soc.* 133: 191–193. (b) Haldón, E., Besora, M., Cano, I. et al. (2014). *Chem. Eur. J.* 20: 3463–3474.

230 Yu, X., Chen, K., Wang, Q. et al. (2018). *Chem. Commun.* 54: 1197–1200.

231 Wu, A. and Sun, J. (2019). *Synlett* 30: 515–518.

232 Zhang, L., Xiao, K., Qiao, Y. et al. (2018). *Eur. J. Org. Chem.* 6913–6918.

233 (a) Wipf, P., Aoyama, Y., and Benedum, T.E.A. (2004). *Org. Lett.* 6: 3593–3595. (b) Hashmi, A.S.K., Weyrauch, J.P., Frey, W., and Bats, J.W. (2004). *Org. Lett.* 6: 4391–4394. (c) Pan, Y.-m., Zheng, F.-j., Lin, H.-x., and Zhan, Z.-p. (2009). *J. Org. Chem.* 74: 3148–3151. (d) Hashmi, A.S.K., Jaimes, M.C.B., Schuster, A.M., and Rominger, F. (2012). *J. Org. Chem.* 77: 6394–6408. (e) Bartoli, G., Cimarelli, C., Cipolletti, R. et al. (2012). *Eur. J. Org. Chem.* 630–636.

234 Pizzo, C. and Mahler, S.G. (2014). *J. Org. Chem.* 79: 1856–1860.

235 Zhang, Z., Guo, B., Zhou, Y., and Hua, R. (2016). *Tetrahedron* 72: 7613–7619.

236 Selected reports, see: (a) Baumann, M. and Baxendale, I.R. (2013). *Beilstein J. Org. Chem.* 9: 2265–2319. (b) Neetha, M., Aneeja, T., Afsina, C.M.A., and Anilkumar, G. (2020). *ChemCatChem* 12: 5330–5358. (c) Zhou, Y. and Hua, R. (2021). *J. Org. Chem.* 86: 8862–8872. and references cited therein.

237 (a) Bönnemann, H. (1985). *Angew. Chem. Int. Ed. Engl.* 24: 248–262. (b) Chelucci, G. (1995). *Tetrahedron: Asymmetry* 6: 811–826. (c) Dzhemilev, U.M., Selimov, F.A., and Tolstikov, G.A. (2001). *ARKIVOC* (ix): 85–116. (d) Varela, J.A. and Saá, C. (2003). *Chem. Rev.* 103: 3787–3801. (e) Nakamura, I. and Yamamoto, Y. (2004). *Chem. Rev.* 104: 2127–2198. (f) Henry, G.D. (2004). *Tetrahedron* 60: 6043–6061. (g) Heller, B. and Hapke, M. (2007). *Chem. Soc. Rev.* 36: 1085–1094. (h) Maryanoff, B.E. and Zhang, H.-C. (2007). *ARKIVOC* (xii): 7–35. (i) Varela, J.A. and Saá, C. (2008). *Synlett* 2571–2578. (j) Hill, M.D. (2010). *Chem. Eur. J.* 16: 12052–12062. (k) Gulevich, A.V., Dudnik, A.S., Chernyak, N., and Gevorgyan, V. (2013). *Chem. Rev.* 113: 3084–3213.

238 (a) Wakatsuki, Y. and Yamazaki, H. (1973). *Tetrahedron Lett.* 36: 3383–3384. (b) Wakatsuki, Y. and Yamazaki, H. (1978). *J. Chem. Soc., Dalton Trans.* 1278–1282.

239 (a) Heller, B. and Oehme, G. (1995). *J. Chem. Soc., Commun. Chem.* 179–180. (b) Heller, B., Sundermann, B., Buschmann, H. et al. (2002). *J. Org. Chem.* 67: 4414–4422.

240 Gutnov, A., Heller, B., Fischer, C. et al. (2004). *Angew. Chem. Int. Ed.* 43: 3795–3797.
241 Bonaga, L.V.R., Zhang, H.-C., and Maryanoff, B.E. (2004). *Chem. Commun.* 2394–2395.
242 (a) Zhou, Y., Porco, J.A., and Snyder, J.K. (2007). *Org. Lett.* 9: 393–396. (b) Zou, Y., Young, D.D., Cruz-Montanez, A., and Deiters, A. (2008). *Org. Lett.* 10: 4661–4664.
243 Young, D.D. and Deiters, A. (2007). *Angew. Chem. Int. Ed.* 46: 5187–5190.
244 Chang, H.-T., Jeganmohan, M., and Cheng, C.-H. (2007). *Org. Lett.* 9: 505–508.
245 Kase, K., Goswami, A., Ohtaki, K. et al. (2007). *Org. Lett.* 9: 931–934.
246 Goswami, A., Ohtaki, K., Kase, K. et al. (2008). *Adv. Synth. Catal.* 350: 143–152.
247 Fatland, A.W. and Eaton, B.E. (2000). *Org. Lett.* 2: 3131–3133.
248 McCormick, M.M., Duong, H.A., Zuo, G., and Louie, J. (2005). *J. Am. Chem. Soc.* 127: 5030–5031.
249 Tekavec, T.N., Zuo, G., Simon, K., and Louie, J. (2006). *J. Org. Chem.* 71: 5834–5836.
250 (a) D'Souza, B.R., Lane, T.K., and Louie, J. (2011). *Org. Lett.* 13: 2936–2939. (b) Lane, T.K., D'Souza, B.R., and Louie, J. (2012). *J. Org. Chem.* 77: 7555–7563.
251 Tanaka, K., Suzuki, N., and Nishida, G. (2006). *Eur. J. Org. Chem.* 3917–3922.
252 (a) Yamomoto, Y., Okuda, S., and Itoh, K. (2001). *Chem. Commun.* 1102–1103. (b) Yamamoto, Y., Ogawa, R., and Itoh, K. (2001). *J. Am. Chem. Soc.* 123: 6189–6190. (c) Varela, J.A., Castedo, L., and Saá, C. (2003). *J. Org. Chem.* 68: 8595–8598. (d) Yamomoto, Y., Kinpara, K., Saigoku, T. et al. (2005). *J. Am. Chem. Soc.* 127: 605–613. (e) Yamomoto, Y., Kinpara, K., Ogawa, R. et al. (2006). *Chem. Eur. J.* 12: 5618–5631. (f) Xu, F., Wang, C., Li, X., and Wan, B. (2012). *ChemSusChem* 5: 854–857.
253 Hrdina, R., Stará, I.G., Dufková, L. et al. (2006). *Tetrahedron* 62: 968–976.
254 Onodera, G.; Shimizu, Y.; Kimura, J.-n.; Kobayashi, J.; Ebihara, Y.; Kondo, K.; Sakata, K.; Takeuchi, R. *J. Am. Chem. Soc.* 2012, 134, 10515–10531.
255 (a) Wang, C., Li, X., Wu, F., and Wan, B. (2011). *Angew. Chem. Int. Ed.* 50: 7162–7166. (b) Wang, C., Wang, D., Xu, F. et al. (2013). *J. Org. Chem.* 78: 3065–3072. (c) Spahn, N.A., Nguyen, M.H., Renner, J. et al. (2017). *J. Org. Chem.* 82: 234–242.
256 Saito, A., Hironaga, M., Oda, S., and Hanzawa, Y. (2007). *Tetrahedron Lett.* 48: 6852–6855.
257 (a) Parthasarathy, K., Jeganmohan, M., and Cheng, C.H. (2008). *Org. Lett.* 10: 325–328. (b) Parthasarathy, K. and Cheng, C.-H. (2009). *Synthesis* 1400–1402.
258 (a) Hyster, T.K. and Rovis, T. (2011). *Chem. Commun.* 47: 11846–11848. (b) Too, P.C., Noji, T., Lim, Y.J. et al. (2011). *Synlett* 2789–2794.
259 Martin, R.M., Bergman, R.G., and Ellman, J.A. (2012). *J. Org. Chem.* 77: 2501–2507.
260 Chen, S., Bergman, R.G., and Ellman, J.A. (2015). *Org. Lett.* 17: 2567–2569.
261 (a) Xu, F., Wang, C., Wang, D. et al. (2013). *Chem. Eur. J.* 19: 2252–2255. (b) Xu, F., Wang, C., Wang, H. et al. (2015). *Green Chem.* 17: 799–803.
262 Yoshida, Y., Kurahashi, T., and Matsubara, S. (2012). *Chem. Lett.* 41: 1498–1499.

263 (a) Neely, J.M. and Rovis, T. (2013). *J. Am. Chem. Soc.* 135: 66–69. (b) Neely, J.M. and Rovis, T. (2014). *J. Am. Chem. Soc.* 136: 2735–2738.
264 Colby, D.A., Bergman, R.G., and Ellman, J.A. (2008). *J. Am. Chem. Soc.* 130: 3645–3651.
265 Kiselyov, A.S. (1995). *Tetrahedron Lett.* 36: 9297–9300.
266 Yamakawa, T. and Yoshikai, N. (2013). *Org. Lett.* 15: 196–199.
267 Wu, J., Xu, W., Yu, Z.-X., and Wang, J. (2015). *J. Am. Chem. Soc.* 137: 9489–9496.
268 Sim, Y.-K., Lee, H., Park, J.-W. et al. (2012). *Chem. Commun.* 48: 11787–11789.
269 Wang, P., Hua, R., and Li, C.-J. (2013). *Curr. Org. Synth.* 10: 655–660.
270 Sasada, T., Sakai, N., and Konakahara, T. (2008). *J. Org. Chem.* 73: 6905–6908.
271 (a) Vessally, E., Hosseinian, A., Edjlali, L. et al. (2016). *RSC Adv.* 6: 71662–71675. (b) Arshadi, S., Vessally, E., Edjlali, L. et al. (2017). *RSC Adv.* 7: 13198–13211.
272 (a) Cacchi, S., Fabrizi, G., and Filisti, E. (2008). *Org. Lett.* 10: 2629–2632. (b) Martins, M.A.P., Rossatto, M., Frizzo, C.P. et al. (2013). *Tetrahedron Lett.* 54: 847–849. (c) Xin, X., Wang, D., Li, X., and Wan, B. (2013). *Tetrahedron* 69: 10245–10248. (d) Karabiyikoglu, S., Kelgokmen, Y., and Zora, M. (2015). *Tetrahedron* 71: 4324–4333. (e) Cheng, G., Weng, Y., Yang, X., and Cui, X. (2015). *Org. Lett.* 17: 3790–3793.
273 (a) Bagley, M.C., Dale, J.W., and Bower, J. (2001). *Synlett* 1149–1151. (b) Bagley, M.C., Dale, J.W., Hughes, D.D. et al. (2001). *Synlett* 1523–1526. (c) Bagley, M.C., Lunn, R., and Xiong, X. (2002). *Tetrahedron Lett.* 43: 8331–8334. (d) Bagley, M.C., Glover, C., and Merritt, E.A. (2007). *Synlett* 2459–2482.
274 Shi, Z. and Loh, T.-P. (2013). *Angew. Chem. Int. Ed.* 52: 8584–8587.
275 Xie, L.-G., Shaaban, S., Chen, X., and Maulide, N. (2016). *Angew. Chem. Int. Ed.* 55: 12864–12867.
276 (a) Janz, G.J. and Wait, S.C. Jr., (1954). *J. Am. Chem. Soc.* 76: 6377–6379. (b) Janz, G.J. and McCulloch, W.J.H. (1955). *J. Am. Chem. Soc.* 77: 3143–3145.
277 Ohashi, M., Takeda, I., Ikawa, M., and Ogoshi, S. (2011). *J. Am. Chem. Soc.* 133: 18018–18021.
278 Chen, Z., Zhu, J., Xie, H. et al. (2010). *Org. Lett.* 12: 4376–4379.
279 Yamamoto, S.-i., Okamoto, K., Murakoso, M. et al. (2012). *Org. Lett.* 14: 3182–3185.
280 Kim, D.-S., Park, J.-W., and Jun, C.-H. (2012). *Chem. Commun.* 48: 11334–11336.
281 He, Z., Dobrovolsky, D., Trinchera, P., and Yudin, A.K. (2013). *Org. Lett.* 15: 334–337.
282 Zhang, Q.-R., Huang, J.-R., Zhang, W., and Dong, L. (2014). *Org. Lett.* 16: 1684–1687.
283 Reddy, C.R., Panda, S.A., and Reddy, M.D. (2015). *Org. Lett.* 17: 896–899.
284 Sha, F. and Huang, X. (2009). *Angew. Chem. Int. Ed.* 48: 3458–3461.
285 Bagley, M.C., Dale, J.W., and Bower, J. (2002). *Chem. Commun.* 1682–1683.
286 Movassaghi, M. and Hill, M.D. (2006). *J. Am. Chem. Soc.* 128: 4592–4593.
287 Shen, J., Cai, D., Kuai, C. et al. (2015). *J. Org. Chem.* 80: 6584–6589.

288 Gati, W., Rammah, M.M., Rammah, M.B. et al. (2012). *J. Am. Chem. Soc.* 134: 9078–9081.
289 Xiong, X., Bagley, M.C., and Chapaneri, K. (2004). *Tetrahedron Lett.* 45: 6121–6124.
290 Sharma, R., Kour, P., and Kumar, A. (2018). *J. Chem. Sci.* 130: 73.
291 Korivi, R.P. and Cheng, C.-H. (2006). *J. Org. Chem.* 71: 7079–7082.
292 Midya, S.P., Sahoo, M.K., Landge, V.G. et al. (2015). *Nat. Commun.* 6: 8591.
293 Xu, X., Yang, Y., Zhang, X., and Yi, W. (2018). *Org. Lett.* 20: 566–569.
294 Phanindrudu, M., Wakade, S.B., Tiwari, D.K. et al. (2018). *J. Org. Chem.* 83: 9137–9143.
295 Kong, L., Yu, S., Zhou, X., and Li, X. (2016). *Org. Lett.* 18: 588–591.
296 Sandelier, M.J. and DeShong, P. (2007). *Org. Lett.* 9: 3209–3212.
297 Huo, Z., Gridnev, I.D., and Yamamoto, Y. (2010). *J. Org. Chem.* 75: 1266–1270.
298 Saunthwal, R.K., Patel, M., and Verma, A.K. (2016). *Org. Lett.* 18: 2200–2203.
299 Zhang, L., Zheng, L., Guo, B., and Hua, R. (2014). *J. Org. Chem.* 79: 11541–11548.
300 Selig, P. and Raven, W. (2014). *Org. Lett.* 16: 5192–5195.
301 (a) Fischer, D., Tomeba, H., Pahadi, N.K. et al. (2007). *Angew. Chem. Int. Ed.* 46: 4764–4766. (b) Fischer, D., Tomeba, H., Pahadi, N.K. et al. (2008). *J. Am. Chem. Soc.* 130: 15720–15725.
302 Niu, Y.-N., Yan, Z.-Y., Gao, G.-L. et al. (2009). *J. Org. Chem.* 74: 2893–2896.
303 Huo, Z. and Yamamoto, Y. (2009). *Tetrahedron Lett.* 50: 3651–3653.
304 Gao, H. and Zhang, J. (2009). *Adv. Synth. Catal.* 351: 85–88.
305 Hwang, S., Lee, Y., Lee, P.H., and Shin, S. (2009). *Tetrahedron Lett.* 50: 2305–2308.
306 He, L., Nie, H., Qiu, G. et al. (2014). *Org. Biomol. Chem.* 12: 9045–9053.
307 (a) Huang, Q., Hunter, J.A., and Larock, R.C. (2001). *Org. Lett.* 3: 2973–2976. (b) Huang, Q., Hunter, J.A., and Larock, R.C. (2002). *J. Org. Chem.* 67: 3437–3444.
308 (a) Dai, G. and Larock, R.C. (2001). *Org. Lett.* 3: 4035–4038. (b) Dai, G. and Larock, R.C. (2003). *J. Org. Chem.* 68: 920–928.
309 Dai, G. and Larock, R.C. (2002). *J. Org. Chem.* 67: 7042–7047.
310 Sakamoto, T., Numata, A., and Kondo, Y. (2000). *Chem. Pharm. Bull.* 48: 669–672.
311 Alfonsi, M., Dell'Acqua, M., Facoetti, D. et al. (2009). *Eur. J. Org. Chem.* 2852–2862.
312 (a) Roesch, K.R., Zhang, H., and Larock, R.C. (1998). *J. Org. Chem.* 63: 5306–5307. (b) Roesch, K.R., Zhang, H., and Larock, R.C. (2001). *J. Org. Chem.* 66: 8042–8051. (c) Roesch, R.C. and Larock, R. (2002). *J. Org. Chem.* 67: 86–94.
313 Korivi, R.P. and Cheng, C.-H. (2005). *Org. Lett.* 7: 5179–5182.
314 Guimond, N. and Fagnou, K. (2009). *J. Am. Chem. Soc.* 131: 12050–12051.
315 Parthasarathy, K. and Cheng, C.-H. (2009). *J. Org. Chem.* 74: 9359–9364.
316 (a) Chinnagolla, R.K., Pimparkar, S., and Jeganmohan, M. (2012). *Org. Lett.* 14: 3032–3035. (b) Kornhaaß, C., Li, J., and Ackermann, L. (2012). *J. Org. Chem.* 77: 9190–9198.

317 Zheng, L., Ju, J., Bin, Y., and Hua, R. (2012). *J. Org. Chem.* 77: 5794–5800.

318 Too, P.C., Wang, Y.-F., and Chiba, S. (2010). *Org. Lett.* 12: 5688–5691.

319 Ju, J. and Hua, R. (2013). *Curr. Org. Synth.* 10: 328–332.

320 Huang, Q. and Larock, R.C. (2003). *J. Org. Chem.* 68: 980–988.

321 Wang, X. and Yu, X. (2014). *J. Org. Chem.* 79: 7854–7860.

322 Zhao, D., Lied, F., and Glorius, F. (2014). *Chem. Sci.* 5: 2869–2873.

323 (a) Ohtaka, M., Nakamura, H., and Yamamoto, Y. (2004). *Tetrahedron Lett.* 45: 7339–7341. (b) Asao, N., Yudha, S., Nogami, T., and Yamamoto, Y. (2005). *Angew. Chem. Int. Ed.* 44: 5526–5528. (c) Asao, N., Iso, K., and Yudha, S. (2006). *Org. Lett.* 8: 4149–4151.

324 Hamama, W.S., Waly, M., El-Hawary, I., and Zoorob, H.H. (2014). *Synth. Commun.* 44: 1730–1759.

325 Duong, H.A., Cross, M.J., and Louie, J. (2004). *J. Am. Chem. Soc.* 126: 11438–11439.

326 Duong, H.A. and Louie, J. (2005). *J. Organomet. Chem.* 690: 5098–5104.

327 Kondo, T., Nomura, M., Ura, Y. et al. (2006). *Tetrahedron Lett.* 47: 7107–7111.

328 Oberg, K.M., Lee, E.E., and Rovis, T. (2009). *Tetrahedron* 65: 5056–5061.

329 (a) Tanaka, K., Wada, A., and Noguchi, K. (2005). *Org. Lett.* 7: 4737–4739. (b) Tanaka, K., Takahashi, Y., Suda, T., and Hirano, M. (2008). *Synlett* 1724–1728.

330 Ghosez, L., Jnoff, E., Bayard, P. et al. (1999). *Tetrahedron* 55: 3387–3400.

331 Su, Y., Zhao, M., Han, K. et al. (2010). *Org. Lett.* 12: 5462–5465.

332 Hyster, T.K. and Rovis, T. (2011). *Chem. Sci.* 2: 1606–1610.

333 Ackermann, L., Lygin, A.V., and Hofmann, N. (2011). *Org. Lett.* 13: 3278–3281.

334 Yu, Y., Huang, L., Wu, W., and Jiang, H. (2014). *Org. Lett.* 16: 2146–2149.

335 Poomathi, N., Perumala, P.T., and Ramakrishna, S. (2017). *Green Chem.* 19: 2524–2529.

336 Bai, H., Sun, R., Chen, X. et al. (2018). *ChemistrySelect* 3: 4635–4638.

337 Shiro, T., Fukaya, T., and Tobe, M. (2015). *Eur. J. Med. Chem.* 97: 397–408.

338 Jia, C., Piao, D., Kitamura, T., and Fujiwara, Y. (2000). *J. Org. Chem.* 65: 7516–7522.

339 Tang, D.-J., Tang, B.-X., and Li, J.-H. (2009). *J. Org. Chem.* 74: 6749–6755.

340 Tang, B.-X., Song, R.-J., Wu, C.-Y. et al. (2011). *Chem. Sci.* 2: 2131–2134.

341 Xie, P., Wang, Z.-Q., Deng, G.-B. et al. (2013). *Adv. Synth. Catal.* 355: 2257–2262.

342 Inamoto, K., Saito, T., Hiroya, K., and Doi, T. (2010). *J. Org. Chem.* 75: 3900–3903.

343 Kadnikov, D.V. and Larock, R.C. (2004). *J. Org. Chem.* 69: 6772–6780.

344 Ferguson, J., Zeng, F., Alwis, N., and Alper, H. (2013). *Org. Lett.* 15: 1998–2001.

345 Manikandan, R. and Jeganmohan, M. (2014). *Org. Lett.* 16: 3568–3571.

346 Wang, W., Peng, X., Qin, X. et al. (2015). *J. Org. Chem.* 80: 2835–2841.

347 Hua, R. (2021). *Catalysts* 11: 620.

348 (a) Hyster, T.K. and Rovis, T. (2010). *J. Am. Chem. Soc.* 132: 10565–10569. (b) Mochida, S., Umeda, N., Hirano, K. et al. (2010). *Chem. Lett.* 39: 744–746.

349 Song, G., Chen, D., Pan, C.-L. et al. (2010). *J. Org. Chem.* 75: 7487–7490.

350 (a) Guimond, N., Gouliaras, C., and Fagnou, K. (2010). *J. Am. Chem. Soc.* 132: 6908–6909. (b) Guimond, N., Gorelsky, S.I., and Fagnou, K. (2011). *J. Am. Chem. Soc.* 133: 6449–6457.
351 Li, B., Feng, H., Xu, S., and Wang, B. (2011). *Chem. Eur. J.* 17: 12573–12577.
352 Too, P.C. and Chiba, S. (2012). *Chem. Commun.* 48: 7634–7636.
353 Zhang, M., Zhang, H.-J., Ruan, W., and Wen, T.-B. (2015). *Eur. J. Org. Chem.* 27: 5914–5918.
354 Khan, D.M. and Hua, R. (2020). *Catalysts* 10: 683.
355 Yang, L. and Hua, R. (2013). *Chem. Lett.* 42 (769): 771.
356 Su, J., Liu, H., and Hua, R. (2015). *Int. J. Mol. Sci.* 16: 3599–3608.
357 (a) Cavalieri, L.F. (1947). *Chem. Rev.* 41: 525–584. (b) Goel, A. and Ram, V.J. (2009). *Tetrahedron* 65: 7865–7913. (c) Lee, J.S. (2015). *Mar. Drugs* 13: 1581–1620.
358 (a) Thakur, A., Singla, R., and Jaitak, V. (2015). *Eur. J. Med. Chem.* 101: 476–495. (b) Medina, F.G., Marrero, J.G., Macías-Alonso, M. et al. (2015). *Nat. Prod. Rep.* 32: 1472–1507. (c) Kubrak, T., Podgórski, R., and Stompor, M. (2017). *Eur. J. Clin. Exp. Med.* 15: 169–175. (d) Stefanachi, A., Leonetti, F., Pisani, L. et al. (2018). *Molecules* 23: 250. (e) Molnar, M., Lončarić, M., and Kovač, M. (2020). *Curr. Org. Chem.* 24: 4–43.
359 (a) Barry, R.D. (1964). *Chem. Rev.* 64: 229–260. (b) Saikia, P. and Gogoi, S. (2018). *Adv. Synth. Catal.* 360: 2063–2075.
360 (a) Gaspar, A., Silva, T., Yáñez, M. et al. (2011). *J. Med. Chem.* 54: 5165–5173. (b) Keri, R.S., Budagumpi, S., Pai, R.K., and Balakrishna, R.G. (2014). *Eur. J. Med. Chem.* 78: 340–374. (c) Gaspar, A., Matos, M.J., Garrido, J. et al. (2014). *Chem. Rev.* 114: 4960–4992. (d) Carlos, F.M., Silva, C.F.M., Batista, V.F. et al. (2018). *Expert Opin. Drug Discov.* 13: 795–798. (e) Kumla, D., Pereira, J.A., Dethoup, T. et al. (2018). *Mar. Drugs* 16: 289. (f) Vavsari, V.F. and Balalaie, S. (2020). *Chem. Heterocycl. Compd.* 56: 404–407.
361 Ryabukhin, D.S. and Vasilyev, A.V. (2016). *Russ. Chem. Rev.* 85: 637–665.
362 Selected reports, see: (a) Nakamura, M., Endo, K., and Nakamura, E. (2003). *J. Am. Chem. Soc.* 125: 13002–13003. (b) Kuninobu, Y., Kawata, A., and Takai, K. (2005). *Org. Lett.* 7: 4823–4825. (c) Hanédanian, M., Loreau, O., Sawicki, M., and Taran, F. (2005). *Tetrahedron* 61: 2287–2294. (d) Endo, K., Hatakeyama, T., Nakamura, M., and Nakamura, E. (2007). *J. Am. Chem. Soc.* 129: 5264–5271. (e) Onodera, G., Kato, M., Kawano, R. et al. (2009). *Org. Lett.* 11: 5038–5041.
363 Zhao, W.-G. and Hua, R. (2007). *Tetrahedron* 63: 11803–11808.
364 Larock, R.C., Han, X., and Doty, M.J. (1998). *Tetrahedron Lett.* 39: 5713–5716.
365 Larock, R.C., Doty, M.J., and Han, X. (1999). *J. Org. Chem.* 64: 8770–8779.
366 Fukuyama, T., Higashibeppu, Y., Yamaura, R., and Ryu, I. (2007). *Org. Lett.* 9: 587–589.
367 Hua, R. and Tanaka, M. (2001). *New J. Chem.* 25: 179–184.
368 Wang, Y. and Burton, D.J. (2006). *J. Org. Chem.* 71: 3859–3862.
369 (a) Ueura, K., Satoh, T., and Miura, M. (2007). *J. Org. Chem.* 72: 5362–5367. (b) Ueura, K., Satoh, T., and Miura, M. (2007). *Org. Lett.* 9: 1407–1409.

370 Shimizu, M., Hirano, K., Satoh, T., and Miura, M. (2009). *J. Org. Chem.* 74: 3478–3483.

371 Mochida, S., Hirano, K., Satoh, T., and Miura, M. (2009). *J. Org. Chem.* 74: 6295–6298.

372 Mo, J., Wang, L., and Cui, X. (2015). *Org. Lett.* 17: 4960–4963.

373 Luo, T., Dai, M., Zheng, S.-L., and Schreiber, S.L. (2011). *Org. Lett.* 13: 2834–2836.

374 Itoh, M., Shimizu, M., Hirano, K. et al. (2013). *J. Org. Chem.* 78: 11427–11432.

375 Selected reviews, see: (a) Braunstein, P., Matt, D., and Nobel, D. (1988). *Chem. Rev.* 88: 747–764. (b) Aresta, M. and Dibeneddetto, A. (2007). *Dalton Trans.* 2975–2992. (c) Sakakura, T., Choi, J.C., and Yasuda, H. (2007). *Chem. Rev.* 107: 2365–2387. (d) Wang, J.L., Miao, C.X., Dou, X.Y. et al. (2011). *Curr. Org. Chem.* 15: 621–646. (e) Tsuji, Y. and Fujihara, T. (2012). *Chem. Commun.* 48: 9956–9964. (f) Yu, B. and He, L.-N. (2015). *ChemSusChem* 8: 52–62. (g) Liu, Q., Wu, L., Jackstell, R., and Beller, M. (2015). *Nat. Commun.* 6: 5933.

376 Inoue, Y., Itoh, Y., and Hashimoto, H. (1977). *Chem. Lett.* 855–856.

377 (a) Tsuda, T., Morikawa, S., Sumiya, R., and Saegusa, T. (1988). *J. Org. Chem.* 53: 3140–3145. (b) Tsuda, T., Morikawa, S., Hasegawa, N., and Saegusa, T. (1990). *J. Org. Chem.* 55: 2978–2981.

378 Louie, J., Gibby, J.E., Farnworth, M.V., and Tekavec, T.N. (2002). *J. Am. Chem. Soc.* 124: 15188–15189.

379 Tekavec, T.N., Arif, A.M., and Louie, J. (2004). *Tetrahedron* 60: 7431–7437.

380 (a) Zhao, Y., Liu, Y., Bi, S., and Liu, Y. (2014). *J. Organomet. Chem.* 758: 45–54. (b) Ishii, M., Mori, F., and Tanaka, K. (2014). *Chem. Eur. J.* 20: 2169–2174. (c) Oliveros-Cruz, S., Arévalo, A., and García, J.J. (2017). *J. Organomet. Chem.* 831: 18–22. (d) Hou, J., Ee, A., Feng, W. et al. (2018). *J. Am. Chem. Soc.* 140: 5257–5263.

381 (a) Tsuda, T., Maruta, K.-i., and Kitaike, Y. (1992). *J. Am. Chem. Soc.* 114: 1499–1500. (b) Tsuda, T. and Maruta, K.-i. (1992). *Macromolecules* 25: 6102–6105. (c) Tsuda, T., Ooi, O., and Maruta, K.-i. (1993). *Macromolecules* 26: 4840–4844. (d) Tsuda, T., Yasukawa, H., and Komori, K. (1996). *Macromolecules* 28: 1356–1359.

382 Reddy, M.S., Thirupathi, N., Babu, M.H., and Puri, S. (2013). *J. Org. Chem.* 78: 5878–5888.

383 Fu, W., Zhu, M., Zou, G. et al. (2015). *J. Org. Chem.* 80: 4766–4770.

384 Zhao, Y., Han, F., Yang, L., and Xia, C. (2015). *Org. Lett.* 17: 1477–1480.

385 Mi, X.; Wang, C.; Huang, M.; Wu, Y.-s.; Wu, Y.-j. *J. Org. Chem.* 2015, 80, 148–155.

386 Yan, K., Yang, D., Wei, W. et al. (2015). *J. Org. Chem.* 80: 1550–1556.

387 Yu, Y., Zhuang, S., Liu, P., and Sun, P. (2016). *J. Org. Chem.* 81: 11489–11495.

388 Ueura, K., Satoh, T., and Miura, M. (2007). *Org. Lett.* 9: 1407–1409.

389 Chinnagolla, R.K. and Jeganmohan, M. (2012). *Chem. Commun.* 48: 2030–2032.

390 Kajita, Y., Kurahashi, T., and Matsubara, S. (2008). *J. Am. Chem. Soc.* 130: 17226–17227.

391 Speraŋca, A., Godoi, B., Pinton, S. et al. (2011). *J. Org. Chem.* 76: 6789–6797.

392 Wang, Y.-H., Qiu, G., Zhou, H. et al. (2019). *Tetrahedron* 75: 3850–3855.

393 Luo, M.-J., Hu, M., Song, R.-J. et al. (2019). *Chem. Commun.* 55: 1124–1127.

394 (a) Saengchantara, S.T. and Wallace, T.W. (1986). *Nat. Prod. Rep.* 3: 465–475. (b) Sosnovskikh, V.Y. (2003). *Russ. Chem. Rev.* 72: 489–516. (c) Keri, R.S., Budagumpi, S., Pai, R.K., and Balakrishna, R.G. (2014). *Eur. J. Med. Chem.* 78: 340–374.

395 Zhou, C., Dubrovsky, A.V., and Larock, R.C. (2006). *J. Org. Chem.* 71: 1626–1632.

396 Shimizu, M., Tsurugi, H., Satoh, T., and Miura, M. (2008). *Chem. Asian J.* 3: 881–886.

397 Yoshida, M., Fujino, Y., and Doi, T. (2011). *Org. Lett.* 13: 4526–4529.

398 Chang, M.-Y., Chen, Y.-H., and Wang, H.-S. (2018). *J. Org. Chem.* 83: 2361–2368.

399 Matsuda, T., Kadowaki, S., and Murakami, M. (2007). *Chem. Commun.* 2627–2629.

400 Matsuda, T., Kadowaki, S., Yamaguchi, Y., and Murakami, M. (2010). *Org. Lett.* 12: 1056–1058.

401 (a) Rhoden, C.R.B. and Zeni, G. (2011). *Org. Biomol. Chem.* 9: 1301–1313. (b) Elsherbini, M., Hamama, W.S., and Zoorob, H.H. (2016). *Coord. Chem. Rev.* 312: 149–177.

402 (a) Alves, D., Luchese, C., Nogueira, C.W., and Zeni, G. (2007). *J. Org. Chem.* 72: 6726–6734. (b) Casola, K.K., Gomes, M.R., Back, D.F., and Zeni, G. (2018). *J. Org. Chem.* 83: 6706–6718.

403 Selected reports, see: (a) Nicolaou, K.C., Pfefferkorn, J.A., Roecker, A.J. et al. (2000). *J. Am. Chem. Soc.* 122: 9939–9953. (b) Pratap, R. and Ram, V.J. (2014). *Chem. Rev.* 114: 10476–10526. (c) Majumdar, N., Paul, N.D., Mandal, S. et al. (2015). *ACS Catal.* 5: 2329–2366. (d) Costa, M., Dias, T.A., Brito, A., and Proença, F. (2016). *Eur. J. Med. Chem.* 123: 487–507.

404 Zeng, H., Ju, J., and Hua, R. (2011). *Tetrahedron Lett.* 52: 3926–3928.

405 Su, J., Ju, J., and Hua, R. (2012). *Curr. Org. Synth.* 9: 273–277.

406 Romanov-Michailidis, F., Ravetz, B.D., Paley, D.W., and Rovis, T. (2018). *J. Am. Chem. Soc.* 140: 5370–5374.

407 Zhu, X.-F., Lan, J., and Kwon, O. (2003). *J. Am. Chem. Soc.* 125: 4716–4717.

408 Wender, P.A., Pedersen, T.M., and Scanio, M.J.C. (2002). *J. Am. Chem. Soc.* 124: 15154–15155.

409 Zhou, M.-B., Song, R.-J., Wang, C.-Y., and Li, J.-H. (2013). *Angew. Chem. Int. Ed.* 52: 10805–10808.

410 Li, J., Yang, Y., Wang, Z. et al. (2017). *Org. Lett.* 19: 3083–3086.

411 Seoane, A., Casanova, N., Quiñones, N. et al. (2014). *J. Am. Chem. Soc.* 136: 834–837.

412 Lee, D.J., Han, H.K., Shin, J., and Yoo, E.J. (2014). *J. Am. Chem. Soc.* 136: 11606–11609.

413 Liang, L., Li, E., Dong, X., and Huang, Y. (2015). *Org. Lett.* 17: 4914–4917.

414 Ni, H., Tang, X., Zheng, W. et al. (2017). *Angew. Chem. Int. Ed.* 56: 14222–14226.

415 Qian Chen, Q., Wang, Y., and Hua, R. (2019). *Molecules* 24: 3773.

9

Carbonyl Compounds from Alkynes and Alkenes

Carbonyl compounds are basic chemicals in organic synthesis and industry process. Carbonyl compounds can be prepared through the hydration of alkynes, the addition reactions of unsaturated hydrocarbons with simple carbonyl compounds, and the carbonylation of a variety of organic compounds with CO [1]. In this chapter, the economic and efficient synthesis of carbonyl compounds through hydration of alkynes, hydroformylation, hydroacylation, hydroamidation, hydrocarboxylation, and cyclocarbonylation of alkynes and/or alkenes are described.

9.1 Hydration of Alkynes

Hydration or hydrolysis of unsaturated hydrocarbons is an important chemical process to produce oxygen-containing organic compounds, and usually the hydration of alkenes and alkynes [2] under different conditions affords alcohols and carbonyl compounds, respectively.

The hydration of alkynes is one of the simple and atom economic ways for the synthesis of ketones and aldehydes with the use of water as a green reagent. Therefore, there have developed a variety of catalyst systems for this type of transformation.

As shown in Scheme 9.1, the hydration of terminal alkyne can theoretically produce either methyl ketone (Markovnikov-type hydration) or aldehyde (*anti*-Markovnikov-type hydration).

It is a great challenge to develop the catalytic systems or the reaction conditions to realize the high regioselectivity to give either of the carbonyl compounds. The formation of methyl ketones via Markovnikov hydration of terminal alkynes has been extensively developed.

For terminal alkyne Markovnikov hydration, mercury(II) salts are the well-used catalysts with long history in the presence of either Brønsted or Lewis acid [3]. Recent studies have disclosed that other metal complexes, such as cobalt [4], copper [5], silver [6], palladium [7], gold [8], ruthenium [9], rhodium [10], tin [11], platinum [12], iridium [13], iron [14], and bismuth [15] are also the efficient catalysts for the regioselective hydration of terminal alkynes approach to methyl ketones.

Recently, Naka and coworker have developed the water-soluble Co (III) porphyrin-catalyzed hydration of terminal alkynes to afford methyl ketones in

Addition Reactions with Unsaturated Hydrocarbons, First Edition. Ruimao Hua.

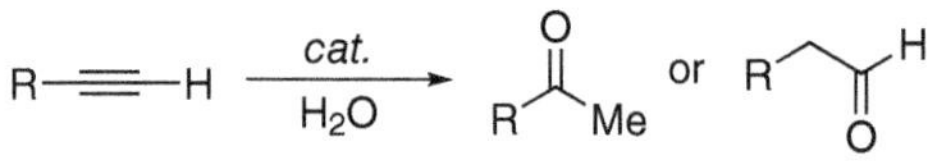

Scheme 9.1 Hydration of terminal alkynes.

good to excellent yields (Scheme 9.2) [16]. The advantages of the procedure are compatible with the presence of acid/base- or redox-sensitive functional groups, such as ethers, carboxylic esters, boronic esters, carboxamides, nitriles, and nitro and acetal groups. In addition, the catalyst could be recovered by aqueous workup. Very interestingly, a water-mediated metal/catalyst/reagent-free hydration of terminal alkynes under temperature and pressure controlled condition afforded methyl ketones in high yields, when alkynes were heated at 150 °C under 11 bar pressure in water–methanol solution [17].

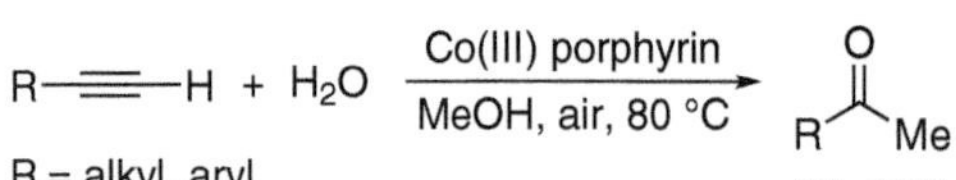

Scheme 9.2 Hydration of terminal alkynes affording methyl ketones catalyzed by water-soluble Co(III) porphyrin complexes.

The microwave-assisted copper salt-catalyzed [18], acid-catalyzed [19], and visible light-promoted CuCl-catalyzed [20] rapid hydration of aromatic terminal alkynes to afford acetophenones have been well developed.

Metal-free Markovnikov-type hydration of terminal alkynes to give methyl ketones under mild conditions has also been investigated. For example, a catalytic amount of TfOH could catalyze the hydration of aromatic/aliphatic terminal alkynes, as well as internal alkynes with excellent regioselectivity in good to excellent yields with the use of CF_3CH_2OH as solvent [21]. Brønsted acidic ionic liquids [22], sulfonated condensed polynuclear aromatic resin in water [23], tropylium ion [24], the combination of a Brønsted acid catalyst and a supramolecular organic capsule [25], as well as other acids and acidic compounds [26] have been used as the catalysts.

In addition, the ketones generated in situ via the hydration of terminal alkynes have been applied for the further transformation of alkynes into another products, such as chiral alcohols [27], α-alkylated ketones [28], and 2-cyclohexenone [29].

α-Halomethyl ketones are the important and versatile intermediates in organic synthesis and many biologically active molecules containing the structure of α-halomethyl ketone moiety [30].

As shown in Scheme 9.3, an alternative synthetic method for α-halomethyl ketones was developed through the hydration of a wide range of alkyl- and aryl-substituted haloalkynes in the presence of $XPhosAuNTf_2$, and the procedure is compatible with a wide range of functional groups [31]. AgF/TFA [32] and $In(OTf)_3$/HOAc [33] have also been confirmed to be the efficient catalysts for the hydration of 1-haloalkynes to α-halomethyl ketones.

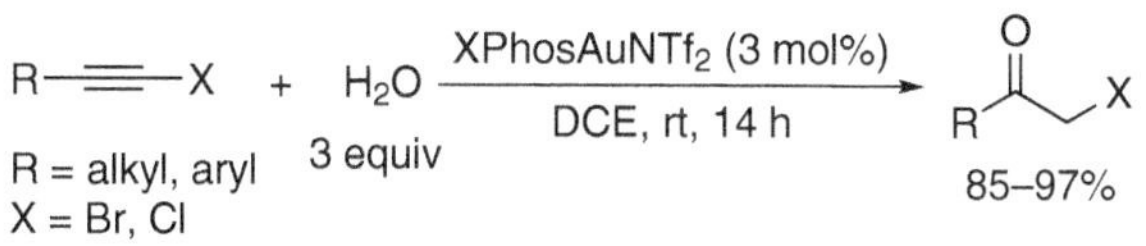

Scheme 9.3 Synthesis of α-halomethyl ketones via hydration of haloalkynes.

With the use of CF_3CH_2OH as solvent, a highly regioselective and efficient metal-free hydration of aromatic haloalkynes to α-halomethyl ketones was then developed by using HBF_4 as catalyst [34].

Compared with the formation of methyl ketones from the hydration of terminal alkynes, there are relative few reports on the regioselective *anti*-Markovnikov hydration of terminal alkynes to afford aldehydes.

The first *anti*-Markovnikov hydration of terminal alkynes was reported by Wakatsuki's group, and $[RuCl_2(C_6H_6)\{PPh_2(C_6F_5)\}]/PPh_2(C_6F_5)$ showed high regioselectivity in 2-propanol [35]. The formation of a ruthenium vinylidene is proposed to be the key intermediate in the *anti*-Markovnikov hydration catalysis.

The ruthenium vinylidene intermediate has also been proposed in the $[Ru_3(dppm)_3Cl_5]PF_6$-catalyzed hydrative cyclization of 1,5-enynes bearing terminal alkyne and Michael acceptor moieties into cyclopentanone derivatives (Scheme 9.4) [36]. The proposed mechanism involves the formation of a ruthenium vinylidene, *anti*-Markovnikov hydration, and intramolecular Michael addition of an acyl ruthenium to the alkene.

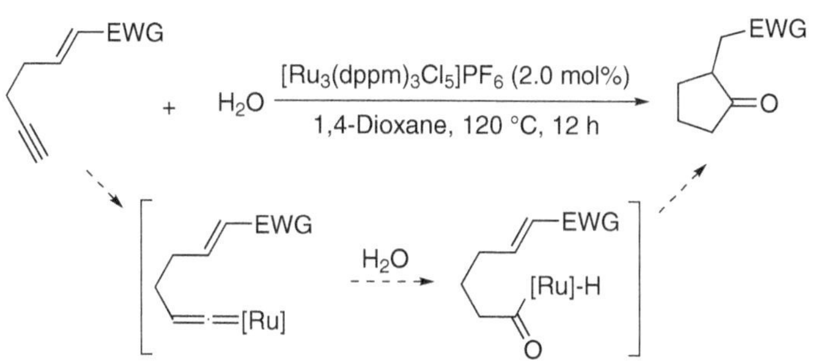

Scheme 9.4 Cyclopentanone formation via an *anti*-Markovnikov hydration step of terminal alkynes.

Other ruthenium complexes with different ligands or with the use of additives have also been confirmed to be the efficient catalysts for the *anti*-Markovnikov hydration of terminal alkynes to produce aldehydes with good regioselectivity [37].

On the other hand, the hydration of internal alkynes is also the useful route to prepare the carbonyl compounds, and the platinum [38], gold complexes [39] have been applied in this transformation. For example, very recently, the synthesis of six-membered 1,4-*P*,*O*-heterocycles has been reported through the hydration/intramolecular cyclization of the internal alkynes of dialkynylphosphine oxides catalyzed by gold(I) complexes [40].

It is also a challenge to establish the efficient catalytic systems for the high regioselective hydration of asymmetric internal alkynes to produce one ketone product. As shown in Scheme 9.5, $PdCl_2$ was found to be an efficient catalyst for the hydration of alkynylphosphonates to give β-ketophosphonates in high yields in 1,4-dioxane without any additive [41].

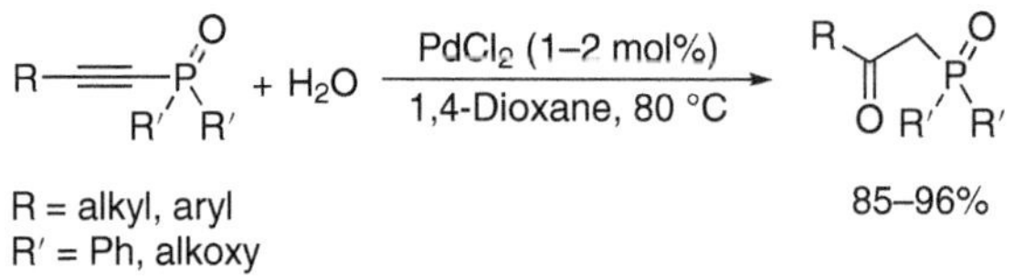

Scheme 9.5 $PdCl_2$-catalyzed hydration of alkynylphosphonate to β-ketophosphonates.

Similarly, the hydration of internal alkynes bearing a strong electron-withdrawing group of CF_3 or SF_5 occurred with excellent regioselectivity to produce the corresponding trifluoromethylated and pentasulfanylated ketones as single regioisomer in the presence of AuCl(JohnPhos)/AgOTf, showing that CF_3 and SF_5 groups are the highly efficient directing groups in this transformation (Scheme 9.6) [42].

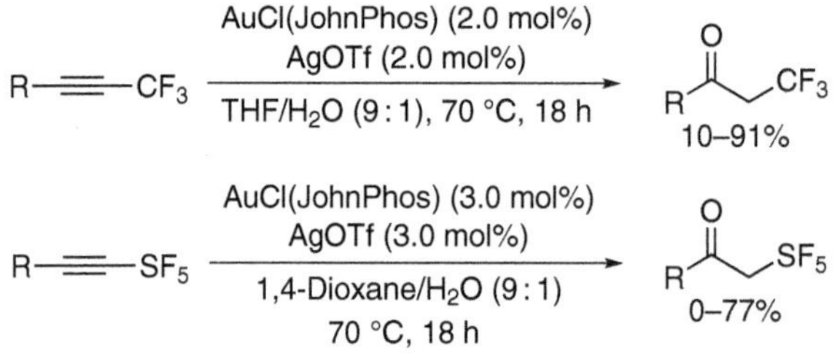

Scheme 9.6 AuCl(JohnPhos)/AgOTf-catalyzed regioselective hydration of internal alkynes bearing CF_3 and SF_5 groups.

Han and coworker have found that CuBr and *p*-fluoroaniline have an excellent synergistic effect in catalyzing the hydration of aromatic internal alkynes without directing groups to α-aryl ketones with regioselectivity up to more than 90% (Scheme 9.7) [43]. It has demonstrated that the formation of α-aryl ketone and aryl ketone is promoted by different catalytic active species, CuBr and CuBr[*p*-fluoroaniline], respectively. CuBr can enlarge the difference of electron population on the two triple-bond carbon atoms, which results in α-aryl ketone formation as the main product.

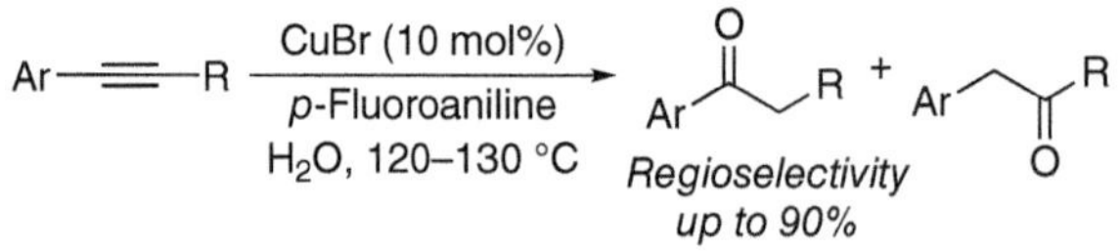

Scheme 9.7 Hydration of aromatic internal alkynes to α-aryl ketones.

Very recently, the carbonyl-directed regioselective hydration of ynones at room temperature has been developed to establish a facile and efficient synthesis of 1,3-diketones in the presence of $AuCl(PPh_3)/AgOTf$ [44]. At the same time, Verma's group developed a metal-free carbonyl-assisted regioselective hydration of *o*-alkynylaldehydes to afford 1,5-dicarbonyls, which could be then used in the construction of pyridine ring (Scheme 9.8) [45].

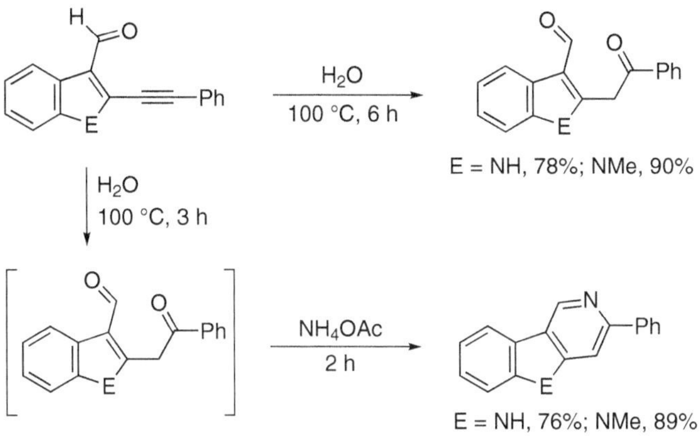

Scheme 9.8 Metal-free carbonyl-assisted hydration of ortho-alkynylaldehydes affording 1,5-dicarbonyls.

$Fe_2(SO_4)_3{\cdot}nH_2O/AcOH$ catalytic system has also been used to develop a highly regioselective hydration of aryl alkyl internal alkynes giving aryl ketones under weak acidic conditions with good functional group compatibility [46].

In addition, a variety of the recoverable heterogeneous catalysts have been applied in the hydration of alkynes, such as TiO_2-supported nanosize gold particles [47]; protonic, Ce^{3+}, and Ca^{2+} exchanged Y zeolites [48]; silver exchanged silicotungstic acid (AgSTA) [49]; encapsulation of *N*-heterocyclic carbine (NHC)-gold (I) catalysts [50], and AgOAc/ionic liquid [51].

9.2 Hydroformylation of Alkynes and Alkenes

Hydroformylation of alkynes and alkenes with syngas (CO/H_2) provides the straight and economic ways for the synthesis of α,β-unsaturated aldehydes and aliphatic aldehydes (Scheme 9.9).

9.2.1 Hydroformylation of Alkynes

Compared to the hydroformylation of alkenes, the reported procedures for alkyne hydroformylation are relatively few, and the early examples of the selective hydroformylation of alkynes to yield unsaturated aldehydes have been summarized in several review papers [52].

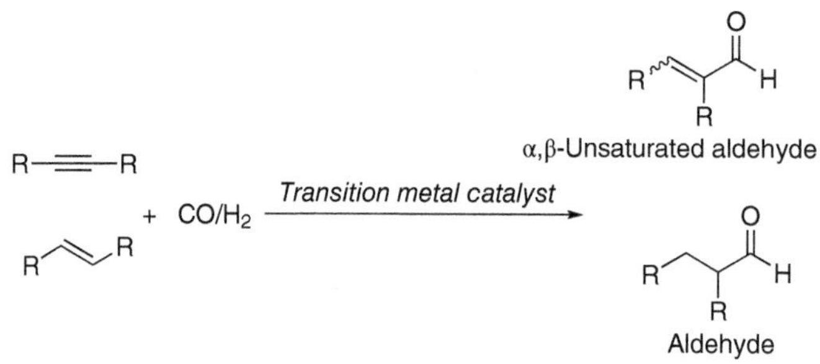

Scheme 9.9 Hydroformylation of alkynes and alkenes.

Rhodium-catalyzed hydroformylation of alkynes usually produces not only the desired α,β-unsaturated aldehydes but also the hydrogenated products of the corresponding saturated aldehyde or/and alkene, as well as other carbonylated compounds, and the chemoselectivity depends greatly on the natures of alkynes and the reaction conditions [53]. A pioneering work on the highly chemoselective rhodium-catalyzed hydroformylation of internal alkynes giving α,β-unsaturated aldehydes was reported by Buchwald's group with the use of $Rh(acac)(CO)_2$/diphosphite ligands [54].

Besides rhodium complexes, under the total initial pressure of 6.0 MPa, at 200 °C, a combination of $Ni(OAc)_2{\cdot}4H_2O$, $CuBr_2$, $MeSO_3H$, and PPh_3 could also catalyze the hydroformylation of acetylene to acrylic acid with 90% conversion of acetylene and 90% selectivity to acrylic acid [55]. The $[Pd(acac)_2]$/bisphosphine/*p*-TsOH catalyst system was reported for selective hydroformylation of alkynes to α,β-unsaturated aldehydes, and the competing hydrogenation side reactions were almost completely suppressed [56].

Hydroformylation of an alkyne in preference to an alkene in enynes was reported. For example, in the presence of $RhH(CO)(PPh_3)_3$ and PPh_3, the reaction of 2-methylhex-1-en-3-yne with CO/H_2 led to the formation of formyl-buta-1,3-dienes [57]. Similarly, the reaction of 1-en-3-ynes with syngas in the presence of zwitterionic rhodium complex and triphenyl phosphite afforded formyl dienes in moderate to good yields with high regio- and stereoselectivities [58]. The same catalyst system was used in the regioselective hydroformylation of acetylenic thiophenes to produce α,β-unsaturated aldehyde with the aldehyde and thiophene attached to the same alkene carbon atom [59].

9.2.2 Hydroformylation of Alkenes

The hydroformylation of terminal alkenes can afford two aldehydes: linear aldehyde and branched aldehyde (Scheme 9.10), which was first developed by Roelen in the 1930s [60]. Afterward, it has been extensively investigated, and has become one of the most important homogeneously transition metal-catalyzed industrial processes [61]. Most often cobalt [62], rhodium complexes [63], and other transition metal complexes have been well applied [64].

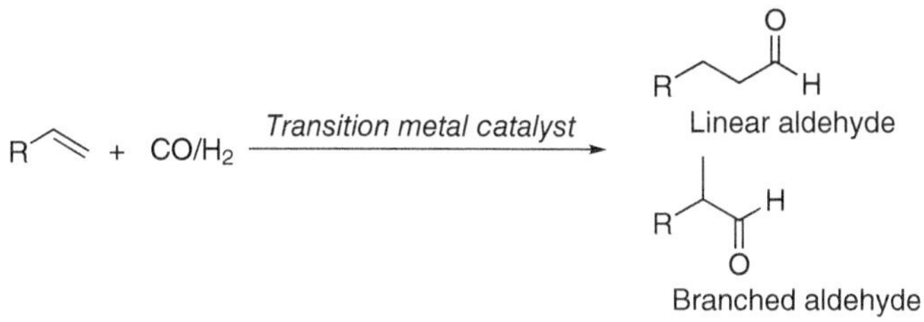

Scheme 9.10 Hydroformylation of alkenes.

Recently, the metal nanoparticles or their supported materials, such as cobalt [65], rhodium [66], oxide-supported rhodium single atom catalysts [67], and simple rhodium black [68] have also been applied in the hydroformylation reactions of alkynes and alkenes.

Very recently, rhodium-catalyzed aqueous biphasic alkene hydroformylation promoted by amphiphilic cyclodextrins [69] and water-soluble rhodium complexes [70] have been developed.

In addition, the hydroformylation of alkenes could also be achieved with the use of paraformaldehyde [71], $HCOOH/Ac_2O$ [72], and alkyl aldehyde [73] as CO/H_2 source.

Moreover, rhodium-catalyzed hydroformylation of 1,1-disubstituted allenes [74], cyclopentadiene [75], and butadiene [76] have been also reported.

On the other hand, as the application examples of alkene hydroformylation, very recently, Chiou and coworker have reported the synthesis of azabicyclic structures via a domino hydroformylation of terminal alkenes and the intramolecular double cyclization strategy to provide a rapid and atom economic access to alkaloid structures under mild conditions with good yields (Scheme 9.11) [77].

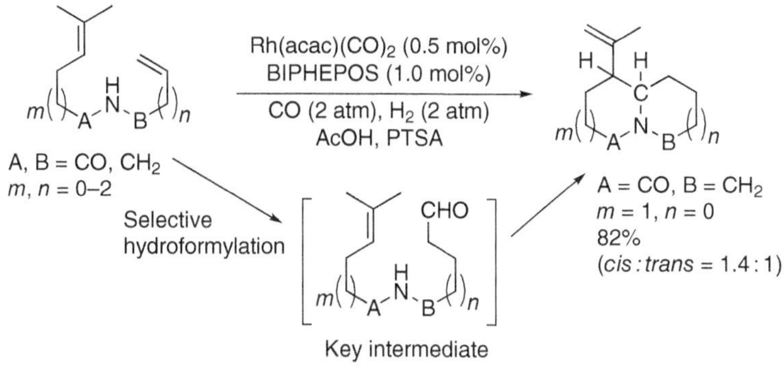

Scheme 9.11 Rhodium-catalyzed hydroformylation-initiated bicyclization affording azabicyclic structures.

The asymmetric hydroformylation of alkenes can afford chiral aldehydes, which has been widely applied in the synthesis of chiral compounds [78].

The asymmetric $Rh(acac)(CO)_2$/chiral ligand-catalyzed *anti*-Markovnikov hydroformylation of 1,1-disubstituted allylic alcohols and amines, and subsequent oxidation with pyridinium chlorochromate (PCC) provided an efficient route to chiral lactones and lactams with good yields and high enantioselectivities (Scheme 9.12) [79].

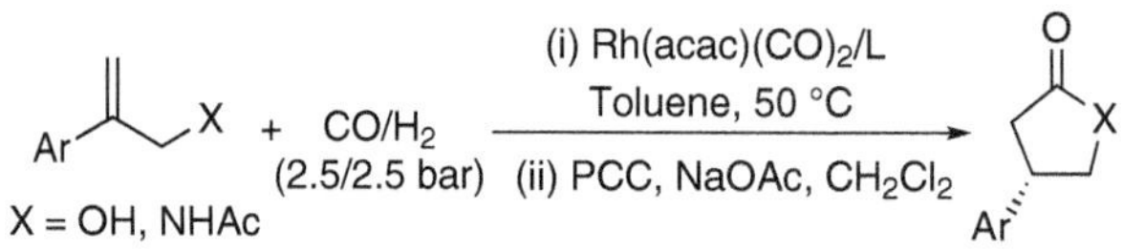

Scheme 9.12 Synthesis of chiral lactones and lactams including *anti*-Markovnikov hydroformylation.

9.3 Hydroacylation of Alkynes and Alkenes

Intermolecular hydroacylation of alkynes and alkenes with aldehydes via direct activation of C—H bond is one of the most efficient synthetic protocols for preparing α,β-unsaturated ketones and aliphatic ketones (Scheme 9.13). The intramolecular hydroacylation of alkynals and alkenals can produce the corresponding cyclic ketones [80].

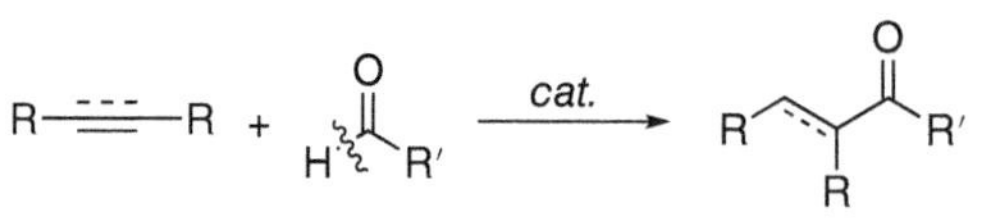

Scheme 9.13 Hydroacylation of alkynes and alkenes via aldehyde addition.

In an early work, a catalyst system of $[RhCl(cod)]_2$/dppf/Na_2CO_3 was used in the hydroacylation of alkynes, alkenes, and allenes with 2-hydroxybenzaldehydes via cleavage of the aldehyde C—H bond [81]. Jun's group then developed a highly regioselective $Rh(PPh_3)_3Cl$-catalyzed Markovnikov intermolecular hydroacylation of terminal alkynes with aldehydes in the presence of the chelation-assisted additives of 2-amino-3-picoline and benzoic acid [82]. Only in the case of 3,3-dimethylbutyne used, the *anti*-Markovnikov addition occurred to give *E*-α,β-enone with excellent stereoselectivity.

Jamison and coworker developed a $Ni(cod)_2/PBu_3/BEt_3$-catalyzed *cis*-intermolecular reductive addition of aldehydes to alkynes, providing an efficient route for the synthesis of allylic alcohols (Scheme 9.14) [83]. The rhodium-catalyzed similar procedure has also been reported by Krische and coworker [84].

Willis's group has developed several rhodium catalytic systems for the hydroacylation of alkynes with the use of β-*S*-substituted aldehydes, chiral aldehydes, β-carbonyl aldehydes, and β-amino-α,β-unsaturated aldehydes [85]. Scheme 9.15 shows a highly selective forming the linear enaminone products as single regioisomer via the enamine-controlled intermolecular hydroacylation of alkynes in the presence of $[Rh(nbd)_2]BF_4$/dppe. Hydrolysis of the obtained products in situ generates α-substituted 1,3-diketone products [86].

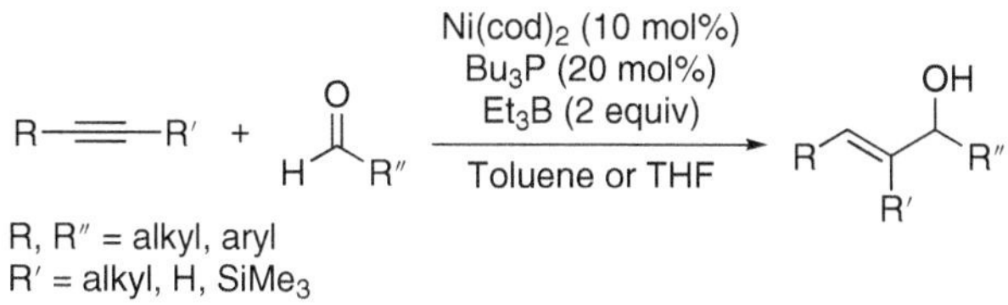

Scheme 9.14 Nickel-catalyzed intermolecular reductive addition of aldehydes to alkynes.

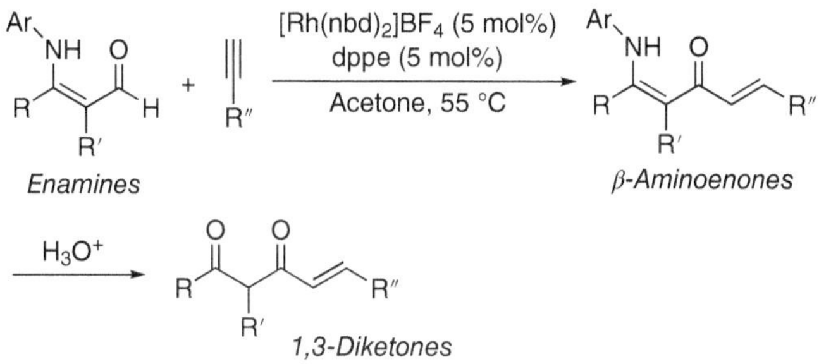

Scheme 9.15 Rhodium-catalyzed hydroacylation of enaminone and its hydrolysis.

The same group has also applied the hydroacylation of alkynes in the synthesis of heterocyclic compounds [87] and enantioselective three-component assembly of β′-aryl enones [88].

A variety of rhodium catalysts have been successfully used in the intermolecular hydroacylation of alkynes [89]. Ruthenium [90], gold [91], Lewis acid of $In(OTf)_3$ [92] also show high catalytic activity for this transformation.

In addition, a variety of catalyst systems have also been developed for the intermolecular hydroacylation of alkenes with various aldehydes, such as rhodium [93], cobalt [94], ruthenium [95], nickel [96], NHC [97], organic photocatalysts [98], and hypervalent iodine (III) reagent under visible light [99]. Particularly, the synthesis of linear ketones via highly regioselective *anti*-Markovnikov hydroacylation of alkenes with aldehydes has been well developed in the presence of rhodium(I) complexes (Scheme 9.16) [100].

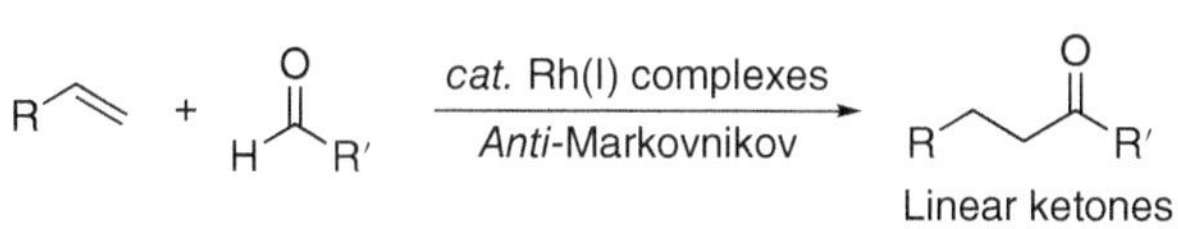

Scheme 9.16 Rhodium(I)-catalyzed synthesis of linear ketones via highly regioselective *anti*-Markovnikov hydroacylation of alkenes with aldehydes.

It has been also reported that under solvent-free conditions, the intermolecular hydroacylation of terminal alkenes with aldehydes is efficiently catalyzed by Wilkinson catalyst under microwave irradiation [101].

In addition, the first intermolecular hydroacylation of 1,3-dienes with aldehydes to give β,γ-unsaturated ketones catalyzed by $Ru(cod)(cot)/PPh_3$ was reported by

Mitsudo's group [102], and then other ruthenium complexes [103] and various catalyst systems with the use of cobalt [104] complexes have been developed for this transformation. Moreover, the stereoselective hydroacylation of bicyclic alkenes has been reported in the presence of iridium complexes [105].

On the other hand, the intramolecular hydroacylation of alkynals and alkenals has been extensively investigated for the construction of different sizes of cyclic ketones, and also rhodium(I) complexes have shown the best choice to catalyze this type of transformation [106]. Nickel complexes [107]; simple organic compounds, such as *N*-heterocyclic carbenes [108]; and phosphines [109] have also been applied as the catalysts in the intramolecular hydroacylation of alkynals.

Nickel [110], cobalt [111], calcium(II) salt or triflimide [112], and *N*-heterocyclic carbenes [113] showed high catalytic activity in the intramolecular hydroacylation of alkenals.

Also rhodium complexes could be used in the intramolecular hydroacylation of allenal [114] and dienals [115].

Scheme 9.17 shows a transition metal-free, thiol-catalyzed intramolecular hydroacylation of alkenals. It involves the thiol-catalyzed generation of acyl radicals and their intramolecular addition to carbon–carbon double bond of alkenals to produce two-substituted five- and six-membered cyclic ketones in good yields. Aldehydes having electron-deficient alkenes cyclizes more easily than those having unactivated alkenes [116].

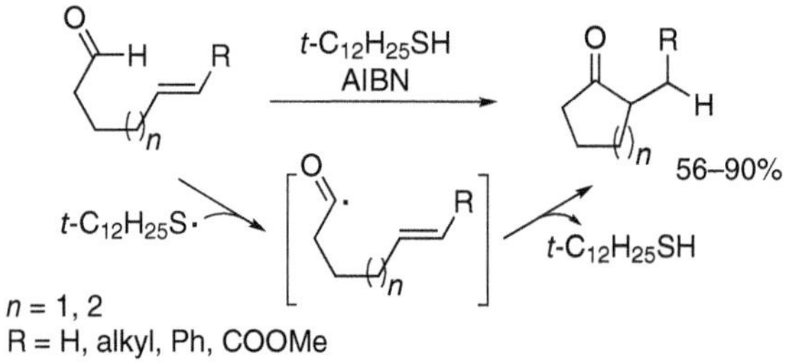

Scheme 9.17 Thiol-catalyzed acyl radical cyclization of alkenals.

Rhodium(I)-catalyzed intramolecular hydroacylation of alkynals could be applied in the annulations with alkynes to furnish unsaturated cyclic ketones, presumably via rhodacycle intermediates [117]. For example, in the presence of [Rh(dppe)]BF_4, 4-alkynals undergo a [4+2] annulations with alkynes giving cyclohexenones (Scheme 9.18) [118].

The intramolecular hydroacylation of ω-alkenals and alkynals containing a sulfur tether atom could be applied in the synthesis of medium-ring sulfur-heterocyclic ketones catalyzed by rhodium complex [119]. The similar protocol has been used in the enantioselective synthesis of seven- and eight-membered heterocyclic ketones with high regio- and enantiocontrol catalyzed by rhodium complexes [120]. As shown in Scheme 9.19, both α- and β-substituted cyclic ketones could be obtained, depending on catalyst choice and substrate structure, and ethers, sulfides, and sulfoxides function as effective directing groups. The medium-ring nitrogen

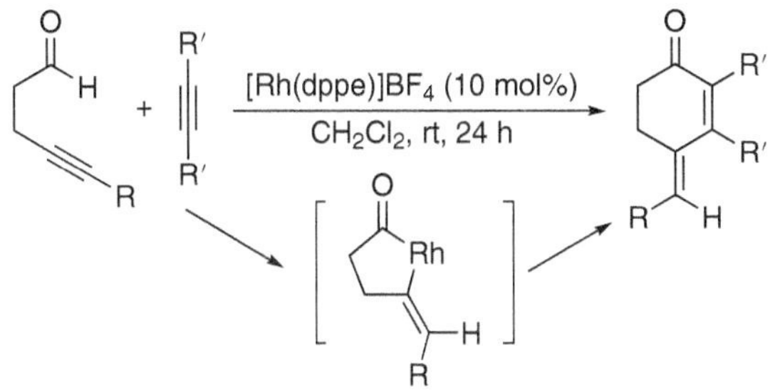

Scheme 9.18 Cyclohexenones from rhodium(I)-catalyzed annulation of alkynals with alkynes.

heterocycles could also be prepared via an amine-directed, rhodium-catalyzed intramolecular hydroacylation of alkenes and alkynes [121].

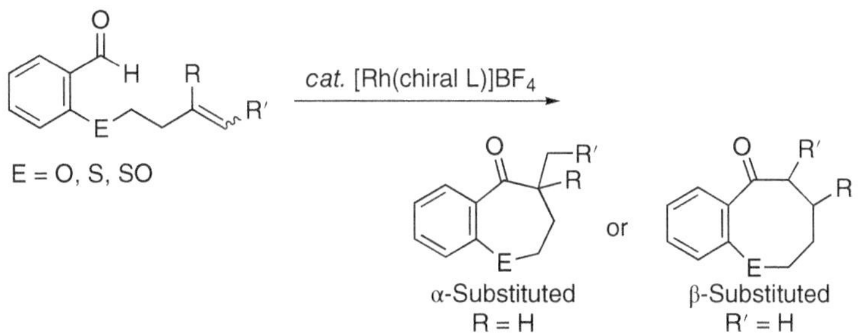

Scheme 9.19 Enantioselective intramolecular hydroacylation of alkenes affording heterocyclic ketones.

The radical intermolecular and intramolecular hydroacylations of alkenes with the use of Di-*t*-butyl peroxide (DTBP) or *t*-Butyl hydroperoxide (TBHP) as the radical initiators have been developed by Lee's group [122]. When 2-(allyloxy) benzaldehydes were treated with TBHP, 4-chromanones could be obtained, and the formation of acyl radical was proposed as the key intermediate (Scheme 9.20).

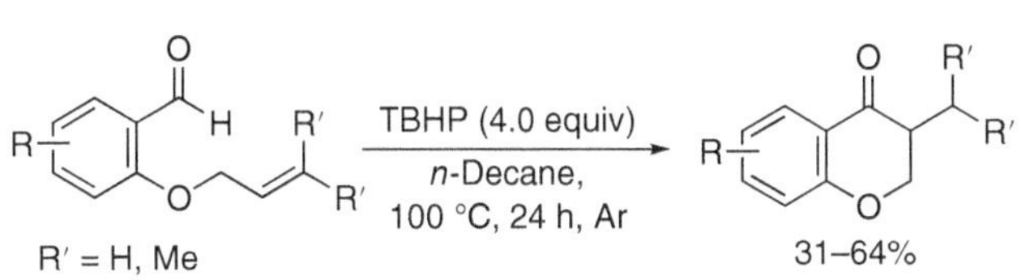

Scheme 9.20 4-Chromanones from TBHP-promoted selectively intramolecular hydroacylation.

On the other hand, the enantioselective intermolecular hydroacylation of alkenes and allenes has been reported to establish the synthetic methods approach to chiral ketones [123]. The enantioselective intramolecular hydroacylation of alkenals or alkenes has become one of the simple and efficient routes for the synthesis of chiral cyclic ketones, and recently, rhodium [124], cobalt [125] complexes with chiral ligands as catalysts, and chiral NHC as organocatalysts [126] have been well studied.

When cyclopropenes were used as the substrates, the enantioselective intermolecular hydroacylation could be realized to give cyclopropylketones with diastereocontrol and excellent enantiomeric excess catalyzed by *N*-heterocyclic carbenes [127] and rhodium complexes [128].

9.4 Hydroamidation of Alkynes and Alkenes

Similar to the hydroacylation of unsaturated hydrocarbons with aldehydes via the activation of RC(O)—H bond, the addition reactions of formamides to alkynes and alkenes via activation of $R_2NC(O)$—H bond are also the atom economic processes for the preparation of α,β-unsaturated amides and aliphatic amides (Scheme 9.21).

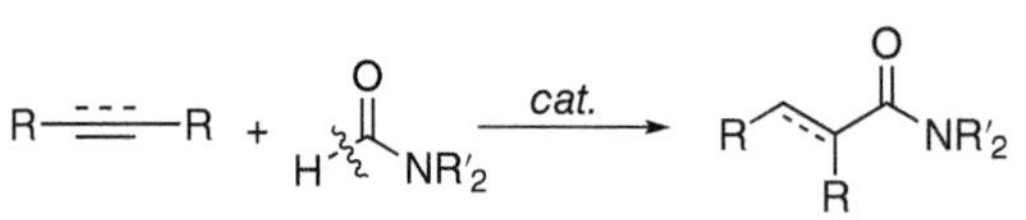

Scheme 9.21 Hydroamidation of alkynes and alkenes via formamide addition.

A pioneering work on the transition metal-catalyzed addition of *N*-substituted formamides to both terminal and internal alkenes was reported by Watanabe's group in 1987, with the use of $Ru_3(CO)_{12}$ as catalyst, under a CO pressure of 20 kg cm^{-2}, the addition of *N*-methylformamide to cyclopentene afforded *N*-methylcyclopentanecarboxamide in 90% yield [129]. As shown in Scheme 9.22, from the same group, Tsuji and coworker then reported a $PdCl_2(PhCN)_2$/phosphine-catalyzed *cis*-hydroamidation of internal alkynes with formamides using acid chloride as an additive to afford (*E*)-α,β-unsaturated amides with high regio- and stereoselectivity, and the same catalyst system shows good catalytic activity for the addition of formamides to terminal alkynes giving the corresponding α,β-unsaturated amides with excellent regioselectivity of Markovnikov adducts [130].

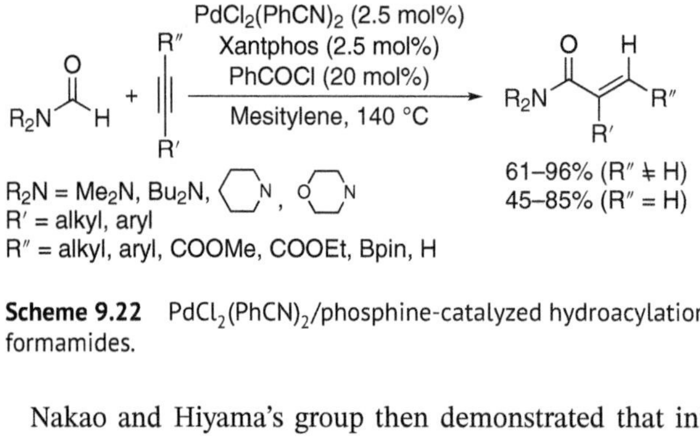

Scheme 9.22 $PdCl_2(PhCN)_2$/phosphine-catalyzed hydroacylation of alkynes with formamides.

Nakao and Hiyama's group then demonstrated that intermolecular hydroamidation of internal alkynes and 1,3-dienes could also be efficiently catalyzed cooperatively by $Ni(cod)_2$/phosphine and Lewis acids access to a range of α,β- and β,γ-unsaturated amides, and the intramolecular hydroamidation of alkenes gave γ-lactam derivatives [131].

Yamamoto and coworker also developed a cyclo-hydroamidation of 1,6-diynes with formamides in dimethylfomamide (DMF) to produce exocyclic-diene-type unsaturated amides with complete stereoselectivity in the presence of $[Cp^*Ru(MeCN)_3]PF_6$ (Scheme 9.23) [132].

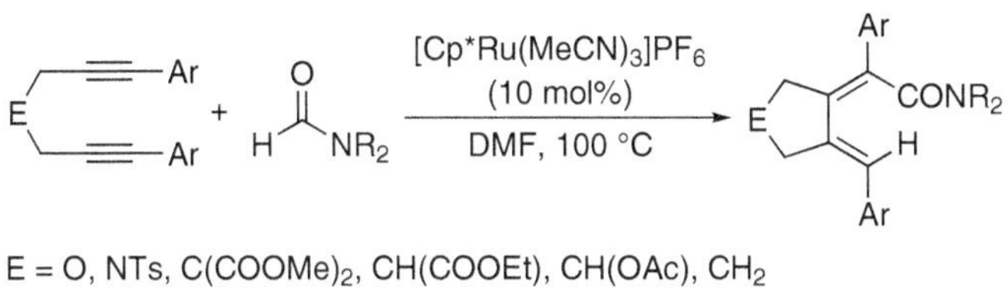

Scheme 9.23 Ruthenium-catalyzed cyclo-hydroamidation 1,6-diynes with formamides.

In the presence of $Rh_4(CO)_{12}$, the intramolecular hydroamidation of formamides having alkynyl group proceeded smoothly to provide an efficient synthetic route for the formation of α-alkylidene-γ-lactams or 3-alkylideneoxindoles in moderate to good yields with high (*E*)-selectivity (Scheme 9.24) [133].

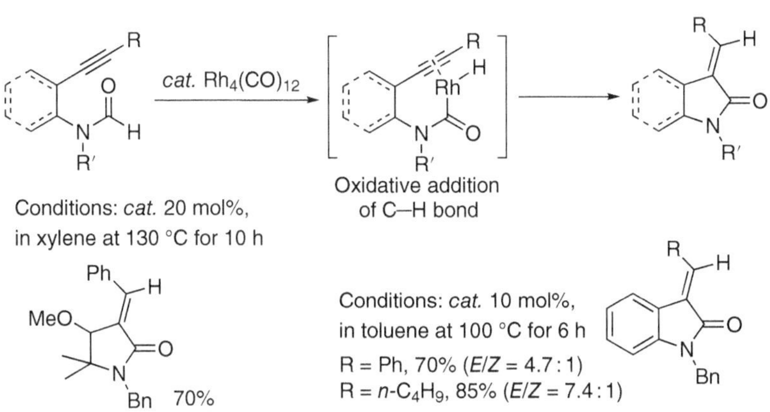

Scheme 9.24 Rhodium-catalyzed intramolecular hydroamidation of alkynylformamides.

On the other hand, for the hydroamidation of alkenes, Kondo and Mitsudo's group have developed a ruthenium-catalyzed hydroamidation of the excess amount of norbornene and ethylene with formamides offering a simple and practical method for synthesizing carboxamides (Scheme 9.25) [134]. $Ru_3(CO)_{12}$ was also used in intermolecular hydroamidation of terminal alkenes and norbornene with chelating formamide of *N*-(2-pyridyl)formamide [135]. In addition, $Ru_3(CO)_{12}$ catalyzes the intramolecular hydroamidation of allylic formamides to develop a convenient access to γ-lactams and chiral γ-lactams with the use of chiral ligand [136].

In a combination of $[Ni(cod)_2]$/NHC and $AlEt_3$, a regioselective hydroamidation of terminal alkenes with variously substituted formamides could also be achieved to afford a range of linear alkanamides with excellent *anti*-Markovnikov addition [137].

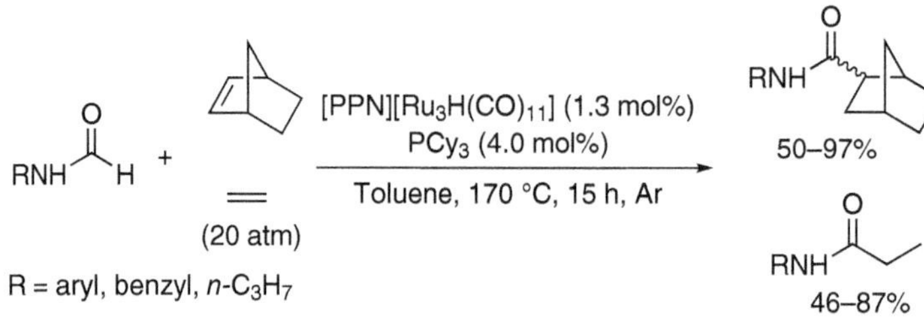

Scheme 9.25 Ruthenium-catalyzed hydroamidation of alkenes with formamides giving carboxamides.

9.5 Hydrocarboxylation of Alkynes and Alkenes

The hydrocarboxylation of alkynes and alkenes is one of the important and efficient protocols for the preparation of acrylic acids and aliphatic acids. Two main routes have been well investigated: (i) addition reaction with formic acid via the cleavage of C(sp^2)—H bond (Scheme 9.26), (ii) hydrocarboxylation with CO_2/H_2 or H_2O and CO/H_2O [138]. In this section, the development of the addition reactions of formic acid and its derivatives to unsaturated hydrocarbons is described.

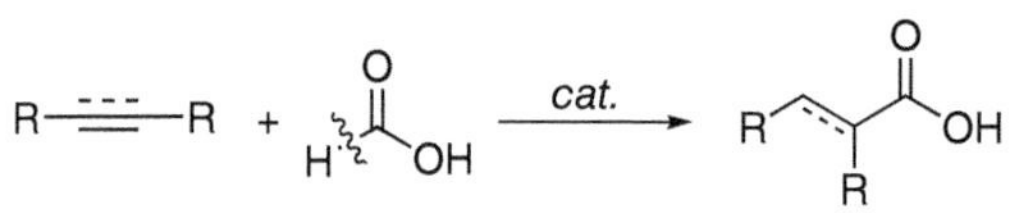

Scheme 9.26 Hydrocarboxylation of alkynes and alkenes via formic acid addition.

The first hydrocarboxylation of alkynes with formic acid was reported by Alper's group [139]. They found that $Pd(OAc)_2$/phosphine catalyzed hydrocarboxylation of alkynes with formic acid under CO atmosphere in 1,2-dimethoxyethane (DME), phenyl alkyne, and a linear alkyl-substituted terminal alkynes underwent the addition with high regioselectivity to give Markovnikov-adduct as a major product. Whereas the bulky group-substituted terminal alkynes, such as 3,3-dimethylbutyne and trimethylsilylacetylene, undergo *anti*-Markovnikov *cis*-addition reaction affording (*E*)-isomer (Scheme 9.27). The *cis*-addition reactions also occurred when internal alkynes were employed with low regioselectivity. Under the similar reaction conditions, the hydrocarboxylation of alkenes also occurred smoothly with the use of Pd/C and phosphine ligand [140]. Other palladium catalyst systems

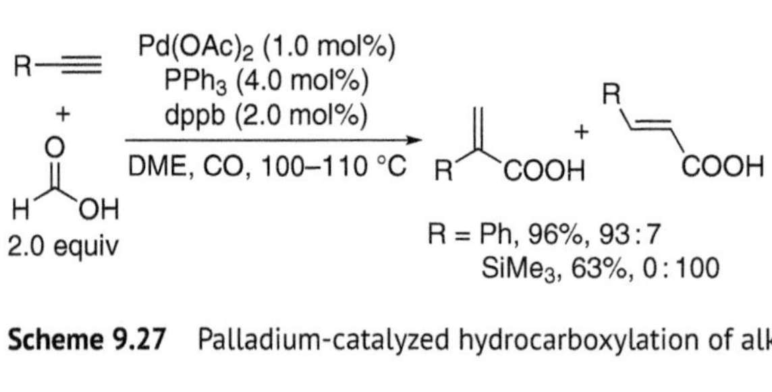

Scheme 9.27 Palladium-catalyzed hydrocarboxylation of alkyne with formic acid.

without the use of CO also showed highly catalytic activity for the same addition reactions [141].

The hydrocarboxylation of alkynes with formic acid catalyzed by nickel complexes [142], and the DFT calculations to elucidate the mechanistic details have also been studied [143].

In addition, the palladium-catalyzed hydrocarboxylation of alkenes with phenyl formate/formic acid, or with formic acid only to give propanoic acid derivatives [144], and hydrocarboxylation of allenes using formate salts as a hydride and a CO_2 source [145] have also been reported.

9.6 Hydroesterification of Alkynes and Alkenes

There are mainly two kinds of hydroesterification of unsaturated hydrocarbons based on the starting materials: (i) three-component coupling reaction of unsaturated hydrocarbon, CO, and alcohol; and (ii) addition reaction of formate esters to unsaturated hydrocarbon via activation of C(sp^2)—H bond (Scheme 9.28).

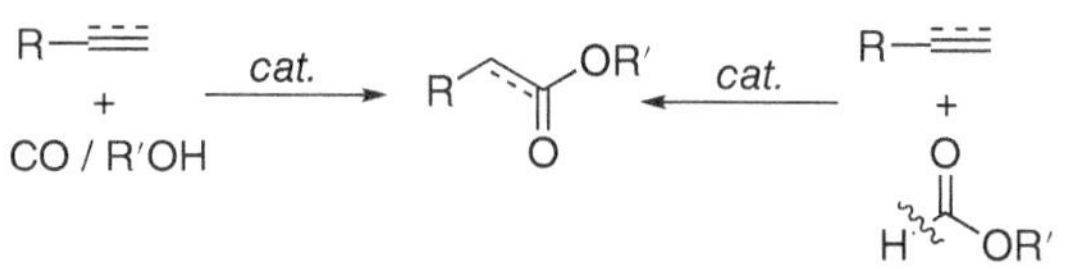

Scheme 9.28 Hydroesterification of alkynes and alkenes.

The pioneering work for the hydroesterification of alkenes and alkynes was reported by Alper and coworker in 1983 [146]. As shown in Scheme 9.29, in the presence of the catalytic amounts of $PdCl_2$ and $CuCl_2$, alkenes and alkynes underwent the hydroesterification at room temperature with CO and CH_3OH under oxygen atmosphere and HCl as the additive. The hydroesterification of 1-decene or *cis*- or *trans*-2-decene afforded methyl 2-methyldecanoate as the only product in

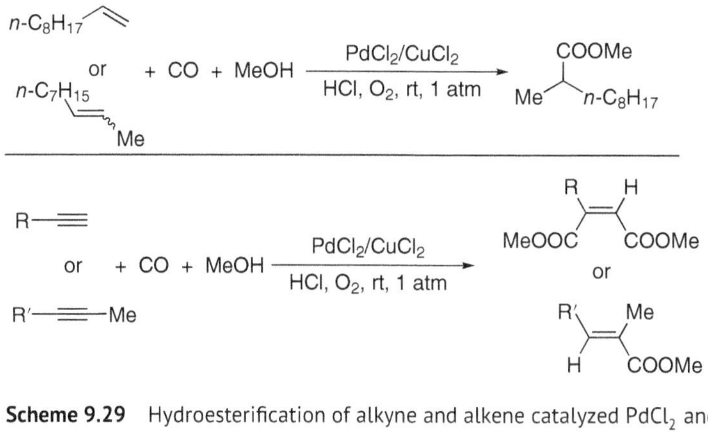

Scheme 9.29 Hydroesterification of alkyne and alkene catalyzed $PdCl_2$ and $CuCl_2$.

excellent yield. In the case of terminal alkynes used, the reactions gave (*Z*)-diesters as the major products with high regioselectivity and stereoselectivity. While the hydroesterification of 2-alkynes, the internal alkynes, gave (*E*)-monoesters, they also developed the hydroesterification of terminal alkenes with CO and alcohols by using other palladium complexes [147].

Since then, the hydroesterification of alkynes [148] and alkenes [149] with CO and alcohol catalyzed by palladium complexes under different conditions in other groups have been well investigated.

In addition, the hydroesterification of alkenes with CO and alcohol took place smoothly catalyzed by $Ru_3(CO)_{12}$/[Bmim]Cl [150] and platinum [151].

On the other hand, paraformaldehyde [152], HCO_2Na [153], and phenyl formate [154] have been used as CO surrogates in the hydroesterification of alkenes.

Alper's group also developed the hydroesterification of terminal and internal alkynes with formate esters to give α,β-unsaturated esters in good yields with the use of the catalytic system of $Pd(OAc)_2$/dppb/PPh_3/*p*-TsOH [155]. The similar hydroesterification of alkynes [156] and alkenes [157] with aryl formats catalyzed by palladium complexes was then reported by other groups.

In addition, $Ru_3(CO)_{12}$ showed high catalytic activity in the hydroesterification of alkyne, alkenes, and 1,3-dienes with 2-pyridylmethyl formate to give a diverse range of α,β-unsaturated esters, alkyl esters, and mono-alkenyl esters [158]. The $Ru_3(CO)_{12}$-catalyzed hydroesterification of allylic and homoallylic alcohols with 2-pyridylmethyl formate gave lactones [159], and hydroesterification of allylic amides involving alkene isomerization afforded δ-amido esters [160]. Other ruthenium complexes were also effective in the hydroesterification of alkenes with 2-pyridylmethyl formate [161].

Moreover, imidazole derivatives were found to be effective ligands in the $Ru_3(CO)_{12}$-catalyzed hydroesterification of alkenes with various benzyl, naphthylmethyl, and aryl formates [162].

9.7 Carbonylation of Alkynes and Alkenes

Transition metal-catalyzed carbonylation and cyclocarbonylation of alkynes and alkenes are the most important and useful reactions for the synthesis of carbonyl compounds, heterocycles, and natural products, as well as in the synthesis of industrial chemicals [163].

Several representative name-called reactions base on the carbonylation of unsaturated compounds have been well-known, such as Pauson–Khand reaction to generate cyclopentenone ring in a formal [2+2+1] cycloaddition of alkyne, alkene and CO, or enyne with CO [164]; Koch synthesis reaction to synthesize highly branched carboxylic acids from alkenes, H_2O, and CO [165]; Reppe carbonylation of alkenes, alkynes, and conjugated dienes with nucleophiles affording α,β-unsaturated carboxylic acids and carboxylic acids, as well as their derivatives of esters and amides [166].

9.7.1 Carbonylation of Alkynes

The outcomes of the carbonylation of alkynes depend greatly on the catalyst systems used and the nature of the alkynes, as well as the reaction conditions. Palladium complexes have been extensively applied in this transformation.

For example, Inoue and coworker developed a palladium-catalyzed hydroesterification of terminal alkynes with CO and methanol in MeCN affording either branched α,β-unsaturated ester or linear α,β-unsaturated ester as major products, depending on the use of phosphine ligand [167]. Other groups reported that the highly chemoselective formation of either methyl 2-phenylacrylate or methyl phenylpropiolate could be controlled under different conditions (Scheme 9.30). In the presence of $MeSO_3H$, $Pd(OAc)_2$/2-pyrimidyldiphenylphosphine could catalyze the carbonylation of phenyl acetylene in *N*-methyl pyrrolidone (NMP) and methanol with CO (60 bar) giving methyl 2-phenylacrylate with high regioselectivity [168], whereas $PdBr_2$ catalyzed the similar reaction in the presence of $CuBr_2$ and base affording methyl phenylpropiolate [169].

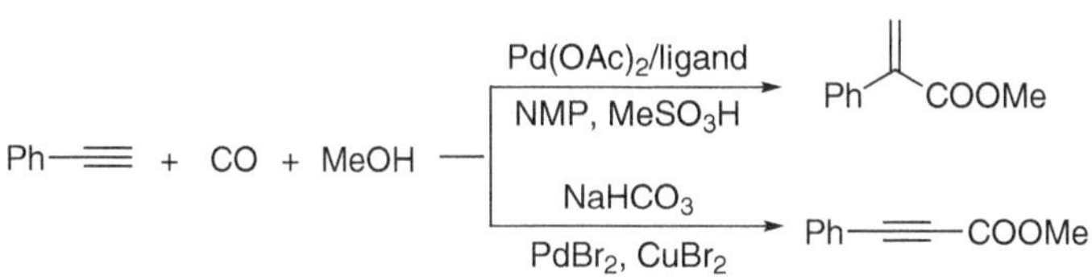

Scheme 9.30 Palladium-catalyzed carbonylation of terminal alkynes.

In addition, using catalytic amount of $Pd(OAc)_2$/dppp and excess amount of Zn, terminal and internal alkynes underwent the hydrophenoxycarbonylation affording α,β-unsaturated arylesters [170].

Chatani's group developed a $Rh_4(CO)_{12}$-catalyzed double hydroesterification of internal alkynes with CO and pyridin-2-yl methanol leading to the formation of 1,4-dicarboxylate esters (Scheme 9.31) [171]. The mechanism studies revealed that the double hydroesterification did not proceed via two consecutive hydroesterifications of alkynes, but the intermediacy of ketene intermediates, and the coordination of the pyridine nitrogen to rhodium was essential for the reaction to take place

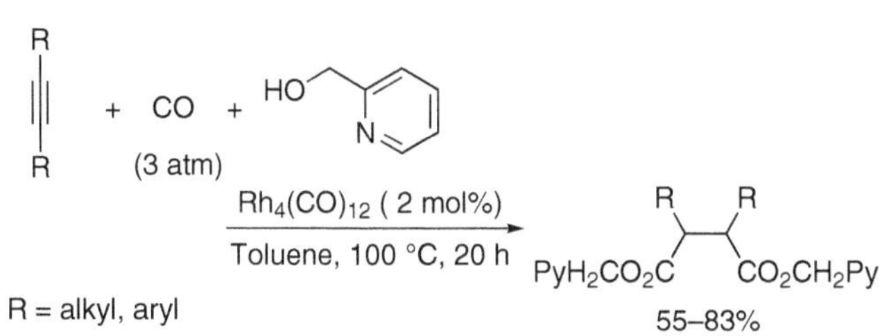

Scheme 9.31 Rhodium-catalyzed double hydroesterification of alkynes to 1,4-dicarboxylate esters.

Very recently, a $Pd(acac)_2$/ligand-catalyzed double hydroesterification of terminal and internal alkynes with methanol, without the use of the special alcohol , has been developed to afford 1,4-dicarboxylic acid diesters through two consecutive hydroesterifications of alkynes [172].

In the presence of a catalytic amount of PdX_2 and excess amount of CuX_2 (X = Cl, Br), the carbonylation of terminal alkynes with CO and various alcohols afforded the corresponding (*Z*)-β-haloacrylates exclusively in moderate to good yields (Scheme 9.32) [173]. In the case of phenyl acetylene used, methyl (*Z*)-β-chlorocinnamate could be obtained in 80%, and the reaction occurred with excellent regioselectivity and stereoselectivity. A proposed mechanism is depicted in Scheme 9.33; it involves the *cis*-addition of alkyne with PdX_2 to form *cis*-halopalladation intermediate, the insertion of CO into C—Pd bond, followed by alcoholysis to afford the products. The active Pd(II) species are regenerated by the oxidation reaction of Pd(0) with CuX_2 to start a new catalytic cycle.

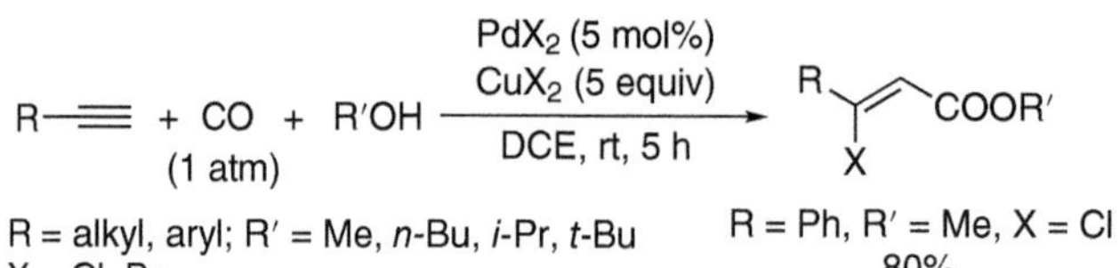

Scheme 9.32 Synthesis of β-haloacrylates via palladium-catalyzed carbonylation of terminal alkynes.

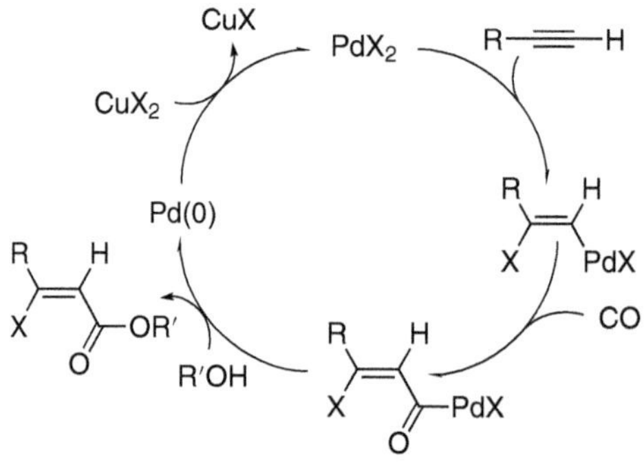

Scheme 9.33 Proposed mechanism for PdX_2-catalyzed carbonylation of terminal alkynes.

Other studies in the same group disclosed that $PdCl_2/CuCl_2$ could catalyze the dicarbonylation of terminal alkynes in H_2O/dioxane to give maleic anhydrides in high to excellent yields (Scheme 9.34) [174].

Yamamoto's group reported the highly selective formation of propiolates by the carbonylation of terminal alkynes catalyzed by $Pd(OAc)_2/PPh_3$ in DMF/MeOH under atmospheric pressure of CO and oxygen [175]. Their further studies revealed that the oxidative tricarbonylation as well as dicarbonylation took place via introducing two or three CO units into alkynes in the presence of iodides, such as NaI and NEt_4I [176].

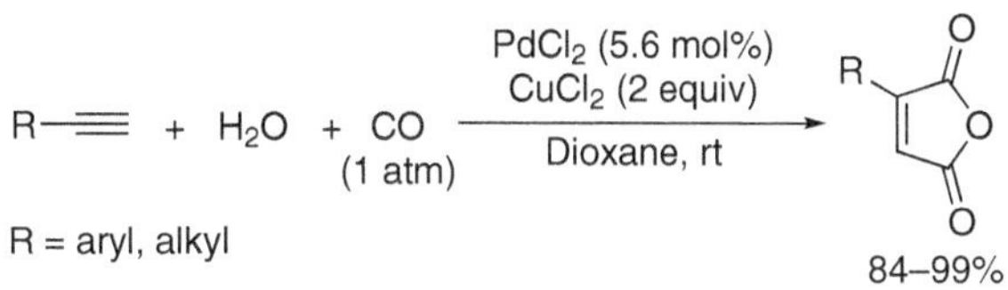

Scheme 9.34 $PdCl_2$-catalyzed dicarbonylation of terminal alkynes giving maleic anhydrides.

Transition metal-catalyzed three-component coupling of terminal alkynes, organo halides and CO represents a straightforward and efficient way for the synthesis of alkynyl ketones [177]. As an example of the recent development, under photoirradiation conditions, in the presence of Et_3N, $PdCl_2(PPh_3)_2$ catalyzed the coupling reaction of terminal alkynes, iodoalkanes, and CO to give alkyl alkynyl ketones in good yields (Scheme 9.35) [178].

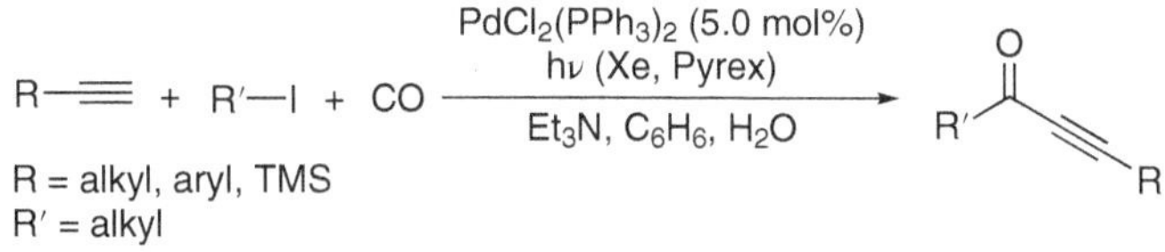

Scheme 9.35 Formation of alkyl alkynyl ketones via $PdCl_2(PPh_3)_2$-catalyzed coupling of terminal alkynes, iodoalkanes, and CO.

With the use of $Pd(PPh_3)Cl_2/P(OPh)_3$ as a catalyst system, the carbonylative coupling of benzyl chlorides with terminal alkynes gave 1,4-diaryl-3-butyn-2-ones [179]. IPrCuCl-catalyzed hydrocarbonylative C—C coupling of terminal alkynes with unactivated alkyl iodides afforded unsymmetrical dialkyl ketones [180].

α,β-Unsaturated ketones can be synthesized from the carbonylative coupling of aromatics, alkynes with CO featuring atom-economic process. Alper's group developed an interesting strategy for the synthesis of α,β-unsaturated ketones via the cationic palladium-catalyzed direct coupling of indoles, alkynes, and CO with linear selectivity (Scheme 9.36) [181].

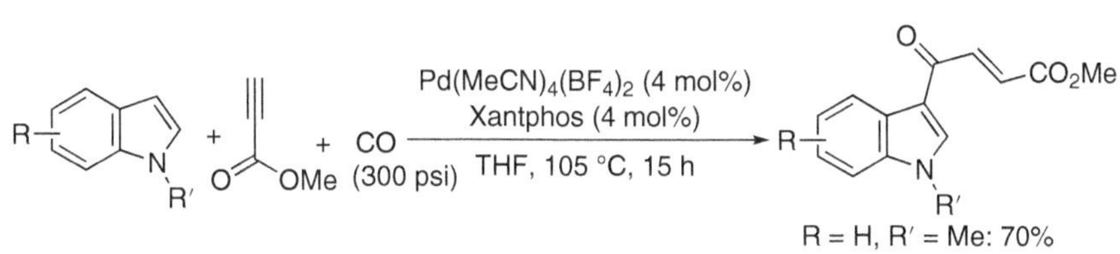

Scheme 9.36 Synthesis of α,β-unsaturated ketones via direct coupling of indoles, alkynes, and CO.

Other palladium complexes-catalyzed alkoxycarbonylation of terminal alkynes have also been reported [182].

9.7.2 Carbonylation of Alkenes

The carbonylative hydroarylation of alkenes via the addition of (hetero)aromatic C—H bond and CO to alkenes has become the atom- and step-economic transformation for the synthesis of aromatic ketones (Scheme 9.37) [183].

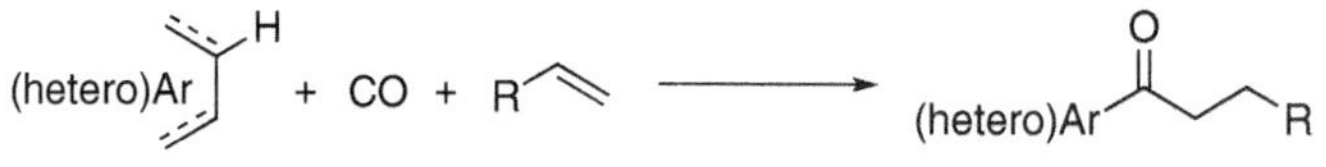

Scheme 9.37 Carbonylative hydroarylation of alkenes affording aromatic ketones.

Beller and coworker studied a carbonylative hydroarylation of alkenes with heteroarenes and arenes in the presence of $PdCl_2$ and various phosphine ligands to produce acylated (hetero)arenes (Scheme 9.38) [184]. In the cases of terminal alkenes used, the addition reactions occurred with excellent *anti*-Markovnikov regioselectivity. This protocol provided a novel type of catalytic acylation reaction complementary to the classic Friedel–Crafts methodologies.

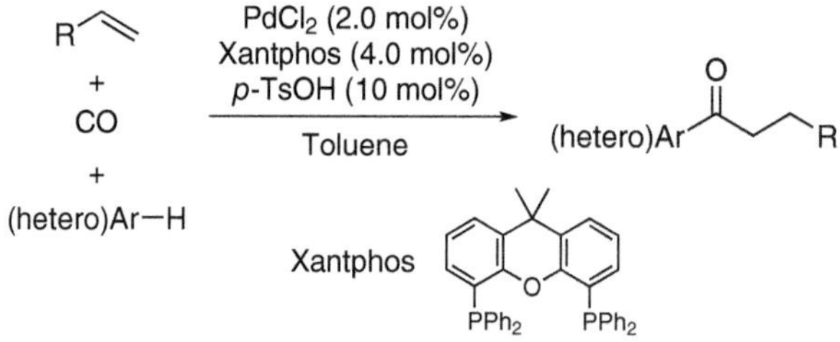

Scheme 9.38 Palladium-catalyzed carbonylative hydroarylation of alkenes.

The carbonylative Heck coupling reactions of aryl alkenes, alkyl iodides, and CO have also been applied in the synthesis of α,β-unsaturated ketones using a palladium/photoirradiation system in the presence of 1,8-diazabicyclo[5.4.0]undec-7-ene (DBU) (Scheme 9.39) [185].

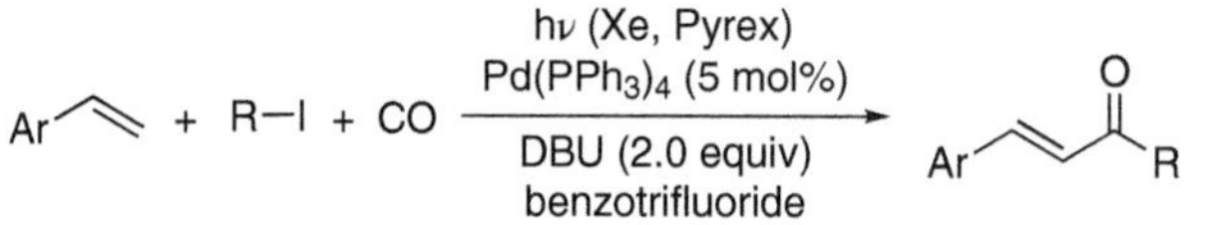

Ar = Ph, R = cyclohexyl: 58% (*E*/*Z* = 57/43)
Ar = Ph, R = *n*-C_8H_{17}: 47% (*E*/*Z* = 50/50)

Scheme 9.39 Synthesis of α,β-unsaturated ketones via carbonylative Heck coupling of aryl alkenes, alkyl iodides, and CO.

9.7.3 Cyclocarbonylation of Alkynes and/or Alkenes

Cyclocarbonylation of unsaturated compounds has become one of the efficient and atom economic processes for the synthesis of cyclocarbonyl compounds [186].

Construction of coumarin ring has been developed by a [3 + 2 + 1] cyclocarbonylation of *o*-iodophenols, alkynes, and CO. As shown in Scheme 9.40, 3,4-disubstituted coumarins could be obtained by palladium-catalyzed reaction of *o*-iodophenols, internal alkynes, and CO [187], and 2-quinolones could be prepared via the similar catalyst system with the use of *o*-iodoanilines to replace *o*-iodophenols [188].

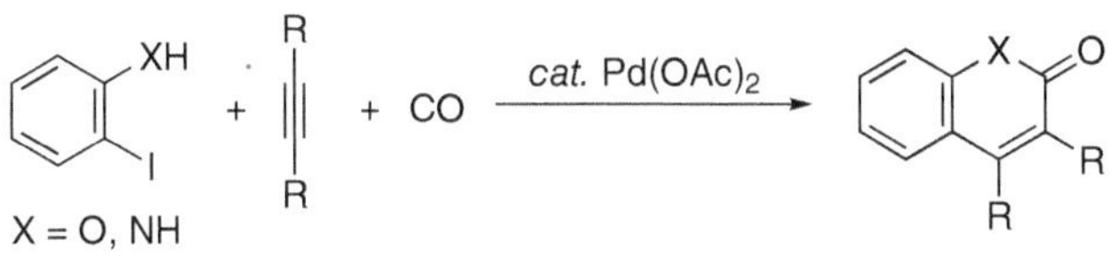

Scheme 9.40 Formation of coumarins or quinolones via [3 + 2 + 1] cyclocarbonylation of *o*-iodophenols or *o*-iodoanilines, alkynes, and CO.

In addition, 3-[(methoxycarbonyl)methyl]coumarins could be synthesized in good to high yields catalyzed by PdI_2 in conjunction with an excess of KI in MeOH as the solvent at room temperature and under 90 atm of CO [189]. Other coumarin derivatives could be obtained from the cyclocarbonylation of 2-alkenylphenols in the presence of oxidants catalyzed by palladium [190], rhodium [191], and cobalt complexes [192]. 2-Quinolones could also be prepared by a rhodium-catalyzed cyclocarbonylation of anilines and alkynes with CO via activation of C—H bond in the presence of oxidants [193].

The [3+2+1] cyclocarbonylation of phenols, alkynes, and CO has also been applied in the synthesis of flavones. As shown in Scheme 9.41, when *o*-halophenols and terminal alkynes were used, the reactions proceeded by means of palladium-catalyzed carbonylative Sonogashira coupling, followed by intramolecular cyclization to give two-substituted flavones [194]. The most ideal protocol for flavone preparation is to start from simple phenols with the activation of C—H bond having atom efficiency, which has been achieved by using iridium complexes with the use of $Cu(OAc)_2$ as an oxidant [195].

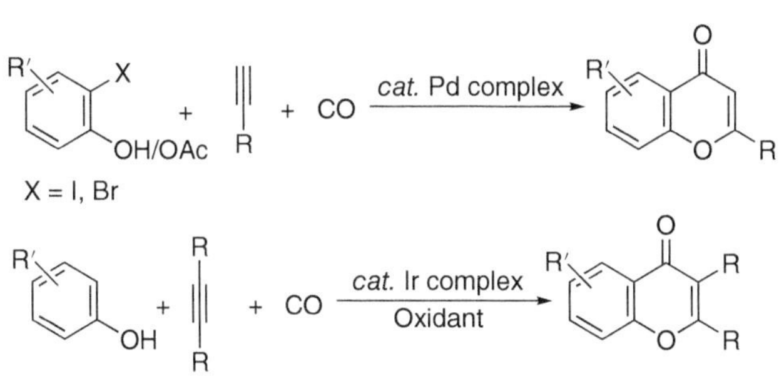

Scheme 9.41 Flavones via [3 + 2 + 1] cyclocarbonylation of phenols, alkynes, and CO.

Interestingly, thiochromenones could be prepared by an efficient $Pd(OAc)_2$-catalyzed one-pot carbonylative cyclization of four-component reaction of 1-fluoro-2-iodobenzenes, terminal alkynes, $Na_2S{\cdot}9H_2O$, and CO (Scheme 9.42) [196]. The reaction is expected to proceed by a carbonylative Sonogashira coupling followed by an aromatic nucleophilic substitution (S_NAr)/conjugate addition tandem reaction.

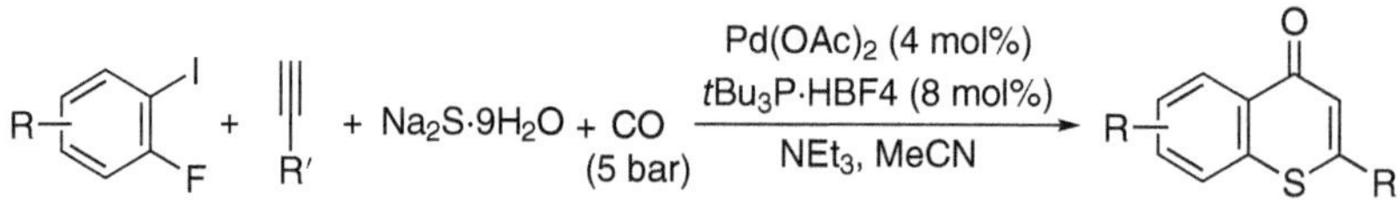

R = H, R′ = Ph: 68%
R = H, R′ = p-$CF_3C_6H_4$: 69%

Scheme 9.42 Formation of thiochromenones via a four-component one-pot $Pd(OAc)_2$-catalyzed carbonylative cyclization.

Mitsudo and coworker have reported the synthesis of functionalized hydroquinones via $[Cp^*RuCl_2]_2$-catalyzed [2+2+1+1] cyclocarbonylation of an alkyne, an alkene, and two molecules of CO (Scheme 9.43) [197]. A variety of electron-deficient alkenes, such as α,β-unsaturated ketones, esters, amides, and nitriles, could be used as an alkene coupling partner to give the corresponding hydroquinones. Their further studies with the use of $Ru_3(CO)_{12}$ as catalyst have developed a carbonylative [3+2+1] cycloaddition of silylacetylenes, α,β-unsaturated ketones, and CO providing an efficient method for the synthesis of tetrasubstituted α-pyrones (Scheme 9.44) [198].

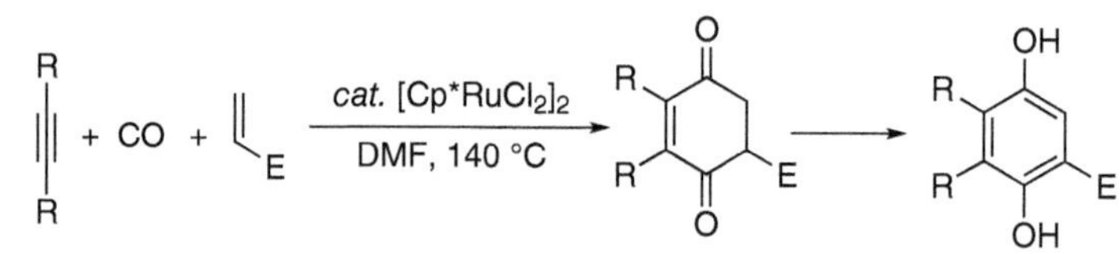

R = alkyl; E = COMe, COEt, CN, COOEt, $CONMe_2$, CHO

Scheme 9.43 $[Cp^*RuCl_2]_2$-catalyzed [2 + 2 + 1 + 1] cyclocarbonylation of alkyne and alkene giving hydroquinones.

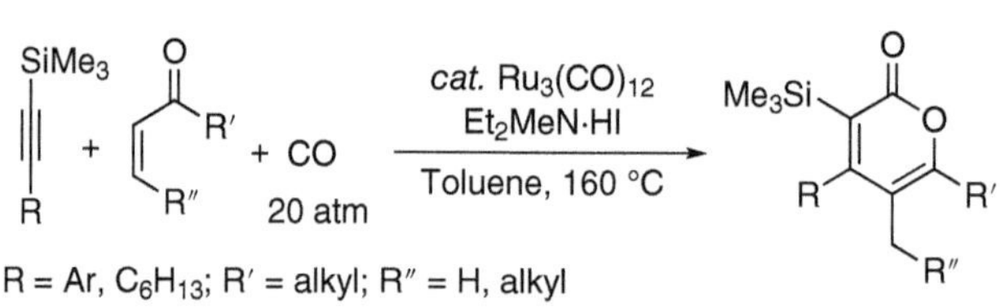

R = Ar, C_6H_{13}; R′ = alkyl; R″ = H, alkyl

Scheme 9.44 $Ru_3(CO)_{12}$-catalyzed carbonylative [3 + 2 + 1] cycloaddition giving α-pyrones.

The same group also found that substituted pyranopyrandiones could be prepared by carbonylative dimerization of cyclopropenones or cross-carbonylation of cyclopropenones with internal alkynes, and polysubstituted maleimides (Scheme 9.45) were synthesized by a [2 + 2 + 1] cyclocarbonylation of isocyanates, internal alkynes, and CO in the presence of $Ru_3(CO)_{12}$ [199].

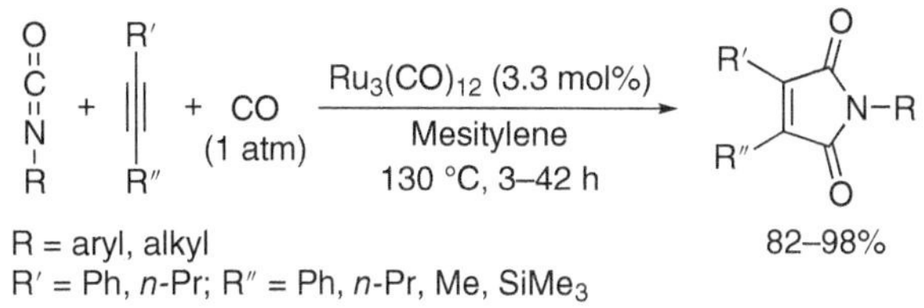

Scheme 9.45 $Ru_3(CO)_{12}$-catalyzed [2 + 2 + 1] cyclocarbonylation of isocyanates and alkynes affording maleimides.

Chatani and coworker investigated the carbonylation of benzyne with CO by using 2-trimethylsilylphenyl trifluoromethansulfonate or its derivatives as the benzyne precursors, and several types of cyclic ketones could be obtained depending upon the use of transition metal complexes under different reaction conditions [200]. For example, the carbonylation of 2-trimethylsilylphenyl trifluoromethansulfonate could produce either anthraquinone or fluorenone as the main products, when $Co_4(CO)_{12}$ or $[RhCl(cod)]_2$ were used, respectively (Scheme 9.46).

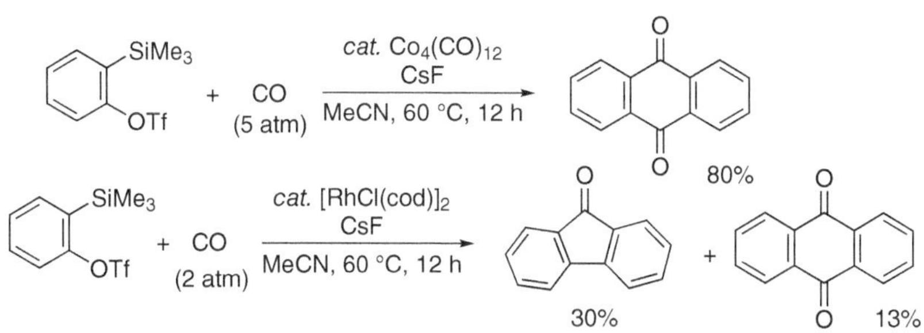

Scheme 9.46 Formation of anthraquinone or fluorenone via carbonylation of benzyne.

$[RhCl(cod)]_2$ was also found to efficiently catalyze the carbonylative cyclization of internal alkynes with 2-bromophenylboronic acids to give indenones reported by the same group [201].

It is well documented that the six-membered carbocyclic ketone of *p*-benzoquinones could be prepared by the reaction of the stoichiometric transition metal carbonyl complexes with alkynes [202], and the formation of *p*-benzoquinones as by-products was also found in a $[RhCl(CO)_2]_2$-catalyzed Pauson–Khand-type reactions [203]. Hua and coworker optimized an alternative catalytic system for the formation of tetrasubstituted *p*-benzoquinones via a direct [2 + 2 + 1 + 1] cyclocarbonylative coupling reaction of two internal alkynes with two CO in the presence of $[RhCl(CO)_2]_2$ (Scheme 9.47) [204]. It was found that the low concentration of CO was the crucial point for the chemoselective formation of *p*-benzoquinones in mild to high yields. Functional groups in R, such as chloro, methoxy, cyano, vinyl, fluoro, and carboxylate, are tolerated under the reaction conditions.

A proposed mechanism involving the formation of rhodacycle intermediates is reported (Scheme 9.48). It involves the formation of five-membered rhodacycle and

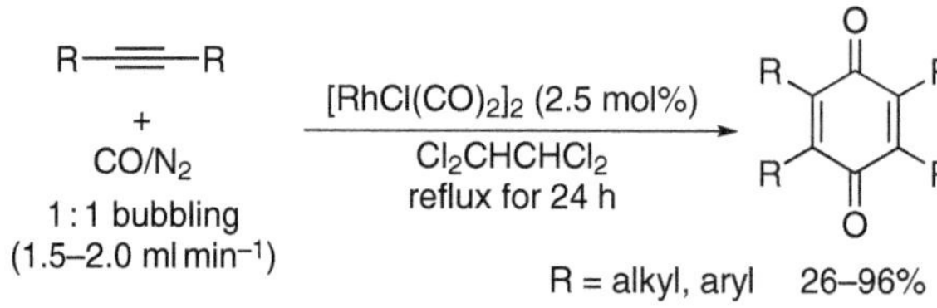

Scheme 9.47 Tetrasubstituted *p*-benzoquinones from rhodium-catalyzed [2 + 2 + 1 + 1] cyclocarbonylative coupling of internal alkynes with CO.

seven-membered rhodacycle as key intermediates. The mechanism is supported by the known procedures for the formation of maleoylmetal complexes from alkynes and two molecules of CO [205].

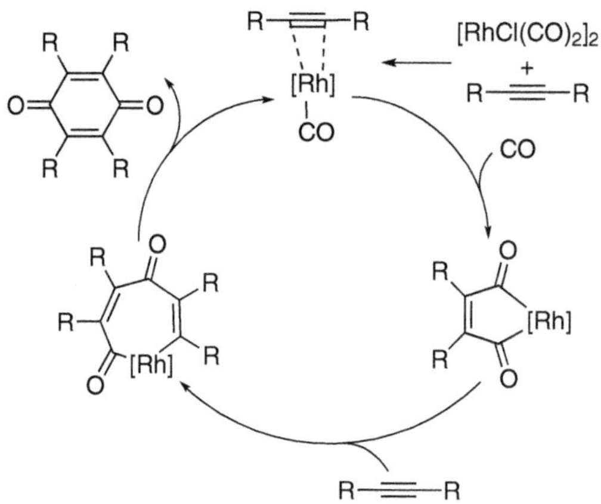

Scheme 9.48 Proposed mechanism for the formation of *p*-benzoquinone.

In the presence of $Pd(PPh_3)_4$, Hua's group also developed a [3 + 2 + 1] cyclocarbonylative coupling of 1,3-cyclohexanediones, terminal alkynes, and CO to provide a straightforward and atom economic process for the synthesis of 3,4,7,8-tetrahydro-2*H*-chromene-2,5(6*H*)-dione derivatives in good to high yields (Scheme 9.49) [206].

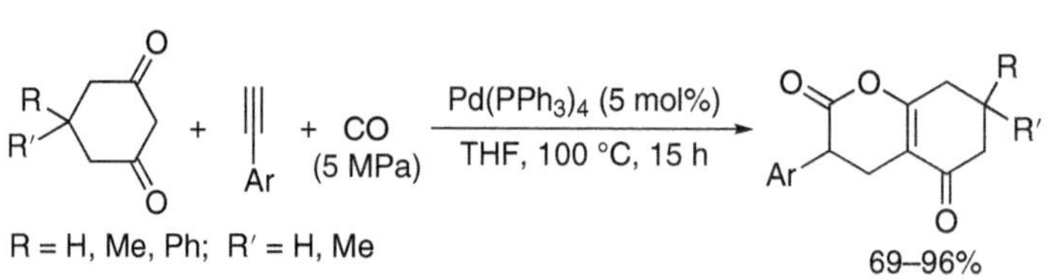

Scheme 9.49 $Pd(PPh_3)_4$-catalyzed [3 + 2 + 1] cyclocarbonylation affording chromene-2,5-diones.

2-Cyclopentenones and indan-1-ones have also been prepared from a $Pd(OAc)_2$-catalyzed [4 + 1] carbonylative cyclization of *o*-iodostyrenes or dienyl triflates, iodides, and bromides with CO in moderate to excellent yields [207]. Scheme 9.50 shows the results from several starting materials.

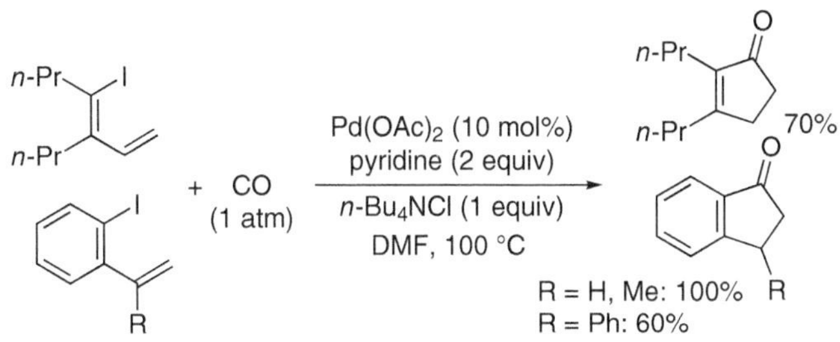

Scheme 9.50 2-Cyclopentenones and indan-1-ones by $Pd(OAc)_2$-catalyzed [4 + 1] carbonylative cyclization.

With the use of $[RhCl(CO)_2]_2$/urea as a catalyst system, Hua's group has developed the efficient synthesis of tetrasubstituted 2-cyclopenten-1-ones by an intermolecular reductive [2 + 2 + 1] cyclocarbonylation of two internal alkynes with CO in the presence of water [208]. As shown in Scheme 9.51, the reductive cyclocarbonylation of internal alkynes with CO to afford three types of cyclic ketones depending on the nature of internal alkynes. In the case of dialkylacetylenes used, the reductive cyclocarbonylation gives *cis*-2-cyclopenten-1-ones with excellent selectivity, whereas the reaction of diarylacetylenes produces both *trans*-2-cyclopenten-1-ones and 5-alkylidenefuran-2(5*H*)-ones, and the electron-deficient diarylacetylenes favors the formation of 5-alkylidenefuran-2(5*H*)-ones. In addition, interestingly, under the similar reaction conditions, as shown in Scheme 9.52, the cyclocarbonylation of diarylacetylenes with *ortho*-substituent on the benzene ring affords selectively indan-1-ones in good yields via a reductive [4 + 1] cyclocarbonylation of one diarylacetylene with CO. An early procedure for the formation of indan-1-one ring was reported by Takahashi's group in 1999, from the cyclocarbonylation of arylated alkynes catalyzed by $Rh_6(CO)_{16}$ in the presence of alcohols [209].

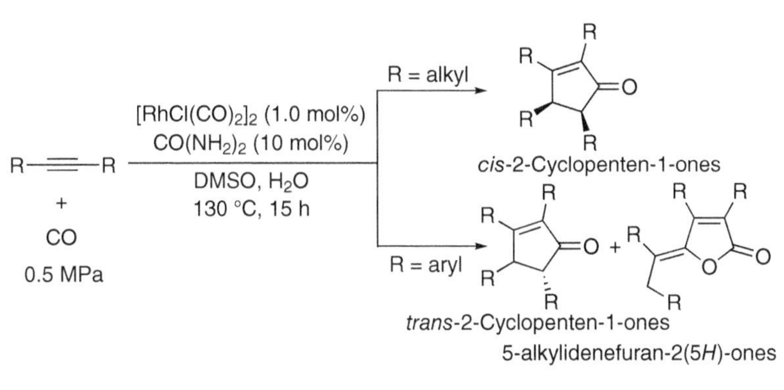

Scheme 9.51 Rhodium(I)-catalyzed reductive [2 + 2 + 1] cyclocarbonylation of internal alkynes with CO.

Several years later, an efficient formation of tetrasubstituted cyclopentadienones was then developed by Jiang's group [210]. As shown in Scheme 9.53, in the presence of $CuCl_2$, $PdCl_2$ could catalyze [2 + 2 + 1] cyclocarbonylation of internal

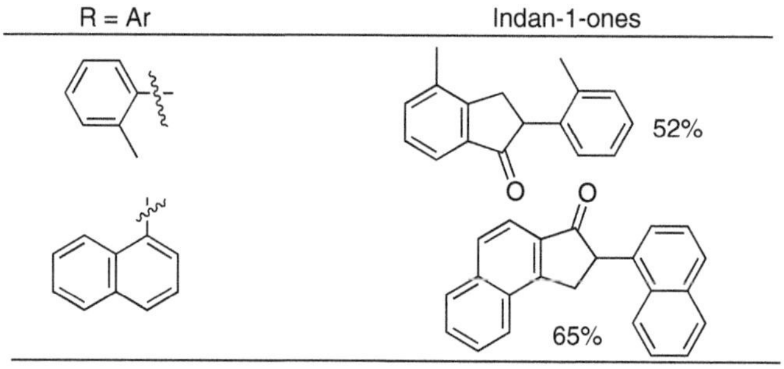

Scheme 9.52 Reductive [4 + 1] cyclocarbonylation of *ortho*-substituted diaryl alkyne with CO.

alkynes under atmospheric pressure of CO and ligand-free conditions producing cyclopentadienones.

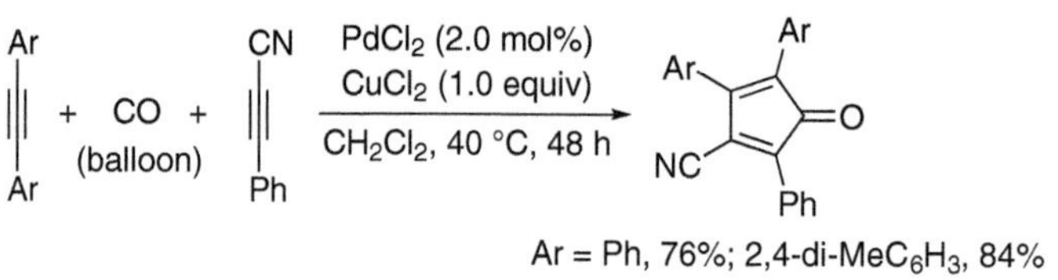

Scheme 9.53 $PdCl_2$-catalyzed [2 + 2 + 1] cyclocarbonylation of internal alkynes giving cyclopentadienones.

Trifluoromethyl-substituted cyclopentadienones could be synthesized by [2 + 2 + 1] cycloaddition of aryl- and trifluoromethyl-substituted internal alkynes with CO catalyzed $PdBr_2$ [211].

The cyclocarbonylation of alkynols is one of the efficient and simple routes for the preparation of lactones, and Pd and platinum complexes have been used as the efficient catalysts [212].

$Pd(OAc)_2$/2-$PyPPh_2$ dissolved in toluene was found to be a simple and efficient catalyst system for the intramolecular alkoxycarbonylation of alkynols with CO to give exclusively five- or six-membered *exo*-methylene lactones (Scheme 9.54) [213]. This catalytic system could also be immobilized in ionic liquids such as 1-butyl-3-methylimidazolium (BMI)·BF_4 and BMI·PF_6 (liquid–liquid biphasic conditions) without any changes in catalytic activity or selectivity.

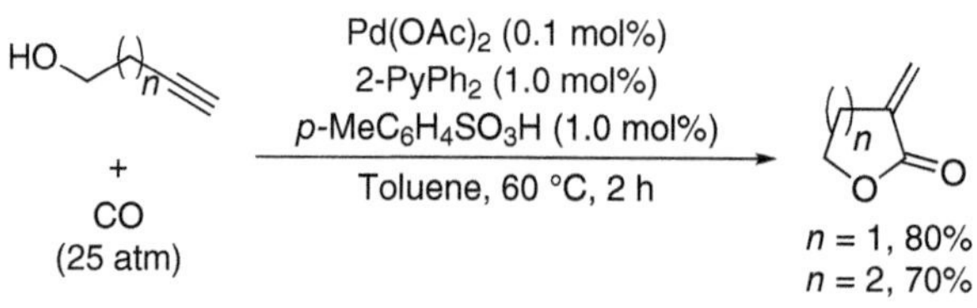

Scheme 9.54 $Pd(OAc)_2$-catalyzed intramolecular alkoxycarbonylation of alkynols.

Allenes have also been well used as the reactants in the cyclocarbonylation to give cyclic ketones. For example, Takahashi and coworker studied the

$Ru_3(CO)_{12}$-catalyzed cyclocarbonylation of allenyl alcohols in the presence of base, providing the selective synthesis of five- and six-membered lactones directly from 4-hydroxybuta-1,2-diene and 5-hydroxypenta-1,2-diene having various functional groups, respectively (Scheme 9.55) [214]. The cyclocarbonylation of allenyl amines could also occur to give lactams.

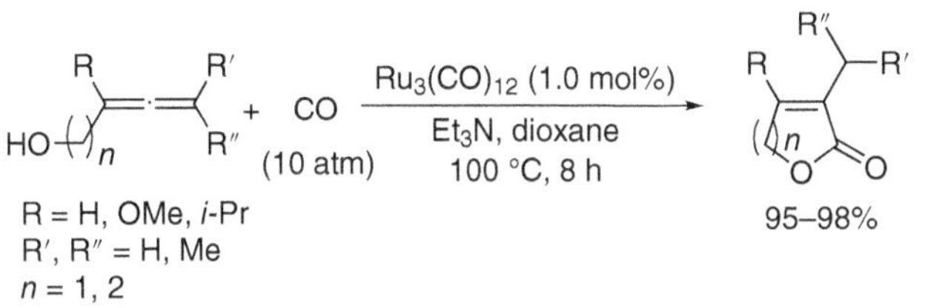

Scheme 9.55 $Ru_3(CO)_{12}$-catalyzed cyclocarbonylation of allenyl alcohols affording lactones.

Furan-2(5*H*)-one ring, a five-membered lactone occurs in many important natural products and pharmaceutical molecules [215]. Furan-2(5*H*)-ones could be selectively prepared directly from a reductive [2 + 1 + 1] cyclocarbonylation of one alkynes with two CO in the presence of water catalyzed by a variety of catalyst systems [216]. $Ru_3(CO)_{12}$ also showed the good catalytic activity in the basic condition under the moderate CO pressure for the reductive cyclocarbonylation of internal alkynes affording 3,4-disubstituted furan-2(5*H*)-ones [217]. As shown in Scheme 9.56, the reaction of 4-octyne with CO (4.0 MPa) in the presence of water in CH_3CN at 140 °C for 15 hours gave 4-dipropyl-furan-2(5*H*)-one in 80% yield.

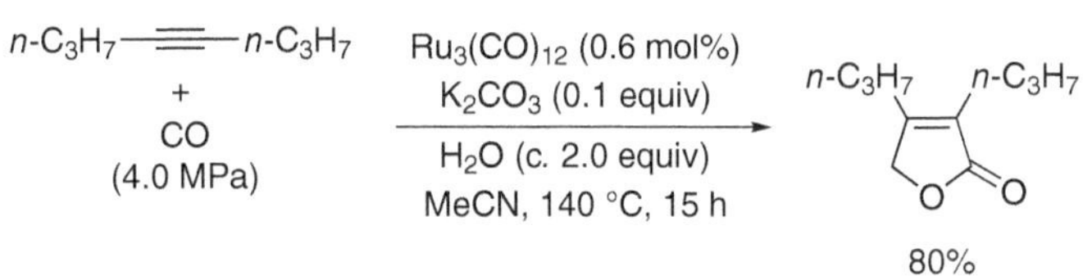

Scheme 9.56 $Ru_3(CO)_{12}$-catalyzed reductive carbonylation of 4-octyne.

In addition, in the presence of Ru/CeO_2, intermolecular [2 + 2 + 1] carbonylative cycloaddition of aldehydes with alkynes could also afford furan-2(5*H*)-ones, which undergo the subsequent oxidation to γ-hydroxybutenolides [218]. Furan-2(5*H*)-ones could be obtained by rhodium-catalyzed cyclohydrocarbonylation of α-keto alkynes [219].

On the other hand, furan-2(3*H*)-ones, the isomer of furan-2(5*H*)-ones could be prepared from 3-alkynoate esters and the corresponding acids via electrophilic cyclization with the use of I_2, ICl, and PhSeCl as the electrophiles [220].

The transition metal-promoted carbonylations of diynes [221] and enyne with CO are the powerful and convergent methods for the construction of bicyclic cyclopentadienones and cyclopentenones, respectively.

Buchwald and coworker developed an early transition metal-catalyzed cyclocarbonylation of enynes to give bicyclic cyclopentenones [222]. For example, in the presence of $Cp_2Ti(CO)_2$, 1,5-enynes underwent the cyclocarbonylation with CO to construct bicyclic cyclopentenone, and the reaction of 1,4-allenyne produces a bicyclic dienones (Scheme 9.57).

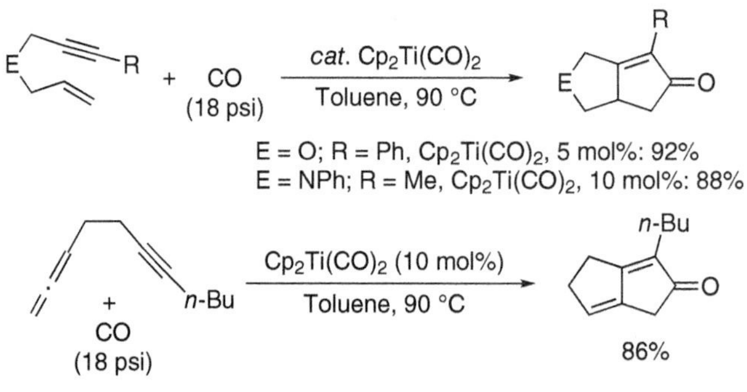

Scheme 9.57 $Cp_2Ti(CO)_2$-catalyzed cyclocarbonylation of enyne and allenyne giving cyclopentenones.

Wender's group has developed several three- and four-component cyclocarbonylation of unsaturated compounds to construct different sizes of cyclic compounds [223]. They have also studied an intramolecular [4+1] or [(2+2)+1] cyclocarbonylation of 1,3-diene-allenes with CO under mild conditions in the presence of $[RhCl(CO)_2]_2$, providing an efficient, selective, and operationally facile method to highly substituted alkylidenyl cyclopentanones (Scheme 9.58) [224].

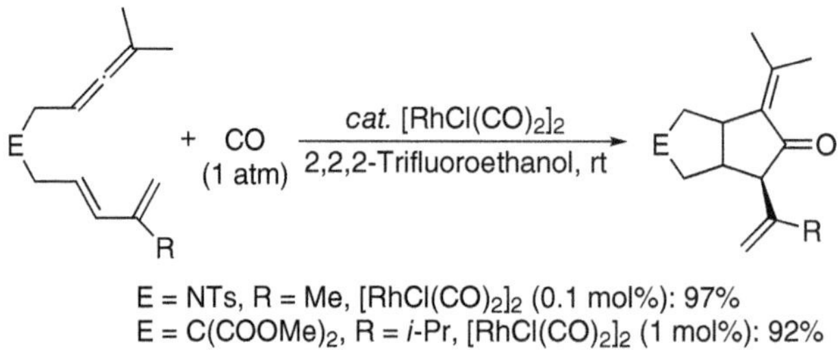

Scheme 9.58 Cyclocarbonylation of 1,3-diene-allenes with CO giving alkylidenyl cyclopentanones.

$[RhCl(CO)_2]_2$ also showed high catalytic activity in an intramolecular cyclocarbonylation of allene-yne tethers with ester, acetonide, and tosylamide functionalities affording high yields of bicyclo[5.3.0]decadienones [225].

It is a challenging work to develop the efficient catalyst systems for the construction of medium-sized cyclic ketones. The medium rings, including methylene eight-, nine-, and ten-membered rings, can be constructed by an intramolecular cyclocarbonylation protocol using palladium-complexed dendrimers on silica (G1-Pd) as recoverable catalyst [226]. Scheme 9.59 shows the examples on the synthesis of methylene eight-, nine-, and ten-membered ring tricyclic lactams via the intramolecular cyclocarbonylation of 2-(2-ethynylphenoxy)anilines, 2-(2-ethynylbenzyloxy)anilines, and 4-amino-3-[2-(2-ethynylphenyl)ethoxy]benzoic

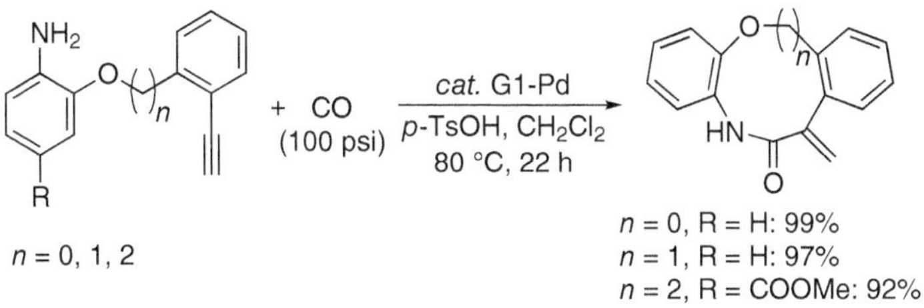

Scheme 9.59 Medium ring formation by an intramolecular cyclocarbonylation protocol.

acid methyl ester, respectively. This reaction can tolerate a wide range of functional groups such as halide, ether, nitrile, ketone, and ester.

Yu's group has extensively studied the Rh(I)-catalyzed cyclocarbonylation of ene-vinylcyclopropanes (ene-VCPs), establishing the simple and easy access to cyclic ketones [227]. For example, in the presence of $[RhCl(CO)_2]_2$, the cyclocarbonylation of nitrogen-tethered ene-VCP with a balloon mixed gas of CO and N_2 (1 : 4, v/v), 81% isolated yield of bicyclic cyclooctenone could be obtained, and carbon- and oxygen-tethered ene-VCPs could also undergo the cyclocarbonylation to give the corresponding eight-membered cyclic ketones in high yields and excellent diastereoselectivities (Scheme 9.60) [228].

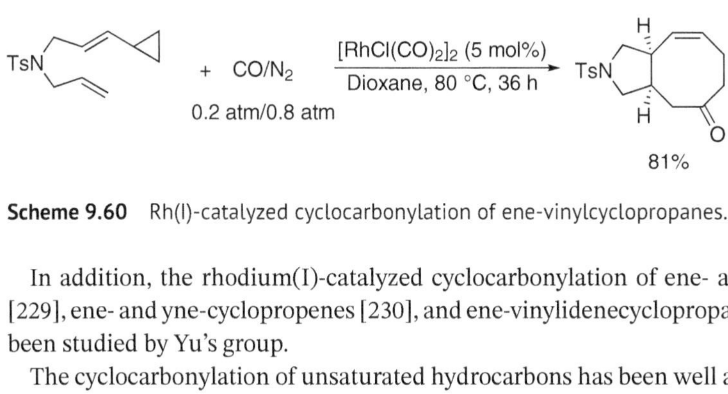

Scheme 9.60 Rh(I)-catalyzed cyclocarbonylation of ene-vinylcyclopropanes.

In addition, the rhodium(I)-catalyzed cyclocarbonylation of ene- and yne-VCPs [229], ene- and yne-cyclopropenes [230], and ene-vinylidenecyclopropanes [231] has been studied by Yu's group.

The cyclocarbonylation of unsaturated hydrocarbons has been well applied as the key step in the total syntheses of natural products [232].

References

1 Selected reviews, see: (a) Wu, X.-F., Neumann, H., and Beller, M. (2013). *Chem. Rev.* 113: 1–35. (b) Wu, L., Liu, Q., Jackstell, R., and Beller, M. (2014). *Angew. Chem. Int. Ed.* 53: 6310–6320. (c) Quintero-Duque, S., Dyballa, K.M., and Fleischer, I. (2015). *Tetrahedron Lett.* 56: 2634–2650. (d) Cao, J., Zheng, Z.-J., Xu, Z., and Xu, L.-W. (2017). *Coord. Chem. Rev.* 336: 43–53.

2 (a) Izumi, Y. (1997). *Catal. Today* 33: 371–409. (b) Chen, L. and Li, C.-J. (2006). *Adv. Synth. Catal.* 348: 1459–1484. (c) Hintermann, L. and Labonne, A. (2007). *Synthesis* 1121–1150. (d) Zuccaccia, D., Del Zotto, A., and Baratta, W. (2019). *Coord. Chem. Rev.* 396: 103–116.

3 Nishizawa, M., Imagawa, H., and Yamamoto, H. (2010). *Org. Biomol. Chem.* 8: 511–521.

4 (a) Wang, S., Miao, C., Wang, W. et al. (2014). *Chin. J. Catal.* 35: 1695–1700. (b) Hou, S., Yang, H., Cheng, B. et al. (2017). *Chem. Commun.* 53: 6926–6929.

5 Hassam, M. and Li, W.-S. (2015). *Tetrahedron* 71: 2719–2723.

6 (a) Thuong, M.B.T., Mann, A., and Wagner, A. (2012). *Chem. Commun.* 48: 434–436. (b) Das, R. and Chakraborty, D. (2012). *Appl. Organomet. Chem.* 26: 722–726. (c) Chen, Z.-W., Ye, D.-N., Qian, Y.-P. et al. (2013). *Tetrahedron* 69: 6116–6120.

7 (a) Xu, C., Du, W., Zeng, Y. et al. (2014). *Org. Lett.* 16: 948–951. (b) Kusakabe, T., Ito, Y., Kamimura, M. et al. (2017). *Org. Chem.* 6: 1086–1090.

8 Selected reports, see: (a) Mizushima, E., Sato, K., Hayashi, T., and Tanaka, M. (2002). *Angew. Chem. Int. Ed.* 41: 4563–4565. (b) Casado, R., Contel, M., Laguna, M. et al. (2003). *J. Am. Chem. Soc.* 125: 11925–11935. (c) Sanz, S., Jones, L.A., Mohr, F., and Laguna, M. (2007). *Organometallics* 26: 952–957. (d) Marion, N., Ramón, R.S., and Nolan, S.P. (2009). *J. Am. Chem. Soc.* 131: 448–449. (e) Leyva, A. and Corma, A. (2009). *J. Org. Chem.* 74: 2067–2074. (f) Almássy, A., Nagy, C.E., Bényei, A.C., and Joó, F. (2010). *Organometallics* 29: 2484–2490. (g) Czégéni, C.E., Papp, G., Kathó, Á., and Joó, F. (2011). *J. Mol. Catal. A* 340: 1–8. (h) Nun, P., Ramón, R.S., Gaillard, S., and Nolan, S.P. (2011). *J. Organomet. Chem.* 696: 7–11. (i) Ghosh, N., Nayak, S., and Sahoo, A.K. (2011). *J. Org. Chem.* 76: 500–511. (j) Goodwin, J.A. and Aponick, A. (2015). *Chem. Commun.* 51: 8730–8741. (k) Xu, Y., Hu, X., Shao, J. et al. (2015). *Green Chem.* 17: 532–537. (l) Li, F., Wang, N., Lu, L., and Zhu, G. (2015). *J. Org. Chem.* 80: 3538–3546. (m) Ebule, R.E., Malhotra, D., Hammond, G.B., and Xu, B. (2016). *Adv. Synth. Catal.* 358: 1478–1481. (n) Chen, T. and Cai, C. (2015). *Catal. Commun.* 65: 102–104. (o) Xu, Y., Hu, X., Zhang, S. et al. (2016). *ChemCatChem* 8: 262–267.

9 Tyagi, D., Rai, R.K., Mobin, S.M. et al. (2017). *Org. Chem.* 6: 1647–1658.

10 Liu, X., Liu, L., Wang, Z., and Fu, X. (2015). *Chem. Commun.* 51: 11896–11898.

11 Chen, D., Wang, D., Wu, W., and Xiao, L. (2015). *Appl. Sci.* 5: 114–121.

12 (a) Lucey, D.W. and Atwood, J.D. (2002). *Organometallics* 21: 2481–2490. (b) Trentin, F., Chapman, A.M., Scarso, A. et al. (2012). *Adv. Synth. Catal.* 354: 1095–1104.

13 Hirabayashi, T., Okimoto, Y., Saito, A. et al. (2006). *Tetrahedron* 62: 2231–2234.

14 (a) Wu, X.-F., Bezier, D., and Darcel, C. (2009). *Adv. Synth. Catal.* 351: 367–370. (b) Cabrero-Antonino, J.R., Leyva-Pérez, A., and Corma, A. (2012). *Chem. Eur. J.* 18: 11107–11114. (c) Bassetti, M., Ciceri, S., Lancia, F., and Pasquini, C. (2014). *Tetrahedron Lett.* 55: 1608–1612.

15 Ötvös, S.B., Szécsényi, Z., and Fülöp, F. (2019). *ACS Sustain. Chem. Eng.* 7: 13286–13293.

16 Tachinami, T., Nishimura, T., Ushimaru, R. et al. (2013). *J. Am. Chem. Soc.* 135: 50–53.

17 Ali, M., Srivastava, A.K., and Joshi, R.K. (2018). *Tetrahedron Lett.* 59: 2075–2078.

18 Jha, M., Shelke, G.M., Pericherla, K., and Kumar, A. (2014). *Tetrahedron Lett.* 55: 4814–4816.

19 (a) Bras, G.L., Provot, O., Peyrat, J.-F. et al. (2006). *Tetrahedron Lett.* 47: 5497–5501. (b) Jacubert, M., Provot, O., Peyrat, J.-F. et al. (2010). *Tetrahedron* 66: 3775–3787.

20 Niu, T.-f., Jiang, D.-y., Li, S.-y. et al. (2017). *Tetrahedron Lett.* 58: 1156–1159.

21 (a) Liang, S., Hammond, G.B., and Xu, B. (2015). *Chem. Commun.* 51: 903–906. (b) Liu, W., Wang, H., and Li, C.-J. (2016). *Org. Lett.* 18: 2184–2187.

22 Kore, R., Kumar, T.J.D., and Srivastava, R. (2012). *J. Mol. Catal. A Chem.* 360: 61–70.

23 Tanemura, K. and Suzuki, T. (2017). *Tetrahedron Lett.* 58: 955–958.

24 Oss, G., Ho, J., and Nguyen, T.V. (2018). *Eur. J. Org. Chem.* 2018: 3974–3981.

25 Sorella, G.L., Sperni, L., Ballester, P. et al. (2016). *Catal. Sci. Technol.* 6: 6031–6036.

26 (a) Deng, T. and Wang, C.-Z. (2016). *Catal. Sci. Technol.* 6: 7029–7032. (b) Xu, S., Yun, Z., Feng, Y. et al. (2016). *RSC Adv.* 6: 69822–69827. (c) Lei, Y., Zhang, M., Li, Q. et al. (2019). *Polymers* 11: 2091. (d) Liu, Y., Wang, B., Kang, L. et al. (2020). *Chem. Mater.* 32: 4375–4382.

27 (a) Li, F., Wang, N., Lu, L., and Zhu, G. (2015). *J. Org. Chem.* 80: 3538–3546. (b) Liu, S., Liu, H., Zhou, H. et al. (2018). *Org. Lett.* 20: 1110–1113.

28 Ma, J., Wang, N., Li, F., and Asian, J. (2014). *Org. Chem.* 3: 940–947.

29 Zhang, C., Cui, D.-M., Yao, L.-Y. et al. (2008). *J. Org. Chem.* 73: 7811–7813.

30 (a) Erian, A., Sherif, S., and Gaber, H. (2003). *Molecules* 8: 793. (b) Conde, S., Pérez, D.I., Martínez, A. et al. (2003). *J. Med. Chem.* 46: 4631–4633. (b) Ostrowski, T., Golankiewicz, B., De Clercq, E. et al. (2009). *Eur. J. Med. Chem.* 44: 3313–3317.

31 Xie, L., Wu, Y., Yi, W. et al. (2013). *J. Org. Chem.* 78: 9190–9195.

32 Chen, Z.-W., Ye, D.-N., Ye, M. et al. (2014). *Tetrahedron Lett.* 55: 1373–1375.

33 Zeng, M., Huang, R.-X., Li, W.-Y. et al. (2016). *Tetrahedron* 72: 3818–3822.

34 Ye, M., Wen, Y., Li, H. et al. (2016). *Tetrahedron Lett.* 57: 4983–4986.

35 Tokunaga, M. and Wakatsuki, Y. (1998). *Angew. Chem. Int. Ed.* 37: 2867–2869.

36 Chen, Y., Ho, D.M., and Lee, C. (2005). *J. Am. Chem. Soc.* 127: 12184–12185.

37 (a) Suzuki, T., Tokunaga, M., and Wakatsuki, Y. (2001). *Org. Lett.* 3: 735–737. (b) Grotjahn, D.B., Incarvito, C.D., and Rheingold, A.L. (2001). *Angew. Chem. Int. Ed.* 40: 3884–3887. (c) Alvarez, P., Bassetti, M., Gimeno, J., and Mancini, G. (2001). *Tetrahedron Lett.* 42: 8467–8470. (d) Tokunaga, M., Suzuki, T., Koga, N. et al. (2001). *J. Am. Chem. Soc.* 123: 11917–11924. (e) Grotjahn, D.B. and Lev, D.A. (2004). *J. Am. Chem. Soc.* 126: 12232–12233. (f) Chevallier, F. and Breit, B. (2006). *Angew. Chem. Int. Ed.* 45: 1599–1602. (g) Labonne, A., Kribber, T., and Hintermann, L. (2006). *Org. Lett.* 8: 5853–5856. (h) Boeck, F., Kribber, T., Xiao, L., and Hintermann, L. (2011). *J. Am. Chem. Soc.* 133: 8138–8141. (i) Li, L., Zeng, M., and Herzon, S.B. (2014). *Angew. Chem. Int. Ed.* 53: 7892–7895.

38 Hartman, J.W., Hiscox, W.C., and Jennings, P.W. (1993). *J. Org. Chem.* 58: 7613–7614.

39 (a) Roembke, P., Schmidbaur, H., Cronje, S., and Raubenheimer, H. (2004). *J. Mol. Catal. A Chem.* 212: 35–42. (b) Rzhevskiy, S.A., Philippova, A.N., Chesnokov, G.A. et al. (2021). *Chem. Commun.* 57: 5686–5689.

40 Zhang, S., Cheng, H., Mo, S. et al. (2019). *Adv. Synth. Catal.* 361: 4227–4231.

41 Li, X., Hu, G., Luo, P. et al. (2012). *Adv. Synth. Catal.* 354: 2427–2432.

42 Cloutier, M., Roudias, M., and Paquin, J.-F. (2019). *Org. Lett.* 21: 3866–3870.

43 Mei, Q., Liu, H., Hou, M. et al. (2017). *New J. Chem.* 41: 6290–6295.

44 Kuang, J., Zhou, T., You, T. et al. (2019). *Org. Biomol. Chem.* 17: 3940–3944.

45 Verma, S., Kumar, M., Mishra, P.K., and Verma, A.K. (2019). *Org. Lett.* 21: 5059–5063.

46 Antenucci, A., Flamini, P., Fornaiolo, M.V. et al. (2019). *Adv. Synth. Catal.* 361: 4517–4526.

47 Liang, S., Jasinski, J., Hammond, G.B., and Xu, B. (2015). *Org. Lett.* 17: 162–165.

48 Finiels, A., Geneste, P., Marichez, F., and Moreau, P. (1989). *Catal. Lett.* 2: 181–184.

49 Rao, K.T.V., Prasad, P.S.S., and Lingaiah, N. (2012). *Green Chem.* 14: 1507–1514.

50 Öztürk, B.Ö., Çetinel, B., and Şehitoğlu, S.K. (2020). *App. Organomet. Chem.* 34: e5686.

51 Li, D., Gong, Y., Du, M. et al. (2020). *ACS Sustain. Chem. Eng.* 8: 8148–8155.

52 Bondžić, B.P. (2015). *J. Mol. Catal. A Chem.* 408: 310–334.

53 (a) Doyama, K., Joh, T., and Takahashi, S. (1986). *Tetrahedron Lett.* 27: 4497–4500. (b) Doyama, K., Joh, T., Shiohara, T., and Takahashi, S. (1988). *Bull. Chem. Soc. Jpn.* 61: 4353–4360. (c) Campi, E.M. and Jackson, W.R. (1989). *Aust. J. Chem.* 42: 471–478.

54 Johnson, J.R., Cuny, G.D., and Buchwald, S.L. (1995). *Angew. Chem. Int. Ed. Engl.* 34: 1760–1761.

55 Tang, C., Zeng, Y., Cao, P. et al. (2009). *Catal. Lett.* 129: 189–193.

56 Fang, X., Zhang, M., Jackstell, R., and Beller, M. (2013). *Angew. Chem. Int. Ed.* 52: 4645–4649.

57 Campi, E.M. and Jackson, W.R. (1989). *Aust. J. Chem.* 42: 471–478.

58 van den Hoven, B.G. and Alper, H. (1999). *J. Org. Chem.* 64: 3964–3968.

59 van den Hoven, B.G. and Alper, H. (1999). *J. Org. Chem.* 64: 9640–9645.

60 (a) Roelen, O. German Patent, DE 849548, 1938. (b) Cornils, B., Herrmann, W.A., and Rasch, M. (1994). *Angew. Chem. Int. Ed. Engl.* 33: 2144–2163.

61 Selected reviews, see: (a) Franke, R., Selent, D., and Börner, A. (2012). *Chem. Rev.* 112: 5675–5732. (b) Sharma, S.K. and Jasra, R.V. (2015). *Catal. Today* 247: 70–81.

62 (a) Hood, D.M., Johnson, R.A., Carpenter, A.E. et al. (2020). *Science* 367: 542–548. (b) MacNeil, C.S., Mendelsohn, L.N., Zhong, H. et al. (2020). *Angew. Chem. Int. Ed.* 59: 8912–8916. and references therein.

63 Selected reports on the application of rhodium complexes in hydroformylation, see: (a) Wang, X. (2015). *J. Flow Chem.* 5: 125–132. (b) Chen, C., Dong, X.-Q., and Zhang, X. (2016). *Chem. Rec.* 16: 2674–2686. (c) Cunyao Li, C., Wang, W., Yan, L., and Ding, Y. (2018). *Front. Chem. Sci. Eng.* 12: 113–123. (d) Bernales, V. and Froese, R.D. (2019). *J. Compt. Chem.* 40: 342–348. Recent reports, see: (e) Iu, L., Fuentes, J.A., Janka, M.E. et al. (2019). *Angew. Chem. Int. Ed.* 58: 2120–2124. (f) Jongkind, L.J., Johannes, A.A.W., Elemans, J.A.A.W., and Reek, J.N.H. (2019). *Angew. Chem. Int. Ed.* 58: 2696–2699. (g) Gamede, N.V., Kapfunde, T.A., Ocansey, E. et al. (2020). *Trans. Met. Chem.* 45: 1–8. (h) Raghuvanshi, K., Zhu, C., Ramezani, M. et al. (2020). *ACS Catal.* 10: 7535–7542. (i) Linnebank, P.R., Ferreira, S.F., Kluwer, A.M., and Reek, J.N.H. (2020). *Chem. Eur. J.* 26: 8214–8219.

64 Selected reports, see: (a) Pospech, J., Fleischer, I., Franke, R. et al. (2013). *Angew. Chem. Int. Ed.* 52: 2852–2872. (b) Tay, D.W.P., Nobbs, J.D., Romain, C. et al. (2020). *ACS Catal.* 10: 663–671.

65 (a) Hertrich, M.F., Scharnagl, F.K., Pews-Davtyan, A. et al. (2019). *Chem. Eur. J.* 25: 5534–5538. (b) Zhao, J., He, Y., Wang, F. et al. (2020). *ACS Catal.* 10: 914–920.

66 (a) Liang, Z., Chen, J., Chen, X. et al. (2019). *Chem. Commun.* 55: 13721–13724. (b) Jia, X., Liang, Z., Chen, J. et al. (2019). *Org. Lett.* 21: 2147–2150. (c) Zhang, J., Sun, P., Gao, G. et al. (2020). *J. Catal.* 387: 196–206. (d) Liu, S., Dai, X., Wang, H. et al. (2020). *Chin. J. Chem.* 38: 139–143. (e) Shi, Y., Ji, G., Hu, Q. et al. (2020). *New J. Chem.* 44: 20–23.

67 Amsler, J., Sarma, B.B., Agostini, G. et al. (2020). *J. Am. Chem. Soc.* 142: 5087–5096.

68 Liu, S., Dai, X., Wang, H. et al. (2019). *Chin. J. Chem.* 37: 139–143.

69 Cocq, A., Bricout, H., Djedaïni-Pilard, F. et al. (2020). *Catalysts* 10: 56.

70 Siangwata, S., Goosen, N.J., and Smith, G.S. (2020). *Appl. Catal. A Gen.* 603: 117736.

71 Rosales, M., González, A., González, B. et al. (2005). *J. Organomet. Chem.* 690: 3095–3098.

72 Ren, W., Chang, W., Dai, J. et al. (2016). *J. Am. Chem. Soc.* 138: 14864–14867.

73 Tan, G., Wu, Y., Shi, Y., and You, J. (2019). *Angew. Chem. Int. Ed.* 58: 7440–7444.

74 Köpfer, A. and Breit, B. (2015). *Angew. Chem. Int. Ed.* 54: 6913–6917. and references therein.

75 Behr, A., Levikov, D., and Vogelsang, D. (2015). *J. Mol. Catal. A Chem.* 406: 114–117.

76 Yu, S.-m., Snavely, W.K., Chaudhari, R.V., and Subramaniam, B. (2020). *Mol. Catal.* 484: 110721.

77 Chiou, W.-H., Hsu, K.-H., and Huang, W.-W. (2020). *ACS Omega* 5: 3717–3724.

78 Selected reviews, see: (a) Gladiali, S., Bayón, J.C., and Claver, C. (1995). *Tetrahedron: Asymmetry* 6: 1453–1474. (b) Agbossou, F., Carpentier, J.F., and Mortreux, A. (1995). *Chem. Rev.* 95: 2485–2506. (c) Diéguez, M., Pàmies, O., and Claver, C. (2004). *Tetrahedron: Asymmetry* 15: 2113–2122. (d) Gual, A., Godard, C., Castillón, S., and Claver, C. (2010). *Tetrahedron: Asymmetry* 21: 1135–1146. (e) Deng, Y., Wang, H., Sun, Y., and Wang, X. (2015). *ACS Catal.* 5: 6828–6837. (f) Brezny, A.C. and Landis, C.R. (2018). *Acc. Chem. Res.* 51: 2344–2354. (g) Bai, S.-T., Kluwer, A.M., and Reek, J.N.H. (2019). *Chem. Commun.* 55: 14151–14154. (h) Jongkind, L.J. and Reek, J.N.H. (2020).

Chem. Asian J. 15: 867–875. (i) Li, S., Li, Z., You, C. et al. (2020). *Org. Lett.* 22: 1108–1112.
79 You, C., Li, S., Li, X. et al. (2019). *ACS Catal.* 9: 8529–8533.
80 Selected reviews, see: (a) Jun, C.-H., Jo, E.-A., and Park, J.-W. (2007). *Eur. J. Org. Chem.* 2007: 1869–1881. (b) Willis, M.C. (2010). *Chem. Rev.* 110: 725–748. (c) González-Rodríguez, C. and Willis, M.C. (2011). *Pure Appl. Chem.* 83: 577–585. (d) Leung, L.C. and Krische, M. (2012). *J. Chem. Sci.* 3: 2202–2209. (e) Ghosh, A., Johnson, K.F., Vickerman, K.L. et al. (2016). *Org. Chem. Front.* 3: 639–644.
81 (a) Kokubo, K., Matsumasa, K., Miura, M., and Nomura, M. (1997). *J. Org. Chem.* 62: 4564–4565. (b) Kokubo, K., Matsumasa, K., Nishinaka, Y. et al. (1999). *Bull. Chem. Soc. Jpn.* 72: 303–311. and references cited therein.
82 Jun, C.-H., Lee, H., Hong, J.-B., and Kwon, B.-I. (2002). *Angew. Chem. Int. Ed.* 41: 2146–2147.
83 Huang, W.-S., Chan, J., and Jamison, T.F. (2000). *Org. Lett.* 2: 4221–4223.
84 (a) Huddleston, R.R., Jang, H.-Y., and Krische, M.J. (2003). *J. Am. Chem. Soc.* 125: 11488–11489. (b) Bausch, C.C., Patman, R.L., Breit, B., and Krische, M.J. (2011). *Angew. Chem. Int. Ed.* 50: 5687–5690.
85 (a) Willis, M.C., Randell-Sly, H.E., Woodward, R.L. et al. (2006). *J. Org. Chem.* 71: 5291–5297. (b) Hooper, J.F., Seo, S., Truscott, F.R. et al. (2016). *J. Am. Chem. Soc.* 138: 1630–1634. (c) Coxon, T.J., Fernández, M., Barwick-Silk, J. et al. (2017). *J. Am. Chem. Soc.* 139: 10142–10149. (d) Barwick-Silk, J., Hardy, S., Willis, M.C., and Weller, A.S. (2018). *J. Am. Chem. Soc.* 140: 7347–7357. (e) Straker, R.N., Formica, M., Lupton, J.D. et al. (2018). *Tetrahedron* 74: 5408–5414.
86 Straker, R.N., Formica, M., Lupton, J.D. et al. (2018). *Tetrahedron* 74: 5408–5414.
87 (a) Lenden, P., Entwistle, D.A., and Willis, M.C. (2011). *Angew. Chem. Int. Ed.* 50: 10657–10660. (b) Castaing, M., Wason, S.L., Estepa, B. et al. (2013). *Angew. Chem. Int. Ed.* 52: 13280–13283. (c) Neuhaus, J.D., Morrow, S.M., Brunavs, M., and Willis, M.C. (2016). *Org. Lett.* 18: 1562–1565. (d) Bouisseau, A., Glancy, J., and Willis, M.C. (2016). *Org. Lett.* 18: 5676–5679.
88 Gao, M. and Willis, M.C. (2017). *Org. Lett.* 19: 2734–2737.
89 (a) Jun, C.-H., Lee, H., Hong, J.-B., and Kwon, B.-I. (2002). *Angew. Chem. Int. Ed.* 41: 2146–2147. (b) Rueda, X.Y. and Castillón, S. (2007). *J. Organomet. Chem.* 692: 1628–1632. (c) Lenden, P., Ylioja, P.M., González-Rodríguez, C. et al. (2011). *Green Chem.* 13: 1980–1982. (d) González-Rodríguez, C., Pawley, R.J., Chaplin, A.B. et al. (2011). *Angew. Chem. Int. Ed.* 50: 5134–5138. (e) Pawley, R.J., Huertos, M.A., Lloyd-Jones, G.C. et al. (2012). *Organometallics* 31: 5650–5659. (f) Chaplin, A.B., Hooper, J.F., Weller, A.S., and Willis, M.C. (2012). *J. Am. Chem. Soc.* 134: 4885–4897. (g) Poingdestre, S.-J., Goodacre, J.D., Weller, A.S., and Willis, M.C. (2012). *Chem. Commun.* 48: 6354–6356.
90 (a) Miura, H., Wada, K., Hosokawa, S., and Inoue, M. (2013). *Chem. Eur. J.* 19: 861–864. (b) Chen, Q.-A., Cruz, F.A., and Dong, V.M. (2015). *J. Am. Chem. Soc.* 137: 3157–3160.

91 Shuai Shi, S., Tao Wang, T., Weingand, V. et al. (2014). *Angew. Chem. Int. Ed.* 53: 1148–1151.

92 Miura, K., Yamamoto, K., Yamanobe, A. et al. (2010). *Chem. Lett.* 39: 766–767.

93 Selected reports, see: (a) Marder, T.B., Roe, D.C., and Milstein, D. (1988). *Organometallics* 7: 1451–1453. (b) Imai, M., Tanaka, M., Tanaka, K. et al. (2004). *J. Org. Chem.* 69: 1144–1150. (c) Tanaka, K., Shibata, Y., Suda, T. et al. (2007). *Org. Lett.* 9: 1215–1218. (d) Roy, A.H., Lenges, C.P., and Brookhart, M. (2007). *J. Am. Chem. Soc.* 129: 2082–2093. (e) Osborne, J.D. and Willis, M.C. (2008). *Chem. Commun.* 5025–5027. (f) Shibata, Y. and Tanaka, K. (2009). *J. Am. Chem. Soc.* 131: 12552–12553. (g) Zhang, H.-J. and Bolm, C. (2011). *Org. Lett.* 13: 3900–3903. (h) Murphy, S.K., Petrone, D.A., Coulter, M.M., and Dong, V.M. (2011). *Org. Lett.* 13: 6216–6219.

94 (a) Lenges, C.P. and Brookhart, M. (1997). *J. Am. Chem. Soc.* 119: 3165–3166. (b) Yang, J., Seto, Y.W., and Yoshikai, N. (2015). *ACS Catal.* 5: 3054–3057.

95 Yi, C.S. and Kim, J. (2016). *ACS Catal.* 6: 3336–3339.

96 Xiao, L.-J., Fu, X.-N., Zhou, M.-J. et al. (2016). *J. Am. Chem. Soc.* 138: 2957–2960.

97 Schedler, M., Wang, D.-S., and Glorius, F. (2013). *Angew. Chem. Int. Ed.* 52: 2585–2589.

98 (a) Papadopoulos, G.N., Voutyritsa, E., Kaplaneris, N., and Kokotos, C.G. (2018). *Chem. Eur. J.* 24: 1726–1731.

99 (a) Moteki, S.A., Usui, A., Selvakumar, S. et al. (2014). *Angew. Chem. Int. Ed.* 53: 11060–11064. (b) Jiang, J., Ramozzi, R., Moteki, S. et al. (2015). *J. Org. Chem.* 80: 9264–9271.

100 Selected reports, see: (a) Willis, M.C., Randell-Sly, H.E., Woodward, R.L., and Currie, G.S. (2005). *Org. Lett.* 7: 2249–2251. (b) Pawley, R.J., Moxham, G.L., Dallanegra, R. et al. (2010). *Organometallics* 29: 1717–1728. (c) Vautravers, N.R., Regent, D.D., and Breit, B. (2011). *Chem. Commun.* 47: 6635–6637.

101 Jun, C.-H., Chung, J.-H., Lee, D.-Y. et al. (2001). *Tetrahedron Lett.* 42: 4803–4805.

102 Kondo, T., Hiraishi, N., Morisaki, Y. et al. (1998). *Organometallics* 17: 2131–2134.

103 Omura, S., Fukuyama, T., Horiguchi, J. et al. (2008). *J. Am. Chem. Soc.* 130: 14094–14095.

104 Chen, Q.-A., Kim, D.K., and Dong, V.M. (2014). *J. Am. Chem. Soc.* 136: 3772–3775.

105 Nagamoto, M. and Nishimura, T. (2015). *Chem. Commun.* 51: 13791–13794.

106 The representative reports, see: (a) Larock, R.C., Oertle, K., and Potter, G.F. (1980). *J. Am. Chem. Soc.* 102: 190–197. (b) Fairlie, D.P. and Bosnich, B. (1988). *Organometallics* 7: 936–945. (c) Aloise, A.D., Layton, M.E., and Shair, M.D. (2000). *J. Am. Chem. Soc.* 122: 12610–12611. (d) Tanaka, K. and Fu, G.C. (2001). *J. Am. Chem. Soc.* 123: 11492–11493. (e) Tanaka, K. and Fu, G.C. (2002). *J. Am. Chem. Soc.* 124: 10296–10297. (f) Takeishi, K., Sugishima, K., Sasaki, K., and Tanaka, K. (2004). *Chem. Eur. J.* 10: 5681–5688. (g) Kundu, K., McCullagh, J.V., and Morehead, A.T. Jr., (2005). *J. Am. Chem. Soc.* 127:

16042–16043. (h) Morgan, J.P., Kundu, K., and Doyle, M.P. (2005). *Chem. Commun.* 3307–3309. (i) Tanaka, K., Hagiwara, Y., and Hirano, M. (2006). *Eur. J. Org. Chem.* 3582–3595. (j) Beletskiy, E.V., Sudheer, C., and Douglas, C.J. (2012). *J. Org. Chem.* 77: 5884–5893. (k) Rastelli, E.J., Truong, N.T., and Coltart, D.M. (2016). *Org. Lett.* 18: 5588–5591.

107 Yang, F., Jin, T., and Yamamoto, Y. (2012). *Tetrahedron* 68: 5223–5228.

108 (a) Biju, A.T., Wurz, N.E., and Glorius, F. (2010). *J. Am. Chem. Soc.* 132: 5970–5971. (b) Vedachalam, S., Wong, Q.-L., Maji, B. et al. (2011). *Adv. Synth. Catal.* 353: 219–225. (c) Wang, Z., Yu, Z., Wang, Y., and Shi, D. (2012). *Synthesis* 1559–1562.

109 Mondal, A., Hazra, R., Grover, J. et al. (2018). *ACS Catal.* 8: 2748–2753.

110 (a) Hoshimoto, Y., Hayashi, Y., Suzuki, H. et al. (2012). *Angew. Chem. Int. Ed.* 51: 10812–10815. (b) Hoshimoto, Y., Hayashi, Y., Ohashi, M., and Ogoshi, S. (2017). *Chem. Asian J.* 12: 278–282.

111 Kim, D.K., Riedel, J., Kim, R.S., and Dong, V.M. (2017). *J. Am. Chem. Soc.* 139: 10208–10211.

112 Qi, C., Yang, S., Gandon, V., and Leboeuf, D. (2019). *Org. Lett.* 21: 7405–7409.

113 (a) Hirano, K., Biju, A.T., Piel, I., and Glorius, F. (2009). *J. Am. Chem. Soc.* 131: 14190–14191. (b) DiRocco, D.A. and Rovis, T. (2011). *Angew. Chem. Int. Ed.* 50: 7982–7983. (c) Janssen-Müller, D., Schedler, M., Fleige, M. et al. (2015). *Angew. Chem. Int. Ed.* 54: 12492–12496.

114 Oonishi, Y., Hosotani, A., and Sato, Y. (2011). *J. Am. Chem. Soc.* 133: 10386–10389.

115 Sato, Y., Oonishi, Y., and Mori, M. (2002). *Angew. Chem. Int. Ed.* 41: 1218–1221.

116 Yoshikai, K., Hayama, T., Nishimura, K. et al. (2005). *J. Org. Chem.* 70: 681–683.

117 Selected reports, see: (a) Tanaka, K., Hagiwara, Y., and Noguchi, K. (2005). *Angew. Chem. Int. Ed.* 44: 7260–7263. (b) Tanaka, K., Hojo, D., Shoji, T. et al. (2007). *Org. Lett.* 9: 2059–2062. (c) Hojo, D. and Tanaka, K. (2012). *Org. Lett.* 14: 1492–1495.

118 Tanaka, K. and Fu, G.C. (2002). *Org. Lett.* 4: 933–935.

119 Bendorf, H.D., Colella, C.M., Dixon, E.C. et al. (2002). *Tetrahedron Lett.* 43: 7031–7034.

120 Coulter, M.M., Dornan, P.K., and Dong, V.M. (2009). *J. Am. Chem. Soc.* 131: 6932–6933.

121 Bendorf, H.D., Ruhl, K.E., Shurer, A.J. et al. (2012). *Tetrahedron Lett.* 53: 1275–1277.

122 Jhuang, H.-S., Reddy, D.M., Chen, T.-H., and Lee, C.-F. (2016). *Asian J. Org. Chem.* 5: 1452–1456.

123 (a) Osborne, J.D., Randell-Sly, H.E., Currie, G.S. et al. (2008). *J. Am. Chem. Soc.* 130: 17232–17233. (b) Coulter, M.M., Kou, K.G.M., Galligan, B., and Dong, V.M. (2010). *J. Am. Chem. Soc.* 132: 16330–16333.

124 Selected reports, see: (a) Murphy, S.K. and Dong, V.M. (2014). *Chem. Commun.* 50: 13645–13649. (b) Johnson, K.F., Schmidt, A.C., and Stanley, L.M. (2015). *Org. Lett.* 17: 4654–4657. (c) Johnson, K.F., Schneider, E.A., Schumacher, B.P.

et al. (2016). *Chem. Eur. J.* 22: 15619–15623. (d) Oonishi, Y., Hosotani, A., Yokoe, T., and Sato, Y. (2019). *Org. Lett.* 21: 4120–4123.

125 (a) Yang, J. and Yoshikai, N. (2014). *J. Am. Chem. Soc.* 136: 16748–16751. (b) Yang, J., Rérat, A., Lim, Y.J. et al. (2017). *Angew. Chem. Int. Ed.* 56: 2449–2453.

126 Selected reports, see: (a) DiRocco, D.A. and Rovis, T. (2011). *Angew. Chem. Int. Ed.* 50: 7982–7983. (b) Walker, J.A. and Stanley, L.M. (2016). *Org. Biomol. Chem.* 14: 9981–9984.

127 Bugaut, X., Liu, F., and Glorius, F. (2011). *J. Am. Chem. Soc.* 133: 8130–8133.

128 Phan, D.H.T., Kou, K.G.M., and Dong, V.M. (2010). *J. Am. Chem. Soc.* 132: 16354–16355.

129 Tsuji, Y., Yoshii, S., Ohsumi, T. et al. (1987). *J. Organomet. Chem.* 331: 379–385.

130 Fujihara, T., Katafuchi, Y., Iwai, T. et al. (2010). *J. Am. Chem. Soc.* 132: 2094–2098.

131 Nakao, Y., Idei, H., Kanyiva, K.S., and Hiyama, T. (2009). *J. Am. Chem. Soc.* 131: 5070–5071.

132 Mori, S., Shibuya, M., and Yamamoto, Y. (2017). *Chem. Lett.* 46: 207–210.

133 Kobayashi, Y., Kamisaki, H., Yanada, K. et al. (2005). *Tetrahedron Lett.* 46: 7549–7552.

134 Kondo, T., Okada, T., and Mitsudo, T.-a. (1999). *Organometallics* 18: 4123–4127.

135 Ko, S., Han, H., and Chang, S. (2003). *Org. Lett.* 5: 2687–2690.

136 Armanino, N. and Carreira, E.M. (2013). *J. Am. Chem. Soc.* 135: 6814–6817.

137 Miyazaki, Y., Yamada, Y., Nakao, Y., and Hiyama, T. (2012). *Chem. Lett.* 41: 298–300.

138 Selected reviews, see: (a) Zhang, Y. and Riduan, S.N. (2011). *Angew. Chem. Int. Ed.* 50: 6210–6212. (b) Liu, Q., Wu, L., Jackstell, R., and Beller, M. (2015). *Nat. Commun.* 6: 5933. (c) Yu, D., Teong, S.P., and Zhang, Y. (2015). *Coord. Chem. Rev.* 293 (294): 279–291. (d) Juliá-Hernández, F., Gaydou, M., Serrano, E. et al. (2016). *Top. Curr. Chem.* 374: 45. (e) Zhang, L. and Hou, Z. (2017). *Curr. Opin. Green Sustain. Chem.* 3: 17–21. (f) Hou, J., Li, J.-S., Wu, J., and Asian, J. (2018). *Org. Chem.* 7: 1439–1447. (g) Yan, S.-S., Fu, Q., Liao, L.L. et al. (2018). *Coord. Chem. Rev.* 374: 439–463. (h) Song, J., Liu, Q., Liu, H., and Jiang, X. (2018). *Eur. J. Org. Chem.* 696–713. (i) Luan, Y.-X. and Ye, M. (2018). *Tetrahedron Lett.* 59: 853–861. (j) Fujihara, T. and Tsuji, Y. (2018). *Beilstein J. Org. Chem.* 14: 2435–2460. (k) Tortajada, A., Juliá-Hernández, F., Bçrjesson, M. et al. (2018). *Angew. Chem. Int. Ed.* 57: 15948–15982. Recent reports: (m) Ostapowicz, T.G., Schmitz, M., Krystof, M. et al. (2013). *Angew. Chem. Int. Ed.* 52: 12119–12123. (n) Gaydou, M., Moragas, T., Juliá-Hernández, F., and Martin, R. (2017). *J. Am. Chem. Soc.* 139: 12161–12164.

139 Zargarian, D. and Alper, H. (1993). *Organometallics* 12: 712–724.

140 El Ali, B., Vasapollo, G., and Alper, H. (1993). *J. Org. Chem.* 58: 4739–4741.

141 Hou, J., Xie, J.-H., and Zhou, Q.-L. (2015). *Angew. Chem. Int. Ed.* 54: 6302–6305.

142 (a) Hou, J., Yuan, M.-L., Xie, J.-H., and Zhou, Q.-L. (2016). *Green Chem.* 18: 2981–2984. (b) Fu, M.-C., Shang, R., Cheng, W.-M., and Fu, Y. (2016). *ACS Catal.* 6: 2501–2505.

143 Jiang, J., Fu, M., Li, C. et al. (2017). *Organometallics* 36: 2818–2825.

144 (a) Wang, Y., Ren, W., Li, J. et al. (2014). *Org. Lett.* 16: 5960–5963. (b) Wang, Y., Ren, W., and Shi, Y. (2015). *Org. Biomol. Chem.* 13: 8416–8419. (c) Liu, W., Ren, W., Li, J. et al. (2017). *Org. Lett.* 19: 1748–1751. (d) Dai, J., Ren, W., Li, J., and Shi, Y. (2018). *Org. Chem. Front.* 5: 561–565.

145 Zhu, C., Takaya, J., and Iwasawa, N. (2015). *Org. Lett.* 17: 1814–1817.

146 (a) Alper, H., Despeyroux, B., and Woell, J.B. (1983). *Tetrahedron Lett.* 24: 5691–5694. (b) Despeyroux, B. and Alper, H. (1983). *Ann. N.Y. Acad. Sci.* 415: 148–151.

147 Reynhardt, J.P.K. and Alper, H. (2003). *J. Org. Chem.* 68: 8353–8360.

148 (a) Akao, M., Sugawara, S., Amino, K., and Inoue, Y. (2000). *J. Mol. Catal. AChem.* 157: 117–122. (b) El Ali, B., Tijani, J., and El-Ghanam, A.M. (2001). *Tetrahedron Lett.* 42: 2385–2387. (c) Akao, M., Sugawara, S., Amino, K., and Inoue, Y. (2000). *J. Mol. Catal. A Chem.* 157: 117–122. (d) Li, J.-H., Tang, S., and Xie, Y.-X. (2005). *J. Org. Chem.* 70: 477–479. (e) Yee, G.M., Hillmyer, M.A., and Tonks, I.A. (2018). *ACS Sustain. Chem. Eng.* 6: 9579–9584. (f) Zhu, L., Liu, L.-J., Jiang, Y.-Y. et al. (2020). *J. Org. Chem.* 85: 7136–7151.

149 (a) Seayad, A., Kelkar, A.A., Toniolo, L., and Chaudhari, R.V. (2000). *J. Mol. Catal. A Chem.* 151: 47–59. (b) Estorach, C.T. and Masdeu-Bulto, A.M. (2008). *Catal. Lett.* 122: 76–79. (c) He, Z., Hou, Z., Luo, Y. et al. (2014). *Catal. Sci. Technol.* 4: 1092–1103.

150 Wu, L., Liu, Q., Jackstell, R., and Beller, M. (2015). *Org. Chem. Front.* 2: 771–774.

151 Yang, J., Liu, J., Ge, Y. et al. (2020). *Chem. Commun.* 56: 5235–5238.

152 Liu, Q., Yuan, K., Arockiam, P.-B. et al. (2015). *Angew. Chem. Int. Ed.* 54: 4493–4497.

153 (a) Park, W.-J., Lee, C.-H., Kim, D.-S., and Jun, C.-H. (2015). *Chem. Commun.* 51: 14667–14670. (b) Kim, D.-S., Park, W.-J., Lee, C.-H., and Jun, C.-H. (2014). *J. Org. Chem.* 79: 12191–12196.

154 (a) Li, J., Chang, W., Ren, W. et al. (2015). *Org. Biomol. Chem.* 13: 10341–10347. (b) Chang, W., Li, J., Ren, W., and Shi, Y. (2016). *Org. Biomol. Chem.* 14: 3047–3052.

155 El Ali, B. and Alper, H. (1995). *J. Mol. Catal. A Chem.* 96: 197–201.

156 Katafuchi, Y., Fujihara, T., Iwai, T. et al. (2011). *Adv. Synth. Catal.* 353: 475–482.

157 Ren, W., Chang, W., Wang, Y. et al. (2015). *Org. Lett.* 17: 3544–3547.

158 (a) Ko, S., Na, Y., and Chang, S. (2002). *J. Am. Chem. Soc.* 124: 750–751. (b) Na, Y., Ko, S., Hwang, L.K., and Chang, S. (2003). *Tetrahedron Lett.* 44: 4475–4478. (c) Park, E.J., Lee, J.M., Han, H., and Chang, S. (2006). *Org. Lett.* 8: 4355–4358.

159 Wang, L. and Floreancig, P.E. (2004). *Org. Lett.* 6: 4207–4210.

160 Armanino, N., Lafrance, M., and Carreira, E.M. (2014). *Org. Lett.* 16: 572–575.

161 Li, B., Lee, S., Shin, K., and Chang, S. (2014). *Org. Lett.* 16: 2010–2013.

162 (a) Konishi, H., Ueda, T., Muto, T., and Manabe, K. (2012). *Org. Lett.* 14: 4722–4725. (b) Konishi, H., Muto, T., Ueda, T. et al. (2015). *ChemCatChem* 7: 836–845.

163 Selected reviews, see: (a) El Ali, B. and Alper, H. (2000). *Synlett* 161–171. (b) Ojima, I., Moralee, A.C., and Vassar, V.C. (2002). *Top. Catal.* 19: 89–99. (c) Chatani, N. (2004). *Top. Organomet. Chem.* 11: 173–195. (d) Brennführer, A., Neumann, H., and Beller, M. (2009). *ChemCatChem* 1: 28–41. (e) Grigg, R. and Mutton, S.P. (2010). *Tetrahedron* 66: 5515–5548. (f) Omae, I. (2011). *Coord. Chem. Rev.* 255: 139–160. (g) Wu, X.-F. and Neumann, H. (2012). *ChemCatChem* 4: 447–458. (h) Wu, X.-F., Fang, X., Wu, L. et al. (2014). *Acc. Chem. Res.* 47: 1041–1053. (i) Kitagaki, S., Inagaki, F., and Mukai, C. (2014). *Chem. Soc. Rev.* 43: 2956–2978. (j) Quintero-Duque, S., Dyballa, K.M., and Fleischer, I. (2015). *Tetrahedron Lett.* 56: 2634–2650. (k) Gautam, P. and Bhanage, B.M. (2015). *Catal. Sci. Technol.* 5: 4663–4702. (l) Song, W., Blaszczyk, S.A., Liu, J. et al. (2017). *Org. Biomol. Chem.* 15: 7490–7504. (m) Shen, C. and Wu, X.-F. (2017). *Chem. Eur. J.* 23: 2973–2987. (n) Bai, Y., Davis, D.C., and Dai, M. (2017). *J. Org. Chem.* 82: 2319–2328. (o) Peng, J.-B., Geng, H.-Q., and Wu, X.-F. (2019). *Chem* 5: 526–552.

164 Selected reviews, see: (a) Chung, Y.K. (1999). *Coord. Chem. Rev.* 188: 297–341. (b) Kobayashi, T., Koga, Y., and Narasaka, K. (2001). *J. Organomet. Chem.* 624: 73–87. (c) Sugihara, T., Yamaguchi, M., and Nishizawa, M. (2001). *Chem. Eur. J.* 7: 1589–1595. (d) Gibson, S.E. and Stevenazzi, A. (2003). *Angew. Chem. Int. Ed.* 42: 1800–1810. (e) Blanco-Urgoiti, J., Añorbe, L., Pérez-Serrano, L. et al. (2004). *Chem. Soc. Rev.* 33: 32–42. (f) Laschat, S., Becheanu, A., Bell, T., and Baro, A. (2005). *Synlett* 2547–2570. (g) Gibson, S.E. and Mainolfi, N. (2005). *Angew. Chem. Int. Ed.* 44: 3022–3037. (h) Shibata, T. (2006). *Adv. Synth. Catal.* 348: 2328–2336. (i) Park, J.H. and Chung, Y.K. (2008). *Dalton Trans.* 2369–2378. (j) Park, J.H., Chang, K.-M., and Chung, Y.K. (2009). *Coord. Chem. Rev.* 253: 2461–2480. (k) Ricker, J.D. and Geary, L.M. (2017). *Top. Catal.* 60: 609–619.

165 Kalck, P. and Urrutigoïty, M. (2015). *Inorg. Chim. Acta* 431: 110–121.

166 Selected reviews, see: (a) Brummonda, K.M. and Kent, J.L. (2000). *Tetrahedron* 56: 3263–3283. (b) Kiss, G. (2001). *Chem. Rev.* 101: 3435–3456.

167 Akao, M., Sugawara, S., Amino, K., and Inoue, Y. (2000). *J. Mol. Catal. A Chem.* 157: 117–122.

168 Reetz, M.T., Demuth, R., and Goddard, R. (1998). *Tetrahedron Lett.* 39: 7089–7092.

169 Li, J., Jiang, H., and Chen, M. (2001). *Synth. Commun.* 31: 199–202.

170 Kuniyasu, H., Yoshizawa, T., and Kambe, N. (2010). *Tetrahedron Lett.* 51: 6818–6821.

171 Inoue, S., Yokota, K., Tatamidani, H. et al. (2006). *Org. Lett.* 8: 2519–2522.

172 Liu, J., Dong, K., Franke, R. et al. (2018). *J. Am. Chem. Soc.* 140: 10282–10288.

173 (a) Li, J., Jiang, H., Feng, A., and Jia, L. (1999). *J. Org. Chem.* 64: 5984–5987. (b) Li, J.-H., Tang, S., and Xie, Y.-X. (2005). *J. Org. Chem.* 70: 477–479.

174 Li, J., Li, G., Jiang, H., and Chen, M. (2001). *Tetrahedron Lett.* 42: 6923–6924.

175 Izawa, Y., Shimizu, I., and Yamamoto, A. (2004). *Bull. Chem. Soc. Jpn.* 77: 2033–2045.

176 Izawa, Y., Shimizu, I., and Yamamoto, A. (2005). *Chem. Lett.* 34: 1060–1061.

177 Selected reports, see: (a) Mohamed Ahmed, M.S. and Mori, A. (2003). *Org. Lett.* 5: 3057–3060. (b) Liang, B., Huang, M., You, Z. et al. (2005). *J. Org. Chem.* 70: 6097–6100. (c) Iizuka, M. and Kondo, Y. (2007). *Eur. J. Org. Chem.* 5180–5182. (d) Liu, J., Chen, J., and Xia, C. (2008). *J. Catal.* 253: 50–56.

178 Fusano, A., Fukuyama, T., Nishitani, S. et al. (2010). *Org. Lett.* 12: 2410–2413.

179 Wu, X.-F., Neumann, H., and Beller, M. (2011). *Org. Biomol. Chem.* 9: 8003–8005.

180 Cheng, L.-J. and Mankad, N.P. (2017). *J. Am. Chem. Soc.* 139: 10200–10203.

181 Zeng, F. and Alper, H. (2013). *Org. Lett.* 15: 2034–2037.

182 (a) Huh, K.-T., Orita, A., and Alper, H. (1993). *J. Org. Chem.* 58: 6956–6957. (b) Jayasree, S., Seayad, A., Gupte, S.P., and Chaudhari, R.V. (1999). *Catal. Lett.* 58: 213–216. (c) Izawa, Y., Shimizu, I., and Yamamoto, A. (2005). *Chem. Lett.* 34: 1060–1061. (d) Kato, K., Motodate, S., Mochida, T. et al. (2009). *Angew. Chem. Int. Ed.* 48: 3326–3328. (e) Qi, H., Huang, Z., Wang, M. et al. (2018). *J. Catal.* 363: 63–68.

183 The pioneering reports, see: (a) Moore, E.J., Pretzer, W.R., Oconnell, T.J. et al. (1992). *J. Am. Chem. Soc.* 114: 5888–5890. (b) Chatani, N., Fukuyama, T., Kakiuchi, F., and Murai, S. (1996). *J. Am. Chem. Soc.* 118: 493–494. (c) Fukuyama, T., Chatani, N., Tatsumi, J. et al. (1998). *J. Am. Chem. Soc.* 120: 11522–11523.

184 (a) Wu, X.-F., Neumann, H., Spannenberg, A. et al. (2010). *J. Am. Chem. Soc.* 132: 14596–14602. (b) Li, H., Dong, K., Jiao, H. et al. (2016). *Nat. Chem.* 8: 1159–1166. (c) Liu, J., Wei, Z., Jiao, H. et al. (2018). *ACS Cent. Sci.* 4: 30–38.

185 Sumino, S., Ui, T., Hamada, Y. et al. (2015). *Org. Lett.* 17: 4952–4955.

186 Selected reviews, see: (a) Gabriele, B., Salerno, G., Costa, M., and Chiusoli, G.P. (2003). *J. Organomet. Chem.* 687: 219–228. (b) Vizer, S.A., Yerzhanov, K.B., Quntar, A.A.A.A., and Dembitsky, V.M. (2004). *Tetrahedron* 60: 5499–5538. (c) Wu, X.-F., Neumann, H., and Beller, M. (2013). *Chem. Rev.* 113: 1–35. (d) Salacz, L., Girard, N., Suffert, J., and Blond, J. (2018). *Monatsh. Chem.* 149: 671–686.

187 (a) Kadnikov, D.V. and Larock, R.C. (2000). *Org. Lett.* 2: 3643–3646. (b) Kadnikov, D.V. and Larock, R.C. (2003). *J. Org. Chem.* 68: 9423–9432. (c) Kadnikov, D.V. and Larock, R.C. (2003). *J. Organomet. Chem.* 687: 425–435.

188 (a) Kadnikov, D.V. and Larock, R.C. (2004). *J. Org. Chem.* 69: 6772–6780.

189 Gabriele, B., Mancuso, R., Salerno, G., and Plastina, P. (2008). *J. Org. Chem.* 73: 756–759.

190 Ferguson, J., Zeng, F., and Alper, H. (2012). *Org. Lett.* 14: 5602–5605.

191 Seoane, A., Casanova, N., Quiñones, N. et al. (2014). *J. Am. Chem. Soc.* 136: 834–837.

192 Liu, X.-G., Zhang, S.-S., Jiang, C.-Y. et al. (2015). *Org. Lett.* 17: 5404–5407.

193 Li, X.-y., Xinwei Li, X.-w., and Jiao, N. (2015). *J. Am. Chem. Soc.* 137: 9246–9249.

194 Selected reports, see: (a) Miao, H. and Yang, Z. (2000). *Org. Lett.* 2: 1765–1768. (b) Liang, B., Huang, M., You, Z. et al. (2005). *J. Org. Chem.* 70: 6097–6100. (c) Awuah, E. and Capretta, A. (2009). *Org. Lett.* 11: 3210–3213. (d) Yang, Q.

and Alper, H. (2010). *J. Org. Chem.* 75: 948–950. (e) Xue, L., Shi, L., Han, Y. et al. (2011). *Dalton Trans.* 40: 7632–7638. (f) Liu, J., Liu, M., Yue, Y. et al. (2013). *Tetrahedron Lett.* 54: 1802–1807. (g) Zhao, H., Cheng, M., Zhang, J., and Cai, M. (2014). *Green Chem.* 16: 2515–2522. (h) Zhu, F., Li, Y., Wang, Z., and Wu, X.-F. (2016). *Catal. Sci. Technol.* 6: 2905–2909.

195 Zhu, F., Wang, Z., Li, Y., and Wu, X.-F. (2017). *Chem. Eur. J.* 23: 3276–3279.

196 Shen, C., Spannenberg, A., and Wu, X.-F. (2016). *Angew. Chem. Int. Ed.* 55: 5067–5070.

197 Fukuyama, T., Higashibeppu, Y., Okamura, T. et al. (2005). *Org. Lett.* 7: 5781–5783.

198 Fukuyama, T., Higashibeppu, Y., Yamaura, R., and Ryu, I. (2007). *Org. Lett.* 9: 587–589.

199 (a) Kondo, T., Kaneko, Y., Taguchi, Y. et al. (2002). *J. Am. Chem. Soc.* 124: 6824–6825. (b) Kondo, T., Nomura, M., Ura, Y. et al. (2006). *J. Am. Chem. Soc.* 128: 14816–14817.

200 Chatani, N., Kamitani, A., Oshita, M. et al. (2001). *J. Am. Chem. Soc.* 123: 12686–12687.

201 Harada, Y., Nakanishi, J., Fujihara, H. et al. (2007). *J. Am. Chem. Soc.* 129: 5766–5771.

202 Selected reports, see: (a) Liebeskind, L.S., Leeds, J.P., Baysdon, S.L., and Iyer, S. (1984). *J. Am. Chem. Soc.* 106: 6451–6453. (b) Jewell, C.F. Jr., and Liebeskind, L.S. (1994). *Inorg. Chim. Acta* 222: 235–246. (c) Rameshkumar, C. and Periasamy, M. (2000). *Organometallics* 19: 2400–2402. (d) Mathur, P., Bhunia, A.K., Mobin, S.M. et al. (2004). *Organometallics* 23: 3694–3700.

203 Kobayashi, T., Koga, Y., and Narasaka, K. (2001). *J. Organomet. Chem.* 624: 73–87.

204 Huang, Q. and Hua, R. (2007). *Chem. Eur. J.* 13: 8333–8337.

205 Selected reports, see: (a) Cheng, M.-H., Syu, H.-G., Lee, G.-H. et al. (1993). *Organometallics* 12: 108–115. (b) Barrow, M., Cromhout, N.L., Manning, A.R., and Gallagher, J.F. (2001). *J. Chem. Soc. Dalton Trans.* 1352–1358. (c) Elarraoui, A., Ros, J., Yanez, R. et al. (2002). *J. Organomet. Chem.* 642: 107–112.

206 Wu, B. and Hua, R. (2010). *Tetrahedron Lett.* 51: 6433–6435.

207 Gagnier, S.V. and Larock, R.C. (2003). *J. Am. Chem. Soc.* 125: 4804–4807.

208 Huang, Q. and Hua, R. (2009). *Chem. Eur. J.* 15: 3817–3822.

209 Yoneda, E., Kaneko, T., Zhang, S.-W. et al. (1999). *Tetrahedron Lett.* 40: 7811–7814.

210 Xu, Y., Zhao, J., Chen, H. et al. (2014). *Chem. Commun.* 50: 2488–2490.

211 Murakami, S., Sonehara, T., Iwakami, K. et al. (2019). *Tetrahedron Lett.* 60: 598–601.

212 The pioneering work, see: (a) Norton, J.R., Shenton, K.E., and Schwartz, J. (1975). *Tetrahedron Lett.* 51–54. (b) Murray, T.F., Samsel, E.G., Varma, V., and Norton, J.R. (1981). *J. Am. Chem. Soc.* 103: 7520–7528. (c) Tsuji, Y., Kondo, T., and Watanabe, Y. (1987). *J. Mol. Catal.* 40: 295–304. (d) Tezuka, K., Ishizaki, Y., and Inoue, Y. (1998). *J. Mol. Catal. AChem.* 129: 199–206.

213 Consorti, C.S., Ebeling, G., and Dupont, J. (2002). *Tetrahedron Lett.* 43: 753–755.

214 (a) Yoneda, E., Kaneko, T., Zhang, S.-W. et al. (2000). *Org. Lett.* 2: 441–443. (b) Yoneda, E., Zhang, S.-W., Zhou, D.-Y. et al. (2003). *J. Org. Chem.* 68: 8571–8576.

215 (a) Ortega, M.J., Zubía, E., Ocaña, J.M. et al. (2000). *Tetrahedron* 56: 3963–3967. (b) Oha, H., Shin, H., Oh, G.-S. et al. (2003). *Phytochemistry* 64: 1113–1118. (c) Bellina, F. and Rossi, R. (2004). *Curr. Org. Chem.* 8: 1089–1103. (d) de Souza, M.V.N. (2005). *Mini-Rev. Org. Chem.* 2: 139–145. (e) Zarghi, A., Rao, P.N.P., and Knaus, E.E. (2007). *J. Pharm. Pharma. Sci.* 10: 159–167. (f) Liu, H., Zhang, T., and Li, Y. (2006). *Chirality* 18: 223–226.

216 (a) Joh, T., Doyama, K., Onitsuka, K. et al. (1991). *Organometallics* 10: 2493–2498. (b) Joh, T., Nagata, H., and Takahashi, S. (1994). *Inorg. Chim. Acta* 220: 45–53. (c) Gabriele, B., Salerno, G., Costa, M., and Chiusoli, G.P. (1999). *Tetrahedron Lett.* 40: 989–990. (d) Chiusoli, G.P., Costa, M., Cucchia, L. et al. (2003). *J. Mol. Catal. A Chem.* 204 (205): 133–142. (e) Carfagna, C., Gatti, G., Mosca, L. et al. (2003). *Organometallics* 22: 3967–3970. (f) Park, K.H., Kim, S.Y., and Chung, Y.K. (2005). *Org. Biomol. Chem.* 3: 395–398.

217 Huang, Q. and Hua, R. (2007). *Catal. Commun.* 8: 1031–1035.

218 Miura, H., Takeuchi, K., and Shishido, T. (2016). *Angew. Chem. Int. Ed.* 55: 278–282.

219 Van den Hoven, B.G., El Ali, B., and Alper, H. (2000). *J. Org. Chem.* 65: 4131–4137.

220 Just, Z.W. and Larock, R.C. (2008). *J. Org. Chem.* 73: 2662–2667.

221 Morimoto, T., Jia, J., Yamaguchi, Y. et al. (2020). *Org. Chem.* 9: 1778–1782.

222 Hicks, F.A., Kablaoui, N.M., and Buchwald, S.L. (1996). *J. Am. Chem. Soc.* 118: 9450–9451.

223 Selected reports, see: (a) Wender, P.A., Gamber, G.G., Hubbard, R.D., and Zhang, L. (2002). *J. Am. Chem. Soc.* 124: 2876–2877. (b) Wender, P.A., Deschamps, N.M., and Gamber, G.G. (2003). *Angew. Chem. Int. Ed.* 42: 1853–1857. (c) Wender, P.A., Deschamps, N.M., and Williams, T.J. (2004). *Angew. Chem. Int. Ed.* 43: 3076–3079. (d) Wender, P.A., Croatt, M.P., and Deschamps, N.M. (2004). *J. Am. Chem. Soc.* 126: 5948–5949. (e) Wender, P.A., Gamber, G.G., Hubbard, R.D. et al. (2005). *J. Am. Chem. Soc.* 127: 2836–2837. (f) Wegner, H.A., de Meijere, A., and Wender, P.A. (2005). *J. Am. Chem. Soc.* 127: 6530–6531.

224 Wender, P.A., Croatt, M.P., and Deschamps, N.M. (2006). *Angew. Chem. Int. Ed.* 45: 2459–2462.

225 Wells, S.M. and Brummond, K.M. (2015). *Tetrahedron Lett.* 56: 3546–3549.

226 Lu, S.-M. and Alper, H. (2008). *J. Am. Chem. Soc.* 130: 6451–6455.

227 Selected reviews, see: (a) Jiao, L. and Yu, Z.-X. (2013). *J. Org. Chem.* 78: 6842–6848. (b) Wang, Y. and Yu, Z.-X. (2015). *Acc. Chem. Res.* 48: 2288–2296.

228 Wang, Y., Wang, J., Su, J. et al. (2007). *J. Am. Chem. Soc.* 129: 10060–10061.

229 Jiao, L., Lin, M., Zhuo, L.-G., and Yu, Z.-X. (2010). *Org. Lett.* 12: 2528–2531.

230 Li, C., Zhang, H., Feng, J. et al. (2010). *Org. Lett.* 12: 3082–3085.

231 Lu, B.-L., Wei, Y., and Shi, M. (2012). *Organometallics* 31: 4601–4609.

232 A recent review, see: Ma, K., Martin, B.S., Yin, X., and Dai, M. (2019). *Nat. Prod. Rep.* 36: 174–219.

10

Natural Product Synthesis via Alkyne Transformation

Addition and cycloaddition reactions of unsaturated hydrocarbons as the atom economic reactions have been well applied as the key steps in the syntheses of natural products, and a number of review papers have appeared [1]. In this section, only a few representative application reports of alkynes or alkynylated intermediates in the total syntheses of natural products are described; it involves the hydrofunctionalization of alkynes, double functionalization of alkynes, cycloaddition of alkynes, and carbonylation of alkynes.

10.1 Hydrofunctionalization of Alkynes in Natural Product Synthesis

A review on the applications of hydrometalation of alkynes, such as hydrosilylation, hydrostannation, and hydrogermylation in the synthesis of natural products, has recently appeared [2].

Catalytic *hydroamination* of alkynes is an important reaction in the C—N bond formation for construction of *N*-heterocyclic structure in the synthesis of natural products. For example, an alternative route for the total synthesis of indolizidine (−)-209D, which is a natural product acting as noncompetitive blocker of neuromuscular transmission, isolated from a single population of dendrobatid frogs [3], has been developed from the commercially available aldehyde including a $Pd(PPh_3)_4$/benzoic acid-catalyzed *intramolecular hydroamination* of ε-alkyne as one of the key reactions with excellent diastereoselectivity (Scheme 10.1) [4].

Monomorine is a kind of natural biological alkaloid, 3,5-disubstituted indolizidine isolated from the cosmopolitan ant *Monomorium pharaoins* L. [5]. Wiest and Helquist's group developed a mild and efficient desymmetrization of diynes via intramolecular hydroamination, which was applied in the synthesis of (±) monomorine starting from the intramolecular hydroamination of diamino-diyne, and the overall yield is 26% with seven steps (Scheme 10.2) [6].

Addition Reactions with Unsaturated Hydrocarbons, First Edition. Ruimao Hua.

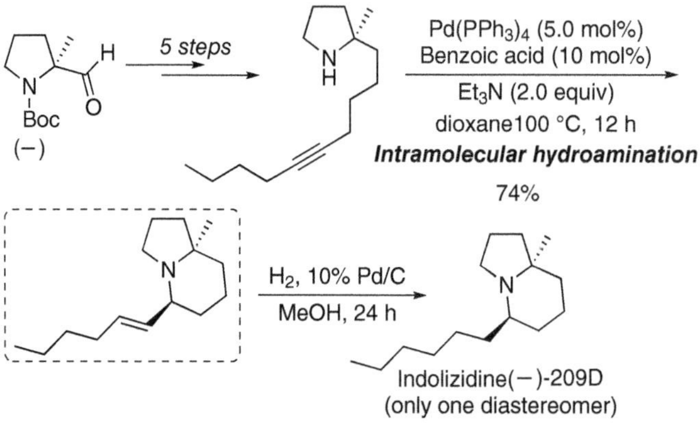

Scheme 10.1 Indolizidine (−)-209D synthesis via diastereoselective intramolecular hydroamination of ε-alkyne as a key reaction.

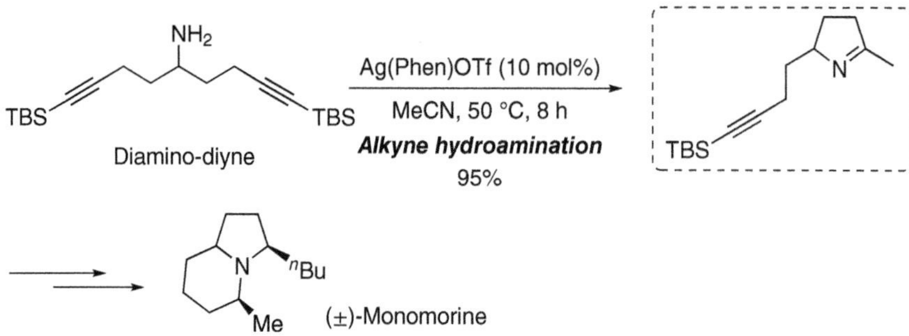

Scheme 10.2 Application of alkyne hydroamination in the synthesis of (±) monomorine.

10.2 Double Functionalization of Alkynes in Natural Product Synthesis

Dihydrojasmone is extracted from the bark of *Koelreuteria paniculata* and the flowers of *Carissa spinarum* L., which has been used in the perfume industry and as fragrances in cosmetics [7]. Therefore, chemists are interested in developing its synthetic method [8]. As described in Section 6.5.1, the regioselective *chloroallylation* of terminal alkynes with allyl chloride as a key step for the formation of carbon chain providing a simple synthetic method for dihydrojasmone via 2,5-undecanedione intermediate was developed by Tsuji and Yasuda (Scheme 10.3) [9].

In addition, a review paper summarizing the application of allylic substitutions in natural product synthesis has been published [10].

10.3 Cycloaddition of Alkynes in Natural Product Synthesis

As described in Chapter 8, the cycloaddition and annulation of alkynes are efficient transformation for access to carbo- and heterocycles with controllable different sizes

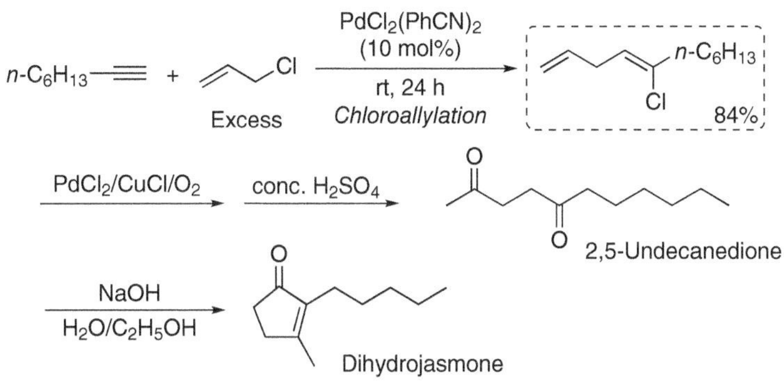

Scheme 10.3 Dihydrojasmone synthesis via regioselective chloroallylation of terminal alkynes with allyl chloride as a key step.

of rings, and with regio- and stereoselectivity, [*m* + 2] cycloaddition has been widely used in natural product synthesis [11].

Gevorgyan and coworker developed a Cu(I)-catalyzed cycloisomerization of alkynyl imines into the pyrrole ring, providing a general and efficient method for the synthesis of pyrroles and pyrrole-containing fused heterocycles [12]. As an example shown in Scheme 10.4, in the presence of CuCl, 2-hexynyl pyridine underwent cycloisomerization at 130 °C to give the substituted indolizine in 91% yield. Therefore, with the utility of this cycloisomerization methodology, they then designed a shortest route for the synthesis of (±) monomorine from the commercially available bromopyridine with three steps in 47% total yield (Scheme 10.5).

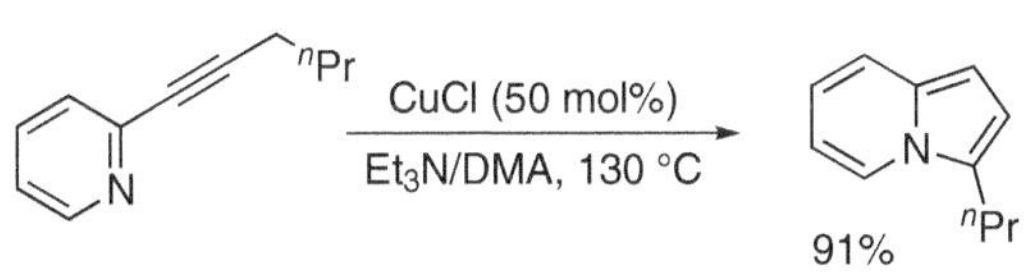

Scheme 10.4 Indolizine formation from cycloisomerization of alkynyl imines.

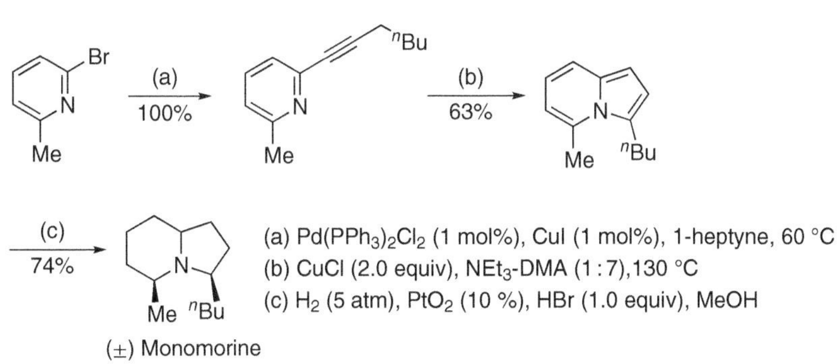

Scheme 10.5 Total synthesis of (±) monomorine from alkynylpyridine intermediate.

As described in Chapter 7, Section 7.3, [2 + 2 + 2] *cyclotrimerization* reactions of alkynes are efficient and versatile reactions for construction of highly substituted benzenes. Transition metal-catalyzed cyclotrimerization of alkynes has been applied in the total synthesis of a few natural products, such as (+)-rubiginone B_2 [13], marine illudalane sesquiterpenoid alcyopterosin E [14], (−)-bruguierol A [15], cannabinoids [16], and sporolide B [17].

Schemes 10.6 and 10.7 show two total synthesis protocols including the [2 + 2 + 2] *cyclotrimerization* reactions of alkynes as the key steps.

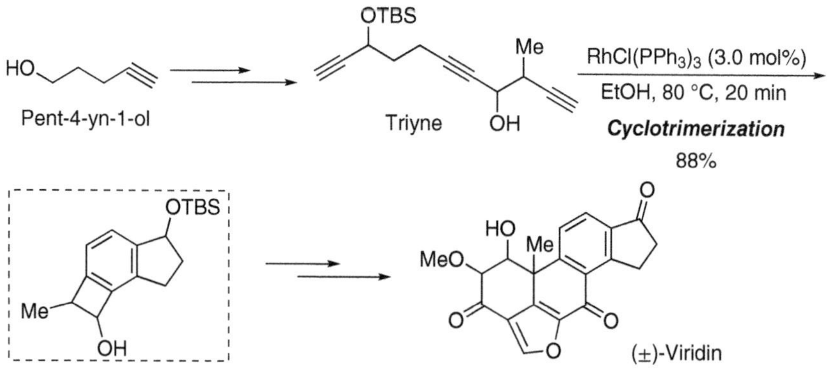

Scheme 10.6 Viridin synthesis via rhodium-catalyzed cyclotrimerization of triyne forming tetrasubstituted aryl cyclobutenol as a key step.

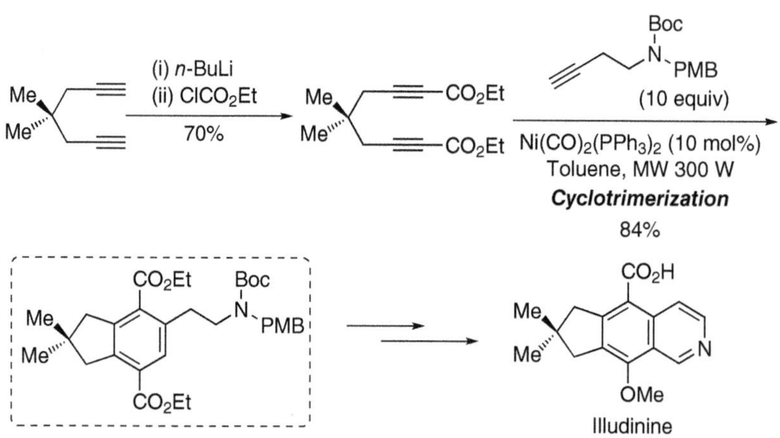

Scheme 10.7 Illudinine synthesis via a microwave-mediated nickel-catalyzed cyclotrimerization.

Viridin is an antifungal metabolite of *Gliocladium virens* [18], which was isolated from *Trichoderma viride* [19], and several total synthesis protocols have been reported [20]. Among them, Sorensen and coworker reported the first total synthesis of racemic viridin starting from pent-4-yn-1-ol, involving a RhCl(PPh)$_3$-catalyzed cyclotrimerization of the prepared triyne in ethanol at 80 °C to form the tetrasubstituted aryl cyclobutenol as one of the key steps to construct the part core structure (Scheme 10.6) [20]. This transformation demonstrated the power of alkyne cyclotrimerization in steroid synthesis.

Illudinine is a sesquiterpene alkaloid, which was isolated as a fungal metabolite from the basidiomycete *Clitocybe illudens* [21]. Deiters and coworker developed a procedure for rapid and efficient [2 + 2 + 2] cyclotrimerization reactions of diynes with alkynes catalyzed by $Ni(CO)_2(PPh_3)_2$ under microwave irradiation, which was employed as the key step in a concise synthesis of the isoquinoline natural product illudinine starting from 1,6-diyne, providing the first example of a nickel-catalyzed cyclotrimerization reaction in total synthesis (Scheme 10.7) [22].

Deiters's group then reported the total synthesis of a tetracyclic terpene natural product of cryptoacetalide, and the microwave-mediated [2 + 2 + 2] cyclotrimerization of triyne was the key step for the formation of the central benzene ring catalyzed by Cp*RuCl(cod) [23].

In addition, [5 + 2] cycloaddition reactions of alkynes have also been well applied in the synthesis of natural product having seven-membered carbocycles [24].

Lactarane natural products usually have a tricyclic scaffold, which have shown interesting and diverse medicinal properties. Although the total synthesis of pyrovellerolactone has been established [25], Evans and coworker designed and developed a concise three-step total synthesis of pyrovellerolactone in 48% overall yield from the commercially available aldehyde, including a regio- and diastereoselective rhodium-catalyzed [5 + 2] *carbocyclization* of an alkenylidenecyclopropane with 4-hydroxybut-2-ynoate as the key step to construct the tricyclic structure (Scheme 10.8) [26].

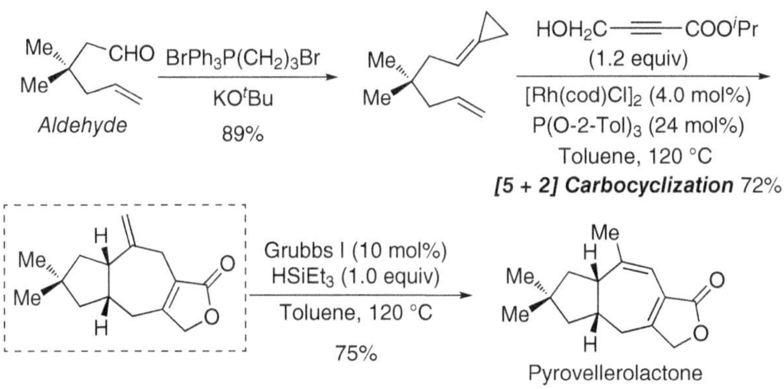

Scheme 10.8 Synthesis of pyrovellerolactone via a regio- and diastereoselective rhodium-catalyzed [5 + 2] carbocyclization as a key step.

10.4 Carbonylation of Alkynes in Natural Product Synthesis

Carbonyl group is one of the important functional groups found in many natural products with impressive biological and medicinal properties, and the carbonylation of alkynes and alkenes in natural product synthesis has been widely applied [27].

Clovan-2,9-dione is one of the clovane-type sesquiterpenes, which was isolated from the gorgonian coral *Rumphella antipathies* [28]. Yu's group has extensively investigated rhodium-catalyzed cyclocarbonylation of unsaturated hydrogens with CO, and they first designed and developed an asymmetric total synthesis of (−)-clovan-2,9-dione starting from the reaction of 3,3-dimethyl-4-pentynal and 2,6-bis(1,1-dimethylethyl)-4-methylphenyl cyclopropanecarboxylate [29]. As shown in Scheme 10.9, the rhodium(I)-catalyzed [(3 + 2) + 1] cyclocarbonylation of the prepared yne-vinylcyclopropane intermediate with CO afforded the key cyclic ketone intermediate for further transformation access to the desired natural product.

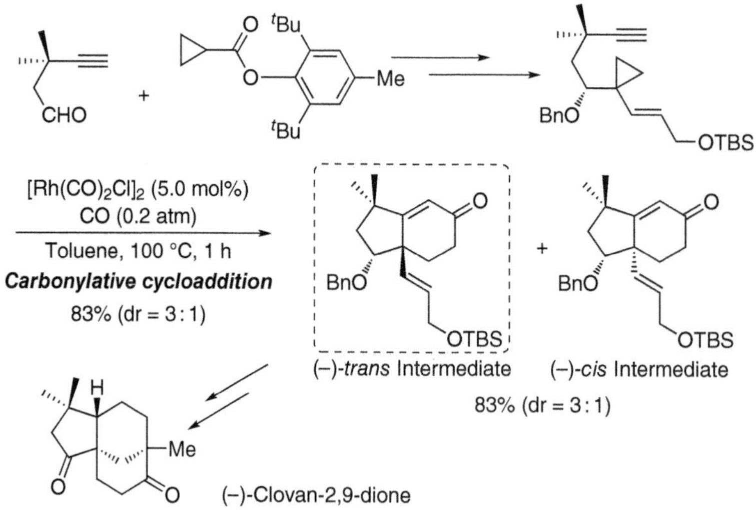

Scheme 10.9 Clovan-2,9-dione synthesis via rhodium-catalyzed carbonylative cycloaddition of yne-vinylcyclopropane.

References

1 Selected reviews, see: (a) Prunet, J. (2003). *Angew. Chem. Int. Ed.* 42: 2826–2830. (b) Fallis, A.G. (2004). *Synlett* 2249–2267. (c) Hsung, R.P., Kurdyumov, A.V., and Sydorenko, N. (2005). *Eur. J. Org. Chem.* 23–44. (d) Nicolaou, K.C., Bulger, P.G., and Sarlah, D. (2005). *Angew. Chem. Int. Ed.* 44: 4442–4489. (e) Takao, K.-i., Munakata, R., and Tadano, K.-i. (2005). *Chem. Rev.* 105: 4779–4807. (f) Evans, P.A., Baum, E.W., Fazal, A.N. et al. (2006). *ARKIVOC* (vii): 338–358. (g) Parenty, A., Moreau, X., and Campagne, J.-M. (2006). *Chem. Rev.* 106: 911–939. (h) Mori, M. (2007). *Adv. Synth. Catal.* 349: 121–135. (i) Touré, B.B. and Hall, D.G. (2009). *Chem. Rev.* 109: 4439–4486. (j) Alcaide, B., Almendros, P., and Alonso, J.M. (2011). *Molecules* 16: 7815–7843. (k) Gampe, C.M. and Carreira, E.M. (2012). *Angew. Chem. Int. Ed.* 51: 3766–3778. (l) Brenzovich, W.E. Jr., (2012). *Angew. Chem. Int. Ed.* 51: 8933–8935. (m) Wilson, M.R. and Taylor, R.E. (2013). *Angew. Chem. Int. Ed.* 52: 4078–4087. (n) Wan, C.Y., Deng, J., Liu, H. et al. (2014). *Sci. China Chem.* 57: 926–929. (o) Lehmann, J., Wright, M.H., and Sieber, S.A. (2016). *Chem. Eur. J.* 22: 4666–4678. (p) Herzon, S.B. and Vanderwal, C.D.

(2017). *Chem. Rev.* 117: 11649–11650. (q) Heravi, M.M., Zadsirjan, V., Hamidi, H., and Amiri, P.H.T. (2017). *RSC Adv.* 7: 24470–24521. (r) Heravi, M.M. and Mohammadkhani, L. (2018). *J. Organomet. Chem.* 869: 106–200. (s) Fürstner, A. (2018). *Angew. Chem. Int. Ed.* 57: 4215–4233. (t) Pflästerer, D., Rudolph, M., and Hashmi, A.S.K. (2018). *Isr. J. Chem.* 58: 622–638. (u) Zheng, L. and Hua, R. (2020). *Front. Chem.* 8: 580355.

2 Frihed, T.G. and Fürstner, A. (2016). *Bull. Chem. Soc. Jpn.* 89: 135–160.

3 Åhman, J. and Somfai, P. (1995). *Tetrahedron Lett.* 36: 303–306.

4 Patil, N.T., Pahadi, N.K., and Yamamoto, Y. (2005). *Tetrahedron Lett.* 46: 2101–2103.

5 Ritter, F.J., Rotgans, I.E.M., Talman, E. et al. (1973). *Experientia* 29: 530–531.

6 Iska, V.B.R., Verdolino, V., Wiest, O., and Helquist, P. (2010). *J. Org. Chem.* 75: 1325–1328.

7 Pinto, Â.V., de Oliveira, J.C., Costa de Medeiros, C.A. et al. (2020). *Lett. Appl. Microbiol.* 72: 292–298. and references cited therein.

8 (a) Ho, T.-L. (1974). *Synth. Commun.* 4: 265–287. (b) Salaün, J. and Almirantis, Y. (1983). *Tetrahedron* 39: 2421–2428. (c) Erickson, J.L.E. and Collins, F.E. Jr., (1965). *J. Org. Chem.* 30: 1050–1052.

9 Tsuji, J. and Yasuda, H. (1978). *Synth. Commun.* 8: 103–107.

10 Helmchen, G., Ernst, M., and Paradies, G. (2004). *Pure Appl. Chem.* 76: 495–506.

11 Selected reviews, see: (a) Maryanoff, B.E. and Zhang, H.-C. (2007). *ARKIVOC* (xii): 7–35. (b) Liu, X., Hu, Y.-J., Fan, J.-H. et al. (2018). *Org. Chem. Front.* 5: 1217–1228. (c) Gao, K., Zhang, Y.-G., Wang, Z., and Ding, H. (2019). *Chem. Commun.* 55: 1859–1878.

12 Kel'in, A.V., Sromek, A.W., and Gevorgyan, V. (2001). *J. Am. Chem. Soc.* 123: 2074–2075.

13 Kalogerakis, A. and Groth, U. (2003). *Synlett* 1886–1888.

14 Witulski, B., Zimmermann, A., and Gowans, N.D. (2002). *Chem. Commun.* 2984–2985.

15 Ramana, C.V., Salian, S.R., and Gonnade, R.G. (2007). *Eur. J. Org. Chem.* 5483–5486.

16 Teske, J.A. and Deiters, A. (2008). *Org. Lett.* 10: 2195–2198.

17 Nicolaou, K.C., Tang, Y., and Wang, J. (2009). *Angew. Chem. Int. Ed.* 48: 3449–3453.

18 Hanson, J.R. (1995). *Nat. Prod. Rep.* 12: 381–384.

19 (a) Brian, P.W. and McGowan, J.C. (1945). *Nature* 156: 144–145. (b) Grove, J.F., McCloskey, P., and Moffatt, J.S. (1966). *J. Chem. Soc. C* 743–746. (c) Hao, J.-J., Geng, C.-D., Xie, W.-J. et al. (1999). *Biol. Chem.* 380: 1243–1245.

20 Selected reports, see: (a) Anderson, E.A., Alexanian, E.J., and Sorensen, E.J. (2004). *Angew. Chem. Int. Ed.* 43: 1998–2001. (b) Bel, M.D., Abela, A.R., Ng, J.D., and Guerrero, C.A. (2017). *J. Am. Chem. Soc.* 139: 6819–6822. (c) Ji, Y., Xin, Z., He, H., and Gao, S. (2019). *J. Am. Chem. Soc.* 141: 16208–16212.

21 Nair, M.S.R., Takeshita, H., McMorris, T.C., and Anchel, M. (1969). *J. Org. Chem.* 34: 240–243.

22 Teske, J.A. and Deiters, A. (2008). *J. Org. Chem.* 73: 342–345.

23 Zou, Y. and Deiters, A. (2010). *J. Org. Chem.* 75: 5355–5358.

24 Selected reviews on the [5 + 2] cycloaddition reactions in organic and natural product synthesis, see: (a) Ylijoki, K.E.O. and Stryker, J.M. (2013). *Chem. Rev.* 113: 2244–2266. (b) Battiste, M.A., Pelphrey, P.M., and Wright, D.L. (2006). *Chem. Eur. J.* 12: 3438–3447. (c) Nguyen, T.V., Hartmann, J.M., and Dieter Enders, D. (2013). *Synthesis* 45: 845–873. (d) Gao, K., Zhang, Y.-G., Wang, Z., and Ding, H. (2019). *Chem. Commun.* 55: 1859–1878.

25 (a) Froborg, J., Magnusson, G., and Thorén, S. (1975). *J. Org. Chem.* 40: 1595–1601. (b) Froborg, J. and Magnusson, G. (1978). *J. Am. Chem. Soc.* 100: 6728–6733.

26 Evans, P.A., Inglesby, P.A., and Kilbride, K. (2013). *Org. Lett.* 15: 1798–1801.

27 Selected reviews, see: (a) Wang, Y. and Yu, Z.-X. (2015). *Acc. Chem. Res.* 48: 2288–2296. (b) Gehrtz, P.H., Hirschbeck, V., Ciszek, B., and Fleischer, I. (2016). *Synthesis* 48: 1573–1596. (c) Bai, Y., Davis, D.C., and Dai, M. (2017). *J. Org. Chem.* 82: 2319–2328. (d) Ma, K., Martin, B.S., Yin, X., and Dai, M. (2019). *Nat. Prod. Rep.* 36: 174–219.

28 Chung, H.-M., Su, J.-H., Hwang, T.-L. et al. (2013). *Tetrahedron* 69: 2740–2744. and references cited therein.

29 Yang, J., Xu, W., Cui, Q. et al. (2017). *Org. Lett.* 19: 6040–6043.

Index

Addition Reactions with Unsaturated Hydrocarbons, First Edition. Ruimao Hua.

b

c

d

i

t